Organization of the sections in the chapters on "Physical data of semiconductors"

The data on the physical properties of semiconductors are generally arranged in subsections of the order given below. Since there is some arbitrariness in the assignment of a property to the seven subsections, physically related properties may appear in different subsections in some cases.

Structure

Static properties of the lattice as structure, space group, lattice parameters, phase transitions, chemical bond. Density, melting point, thermal expansion are sometimes listed under "Further properties". Further static properties are given in "Lattice properties".

Electronic properties

Information and data about electronic and excitonic energy states as well as electron and hole parameters (band structure, density of states, energy gaps, transition energies, effective masses (sometimes also presented in "Transport properties"), g-factors, other band parameters).

Impurity and defect states

Basic data on shallow and deep states (trap levels, nature of defects, point defect thermodynamics, g-factors of defects; for the influence of impurities on other properties, see also the respective subsections).

Lattice properties

Static and dynamic properties of the lattice (phonon dispersion relations, phonon frequencies, sound velocities, elastic and other moduli, Grüneisen parameter, dielectric constants (sometimes also listed under "Optical properties"); for Debye temperature, heat capacity, see also "Further properties"; for structure, density, lattice parameters, chemical bond etc., see subsection "Structure").

Transport properties

Electronic transport parameters (conductivities, carrier concentrations, mobilities, Seebeck coefficient etc.; for thermal conductivity, see also "Further properties").

Optical properties

Optical spectra, optical constants, parameters obtained from optical measurements if not already listed in the subsections "Electronic properties" and "Impurity and defect states" (absorption, reflection, refractive index, dielectric constants, Raman scattering etc.).

Further properties

Thermal, magnetic, thermodynamic properties, data more completely presented in other Landolt-Börnstein volumes (thermal conductivity, thermal expansion, magnetic susceptibility, magnetic transition temperatures, magnetic moments, hardness, melting point, density, heat capacity, Debye temperature etc.).

LANDOLT-BÖRNSTEIN

Numerical Data and Functional Relationships
in Science and Technology

New Series
Editors in Chief: K.-H. Hellwege · O. Madelung

Group III: Crystal and Solid State Physics

Volume 17
Semiconductors

Editors: O. Madelung · M. Schulz · H. Weiss†

Subvolume g

Physics of Non-Tetrahedrally
Bonded Binary Compounds III

J.B. Goodenough · A. Hamnett · G. Huber · F. Hulliger
M. Leiß · S.K. Ramasesha · H. Werheit

Edited by O. Madelung

Springer-Verlag Berlin · Heidelberg · New York · Tokyo 1984

LANDOLT-BÖRNSTEIN

Zahlenwerte und Funktionen
aus Naturwissenschaften und Technik

Neue Serie

Gesamtherausgabe: K.-H. Hellwege · O. Madelung

Gruppe III: Kristall- und Festkörperphysik

Band 17

Halbleiter

Herausgeber: O. Madelung · M. Schulz · H. Weiss †

Teilband g

Physik der nicht-tetraedrisch gebundenen
binären Verbindungen III

J. B. Goodenough · A. Hamnett · G. Huber · F. Hulliger
M. Leiß · S. K. Ramasesha · H. Werheit

Herausgegeben von O. Madelung

Springer-Verlag Berlin · Heidelberg · New York · Tokyo 1984

CIP-Kurztitelaufnahme der Deutschen Bibliothek

Zahlenwerte und Funktionen aus Naturwissenschaften und Technik / Landolt-Börnstein. – Berlin; Heidelberg; New York; Tokyo: Springer. Parallelt.: Numerical data and functional relationships in science and technology. – Teilw. mit d. Erscheinungsorten Berlin, Heidelberg, New York

NE: Landolt, Hans [Begr.]; PT. N.S. Gesamthrsg.: K.-H. Hellwege; O. Madelung. Gruppe 3, Kristall- und Festkörperphysik. Bd. 17. Halbleiter/Hrsg.: O. Madelung... Teilbd. g. Physik der nicht-tetraedrisch gebundenen Verbindungen. – III/ J.B. Goodenough ... Hrsg. von O. Madelung. – 1984.

ISBN 3-540-12744-5 Berlin, Heidelberg, New York, Tokyo
ISBN 0-387-12744-5 New York, Heidelberg, Berlin, Tokyo

NE: Hellwege, Karl-Heinz [Hrsg.]; Madelung, Otfried [Hrsg.]; Goodenough, John B. [Mitverf.]

This work is subject to copyright. All rights are reserved, whether the whole or part of the material is concerned specifically those of translation, reprinting, reuse of illustrations, broadcasting, reproduction by photocopying machine or similar means, and storage in data banks.

Under § 54 of the German Copyright Law where copies are made for other than private use a fee is payable to 'Verwertungsgesellschaft Wort' Munich.

© by Springer-Verlag Berlin-Heidelberg 1984

Printed in Germany

The use of registered names, trademarks, etc. in this publication does not imply, even in the absence of a specific statement, that such names are exempt from the relevant protective laws and regulations and therefore free for general use.

Typesetting: Universitätsdruckerei H. Stürtz AG, Würzburg; printing: Druckhaus Langenscheidt KG, Berlin; bookbinding: Lüderitz & Bauer GmbH, Berlin

2163/3020-543210

Contributors

J.B. Goodenough, Inorganic Chemistry Laboratory, University of Oxford, Oxford OX1 3QR, UK

A. Hamnett, Inorganic Chemistry Laboratory, University of Oxford, Oxford OX1 3QR, UK

G. Huber, Institut für angewandte Physik der Universität Hamburg, 2000 Hamburg, FRG

F. Hulliger, Laboratorium für Festkörperphysik, Eidgenössische Technische Hochschule Zürich, 8093 Zürich, Switzerland

M. Leiß, Institut für angewandte Physik der Universität Hamburg, 2000 Hamburg, FRG

S.K. Ramasesha, Inorganic Chemistry Laboratory, University of Oxford, Oxford OX1 3QR, UK

H. Werheit, Fachbereich 10 (Physik-Technologie) der Universität-Gesamthochschule-Duisburg, 4100 Duisburg, FRG

Vorwort

Der vorliegende Band 17g schließt die in Teilband 17a begonnene und in den Teilbänden 17b, 17e und 17f fortgesetzte Behandlung der physikalischen Eigenschaften binärer Halbleiter ab. Vorgelegt werden Daten über halbleitende Borverbindungen und halbleitende Verbindungen der Übergangsmetalle und der Seltenen Erden.

Die in den früheren Bänden behandelten Familien von Verbindungen (III–V-Verbindungen, II–VI-Verbindungen etc.) enthalten eine überschaubare Anzahl von Substanzen mit wohldefinierten Halbleitereigenschaften, über die in den meisten Fällen zahlreiche Meßdaten vorliegen. Die in diesem Band behandelten Familien bestehen dagegen aus einer extrem großen Zahl von binären Verbindungen, deren Eigenschaften nur zum Teil bekannt sind und bei denen in vielen Fällen Unklarheit besteht, ob sie Halbleiter sind. Dementsprechend war es schwierig, eine sachgemäße Auswahl an Substanzen und Daten zu treffen. Die Autoren versuchten einen Mittelweg zu gehen: über die eindeutig als Halbleiter identifizierten Substanzen hinaus wurden auch solche aufgenommen, bei denen Halbleitereigenschaften zu vermuten sind oder die aus systematischen Gründen einer Gruppe als Halbleiter bekannter Stoffe zuzuordnen sind. Da bei den meisten Substanzen das Datenmaterial wesentlich geringer ist als bei den in früheren Bänden dargestellten Halbleitern, wurde vielfach auf eine detaillierte Untergliederung in strukturelle, elektronische, optische, Gittereigenschaften usw. verzichtet. Unterschieden wurde dann lediglich zwischen „strukturellen" und „physikalischen" Eigenschaften, in einigen Fällen wurde auch auf eine derartige Einteilung verzichtet.

Ein weiterer Unterschied zu den vorhergehenden Bänden muß erwähnt werden: bei den Verbindungen der Übergangsmetalle und Seltenen Erden sind häufig die magnetischen Eigenschaften wichtiger als die halbleitenden Eigenschaften. Solche Verbindungen (wie z.B. die Gruppe der Spinelle) sind schon in anderen Bänden dieser Serie dargestellt. Es erschien hier weder zweckmäßig, lediglich auf diese Bände zu verweisen, noch konnten alle in anderen Bänden mitgeteilten Daten hier wiederholt werden. In diesen Fällen wurden nur solche Daten aufgenommen, die einen Halbleiterphysiker in erster Linie interessieren. Für weitergehende Informationen wurde jeweils auf die anderen Bände verwiesen.

Es ist mir auch hier wieder eine Freude, den Autoren für ihre gründliche und kritische Arbeit und dem Springer-Verlag für die reibungslose Zusammenarbeit danken zu können. Den unermüdlichen Einsatz der Landolt-Börnstein-Redaktion des Verlages, insbesondere von Herrn Dr. W. Polzin, Frau R. Lettmann und Frau I. Lenhart, möchte ich besonders hervorheben.

Marburg, März 1984 **Der Herausgeber**

Preface

In this subvolume 17g, the presentation of the physical properties of binary semiconductors – started in subvolume 17a and continued in subvolumes 17b, 17e and 17f – will be completed. Data on boron compounds and on semiconducting transition-metal and rare earth compounds will be presented.

The families of III–V compounds, II–VI compounds etc. which have been treated in the previous volumes consist of a limited number of compounds with well-defined semiconducting properties. On the other side, the families treated in this volume consist of an extremely large number of compounds which are only partly investigated. For many of these compounds it is not yet clear whether they are semiconductors or not. Accordingly, the choice of substances and data to be presented in this volume was difficult. The authors tried to make a compromise: beyond substances identified definitely as semiconductors, such compounds have been included where semiconducting properties may be expected or where compounds should be added to a group of known semiconductors by systematical reasons. Since considerably less data are available for many of these compounds, the detailed subclassification (structure, electronic, optical, lattice properties etc.) used in previous volumes could frequently not be retained. The data have been then merely grouped under two headings "Structure" and "Physical properties", in some cases the data have not been classified at all.

A further difference to the previous volumes should be mentioned: for several groups of transition-metal and rare earth compounds, magnetic properties are as important as semiconducting properties. Such groups have been already treated in other Landolt-Börnstein volumes. An example is the group of transition-metal compounds with spinel structure. Here it seemed not to be appropriate to refer solely to other volumes. Nor was it possible to reprint all data presented already elsewhere. In these cases, the data have been included to such an extent as a semiconductor physicist may expect to find. For further information, references have been given to the other volumes.

I am glad to take this opportunity to thank the authors for their thorough and critical work and the Springer-Verlag for the excellent cooperation. The competent work of Dr. W. Polzin, Mrs. R. Lettmann and Mrs. I. Lenhart of the Landolt-Börnstein editorial staff is gratefully acknowledged.

Marburg, March 1984 **The Editor**

Table of contents
Semiconductors

Subvolume g: Physics of non-tetrahedrally bonded binary compounds III
(edited by O. MADELUNG)

A Introduction (O. MADELUNG)

		Data
1 List of symbols		1
2 List of abbreviations		5
3 Conversion tables		6

B Physical data of semiconductors V   Data | Figures

9 Non-tetrahedrally bonded binary compounds
(Sections 9.1…9.6, see subvolume III/17e;
sections 9.7…9.13, see subvolume III/17f)

				Data	Figures
9.14		Boron compounds (H. WERHEIT)		9	–
	9.14.0	Introduction, general remarks on structure and chemical bond		9	–
	9.14.1	Compounds with group Ia elements		10	–
		9.14.1.1	Boron-hydrogen alloys	10	345
		9.14.1.2	Boron-lithium compounds	10	–
		9.14.1.3	Boron-sodium compounds	11	345f.
		9.14.1.4	Boron-potassium compounds	13	346f.
	9.14.2	Compounds with group Ib elements		13	347
	9.14.3	Compounds with group IIa elements		14	–
		9.14.3.1	Boron-beryllium compounds	14	348
		9.14.3.2	Beryllium-aluminum-boron compounds	15	348
		9.14.3.3	Boron-magnesium compounds	16	349
		9.14.3.4	Boron-alkaline earth compounds	17	349f.
	9.14.4	Compounds with group IIb elements		19	–
		9.14.4.1	Boron-zinc compounds	19	351
		9.14.4.2	Boron-cadmium compounds	20	–
	9.14.5	Compounds with group IIIa elements		20	352ff.
	9.14.6	Compounds with group IIIb elements		24	–
		9.14.6.1	Boron-scandium compounds	24	–
		9.14.6.2	Boron-yttrium compounds	24	356f.
	9.14.7	Compounds with lanthanides		25	–
		9.14.7.1	Lanthanide hexaborides	25	357ff.
		9.14.7.2	Lanthanide dodecaborides	32	366
		9.14.7.3	Lanthanide borides of the type MB_{66}	33	366f.
	9.14.8	Compounds with actinides		34	–
		9.14.8.1	Boron-thorium compounds	34	–
		9.14.8.2	Boron-uranium compounds	35	–
		9.14.8.3	Boron-neptunium compounds	35	–
		9.14.8.4	Boron-plutonium compounds	35	–
		9.14.8.5	Boron-americium compounds	35	–
	9.14.9	Compounds with group IVa elements		35	–
		9.14.9.1	Boron-carbon compounds	35	367ff.
		9.14.9.2	Boron-silicon compounds	40	376ff.
		9.14.9.3	Boron-germanium compounds	43	–

			Data	Figures
9.14.10	Compounds with group IVb elements		43	–
	9.14.10.1 Boron-titanium compounds		43	–
	9.14.10.2 Boron-zirconium compounds		43	–
	9.14.10.3 Boron-hafnium compounds		43	–
9.14.11	Compounds with group Va elements		44	–
	9.14.11.1 Boron-nitrogen compounds		44	–
	9.14.11.2 Boron-phosphorus compounds		44	379
	9.14.11.3 Boron-arsenic compounds		45	379 f.
9.14.12	Compounds with group Vb elements		46	–
	9.14.12.1 Boron-vanadium compounds		46	–
	9.14.12.2 Boron-niobium compounds		46	–
	9.14.12.3 Boron-tantalum compounds		46	–
9.14.13	Compounds with group VIa elements		47	–
	9.14.13.1 Boron-oxygen compounds		47	–
	9.14.13.2 Boron-sulfur compounds		47	–
	9.14.13.3 Boron-selenium compounds		47	–
9.14.14	Compounds with group VIb elements		48	–
	9.14.14.1 Boron-chromium compounds		48	–
	9.14.14.2 Boron-molybdenum compounds		49	–
	9.14.14.3 Boron-tungsten compounds		49	–
9.14.15	Compounds with group VIIa elements		49	–
9.14.16	Compounds with group VIIb elements		50	–
	9.14.16.1 Boron-manganese compounds		50	–
	9.14.16.2 Boron-technetium compounds		50	–
	9.14.16.3 Boron-rhenium compounds		50	–
9.14.17	Compounds with group VIII elements		51	–
	9.14.17.1 Boron-iron compounds		51	–
	9.14.17.2 Solid solutions of Fe in β-rhombohedral boron		51	380 f.
	9.14.17.3 Boron-cobalt compounds		52	–
	9.14.17.4 Boron-nickel compounds		52	–
	9.14.17.5 Boron-ruthenium compounds		52	–
	9.14.17.6 Boron-rhodium compounds		53	–
	9.14.17.7 Boron-palladium compounds		53	–
	9.14.17.8 Boron-osmium compounds		53	–
	9.14.17.9 Boron-iridium compounds		53	–
	9.14.17.10 Boron-platinum compounds		53	–
9.14.18	References for 9.14		54	–
9.15 Binary transition-metal compounds (F. HULLIGER)			63	–
9.15.1	Compounds with elements of the IIId and IVth groups		63	–
	9.15.1.1 $T_n(III)_{2n-m}$ and $T_n(IV)_{2n-m}$ compounds		63	–
		9.15.1.1.0 Structure, chemical bond	63	381
		9.15.1.1.1 Physical properties of $T(III)_x$ compounds	65	–
		9.15.1.1.2 Physical properties of $Mn_n(IV)_{2n-m}$ compounds	65	381 ff.
		9.15.1.1.3 Doped and ternary Mn_nSi_{2n-m} phases	68	385 f.
		9.15.1.1.4 Physical properties of $Ru_2(IV)_3$ compounds	69	–
		9.15.1.1.5 References for 9.15.1.1	70	–
	9.15.1.2 $T(IV)_2$ compounds		72	–
		9.15.1.2.0 Structures of the TSi_2 phases	72	–
		9.15.1.2.1 Physical properties of $T(IV)_2$ compounds	73	386 ff.
		9.15.1.2.2 References for 9.15.1.2	78	–
	9.15.1.3 $T(V)_2$ compounds		80	–
		9.15.1.3.0 Structure, chemical bond	80	398 f.
		9.15.1.3.1 Physical properties of $T(V)_2$ compounds	85	398 ff.
		9.15.1.3.2 References for 9.15.1.3	106	–

				Data	Figures
	9.15.1.4	$T(V)_3$ compounds		109	–
		9.15.1.4.0	Structure, chemical bond	109	–
		9.15.1.4.1	Physical properties	110	411f.
		9.15.1.4.2	References for 9.15.1.4	112	–
	9.15.1.5	TP_4 compounds		113	–
		9.15.1.5.0	Structure, chemical bond	113	–
		9.15.1.5.1	Physical properties	114	–
		9.15.1.5.2	References for 9.15.1.5	117	–
	9.15.1.6	T-V-VI compounds		117	–
		9.15.1.6.0	Structure, chemical bond	117	–
		9.15.1.6.1	Physical properties	121	–
		9.15.1.6.2	References for 9.15.1.6	128	–
9.15.2	Binary transition-metal oxides (J.B. GOODENOUGH, A. HAMNETT)			129	–
	9.15.2.0	Introduction		129	–
		9.15.2.0.1	Considerations unique to transition-metal compounds	129	–
		9.15.2.0.2	Placement of d bands and formal valence	129	–
		9.15.2.0.3	Character of d electrons	129	–
	9.15.2.1	Titanium oxides		133	–
		9.15.2.1.0	Introduction	133	413
		9.15.2.1.1	Titanium oxide (TiO_2)	133	413ff.
			References for 9.15.2.1.0 and 9.15.2.1.1	148	–
		9.15.2.1.2	Titanium oxide (Ti_2O_3)	151	432ff.
			References for 9.15.2.1.2	154	–
		9.15.2.1.3	Phases between Ti_2O_3 and TiO_2	155	438ff.
			References for 9.15.2.1.3	166	–
	9.15.2.2	Vanadium oxides		167	–
		9.15.2.2.0	Introduction	167	446
		9.15.2.2.1	The shear phases V_nO_{2n-1}	167	447ff.
			References for 9.15.2.2.0 and 9.15.2.2.1	172	–
		9.15.2.2.2	The phases V_nO_{2n+1}	173	455f.
			References for 9.15.2.2.2	175	–
		9.15.2.2.3	Vanadium oxide (V_2O_3)	175	457ff.
			References for 9.15.2.2.3	183	–
		9.15.2.2.4	Vanadium oxide (VO_2)	185	415, 467ff.
			References for 9.15.2.2.4	191	–
		9.15.2.2.5	Vanadium oxide (V_2O_5)	194	486ff.
			References for 9.15.2.2.5	200	–
	9.15.2.3	Monoxides		201	–
		9.15.2.3.1	Manganese oxide (MnO)	201	492ff., 555
			References for 9.15.2.3.1	207	–
		9.15.2.3.2	Iron oxide (FeO and $Fe_{1-x}O$)	208	502ff.
			References for 9.15.2.3.2	211	–
		9.15.2.3.3	Cobalt oxide (CoO)	213	496, 500, 508ff., 555
			References for 9.15.2.3.3	217	
		9.15.2.3.4	Nickel oxide (NiO)	219	517ff., 555
			References for 9.15.2.3.4	225	–
		9.15.2.3.5	Palladium oxide (PdO)	227	529f.
			References for 9.15.2.3.5	228	–
	9.15.2.4	Spinel oxides		229	–
		9.15.2.4.1	Magnetite (Fe_3O_4)	229	530ff., 556
			References for 9.15.2.4.1	234	–
		9.15.2.4.2	Cobalt oxide (Co_3O_4)	236	544ff.
			References for 9.15.2.4.2	237	–

				Data	Figures
		9.15.2.4.3	Manganese oxide (Mn_3O_4)	238	492, 546f.
			References for 9.15.2.4.3	241	–
	9.15.2.5	Sesquioxides and related oxides		242	
		9.15.2.5.1	Oxides of chromium	242	548ff.
			References for 9.15.2.5.1	246	–
		9.15.2.5.2	Hematite (α-Fe_2O_3)	247	552ff.
			References for 9.15.2.5.2	253	–
		9.15.2.5.3	Oxides of rhodium	254	558
			References for 9.15.2.5.3	256	–
		9.15.2.5.4	Oxides of manganese	256	559ff.
			References for 9.15.2.5.4	259	–
	9.15.2.6	Oxides of group V elements (Nb, Ta)		260	
		9.15.2.6.1	Oxides of niobium	260	562ff.
			References for 9.15.2.6.1	273	–
		9.15.2.6.2	Oxides of tantalum (Ta_2O_5)	274	573ff.
			References for 9.15.2.6.2	277	–
	9.15.2.7	Oxides of group VI elements (Mo, W)		277	
		9.15.2.7.1	Molybdenum oxides	277	577ff.
			References for 9.15.2.7.1	281	–
		9.15.2.7.2	Tungsten oxides	282	581ff.
			References for 9.15.2.7.2	289	–
	9.15.3	Binary transition-metal chalcogenides (J.B. GOODENOUGH, S.K. RAMASESHA)		291	–
		9.15.3.1	M_{IV}(=Ti, Zr, Hf)-chalcogenides	291	595ff.
		9.15.3.2	M_V(=V, Nb, Ta)-chalcogenides	295	599f.
		9.15.3.3	M_{VI}(=Cr, Mo, W)-chalcogenides	296	601ff.
		9.15.3.4	M_{VII}(=Mn, Tc, Re)-chalcogenides	301	610f.
		9.15.3.5	M_{VIIIA}(=Fe, Ru, Os)-chalcogenides	305	612ff.
		9.15.3.6	M_{VIIIB}(=Co, Rh, Ir)-chalcogenides	308	–
		9.15.3.7	M_{VIII}(=Ni, Pd, Pt)-chalcogenides	309	615ff.
		9.15.3.8	References for 9.15.3	312	–
9.16	Binary rare earth compounds (G. HUBER, M. LEISS)			317	–
	9.16.1	Hydrides RH_x		317	621ff.
	9.16.2	Borides RB_6		319	628f.
	9.16.3	Monochalcogenides RX		319	629ff.
	9.16.4	Chalcogenides R_xX_y with $x<y$		328	650ff.
	9.16.5	References for 9.16		338	–

A. Introduction

1. List of symbols

In the following list frequently used symbols are specified. The references in the last column refer to the introductory part A in subvolume 17a (cited as 17a/followed by the number for the respective section) or to that section of part B of this subvolume where the quantity is defined or introduced the first time (cited as 17g/followed by the number of the respective section). The units listed in the second-last column are the most frequently used units. In the tables of part B data are generally given in the units of the original paper. To facilitate a conversion from CGS units to SI units or vice versa conversion tables are presented in section 3 below.

Symbol	Property	Unit	Introduced in section
a, b, c	lattice parameters	Å, pm	
a_{O_2}	oxygen activity	–	17g/9.15.2
A	absorbance	–	17g/9.15.2
b	electron-hole mobility ratio (μ_n/μ_p)		
$B(B_S, B_T)$	bulk modulus (adiabatic, isothermal)	bar	17a/2.1.2, eq. (A.17)
B	Nernst coefficient	$cm^2\,K^{-1}\,s^{-1}$	17a/2.2.1
$\mathbf{B}$	magnetic induction	T, G	
B_i	internal magnetic field	T	17g/9.16
c_{lm}	elastic moduli (stiffnesses)	$dyn\,cm^{-2}$	17a/2.1.2, eq. (A.19)
C_A	Curie constant per g-atom	$cm^3\,K\,(g\text{-atom})^{-1}$	17g/9.15.3
C_m	molar Curie constant	$cm^3\,K\,mol^{-1}$	17g/9.14
C_p, C_v	heat capacities	$J\,mol^{-1}\,K^{-1}$	
d	distance, bond length	cm, Å	
d	thickness of a sample	cm	
d	density	$g\,cm^{-3}$	
d_X	X-ray density	$g\,cm^{-3}$	
d_{opt}	optical density ($\log I_0/I$)	–	
$D_{n(p)}$	diffusion coefficient for electrons (holes)	$cm^2\,s^{-1}$	
Dq	crystal field splitting parameter	eV	17g/9.16
e	elementary charge	C	
$e^*_{(T)}$	(transverse) effective ionic charge	e	17a/2.1.2
$\mathbf{e}$	polarization vector	–	
$\mathbf{E}$	electric field strength	$V\,cm^{-1}$	
E	energy	eV	
$E_{0,1,2...}$	energies of critical points in optical spectra	eV	17a/2.1.1
$E(\Gamma_6)...$	energy of band edge of type Γ_6 ...	eV	17a/2.4.1
$E_{[hkl]}$	Young's modulus (measured in [hkl] direction)	$dyn\,cm^{-2}$	17a/2.4.2, eq. (A.100)
$E_{a(d)}$	energy of acceptor (donor) state measured from the respective band edge	eV	17a/2.3.1
E_A	activation energy (of conductivity or other temperature or pressure dependent properties)	eV	
E_b	binding energy	eV	
$E_{c(v)}$	band edge of conduction (valence) band	eV	17a/2.1.1, eq. (A.1)
E_F	Fermi energy	eV	
E_g	energy gap	eV	17a/2.1.1
$E_{g,th}$	energy gap extrapolated to 0 K (thermal energy gap)	eV	17a/2.1.1
$E_{g,dir(ind)}$	direct (indirect) energy gap	eV	17a/2.2.1
E_{kin}	kinetic energy	eV	
$E_{pl(pc)}, E_{peak}$	photoluminescence (photoconductivity) peak energy	eV	

Introduction: 1. List of symbols

Symbol	Property	Unit	Introduced in section
E_t	energy of trap level	eV	
E_{thr}	ionization energy (photothreshold)	eV	17g/9.16
f	frequency	Hz	
f	oscillator strength	–	17g/9.16
$g(E)$	density of states	$cm^{-3} eV^{-1}$	17a/2.1.1, eq. (A.8)
$g_{c(v)}$	density of conduction (valence) band states	$cm^{-3} eV^{-1}$	
g_c, g_n	g-factor of conduction electrons	–	
g_v, g_p	g-factor of holes (valence band)	–	
g_V	Gibbs energy of formation of a neutral cation vacancy	eV	17g/9.15.2
$G_{[hkl]}$	torsional (shear) modulus in [hkl] direction	$dyn\,cm^{-2}$	
ΔG_f^0	standard free energy of formation	$J\,mol^{-1}$	
ΔG_{tr}	Gibbs energy of (phase) transition	$J\,mol^{-1}$	
$H_{(B,K,V)}$	hardness (Brinell, Knoop, Vickers)	$kg\,mm^{-2}$	
$\mathbf{H}$	magnetic field strength	$A\,cm^{-1}$, Oe	
H_A	applied magnetic field strength	A/m	
H_{ext}	external magnetic field strength	A/m	
$\Delta H_{(p-p)}$	(peak to peak) linewidth	A/m	
ΔH_f^0	standard heat of formation	$J\,mol^{-1}$	
ΔH_{tr}	(phase) transformation heat	$J\,mol^{-1}$	
i	current density	$A\,cm^{-2}$	
$I_{(lum, R)}$	intensity (of luminescence, Raman intensity)	$cm^{-2}\,s^{-1}$	
I_{ph}	photocurrent	A	
J	total angular momentum quantum number	–	
J_i	exchange energy (J/k in K)	eV	17g/9.15.2
k	extinction coefficient (absorption index)	–	17a/2.2.2, eq. (A.41)
k	Boltzmann constant	$J\,K^{-1}$	
$\mathbf{k}$	wave vector of electrons	cm^{-1}	17a/2.1.1
K	absorption coefficient	cm^{-1}	17a/2.2.2, eq. (A.38)
K	equilibrium constant (for special meaning, see respective subsection)		17g/9.15.2
K_B	Knight shift	–	
L, l	length	cm	
L	symmetry point in the Brillouin zone	–	17a/2.4, Fig. A.4
$\Delta l/l$	linear thermal elongation	–	
m_0	electron mass	g	
m^*	effective mass	m_0	17a/2.1.1
m^{**}	polaronic mass	m_0	17a/2.1.1, eq. (A.10)
$m_{n(p)}$	effective mass of electrons (holes)	m_0	17a/2.1.1, eq. (A.1)
$m_\parallel, m_l$	longitudinal effective mass	m_0	17a/2.4.1, eq. (A.75)
$m_\perp, m_t$	transverse effective mass	m_0	17a/2.4.1, eq. (A.75)
m_{ω_p}	effective plasma frequency mass	m_0	17a/2.2.2, eq. (A.50)
m_{ω_c}	effective cyclotron resonance mass	m_0	17a/2.2.2, eq. (A.57)
$M_{(0)}$	(saturation) magnetization per unit volume	A/m, Oe	
n	(real) refractive index	–	17a/2.2.2, eq. (A.37)
$n_{a,b,c}$	refractive index in a, b, c direction	–	
Δn	birefringence ($n_\parallel - n_\perp$)	–	
n	electron concentration (also carrier concentration in general)	cm^{-3}	
n_i	intrinsic carrier concentration	cm^{-3}	17a/2.1.3, eq. (A.20)
$n_{a(d)}$	acceptor (donor) concentration	cm^{-3}	
n_H	carrier concentration determined by the Hall effect	cm^{-3}	17g/9.15.2
n_t	trap concentration	cm^{-3}	
n_{eff}, N_{eff}	effective number of electrons contributing to optical properties	–	17a/2.2.2, eq. (A.60)

Introduction: 1. List of symbols

Symbol	Property	Unit	Introduced in section
N	count rate	–	
p	pressure	bar	
p_{tr}	(phase) transition pressure	bar	
p	hole concentration	cm^{-3}	
p_A	magnetic moment per atom (ion)	μ_B	
$p_{A,s}$	saturation magnetic moment per atom (ion)	μ_B	
p_{eff}	effective (paramagnetic) magnetic moment	μ_B	
p_m	magnetic moment per formula unit	μ_B	
P	spontaneous polarization	$C\,m^{-2}$	17g/9.15.2
q	wave vector of phonons	cm^{-1}	
r, R	radius, distance	Å	
R	resistance	Ω	
R	reflectance, reflectivity	–	17a/2.2.2, eq. (A.42)
R_H	normal Hall coefficient	$cm^3\,C^{-1}$	17a/2.2.1
R_s	anomalous Hall coefficient	$cm^3\,C^{-1}$	17g/9.14
S	spin quantum number	–	
S	Seebeck coefficient, thermoelectric power	$V\,K^{-1}$	17a/2.2.1, eq. (A.42)
ΔS_f^0	standard entropy of formation	$J\,mol^{-1}\,K^{-1}$	
ΔS_{tr}	entropy of phase transition	$J\,mol^{-1}\,K^{-1}$	
t	time	s	
t_i	transport numbers (t_+, t_-: for cations, anions)	–	17g/9.15.2
T	transmission	–	
T	temperature	K, °C	
T_C	Curie temperature	K	
T_{dec}	decomposition temperature	K	
T_m	melting temperature	K	
T_N	Néel temperature	K	
T_{perit}	peritectic (decomposition) temperature	K	
T_S	spin flip temperature	K	17g/9.15.2
T_{tr}	transition temperature	K	
T_V	Verwey temperature	K	17g/9.15.2
T_α	critical temperature for cluster formation (see section on FeS)	K	17g/9.15.2
u	position parameter	–	
U	voltage	V	
U	Hubbard-self energy (sometimes electrostatic correlation energy)	eV	17g/9.15.2
U_H	Hall voltage	V	
v_b	bulk velocity	$cm\,s^{-1}$	17g/9.14
v_s	shear sound velocity	$cm\,s^{-1}$	17g/9.14
$v_{l(t)}, v_{L(T)}$	velocity of longitudinal (transverse) waves	$cm\,s^{-1}$	17a/2.1.2
$V_{(m)}$	(molar) volume	$cm^3\,(mol^{-1})$	
W	width of valence or conduction bands	eV	
x, y, z	fractional coordinates of atoms in unit cell	–	
X	symmetry point in the Brillouin zone	–	17a/2.4.1, Fig. A.4
X_{ik}	stress tensor (6×6) (in the literature often labeled T_{ij})	bar	17a/2.1.2
X_k	stress vector (6-component)	bar	17a/2.1.2
Y	photoyield	–	17g/9.16
Z	coordination number	–	
$\alpha(\alpha_F)$	Fröhlich polaron coupling constant	–	17a/2.1.1, eq. (A.11)
α	linear thermal expansion coefficient	K^{-1}	17a/2.1.2, eq. (A.17)
$\alpha_{a,b,c}$	linear thermal expansion coefficient in a, b, c direction	K^{-1}	

Introduction: 1. List of symbols

Symbol	Property	Unit	Introduced in section
β	volume thermal expansion coefficient (3α)	K^{-1}	17a/2.1.2, eq. (A.17)
γ	Grüneisen constant	–	17a/2.1.2, eq. (A.18)
Γ	linewidth (e.g. of phonon wavenumber)	cm^{-1}	
Γ	center of Brillouin zone	–	17a/2.4.1, Fig. A.4
Δ	[100]-axis in k-space	–	17a/2.4.1, Fig. A.4
Δ_{cf}	crystal field splitting energy	eV	17g/9.15.2
Δ_{el}	electronic stabilization energy	eV	17g/9.15.2
Δ_{ex}	exchange splitting parameter	eV	17g/9.15.1
$\tan \delta$	dielectric loss tangent ($=\varepsilon_2/\varepsilon_1$)	–	
ε_0	permittivity of free space	$F\,cm^{-1}$	
ε	dielectric constant	–	
$\varepsilon_{1(2)}$	real (imaginary) part of dielectric constant	–	17a/2.2.2, eq. (A.39)
$\varepsilon(0)$	low frequency dielectric constant	–	17a/2.2.2, eq. (A.53)
$\varepsilon(\infty)$	high frequency dielectric constant	–	17a/2.2.2, eq. (A.53)
ζ	reduced wave vector coordinate	–	
Θ_a	asymptotic Curie temperature	K	
Θ_p	paramagnetic Curie temperature	K	
Θ_D	Debye temperature	K	17a/2.1.2
$\kappa(\kappa_{L,\,el})$	thermal conductivity (lattice, electronic contribution)	$W\,cm^{-1}\,K^{-1}$	17a/2.2.1, eq. (A.36)
κ	compressibility ($=1/$bulk modulus)	$cm^2\,dyn^{-1}$	
$\kappa_{v(l)}$	volume (linear) compressibility	$cm^2\,dyn^{-1}$	
λ	wavelength	cm	
λ_L	wavelength of laser light	cm	
Λ	phonon mean free path	cm	17g/9.14
$\mu_{n(p)}$	electron (hole) mobility	$cm^2\,V^{-1}\,s^{-1}$	17a/2.2.1, eq. (A.25)
μ_{dr}	drift mobility	$cm^2\,V^{-1}\,s^{-1}$	17a/2.2.1
μ_H	Hall mobility	$cm^2\,V^{-1}\,s^{-1}$	17a/2.2.1, eq. (A.27)
$\mu_{a,b,c}$	mobility in a, b, c direction	$cm^2\,V^{-1}\,s^{-1}$	
ν	Poisson's ratio	–	17a/2.4.2, eq. (A.100)
ν	frequency	s^{-1}	
$\bar{\nu}$	wavenumber	cm^{-1}	
$\bar{\nu}_p$	plasma wavenumber	cm^{-1}	
ν_R	Raman wavenumber	cm^{-1}	
π_{ik}	piezoresistance coefficients	$cm^2\,dyn^{-1}$	17a/2.1.1, eq. (A.32)
ϱ	resistivity	$\Omega\,cm$	
$\varrho_{a,b,c}$	resistivity in a, b, c direction	$\Omega\,cm$	
$\Delta\varrho/\varrho_0$	magnetoresistance	–	17a/2.2.1, eq. (A.30)
$\sigma_{(i)}$	(intrinsic) conductivity	$\Omega^{-1}\,cm^{-1}$	17a/2.2.1
$\sigma_{n(p)}$	conductivity of electrons (holes)	$\Omega^{-1}\,cm^{-1}$	
$\sigma_{a,b,c}$	conductivity in a, b, c direction	$\Omega^{-1}\,cm^{-1}$	
σ_d	dark conductivity	$\Omega^{-1}\,cm^{-1}$	
σ_{ph}	photoconductivity	$\Omega^{-1}\,cm^{-1}$	
$\tau_{(n,p)}$	relaxation time, decay time, rise time, lifetime of carriers	s	
Φ	work function	eV	
χ_v	magnetic volume susceptibility	–	
χ_g, χ	magnetic mass susceptibility	$cm^3\,g^{-1}$	
χ_m	magnetic molar susceptibility	$cm^3\,mol^{-1}$	
ω	circular frequency	$rad\,s^{-1}$	
$\hbar\omega$	photon energy	eV	
ω_p	plasma resonance frequency	s^{-1}	17a/2.2.2, eq. (A.49)

2. List of abbreviations

a	acceptor
ac	alternating current
arb.	arbitrary
av	average
AF	antiferromagnetically ordered spin system
APB	antiphase boundary
APW	augmented plane wave method
bcc	body centered cubic
bct	base centered tetragonal
BZ	Brillouin zone
calc	calculated
cub	cubic
crit	critical
CB	conduction band
CCDW	commensurate charge density wave
CS	crystallographic shear (plane)
CVD	chemical vapor deposition
d	donor
dc	direct current
dir	direct
DOS	density of states
DTA	differential thermal analysis
e	electron
eff	effective
epr, EPR	electron paramagnetic resonance
esr, ESR	electron spin resonance
exp	experimental
EDC	energy distribution curve
EELS, ELS	electron energy loss spectroscopy
EMF	electromotive force
ESCA	electron spectroscopy for chemical analysis
fcc	face centered cubic
fir, FIR	far infrared
F	ferromagnetic ordered spin system
h	hole
hex	hexagonal
HT	high temperature
i	intrinsic; sometimes used for interstitial
ind	indirect
ion	ionic, ionization
ir, IR	infrared
I	insulator
ICDW	incommensurate charge density wave
l, L	longitudinal (sub- or superscript); sometimes abbreviation for "laser"
l, L, liq	liquid
LA	longitudinal acoustic
LO	longitudinal optical
LCAO	linear combination of atomic orbitals
Ln	lanthanides
LT	low temperature
m	monoclinic (mostly subscript)
magn	magnetic
max	maximum
min	minimum
M	metal
$Mn_{Mn}^{\cdot}$	Mn ion on Mn site, positively charged
MO	molecular orbital

MSC	magnetically stimulated current
NMR	nuclear magnetic resonance
obs	observed
oct	octahedral
opt	optical
O_d	oxygen deficiency (concentration of deficient oxygen atoms)
OPW	orthogonalized plane wave method
ph	photo-, phonon- (subscript)
P	paramagnetic
PE	photoelectron spectrum
rel	relative
res	resolution
rh	rhombohedral
R	rutile (mostly subscript)
RE, R	rare earth
RPA	random phase approximation
RT	room temperature
s	surface
SICF	single ion in a crystal field
SXS	soft X-ray spectrum
tetr	tetragonal
th	thermal; sometimes for theoretical
tot	total
tr	transition (subscript for phase transition parameters)
T	transition metal
TA	transverse acoustic
TO	transverse optical
TIP	temperature independent paramagnet
UPS, UPE	uv photoelectron spectroscopy
uv, UV	ultraviolet
vac	vacuum, sometimes abbreviation for vacancy
V_{Ga}	vacancy on Ga site
V'_{Mn}	Mn vacancy, negatively charged
VB	valence band
VT	vapor transport
X	anion (e.g. S, Se, Te)
XPS, XPE	X-ray photoelectron spectroscopy
XRD	X-ray diffraction
$\perp, \parallel$	perpendicular, parallel to a crystallographic axis

3. Conversion tables

A. Conversion factors from the SIU system to the CGS-esu and the CGS-emu systems.

Quantities	Symbols	SIU	CGS-esu	CGS-emu
			(non-rationalized)	
bulk modulus	B	$Pa (= N m^{-2})$	$10\ dyn\ cm^{-2}$	$10\ dyn\ cm^{-2}$
magnetic induction	B	$T (= Wb\ m^{-2})$ $= V s m^{-2}$	$10^{-6}/3$ esu	10^4 G
molar heat capacity at const. pressure	C_p	$J K^{-1} mol^{-1}$	$10^7\ erg\ K^{-1} mol^{-1}$ $(= 0.239\ cal\ K^{-1} mol^{-1})$	$10^7\ erg\ K^{-1} mol^{-1}$ $(= 0.239\ cal\ K^{-1} mol^{-1})$
elastic moduli (stiffnesses)	c_{lm}	$N m^{-2} (= Pa)$	$10\ dyn\ cm^{-2}$	$10\ dyn\ cm^{-2}$
density	d	$kg m^{-3}$	$10^{-3}\ g\ cm^{-3}$	$10^{-3}\ g\ cm^{-3}$
piezoelectric strain coefficient	d_{ik}	$C N^{-1}$ $(= m V^{-1})$	$3 \cdot 10^4$ esu	10^{-6} emu

(continued)

Introduction: 3. Conversion tables

Quantities	Symbols	SIU	CGS-esu	CGS-emu
			(non-rationalized)	
Conversion factors (continued)				
strain tensor	e_{ik}	$m \cdot m^{-1}$	$1\ cm \cdot cm^{-1}$	$1\ cm \cdot cm^{-1}$
piezoelectric stress coefficients	e_{ik}	$C\,m^{-2}$	$3 \cdot 10^5$ esu	10^{-5} emu
Young's modulus	E	$N\,m^{-2}(=Pa)$	$10\ dyn\,cm^{-2}$	$10\ dyn\,cm^{-2}$
electric field strength	E	$V\,m^{-1}$	$10^{-4}/3$ esu	10^6 emu
piezoelectric strain coefficients	g_{ik}	$m^2\,C^{-1}$	$10^{-5}/3$ esu	10^5 emu
molar free energy change	ΔG	$J\,mol^{-1}$	$10^7\ erg\,mol^{-1}$ $(=0.239\ cal\,mol^{-1})$	$10^7\ erg\,mol^{-1}$ $(=0.239\ cal\,mol^{-1})$
piezoelectric stress coefficient	h_{ik}	$V\,m^{-1}(=N\,C^{-1})$	$10^{-4}/3$ esu	10^6 emu
hardness	H	$N\,m^{-2}(=Pa)$	$10\ dyn\,cm^{-2}$	$10\ dyn\,cm^{-2}$
magnetic field strength	H	$A\,m^{-1}$	$12\pi \cdot 10^7$ esu	$4\pi \cdot 10^{-3}$ Oe
molar enthalpy change	ΔH	$J\,mol^{-1}$	$10^7\ erg\,mol^{-1}$ $(=0.239\ cal\,mol^{-1})$	$10^7\ erg\,mol^{-1}$ $(=0.239\ cal\,mol^{-1})$
current density	i	$A\,m^{-2}$	$3 \cdot 10^5$ esu	10^{-5} emu
elastoresistance coefficients	m_{ik}	dimensionless	1 (dimensionless)	1 (dimensionless)
pressure	p	$Pa(=10^{-5}\ bar)$	$10\ dyn\,cm^{-2}$ $(=1.019 \cdot 10^{-5}\ kg\,cm^{-2}$ $=7.5 \cdot 10^{-3}\ Torr)$	$10\ dyn\,cm^{-2}$ $(=1.019 \cdot 10^{-5}\ kg\,cm^{-2}$ $=7.5 \cdot 10^{-3}\ Torr)$
dielectric polarization	P	$C\,m^{-2}$	$3 \cdot 10^5$ esu	10^{-5} emu
pyroelectric coefficient	p_i	$C\,m^{-2}\,K^{-1}$	$3 \cdot 10^5$ esu K^{-1}	10^{-5} emu K^{-1}
elastooptic constant (in cubic crystals)	p_{ij}	dimensionless	1 (dimensionless)	1 (dimensionless)
piezooptic constant (in cubic crystals)	q_{ij}	dimensionless	1 (dimensionless)	1 (dimensionless)
Hall coefficient	R_H	$m^3\,C^{-1}$	$10^{-3}/3$ esu	10^7 emu
linear electrooptic constant	r_{ij}	$m\,V^{-1}$	$3 \cdot 10^4$ esu	10^{-6} emu
elastic compliances	s_{ml}	$m^2\,N^{-1}$	$10^{-1}\ cm^2\,dyn^{-1}$	$10^{-1}\ cm^2\,dyn^{-1}$
molar entropy change	ΔS	$J\,K^{-1}\,mol^{-1}$	$10^7\ erg\,K^{-1}\,mol^{-1}$ $(=0.239\ cal\,mol^{-1})$	$10^7\ erg\,K^{-1}\,mol^{-1}$ $(=0.239\ cal\,mol^{-1})$
stress tensor	X_{ij}	$N\,m^{-2}$	$10\ dyn\,cm^{-2}$	$10\ dyn\,cm^{-2}$
thermal conductivity	κ	$W\,m^{-1}\,K^{-1}$ $(=J\,m^{-1}\,s^{-1}\,K^{-1})$	$10^5\ erg\,cm^{-1}\,s^{-1}\,K^{-1}$	$10^5\ erg\,cm^{-1}\,s^{-1}\,K^{-1}$
dielectric constant	ε	dimensionless	1 (dimensionless)	1 (dimensionless)
piezoresistance tensor coefficients	π_{ik}	$m^2\,N^{-1}$	$10^{-1}\ cm^2\,dyn^{-1}$	$10^{-1}\ cm^2\,dyn^{-1}$
piezooptic constant (in cubic crystals)	π_{ik}	$m^2\,N^{-1}$	$10^{-1}\ cm^2\,dyn^{-1}$	$10^{-1}\ cm^2\,dyn^{-1}$
resistivity	ϱ	$\Omega\,m$	$10^{-9}/9$ esu	10^{11} emu
conductivity	σ	$\Omega^{-1}\,m^{-1}$	$9 \cdot 10^9$ esu	10^{-11} emu
magnetic volume susceptibility	χ_v	dimensionless	$\frac{1}{4\pi}$ dimensionless	$\frac{1}{4\pi}$ dimensionless
magnetic mass susceptibility	χ_g	$m^3\,kg^{-1}$	$10^3/4\pi$ esu $cm^3\,g^{-1}$	$10^3/4\pi$ emu g^{-1} $(=10^3/4\pi\ cm^3\,g^{-1})$
magnetic molar susceptibility	χ_m	$m^3\,mol^{-1}$	$10^6/4\pi$ esu $cm^3\,mol^{-1}$	$10^6/4\pi$ emu mol^{-1} $(=10^6/4\pi\ cm^3\,mol^{-1})$

Lattice parameters a, b, c are mostly given in Å; $1\,\text{Å} = (1 \pm 5 \cdot 10^{-7}) \cdot 10^{-10}$ m.

B. Energy conversion

Energy: $E = eV = h\nu = hc\bar{\nu}$ $1\,\text{VAs} = 1\,\text{J} = 10^7\,\text{erg} = 2.38845 \cdot 10^{-4}\,\text{kcal}$

Energy and equivalent quantities	E	V	ν	$\bar{\nu}$
	J	V	Hz, s^{-1}	cm^{-1}
1 J ≙	1	$6.2415 \cdot 10^{18}$	$1.50916 \cdot 10^{33}$	$5.03403 \cdot 10^{22}$
1 V ≙	$1.60219 \cdot 10^{-19}$	1	$2.41797 \cdot 10^{14}$	$8.06547 \cdot 10^{3}$
1 s^{-1} ≙ (=1 Hz)	$6.62619 \cdot 10^{-34}$	$4.13550 \cdot 10^{-15}$	1	$3.33564 \cdot 10^{-11}$
1 cm^{-1} ≙	$1.98648 \cdot 10^{-23}$	$1.23979 \cdot 10^{-4}$	$2.99792 \cdot 10^{10}$	1

Error: In this volume, experimental errors are given in parentheses referring to the last decimal places. For example, 1.352 (12) stands for 1.352 ± 0.012 and 342.5 (21) stands for 342.5 ± 2.1.

B. Physical data of semiconductors V
9 Non-tetrahedrally bonded binary compounds

(Sections 9.1 ··· 9.6, see subvolume III/17e; sections 9.7 ··· 9.13, see subvolume III/17f)

9.14 Boron compounds

9.14.0 Introduction, general remarks on structure and chemical bond

In order to facilitate the locating of the various boron compounds the systematic arrangement of the binary compounds in this chapter follows consequently the periodic table of elements in the sequence of its groups, in spite of the fact that a classification in terms of boron groups and size of the other atoms in the compounds often appears to be more meaningful with regard to the relationship of their solid state properties. Ternary compounds are only considered, if their structure is immediately related to a semiconducting binary compound and when their semiconducting behavior has been proofed. They are arranged according to the atom of the compound which belongs to the lower group of the periodic table.

Compounds of the following types remain generally unconsidered, since they are treated in other chapters of this book, they are no semiconductors, their semiconducting properties seem to be of low interest, or their semiconducting properties are not essentially influenced by their low boron content:

1. $A_{III}B_V$ compounds (see vol. 17a, p. 148 ff.).
2. Boranes.
3. Borate glasses.
4. Metglasses containing boron.
5. Boron ferrites.

Otherwise completeness has been aimed at. Of course semiconductors are of primary interest, but in many cases an unambiguous separation between semiconducting and metallic boron compounds is not possible for the following reasons:

1. The theoretical considerations and calculations regarding band structures, electronic structures, nature of bonding, atomic coordination and general nature of the formation of boron compounds are quantitatively or even qualitatively uncertain in some cases at present.

2. The conductivity character of the compounds has been determined experimentally not at all or not with sufficient certainty.

3. In many boron compounds impurity atoms decisively influence the conductivity character. Therefore in cases where the purity of the material is not guaranteed a revision of the apparently clarified conductivity character may be useful.

Hence, with the restrictions cited above, all binary boron compounds, whose semiconducting properties are proofed, are considered in detail. For the other compounds the probable conductivity character is mentioned if possible, and references, largely restricted to recent review articles, are given to facilitate the access to further information.

General remarks on structure and chemical bond of boron compounds

The boron compounds and especially the borides show a wide variety of structures and compositions which include single atoms, connected pairs, multiple chains, planar networks and polyhedral groupings which are preferably based on octahedra, icosahedra or even more complex polyhedral structures. This variety is the consequence of the peculiar electronic structure of boron atoms characterized by an unpaired electron in the $2s^2 2p^1$ configuration of the outer electron shell of the isolated boron atom, and leading to the various combinations of s^2p, sp, sp^2, and sp^3 electron configurations by the electron interchange between boron atoms or by the interaction with electrons of other atoms.

From this point of view according to Serebryakova [79S1] the borides can be divided into three classes:

1. Borides formed by elements with s outer electrons and with the deeper electron shells completely occupied or completely unoccupied (borides of the alkaline metals, beryllium, magnesium, alkaline earth metals).

2. Borides formed by elements with s outer electrons and incomplete d or fd subshells (borides of the IV to VIII group elements (the transition metals, the rare earths metals and the actinides).

3. Borides formed by elements with s, p outer electrons (e.g. BN, BP).

Each of these classes of borides can be divided into two groups:

a) Metal-rich borides (compositions like Me_4B, Me_3B, Me_2B, MeB, Me_3B_4).

b) Boron-rich borides (compositions like MeB_2, MeB_6, MeB_{12}, MeB_{41}, $MeB_{66\cdots100}$).

From lower to higher borides the tendency of the formation of direct B–B bonds with covalent bond character increases leading to the change from metallic conduction of lower borides to semiconduction of higher borides. Nevertheless no general limit between both types of solids with regard to the composition can be indicated [77S1, 77S2, 79S1]. On the binding in borides, see also [77S6].

Besides, another type of boron compound has recently become rather important because of the possibility to influence the transport mechanism in a definite way.:

The crystal structures of β-rhombohedral boron (vol. 17e, p.9ff) consists of a complicated three-dimensional boron network based on B_{12} icosahedra and related aggregates. This network contains a number of holes which are large enough to accommodate transition metal atoms. The distribution of the transition metal atoms among the different holes and in some cases substitutionally on B sites depends on the kind of atoms dissolved [70A1, 74A, 76L1, 74L, 76H, 76C, 81W1, 82L].

Physical property	Numerical value	Experimental conditions	Experimental method, remarks	Ref.

9.14.1 Compounds with group Ia elements

9.14.1.1 Boron-hydrogen alloys

energy gap (see also Fig. 1):

E_g	1.10(5) eV	$T=300$ K	9 at% H; ellipsometry (see Fig. 1)	80B
	1.15(5) eV		11 at% H; ellipsometry (see Fig. 1)	
	1.7(1) eV		24 at% H; ellipsometry (see Fig. 1)	

phonon wavenumbers:

$\bar{\nu}$	2560 cm^{-1}	$T=300$ K	B–H stretching mode	80B,
	1108 cm^{-1}		B–H bending mode	79B2
	1900$\cdots$2000 cm^{-1}		hydrogen three-center bridge bond	79T1

dielectric constant: see Fig. 2

9.14.1.2 Boron-lithium compounds

Li$_5$B$_4$:

metallic conduction; crystalline structure: [78W], electronic structure: [79L2]

LiB$_6$:

metallic conduction(?); preparation: [75S, 77N1], crystalline structure: [77N1], CESR (conduction electron spin resonance): [74R]

LiB alloys (0$\cdots$60 at% B):

metallic conduction; preparation: [78M], electrical conductivity: [78M]

solid solution of Li in β-rhombohedral B:

semiconductor (?), preparation: [77N1]

Physical property	Numerical value	Experimental conditions	Experimental method, remarks	Ref.

LiAlB$_{14}$:

Structure

The boron framework contains four B$_{12}$ icosahedra and eight isolated B atoms in the unit cell. The icosahedra are centered at 0,0,0; 1/2,1/2,0; 0,0,1/2; 1/2,1/2,1/2 being oriented with one of their mirror planes parallel to the (100) plane. There are two kinds of intericosahedral B–B bonds. One of them is parallel to the c-axis and the other one parallel to the (110) planes thus leading to infinite chains of B$_{12}$ icosahedra. All icosahedra are crystallographically equivalent, the isolated B atoms as well. Six apical atoms of an icosahedron are linked to those of neighboring icosahedra, the remaining six are linked via the isolated B atoms. The Li and Al atoms are accommodated in the large holes between the icosahedra.

lattice: orthorhombic; space group: D_{2h}^{28} – Imam [81H].

lattice parameters:

a	5.8469 (9) Å	$T=300$ K	Weißenberg and precession method	81H
b	8.1429 (8) Å			
c	10.3542 (6) Å			

Physical properties

energy gap:

$E_{g, ind}$	1.82 eV	$T=300$ K	single crystals (orientation not defined); optical absorption	81W2
	2.05 eV	300 K	maximum of photoconduction	81W2

density:

d	2.46 (2) g cm^{-3}	$T=300$ K		81H

melting point:

T_m	< 1500 °C			81H

microhardness:

H_K	2620 kg mm^{-2}	$T=300$ K	1N load	81W2

9.14.1.3 Boron-sodium compounds

NaB$_6$:

Structure

There is some controversy, whether NaB$_6$ cristallizes in the CaB$_6$ structure or not. Experimental lattice parameters are not reported so far [63H, 74R, 75S, 77N1].

Physical properties

Based on the CaB$_6$ structure the calculated band structure is basically similar in topology to that of CaB$_6$. But because of the lower number of electrons introduced by the monovalent metal the position of the Fermi level is expected at -9.1 eV, i.e. within the valence band [76P1]. This contrasts with the high electrical resistance (see below) which indicates semiconducting behavior.

band structure: Fig. 3

g-factor:

g	2.0073 (4)	$T=295$ K	linewidth, $B=76$ G	74R
Δg	5 (4) · 10^{-3}	295 to 1.4 K	positive temperature independent g-shift $\Delta g = g - 2.0023$	

Physical property	Numerical value	Experimental conditions	Experimental method, remarks	Ref.
resistivity:				
ϱ	10^6 Ω cm	$T=300$ K	sintered sample	77N1
stability limit:				
T	650 °C		high vacuum	77N1
density:				
d	2.12 g cm^{-3}	$T=300$ K	pycnometric	63H, 67N

NaB$_{15}$:

Structure

The elementary cell contains four formula groups. The structure can be described as a stacking of rather compact planes of quasi-spherical icosahedral clusters of boron atoms with a packing diameter of about 5.1 Å (see Fig. 4a). Otherwise it can be described as an alignment of icosahedron chains parallel to the c-axis, which are centered at ($x=0$; $y=0$) and ($x=1/2$; $y=1/2$) (see Fig. 4b). Two types of isolated boron atoms are present in the structure. The covalent skeleton formed by the boron atoms has two types of interstitial sites, one of them formed by 16, the other one by 12 boron atoms. The sodium atoms obtain these interstitial sites and can be assumed as at least partially ionized and to act as electron donors [70N, 76N, 77N1].

lattice: orthorhombic; space group: D_{2h}^{28} – Imam.

lattice parameters:

a	5.847 Å			70N,
b	8.415 Å			76N,
c	10.248 Å			77N1, 77M2, 77N2

Physical properties

energy gap:

E_g	0.32 eV	$T \gtrsim 300$ K	electrical conductivity (see Fig. 5)	77N1

impurity and defects:

Because of the high volatility of sodium, deviations from stoichiometry are possible which lead to Na$_x$B$_{15}$ phases ($0 \leq x < 1$). Limit phase at 850 °C: Na$_{0.3}$B$_{15}$. Above 950 °C the last traces of sodium can be removed [76N, 77N1].

conductivity:

σ	$4 \cdot 10^{-4}$ Ω^{-1} cm^{-1}	$T=300$ K	polycrystalline sample, sintered at 650 °C, pressure 50 kbar; see Fig. 5	77N1

densities:

d	2.428 g cm^{-3}		calculated	77N1
	2.44 (2) g cm^{-3}		NaB$_{0.84}$B$_{14}$, experimental	76N
	2.398 g cm^{-3}		NaB$_{0.84}$B$_{14}$, calculated	

Na$_x$Ba$_{1-x}$B$_6$:

semiconductor (?); preparation [77N1, 52B]; crystalline structure CaB$_6$-type [77N1, 54B]

Na$_x$Th$_{1-x}$B$_6$:

semiconductor (?); preparation [77N1, 52B]; crystalline structure CaB$_6$-type [77N1, 54B]

| Physical property | Numerical value | Experimental conditions | Experimental method, remarks | Ref. |

9.14.1.4 Boron-potassium compounds

KB$_6$:

Structure

KB$_6$ is isostructural with CaB$_6$.
lattice: cubic; space group: O_h^1 – Pm3m.

lattice parameter:

a	4.232 Å	$T = 300$ K	X-ray diffraction	66N, 77N1

Physical properties

electronic structure:

The calculated band structure is basically similar in topology to that of CaB$_6$ with the empty d orbitals of potassium forming a narrow set of bands in the accessible conduction region. The band gap (at X) is reduced by the d-orbital participation. The Fermi level (at -8.1 eV) is very close to the lower edge of the energy gap; only a few holes are left vacant [76P1, 77P1].

energy gap:

E_g	0.29 eV	$T \gtrsim 300$ K	electrical conductivity (see Fig. 6)	77N1

impurity and defects:

Because of the high volatility of potassium deviations from stoichiometry are possible. It is assumed that beyond a limiting value the B$_6$-sublattice is no longer electronically stable and collapses to amorphous boron. Thus a mixture of amorphous boron and potassium deficient hexaboride is expected [77N1].

conductivity:

σ	$10^{-4}\,\Omega^{-1}\,\text{cm}^{-1}$	$T = 300$ K	For temperature dependence, see Fig. 6	77N1

activation energy:

E_A	0.008 eV	$T < 120$ K	electrical conductivity	77N1

magnetic susceptibility: see Fig. 7

density:

d	2.99 g cm^{-3}	$T = 300$ K	pycnometric	66N, 75S

9.14.2 Compounds with group Ib elements

The ability of the group Ib elements to form real boron compounds remains questioned. Only in the case of Cu the existence of solid solutions has been proofed. The existence of some compounds containing Ag and Au is discussed controversely (see e.g. [77G3, 60S] and references therein).

Solid solutions of Cu in β-rhombohedral boron:

Structure

The crystal structure of β-rhombohedral boron contains a number of holes which are large enough to accommodate transition metal atoms [70A1, 73A1, 74A, 74H2, 74L, 76L1, 76H, 82L]. The Cu atoms occupy a maximum of four non-equivalent crystallographic positions (Fig. 8), by which the unit cell is expanded (Fig. 9) [76C].

limits of solubility [74P1]:

1700 °C: CuB$_{26.4}$
1500 °C: CuB$_{27.8}$
1300 °C: CuB$_{30.5}$

positions of the Cu atoms in the unit cell [76C]:

Atom	Position	Coordinates		
		x	y	z
Cu_1	6 (c)	0	0	0.13654 (13)
$Cu_{2'}$	18 (h)	0.19638 (92)	0.39277 (92)	0.17791 (45)
$Cu_{2''}$	36 (i)	0.2043 (20)	0.4317 (18)	0.16996 (42)
Cu_3	6 (c)	0	0	0.22554 (8)

Physical properties

energy gap:

E_g	0.54 eV	$T = 300 \cdots 1000$ K CuB_{24}	electrical conductivity (see Fig. 10),	76G1, 79G2

conductivity:

p-type conductivity for $CuB_{42.18}$ [81W3]

σ	$10^{-2} \Omega^{-1} cm^{-1}$	$T = 300$ K CuB_{24}	For temperature dependence, see Figs. 10 for CuB_{24} and 11 for $CuB_{42.18}$	76G1, 79G2, 81W3

9.14.3 Compounds with group IIa elements

9.14.3.1 Boron-beryllium compounds

Be_5B:

metallic conduction; preparation [75S, 63B]; crystalline structure [63B], electrical conductivity [75S]

Be_2B:

metallic conduction; preparation [75S]; crystalline structure [63B]; electrical conductivity [75S]

BeB_2:

semimetallic behavior (?); preparation [75S, 61S]; electronic structure [77P1, 77S2]; electrical conductivity [75S]

BeB_3:

preparation [61S, 76M1]; crystalline structure [76M1]

BeB_4:
Structure:

lattice: tetragonal [75S], similarity to UB_4 assumed.
No lattice parameters available.

Physical properties
resistivity:

ϱ	$1.1 \cdot 10^7 \Omega$ cm	$T = 300$ K	polycrystalline material	75S

density:

d	2.57 g cm^{-3}	$T = 300$ K	pycnometric	75S

melting point:

T_m	> 2000 °C			75S

BeB_6:
Structure

lattice: tetragonal (related to AlB_{12}) [61H; 61S; 77M2].

Physical property	Numerical value	Experimental conditions	Experimental method, remarks	Ref.
lattice parameters:				
a	10.16 Å	$T = 300$ K	X-ray diffraction	61H
c	14.28 Å			

Physical properties

For electronic structure, cp. [77E, 77P1].

resistivity:				
ϱ	$1.1 \cdot 10^7$ Ω cm	$T = 300$ K	polycrystalline material	75S
density:				
d	2.33 g cm^{-3}	$T = 300$ K	X-ray	75S
	2.35 g cm^{-3}	300 K	pycnometric	75S
melting point:				
T_m	> 2000 °C			75S

BeB$_9$:

semiconductor; preparation [61S, 75S]; electrical conductivity [75S]

BeB$_{12}$:

Structure

The crystal structure is directly related to α-tetragonal boron (tetr. I; see section 8.1, vol. 17e). The unit cell contains four B$_{12}$ icosahedra, two beryllium atoms at the sites of the bisphenoidally coordinated boron atoms (position 2(b)) and two beryllium atoms at the interstitial voids (position 2(a)).

lattice: tetragonal; space group: $D_4^4 - P4_12_12$; 52 atoms per unit cell [77B1, 77M3].

lattice parameters:				
a	8.80 Å	$T = 300$ K		77B1
c	5.08 Å			

Physical properties

electronic band structure: cp. [77P1]
optical transmission: Fig. 12

density:				
d	2.36 g cm^{-3}	$T = 300$ K		77B1
melting point:				
T_m	> 2000 °C			75S

9.14.3.2 Beryllium-aluminum-boron compounds

The compound Al$_{1.0}$Be$_{0.7}$B$_{22}$ [79H2] and the compound (Be, Al)B$_{12}$ with the "believed" composition Be$_{0.1}$Al$_{0.9}$B$_{12}$ [79G1] are subsumed in this section (see also [68B, 75B2, 72K2]).

Structure

The crystal structure is stongly related to that of α-AlB$_{12}$, but the metal distribution is different. The beryllium atoms are accommodated in the two types of truncated tetrahedral holes which are vacant in α-AlB$_{12}$ [79H2].

Physical property	Numerical value	Experimental conditions	Experimental method, remarks	Ref.

position and occupancy of the metal atoms [79H2]:

Atom	Position	Occupancy [%]	
		$Al_{1.0}Be_{0.7}B_{22}$	α-AlB_{12}
Al_1	8 (b)	35.1 (6)	71.7 (7)
Al_2	8 (b)	29.1 (7)	49.1 (3)
Al_3	8 (b)	38.7 (7)	24.0 (6)
Al_4	8 (b)	0	15.0 (3)
Al_5	8 (b)	0	2.1 (5)
Be_1	8 (b)	46 (3)	0
Be_2	4 (a)	42 (6)	0

space group: D_4^4–$P4_12_12$ or D_4^8–$P4_32_12$.

lattice parameters:

a	10.180 (2) Å			79H2
c	14.257 (2)			

Physical properties

energy gaps:

E_g	2.12 eV	$T = 300$ K	optical absorption	79G1
$E_{g,th}$	2.1 eV	$T \gtrsim 800$ K	electrical conductivity (see Fig. 13)	

activation energies:

E_A	0.7 eV	$T = 300 \cdots 500$ K	electrical conductivity (see Fig. 13)	79G1
	0.55 eV	$200 \cdots 300$ K		

trap density:

n_t	$4 \cdot 10^{13}$ cm^{-3}	I–U^2 region	I–U characteristic	79G1

electrical conductivity: Fig. 13, I–U characteristic: Fig. 14
lattice absorption spectrum: Fig. 15

dielectric constants:

$\varepsilon(0)$	18	$T = 300$ K	improbable value (cp. comment to Fig. 15)	79G1
$\varepsilon(\infty)$	8.3	300 K	interferences, platelets $d = 50 \cdots 100$ μm; $\lambda = 2 \cdots 9$ μm	79G1

9.14.3.3 Boron-magnesium compounds

MgB_2:
metallic conduction (?); preparation [75S], crystalline structure [75S, 77C1], electronic structure [77P1, 77C1]

MgB_4:
preparation [75S, 72G, 77N2]; crystalline structure [72G, 77N2]

MgB_6:
semiconductor (?); preparation [75S]; crystalline structure [75S], electronic structure [77P1, 77E]

MgB_{12}:
preparation [75S]; crystalline structure [75S]

Mg_2B_{14}:
semiconductor (?); preparation [81G1, 70M1]; crystalline structure [81G1, 76N, 70M1, 70E1]

Physical property	Numerical value	Experimental conditions	Experimental method, remarks	Ref.

$MgAlB_{14}$:

Structure (Fig. 16)

The structure can be considered as being constituted of distorted, closest-packed layers of quasi-spherical icosahedra stacked directly one above the other in the c-direction [65M1, 70M1, 77M2, 77M3]. Another description is based on chains of B_{12} icosahedra running in the c-direction. In a chain, the icosahedra are orientated in such a way that one of their mirror planes is parallel to the bc-plane, one of their pseudo fivefold axes being alternately inclined 7.88° to the c-axis [76N]. The chains are linked either by direct inter-icosahedral B–B bonds or by bridges involving isolated B atoms. The metal atoms partially occupy interstitial holes in the boron network.

space group: D_{2h}^{28} – Imam.

lattice parameters

a	5.858 Å	$Mg_{0.5}AlB_{14}$	70M1,
b	8.115 Å		76N,
c	10.313 Å		77M2

occupancies and positions of the metal atoms [77M3]:

Atoms	Position	Coordinates			Occupancies
		x	y	z	%
M_1	4 (e)	−0.359	1/4	1.205	Al: 25; Mg 50
M_2	4 (d)	1/4	1/4	1/4	Al: 75

Physical properties

electrical conductivities (polycryst. samples):

σ	$2 \cdot 10^{-1} \, \Omega^{-1} \, cm^{-1}$	$T = 300$ K	pure $MgAlB_{14}$ (see also Fig. 17)	79B1
	$5 \cdot 10^{-2} \, \Omega^{-1} \, cm^{-1}$		Ni-alloyed	79B1

thermoelectric powers (polycryst. samples):

S	260 μV K^{-1}	$T = 400$ K	pure $MgAlB_{14}$ (see also Fig. 18)	79B1
	230 μV K^{-1}		Ni-alloyed	79B1

density:

d	2.585 g cm^{-3}	$Mg_{0.5}AlB_{14}$, calculated	70M1
	2.60 g cm^{-3}	experimental	76N

9.14.3.4 Boron-alkaline earth compounds

Alkaline earth hexaborides CaB_6, SeB_6, BaB_6:

Structure

The CaB_6-type lattice structure, the same for all alkaline earth hexaborides, is characterized by a three-dimensional skeleton of B_6 octahedra, whose interstices are occupied by metal atoms. Hence the crystal has cubic symmetry [66N, 69E, 60M, 75S, 77E]. The covalent boron sublattice is electron deficient and an electron transfer of at least one per atom from the metal sublattice to that of boron is required for stabilization [66N, 54B, 54E, 60L, 77E, 77P1, 79A3, 77S1].

space group: O_h^1 – Pm3m.

lattice structure: Fig. 19

For linear thermal expansion, see below.

lattice parameters (in Å):

CaB_6:

a	4.1520	$T = 300$ K	X-ray diffraction	70E2, 77E
	4.148			75S

(continued)

Physical property	Numerical value	Experimental conditions	Experimental method, remarks	Ref.
lattice parameters (continued)				
SrB_6:				
a	4.1981	$T = 300$ K	X-ray diffraction	70E2, 77E
	4.190			75S
BaB_6:				
a	4.2706	$T = 300$ K	X-ray diffraction	61J, 77E
	4.280			75S

Electronic properties

electron density contours: Fig. 20, band structure and density of states of CaB_6: Fig. 21, ESCA spectrum: Fig. 21 α, Brillouin zone: Fig. 22

energy gaps (in eV):				
CaB_6:				
E_g	2.11		calculated	77P1
	0.3		calculated (see Fig. 21)	79H1
	0.4	$T > 820$ K	electrical resistance	63J
	0.2	$T > 500$ K	electrical conductivity	70P1
SrB_6:				
E_g	3.68		calculated	77P1
	0.38	$T > 1250$ K	electrical resistance	63J
	0.45	$T > 700$ K		60E
BaB_6:				
E_g	2.64		calculated	77P1
	0.12	$T > 700$ K	electrical resistance	63J
	0.15	$T > 300$ K	electrical conductivity	70P1

Transport properties

resistivity (in Ω cm):				
CaB_6:				
ϱ	0.1	$T = 300$ K	polycrystal	63J
	0.43	300 K		70P1
	0.16	600 K		
SrB_6:				
ϱ	0.216	$T = 300$ K	single crystal	63J
BaB_6:				
ϱ	0.07	$T = 300$ K		70P1
	0.01	600 K		

For resistivity, see also [75S].

Further properties

densities (in g cm^{-3}):				
CaB_6:				
d	2.45		calculated	75S
	2.49		experimental	

(continued)

Physical property	Numerical value	Experimental conditions	Experimental method, remarks	Ref.
densities (continued)				
SrB_6:				
d	3.42		calculated	75S
	3.28		experimental	
BaB_6:				
d	4.25		calculated	75S
	4.26		experimental	
melting points (in °C):				
CaB_6:				
T_m	2230			74S1
	2235			51L
SrB_6:				
T_m	2235			51L
BaB_6:				
T_m	2270			51L
coefficients of linear thermal expansion (in 10^{-6} K^{-1}):				
CaB_6:				
$\bar{\alpha}$	5.64			75S
SrB_6:				
$\bar{\alpha}$	6.7			75S
BaB_6:				
$\bar{\alpha}$	6.8			75S
work function:				
SrB_6:				
Φ	4.24···4.38 eV	$A = 29···120$ A cm^{-2} K^{-2}	A: emission constant, defined by $i_s = AT^2 \exp(-\Phi/kT)$ where i_s is the saturation current density	78R

Ternary compounds:
System $CaB_6 - SmB_6$: see [79S7; 77S3; 79R2].

9.14.4 Compounds with group IIb elements

Only few investigations have been performed on compounds of boron with group IIb elements (see [77G3] and references therein).

9.14.4.1 Boron-zinc compounds

For solid solutions of Zn in boron, see Fig. 23.

ZnB_{22}:
Structure
Since the crystalline structure deviates only insignificantly from that one of β-rhombohedral boron, the accommodation of the Zn atoms in holes of the β-rhombohedral boron structure seems probable, but the possibility of a substitution of boron atoms cannot be excluded [77G3, 82L].

Physical property	Numerical value	Experimental conditions	Experimental method, remarks	Ref.
Physical properties				
resistivity:				
ϱ	$2.5 \cdot 10^3$ Ω cm	$T = 298$ K	crystalline (see Fig. 24), p-type conductivity	74K, 77G3
	$2.5 \cdot 10^6$ Ω cm	298 K	powder	
optical transmission: Fig. 25				
magnetic susceptibility:				
χ_g	$-0.64\,(2)$ $\cdot 10^{-6}$ cm^3 g^{-1}	$T = 298$ K	powder (χ_g in CGS-emu)	74K, 77G3
density:				
d	2.86 (4) g cm^{-3}	$T = 298$ K	powder	74K, 77G3
microhardness:				
H	3510 (210) kg mm^{-2}		load 30 g, hardness not specified	77G3

9.14.4.2 Boron-cadmium compounds

Amorphous boron reacts with molten Cd at $T = 800\,°\text{C}$ to a blackish-brown powder containing about 5 wt% Cd. The chemical properties are similar to ZnB$_{22}$, but X-ray powder patterns and IR absorption indicate amorphous structure [77G3, 74K].

9.14.5 Compounds with group IIIa elements

Boron-aluminum compounds:

AlB$_2$:

semiconducting or metallic (?); preparation [56F, 79S2], crystalline structure [75S, 76S1, 77L], electronic structure [77S1, 77P1], thermal and electrical conductivity [77C1, 75S]

AlB$_4$:

preparation [75S]

AlB$_{10}$:

preparation [60K1, 79S2], crystalline structure [60K1, 79S2, 70W1, 77M3], thermal properties [75S]

α-AlB$_{12}$:

Structure

The tetragonal structure is based on a three-dimensional framework consisting of B$_{12}$ icosahedra, B$_{19}$ units, and single B atoms. The B$_{19}$ unit is a twinned icosahedron with a triangular composition plane and a vacant apex on each side. The Al atoms are distributed statistically over five sites in the boron framework. The unit cell contains four times the chemical unit Al$_{3.2}$ · 2B$_{12}$ · B · B$_{19}$.

space group: D_4^4–P4$_1$2$_1$2 or D_4^8–P4$_3$2$_1$2 [77H1, 77K1].

lattice parameters:

a	10.158 (2) Å	$T = 300$ K	single crystal, X-ray diffraction	77H1
c	14.270 (5) Å			

B$_{12}$ icosahedral arrangement: Fig. 26; arrangement of B$_{19}$ units: Fig. 27; nature of linkages between B$_{12}$ and B$_{19}$: Fig. 28; nature of linkages between B$_{12}$ and B$_{19}$ through B: Fig. 29.

Physical property	Numerical value	Experimental conditions	Experimental method, remarks	Ref.

position and occupancies of the Al atoms [77H1, 77K1]:

Atoms	Position	Coordinates			Occupancies
		x	y	z	%
Al_1	8 (b)	0.3021 (1)	0.3688 (1)	0.2589 (2)	71.7 (7)
Al_2	8 (b)	0.0823 (1)	0.0118 (1)	0.3086 (1)	49.1 (3)
Al_3	8 (b)	0.3130 (3)	0.3932 (3)	0.3419 (4)	24.0 (6)
Al_4	8 (b)	0.2873 (4)	0.4785 (4)	0.1224 (3)	15.0 (3)
Al_5	8 (b)	0.3037 (35)	0.3831 (43)	0.2958 (91)	2.1 (5)

Electronic properties

For valence electron distribution, see Fig. 30.

energy gaps:

E_g	1.9 eV	$T = 300$ K	optical absorption	74G1
	1.96 eV	300 K	optical absorption	79G1
$E_{g,th}$	2.2 eV	$T \gtrsim 1000$ K	electrical conductivity (intrinsic)	77B1, 74B3

Impurity and defects

activation energies:

E_A	0.26···0.33 eV	$T = 300$ K	optical absorption	74G1
	0.2···0.3 eV		electrical conductivity	73B1

Lattice properties

For lattice parameters, see above ("Structure").

IR active phonon wavenumbers (in cm^{-1}):

$\bar{v}$	430	$T = 300$ K	optical absorption (see Fig. 31), polycrystalline samples	74G1
	480			
	565			
	600			
	650			
	710			
	785			
	905			
	1040			
	1180			

Transport properties

See also Figs. 32···36.

conductivities (in Ω^{-1} cm^{-1}):

		T [K]		
σ	10^{-2}···10^{-7}	300	polycrystalline samples; see also Figs. 32, 33, 34	79B1, 74B3, 77B1
	10^{-1}	600	0.5 at% Mg doped, polycrystal	77B2
	$7 \cdot 10^{-2}$	600	0.5 at% Zr doped, polycrystal	
	$3 \cdot 10^{-3}$	600	0.5 at% Co doped, polycrystal	

For current-voltage characteristics, Hall coefficient and temperature dependence of resistance, see Figs. 33, 35, 36.

Physical property	Numerical value	Experimental conditions	Experimental method, remarks	Ref.
thermoelectric powers (in µV K^{-1}):		T [K]		
S	300…600	300	polycrystalline sample; see also Figs. 33, 34	74B3, 77B1, 79B1
	260	1000	0.5 at% Mg doped, polycrystal	77B2
	330	1000	0.5 at% Zr doped, polycrystal	
	400	1000	0.5 at% Co doped, polycrystal	
Optical properties				
dielectric constants:				
$\varepsilon(0)$	≈19	$T = 300$ K	questionable result, derived from reflectivity, see comment at Fig. 31	73B1, 77B1
$\varepsilon(\infty)$	≈11	300 K		
Further properties				
thermal conductivity:				
κ	0.04 W cm^{-1} K^{-1}	$T = 300$ K	polycrystalline material. For temperature dependence, see Fig. 37	74B3
density:				
d	2.557…2.660 g cm^{-3}	$T = 300$ K	experimental	75S
melting point:				
T_m	2163 °C			60G1
	2070 °C			75S
microhardness:				
H_K	2515 kg mm^{-2}	$T = 300$ K	100 g load, Knoop hardness	60G1

β-AlB$_{12}$:

Structure

For details of the lattice structure, see [64P]. Possibly non-metal impurity atoms are necessary to stabilize the lattice [70E1].

lattice: orthorhombic.

lattice parameters:

a	12.34 Å	$T = 300$ K		75S
b	12.631 Å			
c	10.161 Å			

Other indication of the lattice structure:
lattice: tetragonal [74G3].

lattice parameters:

a	8.82 Å	$T = 300$ K		74G3
c	5.09 Å			

Physical properties

energy gaps:

E_g	2.5 eV	$T = 300$ K	optical absorption	74G1
$E_{g,th}$	2.5 eV	$T \gtrsim 700$ K	electrical conductivity (see Fig. 39)	76G1, 79G2

For lattice parameters, see above (Structure).

Physical property	Numerical value	Experimental conditions	Experimental method, remarks	Ref.
IR-active phonon wavenumbers (in cm^{-1}):				
$\bar{\nu}$	450	$T = 300$ K	optical absorption (see Fig. 38)	76G1
	515			
	570			
	615			
	690			
	745			
	810			
	875			
	950			
	1030			
	1080			
electrical conductivity:				
σ	$5 \cdot 10^{-9}$ Ω^{-1} cm^{-1}	$T = 300$ K	polycrystal; see Fig. 39	76G1
thermoelectric power:				
S	180 µV K^{-1}	$T = 300$ K	polycrystal; see Fig. 40	79G2
thermal conductivity:				
$\varkappa$	0.2 W K^{-1} cm^{-1}	$T = 300$ K	polycrystalline sample; for temperature dependence, see references	74G2, 79G2
	0.105 W K^{-1} cm^{-1}	300 K	single crystal; for temperature dependence, see reference	74G3
Debye temperature:				
Θ_D	≈ 1200 K			74G3
microhardness:				
H_K	2380 kg mm^{-2}	$T = 300$ K	100 g load, Knoop hardness	60G1
density:				
d	2.60 g cm^{-3}	$T = 300$ K	pycnometric	75S
melting point:				
T_m	2214 °C			60G1

γ-AlB$_{12}$:
preparation [77S1, 77H2, 60K1], crystalline structure [77K2, 77H2, 60K1, 75S, 77M2, 77M3]

Al$_3$B$_{32}$:
preparation [77M3], crystalline structure [77M3, 70M3]

AlC$_2$B$_{12}$:
preparation [66L2], crystalline structure [77M2, 77M3]

AlC$_4$B$_{24\cdots26}$
preparation [79S2], crystalline structure [70W1, 77M2, 77M3]

AlC$_4$B$_{40}$:
preparation [61M], crystalline structure [77M3]

Al$_3$C$_2$B$_{48}$:
preparation [79S2, 77M3], crystalline structure [77M3, 70M3]

Al$_{2.1}$C$_8$B$_{51}$:
preparation [69P1], crystalline structure [70W1]

AlMoB:
preparation [76S2]

Physical property	Numerical value	Experimental conditions	Experimental method, remarks	Ref.

9.14.6 Compounds with group IIIb elements

9.14.6.1 Boron-scandium compounds

ScB$_2$:

metallic conduction (?); preparation [70P2, 75S], crystalline structure [70P2, 75S], electronic structure [81A1, 79A3]

ScB$_4$:

preparation [79S3]

ScB$_{12}$:

preparation [70P2, 75S], crystalline structure [70P2, 75S], electronic structure [77P1]

LaB$_6$– "ScB$_6$"-system:

preparation [79S3]

9.14.6.2 Boron-yttrium compounds

YB$_2$:

preparation [75S], crystalline structure [75S]

YB$_4$:

preparation [75S], crystalline structure [63R], electrical and thermal conductivity [74S2]

YB$_6$:

superconducting, semimetal?; preparation [75S, 76F, 77P3], crystalline structure [75S, 76F], magnetic susceptibility [79S4], electrical conductivity [76F], electronic transport properties [71G3], ESCA spectrum [76A1]

YB$_{12}$:

preparation [65M2, 75S, 65M4], crystalline structure [65M2, 65M4, 77M2, 60B], magnetic susceptibility [79M1, 73O]

YB$_{66}$:
Structure

The most boron-rich yttrium-boron compound is the cubic structure YB$_n$ ($100 > n > 20$). Commonly n is considered to be 66 [69R, 75B4, 76K] comprising 104 icosahedra or 1248 atoms in a unit cell; the stoichiometric composition is at n = 68 [77S4].
lattice: cubic; space group: O_h^6 – Fm3c [72S1].
  A simplified schematic representation of the structure is shown in Fig. 41.

lattice parameter:

a	23.4···23.5 Å	$T = 300$ K	X-ray diffraction, electron microscopy	76K, 77S4

Physical properties
energy gap:

E_g	0.8 eV	$T > 500$ K	electrical conductivity (estimated from Fig. 42; YB$_{61.5}$)	77S4

elastic moduli (in 10^{12} dyn cm^{-2}):

c_{11}	3.80 (3)	$T = 300$ K,	from longitudinal-wave sound	77S4
c_{44}	1.60 (8)	$v = 60$ MHz	velocity	
c_{12}	0.4 (5)		derived value	

sound velocity:

v_l	$1.18 \cdot 10^6$ cm s^{-1}	$T = 300$ K		77S4

Physical property	Numerical value	Experimental conditions	Experimental method, remarks	Ref.
resistivity:				
ϱ	$3.6 \cdot 10^2$ Ω cm	$T = 300$ K	For temperature dependence, see Fig. 42	77S4
far infrared optical absorption: see Fig. 43				
thermal conductivity:				
$\varkappa$	$\approx 20 \cdot 10^{-3}$ W cm^{-1} K^{-1}	$T = 300$ K		71S3
	0.58 W cm^{-1} K^{-1}		computed value for $T = 0$ K. For temperature dependence, see Fig. 44	71S3
phonon mean free path:				
Λ	$23.44 \cdot 10^{-8}$ cm	$T = 300$ K	For temperature dependence, see Fig. 45	71S3
acoustic attenuation: see Fig. 46.				
Debye temperature:				
Θ_D	1300 (50) K	$T = 4 \cdots 300$ K	calculated from elastic constants	71S3, 77S4
	1200 K		heat capacity (preliminary measurements)	77S4
density:				
d	2.5687 (50) g cm^{-3}	$T = 300$ K	pycnometric	77S4, 71S3
	2.482 g cm^{-3}	300 K	X-ray	71S3
melting point:				
T_m	2100 °C			77S4

9.14.7 Compounds with lanthanides

The diborides and the tetraborides of the lanthanides seem all to be metallic (cf. [75S, 77E, 77S1, 77S2, 81G2, 81W4, 79W1, 79K1, 79E, 79P2]).

9.14.7.1 Lanthanide hexaborides

The conductivity character of the lanthanide hexaborides depends on the lattice parameters and on the atomic radii of the metal atoms, respectively (see Fig. 47). Only EuB$_6$ and YbB$_6$ (Eu and Yb in +II oxidation state) are semiconductors at normal temperatures, SmB$_6$ behaves as a semiconductor at low temperatures and hence only these lanthanide hexaborides are considered here.

Structure

Cubic lattice structure of CaB$_6$ type, space group $O_h^1 - Pm3m$.

lattice parameters:

SmB$_6$:

a	4.1304 Å	$T = 300$ K	see also [56P]	71N1, 77E, 79G3

EuB$_6$:

a	4.1780 Å	$T = 300$ K	see also [56P] and [74M1]	73S1, 77E, 79G3
	4.186 Å		see Fig. 51, see also [74M1]	78K1, 79K3

Physical property	Numerical value	Experimental conditions	Experimental method, remarks	Ref.
YbB_6:				
a	4.1478 Å	$T = 300$ K	see also [56P]	70E3, 77E, 79G3

SmB_6:

For NMR studies of valence fluctuations, see [81T4]; for NMR studies and spin/charge fluctuations, see [81P].

EuB_6:

transverse nuclear spin-spin relaxation time (in µs):

τ_t	4.5 (4)	$T = 1.7$ K; $B = 0$		77B4
	4.5 (4)	$B = 0.15$ T		
	5.7 (6)	$B = 0.3$		
	9.6 (10)	$B = 0.6$		
	11.0 (10)	$B = 0.75$		
	11.0 (10)	$B = 1$ T		
	4.5 (4)	$T = 4.2$ K; $B = 0$		
	8.9 (9)	$B = 1$ T		

longitudinal nuclear spin-lattice relaxation time:

τ_l	54 (5) µs	$T = 1.7$ K; $B = 1$ T		77B4

Knight shift:

K_B	−0.059%		commercial EuB_6	63M
	−0.133%		EuB_6 reacted with boron	

Electronic properties

General considerations, also on alloys: see [74H3].

SmB_6:

In the electronic structure of SmB_6 a narrow f band with large Coulomb interaction, a wide d band with negligible interactions, f–d hybridization, Coulomb, and exchange interactions are assumed. Since in the case of SmB_6 the two bands contain an even integral number of electrons per rare earth ion, intersite correlations profoundly affect states near the Fermi energy and lead to a small insulating gap [79M3]. See also [70C3, 71N2, 73M2, 77O].

Calculated densities of states: see Fig. 48.

For $SmM_{4,5}$ spectra of SmB_6, see [82A]; for Sm $L\gamma_1$ emission spectra in SmB_6, see [82T]; for B K spectra in SmB_6, see [82O].

EuB_6:

EuB_6 is a magnetic semiconductor. It has been assumed that the Eu^{2+}: $4f^7$ level is located in the band gap. Eu vacancies would introduce holes in the $4f^7$ configuration. These holes are small polarons, i.e. Eu^{3+} ions trapped by electrostatic forces on a site near-neighbor to the vacancy (acceptor density n_a). The negative Seebeck coefficient indicates that electron donor sites with $n_d > n_a$ are also present. These are rare earth atoms as substitutional impurities, represented in Fig. 49 by $5d^1$ donor levels of Ln^{2+}: $4f^{n-1} 5d^1$ centers. When the density of the impurity atoms is sufficient, a narrow, partly occupied $5d^1$ impurity band is formed [77E, 73G]. Some recent experimental results seem not to agree satisfactorily with this band scheme [77I, 79F1, 79W2].

Physical property	Numerical value	Experimental conditions	Experimental method, remarks	Ref.

The semiconducting behavior in the paramagnetic region changes to semimetallic behavior at the ferromagnetic ordering temperature $T_C = 13.7$ K [81T1; 80G1].

An energy diagram for EuB_6 is shown in Fig. 49; for band structure calculation, cf. [75P, 77P1].

energy gaps*) and g-factors:

SmB_6:

E_g	2.3 meV		electrical conductivity	69M1
	5 meV		electrical conductivity, see Fig. 54	77E

EuB_6:

E_g	0.38 eV		electrical conductivity	69F
	0.30 eV		electrical conductivity	73G, 77E
	0.032 eV		electrical conductivity (cp. remark below (impurity and defects))	80L
	0.05···0.1 eV		electrical conductivity	81T1
g	2.004 (1)		ESR	72H1
	2.000 (2)		ESR	76G2, 76G3
	1.995 (10)		ESR	79O

For temperature dependence of ESR resonance points and linewidth, see Fig. 50; for further ESR investigations, see [73S3, 73S4, 75G1, 75G2, 76G3, 76G4, 76G5, 79M2, 79K2].

YbB_6:

E_g	0.14 eV		electrical conductivity	77E, 73G
	0.081 eV			80L

pressure dependence of the energy gap:

EuB_6:

$\dfrac{dE_g}{dp}$	-1.5 meV/kbar	$T = 300$ K		80L

YbB_6:

$\dfrac{dE_g}{dp}$	-3.7 meV/kbar	$T = 300$ K		80L

Impurity and defects

SmB_6:

The influence of Sm vacancies is discussed in [77K3] and [78A].

EuB_6:

The electronic properties of EuB_6 are strongly influenced by carbon. Even very small C contents (C substitutional on B sites) lead to a degenerate semiconductor, in which carrier concentration and antiferromagnetic interaction increase with C content [81T1]. Possibly many of the experimental results available are influenced in this way.

For effects of carbon on lattice parameter, paramagnetic Curie temperature and resistivity, see Figs. 51···53; for effects on the magnetism of EuB_6, see [79K3].

---

*) or activation energy for conductivity (cf. section 9.16.2)

Physical property	Numerical value	Experimental conditions	Experimental method, remarks	Ref.
Transport properties				
LaB$_6$:				
resistivity:				
ϱ	$1.5 \cdot 10^{-5}$ Ω cm	$T = 300$ K		80L
SmB$_6$:				
resistivity (in Ω cm):				
		T [K]		
ϱ	$3 \cdots 10 \cdot 10^{-4}$	300	see Figs. 54, 55; see also [70Y]	75A1, 78A, 75S, 69P2, 80L
	$1.2 \cdot 10^{-3}$	300, 230 Hz		79A6
	$1.7 \cdot 10^{-4}$	10		77E
	$3 \cdot 10^{-2}$	10		71N2
	3	4, 230 Hz	For temperature dependence, see Fig. 55α	76A6
temperature coefficient of ϱ:				
$\dfrac{1}{\varrho}\dfrac{d\varrho}{dT}$	$-42 \cdot 10^{-2}$ K^{-1}	$T = 300$ K		75S
pressure dependence of ϱ:				
$\left(\dfrac{d \log \varrho}{dp}\right)_{p=0}$	$-25 \cdot 10^{-3}$ kbar^{-1}	$T = 300$ K		80L
mobility:				
μ_H	-20 cm^2 V^{-1} s^{-1}	$T = 10$ K	see Fig. 55	71N2
thermoelectric power:				
S	-320 μV K^{-1}	$T \approx 10$ K	see Fig. 56	77E
	$+8.4$ μV K^{-1}	300 K		69P2, 75S, 62S

For Hall coefficient, see Fig. 55β, see also [71N2].
For electrical properties of SmB$_6$, see also [81K].

EuB$_6$:				
resistivity (in Ω cm):				
		T [K]		
ϱ	$6.16 \cdot 10^{-4}$	300	see Figs. 57, 58	80G1
	$8.47 \cdot 10^{-5}$	300		75S
	$1 \cdot 10^{-2}$	300	see Fig. 72	77E1, 74M1
	2	300	possibly carbon-doped	69P2, 75S
	$5 \cdot 10^{-4}$	300	semimetal	79F1
pressure dependence of ϱ:				
$\left(\dfrac{d \log \varrho}{dp}\right)_{p=0}$	$-27 \cdot 10^{-3}$ kbar^{-1}	$T = 300$ K		80L

For pressure dependence of resistivity (samples of different composition), see Fig. 61.

temperature coefficient of resistivity:				
$\dfrac{1}{\varrho}\dfrac{d\varrho}{dT}$	$+0.09 \cdot 10^{-2}$ K^{-1}	$T = 300$ K		75S
	30 Ω^{-1} cm^{-1}			77E

Physical property	Numerical value	Experimental conditions	Experimental method, remarks	Ref.
electron concentration:				
n	$1.7 \cdot 10^{20}$ cm^{-3}	$T = 4.2$ K	electrical conductivity and Hall effect, see Fig. 59	80G1
	$3.5 \cdot 10^{19}$ cm^{-3}	100···150 K		
	$3 \cdot 10^{19}$ cm^{-3}	4.2···300 K		79W2
electron mobility (in cm^2 V^{-1} s^{-1}):		T [K]		
μ_n	2560	4.2	electrical conductivity and Hall effect	80G1
	300	100···150		
	2000	4.2		79W2
	325	77		
	100	290		
pressure dependence of galvanometric effects:				
$\dfrac{\mathrm{d}\log\mu_n}{\mathrm{d}p}$	$10(4) \cdot 10^{-6}$ bar^{-1}	$T = 4.2$ K		79W2
	$25(15) \cdot 10^{-6}$ bar^{-1}	77 K		
	≈ 0	290 K		

For pressure dependence of ϱ, see Fig. 60, of magnetoresistance, see Fig. 61.
For electrical conductivity, Hall effect of EuB$_6$, see also [73G, 73M1, 76M2, 77E, 81K]; for conductivity, carrier concentration, mobility of EuB$_6$ with deviations from exact stoichiometry, see [79G3, 79C1].

thermoelectric power:				
S	-75 μV K^{-1}	$T = 300$ K	see Fig. 62; see also [73M1]	73G
	-17.7 μV K^{-1}	300 K		75S
	-38.5 μV K^{-1}	300 K		69P2

For Hall coefficient, see [62S].

YbB$_6$:

conductivity:				
σ	40 Ω^{-1} cm^{-1}	$T = 300$ K	see Fig. 58	73G, 77E
	30 Ω^{-1} cm^{-1}			73M1
temperature coefficient of ϱ:				
$\dfrac{1}{\varrho}\dfrac{\mathrm{d}\varrho}{\mathrm{d}T}$	$+0.234 \cdot 10^{-2}$ K^{-1}	$T = 300$ K		75S
pressure dependence of ϱ:				
$\left(\dfrac{\mathrm{d}\log\varrho}{\mathrm{d}p}\right)_{p=0}$	$-65 \cdot 10^{-3}$ kbar^{-1}	$T = 300$ K		80L
thermoelectric power:				
S	-190 μV K^{-1}	$T = 300$ K	see Fig. 62; see also [73M1]	73G
	-25.5 μV K^{-1}			75S
	$+18.4$ μV K^{-1}			69P2

For Hall coefficient, see [62S].
For electrical properties of YbB$_6$, see also [70P1, 81K].

Optical properties

Raman spectra of SmB$_6$ and LaB$_6$ are shown in Fig. 62a. For Raman frequencies of SmB$_6$, EuB$_6$, YbB$_6$, see Fig. 63. The reflectivity and dielectric function of SmB$_6$ is shown in Fig. 64. The reflectivity of EuB$_6$ and LaB$_6$ is given in Fig. 64a. A Raman spectrum of EuB$_6$ is presented in Fig. 65. Reflectivity spectra between 0.05 eV and 18 eV for EuB$_6$ are announced in [81S1].

Physical property	Numerical value	Experimental conditions	Experimental method, remarks	Ref.
Further properties				
thermal conductivity:				
EuB_6:				
κ	19.26 W m^{-1} K^{-1}	$T = 90$ °C		73S2
	19 W m^{-1} K^{-1}	90 °C		81S2
heat capacity:				
SmB_6: see Fig. 66				
EuB_6: see Fig. 67				
densities (all values at $T = 300$ K and in g cm^{-3}):				
SmB_6:				
d	5.076		X-ray	75S
	4.79		pycnometric	75S
EuB_6:				
d	4.938		X-ray	75S
	4.94		pycnometric	73S2
	4.91		pycnometric	81S2
YbB_6:				
d	5.556		X-ray	75S
	5.53		X-ray	69M2
	4.37		pycnometric	75S
	5.45		pycnometric	69M2
melting points (in °C):				
SmB_6:				
T_m	2540			75S
	1810			
	2400			
	>2500			80H
EuB_6:				
T_m	2150			75S
	2580			73S2
	>2500			80H
YbB_6:				
T_m	>2500			80H
linear thermal expansion coefficients (in 10^{-6} K^{-1}):				
SmB_6:		T [K]		
α	6.5	300		75S
EuB_6:				
α	6.86	300		75S
	7.4	293···1273		73S2
YbB_6:				
α	5.85	300		75S

9.14.7 Boron compounds with lanthanides

Physical property	Numerical value	Experimental conditions	Experimental method, remarks	Ref.
work functions (in eV):				
SmB_6:				
Φ	4.4			75S
	4.30(5)	$T = 300$ K	For temperature dependence at high temperatures, see original paper.	79S8
	4.24(7)			80F2
EuB_6:				
Φ	4.9			75S
	3.95$\cdots$4.03	$A = 29\cdots 120$ A cm^{-2} K^{-2}	A is defined by $i_s = AT^2 \exp(-\Phi/kT)$, where i_s is the saturation current density	78B
	4.32(7)			80F2
YbB_6:				
Φ	3.13			75S
emissivities at 655 nm:				
SmB_6:				
ε	0.68			75S
EuB_6:				
ε	0.83			75S
YbB_6:				
ε	0.7			75S
microhardnesses:				
SmB_6:				
H	2500 kg mm^{-2}	$T = 300$ K	hardness not specified	75S
EuB_6:				
H_K	1840 kg mm^{-2}	$T = 300$ K	Knoop hardness; load 1 N (see also [79F2])	81S2
YbB_6:				
H	3080 kg mm^{-2}	$T = 300$ K	hardness not specified	75S
magnetic moments (in Bohr magnetons μ_B):				
SmB_6:				
p_{eff}	2.52		effective magnetic moment	75S
EuB_6:				
p_{eff}	8.1		effective magnetic moment	67P
	7.90 (1)			79V
	7.76			78F1
p_{calc}	7.94		theoretical (divalent Eu ion)	79V
YbB_6:				
p_{eff}	4.58		effective magnetic moment	67P

Physical property	Numerical value	Experimental conditions	Experimental method, remarks	Ref.
magnetic susceptibilities:				
SmB_6:				
χ_m	$1810 \cdot 10^{-6}$ cm^3 mol^{-1}	$T = 300$ K	χ_m in CGS-emu	75S
YbB_6:				
χ_m	$8740 \cdot 10^{-6}$ cm^3 mol^{-1}	$T = 300$ K	χ_m in CGS-emu	75S

molar susceptibility of SmB_6: Fig. 68; low field magnetic susceptibility of $Eu_{1-x}B_6$: Fig. 69

ferromagnetic Curie temperatures:

EuB_6:				
T_C	8.5 K		polycrystal (?)	77B5
	13.7 K		$Eu_{1-x}B_6$, single crystal	78F1
	14 K		single crystal	79V
			polycrystal	77I, 68G

asymptotic (paramagnetic) Curie temperatures:

EuB_6:				
Θ_a	+9 K			77B5, 67P
YbB_6:				
Θ_a	−161 K			77B5

Temperature dependence of the exchange field (EuB_6): Fig. 70; magnetic phase diagram ($Eu_{1-x}B_6$): Fig. 71.

For further investigations on ferromagnetism in EuB_6, see [71F1, 71W1, 73K, 74S3, 75G, 76G2, 76G3, 76L2, 77B5, 79V].

Ternary Compounds

$La_xEu_{1-x}B_6$:

In $M^{II}_{1-x}M^{III}_xB_6$ ternary compounds metals in the +II and the +III oxidation state are present simultaneously. So the development of their properties from semiconductor ($M^{II}B_6$) to metallic ($M^{III}B_6$) behavior can be observed [74M1, 76M2, 77E, 79A4].

A composition-induced metal-insulator transition occurs at $0 < x < 0.01$ [76M2].

For work function, see [78B]; electrical resistivity: Fig. 72; thermoelectric power: Fig. 73.

For further ternary compounds, see [74S3, 79K1, 79M2, 74H1, 77S3, 78B].

9.14.7.2 Lanthanide dodecaborides

Dodecaborides of the lanthanides Tb, Dy, Ho, Er, Tm, Yb, Lu have been obtained. Additionally a number of alloys of these dodecaborides with dodecaborides of other elements is known. Some of these alloying dodecaborides are not synthetized in pure form [72S4, 81S2, 77S5].

Structure

The crystalline structure can be described in two ways:
a) Metal atoms and cubooctahedral B_{12} units are arranged in a cubic face-centered lattice (Fig. 74a).
b) B_{24} cages surround the metal atoms which are arranged in a cubic lattice (Fig. 74b) [72S4, 75S, 76C, 77M2, 81S2, 77S1, 77S2].
lattice parameters: Fig. 75

Physical properties

Systematic investigations on the semiconductor properties are not known. But the distinct coloring of most of the dodecaborides (see [72S4, 81S2]) indicate band gaps up to some eV.

For transport properties, see [71G3, 71O1, 71O2, 71P2, 72O].

magnetic susceptibilities χ_m, Curie-Weiss constants C_m, effective magnetic moments p_{eff}, and paramagnetic Curie points Θ_p:

Physical property	Numerical value	Experimental conditions	Experimental method, remarks	Ref.
TbB_{12}:				
χ_m	7770 cm^3/mol	$T = 300$ K	CGS-emu	73O
C_m	7.4 cm^3 K/mol			
p_{eff}	7.70 μ_B			
Θ_p	-652 K			
DyB_{12}:				
χ_m	38300 cm^3/mol	$T = 300$ K	CGS-emu	73O
C_m	12.3 cm^3 K/mol			
p_{eff}	9.90 μ_B			
Θ_p	-21 K			
HoB_{12}:				
χ_m	44370 cm^3/mol	$T = 300$ K	CGS-emu	73O
C_m	13.8 cm^3 K/mol			
p_{eff}	10.50 μ_B			
Θ_p	-11 K			
ErB_{12}:				
χ_m	38550 cm^3/mol	$T = 300$ K	CGS-emu	73O
C_m	12.8 cm^3 K/mol			
p_{eff}	10.10 μ_B			
Θ_p	-32 K			
TmB_{12}:				
χ_m	25850 cm^3 K/mol	$T = 300$ K	CGS-emu	73O
C_m	9.0 cm^3 K/mol			
p_{eff}	8.50 μ_B			
Θ_p	-48 K			
YbB_{12}:				
χ_m	6820 cm^3/mol	$T = 300$ K	CGS-emu	73O
C_m	5.6 cm^3 K/mol			
p_{eff}	6.70 μ_B			
Θ_p	-520 K			
LuB_{12}:				
χ_m	2896 cm^3/mol	$T = 300$ K	CGS-emu	73O

On magnetic properties, see also [77B5, 79M1].

9.14.7.3 Lanthanide borides of the type MB_{66}

Many of the lanthanide metals form boron-rich compounds of the composition MB_{66}. The compounds NdB_{66}, SmB_{66}, GdB_{66}, TbB_{66}, DyB_{66}, HoB_{66}, ErB_{66}, TmB_{66}, YbB_{66}, LuB_{66} are known at now. They exhibit the same lattice structure as YB_{66} (see 9.14.6.2) with 104 icosahedra or 1248 atoms in a unit cell [72S1, 72S2, 72S5, 76K, 76S4, 77N2, 77S5, 75B4, 81S2].

Investigations of the physical properties have been performed only on SmB_{66}, GdB_{66}, and YbB_{66}.

Physical property	Numerical value	Experimental conditions	Experimental method, remarks	Ref.

Electronic properties

Band tails with localized states caused by the complex lattice structure are assumed to explain the transport properties [81G3].

energy gaps and activation energies (in eV):

SmB_{66}:

E_g	0.80		electrical conductivity, see Fig. 76	81G3
E_A	0.15		distance between mobility edge and E_F	

GdB_{66}:

E_g	0.73	$T = 300$ K	optical absorption	81G3
	0.87		electrical conductivity	
E_A	0.20		distance between mobility edge and E_F	

YbB_{66}:

E_g	1.27		electrical conductivity	81G3
E_A	0.10		distance between mobility edge and E_F	

Transport properties (see also Figs. 76···78)

carrier concentration p, **Seebeck coefficient** S, **and Hall mobility** μ_H (at $T = 300$ K):

SmB_{66}:

p	$2.7 \cdot 10^{16}$ cm^{-3}		derived from Hall effect	81G3
S	$+100$ μV K^{-1}			
$\mu_{H,p}$	15 cm^2 V^{-1} s^{-1}			

GdB_{66}:

p	$1.5 \cdot 10^{15}$ cm^{-3}		derived from Hall effect	81G3
S	$+390$ μV K^{-1}			
$\mu_{H,p}$	15 cm^2 V^{-1} s^{-1}			

YbB_{66}:

p	$1.2 \cdot 10^{16}$ cm^{-3}		derived from Hall effect	81G3
S	$+170$ μV K^{-1}			
$\mu_{H,p}$	5 cm^2 V^{-1} s^{-1}			

9.14.8 Compounds with actinides

9.14.8.1 Boron-thorium compounds

ThB_4:

preparation [77P3, 75S, 75B4], crystalline structure [77E, 75S], electronic structure [75S], magnetic properties [77B5], ESR spectra [73T]

ThB_6:

preparation [71E, 75S], crystalline structure [77E], magnetic properties [77B5], ESR spectra [73T]

ThB_{66}:

semicond.?; preparation [75B4], crystalline structure (YB_{66}-type) [75B4, 81S2, 76K]

ThB_{76}:

semicond.?; preparation [75S], crystalline structure [75S]

9.14.8.2 Boron-uranium compounds

UB_2:
preparation [75S, 77P3, 75M], crystalline structure [75S, 77C1], magnetic properties [77B5]

UB_4:
preparation [75S, 77P3, 75M, 75B4], crystalline structure [75S], magnetic properties [77B5, 75F]

UB_{12}:
semicond.?; preparation [75S, 75M], crystalline structure [75S, 77M2], magnetic properties [77B5, 73O]

9.14.8.3 Boron-neptunium compounds

NpB_2:
preparation [81S2], crystalline structure [81S2], magnetic properties [74S4]

NpB_4:
preparation [75B4, 81S2], crystalline structure [75B4, 81S2]

NpB_6:
preparation [69E], crystalline structure [69E, 77E], magnetic properties [74S4]

NpB_{12}:
preparation [75B4], magnetic properties [74S4]

9.14.8.4 Boron-plutonium compounds

PuB:
preparation [75S], crystalline structure [75S, 60M]

PuB_2:
preparation [75S, 81S2], crystalline structure [75S, 60M, 81S2]

PuB_4:
preparation [75B4, 75S, 60M], crystalline structure [77N2, 75S, 60M]

PuB_6:
preparation [75B4, 75S], crystalline structure [77E, 60M, 77N2, 75S], electronic structure [77P1]

PuB_{12}:
preparation [75B4]

PuB_{66}:
semicond.?; preparation [75B4], crystalline structure (YB_{66}-type) [81S2]

9.14.8.5 Boron-americium compounds

AmB_4:
preparation [75B4]

AmB_6:
preparation [69E], crystalline structure [69E, 77E]

9.14.9 Compounds with group IVa elements

9.14.9.1 Boron-carbon compounds

Besides the boron compounds with carbon explicitly described below, the compounds B_8C, $B_{12}C$, $B_{49}C_3$, $B_{25}C$, $B_{51}C$ have been obtained, which are possibly not thermodynamically stable [61Z, 74P1, 74P2]. – For phase diagrams and limits of the homogeneity range, see [81B3]. For chemical and related properties, see [81S3].

| Physical property | Numerical value | Experimental conditions | Experimental method, remarks | Ref. |

Boron carbide:

Structure

Boron carbide exhibits a wide homogeneity range from the limit composition $B_{12}C_3$ at the carbon-rich side [81B2, 81W2] to a limit composition of about $B_{10}C$ at the boron-rich side [76M3, 81B1, 81B2]. The lattice structure of boron carbide is strongly related to the lattice structures of the rhombohedral modifications of crystalline boron (cp., vol. 17e, section 8.1) [79B4].

The recent structure conceptions are based on a rhombohedral lattice with B_{12} icosahedra at the corners of the unit cell, and additionally a linear C−B−C chain arranged along the trigonal axis. In addition to these two C atoms per unit cell a substitutional occupation of B sites in the icosahedra is possible, which is strongly limited to one C atom per icosahedron (cf. [81W1]). In boron-rich boron carbide the C−B−C chains are partly replaced by planar B_4 groups [75Y].

lattice: rhombohedral; space group: $D_{3d}^5 - R\bar{3}m$; 15 atoms per unit cell
[59S, 66S, 70A3, 71H, 75Y, 73Y, 77M3, 77M1, 76M3, 71K1, 76W1, 76W2, 78H, 81B1, 81B2, 81W1, 79K7, 79K8, 80K].

unit cell: Fig. 79, C−B−C chain: Fig. 80a, B_4 plane: Fig. 80b, icosahedron: Fig. 80c.

For morphology of boron carbide single crystals, see [71S4]. For dislocation nodes, see [72A, 73A2, 75A3, 75L3].

In many cases the composition of boron carbide is not specified in the original literature. But usually then the investigations have been performed on technical boron carbide, and the composition will not deviate significantly from B_4C.

lattice parameters ($B_{13}C_2$):

rhombohedral presentation:

a	5.198 (2) Å	$T = 300$ K	X-ray diffraction	79K7
α	65.62°			

hexagonal presentation:

a	5.633 (1) Å	$T = 300$ K	X-ray diffraction	79K7
c	12.164 (2) Å			

Dependence of lattice parameters on C content: Fig. 81.

quadrupole coupling constants:

ν_{Qa}	520 (30) kHz		NMR	71H
ν_{Qb}	30 kHz		NMR	

Electronic properties

Detailed band structure calculations have been performed on $B_{12}C_3$ in [81A1, 83A]. The valence band structure exhibits a series of flat, non-free-electron-like bands, with little variation in energy for any one band throughout the k-space (Fig. 81 β). According to these calculations the Fermi level is expected within the valence band near its upper limit (Figs. 81 β, γ). The metallic behavior expected according to these calculations does not agree with the experimental results which are available at present. Besides, the calculated band gap of about 2.5 eV disagrees with the experiments (cf. [71W2, 74W, 81W1]).

energy gaps:

$E_{g, ind}$	0.48 eV	$T = 300$ K	optical absorption (see Fig. 92)	71W2, 74W
E_g	$\gtrsim 4$ eV		calculated	81A1, 83A
	0.5···1 eV		calculated (tight binding approximation)	57Y

Physical property	Numerical value	Experimental conditions	Experimental method, remarks	Ref.

g-values (± 0.0002 mostly) **and line widths** ΔB ($\cdot 10^{-4}$ T) of the single, isotropic, symmetrical EPR line in boron carbide [77G4]:

	single crystal [64G]		polycrystal [64G, 70G2]		powder, powder +5 wt% C [70K1]		crystalline, also with Be [70K2]		sintered (asymm. EPR line) [80V]	
T [K]	g	ΔB	g	ΔB	g	ΔB	g	ΔB	g	ΔB
1.7	2.0030	14	2.0035	10	–	–	–	–	2.0038 (3)	12.6
77	–	5	2.001	8	–	4	–	6	2.0038 (3)	6.2
300	–	8	2.000		2.0032	9	2.0030 (asymp.)	–	2.0028 (3)	5.3

For EPR investigation on various boron-carbon compounds, see [68K1, 68K2].

effective hole mass:

m_p	$3 \cdots 5 \, m_o$	$T = 300$ K	plasma resonance	71W2, 79B4

According to the experimental results a metal-insulator transition appears near the composition $B_{12}C_3$ accompanied with a change in lattice structure, which is indicated by the shift of an optical phonon [80W, 81W6]. The transition is assumed to occur within a split band induced by the electronic states of the C atoms in a level at about 0.17 eV above the valence band edge, respectively above the mobility edge of the valence band.

Figures:

Brillouin zone: Fig. 81α, band structure: Fig. 81β, total density of states: Fig. 81γ, electron paramagnetic resonance: Fig. 82; valence electron density distribution: Fig. 83; position of Fermi level: Fig. 84; density of states and hopping energy: Fig. 85

Lattice properties

For lattice parameters, see above (Structure) and further below (Further properties).
For phonon wavenumbers, see next page.

longitudinal sound velocity:

v_l	13780 m s^{-1}	$T = 300$ K		71G2, 74M2
	13800 m s^{-1}	300 K		82S

shear sound velocity:

v_s	8540 m s^{-1}	$T = 300$ K		71G2
	8900 m s^{-1}	300 K		82S

bulk velocity:

v_b	9630 m s^{-1}	$T = 300$ K		71G2

Young's modulus (in 10^{11} Pa):

E	4.34	$T = 300$ K		71G2
	4.80	300 K	depends on density	82S
	4.48	300 K		77M4

shear modulus (in 10^{11} Pa):

G	1.82	$T = 300$ K		71G2
	2.00	300 K	depends on density	82S
	1.86	300 K		77M4

Physical property	Numerical value	Experimental conditions	Experimental method, remarks	Ref.

Poisson's ratio:

ν	0.21	$T = 300$ K		77M4

IR-active optical phonon wavenumbers of boron carbide and their medium oscillator-strengths $\Delta\varepsilon$ (uncertain values in brackets) [79B4]:

The complex dielectric constant can be described by

$$\varepsilon(\bar{\nu}) = \varepsilon(\infty) + \sum_j \frac{\Delta\varepsilon_j}{1-(\bar{\nu}/\bar{\nu}_{j_0})^2 - i(\gamma/\bar{\nu}_{j_0})(\bar{\nu}/\bar{\nu}_{j_0})}$$

$\varepsilon(\infty)$: optical dielectric constant (contribution of bound electrons)
$\Delta\varepsilon_j$: oscillator strength (contribution of the j-th oscillator)
$\bar{\nu}$: wavenumbers (in cm^{-1})
$\bar{\nu}_j$: resonance wavenumber of the sustained oscillators
γ: frequency independent damping constant (in cm^{-1})

$\bar{\nu}$ cm^{-1}	$\Delta\varepsilon$ Polycryst. material	$\Delta\varepsilon$ Single crystals $E \parallel c$	$\Delta\varepsilon$ Single crystals $E \perp c$	Prevailing symmetry type
1580	0.15	0.45	0.12	A_{2u}
1090	0.35	0.25	0.5	E_u
(980)				
870	0.02	0.08	0.06	
873	0.01	0.04	0.02	
845	0.02	0.03	0.01	A_{2u}
700	0.07	0.05	0.04	
(660)				
582	0.06		0.20	E_u
(530)				
475				
390	0.6	0.35	0.60	E_u

For the dependence of $\bar{\nu}$ and $\Delta\varepsilon$ on the C content of boron carbide, see Fig. 98.

Transport properties (see also Figs. 86, 87, 90, 91)

electrical conductivity:

σ	$1.4 \cdots 12\ \Omega^{-1}$ cm^{-1}	$T = 300$ K	value depends on C content (cf. Fig. 86)	80W, 81W6
	$8 \cdot 10^{-2} \cdots 5\ \Omega^{-1}$ cm^{-1}	100 K	value depends on C content	

See also [70G2, 79B6, 79T3, 79K9].

For temperature dependence, see Fig. 87.
Activation energy of the electrical conductivity and Peltier energy: Figs. 88, 89.

thermoelectric power:

S	$15 \cdots 160\ \mu$V K^{-1}	$T = 300$ K	value depends on C content, cf. Fig. 94	80W, 81W6

For temperature dependence, see Fig. 90.

Hall coefficient:

R_H	$(-4 \cdots +19)$ $\cdot 10^{-2}$ cm^3 A^{-1} s^{-1}	$T = 300$ K	value depends on C content, see Fig. 91	80W

Physical property	Numerical value	Experimental conditions	Experimental method, remarks	Ref.
hole concentration:				
p	$(4 \cdots 9) \cdot 10^{19}$ cm^{-3}	$T = 300$ K	Hall effect, ESR	70G2, 77G4
hole mobility:				
μ_p	$0.2 \cdots 1$ cm^2 V^{-1} s^{-1}	$T = 300$ K	Hall effect, ESR, conductivity	70G2, 77G4

Optical properties

absorption: Fig. 92, photo-absorption: Fig. 93, occupation of traps by photoexcitation: Fig. 94, reflectivity: Fig. 95, dependence of optical parameters on C content: Figs. 96···98.
For Auger investigations, see [77D].

Further properties
(Review articles, see [65L, 79B6, 79T3].)

thermal conductivity (in W m^{-1} K^{-1}):

		T [°C]		
κ	14.6	100	depends on density	79B3
	26	100		82S
	11.7	500	depends on density	79B3
	19	500		82S
	9.3	1000	depends on density	79B3
	17	1000		82S

For thermal conductivity of theoretically dense boron carbide, see [77M4].
heat capacity: see Fig. 100

densities (in g cm^{-3}):

d	2.52		calculated	71G2
	2.456	$T = 300$ K	experimental	75Y
	2.47···2.51	300 K	exp., dependent on composition ($B_7C - B_4C$)	53A2
	2.465···2.52	300 K	exp., dependent on composition (8.8···20.0 at% C)	81B1

melting point:

T_m	2490 °C		For dependence on composition, see Fig. 99.	71K1

linear thermal expansion coefficients:

α	$3.016 \cdot 10^{-6} + 4.30 \cdot 10^{-9} T - 9.18 \cdot 10^{-13} T^2$ (T in °C)			77M4
	$2.6 \cdots 4.5 \cdot 10^{-6}$ K^{-1}	$T = 25 \cdots 800$ °C	polycrystals	77M1, 79B3
	$5.65(5) \cdot 10^{-6}$ K^{-1}	0···1000 °C	carbon-rich boron carbide	73Y
	$5.87(4) \cdot 10^{-6}$ K^{-1}	0···1000 °C	boron-rich boron carbide	73Y

For temperature dependence, see literature.

microhardnesses:

H_K	2970 (150) kg mm^{-2}	$T = 300$ K	B_4C, 1N load, Knoop hardness	75L2, 82S
H_V	3200 (150) kg mm^{-2}	300 K	B_4C, 1N load, Vickers hardness	75L2, 82S

For dependence on composition, see [75L2] and [79B7]; for dependence on temperature, see [79B3].
For metallography of boron carbide, see [76C].
For properties of irradiated boron carbide, see [77M4].
For mechanical properties of hot-pressed $B - B_4C$ materials, see [79C2].

Physical property	Numerical value	Experimental conditions	Experimental method, remarks	Ref.

Carbon saturated β-rhombohedral boron:
Structure and physical properties comparable to SiB_{14} are probable.

lattice parameters (hexagonal presentation):

a	10.929 Å	$T = 300$ K		81B2
c	23.921 Å	300 K		

Ternary system boron-carbon-silicon
For chemical, structural, and technological properties, see [72K1].

9.14.9.2 Boron-silicon compounds

general literature: [70E4, 79L1]

$Si_{11}B_{31}$:
preparation [79L1], crystalline structure [79L1]

SiB_3:
preparation [77M1, 70E4], crystalline structure [77M1, 77M3]

SiB_4:
lattice: rhombohedral (boron-carbide type) [70E4; 77M1]; space group: $D_{3d}^5 - R\bar{3}m$.

lattice parameters:
(hexagonal presentation)

a	6.35 Å	$T = 300$ K	X-ray diffraction	61C
c	12.69 Å			

resistivity:

ϱ	1.75 Ω cm	$T = 293$ K	polycrystal	65M5, 77M1

thermal conductivity:

κ	5.9 (5) W m^{-1} K^{-1}	$T = 100$ °C	polycrystal	79B3
	4.7 (4) W m^{-1} K^{-1}	500 °C		
	3.5 (3) W m^{-1} K^{-1}	1000 °C		

density:

d	2.41 g cm^{-3}	$T = 300$ K	X-ray	61C
	2.44 g cm^{-3}	300 K	pycnometric	77M1

melting point:

T_m	1269 °C			70E5, 77M1

coefficient of linear thermal expansion:

$\bar{\alpha}$	$6 \cdot 10^{-6}$ K^{-1}	$T = 20 \cdots 1000$ °C	polycrystal, $\bar{\alpha}$: average value	79B3

microhardness:

H_K	2000 $\cdots$ 2500 kg mm^{-2}	$T = 300$ K	Knoop hardness	64S
	1870 $\cdots$ 2290 kg mm^{-2}	300 K	Knoop hardness	60R, 77M1

For temperature dependence, see [79B3].

Young's modulus:

E	280 GPa	$T = 300$ K		79B3

Physical property	Numerical value	Experimental conditions	Experimental method, remarks	Ref.
SiB$_6$:				
lattice: orthorhombic [70E4, 77M1, 81A4]				
lattice parameters:				
a	14.392 Å	$T = 300$ K	X-ray diffraction	58A,
b	18.267 Å			77M1,
c	9.8852 Å			81A4
For dependence on composition and temperature, see [79L1].				
resistivity:				
ϱ	0.2 Ω cm	$T = 300$ K	polycrystal	61C, 77M1
density:				
d	2.39 g cm^{-3}	$T = 300$ K	X-ray	58A
	2.43 g cm^{-3}	300 K	pycnometric	77M1
melting point:				
T_m	1864 °C			70E5, 77M1
volume expansion coefficient:				
β	$18.15 \cdot 10^{-6}$ K^{-1}	$T = 20 \cdots 1190$ °C	a decreases, b and c increase with increasing temperature	79L1
microhardness:				
H_K	$3200 \cdots 3500$ kg mm^{-2}	$T = 300$ K	Knoop hardness	64S
	$2520 \cdots 2870$ kg mm^{-2}			60R, 77M1
SiB$_{14}$:				
Structure				

The crystalline structure has been suggested to be isotypic with β-rhombohedral boron (see vol 17e, section 8.1). It is based on a nearly cubic closest-packed arrangement of B$_{84}$ units. Additionally a (B$_7$Si$_3$)–Si–(B$_7$Si$_3$) group is expected to be arranged along the trigonal axis of the rhombohedral cell (i.e. the crystallographic c-axis) [70M2, 70M3, 77M3].

Physical property	Numerical value	Experimental conditions	Experimental method, remarks	Ref.
lattice parameters: (hexagonal presentation)				
a	11.13 Å	$T = 300$ K	X-ray diffraction	65G,
c	23.83 Å			70M2
Physical properties				
conductivity:				
σ	$6 \cdot 10^{-6}$ Ω^{-1} cm^{-1}	$T = 300$ K	For temperature dependence, see Fig. 101.	78D1, 78P2, 78P3, 79P1
thermoelectric power:				
S	$+400$ μV K^{-1}	$T = 300$ K	For temperature dependence, see Fig. 102.	78D1, 78P3, 79P1
mobility:				
$\mu_{H,p}$	<1 cm^2 V^{-1} s^{-1}	$T = 300$ K	Hall mobility	76A2

p-type conductivity character of pure material [79P1, 76A2].

Physical property	Numerical value	Experimental conditions	Experimental method, remarks	Ref.
\multicolumn{5}{l}{A dominating trapping level similar to β-rhombohedral boron exists:}				
E_t	0.38 eV		trapping level	78P2, 79P1
\multicolumn{5}{l}{For dielectric constants, see Figs. 103···105.}				
thermal conductivity:				
κ	0.015···0.03 W cm^{-1} K^{-1}	$T = 300$ K		76A2
density:				
d	2.48 g cm^{-3}	$T = 300$ K	experimental	76A2
	2.51 g cm^{-3}	300 K	X-ray	76A2
\multicolumn{5}{l}{The transport properties are strongly influenced by impurities:}				
resistivity:				
ϱ	60···7.6·10^4 Ω cm	$T = 300$ K	For temperature dependence, cf. Fig. 106.	76A2
thermoelectric power:				
S	40···340 μV K^{-1}	$T = 300$ K	For temperature dependence, cf. Fig. 107.	76A2

(cp. below: general information on doped B−Si compounds with undefined structure)

SiB$_{12}$:

preparation [77M1, 70E4], crystalline structure [77M1]

SiB$_{\approx 36}$:

Solid solution of Si in β-rhombohedral boron; space group $D_{3d}^5 - R\bar{3}m$. The Si atoms are distributed in the lattice by the way that two Si atoms occupy interstitial holes and a third Si atom substitutes partially a framework boron atom [81A4, 81V, 82L].

lattice parameters:
(hexagonal presentation)

a	11.01 (1) Å	$T = 300$ K	X-ray diffraction	81V
c	23.90 (2) Å	300 K		

B−Si compounds of undefined structure (possibly SiB$_{14}$ structure):

Compositions Si$_x$M$_{1-x}$B$_{\approx 30}$ (M = Fe, Co, Ni); (cp. FeB$_{29.5}$).

resistivity: Fig. 106, thermoelectric power: Fig. 107.

n-type conductivity character at low, p-type at high temperatures [81D].

Amorphous Si−B alloys:

The amorphous films are prepared by radio frequency plasma decomposition of silane-diborane gas mixtures [79T1]. The films are hydrogenated.

Physical properties

The energy gaps of films with zero Si content and zero B content (Fig. 108) are considerably greater than those of sputtered a-Si (1.34 eV) and a-B (1.1 eV). This is attributed to the high Si−H and B−H bond energy.

Optical gap and electrical activation energy: Fig. 108, g-value: Fig. 109, electrical conductivity: Fig. 110, spin density: Fig. 111, absorption edge: Fig. 112, IR transmission: Fig. 113.

9.14.9.3 Boron-germanium compounds

For a general phase diagram of the B−Ge system, see [77M1, 70B].

GeB$_{\approx 90}$:

Solid solution of Ge in β-rhombohedral boron; space group: $D_{3d}^5-R\bar{3}m$. The Ge atoms partially occupy seven crystallographic positions, of which all except one are interstitial sites [81T3].

lattice parameters:
(hexagonal presentation)

a	10.9588 (8) Å	$T=300$ K	X-ray diffraction	81T3
c	23.8622 (11) Å			

9.14.10 Compounds with group IVb elements

9.14.10.1 Boron-titanium compounds

TiB:

metallic; preparation [69L, 75S, 77L], crystalline structure [69L, 75S, 77L], electronic structure [79P3]

Ti$_3$B$_4$:

preparation [75S, 81N2], crystalline structure [75S]

TiB$_2$:

metallic; preparation [69L, 75S, 77L, 77G1, 79F3, 79S5, 75V], crystalline structure [69L, 75S, 76S2], electronic structure [81A2, 81W5, 77S1, 79P3, 76P1, 76S3], X-ray photoelectron spectra [79A3], electrical conductivity [79L4]

Ti$_2$B$_5$:

preparation [75S], crystalline structure [75S]

(B$_{12}$)$_4$B$_2$Ti$_{1.3-2.0}$:

preparation [75A2], crystalline structure [75A2, 76A3, 76P2]

Ti−Cr−B system:

preparation [79K5], crystalline structure [79K5]

TiB$_x$N$_{1-x}$:

(x = 0.17, 0.35, 0.78), preparation [81L]

9.14.10.2 Boron-zirconium compounds

ZrB$_2$:

metallic; preparation [75S, 52P, 77G1, 77L], crystalline structure [75S, 77G1], electronic structure [77S1, 77S2, 77P1, 76S3], electrical conductivity [79L4], electrical transport [77C1].

ZrB$_{12}$:

superconducting; preparation [75S], crystalline structure [75S], Curie temperature [71F2], magnetic properties [73O]

ZrB$_{51}$:

semiconductor?; preparation [82L], crystalline structure [82L], solid solution of Zr in β-rhombohedral B

9.14.10.3 Boron-hafnium compounds

HfB$_2$:

metallic; preparation [75S], crystalline structure [75S], electronic structure [76S3]
   For ternary compounds with Zr and Hf, see [79R1].

9.14.11 Compounds with group Va elements

9.14.11.1 Boron-nitrogen compounds

BN: see vol. 17a, section 2.1, p. 148

$B_{50}N_2$:

semiconductor (?); preparation [72P], crystalline structure [76W1]

9.14.11.2 Boron-phosphorus compounds

BP: see vol. 17a, section 2.2, p. 153

B_5P_3:

preparation [1891M], no further information available

B_6P ($B_{12}P_2$); $B_{13}P_2$:

Structure

Boron-carbide type structure [58P, 60W, 61L, 61M, 66S, 70E1, 77M2, 77M3].
lattice: rhombohedral; space group: $D_{3d}^5 - R\bar{3}m$.

lattice parameters:
(hexagonal presentation)

a	5.984 Å	$T = 300$ K	X-ray diffraction	61L, 61M, 77M2, 77M3
c	11.850 Å			

Physical properties
energy gap:

$E_{g, ind}$	3.3 eV	$T = 300$ K, $E \perp c$	optical absorption	65B, 67B

For X-ray emission, see [76D].

resistivity (in Ω cm):

		T [K]		
ϱ	$3 \cdot 10^2$	300	polycrystals (temperature dependence: Fig. 114)	59G2, 60G2
	10^8	300	epitaxial grown film	73T, 74T
	10^6	295	non-colored crystals, $p = 10^{16}$ cm^{-3}	64P
	$10 \cdots 100$	295	blue-black crystals	64P

p-type conductivity [59G2, 64P].
current voltage characteristic: Fig. 115.

thermal conductivity:

κ	$3.8 \cdot 10^{-1}$ W cm^{-1} K^{-1}	$T = 300$ K	polycrystal For temperature dependence, see Fig. 116.	71S3

phonon mean free path:

Λ	$2 \cdot 10^{-3}$ cm	$T < 10$ K		71S3

Debye temperature:

Θ_D	1160 K		calculated from sound velocities	71S3, 74G3

Physical property	Numerical value	Experimental conditions	Experimental method, remarks	Ref.
density:				
d	2.597 g cm^{-3}	$T = 300$ K	X-ray	71S3
	2.599 (5) g cm^{-3}		pycnometric	71S3
melting point:				
T_m	> 2000 °C			64P

9.14.11.3 Boron-arsenic compounds

BAs: see vol. 17a, section 2.3, p. 156

B$_6$As (B$_{12}$As$_2$); B$_{13}$As$_2$:

Structure

Boron carbide type structure [61L, 61M, 66S, 70E1, 73H, 77M2, 77M3].

lattice: rhombohedral; space group: $D_{3d}^5 - R\bar{3}m$.

lattice parameters:
(hexagonal presentation)

a	6.142 Å			61L, 61M, 77M2, 77M3
c	11.892 Å			

(rhombohedral presentation)

a	5.3 Å			61L, 61M, 77M2, 77M3
α	70°30′			

Other indication of the lattice structure:
lattice: orthorhombic [74G3].

lattice parameters:

a	9.709 Å	$T = 300$ K		74G3
b	4.343 Å			
c	3.069 Å			

Physical properties

For X-ray emission, see [76D].
(The band gap of 1.45 eV, which has been ascribed to B$_{13}$As$_2$ in [79G2], belongs in fact to BAs.)

thermal conductivity:				
κ	1.2 W cm^{-1} K^{-1}	$T = 300$ K	crystal orientation not reported Temperature dependence: Fig. 117.	71S3
phonon mean free path:				
Λ	≈ 10^{-1} cm	$T < 10$ K	Temperature dependence: Fig. 118.	71S3
Debye temperature:				
Θ_D	940 K		calculated from sound velocities	71S3, 74G3
density:				
d	3.583 g cm^{-3}	$T = 300$ K	X-ray	71S3
	3.59 (5) g cm^{-3}	300 K	pycnometric	71S3

9.14.12 Compounds with group V b elements
9.14.12.1 Boron-vanadium compounds

V_3B_2:
preparation [69S2, 75S]

VB:
metallic; preparation [69S2, 75S, 77L], crystalline structure [77L], electronic structure [79P3]

V_5B_6:
preparation [69S2, 75S, 77L], crystalline structure [77L]

V_3B_4:
preparation [69S, 75S, 77L], crystalline structure [77L]

V_2B_3:
preparation [69S2, 75S, 77L], crystalline structure [77L]

VB_2:
metallic; preparation [69S2, 75S, 77G1], crystalline structure [77C1, 77G1, 81N1], electronic structure [79P3, 76S3, 81N1], electrical conductivity [79L4], electronic transport, thermal properties [77C1]

9.14.12.2 Boron-niobium compounds

NB_3B_2:
preparation [75S]

NbB:
preparation [75S, 77L], electronic structure [76S3]

Nb_3B_4:
preparation [75S, 77L]

NbB_2:
metallic; preparation [75S, 77L], crystalline structure [81N1], electronic structure [76S3, 79P3], electronic transport, thermal properties [77C1]

9.14.12.3 Boron-tantalum compounds

Ta_2B:
preparation [75S, 77L], crystalline structure [75S, 77G1], electronic structure [76S3, 79P3]

Ta_3B_2:
preparation [75S], crystalline structure [75S, 77G1]

TaB:
preparation [75S, 77L], crystalline structure [75S, 77G1], electronic structure [76S3, 79P3]

Ta_3B_4:
preparation [75S, 77L], crystalline structure [75S, 77G1]

TaB_2:
metallic; preparation [75S, 77L], crystalline structure [75S, 77G1], electronic structure [76S3, 79P3], electronic transport, thermal properties [77C1]

Physical property	Numerical value	Experimental conditions	Experimental method, remarks	Ref.

9.14.13 Compounds with group VIa elements

9.14.13.1 Boron-oxygen compounds

$(BO)_x$:

preparation [77G2], crystalline structure [77G2]

B_2O_3:

preparation [77G2], crystalline structure [77G2, 71S2, 70G1, 68S], Raman and IR spectra [71B1, 71B2, 71S1, 60P], stimulated Brillouin scattering [71G1], nuclear magnetic resonance [76R]

B_2O:

preparation [77G2], crystalline structure [77G2, 65H]

B_6O, $B_{12}O_2$, B_7O:

semiconductor (?); preparation [77G2], crystalline structure: B_4C-type [61L, 66L1, 65M2, 76B1], thermal conductivity, physicochemical properties [79B3], X-ray emission [71P1], thermal properties [73P], electrical conductivity [70W2]

9.14.13.2 Boron-sulfur compounds

$(BS_2)_n$:

preparation [77G2]

B_2S_3:

preparation [77G2], crystalline structure [70C1, 77G2], IR emission and absorption spectra [59G1]

BS, B_2S_2, $(B_2S_3)_n$:

preparation [77G2]

B_4S:

semiconductor (?); preparation [77G2]

$B_{12}S$:

semiconductor (?); preparation [77G2], crystalline structure: B_4C-type [77G2]

9.14.13.3 Boron-selenium compounds

B_2Se_3:

Structure

crystalline structure: monoclinic [72H1, 77G2]
space group: P_2P_m (?) or $C_{2h}^1-P_2/m$ [77G2]

lattice parameters:

a	4.06 (1)	$T = 300$ K	X-ray diffraction	72H2
b	38.6 (1)			
c	10.43 (2)			
β	90°			

Physical properties

No detailed investigations on the electronic properties are available. But the obtained materials, either in crystalline or glassy form are reported to be colored in red, orange or yellow [62H, 67C, 70G3, 71D, 72D, 72H1, 77G2]. Therefore semiconducting behavior can be expected.

Physical property	Numerical value	Experimental conditions	Experimental method, remarks	Ref.
IR spectra absorption maxima wavenumbers (in cm^{-1}):				
B_2Se_3 (polycrystal)				
$\bar{\nu}$	768	$T = 300$ K	IR absorption	72H2, 77G2
	789			
	804			
	831			
	868			
	883			
B_2Se_3 (glass)				
$\bar{\nu}$	775	300 K	IR absorption	72H2, 77G2
	838			
	861			

Deviating results on IR spectra are reported in [62H].

density:

d	2.10 g cm^{-3}		calculated	77G2

melting point:

T_m	753 K			69B, 72H2

glass softening temperature:

T_{gs}	325 °C		beginning of glass softening	71D, 72D

9.14.14 Compounds with group VIb elements

9.14.14.1 Boron-chromium compounds

Cr_2B:

metallic; preparation [75S, 77P2, 77P3], crystalline structure [75S], electronic structure [79P3], X-ray photoelectron spectra [79A3], transport [75S], electrical properties [76G6]

Cr_5B_3:

metallic; preparation [75S], crystalline structure [75S, 76C], thermal and magnetic properties [79L3], transport [75S]

CrB:

metallic; preparation [75S, 77P2, 77P3], crystalline structure [75S, 76C], electronic structure [79P1, 79P3], X-ray photoelectron spectra [79A3], thermal and magnetic properties [79L3]

Cr_3B_4:

metallic; preparation [75S], crystalline structure [75S, 76C], transport [75S], electronic properties [76G6]

CrB_2:

metallic or semimetallic; preparation [75S, 77P2], crystalline structure [77C1, 75S], electronic structure [75L1, 77S1, 77P1, 79P3, 81A2], transport [77C1, 77S2], thermal properties [77C1], X-ray photoelectron spectra [79A3], magnetic and thermal properties [79L3], electronic conduction [79L4], electronic properties [76G6]

CrB_4:

metallic; preparation [75S], crystalline structure [75S]

CrB_{41}:

semiconducting?; preparation [70A1, 68A], solid solutions in β-rhombohedral boron [70A1, 82L], hardness [70C2]

9.14.14.2 Boron-molybdenum compounds

Mo$_2$B:

metallic; preparation [75S], crystalline structure [77L], electronic structure [79P3], solid state properties [75S]

α-MoB:

metallic; preparation [75S, 77G1], crystalline structure [76C, 77G1], electronic structure [79P3], solid state properties [75S], electrical conductivity [79L4]

β-MoB:

metallic; preparation [75S, 77G1], crystalline structure [76C, 77G1], electronic structure [79P3], solid state properties [75S], electrical conductivity [79L4]

MoB$_2$:

metallic; preparation [77L], crystalline structure [77L], electronic structure [79P3], electronic transport [79K4]

Mo$_2$B$_5$:

metallic; preparation [75S], crystalline structure [76S2, 76C, 77L], electronic transport [79K4]

Mo$_{1-x}$B$_3$:

preparation [73L], crystalline structure [73L]

MoB$_4$:

metallic; preparation [75S], crystalline structure [75S], electronic transport [79K4]

9.14.14.3 Boron-tungsten compounds

W$_2$B:

metallic?; preparation [75S], crystalline structure [75S, 77L], electronic structure [79P3]

α-WB:

metallic?; preparation [75S, 77G1], crystalline structure [75S, 77L, 77P3], electronic structure [79P3]

β-WB:

metallic?; preparation [75S, 77G1], crystalline structure [75S, 77L, 77P3], electronic structure [79P3]

α-W$_2$B$_5$:

metallic; preparation [75S, 77G1], crystalline structure [75S, 77L, 76S2, 76P2, 77P3], electronic transport [79K4]

β-W$_2$B$_5$:

metallic; preparation [75S, 77G1], crystalline structure [75S, 77L, 76S2, 76P2, 77P3], electronic transport [75S]

WB$_2$:

preparation [68L], crystalline structure [68L, 76C]

WB$_4$:

metallic; preparation [75S], crystalline structure [75S, 77L], electronic transport [79K4]

WB$_{12}$:

semicond.; preparation [79A5], solid state properties [79A5]

9.14.15 Compounds with group VIIa elements

BF$_3$:

insulators (molecular crystals); crystalline structure [59D], IR and Raman spectra [74B1, 72C, 74B4]

BCl$_3$:

insulators (molecular crystals); crystalline structure [61D, 57A, 57S], IR and Raman spectra [74B1, 72C, 75B1]

BBr₃:

insulators (molecular crystals); crystalline structure [47R, 62R], IR and Raman spectra [74B1, 72C]

BJ₃:

insulators (molecular crystals); crystalline structure [58W]

B₄Cl₄:

insulators (molecular crystals); crystalline structure [53A1], electronic structure [65M3], IR and Raman spectra [76B2]

9.14.16 Compounds with group VIIb elements

9.14.16.1 Boron-manganese compounds

Mn₄B:

preparation [75S, 77L], crystalline structure [81T2], magnetic properties [77B5], X-ray photoelectron spectra [79A3]

Mn₂B:

preparation [75S, 77L], crystalline structure [81T2], magnetic properties [77B5], X-ray photoelectron spectra [79A3]

MnB:

preparation [75S, 77L], electronic structure [81A3], magnetic properties [77B5], X-ray photoelectron spectra [79A3]

Mn₃B₄:

preparation [75S, 77L], magnetic properties [77B5], X-ray photoelectron spectra [79A3]

MnB₂:

metallic?; preparation [75S, 77L], electronic structure [81A2], magnetic properties [77B5], X-ray photoelectron spectra [79A3], electronic and thermal properties [77C1]

MnB₄:

preparation [75S, 77L], crystalline structure [77L], magnetic properties [77B5]

MnB₂₃:

semiconducting?; preparation [74A, 70A2], crystalline structure [77N2, 70A2, 73A1, 82L], hardness [70C2]

9.14.16.2 Boron-technetium compounds

Tc₃B:

metallic?; preparation [75S, 77L]

Tc₇B₃:

preparation [75S, 77L]

TcB₂:

semiconducting?; preparation [75S, 77L, 77C2], electronic structure [77C2], X-ray photoelectron spectra [79A3]

9.14.16.3 Boron-rhenium compounds

Re₃B:

preparation [75S, 77L]

Re₇B₃:

preparation [75S, 77L]

ReB₂:

semiconducting?; preparation [75S, 77L, 77C2], electronic structure [77C2], X-ray photoelectron spectra [79A3]

| Physical property | Numerical value | Experimental conditions | Experimental method, remarks | Ref. |

9.14.17 Compounds with group VIII elements

9.14.17.1 Boron-iron compounds

Fe_2B:

metallic; preparation [75S, 77L], crystalline structure [75S, 77L], electronic structure [77S1], electronic transport [75S], X-ray photoelectron spectra [79A3], magnetic properties [77B5]

FeB:

metallic?, semiconducting?; preparation [75S, 77L], crystalline structure [75S, 77L], electronic structure [77S1, 81A3], electronic transport [75S], X-ray photoelectron spectra [79A3], magnetic properties [73B2, 77B5], electrical resistance, thermoelectric power [75B3]

amorphous (glassy) Fe–B alloys:

metallic (Fe-rich); preparation [78F3, 79G4, 79T2], crystalline structure [78F3, 79G4, 79T2], magnetic properties [78J, 78O, 78E, 78K2, 79S6]

9.14.17.2 Solid solutions of Fe in β-rhombohedral boron

Structure

The crystal structure of β-rhombohedral boron contains a number of holes, which are large enough to accommodate transition metal atoms [70A1, 74A, 74H2, 74L, 76L1, 76H, 82L]. The solubility limit is at the composition $FeB_{29.5}$ [81W1].

position and occupancy of the Fe sites in FeB_{49} [76C]:

position	approximate coordinates			occupancy
	x	y	z	%
6 (c)	0	0	0.135	50.7 (3)
18 (h)	0.203	0.406	0.175	18.5 (2)

The same atomic positions are occupied in $FeB_{29.5}$ [76C, 81W1]; both occupancies are higher, roughly proportional to the Fe content.

lattice parameters (hexagonal presentation):

pure β-rhombohedral boron: $a = 10.9251\,(2)$ Å, $c = 23.8143\,(8)$ Å [77C3]
FeB_{49}: $a = 10.9514\,(8)$ Å, $c = 23.8609\,(16)$ Å [81W1]
$FeB_{29.5}$: $a = 10.9700\,(4)$ Å, $c = 23.880\,(1)$ Å [81W1]
For Mössbauer investigations of Fe in B, see [72W, 72S3].

Physical properties

In $FeB_{29.5}$ the acceptor-like level 0.19 eV above the valence band (E_v) of pure β-rhombohedral boron (see vol. 17e, section 8.1) is overcompensated by electrons originating from the Fe atoms. From the results available at present it cannot be decided whether the transport in $FeB_{29.5}$ (n-type) occurs within the level 0.4 eV below the conduction band edge (E_c) of pure β-rhombohedral boron or within an additional level evoked by the Fe atoms. Activation energy: Fig. 119.

conductivity:

σ	$0.7\,\Omega^{-1}\,cm^{-1}$	$T = 300$ K	$FeB_{29.5}$; for temperature dependence, see Fig. 120	81W1

carrier concentration:

n	$7.6 \cdot 10^{18}\,cm^{-1}$		formally calculated	81W1

Hall coefficient:

R_H	$-0.81\,cm^3\,C^{-1}$	$T = 300$ K	$FeB_{29.5}$	81W1

mobility:

μ_H	$0.58\,cm^2\,V^{-1}\,s^{-1}$		formally calculated	81W1

Physical property	Numerical value	Experimental conditions	Experimental method, remarks	Ref.
thermoelectric power:				
S	$-19\ \mu V\ K^{-1}$	$T = 300$ K	$FeB_{29.5}$; for temperature dependence, see Fig. 121	81W1
	$-23\ \mu V\ K^{-1}$		FeB_{30}	81D
reflectivity: for relative difference of spectra for B and $FeB_{29.5}$, see Fig. 122				
hardness:				
H	4010 kg mm^{-2}	$T = 300$ K	50 g load; hardness not specified	70C2

9.14.17.3 Boron-cobalt compounds

Co_3B, Co_2B:
metallic; preparation [75S], crystalline structure [75S, 77L], electronic structure [77S1], electronic transport [75S], magnetic properties [77B5]

CoB:
semiconducting?; preparation [75S], crystalline structure [77S1], electronic transport [75B3], magnetic properties [73B2]

CoB_{12}:
semiconducting; preparation [79A5], electronic transport [79A5]

9.14.17.4 Boron-nickel compounds

Ni_3B:
metallic; preparation [75S], crystalline structure [77L, 75S], electronic transport [75S], magnetic properties [77B5]

Ni_2B:
metallic; preparation [75S], crystalline structure [77L, 75S], electronic transport [75S], magnetic properties [77B5]

Ni_4B_3 (o):
crystalline structure [77L]

Ni_4B_3 (m):
crystalline structure [77L]

NiB:
metallic; preparation [75S], crystalline structure [77L, 75S], electronic transport [75S, 73B2], magnetic properties [73B2, 77B5]

NiB_{25}:
semiconducting?; preparation [60D], crystalline structure [77M2, 60D, 77M3, 70M3]

amorphous (glassy) Ni−B alloys:
metallic; preparation [78F2, 79M4], crystalline structure [79M4, 79K6], electrical conductivity [78F2, 79M4]

9.14.17.5 Boron-ruthenium compounds

Ru_7B_3:
crystalline structure [77L]

$Ru_{11}B_8$:
crystalline structure [77L]

$RuB_{1.1}$:
crystalline structure [77L]

Ru$_2$B$_3$:
crystalline structure [77L]

RuB$_2$:
semiconducting?; preparation [77C2], crystalline structure [77L, 77C2], electronic structure [77C2]

9.14.17.6 Boron-rhodium compounds

RhB$_{\approx 1.1}$:
semiconducting?; preparation [75S], crystalline structure [75S, 77L], electrical conductivity [75S]

Rh$_7$B$_3$:
semiconducting?; preparation [75S], crystalline structure [75S, 77L], electrical conductivity [75S]

Rh$_5$B$_4$:
preparation [81T3], crystalline structure [81S2, 81T3]

9.14.17.7 Boron-palladium compounds

Pd$_3$B:
metallic; preparation [75S], crystalline structure [75S, 77L], electrical conductivity [75S]

Pd$_5$B$_2$:
metallic; preparation [75S], crystalline structure [75S, 77L], electrical conductivity [75S]

9.14.17.8 Boron-osmium compounds

OsB$_{1.2}$:
crystalline structure [77L]

OsB$_{1.6}$:
crystalline structure [81S2]

Os$_2$B$_3$:
crystalline structure [77L]

OsB$_2$:
semiconducting?; crystalline structure [77C2], electronic structure [77C2]

9.14.17.9 Boron-iridium compounds

IrB$_{0.9}$:
crystalline structure [81S2]

IrB$_{1.15}$:
metallic?; preparation [75S], crystalline structure [75S, 77L], electrical conductivity [75S]

IrB$_{1.35}$:
preparation [75S], crystalline structure [75S]

IrB$_{3-x}$:
crystalline structure [77L]

9.14.17.10 Boron-platinum compounds

Pt$_4$B:
crystalline structure [81S2]

Pt$_2$B:
crystalline structure [81S2]

Pt$_3$B:
crystalline structure [81S2]

PtB:
crystalline structure [77L]

9.14.18 References for 9.14

1891M	Moissan H.: C.R. Acad. Sci. (Paris) **113** (1891) 726.
47R	Rollier, M.A., Riva, A.: Gazz. Chim. Ital. **77** (1947) 361.
51L	Lafferty, J.: J. Appl. Phys. **22** (1951) 299.
52B	Bertaut, F., Blum, P.: C.R. Acad. Sci. (Paris) **234** (1952) 2621.
52P	Post, B., Glaser, F.: J. Chem. Phys. **20** (1952) 1050.
53A1	Atoji, M., Lipscomb, W.N.: J. Chem. Phys. **21** (1953) 172.
53A2	Allen, R.D.: J. Am. Chem. Soc. **75** (1953) 3582.
54B	Blum, P., Bertraut, F.: Acta Crystallogr. **7** (1954) 81.
54E	Eberhardt, W.H., Crawford, J.B., Lipscomb, W.N.: J. Chem. Phys. **22** (1954) 989.
56F	Felten, E.J.: J. Am. Chem. Soc. **78** (1956) 5977.
56P	Post, B., Moskowitz, D., Glaser, F.W.: J. Am. Chem. Soc. **78** (1956) 1800.
56S	Samsonov, G.V.: Metallphys. Metallkunde SSSR Ural Filial **3** (1956) 309.
57A	Atoji, M., Lipscomb, W.N.: J. Chem. Phys. **27** (1957) 195.
57S	Spencer, C., Lipscomb, W.N.: J. Chem. Phys. **27** (1957) 355.
57Y	Yamazaki, M.: J. Chem. Phys. **27** (1957) 746.
58A	Adamsky, R.: Acta Crystallogr. **11** (1958) 744.
58P	Perri, J.A., La Placa, S., Post, B.: Acta Crystallogr. **11** (1958) 310.
58W	Wentink, T., Tiensun, V.H.: J. Chem. Phys. **28** (1958) 826.
59D	Dows, D.A.: J. Chem. Phys. **31** (1959) 1637.
59G1	Greene, F.T., Margrave, J.L.: J. Am. Chem. Soc. **81** (1959) 5555.
59G2	Greiner, E.S., Gutowski, J.A.: J. Appl. Phys. **30** (1959) 1842.
59S	Silver, A.H., Bray, P.J.: J. Chem. Phys. **31** (1959) 247.
60B	Binder, I., La Placa, S., Post, B., in: Boron I, J.A. Kohn, W.F. Nye, G.K. Gaulé, eds., Plenum Press: New York, **1960** p. 86.
60D	Decker, B.F., Kasper, J.S.: Acta Crystallogr. **13** (1960) 1030.
60E	Eubank, W.R., Pruitt, L.E., Thurnauer, H.: see [60B], p. 116.
60G1	Giardini, A.A., Kohn, J.A., Toman, L., Eckart, D.W.: see [60E1], p. 140.
60G2	Greiner, E.S.: see [60B1], p. 105.
60K1	Kohn, J.A.: see [60B1], p. 75.
60K2	Kelley: Bur. Mines. Bull. **12** (1960) 584.
60L	Lipscomb, W.N., Britton, D.: J. Chem. Phys. **33** (1960) 275.
60M	McDonald, B.J., Stuart, W.I.: Acta Crystallogr. **13** (1960) 447.
60P	Parsons, J.L., Milberg, M.E.: J. Am. Ceram. Soc. **43** (1960) 326.
60R	Rizzo, M.F., Bidwell, L.R.: J. Am. Ceram. Soc. **43** (1960) 550.
60S	Samsonov, G.V., Markowsky, L.Ya., Zhigach, A.F., Valyashko, M.G.: Bor, ego soedineniya i splavy, Kiev Izd. Akad. NAUK Ukr. SSR, **1960**.
60W	Williams, F.V., Ruehrwein, R.A.: J. Am. Chem. Soc. **82** (1960) 1330.
61C	Colton, E.: Mater. Des. Eng. **53** (1961) 9.
61D	Dows, D.A., Bottger, G.: J. Chem. Phys. **34** (1961) 689.
61E	Elliot, R.P.: IIT Research Inst., ARF–2200–12, Final Rep. US. At. Energy Comm. Contract At. (11-1) 578, Project Agreement No. 4, **1961**, 44 S.
61H	Hoenig, C.L., Cline, C.F.: J. Am. Ceram. Soc. **44** (1961) 385.
61J	Johnson, R.W., Daane, A.H.: J. Phys. Chem. **65** (1961) 909.
61L	La Placa, S., Post, B.: Planseeber. Pulvermetall. **9** (1961) 109.
61M	Matkovich, V.I.: Acta Crystallogr. **14** (1961) 93.
61S	Sands, D., Cline, G.F., Zalkin, A., Hoenig, C.L.: Acta Crystallogr. **14** (1961) 309.
61Z	Zhuravlev, N.N., Makarenko, G., Samsonov, G.: Izv. Akad. Nauk SSSR OTN **1** (1961) 133.
62H	Hutchinson, W.E., Eick, H.A.: Inorg. Chem. **1** (1962) 434.
62R	Ring, M.A., Donnay, J.D.H., Koski, W.S.: Inorg. Chem. **1** (1962) 109.
62S	Samsonov, G.V., Vainshtein, E.Y., Paderno, Y.B.: Fiz. Metal. Metallov. **13** (1962) 764.
63B	Becher, H.I., Schäfer, A.: Z. Anorg. Allg. Chem. **321** (1962) 217.
63H	Hagenmuller, P., Naslain, R.: C.R. Acad. Sci. (Paris) **257** (1963) 1294.
63J	Johnson, R.W., Daane, H.H.: J. Chem. Phys. **38** (1963) 425.
63M	McNiff, E.J., Shapiro, S.: J. Phys. Chem. Solids **24** (1963) 939.
63R	Rudman, R., La Placa, S., Post, B.: Acta Crystallogr. **16** (1963) 29.
64B	Becher, H.J. in: Boron Vol. 2 Ed. G.K. Gaulé, Plenum Press: New York, **1964** p. 89.

9.14.18 References for 9.14

64G	Geist, D.: Int. Conf. Phys. Semicond., Paris (M. Hulin, ed.) **1964** p. 767.
64P	Peret, J.L.: J. Am. Ceram. Soc. **47** (1964) 44.
64S	Samsonov, G.V., Sleptsov, V.M.: Proshk. Metall. **6** (1964) 58.
65B	Burmeister, R.A. Jr., Greene, P.E.: Bull. Am. Phys. Soc. Ser. II; **10** (1965) 1184.
65E	Eick, H.A.: Inorg. Chem. **4** (1965) 1237.
65G	Giese, R.F. Jr., Economy, J., Matkovich, V.I.: Z. Kristallogr. **122** (1965) 144.
65H	Hale, H.T., Compton, L.A.: Inorg. Chem. **4** (1965) 1213.
65L	Lipp, A.: Tech. Rdsch. 57, No. **14** (1965) 5; 57, No. **28** (1965) 14; 57, No. **33** (1965) 3; 58, No. **7** (1966) 3.
65M1	Matkovich, V.I., Giese, R.F., Economy, J.: Z. Kristallogr. **122** (1965) 66.
65M2	Matkovich, V.I., Giese, R.F., Economy, J.: Z. Kristallogr. **122** (1965) 116.
65M3	Massey, A.G., Urch, D.S.: J. Chem. Soc. Am. (1965) 6180.
65M4	Matkovich, V.I., Economy, J., Giese, R.F., Barret, R.: Acta Crystallogr. **19** (1965) 1056.
65M5	Meerson, G.A., Kiparisov, S.S., Gurevich, M.A.: Proshk. Metall. **3** (1965) 32.
66L1	Lipp, A.: Ber. Dtsch. Keram. Ges. **43** (1966) 60.
66L2	Lipp, A., Roeder, M.: Z. Anorg. Allg. Chem. **344** (1966) 225.
66N	Naslain, R., Etourneau, J.: C.R. Acad. Sci. (Paris) **263** (1966) 484.
66S	Sullenger, B., Kennard, Ch.L.: Sci. Am. 215 No. **7** (1966) 96.
67B	Burmeister, R.A.Jr., Greene, P.E.: Trans. MS. AIME **239** (1967) 408.
67C	Cueilleron, J., Hillel, R.: Bull. Soc. Chim. Fr. N **8** (1967) 2973.
67N	Naslain, R.: Thèse No. 188 Docteurs Sciences Physiques, Paris, **1967**.
67P	Paderno, Yu.B., Pokrzywnicki, S., Stalinski, B.: Phys. Status Solidi **24** (1967) K 73.
68A	Andersson, S., Lundström, T.: Acta Chem. Scand. **22** (1968) 3103.
68B	Becher, H.J., Neidhard, H.: Acta Crystallogr. B **24** (1968) 280.
68G	Geballe, T.H., Matthias, B.H., Andres, K., Maita, I.P., Copper, A.S., Corenzwit, E.: Science **160** (1968) 1443.
68K1	Koulmann, J.J., Perol, N., Taglang, P.: C.R. Acad. Sci. (Paris) B **267** (1968) 15.
68K2	Koulmann, J.J., Taglang, P.: C.R. Acad. Sci. (Paris) B **266** (1968) 759.
68L	Lundström, T.: Ark. Kemi **30** (1968) 115.
68S	Strong, S.L., Kaplow, R.: Acta Crystallogr. **324** (1968) 1032.
69B	Boryakova, V.A., Greenberg, Ya.H., Zhukov, E.G. Loryazhkin, V.A., Medvedeva, Z.S.: Izv. AN USSR, Ser. Neorg. Mater. **5** (1969) 477.
69E	Eick, H.A., Mulford, R.N.R.: J. Inorg. Nucl. Chem. **31** (1969) 371.
69F	Fisk, Z.: Ph. D. Thesis, Univ. California, **1969**.
69L	Lundström, T.: Ark. Kemi **31** (1969) 227.
69M1	Menth, H., Buehler, E., Geballe, T.H.: Phys. Rev. Lett. **22** (1969) 295.
69M2	Mercurio, J.P.: C.R. Acad. Sci. colon **268** (1969) 1766.
69P1	Perotta, A.J., Townes, W.D., Potenza, J.A.: Acta Crystallogr. B **25** (1969) 1223.
69P2	Paderno, Yu.B., et al.: Poroshk. Metall. Acad. NAUK Ukr. SSR **10** (1969) 55.
69R	Richards, S.M., Kasper, J.S.: Acta Crystallogr. B **25** (1969) 237.
69S1	Sullenger, D.B., Phipps, K.D., Seabaugh, P.W., Hudgens, C.R., Sands, D.E., Cantrell, J.S.: Science **163** (1969) 93f.
69S2	Spear, K.E., Gilles, P.W.: High Temp. Sci. **1** (1969) 86.
70A1	Andersson, S., Lundström, T.: J. Solid State Chem. **2** (1970) 603.
70A2	Andersson, S., Carlsson, J.O.: Acta Chem. Scand. **24** (1970) 1791
70A3	Amberger, E., Druminski, M., Ploog, K.: J. Less-Common Met. **23** (1971) 43.
70B	Bidwell, L.R.: J. Less-Common Met. **20** (1970) 19.
70C1	Chen, H., Conrad, B.R., Gilles, P.W.: Inorg. Chem. **9** (1970) 1776.
70C2	Carlsson, J.O., Lundström, R.: J. Less-Common Met. **22** (1970) 317.
70C3	Cohen, R.L., Eibschutz, M., West, K.: Phys. Rev. Lett. **24** (1970) 383.
70E1	Economy, J., Matkovich, V.I.: Boron 3, T. Niemyski, ed., PWN Warsaw, **1970** p. 159.
70E2	Etourneau, J., Thèse de Doct. Sci. Phys. Univ. Bordeaux I, **1970**.
70E3	Etourneau, J., Mercurio, J.P., Naslain, R., Hagenmuller, P.: J. Solid-State Chem. **2** (1970) 332.
70E4	Ettmayer, P., Horn, H.C., Schwetz, K.A.: Microchim. Acta Suppl. **IV** (1970) 87.
70E5	Elliot, R.P.: Metallurgia, **1970**.
70G1	Gurr, G.E., Montgomery, P.W., Knutson, C.D., Gorres, B.T.: Acta Crystallogr. B **26** (1970) 906.
70G2	Geist, D., Meyer, J., Peußner, H.: see [70E1] p. 207.
70G3	Greenberg, Ya.H., Zhukov, E.G., Koryazhkin, V.A.: Dokl. AN USSR **191** (1970) 589.

9.14.18 References for 9.14

70K1	Koulmann, J.J., Kappel, H., Taglang, P.: C.R. Acad. Sci. (Paris) **270** B (1970) 445.
70K2	Khusidman, M.B., Neshpor, V.S.: Poroshk. Metall. **9** (1970) 67.
70M1	Matkovich, V.I., Economy, J.: Acta Crystallogr. B **26** (1970) 616.
70M2	Matkovich, V.I., Economy, J.: see [70E1] p. 167.
70M3	Matkovich, V.I., Economy, J.: see [70E1] p. 159.
70N	Naslain, R., Kasper, J.S.: J. Solid State Chem. **1** (1970) 150.
70P1	Paderno, Yu.B., Goryachev, Yu.M., Garf, E.S.: see [70E1] p. 175.
70P2	Peshev, P., Etourneau, J., Naslain, R.: Mater. Res. Bull. **5** (1970) 319.
70W1	Will, G.: see [70E1], p. 119.
70W2	Werheit, H., Runow, P., Leis, H.G.: Phys. Status Solidi (a) **2** (1970) K 125.
70Y	Yajima, S., Niihara, K.: Proc. 9th Rare Earth Research Conf., US Dept. of Commerce, **1970**, p. 598.
71A	Armstrong, D.R., Perkins, P.G., Stewart, J.J.P.: J. Chem. Soc. Sect. A (1971) 3674.
71B1	Broadhead, P., Newman, G.A.: Inorg. Macromol. Rev. **1** (1971) 191.
71B2	Broadhead, P., Newman, G.A.: J. Mol. Struct. **10** (1971) 157.
71D	Dembovsky, S.A., Kirilenko, V.V., Buslaev, Yu.A.: Izv. AN, USSR, Ser. Neorg. Materiali **7** (1971) 510.
71E	Etourneau, J., Naslain, R., La Placa, S.: J. Less-Common Met. **24** (1971) 183.
71F1	Fisk, Z.: Phys. Lett. **34** A (1971) 261.
71F2	Fisk, Z., Lawson, A.C., Matthias, B.T., Corenzwit, E.: Phys. Lett. **37** A (1971) 251.
71G1	Goldblatt, N., Figgins, R., Montrose, C.J., Macedo, P.B.: Phys. Chem. Glasses **12** (1971) 15.
71G2	Gust, W.H., Royce, E.B.: J. Appl. Phys. **42** (1971) 276.
71G3	Gorjachev, Yu.M., Odintsov, V.V., Paderno, Yu.B.: Metallofizika, Kiev (USSR) **37** (1971) 29.
71H	Hynes, T.V., Alexander, M.N.: J. Chem. Phys. **54** (1971) 5296.
71K1	Kieffer, R., Gugel, E., Leimer, G., Ettmayer, P.: Ber. Dtsch. Keram. Ges. **48** (1971) 385.
71K2	Krishman, R.S.: Indian J. Pure Appl. Phys. **9** (1971) 916.
71N1	Niihara, K.: Bull Chem. Soc. Jpn. **44** (1971) 963.
71N2	Nickerson, J.C., White, R.M., Lee, K.N., Bachmann, R., Geballe, T.H., Hull, G.W.Jr.: Phys. Rev. B **3** (1971) 2030.
71O1	Odintsov, V.V., Paderno, Yu.B.: Izv. Akad. Nauk SSSR, Neorg. Mater. **7** (1971) 343.
71O2	Odintsov, V.V., Paderno, Yu.B.: At. Energ. **30** (1971) 453.
71P1	Petrak, D.R., Ruh, R., Goosey, B.F.: Conf. Gaithersburg M.D. USA, **1971**, 93.
71P2	Paderno, Yu.B., Odintsov, V.V., Timofeeva, I.I., Klochkov, L.A.: Teplofiz. Vys. Temp. **9** (1971) 216.
71S1	Shuker, R., Gammon, R.W.: J. Chem. Phys. **55** (1971) 4784.
71S2	Strong, S.L., Wells, A.F., Kaplow, R.: Acta Crystallogr. B **27** (1971) 166.
71S3	Slack, G.A., Oliver, D.W., Horn, E.H.: Phys. Rev. B **4** (1971) 1714.
71S4	Sugaya, T., Watanabe, O.: J. Less-Common Met. **26** (1972) 25.
71W1	Wood, V.E.: Phys. Lett. **37** A (1971) 357.
71W2	Werheit, H., Binnenbruck, H., Hausen, A.: Phys. Status Solidi (b) **47** (1971) 153.
72A	Ashbee, K.H.G., DuBose, C.K.H.: Acta Metall. **20** (1972) 241.
72C	Clark, R.J.H., Mitchell, P.D.: J. Chem. Phys. **56** (1972) 2225.
72D	Dembovsky, S.A., Kirilenko, V.V., Buslaev, Yu.A.: Izo. AN USSR Ser. Neorg. Mater. **8** (1972) 237.
72G	Guette, A., Naslain, R., Gally, J.: C.R. Acad. Sci. Colon. Ser. C **275** (1972) 41.
72H1	Hacker, H.: J. Magn. Reson. **8** (1972) 175.
72H2	Hillel, R., Cueilleron, J.: Bull. Soc. Chim. France N **1** (1972) 98.
72K1	Kieffer, R., Gugel, E., Leimer, G., Ettmayer, P.: Ber. Dtsch. Keram. Ges. **49** (1972) 41.
72K2	Krogmann, K., Becher, J.H.: Z. Anorg. Allg. Chem. **392** (1972) 197.
72O	Odintsov, V.V., Paderno, Yu.B., in: Electronic Structure and Physical properties of Solids (in Russian), Naukova dumka, Kiev, **1972**, p. 112.
72P	Ploog, K., Schmidt, H., Amberger, E., Will, G., Kossobutzki, K.H.: J. Less-Common Met. **29** (1972) 161.
72S1	Schwetz, K., Ettmayer, P., Kieffer, R., Lipp, A.: J. Less-Common Met. **26** (1972) 99.
72S2	Spear, K.E., Solovyev, G.J. in: Solid State Chemistry, (Ed.: Roth R.S., Schneider S.J.), Nat. Bur. Stand. US Spec.-Publ. **364** (1972) 597.
72S3	Stanke, F., Parak, F.: Phys. Status Solidi (b) **52** (1972) 69.
72S4	Schwetz, K., Ettmayer, P., Kieffer, R., Lipp, A.: Radex-Rundschau **3/4** (1972) 257.

9.14.18 References for 9.14

72S5	Solovyev, G.J., Spear, K.E.: J. Am. Ceram. Soc. **55** (1972) 475.
72V	Sheindlin, A.E., Belevich, I.S., Kozhevnikov, I.G.: Teplofizika Vysokikh Temp. **10** (1972) 421; High Temp. USSR (English Transl.) **10** (1972) 369.
72W	Wäppling, R., Häggström, L., Devanarayanan, S.: Phys. Scr. **5** (1972) 97.
73A1	Andersson, S., Callmer, B.: J. Solid State Chem. **10** (1974) 219.
73A2	Ashbee, K.H.G., Frank, F.C., DuBose, C.K.H.: J. Nucl. Mater. **48** (1973) 193.
73B1	Berezin, A.A., Golikova, O.A., Zaitsev, V.K., Kazanin, M.M., Orlov, V.M., Stil'bans, L.S., Tkalenko, E.N.: Sov. Phys. Solid State **15** (1973) 664.
73B2	Budozhapov, V.D., Zelenin, L.P., Chemerinskaya, L.S., Sidorenko, F.A., Gel'd, P.V.: Izv. Akad. NAUK SSSR Neorg. Mater. **9** (1973) 1447.
73G	Goodenough, J.B., Mercurio, J.P., Etourneau, J., Naslain, R., Hagenmuller, P.: C.R. Acad. Sci. (Paris) **277** (1973) 1239.
73H	Hirayama, M., Shohno, K.: Jpn. J. Appl. Phys. **12** (1973) 1960.
73K	Krause, J.L., Sienko, M.J.: J. Solid State Chem. **6** (1973) 590.
73L	Lundström, T., Rosenberg, I.: J. Solid State Chem. **6** (1973) 299.
73M1	Mercurio, J.P., Etourneau, J., Naslain, R., Hagenmuller, P.: Mater. Res. Bull **8** (1973) 837.
73M2	Mott, N.F.: Philos. Mag. **30** (1973) 403.
73O	Odintsov, V.V., Kostetskii, I.I., L'vov, S.N.: Izv. Akad. Nauk, Neorg. Mater. **9** (1973) 944; (engl. transl. in Inorganic materials).
73P	Plumkett, J.D.: Am. Ceram. Soc. Bull. **52** (1973) 341.
73S1	Schwetz, K., Lipp, A.: J. Less-Common Met. **33** (1973) 295.
73S2	Schwetz, K., Lipp, A.: Atomwirtsch. Atomtechn. **18** (1973) 531.
73S3	Sperlich, G., Buschow, K.H.J.: Proc. 4th. Int. Conf. Solid Comp. Trans. Elements, Geneva, **1973**, p. 216.
73S4	Sperlich, G.: Int. J. Magn. **5** (1973) 125.
73S5	Stalinski, B., Drulis, H.: Proc. 4th Int. Conf. Solid Comp. Trans. Elements, Geneva, **1973** p. 213.
73T	Takigawa, M., Hirayama, M., Shohno, K.: Jpn J. Appl. Phys. **12** (1973) 1504.
73Y	Yahel, H.L.: J. Appl. Crystallogr. **6** (1973) 471.
74A	Andersson, S., Callmer, B.: J. Solid State Chem. **10** (1974) 219.
74B1	Binbrek, O.S., Krishnamurthy, N., Anderson, A.: J. Chem. Phys. **60** (1974) 4400.
74B2	Berezin, A.A., Golikova, O.A., Kazanin, M.M., Khomidov, T., Mirlin, D.N., Petrov, A.V., Umarov, A.S., Zaitsev, V.K.: J. Non-Cryst. Solids **16** (1974) 237.
74B3	Berezin, A.A., Golikova, O.A., Zaitsev, V.K., Kazanin, M.M., Orlov, V.M., Tkalenko, E.N.: Proc. 12th Int. Conf. Phys. Semicond. **1974**, p. 291.
74B4	Binbrek, O.S., Brandon, J.K., Anderson, A.: Can. J. Spectrosc. **20** (1974) 52.
74G1	Golikova, O.A., Drabkin, I.A., Zaitsev, V.K., Kazanin, M.M., Mirlin, D.N., Nelvson, I.V., Tkalenko, E.N., Chomidov, T., in: Bor, Poluchenie, Struktura i Svoistva; Ed.: Tavadze F.N., Mecniereba, Tbilisi, **1974**, p. 44.
74G2	Golikova, O.A., Zaitsev, V.K., Orlov, V.M., Petrov, A.V., Stilbans, L.S., Tkalenko, E.N.: Phys. Status Solidi (a) **21** (1974) 405.
74G3	Golikova, O.A., Zaitsev, V.K., Orlov, V.M., Petrov, A.V., Tkalenko, E.N.: see [74G1], p. 25.
74H1	Higashi, J., Tagahashi, Y., Atoda, T.: J. Less-Common Met. **37** (1974) 283.
74H2	Higashi, J., Tagahashi, Y., Atoda, T.: J. Less-Common Met. **37** (1974) 199.
74H3	Hulliger, F.: Solid State Commun. **15** (1974) 933.
74K	Korsukova, M.M., Gurin, V.N., Sorokin, V.N., Yusov, Yu.P., Terent'eva, S.P., in: Bor, Poluchenie, Struktura i Svoistva, Ed.: Tavadze F.N., Moscow Nauka, **1974**, p. 235.
74L	Lundström, T.: see [74K], p. 44.
74M1	Mercurio, J.P., Etourneau, J., Naslain, R., Hagenmuller, P., Goodenough, J.B.: J. Solid State Chem. **9** (1974) 37.
74M2	Matuschka, A. v.: Chem. Ztg. **98** (1974) 504.
74P1	Ploog, K.: J. Cryst. Growth **24/25** (1974) 197.
74P2	Ploog, K., Druminski, M.: Krist. Techn. **9** (1974) 25.
74R	Rupp, L.W. Jr., Hodges, D.J.: J. Phys. Chem. Solids **35** (1974) 617.
74S1	Samsonov, G.V., Goryachev, Yu.M., Kovenskaya, B.A., Arabej, B.G.: see [74K], p. 171.
74S2	Severyanina, E.N., Dudnik, E.M., Paderno, Yu.B.: Poroshk. Metall. **13** (1974) 83.
74S3	Sperlich, G., Jansen, K.: Solid State Commun. **15** (1974) 1105.
74S4	Smith, J.L., Hill, H.H.: AIP Conf. Proc. **24** (1974) 382.
74T	Takigawa, M., Hirayama, M., Shohno, K.: Jpn. J. Appl. Phys. **13** (1974) 411.

9.14.18 References for 9.14

74W	Werheit, H., Binnenbruck, H.: see [74K], p. 110.
75A1	Andres, K., Graebner, J.E., Ott, H.R.: Phys. Rev. Lett. **35** (1975) 1779.
75A2	Amberger, E., Polborn, K.: Acta Crystallogr. B **31** (1975) 949.
75A3	Ashbee, K.H.G., Frank, F.C.: J. Nucl. Mater. **55** (1975) 116.
75B1	Binbrek, O.S., Branon, J.K., Anderson, A.: Can. J. Spectrosc. **20** (1975) 52.
75B2	Becker, H.J., Rethfeld, H., Mattes, R.: Z. Anorg. Allg. Chem. **414** (1975) 203.
75B3	Budozhapov, V.D., Sidorenko, F.A., Zelenin, L.P., Gel'd, P.V., Chemerinskaya, L.S.: Izv. Akad. Nauk SSSR Neorg. Mater. **11** (1975) 173.
75B4	Benesovsky, F., in: Ullmanns Enzyklopädie der techn. Chemie **8** (1975) 657.
75F	Fukushima, E., Strubeing, V.O., Hill, H.H.: J. Phys. Soc. Jpn. **39** (1975) 921.
75G1	Glaunsinger, W.S.: Phys. Status Solidi (b) **70** (1975) K 151.
75G2	Glaunsinger, W.S.: J. Magn. Reson. **18** (1975) 265.
75L1	Liu, S.H., Kopp, L., England, W.B., Myron, H.W.: Phys. Rev. B **11** (1975) 3463.
75L2	Lipp, A., Schwetz, K.: Ber. Dtsch. Keram. Ges. **52** (1975) 335.
75L3	Leitnaker, J.M., Stiegler, J.O.: J. Nucl. Mater. **55** (1975) 113.
75M	Mar, R.W.: J. Am. Ceram. Soc. **58** (1975) 145.
75P	Perkins, P.G., Armstrong, D.R., Breeze, A.: J. Phys. C **8** (1975) 3558.
75S	Samsonov, G.V., Serebryakova, T.I., Neronov, V.A.: Boridy, Moskva Atomizdat, **1975**.
75V	Voitovich, R.F., Pugach, E.A.: Poroshk. Metall. **14** (1975) 57.
75Y	Yakel, H.L.: Acta Crystallogr. B **31** (1975) 1797.
75Z	Zaitsev, V.K., Golikova, O.A., Kazanin, M.M., Orlov, V.M., Tkalenko, E.N.: Sov. Phys. Semicond. **9** (1976) 1372.
76A1	Aono, M., Kawai, S., Kono, S., Okusawa, M., Sagawa, T., Takehana, Y.: J. Phys. Chem. Solids **37** (1976) 215.
76A2	Armas, B., Combescure, C., Dusseau, J.M., Lepetre, T.P., Robert, J.L., Pistoulet, B.: J. Less-Common Met. **47** (1976) 135.
76A3	Amberger, E., Polborn, K.: Acta Crystallogr. B **32** (1976) 981.
76A4	Armstrong, D.R., Breeze, A., Perkins, P.G.: J. Phys. C **8** (1976) 3558.
76B1	Bills, P.M., Lewis, D.: J. Less-Common Met. **45** (1976) 343.
76B2	Brown, F.R., Miller, F.A., Sourisseau, C.: Spectrochim Acta A **32** (1976) 125.
76C	Champagne, B., Beauvy, M., Angers, R.: Metallography **9** (1976) 357.
76D	Domashevskaya, E.P., Solovjev, N.E., Terechov, V.A., Ugai, Ya.A.: J. Less-Common Met. **47** (1976) 189.
76F	Fisk, Z., Schmidt, P.H., Longinotti, L.D.: Mater. Res. Bull. **11** (1976) 1019.
76G1	Golikova, O.A., Kazanin, M.M., Lutsenko, E.L., Orlov, V.M., Tkalenko, E.N., Zaitsev, V.K.: Proc. 13th Int. Conf. Phys. Semicond., Rome (Ed. Fumi F.G.) Tipografia Marves, Rome, **1976**, p. 497.
76G2	Glaunsinger, W.S.: AIP Conf. Proc. **1976** (Magn. Mat. Ann. Conf. 21th, 1975) 412.
76G3	Glaunsinger, W.S.: Phys. Status Solidi (b) **74** (1976) 443.
76G4	Glaunsinger, W.S.: J. Phys. Chem. Solids **37** (1976) 51.
76G5	Glaunsinger, W.S.: J. Magn. Reson. **21** (1976) 147.
76G6	Guy, C.N.: J. Phys. Chem. Solids **37** (1976) 1005.
76H	Higashi, J., Sakurai, T., Atoda, T.: J. Less-Common Met. **45** (1976) 283.
76I	Ishii, M., Aono, M., Muranaka, S., Kawai, S.: Solid State Commun. **20** (1976) 437.
76K	Kasper, J.S.: J. Less-Common Met. **47** (1976) 17.
76L1	Lundström, T., Tergenius, L.E.: Less-Common Met. **47** (1976) 23.
76L2	Lalanne, M., Georges, R., Mercurio, J.P., Chevalier, B., Etourneau, J.: J. Less-Common Met. **46** (1976) 39.
76M1	Mattes, R., Tebbe, K.F., Neidhard, H., Rethfeld, H.: J. Less-Common Met. **47** (1976) 29.
76M2	Mercurio, J.P., Etourneau, J., Naslain, P., Hagenmuller, P.: J. Less-Common Met. **47** (1976) 175.
76M3	Matkovich, V.I.: J. Less-Common Met. **46** (1976) 39.
76N	Naslain, P., Guette, A., Hagenmuller, P.: J. Less-Common Met. **47** (1976) 1.
76P1	Perkins, P.G., Sweeney, A.V.J.: J. Less-Common Met. **47** (1976) 165.
76P2	Pauling, L.: Acta Crystallogr. B. **32** (1976) 3359.
76R	Rubinstein, M., Resing, H.A.: Phys. Rev. B **13** (1976) 959.
76S1	Spear, K.E.: J. Less-Common Met. **47** (1976) 197.
76S2	Sinel'nikova, V.S., Gurin, V.N., Pilyankevich, A.N., Strachinskaya, L.V., Korsukova, M.M.: J. Less-Common Met. **47** (1976) 265.

9.14.18 References for 9.14

76S3	Samsonov, G.V., Goryachev, Yu.M., Kovenskaya, B.A.: J. Less-Common Met. **47** (1976) 147.
76S4	Spear, K.E., in: Phase Diagrams, Materials Science and Technology, Vol. 4, Acad. Press: New York, **1976**.
76W1	Will, G., Kossobutzki, K.H.: J. Less-Common Met. **47** (1976) 33.
76W2	Will, G., Kossobutzki, K.H.: J. Less-Common Met. **47** (1976) 43.
77A	Allen, J.W., in: Valence Instab. Relat. Narrow Band Phenom. (Proc. Int. Conf.), Parks R.D., ed., Plenum: New York, **1977**, p. 533.
77A2	Armstrong, D.R., Breeze, A., Perkins, P.G.: J. Chem. Soc. Faraday Trans. II **73** (1977) 952.
77B1	Berezin, A.A., Golikova, O.A., Zaitsev, V.R., Kazanin, M.M., Orlov, V.M., Tkalenko, E.N., in: Boron and Refractory Borides, (Matkovich V.I., ed.) Springer: Berlin, Heidelberg, New York **1977**, p. 52.
77B2	Bairamashvili, I.A., Golikova, O.A., Kekelidze, L.I., Orlov, V.M.: Sov. Phys. Semicond. **11** (1977) 451.
77B3	Becher, H.J., Mattes, R.: see [77B1], p. 107.
77B4	Bajaj, M.M., Kasaya, M.: Pramana **9** (1977) 297.
77B5	Buschow, K.H.J.: see [77B1], p. 494.
77C1	Castaing, J., Costa, P.: see [77B1], p. 390.
77C2	Chapnik, I.M.: Phys. Status Solidi (a) **41** (1977) K 71.
77C3	Callmer, B.: Acta Crystallogr. B **33** (1977) 1951.
77D	Dagourry, G., Vigner, D.: Le Vide **187** (1977) 51.
77E	Etourneau, J., Mercurio, J.P., Hagenmuller, P.: see [77B1], p. 115.
77G1	Gurin, V.N., Sinelnikova, V.S.: see [77B1], p. 377.
77G2	Gal'chenko, G.L., Lavut, E.G., Lavut, E.A., Vidavsky, L.M.: see [77B1], p. 331.
77G3	Gurin, V.N., Korsukova, M.M.: see [77B1], p. 293.
77G4	Geist, D.: see [77B1], p. 65.
77H1	Higashi, I., Sakurai, T., Atoda, T.: J. Solid State Chem. **20** (1977) 67.
77H2	Hughes, R.E., Leonovicz, M.E., Lemley, J.T., Leung-Tak Tai, J. Am. Chem. Soc. **99** (1977) 5507.
77I	Isikawa, Y., Bajaj, M.M., Kusaya, M., Tanaka, T., Bannay, E.: Solid State Commun. **22** (1977) 573.
77K1	Kasper, J.S., Vlasse, M., Naslain, R.: J. Solid State Chem. **20** (1977) 281.
77K2	Kobayashi, T., Iiono, K., Hiraoka, T.: J. Am. Chem. Soc. **99** (1977) 5507.
77K3	Kasuya, T., Kojima, K., Kasaya, M.: see [77A], p. 137.
77L	Lundström, T.: see [77B1], p. 351.
77M1	Makarenko, G.N.: see [77B1], p. 310.
77M2	Matkovich, V.I., Economy, J.: see [77B1], p. 78.
77M3	Matkovich, V.I., Economy, J.: see [77B1], p. 96.
77M4	Murgatroyd, R.A., Kelly, B.T.: At. Energy Rev. **15** (1977) 3.
77N1	Naslain, R., Etourneau, J., Hagenmuller, P.: see [77B1], p. 262.
77N2	Naslain, R.: see [77B1], p. 139.
77O	Oudet, X.: see [77B1], p. 525.
77P1	Perkins, P.G.: see [77B1], p. 31.
77P2	Pastor, H.: see [77B1], p. 257.
77P3	Pastor, H.: see [77B1], p. 457.
77S1	Samsonov, G.V., Kovenskaya, B.A.: see [77B1], p. 5.
77S2	Samsonov, G.V., Kovenskaya, B.A.: see [77B1], p. 19.
77S3	Samsonov, G.V., Rud', B.M., Shulishova, O.I., Konovalova, E.S., Mudrolyubov, Yu.M., Budanova, I.G.: Izd. Akad. Nauk SSSR Neorg. Mater. **13** (1977) 1314.
77S4	Slack, G.A., Oliver, D.W., Brower, G.D., Young, J.D.: J. Phys. Chem. Solids **38** (1977) 45.
77S5	Spear, K.E.: see [77B1], p. 439.
77S6	Schubert, K.: Chem. Scr. **12** (1977) 109.
78A	Allen, J.W., Martin, R.M., Batlogg, B., Wachter, P.: J. Appl. Phys. **49** (1978) 2078.
78B	Berrada, H., Mercurio, J.P., Etourneau, J., Hagenmuller, P., Shroff, A.M.: J. Less-Common Met. **59** (1978) 7.
78D1	Dusseau, J.M., Ensuque, L., Im-Sareoun, Lepetre, T.P.: Phys. Status Solidi (a) **47** (1978) K 11.
78E	Egami, T., Dahlgren, S.D.: J. Appl. Phys. **49** (1978) 1703.
78F1	Fisk, Z., Johnston, D.C., Cornut, B., v. Molnar, S., Oseroff, S., Calvo, R.: J. Appl. Phys. **50** (1979) 1911.

9.14.18 References for 9.14

78F2	Flechon, J., Kuhnast, F. A., Machizaud, F.: Thin Solid Films **52** (1978) 89.
78F3	Fujita, F. E.: J. Phys. Colloq. C 2, **40** (1979) 120.
78H	Haworth, D. T., Wilkie, C. A.: J. Inorg. Nucl. Chem. **40** (1978) 1689.
78J	Jagielinski, T., Arai, K. I., Tsuja, N., Fukamichi, K.: Phys. Status Solidi (a) **50** (1978) K 25.
78K1	Kasaya, M., Tarascon, J. M., Etourneau, J., Hagenmuller, P.: Mater. Res. Bull. **13** (1978) 751.
78K2	Kikuchi, M., Fukamichi, K., Masumoto, T., Jagielinski, T., Arai, K. I., Tsuya, N.: Phys. Status Solidi (a) **48** (1978) 175.
78M	Mitchell, M. A., Sutula, R. A.: J. Less-Common Met. **57** (1978) 161.
78O	Onodera, H., Yamamoto, H., Watanabe, H. J.: J. Phys. Colloq. C 2, **40** (1979) 142.
78P1	Pistoulet, B., Robert, J. L., Dusseau, J. M., Roche, F. M., Girard, P., in: Physics of Semiconductors (Edinburgh Conf.), **1978**, p. 793.
78P2	Pistoulet, B., Robert, J. L., Dusseau, J. M., Ensuque, L.: J. Non-Cryst. Solids **29** (1978) 29.
78P3	Pistoulet, B., Robert, J. L., Dusseau, J. M.: see [78P1], p. 1009.
78S	Stepanov, N. N., Zynzin, A. Yu., Shul'man, S. G., Gurin, V. N., Korsukova, M. M., Nikanorov, S. P., Smirnov, I. A.: Sov. Phys. Solid State **20** (1978) 542.
78W	Wang, F. E., Mitchell, M. A., Sutula, R. A., Holden, J. R., Bennet, L. H.: J. Less-Common Met. **61** (1978) 237.
79A1	Aida, T., Honda, Y., Yamamoto, S., Kawabe, U.: J. Jpn. Inst. Met. **43** (1979) 901.
79A2	Armas, B., Combescure, C.: Opt. Pura Appl. **12** (1979) 155.
79A3	Aleshin, V. G., Kosolapova, T. Ya., Nemoshkalenko, V. V., Serebryakova, T. I., Chudimov, N. G.: J. Less-Common Met. **67** (1979) 173.
79A4	Aivazov, M. I., Bashilov, V. A., Zinchenko, K. A., Kagramanova, R. R.: Poroshk. Metall. **18** (1979) 46.
79A5	Avlokhashvili, J. A., Tavadze, F. N., Tavadze, G. F., Tsikaridze, D. N., Gabunia, D. L., Tsomaya, K. P.: J. Less-Common Met. **67** (1979) 367.
79A6	Allen, J. W., Batlogg, B., Wachter, P.: Phys. Rev. **20B** (1979) 4807.
79B1	Bairamashvili, I. A., Kekelidze, L. I., Golikova, O. A., Orlov, V. M.: J. Less-Common Met. **67** (1979) 461.
79B2	Bagley, B. G., Aspnes, D. E., Benenson, R. E., Adams, A. C., Alexander, F. B., Mogab, C. J.: Bull. Am. Chem. Soc. **24** (1979) 399.
79B3	Bairamashvili, I. A., Kalandadze, G. I., Eristavi, A. M., Jobava, J. Sh., Chotulidi, V. V., Saloev, Yu. I.: J. Less-Common Met. **67** (1979) 455.
79B4	Binnenbruck, H., Werheit, H.: Z. Naturforsch. **34a** (1979) 787.
79B5	Bouchacourt, M., Thevenot, F., Ruste, J.: J. Microsc. Spectrosc. Electron. **4** (1979) 143.
79B6	Beauvy, M., Thevenot, F.: L'Industrie Ceramique **732** (1979) 734.
79B7	Bouchacourt, M., Thevenot, F.: J. Less-Common Met. **67** (1979) 327.
79C1	Coey, J. M. D., Massenet, O., Kasaya, M., Etourneau, J.: J. Phys. Coloq. C 2, **40** (1979) 333.
79C2	Champagne, B., Angers, R.: J. Am. Ceram. Soc. **62** (1979) 149.
79E	Etourneau, J., Mercurio, J. P., Berrada, A., Hagenmuller, P., Georges, R., Bourezg, R., Gianduzzo, J. C.: J. Less-Common Met. **67** (1979) 531.
79F1	Fisk, Z., Johnston, D. C., Cornut, B., v. Molnar, S., Oserov, S., Calvo, R.: J. Appl. Phys. **50** (1979) 1911.
79F2	Futamoto, M., Aita, T., Kawabe, U.: Mater. Res. Bull. **14** (1979) 1329.
79F3	Feurer, R., Constant, G., Bernard, C.: J. Less-Common Met. **67** (1979) 107.
79G1	Golikova, O. A., Kazanin, M. M., Orlov, V. M., Tkalenko, E. N., Fedorov, M. I.: J. Less-Common Met. **67** (1979) 363.
79G2	Golikova, O. A.: Phys. Status Solidi (a) **51** (1979) 11.
79G3	Gurin, V. N., Korsukova, M. M., Nikanorov, S. P., Smirnov, I. A., Stepanov, N. N., Shulman, S. G.: J. Less-Common Met. **67** (1979) 115.
79G4	Garibashvili, V. I., Zoidze, N. A., Nakaidze, Sh. G., Tavadze, F. N.: Izv. Akad. Nauk SSSR Met. Co. **1** (1979) 118.
79H1	Hasegawa, A., Yanase, A.: J. Phys. C **12** (1979) 5431.
79H2	Higashi, I.: J. Less-Common Met. **67** (1979) 7.
79K1	Kuz'ma, Yu. B., Bilinozhko, N. S., Mykhalenko, S. I., Stepanchikova, G. F., Chaban, N. E.: J. Less-Common Met. **67** (1979) 51.
79K2	Kunii, S., Kasuya, T.: J. Phys. Soc. Jpn. **46** (1979) 13.
79K3	Kasaya, M., Tarascon, J. M., Etourneau, J., Hagenmuller, P., Coey, J. M. D.: J. Phys. Colloq. **40** (1979) 393.

9.14.18 References for 9.14

79K4	Koval'chenko, M.S., Bodrova, L.G., Nemchenko, V.F., Kolotun, V.F.: J. Less-Common Met. **67** (1979) 357.
79K5	Knychev, E.A., Novgorodtzev, V.M., Svistunov, V.V., Beketov, A.R., Plyshevskii, Yu.S., Obabkov, N.V.: J. Less-Common Met. **67** (1979) 347.
79K6	Kuhnast, F.A., Machizand, F., Vangelisti, R., Flechon, J.: J. Microsc. Spectrosc. Electron. **4** (1979) 553.
79K7	Kirfel, A., Gupta, A., Will, G.: Acta Crystallogr. B **35** (1979) 1052.
79K8	Kirfel, A., Gupta, A., Will, G.: Acta Crystallogr. B **35** (1979) 2291.
79K9	Kuzenkova, M.A., Kishyi, P.S., Grabchuk, B.L., Bodnaruk, N.I.: J. Less-Common Met. **67** (1979) 217.
79L1	Lugscheider, E., Reimann, H., Quadakkers, W.J.: Ber. Dtsch. Keram. Ges. **56** (1979) 301.
79L2	Letelier, J.R., Chiu, Y.N., Wang, F.E.: J. Less-Common Met. **67** (1979) 179.
79L3	Leyarovska, I., Leyrarovski, E., Popov, Chr., Midlarz, T.: J. Less-Common Met. **67** (1979) 389.
79L4	Leyarovska, I., Leyarovski, E.: J. Less-Common Met. **67** (1979) 249.
79M1	Moiseenko, L.L., Odintsov, V.V.: J. Less-Common Met. **67** (1979) 237.
79M2	Mercurio, J.P., Angelov, S., Etourneau, J.: J. Less-Common Met. **67** (1979) 237.
79M3	Martin, R.M., Allen, J.W.: J. Appl. Phys. **50** (1979) 7561.
79M4	Machizand, F., Kuhnast, F.A., Flechon, J.: Thin Solid Films **61** (1979) 327.
79O	Oseroff, S., Calvo, R., Stankiewicz, J., Fisk, Z., Johnston, D.C.: Phys. Status Solidi (b) **94** (1979) K 133.
79P1	Pistoulet, B., Robert, J.L., Dusseau, J.M., Roche, F., Girard, P., Ensuque, L.: J. Less-Common Met. **67** (1979) 131.
79P2	Paderno, Yu.B., Konovalova, E.S.: J. Less-Common Met. **67** (1979) 185.
79P3	Povzner, A.A., Zilichiklis, A.L., Abel'skii, Sh.Sh., Borukhovich, A.S., Gel'd, P.V., Knyshev, E.A.: J. Less-Common Met. **67** (1979) 211.
79R1	Rogl, P., Nowotny, H.: J. Less-Common Met. **67** (1979) 41.
79R2	Rud', B.M., Levandovskii, V.D., Skulishova, O.I.: Izv. Akad. Nauk, Neorg. Mater. **15** (1979) 709; Inorg. Mater. (English Transl.) **15** (1979) 553.
79S1	Serebryakova, T.: J. Less-Common Met. **67** (1979) 499.
79S2	Samsonov, G.V., Neronov, V.A., Lamikhov, L.K.: J. Less-Common Met. **67** (1979) 291.
79S3	Samsonov, G.V., Kondrashov, A.I., Okhremchuk, L.N., Podchernyaeva, I.A., Siman, N.I., Fomenko, V.S.: J. Less-Common Met. **67** (1979) 415.
79S4	Sobczak, R.J., Sienko, M.J.: J. Less-Common Met. **67** (1979) 167.
79S5	Svistunov, V., Beketov, A., Knyshev, E., Obabkov, N., Novgorodtsev, V.: J. Less-Common Met. **67** (1979) 287.
79S6	Shimada, Y., Kojima, H.: J. Appl. Phys. **50** (1979) 1541.
79S7	Rud', B.M., Levandovskii, V.D., Shulishova, O.I.: Izv. Akad. Nauk SSSR Neorg. Mater. **15** (1979) 709.
79S8	Swanson, L.W., Mc Neely, D.R.: Surf. Sci. **83** (1979) 11.
79T1	Tsai, C.C.: Phys. Rev. B. **19** (1979) 2041.
79T2	Tagahashi, M., Koshimura, M., Suzuki, T.: J. Phys. Colloq. C 2 **40** (1979) 144.
79T3	Thevenot, F., Bouchacourt, M.: L'Industrie Ceramique **732** (1979) 655.
79V	Volkonskaya, T.I., Kizhaev, S.A., Smirnov, I.A., Korsukova, M.M., Gurin, V.N.: Sov. Phys. Solid State **21** (1979) 375.
79W1	Will, G., Schäfer, W.: J. Less-Common Met. **67** (1979) 31.
79W2	Weill, G., Smirnov, I.A., Gurin, V.N.: Phys. Status Solidi (a) **53** (1979) K 119.
79W3	Werheit, H.: J. Less-Common Met. **67** (1979) 143.
80B	Bagley, B.G., Aspnes, D.E., Adams, A.C., Benenson, R.E., J. Non-Cryst. Solids **35–36** Pt. 1 (1980) 441.
80F1	Fujita, T., Suzuki, M., Isikawa, Y.: Solid State Commun. **33** (1980) 947.
80F2	Futamoto, M., Nakazawa, M., Kawabe, U.: Surf. Sci. **100** (1980) 470.
80G1	Guy, C.N., v. Molnar, S., Etourneau, J., Fisk, Z.: Solid State Commun. **33** (1980) 1055.
80G2	Gurin, V.N., Korsukova, M.M., Karin, M.G., Sidorin, K.K., Smirnov, I.A., Shelikh, A.I.: Sov. Phys. Solid. State **22** (1980) 418.
80H	Holtzberg, F., v. Molnar, S., Coly, J.M.D.: Handbook on Semicond., Vol. 3 (ed. Keller, S.P.) North Holland: Amsterdam, **1980**, p. 803.
80K	Kirfel, A., Will, G.: Acta Crystallogr. B **36** (1980) 1311.

9.14.18 References for 9.14

80T	Tarascon, J.M., Etourneau, J., Dordor, P., Hagenmuller, P., Kasaya, M., Coey, J.M.D.: J. Appl. Phys. **51** (1980) 574.
80V	Vlasova, M.V., Kakazey, N.G., Kosolapova, T.Y., Makarenko, G.N., Marek, E.V., Uskoković, D., Ristić, M.M.: J. Mater. Sci. **15** (1980) 1041.
80W	Werheit, H., de Groot, K.: Phys. Status Solidi (b) **97** (1980) 229.
81A1	Armstrong, D.R.: Proc. 7th Int. Symp. Boron, Borides and Related Compounds, Uppsala, Sweden, **1981**; spec. issue of J. Less-Common Met. **82** (1981) 357.
81A2	Armstrong, D.R., Bolland, J., Perkins, P.G. (abstract only): see [81A1], p. 357.
81A3	Armstrong, D.R., Cetina, E., Perkins, P.G. (abstract only): see [81A1], p. 358.
81A4	Armas, B., Malé, G., Salanoubat, D., Chatillon, C., Allibert, M.: see [81A1], p. 245.
81B1	Bouchacourt, M., Thevenot, F.: see [81A1], p. 219.
81B2	Bouchacourt, M., Thevenot, F.: see [81A1], p. 227.
81B3	Beauvy, M.: see [81A1], p. 359.
81D	Dusseau, J.M., Robert, J.L., Armas, B., Combescure, C.: see [81A1], p. 137.
81G1	Guette, A., Barret, M., Naslain, R., Hagenmuller, P., Tergenius, L.-E., Lundström, T.: see [81A1], p. 325.
81G2	Gianduzzo, J.C., Georges, R., Chevalier, B., Etourneau, J., Hagenmuller, P., Will, G., Schäfer, W., Elf, F.: see [81A1], p. 29.
81G3	Golikova, O.A., Tadzhiev, A. (abstract only): see [81A1], p. 169.
81H	Higashi, I.: see [81A1], p. 317.
81K	Korsukova, M.M., Stepanov, N.N., Gontcharova, E.V., Gurin, V.N., Nikanorov, S.P., Smirnov, I.A. (abstract only): see [81A1], p. 211.
81L	Lebugle, A., Nyholm, R., Mårtenson, N.: see [81A1], p. 269.
81M	Mörke, I, Dvorak, V., Wachter, P.: Solid State Commun. **40** (1981) 331.
81N1	Nakano, K., Doi, M., Kuwayama, K., Imura, T.: see [81A1], p. 309.
81N2	Neronov, V.A., Aleksandrov, V.V., Korchagin, M.A., Gusenko, S.N.: see [81A1], p. 125.
81P	Peña, O., MacLaughlin, D.E., Lysak, M.: J. Appl. Phys. **52** (1981) 2152.
81S1	Shelich, A.I., Sidorin, K.K., Karin, M.G., Bobrikov, V.N., Korsukova, M.M., Gurin, V.N., Smirnov, I.A. (abstract only): see [81A1], p. 291.
81S2	Schwetz, K.A., Reinmuth, K., Lipp, A.: Radex-Rundschau, **1981**, 568.
81S3	Stumpf, W.: in Gmelin, Handbook of Inorganic Chemistry; Supplement Vol. 2, Springer: Berlin, Heidelberg, New York, **1981**, p. 117.
81T1	Tarascon, J.M., Etourneau, J., Dance, J.M., Hagenmuller, P., Georges, R., Angelov, S., v. Molnar, S.: see [81A1], p. 277.
81T2	Tergenius, L.E.: see [81A1], p. 335.
81T3	Tergenius, L.E., Lundström, T.: see [81W1], p. 341.
81T4	Takigawa, M., Yasuoka, H., Kitaoka, Y., Tanaka, T., Nozaki, H., Ishizawa, Y.: J. Phys. Soc. Jpn. **50** (1981) 2525.
81V	Vlasse, M., Viala, J.L.: see [81A1], p. 369.
81W1	Werheit, H., de Groot, K., Malkemper, W., Lundström, T.: see [81A1], p. 163.
81W2	Werheit, H., Higashi, I., preliminary results.
81W3	Werheit, H., Lundström, T., unpublished results.
81W4	Will, G., Schäfer, W., Pfeiffer, F., Elf, F., Etourneau, J.: see [81A1], p. 349.
81W5	Will, G., Kirfel, A., Josten, B., Fröhlich, T.: see [81A1], p. 255.
81W6	Werheit, H., de Groot, K., Malkemper, W.: see [81A1], p. 153.
82A	Aita, O., Watanabe, T., Fujimoto, Y., Tsutsumi, K.: J. Phys. Soc. Jpn. **51** (1982) 483.
82L	Lundström, T.: The Formation of the Bonds to the Group IIIb Elements in Inorganic Reactions and Methods, (Ed.: J.J. Zuckerman) Verlag Chemie: Weinheim, **1982**.
82O	Okusawa, M., Ichikawa, K., Matsumoto, T., Tsutsumi, K.: J. Phys. Soc. Jpn. **51** (1982) 1921.
82S	Schwetz, K.A.: unpublished results.
82T	Tsutsumi, K., Aita, O., Watanabe, T.: Phys. Rev. **25B** (1982) 5415.
82W	Werheit, H.: Colloque France-Allemand Ceramiques Techniques Lyon, March 15–17; **1983**, Nr. 3–4-p.
83A	Armstrong, D.R., Bolland, J., Perkins, P.G., Will, G., Kirfel, A.: Acta Crystallogr. **B39** (1983) 324.

9.15 Binary transition-metal compounds
9.15.1 Compounds with elements of the IIId and IVth groups
9.15.1.1 $T_n(III)_{2n-m}$ and $T_n(IV)_{2n-m}$ compounds

$RuAl_2$, $Mn_{11}Si_{19}$, $Mn_{26}Si_{45}$, $Mn_{15}Si_{26}$, $Mn_{27}Si_{47}$, Mn_4Si_7, Ru_2Si_3, Ru_2Ge_3, Ru_2Sn_3, Os_2Si_3, Os_2Ge_3

9.15.1.1.0 Structure, chemical bond (including lattice properties)

The structures of the Nowotny chimney-ladder compounds $T_n(III)_{2n-m}$ and $T_n(IV)_{2n-m}$ (T: transition element) are derived from the $TiSi_2$ type. The Ti atoms in orthorhombic $TiSi_2$ and the T atoms in tetragonal $T_n(III)_{2n-m}$ and $T_n(IV)_{2n-m}$ occupy a β-Sn-like array of sites with a strongly increased axial ratio c/a. The number n in the formula corresponds to the number of white-tin-like pseudocells stacked along c. According to the (III)- or (IV)-deficiency the distribution of the (III)- or (IV)-atoms is stretched up along c as compared with the $TiSi_2$ structure [70N2, 71D, 72P, 74B, 76D].

The electronegativity of the transition element is larger than that of the group III and group IV element. Thus the T atoms act as anions so that the bonding is similar to that in the transition-element carbonyls. According to Jeitschko and Parthé [67J, 69P, 70N2, 77J] semiconductor behavior and hence filled energy bands occur with 14 valence electrons per T atom.

In Ru_2Si_3 Si has no like neighbors within bonding distance in contrast to the Mn_nSi_{2n-m} phases, where Si chain fragments are formed even in Mn_4Si_7, thus reducing the number of valence electrons available for Mn–Si bonding to about 5 per Mn atom. Unfortunately, there is no clear-cut separation between bonding and nonbonding Si–Si distances.

In ternary silicides $(Mn_{1-x}T_x)_nSi_{2n-m}$ the silicon content depends specifically upon the second metal atom and its concentration (Fig. 1).

The tetragonal high-temperature structures of Ru_2Si_3, Ru_2Ge_3 and Ru_2Sn_3 undergo second-order transitions to orthorhombic symmetry (D_{2h}^{14} – Pbcn; $Z=8$) at 1300 K (Ru_2Si_3), 800 K (Ru_2Ge_3), and < 100 K (Ru_2Sn_3). Os_2Si_3 and Os_2Ge_3 are known only in the latter structure.

Abbreviations: r: room temperature modification, l: low-temperature modification, h: high-temperature modification.

Crystallographic data of the $TiSi_2$ derivatives (RT values if not otherwise noted)

Compound (type, space group, coordination number)	a [Å]	b [Å]	c [Å]	Ref.
$RuAl_2$ ($TiSi_2$ type; D_{2h}^{24} – Fddd)	4.715	8.015	8.715	63S
$RuGa_2$ ($TiSi_2$ type; $Z=24$)	4.749	8.184	8.696	63J
Ir_3Ga_5 (D_{2d}^8 – P$\bar{4}$n2; $Z=4$)	5.823		14.20	67V1
$Mn_{11}Si_{19}=MnSi_{1.727}$ (D_{2d}^8 – P$\bar{4}$n2)	5.530		47.90	67K
	5.518		48.136	64S, 67F, 70N2
$Mn_{26}Si_{45}=MnSi_{1.730}$ (D_{2d}^2 – P$\bar{4}$c2)	5.515		113.36	67F, 70N2
$Mn_{15}Si_{26}=MnSi_{1.733}$ (D_{2d}^{12} – I$\bar{4}$2d)	5.531		65.44	67K
	5.525		65.55	67F, 70N2
$Mn_{27}Si_{47}=MnSi_{1.741}$ (D_{2d}^8 – P$\bar{4}$n2)	5.53		117.9	71Z, 73Z
	5.530		117.94	70N2
$Mn_4Si_7=MnSi_{1.75}$ (D_{2d}^2 – P$\bar{4}$c2; $Z=4$)	5.532		17.54	67K
	5.525		17.463	69K, 72K
	5.526		17.455	71Z
$(Mn_{0.93}Re_{0.07})_nSi_{2n-m}$ (solubility limit)				69E
$(Mn_{0.95}Cr_{0.05})_{19}Si_{33}=MSi_{1.737}$ (D_{2d}^8 – P$\bar{4}$n2)	5.533		83.068	70N2, 68F
$(Mn_{0.85}Cr_{0.15})_4Si_7=MSi_{1.75}$ (D_{2d}^2 – P$\bar{4}$c2)	5.537		17.532	70N2, 68F
$(Mn_{0.75}Cr_{0.25})_{17}Si_{30}=MSi_{1.765}$ (D_{2d}^{12} – I$\bar{4}$2d)	5.552		74.664	70N2, 68F
$(Mn_{0.9}Fe_{0.1})_7Si_{12}=MSi_{1.714}$ (D_{2d}^{12} – I$\bar{4}$2d)	5.510		30.464	70N2, 68F
$(Mn_{0.8}Fe_{0.2})_{23}Si_{39}=MSi_{1.696}$ (D_{2d}^8 – P$\bar{4}$n2)	5.496		99.797	70N2, 68F
$(Mn_{0.7}Fe_{0.3})_{22}Si_{37}=MSi_{1.682}$ (D_{2d}^2 – P$\bar{4}$c2)	5.483		95.172	70N2, 68F
$(Mn_{0.95}Co_{0.05})_{25}Si_{43}=MSi_{1.720}$ (D_{2d}^8 – P$\bar{4}$n2)	5.516		108.950	68F

(continued)

Crystallographic data of the TiSi$_2$ derivatives (continued)

Compound (type, space group, coordination number)	a [Å]	b [Å]	c [Å]	Ref.
Tc$_4$Si$_7$ (D$_{2d}^2$–P$\bar{4}$c2; Z=4)	5.737		18.10	65W, 67V1
Ru$_2$Si$_3$(h) (Ru$_2$Sn$_3$ type)				75P, 80S
Ru$_2$Ge$_3$(h) (Ru$_2$Sn$_3$ type), 800°C	5.739		9.952	75P
Ru$_2$Sn$_3$ (D$_{2d}^2$–P$\bar{4}$c2; Z=4)	6.172		9.915	65W
	6.178		9.916	75P
Ir$_4$Ge$_5$ (D$_{2d}^2$–P$\bar{4}$c2; Z=4)	5.6123		18.361	67P
	5.616		18.35	67V1

Coordination distances (in Å) in Mn$_n$Si$_{2n-m}$ phases (Si–Si < 2.70 Å)

Mn$_{11}$Si$_{19}$ [64S]:

Mn –4 Mn	2.97	Si(2)–Si(3)	2.50	Si(6) –Si(5)	2.66
Mn(1) –8 Si	2.37 and 2.46	Si(4)	2.64	Si(7) –Si(8)	2.63
Mn(2) –8 Si	2.33···2.53	Si(3)–Si(4)	2.39	Si(8) –Si(9)	2.61
Mn(3) –8 Si	2.30···2.55	Si(2)	2.50	Si(7)	2.63
Mn(4) –8 Si	2.29···2.74	Si(3)	2.65		
Mn(5) –8 Si	2.35···2.71	Si(5)	2.69	Si(9) –Si(10)	2.45
Mn(6) –8 Si	2.30···2.60			Si(9)	2.60
Mn(7) –8 Si	2.27···2.55	Si(4)–Si(3)	2.39	Si(8)	2.61
Mn(8) –8 Si	2.35···2.44	Si(5)	2.47	Si(9)	2.69
Mn(9) –8 Si	2.33···2.53	Si(4)	2.62		
Mn(10)–8 Si	2.36···2.57	Si(2)	2.64	Si(10)–2 Si(9)	2.45
Mn(11)–8 Si	2.31···2.61	Si(5)–Si(4)	2.47	Si(10)	2.67
Mn(12)–6 Si	2.27···2.57	Si(6)	2.66		
		Si(3)	2.69		

Mn$_{15}$Si$_{26}$ [67K]:

Mn –4 Mn	2.93···3.03	Si(1)–Si(2)	2.54	Si(4)–Si(5)	2.40
Mn(1)–8 Si	2.38···2.42	Si(3)	2.62	Si(4)	2.52
Mn(2)–8 Si	2.28···2.66	Si(2)–Si(3)	2.39	Si(3)	2.53
Mn(3)–8 Si	2.30···2.52	Si(1)	2.54	Si(5)–Si(6)	2.40
Mn(4)–8 Si	2.32···2.63	Si(2)	2.59	Si(3)	2.68
Mn(5)–8 Si	2.38···2.50	Si(3)–Si(2)	2.39	Si(6)–Si(7)	2.63
Mn(6)–8 Si	2.32···2.65	Si(4)	2.53	Si(6)	2.67
Mn(7)–8 Si	2.34···2.51	Si(1)	2.62	Si(7)–2 Si(6)	2.63
Mn(8)–8 Si	2.27···2.65	Si(5)	2.68		

Mn$_4$Si$_7$ [69K, 72K]:

Mn – 4 Mn	2.970	Si(1)–Si(2)	2.578	Si(4)–2 Si(3)	2.443
Mn(1)–10 Si	2.320···2.701	Si(2)–Si(3)	2.488	Si(4)	2.605
Mn(2)– 8 Si	2.387···2.392	Si(1)	2.578		
Mn(3)– 8 Si	2.272···2.578	Si(3)–Si(4)	2.443		
Mn(4)– 8 Si	2.375···2.471	Si(2)	2.488		
Mn(5)– 8 Si	2.295···2.650				

9.15.1.1 Transition-metal compounds $T_n(III)_{2n-m}$ and $T_n(IV)_{2n-m}$

Crystallographic data of the orthorhombic Ru_2Ge_3-type representatives (Space group D_{2h}^{14} – Pbcn, $Z=8$)

Compound	a [Å]	b [Å]	c [Å]	d_X [g cm^{-3}]	Ref.
Ru_2Si_3(r)	11.057	8.934	5.533	6.96	74P
	11.060	8.952	5.530	6.95	74I
Ru_2Ge_3(r)	11.436	9.238	5.716	9.23	74P
	11.436	9.240	5.718		74V
Ru_2Sn_3(l)					75P
Os_2Si_3(r)	11.124	8.932	5.570	11.15	74P
	11.158	8.962	5.579		74V
Os_2Ge_3(r)	11.544	9.281	5.783	12.82	74P
	11.511	9.267	5.755		74V
$Ir_2(Ga_{0.6}Ge_{0.4})_3$	11.444	9.360	5.722		74V

Coordination distances (in Å) of the orthorhombic Ru_2Ge_3 representatives [74P]
(Ru–Si < 2.99 Å, Ru–Ge < 3.08 Å, Si–Si < 2.95 Å, Ge–Ge < 3.02 Å, Ru–Ru < 3.5 Å)

		Ru_2Si_3	Ru_2Ge_3			Ru_2Si_3	Ru_2Ge_3
Ru(1)–	X(2)	2.375	2.424	X(1)–Ru(3)		2.326	2.417
	X(3)	2.418	2.495		Ru(2)	2.379	2.443
	X(3)	2.428	2.516		Ru(1)	2.477	2.550
	X(1)	2.477	2.550		Ru(1)	2.512	2.557
	X(1)	2.512	2.557		Ru(2)	2.711	2.910
	X(2)	2.517	2.598		X(2)	2.659	2.817
	X(2)	2.519	2.664		X(1)	2.766	2.992
	Ru(3)	2.968	3.076		X(2)	2.859	2.991
	2 Ru(1)	2.991	3.087	X(2)–Ru(2)		2.334	2.416
	Ru(2)	3.011	3.111		Ru(1)	2.375	2.424
Ru(2)–	2 X(2)	2.334	2.416		Ru(1)	2.517	2.598
	2 X(1)	2.379	2.443		Ru(1)	2.519	2.664
	2 X(2)	2.715	2.757		Ru(2)	2.715	2.757
	2 X(1)	2.711	2.910		X(1)	2.659	2.817
	2 Ru(2)	2.882	2.990		X(3)	2.689	2.837
	2 Ru(1)	3.011	3.111		X(3)	2.784	2.887
Ru(3)–	2 X(3)	2.319	2.383		X(1)	2.859	2.991
	2 X(1)	2.326	2.417	X(3)–Ru(3)		2.319	2.383
	2 X(3)	2.490	2.557		Ru(1)	2.418	2.495
	2 Ru(1)	2.968	3.076		Ru(1)	2.428	2.516
	2 Ru(3)	3.072	3.152		Ru(3)	2.490	2.557
					X(2)	2.689	2.837
					X(2)	2.784	2.887

9.15.1.1.1 Physical properties of $T(III)_X$ compounds

$RuAl_2$:

Semiconductivity was deduced from resistivity measurements (cited by [77J]). Questionable.

9.15.1.1.2 Physical properties of $Mn_n(IV)_{2n-m}$ compounds

Mn_nSi_{2n-m}:

The Mn_nSi_{2n-m} phases form by a peritectic reaction around 1550°C [66M] (earlier interpreted as congruent melting [64F]). In most cases the exact stoichiometry of the samples was not determined. Many samples probably contained a mixture of different Mn_nSi_{2n-m} phases as well as traces of metallic and paramagnetic MnSi [73L]. Mn_4Si_7 is metastable below 1125°C. On slow cooling it decomposes into $MnSi_{1.73}$ + interstitial Si. If all valence electrons are bonded in Mn_4Si_7, the interstitial silicon in the low-temperature phase should act as an acceptor. This assumption is supported by the fact that the experimental hole density ($4.6 \cdot 10^{20}$ cm^{-3}) is roughly equal to the number of Frenkel defects ($6.5 \cdot 10^{20}$ cm^{-3}) [72U].

Reports on the magnetic properties of the Mn_nSi_{2n-m} phases are contradictory. According to [70M, 73L] the pure Mn_nSi_{2n-m} phases (all?) should be diamagnetic, but other authors report on antiferromagnetism [66S, 80N] and Curie-Weiss behavior [65R].

Electrical conductivity, Seebeck coefficient and thermal conductivity for different Si concentrations: Figs. 2···5.

$Mn_{11}Si_{19}$ and $Mn_{26}Si_{45}$:

energy gaps (in eV):

Physical property	Numerical value	Experimental conditions T [K]	Experimental method, remarks	Ref.
$E_{g,th}$	≈0.5	0	from $\log \varrho \propto E_g/2kT$	69I, 69N1, 72A, 72U
	0.8	0		69N2
	≈1	0	from $\log \varrho \propto E_g/2kT$ and $S(T)$	74A1
$E_{g,dir}$	0.66	RT	from optical absorption	79Z

Near the edge of the valence band there is a narrow d-band whose effective electron mass is 100 times greater than the effective mass of holes [74A1].

effective masses (in units of m_0):

		T [K]		
m_n	>10^3	300		69N2
m_p	10	300		69N2
	3.3	300		74A1
	2.7	300	density of states mass	69N1
	3.2	650		

resistivity $\varrho(T)$, Hall coefficient $R_H(T)$, thermopower $S(T)$ and optical absorption: Figs. 6···12.

mobilities (in $cm^2 V^{-1} s^{-1}$):

		T [K]		
μ_p	230	300	Hall mobility	69I
	16	800		
	5···8	300		74A1
$\mu_{p\parallel}$	1.7	300		74A1
$\mu_{p\perp}$	7.9	300		74A1

mobility ratio:

b	≈0.03	$T=300$ K	estimated from E_g and dS/dT	74A1

carrier concentration (in 10^{19} cm^{-3}): (see Fig. 13)

		T [K]		
p	2	300	estimated from $R_H(T)$	69I
	12	800		
	4	300		69N1
	73	300		74A1

RT data on conductivity (in Ω^{-1} cm^{-1}), **thermopower** (in μV K^{-1}) and **thermal conductivity** (in W cm^{-1} K^{-1}) **in the extrinsic region:** (temperature dependence: Fig. 8)

σ	160	along [001]		69I
	200			74A1
	800	along [100]		69I
	932			74A1
S	+170	along [001]		69I
	+134			74A1
	+105	along [100]		69I
	+81			74A1

(continued)

Physical property	Numerical value	Experimental conditions	Experimental method, remarks	Ref.

RT data on conductivity, thermopower and thermal conductivity in the extrinsic region (continued)

κ	0.021	along [001]		69I
	0.018			74A1
	0.043	along [100]		69I
	0.037			74A1

$Mn_{15}Si_{26}$:

Single crystal boules of $MnSi_{\approx 1.73}$ grown by the Bridgman technique contained about 2 vol.% of plate-like MnSi precipitates parallel to [001] of the matrix $Mn_{15}Si_{26}$ (stoichiometry deduced from X-ray diffraction patterns; electron microprobe X-ray analysis yielded $MnSi_{1.717}$ which is closer to $Mn_{11}Si_{19}$). Electrical measurements were corrected for the contribution of the metallic MnSi.

energy gap (in eV):

E_g	0.702	$T=0$ K	from $\log \varrho \propto E_g/2kT$ measured along [001] and along [100]	81K
	0.70	900···1200 K	from $\log(R_H T^{3/2}) \propto E_g/2kT$	81K

effective masses (in units of m_0):

m_p	11	along [100]	from thermoelectric power near 600K	81K
	15	along [001]		

hole mobilities (in $cm^2 V^{-1} s^{-1}$):

μ_p	$1.2\ T^{-3/2}$	along [100], $T>200$ K	Hall mobility; temperature dependence shows that acoustic phonon scattering is dominant (Fig. 14)	81K
	$0.62\ T^{-3/2}$	along [001] $T>600$ K		

mobility ratios:

b	0.02		from $\varrho(T)$ between extrinsic and intrinsic region with Hunter's method, confirmed by thermopower	81K
	0.017	from $\varrho^{[100]}$		
	0.023	from $\varrho^{[001]}$		

carrier concentrations from resistivity and Hall coefficient: Fig. 15; $n \approx p$ above 1000 K [81K], electrical resistivity: Fig. 16, Hall coefficient is isotropic at 77···1200 K: Fig. 17, thermoelectric power: Fig. 18.

anisotropy of intrinsic resistivity (in Ω cm):

ϱ_i	$4.42 \cdot 10^{-5} \exp(4073/T)$	along [100]		81K
	$1.09 \cdot 10^{-4} \exp(4073/T)$	along [001]		

On a sample with 47 wt% Si ($=MnSi_{1.735}$) a carrier concentration $n=9 \cdot 10^{19}$ cm^{-3} was found to be constant in the range 300···700 K. From the resistivity in the range 300···700 K an energy gap $E_g \approx 0.6$ eV was derived [61K]. A very similar behavior was observed on a sample containing 47.5 wt% Si ($=MnSi_{1.770}$) (Fig. 5). Curie-Weiss-type paramagnetism was reported for alloys with 46.0, 46.3, 46.5, 46.8, 47.0 and 47.3 wt% Si [65R].

$Mn_{27}Si_{47}$:

conductivities (in Ω^{-1} cm^{-1}):

σ	220	$\parallel c$, RT		67V2
	218			73L
	420			74Z
	760	$\perp c$, RT		67V2
	1060			73L
	900			74Z

9.15.1.1 Transition-metal compounds $T_n(III)_{2n-m}$ and $T_n(IV)_{2n-m}$

Physical property	Numerical value	Experimental conditions	Experimental method, remarks	Ref.
thermoelectric powers (in $\mu V\,K^{-1}$):				
S	160	$\|c$, RT		67V2
	170			73L
	100	$\perp c$, RT		67V2
	105			73L
thermal conductivities (in $W\,cm^{-1}\,K^{-1}$):				
κ	0.029	$\|c$, RT		67V2
	0.030			73L
	0.046	$\perp c$, RT		67V2
	0.053			73L

From 77 to 300 K $\varrho(T)$ of $Mn_{27}Si_{47}$ is fairly well represented by a straight line: $\varrho_\perp = aT$ and $\varrho_\| = aT + b$ with $a \approx 8.3 \cdot 10^{-6}\,\Omega\,cm\,K^{-1}$ and $b \approx -2.4 \cdot 10^{-4}\,\Omega\,cm$ [74Z].

Magnetic properties
(second-phase effects?)

Itinerant antiferromagnet with a spiral spin structure of incommensurate periodicity $l = 163$ Å along the c axis (from small-angle neutron scattering). Angular change between two consecutive Mn planes (which are separated by 1.09 Å) is 2.4°. Easy plane [001] [80N].

Néel temperature:

T_N	42 K		temperature of discontinuous change of χ^{-1}	80N

paramagnetic Curie temperature:

Θ_p	$+38$ K		from Curie-Weiss law, $T = 50\cdots110$ K	80N
	$+5$ K		from Curie-Weiss law, $T = 130\cdots200$ K	

magnetic moment per ion:

p_A	0.056 μ_B/Mn	$T = 4$ K	from neutron diffraction intensities	80N

saturation magnetic moment per ion:

$p_{A,s}$	0.0135 μ_B/Mn	$T = 4$ K	magnetization at $B = 1$ T	80N

paramagnetic moment:

p_{eff}	0.09 μ_B/Mn	$T > T_N$	from Curie-Weiss law	80N

9.15.1.1.3 Doped and ternary Mn_nSi_{2n-m} phases

Solubility limits:

$MnSi_{1.73-x}B_x$:	$x < 0.06$	[70N1]
$MnSi_{1.73-x}Al_x$:	$x \leq 0.006$	[70N1]
$MnSi_{1.73-x}Ge_x$:	$x \leq 0.015$	[70N1, 72A]
$(Mn_{1-x}Cr_x)_nSi_{2n-m}$:	$x < 0.05$	[70U]
	$x \approx 0.09$	at 1000 °C [72A]
$(Mn_{1-x}Re_x)_nSi_{2n-m}$:	$x < 0.07$	[69E, 70U]
$(Mn_{1-x}Fe_x)_nSi_{2n-m}$:	$x = 0.25$	[74A1]
	$x \leq 0.30$	[70N1]

The solubility of B and Al in $Mn_{11}Si_{19}$ is increased by at least an order of magnitude by the presence of Ge. Doping with Ge slightly raises the conductivity, but does not affect the thermoelectric power. Alloying with boron increases the concentration of charge carriers without markedly reducing their mobility. Partial substitution of Si by Al results in a greater high-temperature conductivity.

Physical property	Numerical value	Experimental conditions	Experimental method, remarks	Ref.

Cr substitution for Mn, which reduces the valence-electron concentration but increases the (intrinsic?) conductivity and the slope of $\log \sigma$ vs. T^{-1} [72A], leads to a smaller defect, i.e. to a smaller value of $(2n-m)/n$, while the reverse holds for Fe substitution [70N1]. Fe substitution for Mn increases the low-temperature resistivity, the carrier concentration and the effective mass m_p, and reduces the mobility and the thermal conductivity in [100] direction. The anisotropy of S and σ decreases with temperature and vanishes toward 1000 K [74A1] (Figs. 19···23).

9.15.1.1.4 Physical properties of Ru$_2$(IV)$_3$ compounds

Ru$_2$Si$_3$(r) (room-temperature modification):

energy gap:

$E_{g,th}$	≈ 0.7 eV		from resistivity measurement assuming $\varrho \propto e^{E_g/2kT}$, $T = 1050···1300$ K, four-probe method, polycrystalline bar, pressed Mo contacts	80S

magnetic susceptibility:

χ_g	$-0.52 \cdot 10^{-6}$ cm^3 g^{-1}	RT, $B = 5···10$ kG	Faraday method, powder sample, χ_g in CGS-emu	80S

transition temperature:

T_{tr}	≈ 1300 K		diffusionless phase transition	80S

Ru$_2$Si$_3$(h) (high-temperature modification):

energy gap:

$E_{g,th}$	0.44 eV		from $\varrho(T)$, $T = 1400···1700$ K	80S

melting point:

T_m	≈ 2070 K		congruent	75P
	≈ 1983 K		congruent	65O

Ru$_2$(Si,Ge)$_3$

The energy gap of the h-modification, as well as the transition temperature decrease with growing Ge concentration. For all concentrations $E_g(r) > E_g(h)$ [80S].

Ru$_2$Ge$_3$(r) (room-temperature modification):

energy gap:

$E_{g,th}$	0.52 eV		from $\varrho \propto e^{E_g/2kT}$, $T = 550···700$ K, polycrystalline bar	80S

electron concentration (in 10^{18} cm^{-3}):

		T [K]		
n	0.7	1.5	estimated from $R_H = -8.62$ cm^3 A^{-1} s^{-1}	80S
	0.9	41	estimated from $R_H = -7.14$ cm^3 A^{-1} s^{-1}	
	1.7	175	estimated from $R_H = -3.63$ cm^3 A^{-1} s^{-1}	
	≤ 7.5	1.5···5	free-electron concentration derived from $\gamma = 29 \cdot 10^{-6}$ J K^{-2} (g-atom)$^{-1}$ of the heat capacity	

Debye temperature:

Θ_D	435 K	$T = 1.5···5$ K	from heat-capacity measurements	80S

9.15.1.1.5 References for 9.15.1.1

Physical property	Numerical value	Experimental conditions	Experimental method, remarks	Ref.
magnetic susceptibility:				
χ_g	$-0.43 \cdot 10^{-6}$ cm^3 g^{-1}	$T = 50 \cdots 900$ K, $B = 5 \cdots 10$ kG	Faraday method, powder sample; χ roughly temperature independent; χ in CGS-emu	80S
transition temperature:				
T_{tr}	≈ 800 K		from X-ray diffraction and resistivity	75P, 80S
Ru$_2$Ge$_3$(h) (high-temperature modification):				
energy gap:				
$E_{g,th}$	0.34 eV		from resistivity measurements, $T = 800 \cdots 1100$ K	80S
magnetic susceptibility (in 10^{-6} cm^3 g^{-1}):				
χ_g	-0.43	$T = 900$ K, $B = 5 \cdots 10$ kG	linearly increasing above 900 K, Faraday method, powder sample, χ in CGS-emu	80S
	-0.39			
peritectic temperature:				
T_{perit}	≈ 1770 K			80S

Ru$_2$(Ge,Sn)$_3$:
Roughly linear decrease of E_g of the high-temperature modification with Sn concentration [80S].

Ru$_2$Sn$_3$(r) (room-temperature modification):
Probably semiconducting with a very small energy gap.
Extrapolation of E_g of Ru$_2$(Ge,Sn)$_3$(h) yields $E_g \approx 0.1$ eV for Ru$_2$Sn$_3$. For pure Ru$_2$Sn$_3$(h) $\varrho(T)$ showed a linear increase between 750 and 1100 K [80S].

Physical property	Numerical value	Experimental conditions	Experimental method, remarks	Ref.
magnetic susceptibility (in 10^{-6} cm^3 g^{-1}):		T [K]		
χ_g	-0.30	4, $B = 5 \cdots 10$ kG	Faraday method, powder sample, χ in CGS-emu	80S
	-0.33	200		
	-0.35	400, on heating		
	-0.34	400, on cooling from 700 K		
transition temperatures (in K):				
T_{tr}	<100		diffusionless phase transition to the orthorhombic modification	75P
	$300 \cdots 500$		first-order(?) transition to an unknown structure (this phase was not detected on high-temperature X-ray patterns up to 800 °C [75P])	80S
T_{dec}	≈ 1500		peritectoidic reaction	80S

9.15.1.1.5 References for 9.15.1.1

61K Korshunov, V.A., Gel'd, P.V.: Fiz. Metal. Metalloved. **11** (1961) 945 (translation: Phys. Met. Metallogr. **11**, Nr. 6 (1961) 118).
62R Raub, E., Fritzsche, W.: Z. Metallkde. **53** (1962) 779.
63J Jeitschko, W., Holleck, H., Nowotny, H., Benesovsky, F.: Monatsh. Chem. **94** (1963) 838.
63S Schwomma, O., Nowotny, H., Wittmann, A.: Monatsh. Chem. **94** (1963) 924.

9.15.1.1.5 References for 9.15.1.1

64F	Fujino, Y., Shinoda, D., Asanabe, S., Sasaki, Y.: Jap. J. Appl. Phys. **3** (1964) 431.
64S	Schwomma, O., Preisinger, A., Nowotny, H., Wittmann, A.: Monatsh. Chem. **95** (1964) 1527.
65O	Obrowski, W.: Metall **19** (1965) 741.
65R	Radovskiy, I.Z., Sidorenko, F.A., Gel'd, P.V.: Fiz. Metal. Metalloved. **19** (1965) 514 (translation: Phys. Met. Metallogr. **19**, Nr. 4 (1965) 30).
65W	Wittmann, A., Nowotny, H.: J. Less-Common Met. **9** (1965) 303.
66M	Morokhovets, M.A., Elagina, E.I., Abrikosov, N.Kh.: Izv. Akad. Nauk SSSR, Neorg. Mater. **2** (1966) 650 (translation: Inorg. Mater. **2** (1966) 561).
66S	Shinoda, D., Asanabe, S.: J. Phys. Soc. Jpn. **21** (1966) 555.
67F	Flieher, G., Völlenkle, H., Nowotny, H.: Monatsh. Chem. **98** (1967) 2173.
67J	Jeitschko, W., Parthé, E.: Acta Crystallogr. **22** (1967) 417.
67K	Knott, H.W., Mueller, M.H., Heaton, L.: Acta Crystallogr. **23** (1967) 549.
67P	Panday, P.K., Singh, G.S.P., Schubert, K.: Z. Krist. **125** (1967) 274.
67V1	Völlenkle, H., Preisinger, A., Nowotny, H., Wittmann, A.: Z. Krist. **124** (1967) 9.
67V2	Voronov, B.K., Dudkin, L.D., Trusova, N.N.: Kristallografiya **12** (1967) 519 (translation: Sov. Phys. Cryst. **12** (1967) 448).
68F	Flieher, G., Völlenkle, H., Nowotny, H.: Monatsh. Chem. **99** (1968) 2408.
69E	Elagina, E.I., Abrikosov, N.Kh.: Izv. Akad. Nauk SSSR, Neorg. Mater. **5** (1969) 1637 (translation: Inorg. Mater. **5** (1969) 1386).
69I	Ivanova, L.D., Abrikosov, N.Kh., Elagina, E.I., Khvostikova, V.D.: Izv. Akad. Nauk SSSR, Neorg. Mater. **5** (1969) 1933 (translation: Inorg. Mater. **5** (1969) 1645).
69K	Karpinskii, O.G., Evseev, B.A.: Izv. Akad. Nauk SSSR, Neorg. Mater. **5** (1969) 525 (translation: Inorg. Mater. **5** (1969) 483).
69N1	Nikitin, E.N., Tarasov, V.I., Tamarin, P.V.: Fiz. Tverd. Tela **11** (1969) 234 (translation: Sov. Phys. Solid State **11** (1969) 187).
69N2	Nikitin, E.N., Tarasov, V.I., Andreev, A.A., Shumilova, L.N.: Fiz. Tverd. Tela **11** (1969) 3389 (translation: Sov. Phys. Solid State **11** (1969/70) 2757).
69P	Parthé, E.: in "Developments in the Structural Chemistry of Alloy Phases" (B.C. Giessen, ed.), Plenum New York **1969**, p. 49.
70M	Mager, T., Wachtel, E.: Z. Metallkde. **61** (1970) 853.
70N1	Nikitin, E.N., Sidorov, A.F., Tarasov, V.I., Zaslavskii, A.I.: Izv. Akad. Nauk SSSR, Neorg. Mater. **6** (1970) 604 (translation: Inorg. Mater. **6** (1970) 537).
70N2	Nowotny, H.: in "The Chemistry of Extended Defects in Non-Metallic Solids", North Holland **1970**, p. 223.
70P	Pearson, W.B.: Acta Crystallogr. **B26** (1970) 1044.
70U	Ugai, Ya.A., Anokhin, V.Z., Ivanova, T.V.: Radiofiz. Mikroelektron. (1970) 75 (Chem. Abstracts **76** (1972) 146, 581f.).
71A	Abrikosov, N.Kh., Elagina, E.I., Ledenkova, L.I.: Izv. Akad. Nauk SSSR, Neorg. Mater. **7** (1971) 870 (translation: Inorg. Mater. **7** (1971) 763).
71D	De Ridder, R., Amelinckx, S.: Mat. Res. Bull. **6** (1971) 1223.
71N	Nikitin, E.N., Tarasov, V.I.: Fiz. Tverd. Tela **13** (1971) 3473 (translation: Sov. Phys. Solid State **13** (1972) 2938).
71Z	Zwilling, G., Nowotny, H.: Monatsh. Chem. **102** (1971) 672.
72A	Abrikosov, N.Kh., Ivanova, L.D., Murav'ev, V.G.: Izv. Akad. Nauk SSSR, Neorg. Mater. **8** (1972) 1194 (translation: Inorg. Mater. **8** (1972) 1049).
72K	Karpinskii, O.G., Evseev, B.A.: in "Chemical Bonds in Solids" Vol. **4** (ed. N.N. Sirota) Consultants Bureau New York **1972**, p. 3.
72P	Pearson, W.B.: "The Crystal Chemistry and Physics of Metals and Alloys"; Wiley, New York **1972**.
72U	Ugai, Ya.A., Anokhin, V.Z., Averbakh, E.M.: in "Chemical Bonds in Solids", Vol. **4** (ed. N.N. Sirota) Consultants Bureau, New York **1972**, p. 16.
73L	Levinson, L.M.: J. Solid State Chem. **6** (1973) 126.
73Z	Zwilling, G., Nowotny, H.: Monatsh. Chem. **104** (1973) 668.
74A1	Abrikosov, N.Kh., Ivanova, L.D.: Izv. Akad. Nauk SSSR, Neorg. Mater. **10** (1974) 1016 (translation: Inorg. Mater. **10** (1974) 873).
74A2	Abrikosov, N.Kh., Ivanova, L.D., Petrova, L.I.: Izv. Akad. Nauk SSSR, Neorg. Mater. **10** (1974) 2226 (translation: Inorg. Mater. **10** (1974) 1907).
74B	Boller, H.: Monatsh. Chem. **105** (1974) 934.

74I	Israiloff, P., Völlenkle, H.: Monatsh. Chem. **105** (1974) 1313.
74P	Poutcharovsky, D.J., Parthé, E.: Acta Crystallogr. **B30** (1974) 2692.
74V	Völlenkle, H.: Monatsh. Chem. **105** (1974) 1217.
74Z	Zwilling, G., Nowotny, H.: Monatsh. Chem. **105** (1974) 666.
75P	Poutcharovsky, D.J., Yvon, K., Parthé, E.: J. Less-Common Met. **40** (1975) 139.
75Z	Zaitsev, V.K., Tarasov, V.I., Adilbekov, A.A.: Fiz. Tverd. Tela **17** (1975) 581 (translation: Sov. Phys. Solid State **17** (1975) 370).
76D	De Ridder, R., van Tendeloo, G., Amelinckx, S.: Phys. Status Solidi (a) **33** (1976) 383.
77J	Jeitschko, W.: Acta Crystallogr. **B33** (1977) 2347.
79Z	Zaitsev, V.K., Ordin, S.V., Tarasov, V.I., Fedorov, M.I.: Fiz. Tverd. Tela **21** (1979) 2517 (translation: Sov. Phys. Solid State **21** (1979) 1454).
80N	Nakajima, T., Schelten, J.: J. Mag. Magn. Mater. **21** (1980) 157.
80S	Susz, C.P., Muller, J., Yvon, K., Parthé, E.: J. Less-Common Met. **71** (1980) P1.
81K	Kawasumi, I., Sakata, M., Nishida, I., Masumoto, K.: J. Mater. Sci. **16** (1981) 355.

9.15.1.2 T(IV)$_2$ compounds

CrSi$_2$, ReSi$_2$, FeSi$_2$, (RuSi$_2$) OsSi$_2$

9.15.1.2.0 Structures of the TSi$_2$ phases (including lattice properties)

The hexagonal CrSi$_2$ structure and the tetragonal MoSi$_2$ type structure belong to the same family of polytypic structures as does the orthorhombic TiSi$_2$ structure. These three structure types can be generated from close-packed layers in bcc [110] stacking [72P]. Of the four possible stacking positions ABCD two (AB) are involved in tetragonal MoSi$_2$, three (ABC) in hexagonal CrSi$_2$ and four (ACBD) in orthorhombic TiSi$_2$. The MoSi$_2$ cell contains three bc tetragonal ($c/a < 1$) pseudo-cells. Thus, Si pairs in c direction result which however are connected with four Si atoms of the adjacent pseudo-cell at virtually the same distance. The metal atom is coordinated to 10 Si atoms. ReSi$_2$ crystallizes in a slightly distorted MoSi$_2$ structure, isopuntal with orthorhombic MoPt$_2$. In CrSi$_2$ Si has 2+3 close Cr neighbors and 2+3 Si neighbors at 2.48 and 2.56 Å. FeSi$_2$ (so-called β phase) is stable only below 950⋯970°C. At this temperature it decomposes into FeSi + Fe$_{\approx 0.8}$Si$_2$ (so-called α-FeSi$_2$). The structure of FeSi$_2$ is of the orthorhombic OsSi$_2$ type [71D] which can be described as a distorted CaF$_2$ type derivative, i.e. a distorted arrangement of Si cubes half of which are occupied by Fe atoms. The deformation leads to distorted Fe tetrahedrons around each Si atom. At the same time the Fe atoms as well as the Si atoms approach each other. Fe atoms cluster to a square with distances of 2.97 Å and the Si atoms acquire 5 Si neighbors as close as 2.45⋯2.59 Å.

Crystallographic data of the TSi$_2$ phases (RT values)

Compound (Structure type, space group, coordination number)	a [Å]	b [Å]	c [Å]	Ref.
CrSi$_2$ (C 40 type, $D_6^4 - P6_222$; $Z=3$)	4.428		6.364	78N
	4.431		6.364	64S
	4.428		6.363	67P
(Cr$_{33.58}$Si$_{66.42}$)	4.424		6.347	72N
Cr$_{0.775}$Mn$_{0.225}$Si$_2$ (CrSi$_2$ type, solubility limit)	4.412		6.345	78N, 62W
MoSi$_2$(h) (CrSi$_2$ type, stable >2170 K)	4.605		6.559	65A
	4.642		6.529	69S
ReSi$_2$ (MoSi$_2$ type, $D_{4h}^{17} - I4/mmm$; $Z=2$; (h)?)	3.131		7.676	67P
(distorted MoSi$_2$ type, $D_{2h}^{25} - Immm$; $Z=2$)	3.128	3.144	7.677	83S
FeSi$_2$ (OsSi$_2$ type)	9.863	7.791	7.833	71D
Fe$_{0.88}$Co$_{0.12}$Si$_2$ (OsSi$_2$ type, solubility limit)	9.918	7.798	7.840	70H
Fe$_{0.8}$Mn$_{0.2}$Si$_2$ (OsSi$_2$ type, solubility limit)				74A
RuSi$_2$ (OsSi$_2$ type (?), existence questioned [63S2])[a]				60W
RuGe$_2$ (OsSi$_2$ type ?)[a]				60W
OsSi$_2$ ($D_{2h}^{18} - Cmca$; $Z=16$)	10.1496	8.1168	8.2230	70E

[a] "RuSi$_2$" and "RuGe$_2$" of [60W] probably correspond to Ru$_2$Si$_3$ and Ru$_2$Ge$_3$, isostructural with Os$_2$Si$_3$. Above 1000 K the Ru−Ge phase diagram contains only RuGe and Ru$_2$Ge$_3$ [62R].

9.15.1.2 Transition-metal compounds T(IV)$_2$

Physical property	Numerical value	Experimental conditions	Experimental method, remarks	Ref.

Coordination of the atoms in orthorhombic FeSi$_2$ [71D] and OsSi$_2$ [70E] (distances in Å)

	FeSi$_2$	OsSi$_2$			
T(1) — 2 Si(1)	2.338	2.439			
2 Si(2)	2.338	2.462			
2 Si(1)	2.376	2.471			
2 Si(2)	2.385	2.486			
2 T(2)	2.967	3.046			
T(2) — 2 Si(1)	2.333	2.492			
2 Si(2)	2.335	2.474			
2 Si(2)	2.429	2.495			
2 Si(1)	2.437	2.521			
2 T(1)	2.967	3.046			
Si(1) — 4 T	2.333···2.437	2.439···2.521			
Si(2)	2.499	2.548			
Si(2)	2.512	2.609			
Si(1)	2.529	2.640			
Si(1)	2.561	2.667			
Si(2)	2.587	2.708			
Si(2) — 4 T	2.335···2.429	2.462···2.495			
Si(2)	2.449	2.509			
Si(1)	2.499	2.548			
Si(2)	2.511	2.599			
Si(1)	2.512	2.609			
Si(1)	2.587	2.708			

9.15.1.2.1 Physical properties of T(IV)$_2$ compounds

CrSi$_2$:

Electronic properties

energy gap (in eV):

$E_{g,th}$	0.27		from $\varrho(T)$ above 600 K	63S1
	0.35		from $\varrho(T)$ and $R_H(T)$	64S
	0.32		from $\varrho(T)$, $R_H(T)$, $S(T)$ along [100] and [001]	72N
	0.30		from $\varrho(T)$, $S(T)$	73N1
	0.29···0.32		from $\log \varrho$ vs. T^{-1}	78N
	0.30		from $\log(T^{3/2} R_H)$ vs. T^{-1}	78N

band overlap ≈ 0.1 eV at 100 K; overlap disappears at ≈ 400 K [70K].

effective masses (∥: along c; ⊥: within the (a_1, a_2)-plane) (in units of m_o):

m_n	≈ 7		density of states mass	64S
	12			73N1
	20.2			78N
m_p	≈ 5		density of states mass	64S
	3.2			73N1, 78N
$m_{p\parallel}$	5			72N
$m_{p\perp}$	3			72N

m_p independent of x in Cr$_{1-x}$Mn$_x$Si$_2$, whereas m_n decreases to 7.5 m_o for x = 0.182 [78N].

Physical property	Numerical value	Experimental conditions	Experimental method, remarks	Ref.
Transport properties				
resistivities (in $10^{-4}\,\Omega\,\text{cm}$): (see Figs. 1 and 2)		T [K]		
ϱ	≈ 10	RT	highly degenerate p-type region	64S, 72N, 78N
	$1.9\exp(1740/T)$	>700	hot-pressed bar, intrinsic region	78N
$\varrho_\parallel$	$3.0\exp(1850/T)$	>700, along [001]	intrinsic region; dc method	72N
$\varrho_\perp$	$1.6\exp(1850/T)$	>700	intrinsic region	72N
Hall effect, thermoelectric power, Nernst-Ettingshausen effect and thermal conductivity: Figs. 3···9.				
carrier mobilities (in $\text{cm}^2\,\text{V}^{-1}\,\text{s}^{-1}$): (see also Figs. 11···13)				
μ_n	0.27	RT		73N1
μ_p	8	RT		73N1
	$7.0\cdot 10^4\,T^{-3/2}$	$T=300\cdots 1200$ K	Hall mobility R_H/ϱ	78N
mobility ratio:				
b	0.035	RT		73N1
	0.01		derived from extrinsic/intrinsic resistivities	78N
thermal conductivity:				
κ	$0.071\,\text{W}\,\text{cm}^{-1}\,\text{K}^{-1}$	RT		60N2
anisotropy of resistivity, mobility, thermoelectric power and thermal conductivity:				
$\varrho_\parallel/\varrho_\perp$	1.9		roughly constant below 700 K, decreases at higher temperature	72N
	1.5···1.7	RT		67V
no anisotropy of the Hall coefficient at 100···500 K [70K], 85···1100 K [72N].				
$S_\parallel/S_\perp$	1.7	RT	for the temperature range 100···1100 K, see Figs. 7, 8. No variation of $S_\parallel - S_\perp$ with carrier concentration for samples with $\varrho=0.09\cdots 0.3$ mΩ cm [67V].	67V
	1.3			
$\kappa_\parallel/\kappa_\perp$	1.2	RT	for the temperature range 300···950 K, see Fig. 10	67V
anisotropy of Hall mobility: Fig. 13, mobility ratio $b_\parallel \ll b_\perp = 0.01$ [72N].				
hole concentration (in $10^{20}\,\text{cm}^{-3}$):				
p	6.2	$T=85\cdots 600$ K	from Hall effect in 0.5···6.5 kG (degenerate semiconductor)	72N
	≈ 4	90···500 K, undoped sample	from Hall effect in 5 kG, p decreases with excess Si (compare Fig. 3)	64S
	7.7	RT, pure CrSi$_2$		78N
Further properties				
magnetic susceptibility:				
χ_g	$-0.5\cdot 10^{-6}\,\text{cm}^3\,\text{g}^{-1}$	RT	χ_g in CGS-emu	66S
	<0	$T=300\cdots 1100$ K	for compositions within the homogeneity range (66.5···66.9% Si [68V])	65R

CrSi$_2$ is said to undergo a semimetal→semiconductor transition near 400 K [70K].

Physical property	Numerical value	Experimental conditions	Experimental method, remarks	Ref.
density (in g cm^{-3}):				
d	4.99	RT	X-ray density	68V
	4.93	RT	experimental	68V
melting point:				
T_m	1823 or 1748 K			71A

CrSi$_2$-type ternary alloys

Cr$_{1-x}$M$_x$Si$_2$ homogeneity ranges:
Ti: $x \leq 0.9$ at 1570 K [58P]; Zr: $x \approx 0$ at 1570 K [58P];
V: $x = 0 \cdots 1$ [62W]; Ta: $x \leq 0.12$ at 1570 K [58P];
Mo: $x \leq 0.3$ at 1570 K [58P, 65N]; W: $x \leq 0.15$ at 1570 K [58P];
Mn: $x \leq 0.225$ [78N, 62W]; Re: $x \leq 0.1$ at room temperature [65N];
Fe: $x \leq \approx 0.01$ [62W]; Co: $x \leq 0.02$ [71N], $x \approx 0$ [62W]; Ni: $x \approx 0$ [62W].

Cr$_{1-x}$V$_x$Si$_2$: non-metallic limit $x = 0.1$ [63S1]. RT values of electrical conductivity $1/\varrho$ and Seebeck coefficient S: Fig. 15.

Cr$_{1-x}$Mo$_x$Si$_2$, Cr$_{1-x}$W$_x$Si$_2$: resistivity ϱ and thermoelectric power S decrease with increasing x. Semiconductor properties of Cr$_{0.9}$Mo$_{0.1}$Si$_2$ (Cr$_{0.25}$Mo$_{0.1}$Si$_{0.65}$?) deduced from $\varrho(T)$ and $S(T)$ up to ≈ 1800 K [70T].

Cr$_{1-x}$Mn$_x$Si$_2$: resistivity ϱ: Figs. 1, 2, 15; Hall effect: Figs. 3, 4; Seebeck coefficient S: Figs. 5, 6, 15; mobility: Figs. 11, 12; carrier concentration: Fig. 14.

ReSi$_2$:

energy gap (in eV):				
$E_{g,th}$	0.13	$T = 0$ K	from $\varrho(T)$ above 500 K assuming $\log \varrho \propto E_g/2kT$	65N, 61N, 60N1
	0.21$\cdots$0.22	0 K	from $\varrho(T)$ in the range 500$\cdots$1200 K	83S
	0.195	RT	from optical reflectivity	83S
resistivity (in Ω cm):				
(for temperature dependence, see Fig. 16)				
ϱ	≈ 0.01	RT	impurity conduction	61N
	0.016	$T = 295$ K	extrapolated from intrinsic range	61N
	0.003$\cdots$0.01	RT	four-probe ac method, polycrystalline sample	83S
thermoelectric power (in μV K^{-1}):				
(see Fig. 17)				
S	$-90 \cdots -130$	$T \approx 310$ K	polycrystalline sample	83S
	$+80$	≈ 300 K	polycrystal; p-type conduction confirmed by Hall effect; impurity concentration $\approx 10^{18}$ cm^{-3}	65N
magnetic susceptibility:				
χ_m	$-75.3 + 0.0092\,T$ [10^{-6} cm^3 mol^{-1}]	$T = 300 \cdots 1000$ K	polycrystalline bar, Faraday method, $B = 4$ kG; χ in CGS-emu	83S
dielectric constant:				
$\varepsilon(0)$	≈ 70	$T = 0$ K	extrapolated from RT reflectivity by subtracting the free-carrier contribution	83S
melting point:				
T_m	2253 K			61N

Physical property	Numerical value	Experimental conditions	Experimental method, remarks	Ref.
density (in g cm^{-3}):				
d	10.8	$T=295$ K		61N
	10.70	295 K	X-ray density	

Nonmetallic character of ReSi$_2$ questioned by [80C] based on the high atomic-volume contraction on compound formation.

Re$_{1-x}$Ti$_x$Si$_2$: solubility x$\leq$0.15 at 1370 K [64D].
Re$_{1-x}$Cr$_x$Si$_2$: solubility limit x=0.2 [65N]; decrease of ϱ, S and κ with x [65N].
Re$_{1-x}$Mo$_x$Si$_2$: complete solid solubility [64N]; smooth decrease of ϱ at room temperature (i.e. in the extrinsic region) [64N].
Re$_{1-x}$Mn$_x$Si$_2$: solubility x$\leq$0.03 at 1370 K [69E].

FeSi$_2$:
Electronic properties
energy gaps (in eV):

		T [K]		
$E_{g,\text{th}}$	0.85	0	calculated from the transition temperature 1210 K	69B2
	0.88	0	from $\log \varrho \propto E_g/2\,kT$ in the intrinsic range	64W
	0.9$\cdots$1.0	0	from $\log \varrho \propto E_g/2\,kT$ in the intrinsic range (≈ 0.9 eV for Al-doped, ≈ 1.0 eV for Co-doped samples)	68B1
	0.95	0	from $\log \varrho \propto E_g/2\,kT$	73N2
E_g	1.0	295	from optical absorption	70B2

E_g decreases with x in Fe$_{1-x}$Mn$_x$Si$_2$: 0.92 eV for x=0.03, 0.80 eV for x=0.06 [73N2].
E_g increases (?) with x in Fe$_{1-x}$Co$_x$Si$_2$ (x=0: ≈ 1.14 eV, x=0.04: ≈ 1.3 eV) [71U].

dE_g/dT	-0.45 meV K^{-1}	$T=700\cdots1200$ K	from thermal conductivity	73W

density of states (in 10^{22} cm^{-3}):

g	5.2	T_{tr}	from magnetic susceptibility (in CGS-emu) $\chi_g = \dfrac{2g\mu_B^2}{d_x kT} e^{-E_g(T)/2kT}$; see also Fig. 44	69B1
	4.9	T_{tr}	from electrical conductivity	69B2
	7.46			73N2

Various energy band models for FeSi$_2$ are discussed in [68B1, 69B2, 69H, 70B1, 73B] (Figs. 18, 19).

Transport properties

resistivity, Hall effect, thermoelectric power and thermal conductivity of pure and Mn-, Co- and Al-doped samples: Figs. 20$\cdots$36. (See also Figs. 46, 47.)

absolute thermoelectric power in the intrinsic range (in µV K^{-1}):

S	$\dfrac{k}{e}\left[\dfrac{620}{T}-0.00084\right]$	$T=600\cdots900$ K	polycrystalline samples, annealed at 1173 K, $R_H < 1$ cm^3 C^{-1}	64W

According to [68B1], however, $S(T)$ does not follow a T^{-1}-law as the mobility ratio b depends strongly on temperature.

Physical property	Numerical value	Experimental conditions	Experimental method, remarks	Ref.
mobilities of charge carriers (in $cm^2 V^{-1} s^{-1}$): Fig. 37				
μ_n	0.26	$T = 300$ K	assuming small polarons in n-type $FeSi_2$ (activation energy of the mobility 0.06 eV)	68B1
μ_p	2···4	300 K	$\mu_p \propto T^{1/2}$ in the extrinsic range	68B1
$\mu_p + \mu_n$	≈ 1	1210 K		69B2
mobility ratio ($b = \mu_n/\mu_p$):				
b	0.8	$T = 600···850$ K	from $S(T)$ (Fig. 26) assuming a simple 2-band semiconductor in the intrinsic range: $$S = -\frac{k}{e}\left[\frac{b-1}{b+1}\frac{E_g}{2kT} + A\right],$$ with E_g from $\varrho(T)$	64W
	0.33	1260 K	(the decomposition temperature of $FeSi_2$)	73N2

free-carrier contribution to the thermal conductivity: Fig. 35. Doping with Co and Ni produces n-type material, whereas doping with Mn or substitution of a Group III element for Si leads to p-type material [64W, 73N2].

Optimum properties for thermoelectric generators were obtained with Co for n-type doping and Al for p-type doping [64W].

Optical and further properties

reflectivity and optical constants of orthorhombic $FeSi_2$: Figs. 38···43.

The reflectivity of p-type $FeSi_2$ can be interpreted with Drude's theory of free-carrier absorption in contrast to the case of n-type $FeSi_2$. A consistent interpretation of the optical properties of n-type $FeSi_2$ is possible in terms of the small-polaron picture [70B2, 72B].

dielectric constants:

$\varepsilon(0)$	61.6	RT, polycryst.	from a Kramers-Kronig analysis of reflectivity in the range 1···50 μm	68B2, 73W
$\varepsilon(\infty)$	27.6	RT, polycryst.		68B2

magnetic susceptibility (in 10^{-8} cm^3 g^{-1}): (Figs. 44, 45)

χ_g	-4	$T = 300$ K	χ_g in CGS-emu	69B1
	$+20$	T_{dec}		

magnetic susceptibility $\chi(T)$ in the intrinsic conduction range: Fig. 44.

phase transition: Figs. 45···47.

decomposition temperatures (peritectic temperature) (in K):
(compare phase diagram, Fig. 48)

T_{dec}	1223			59S, 60S
	1200			68B1
	1259			68P
	~1250			71U
	1260		from resistivity jump	73N2
	1253		for $Fe_{0.97}Mn_{0.03}Si_2$	
	1218		for $Fe_{0.94}Mn_{0.06}Si_2$	

Figs. p. 397

9.15.1.2.2 References for 9.15.1.2

Physical property	Numerical value	Experimental conditions	Experimental method remarks	Ref.
Debye temperature:				
Θ_D	630 K	$T = 0$ K	from heat capacity $c_p(T)$	73W
heat capacity: Fig. 49.				
density (in g cm^{-3}):				
d	4.93	RT	pycnometric	71D
	4.94	RT	X-ray density	71D
OsSi$_2$:				
energy gaps (in eV):				
$E_{g,th}$	≥ 0.26		from $\log \varrho \propto E_g/2kT$; $T = 400 \cdots 700$ K, sintered sample	82H
	1.8		from $\log \varrho \propto E_g/2kT$; $T = 1000 \cdots 1370$ K, p-type sintered sample containing 1.5 at% Al and 0.2 at% Fe; dc four-probe method with Mo contacts	83M
E_g	≈ 2	RT	on sintered sample as above optical reflectivity similar to that of FeSi$_2$; thermoelectric power ($+100$ μVK^{-1} at 300 K, $\approx +400$ μVK^{-1} at 900 K) decreasing above 1000 K	

9.15.1.2.2 References for 9.15.1.2

58P	Pearson, W.B.: "A Handbook of Lattice Spacings and Structures of Metals and Alloys", Vol. 1, Pergamon Press **1958**, reprinted 1964.
59G	Gel'd, P.V.: Zhur. Tekh. Fiz. **27** (1957) 113 (translation: Sov. Phys. Tech. Phys. **2** (1957) 95).
59S	Sidorenko, F.A., Gel'd, P.V., Dubrovskaya, L.B.: Fiz. Metal. Metalloved. **8** (1959) 735 (translation: Phys. Met. Metallogr. **8**, Nr. 5 (1959) 88).
60N1	Neshpor, V.S., Samsonov, G.V.: Fiz. Tverd. Tela **2** (1960) 2202 (translation: Sov. Phys. Solid State **2** (1960) 1966).
60N2	Nikitin, E.N.: Fiz. Tverd. Tela **2** (1960) 2685 (translation: Sov. Phys. Solid State **2** (1961) 2389).
60S	Sidorenko, F.A., Gel'd, P.V., Shumilov, M.A.: Fiz. Metal. Metalloved. **9** (1960) 861 (translation: Phys. Met. Metallogr. **9**, Nr. 6 (1960) 56).
60W	Weitz, G., Born, L., Hellner, E.: Z. Metallkde. **51** (1960) 238.
61N	Neshpor, V.S., Samsonov, G.V.: Fiz. Metal. Metalloved. **11** (1961) 638 (translation: Phys. Met. Metallogr. **11**, Nr. 4 (1961) 146).
62R	Raub, E., Fritzsche, W.: Z. Metallkde. **53** (1962) 779.
62W	Wittmann, A., Burger, K.O., Nowotny, H.: Monatsh. Chem. **93** (1962) 674.
63S1	Sakata, T., Tokushima, T.: Trans. Ntl. Res. Inst. Met. **5** (1963) 34.
63S2	Schwomma, O., Nowotny, H., Wittmann, A.: Monatsh. Chem. **94** (1963) 681.
64D	Duffin, W.J., Parthé, E., Norton, J.T.: Acta Crystallogr. **17** (1964) 450.
64N	Neshpor, V.S., Samsonov, G.V.: Fiz. Metal. Metalloved. **18** (1964) 187 (translation: Phys. Met. Metallogr. **18**, Nr. 2 (1964) 31).
64S	Shinoda, D., Asanabe, S., Sasaki, Y.: J. Phys. Soc. Jpn. **19** (1964) 269.
64W	Ware, R.M., Mc Neill, D.J.: Proc. IEE **111** (1964) 178.
65A	Aubry, J., Duval, R., Roques, B.: C.R. Acad. Sci., Paris **261** (1965) 2665.
65N	Neshpor, V.S., Samsonov, G.V.: Izv. Akad. Nauk SSSR, Neorg. Mater. **1** (1965) 665 (translation: Inorg. Mater. **1** (1965) 599).
65R	Radovskiy, I.Z., Sidorenko, F.A., Gel'd, P.V.: Fiz. Metal. Metalloved. **19** (1965) 915 (translation: Phys. Met. Metallogr. **19**, Nr. 6 (1965) 104).
66S	Shinoda, D., Asanabe, S.: J. Phys. Soc. Jpn. **21** (1966) 555.

9.15.1.2.2 References for 9.15.1.2

67P	Pearson, W.B.: "A Handbook of Lattice Spacings and Structures of Metals and Alloys", Vol. 2, Pergamon Press **1967**.
67V	Voronov, B.K., Dudkin, L.D., Trusova, N.N.: Kristallografiya **12** (1967) 519 (translation: Sov. Phys. Cryst. **12** (1967) 448).
68B1	Birkholz, U., Schelm, J.: Phys. Status Solidi **27** (1968) 413.
68B2	Birkholz, U., Finkenrath, H., Naegele, J., Uhle, N.: Phys. Status Solidi **30** (1968) K81.
68P	Piton, J.P., Fay, M.F.: C.R. Acad. Sci., Paris **266C** (1968) 514.
68V	Voronov, B.K., Dudkin, L.D., Kiryukhina, N.I., Trusova, N.N.: Izv. Akad. Nauk SSSR, Neorg. Mater. **4** (1968) 325 (translation: Inorg. Mater. **4** (1968) 271).
69B1	Birkholz, U., Frühauf, A.: Phys. Status Solidi **34** (1969) K181.
69B2	Birkholz, U., Schelm, J.: Phys. Status Solidi **34** (1969) K177.
69E	Elagina, E.I., Abrikosov, N.Kh.: Izv. Akad. Nauk SSSR, Neorg. Mater. **5** (1969) 1637 (translation: Inorg. Mater. **5** (1969) 1386).
69H	Hesse, J.: Z. Metallkde. **60** (1969) 652.
69S	Svechnikov, V.N., Kocherzhinskii, Yu.A., Yupko, L.M.: Dokl. Akad. Nauk SSSR **182** (1968) 1313 (translation: Sov. Phys. Doklady **13** (1969) 1072).
70B1	Birkholz, U., Frühauf, A., Schelm, J.: Proc. Tenth Int. Conf. Phys. Semicond., Cambridge, Mass. (1970) 311.
70B2	Birkholz, U., Naegele, J.: Phys. Status Solidi **39** (1970) 197.
70E	Engström, I.: Acta Chem. Scand. **24** (1970) 2117.
70H	Hesse, J., Bucksch, R.: J. Mater. Sci. **5** (1970) 272.
70K	Kaidanov, V.I., Tselishchev, V.A., Usov, A.P., Dudkin, L.D., Voronov, B.K., Trusova, N.N.: Fiz. Tekh. Poluprovod. **4** (1970) 1338 (translation: Sov. Phys. Semicond. **4** (1971) 1135).
70T	Terekhov, G.I., Tagirova, R.Kh., Ivanov, O.S.: Izv. Akad. Nauk SSSR, Metal. (1970) 177 (translation: Russian Metallurgy (1970) 126).
71A	Abrikosov, N.Kh., Elagina, E.I., Ledenkova, L.I.: Izv. Akad. Nauk SSSR, Neorg. Mater. **7** (1971) 870 (translation: Inorg. Mater. **7** (1971) 763).
71D	Dusausoy, Y., Protas, J., Wandji, R., Roques, B.: Acta Crystallogr. B **27** (1971) 1209.
71N	Nikitin, E.N., Tarasov, V.I.: Kristallografiya **16** (1971) 372 (translation: Sov. Phys. Cryst. **16** (1971) 305).
71S	Suchet, J.P.: "Crystal Chemistry and Semiconduction in Transition Metal Binary Compounds". Academic Press, New York **1971**.
71U	Ugai, Ya.A., Ivanova, T.V., Inozemtseva, V.P.: Izv. Akad. Nauk SSSR, Neorg. Mater. **7** (1971) 1983 (translation: Inorg. Mater. **7** (1971) 1766).
72B	Baltz, R.v., Birkholz, U.: Festkörperprobleme XII, Pergamon/Vieweg **1972**, 233.
72N	Nishida, I.: J. Mater. Sci. **7** (1972) 1119.
72P	Pearson, W.B.: "The Crystal Chemistry and Physics of Metals and Alloys", Wiley, New York **1972**.
72V	Voronov, B.K., Dudkin, L.D., Trusova, N.N.: in "Chemical Bonds in Solids" Vol. 4 (ed. N.N. Sirota) Consultants Bureau, New York **1972**, 21.
73B	Blaauw, C., van der Woude, F., Sawatzki, G.A.: J. Phys. C**6** (1973) 2371.
73N1	Nikitin, E.N., Tarasov, V.I., Zaitsev, V.K.: Fiz. Tverd. Tela **15** (1973) 1254 (translation: Sov. Phys. Solid State **15** (1973) 846).
73N2	Nishida, I.: Phys. Rev. B**17** (1973) 2710.
73W	Waldecker, G., Meinhold, H., Birkholz, U.: Phys. Status Solidi (a) **15** (1973) 143.
74A	Abrikosov, N.Kh., Ivanova, L.D., Petrova, L.I.: Izv. Akad. Nauk SSSR, Neorg. Mater. **10** (1974) 2226 (translation: Inorg. Mater. **10** (1974) 1907).
78N	Nishida, I., Sakata, T.: J. Phys. Chem. Solids **39** (1978) 499.
78S	Sakata, T., Sakai, Y., Yoshino, H., Fujii, H.: J. Less-Common Met. **61** (1978) 301.
80C	Chapnik, I.M.: Phys. Status Solidi (a) **58** (1980) K193.
82H	Hulliger, F.: unpublished.
83M	Mason, K., Müller-Vogt, G.: J. Cryst. Growth **63** (1983) 34.
83S	Siegrist, T., Hulliger, F., Travaglini, G.: J. Less-Common Met. **92** (1983) 119.

9.15.1.3 T(V)$_2$ compounds

T = Cr; Fe, Ru, Os; Co, Rh, Ir; Ni, Pd, Pt
V = P, As, Sb, (Bi)

9.15.1.3.0 Structure, chemical bond (including lattice properties)

Crystallographic data: tables below and Figs. 1···6.
Energy-band schemes for CrSb$_2$, FeAs$_2$ and CoAs$_2$: Figs. 7···9.

In nonmetallic transition-element pnictides not more than 4 cation valence electrons are engaged in the cation-anion bonds, therefore the semiconducting compounds TV$_n$ (n ≥ 2) as well as the chalcopnictides T–V–VI all contain polyanions. The most stable configuration of the cation d electrons is nonmagnetic d^6 and thus octahedral (although distorted) coordination ($\sim t_{2g}^6$) is predominant. Structures that match this requirement for TX$_2$ compositions are the pyrite and the marcasite structure as well as the pararammelsbergite structure. The pyrite structure is a rocksalt structure where the single anions are substituted by polyanions [X$_2$], oriented along the body diagonals. Each anion X thus acquires 1X + 3T neighbors which together form a distorted tetrahedron. The marcasite structure is closely related to the rutile TiO$_2$ structure and can be derived from a cation-deficient NiAs structure, i.e. from a distorted hexagonal anion close packing. Whereas in pyrite the [TX$_6$] octahedra are connected via corners they share two opposite edges in marcasite. In the marcasite structure the deformation of the [TX$_6$] octahedron is more pronounced than in the pyrite structure. Frequently the marcasite type is met in the low-temperature modification while the pyrite type occurs in the high-temperature or high-pressure modification. The pararammelsbergite structure, which is isopuntal with brookite TiO$_2$, represents a combination of pyrite and marcasite.

In the compounds with low-spin d^6 cations no distortion of the [TX$_6$] octahedron is required in order to warrant nonmetallic properties. The distortions found may be due to geometrical constraints only. With a cation in octahedral coordination carrying less than six excess valence electrons, however, the t$_{2g}$ band would be incompletely filled. Metallic behavior can be avoided only by splitting this d-band by means of additional distortions or cation-cation bonds. The loellingite branch of the marcasite family indeed reveals such strong structural distortions which at the same time lead to distinctly shortened metal-metal distances within the metal chains along the c axis (T–T = c). Loellingites form with d^4 and d^2 cations. The case of the antiferromagnetic semiconductor CrSb$_2$ raises doubts that metal-metal bonding is the driving force for the loellingite-type contractions of the marcasite structure [68H3, 72G, 74K3, 79K1].

Some kind of metal-metal bonding appears to occur in the low-spin d^5 compounds TX$_2$ and TXY which crystallize in the arsenopyrite structure. The metal atoms, which are equidistant in the chains of the loellingites, are shifted out of the centers of the [TX$_6$] octahedra such as to form T–T pairs separated from the neighboring pairs. The bonding within these T–T pairs is gradually weakened at higher temperatures and in CoAs$_2$, CoSb$_2$ and RhSb$_2$ a second-order transition to the loellingite structure is observed [77K3].

Various models for the chemical bonding in pyrites, marcasites and arsenopyrites have been proposed [65H, 65P, 70B2, 71K2, 72G, 72H, 74K2, 74K3, 75F, 75J]. Recent molecular-orbital calculations [81T] support Goodenough's model [72G] which ignores T–T interactions. Attention is focused upon the T–X–T angles which differ substantially among the three structure types. Direct T–T d orbital interaction, on the other hand, is argued to be small.

The PdP$_2$ structure owes its existence to the particular stability of the diamagnetic d^8 configuration of Ni^{2+} and Pd^{2+}. In this structure type anion chains spiralling along the a axis are connected by cations in such a way that the latter acquire a square-planar coordination while the anions see four neighbors (2P + 2T) at the apices of a strongly deformed tetrahedron.

Abbreviations: r: room temperature modification, h: high-temperature modification, l: low-temperature modification, p: high-pressure phase.

9.15.1.3 Transition-metal compounds T(V)$_2$

Crystallographic data of transition-element dipnictides with octahedrally coordinated cations (RT values)

Cation configuration	Compound (Space group, coordination number)	a [Å]	b [Å]	c [Å]	β [°]	d [g cm^{-3}]	Ref.
d^6	Pyrites (T$_h^6$–Pa3; Z=4)						
	NiP$_2$(p)*)	5.4831				4.86	68D
	NiAs$_2$(p)*)	5.7634				7.24	68D
	Ni$_{0.5}$Pd$_{0.5}$As$_2$*)	5.877				7.60	66B
	PdAs$_2$*)	5.9845				7.94	66B
		5.9855				7.935	65F
	PdSb$_2$*)	6.459				8.625	63M
		6.4584					65F
	PtP$_2$	5.6956				9.238	60K1, 65F, 69D, 74B2
	PtPAs	5.842					63H2
	PtAs$_2$	5.968					65M, 66B, 68S
		5.9665				10.79	65F, 60K1
	PtSb$_2$	6.428				10.964	65D
		6.4400					60K1, 65F
	PtSbBi	6.570					63H2
	PtBi$_2$(h$_1$)*)	6.7022				13.526	60K1, 65F
d^6	Marcasites (D$_{2h}^{12}$–Pnnm; Z=2)						
	NiAs$_2$(h)^{b,d}) (β-NiAs$_2$)	4.759	5.797	3.539		7.09	60S
		4.757	5.793	3.544			61Y, 66B
		4.7583	5.7954	3.5449		7.088	68H2, 74K2, 79K1
	NiSb$_2$^{b,d})	5.178	6.319	3.832		8.01	70C
		5.1823	6.3168	3.8403		7.984	74K2
		5.1837	6.3184	3.8408		7.979	68H2, 79K1
d^6	Pararammelsbergites (D$_{2h}^{15}$–Pbca; Z=8)						
	NiAs$_2$(r) (α-NiAs$_2$)	5.75	5.82	11.43			60H
		5.770	5.838	11.419		7.20	66B, 68S
		5.772	5.834	11.420			77K3, 68H2
	Ni$_{0.94}$Co$_{0.06}$As$_{1.86}$S$_{0.14}$	5.753	5.799	11.407			72F2
	PtBi$_2$(r)	6.732	6.794	13.346		13.34	80B
d^5	Arsenopyrites (C$_{2h}^5$–P2$_1$/c; Z=4)						
	CoP$_2$	5.610	5.591	5.643	116.82	5.08	72D2
	CoAs$_2$^{b})	5.918	5.876	5.964	116.45		62Q, 67P, 66B
		5.916	5.872	5.960	116.45	7.480	66S
		5.9106	5.8680	5.9587	116.432	7.49	71K1
	CoSb$_2$^{b,c})	6.52	6.38	6.55	118.2	8.37	61Z
		6.5077	6.3879	6.5430	117.660	8.34	71K1
	RhP$_2$	5.7429	5.7942	5.8370	112.92	6.053	61R
		5.7417	5.7951	5.8389	112.911	6.12	71K1
	RhAs$_2$^{b,c})	6.041	6.082	6.126	114.33	8.127	62Q, 66B
		6.0629	6.0816	6.1498	114.707	8.15	71K1
	RhAsSb^a)	6.36	6.28	6.41	115.4	8.55	63H3
	RhSb$_2$^{b,c})	6.604	6.557	6.668	116.63	8.867	62Q, 61Z
		6.6156	6.5596	6.6858	116.821	8.89	71K1

For footnotes, see page 83.

(continued)

Crystallographic data of transition-element dipnictides with octahedrally coordinated cations (continued)

Cation configuration	Compound (Space group, coordination number)	a [Å]	b [Å]	c [Å]	β [°]	d [g cm^{-3}]	Ref.
d^5	Arsenopyrites						
	RhBi$_2$(r)[b]) (α-RhBi$_2$)	6.96	6.83	7.01	118.2	11.79	61K, 61Z
		6.9207	6.7945	6.9613	117.735	11.86	71K1
	IrP$_2$	5.746	5.791	5.851	111.60	9.32	77K1
		5.7453	5.7915	5.8494	111.575	9.25	71K1
		5.7457	5.7906	5.8500	111.60	9.33	61R
	IrAs$_2$[b])	6.060	6.071	6.158	113.27	10.92	62Q, 66B
		6.0549	6.0717	6.1587	113.197	10.83	71K1
	IrAsSb[a])	6.35	6.30	6.41	114.2	11.04	63H3
	IrSb$_2$[b])	6.58	6.53	6.68	115.5	11.17	61Z, 65Z
		6.5945	6.5492	6.6951	115.158	10.98	71K1
	IrBi$_2$(?)	7.0	6.9	7.1	117	13.3	65Z
d^4	Loellingites (D_{2h}^{12}–Pnnm; $Z=2$)						
	FeP$_2$	4.9732	5.6570	2.7235		5.11	68H2, 68H3
		4.9729	5.6568	2.7230		5.107	69D
	FeAs$_2$[b])	5.3007	5.9792	2.8816			63R
		5.3013	5.9859	2.8822		7.47	68H2, 74K2, 79K1
		5.291	5.981	2.880			68R1
		5.300	5.984	2.881		7.48	72R, 60S, 60H
	FeSb$_2$[b, d])	5.830	6.535	3.197		8.16	72R, 72F1
		5.8328	6.5376	3.1973		8.154	68H2, 68H3
		5.819	6.520	3.189			53R
		5.830	6.534	3.195			77G
	RuP$_2$	5.1169	5.8915	2.8709		6.256	77K2
		5.1173	5.8932	2.8711			68H2, 68H3
		5.115	5.888	2.870		6.26	77K1
	RuPAs	5.26	6.04	2.92		7.41	63H2
	RuAs$_2$	5.4279	6.1834	2.9685		8.364	77K2
		5.4302	6.1834	2.9714			68H2, 68H3
	RuAsSb	5.76	6.48	3.10		8.45	63H2
	RuSb$_2$	5.9514	6.6743	3.1790		9.063	77K2
		5.9524	6.6737	3.1803			68H2, 68H3
		5.930	6.637	3.168		9.18	60K2
	OsP$_2$	5.1012	5.9022	2.9183		9.531	77K2
		5.1001	5.9012	2.9182			68H2, 68H3
	OsAs$_2$	5.4115	6.1900	3.0127		11.191	77K2
		5.4129	6.1910	3.0126			68H2, 68H3
	OsSb$_2$	5.937	6.684	3.210		11.31	61J
		5.9409	6.6880	3.2112			68H2, 68H3
		5.9411	6.6873	3.2109		11.291	77K2
		5.912	6.653	3.196		11.45	60K2
d^2	CrSb$_2$	6.020	6.877	3.275		7.24	67P
		6.025	6.873	3.266			78A2
		6.0275	6.8738	3.2715		7.24	70H, 68H2
	at 80 K	6.0183	6.8736	3.2704		7.25	70H
d^4, d^6	Fe$_{0.5}$Ni$_{0.5}$As$_2$[b])	5.1377	5.9205	3.1077		7.28	79K1
	Fe$_{0.5}$Ni$_{0.5}$Sb$_2$[b])	5.6417	6.4402	3.3855		8.12	79K1
d^3, d^5(?)	Mn$_{0.5}$Co$_{0.5}$P$_2$*) (?)	4.996	5.685	2.745		5.06	72D2

For footnotes, see next page.

*) Metallic.
a) Anions probably disordered.
b) Temperature dependence of the lattice parameters: [77K3] and Figs. 3, 4, 5, 6.
c) Arsenopyrite→loellingite transformation at high temperatures [77K3].
d) Marcasite structure persists down to 4 K [70H].

Interatomic distances in pyrite- and pararammelsbergite-type compounds (RT values)

Compound	T–6X [Å]	X–X [Å]	Ref.
$PdAs_2$*)	2.498	2.426	65F
$PdSb_2$*)	2.671	2.89	65F
PtP_2	2.401	2.09	65F
	2.386	2.22	74B2
	2.391	2.172	69D
$PtAs_2$	2.490	2.42	65F
	2.494	2.38	68S
$PtSb_2$	2.669	2.79	65F
$PtBi_2(h_1)$*)	2.771	3.00	65F
$NiAs_2(r)$	2.25···2.46 (av. 2.38)	2.46	68S
	2.355···2.415 (av. 2.379)	2.449	79K2
$Ni_{0.94}Co_{0.06}As_{1.86}S_{0.14}$	2.359···2.415 (av. 2.372)	2.432	72F2
$PtBi_2(r)$	2.742···2.771 (av. 2.760)	3.018	80B

*) Metallic.

Interatomic distances in marcasite- and loellingite-type compounds ($T-2T$ and $X-2X = c$) (RT values)

d^n	Compound	T–4X [Å]	T–2X [Å]	T–2T [Å]	X–X [Å]	Ref.
d^6	$NiAs_2(h)$	2.345	2.394	3.545	2.447	79K1
		2.341	2.392	3.545	2.464	68H3
	$NiSb_2$	2.538	2.569	3.841	2.883	79K1
		2.55	2.56	3.841	2.86	68H3
d^6, d^4	$Fe_{0.5}Ni_{0.5}As_2$	2.367	2.379	3.108	2.475	79K1
	$Fe_{0.5}Ni_{0.5}Sb_2$	2.568	2.572	3.386	2.859	79K1
d^4	FeP_2	2.29	2.20	2.724	2.27	68H3
		2.264	2.248	2.723	2.237	69D
	$FeAs_2$	2.362	2.387	2.882	2.492	79K1
	$FeSb_2$	2.575	2.597	3.197	2.894	79K1, 68H3
	RuP_2	2.371	2.347	2.871	2.234	77K2
	$RuAs_2$	2.468	2.448	2.969	2.475	77K2
	$RuSb_2$	2.648	2.628	3.179	2.863	77K2
		2.656	2.618	3.180	2.86	68H3
	OsP_2	2.376	2.350	2.918	2.248	77K2
	$OsAs_2$	2.477	2.452	3.013	2.469	77K2
	$OsSb_2$	2.639	2.644	3.211	2.889	77K2
d^2	$CrSb_2$	2.717	2.712	3.272	2.833	79K1

9.15.1.3 Transition-metal compounds T(V)$_2$ [Ref. p. 106]

Interatomic distances for binary arsenopyrite-type phases. (The shortest non-bonding T−T and X−X distances are added for comparison (RT values).)

Compound	T−6X [Å]	T−T [Å]	X−X [Å]	Ref.
CoAs$_2$	2.29···2.42	2.77 3.47	2.45 3.00	66D
	2.310···2.414	2.780 3.476	2.462 3.146	71K1
CoSb$_2$	2.48···2.66	3.04 3.72	2.82 3.55	61Z (71K1)
	2.505···2.607	3.01 3.74	2.857 3.282	81S
RhP$_2$	2.30···2.44	2.68 3.72	2.24 3.00	71K1
RhAs$_2$	2.39···2.52	2.83 3.77	2.49 3.21	71K1
RhSb$_2$	2.53···2.73	3.03 3.91	2.84 3.57	61Z (71K1)
α-RhBi$_2$	2.64···2.83	3.23 3.95	3.00 3.77	61Z (71K1)
IrP$_2$	2.29···2.45	2.75 3.78	2.25 2.99	71K1
IrAs$_2$	2.39···2.53	2.86 3.87	2.48 3.20	71K1
IrSb$_2$	2.53···2.74	3.06 4.07	2.81 3.54	61Z (71K1)

Crystallographic data for the PdP$_2$-type compounds with square-planar cation coordination (C_{2h}^6 − C2/c; Z = 4) (RT values)

Compound	a [Å]	b [Å]	c [Å]	β [°]	Ref.
NiP$_2$	6.3659	5.6152	6.0715	126.224	61R
PdP$_2$	6.7771	5.8563	6.2063	126.427	61R
	6.779 [a]	5.857	6.207	126.4	63Z
PdPAs	6.88	6.03	6.41	126.1	63H2

[a]) Data given for space group I2/a (a', c') were transformed to C2/c (a, c): $a = a' + c'$; $c = -a$.

Interatomic distances and bond angles for the PdP$_2$-type compounds (RT values)

Compound	T−4P [Å]	T−2T [Å]	P−2P [Å]	P−P−P [°]	P−T−P [°]	T−P−T [°]	Ref.
NiP$_2$	2.205 (2×) 2.208 (2×)	3.036	2.219 2.221	99.66	89.18 90.82	128.58	64L
PdP$_2$	2.335 (2×) 2.341 (2×)	3.104	2.201 2.224	102.0	88.9 91.1	125.0	63Z

9.15.1.3.1 Physical properties of T(V)$_2$ compounds

Physical property	Numerical value	Experimental conditions	Experimental method remarks	Ref.

CrSb$_2$

energy level scheme: Fig. 7.

energy gap (in eV):

$E_{g,th}$	0.16		from $\log \varrho \propto E_g/2kT$, $T = 300 \cdots 550$ K	56A, 57A
	(0.32?)		from $\log \varrho \propto E_g/2kT$, $T = 620 \cdots 710$ K (the values given for σ at 669 and 710 K yield 0.45 eV. These large E_g figures may not be reliable since no kink is detectable on the thermopower curve, Fig. 11. Reproducibility of the measurements probably not checked on cooling)	56A, 57A
	0.14		from $\log \varrho \propto E_g/2kT$, $T = 200 \cdots 260$ K	69A
	0.07		from $\log \varrho \propto E_g/2kT$, $T < 200$ K (see Fig. 10)	69A

resistivity, temperature dependence: Figs. 10, 11.
thermoelectric power; temperature dependence: Fig. 11.

Debye temperature:

Θ_D	290 K	$T = 90$ K	maximum of Θ_D in the low-temperature region, from $C_p(T)$	78A2

magnetic susceptibility: Curie-Weiss law, Fig. 12.
$\chi(T)$ curves in [68H] and [69A] falsified by oxidation of the samples [79K1].

Néel temperature:

T_N	273 K		from neutron-diffraction intensities	70H, 79K1
	274 K		from heat capacity and magnetic susceptibility	78A2, 79K1

antiferromagnetic order: simple uniaxial type ($a_{magn} = a$, $b_{magn} = 2b$, $c_{magn} = 2c$), moments coupled ferromagnetically within (011) of the chemical cell, the moments in adjacent planes being antiparallel; moments perpendicular to b_{magn} [79K1].

magnetic moments:

p_A	1.94 μ_B	$T = 4$ K	from neutron diffraction (2 unpaired electrons/Cr, hence no Cr–Cr bonds)	70H, 79K1
p_{eff}	3.1 μ_B	$300 \cdots 800$ K	from Curie-Weiss behavior with $\Theta_p = -500$ K	79K1

peritectic temperature:

T_{perit}	949 K			57A
	991.3 K		from heat capacity measurements	78A2

heat capacity: temperature dependence: Figs. 13 and 14.

Physical property	Numerical value	Experimental conditions	Experimental method remarks	Ref.
$Cr_{1-x}Fe_xSb_2$				
$x = 0 \cdots 1$, linear variation of the lattice constants b and c ($= T-T$) with x [70B1, 79K1].				
Néel temperatures (in K):				
T_N	246	$x = 0.07$	the Fe atoms carry no magnetic moment	79K1
	210	0.12		
	127	0.28		
	60	0.40		
	0	≈ 0.50 (extrapolated)		
FeP_2				
Energy level scheme is similar to that given for $FeAs_2$ (Fig. 8).				
energy gap (in eV):				
$E_{g,th}$	≈ 0.4		from $\log \varrho \propto E_g/2kT$, $T = 550 \cdots 800$ K, sintered sample	59H
	0.37		from $\log \varrho \propto E_g/2kT$, $T = 400 \cdots 550$ K, single crystal	71B
electrical resistivity and Hall coefficient $5 \cdots 300$ K: Fig. 15.				
Hall voltage proportional to the magnetic field up to 0.7 T [71B].				
carrier concentration:				
n	$\approx 10^{17}$ cm^{-3}	$T \approx 5$ K	from Hall effect on iodine-transported single crystal, n-type conduction according to R_H and S	71B
mobility of charge carriers: Fig. 16.				
			below 60 K $\mu_H \propto T^{0.8}$ above 300 K $\mu_H \propto T^{-2 \cdots -2.5}$	71B
magnetic susceptibility (in 10^{-6} cm^3 g^{-1}):				
χ_g	-0.14	$T = 300$ K	slightly increasing below 220 K	71B
	-0.12	115 K	Faraday balance, three single crystals of total mass 2.4 mg	
far-infrared absorption: for spectrum in the range $180 \cdots 500$ cm^{-1}, see [77L].				
$FeAs_2$				
energy level scheme: Fig. 8.				
energy gap (in eV):				
$E_{g,th}$	0.2		from $\log \varrho \propto E_g/2kT$, $T = 300 \cdots 550$ K	59H, 65J
	0.22		from $\log \varrho \propto E_g/2kT$, intrinsic above 170 K	72F1
resistivity (in Ω cm):				
ϱ	0.012	RT	single crystal grown by chlorine transport	72F1
	≈ 0.05		sintered sample	59H
	≈ 0.002		natural crystal	65J
For temperature dependence, see Fig. 19.				

Physical property	Numerical value	Experimental conditions	Experimental method remarks	Ref.
Hall coefficient (in cm^3 C^{-1}):				
R_H	-10	$T = 60\cdots170$ K	exponential function of T above 170 K, single crystal of unknown orientation	72F1
	-100	9 K		
For temperature dependence, see Fig. 19.				
carrier concentration:				
n	$\approx 5 \cdot 10^{19}$ cm^{-3}	$T = 60\cdots170$ K	from Hall coefficient	74B1, 72F1
mobility ratio:				
b	<1	high T	from sign change of $S(T)$ at 800 K	65J
thermoelectric power (in µV K^{-1}):				
S	≈ -200	$T \approx 285$ K	single crystal	72F1
	-120	$T \approx 300$ K	polycrystalline natural sample	65J
For temperature dependence in the range 77$\cdots$900 K, see Fig. 18; $S>0$ above 800 K [65J].				
magnetic susceptibility:				
χ_m	$-14.0 \cdot 10^{-6}$ cm^3 mol^{-1}	$T = 80\cdots500$ K (?)	temperature independent, single crystal of unknown orientation; χ in CGS-emu	72F1
far infrared absorption: for spectrum in the range 50$\cdots$400 cm^{-1}, see [83L].				
decomposition temperature:				
T_{dec}	1287 K		loses As in vacuum at 800 K	77K3, 79K1, 72R
lattice parameters $a(T)$, $b(T)$ linear with T (300$\cdots$1280 K), $c(T)$ linear above 600 K [77K3].				
volume expansion coefficient (in 10^{-5} K^{-1}):				
β	≈ 3.7	$T = 300$ K	increases up to ≈ 600 K, taken from graphical representation	77K3
	6.3	800 K	$T = 600\cdots1280$ K	

FeAs$_{1.94}$Se$_{0.06}$

metallic behavior for $T = 77\cdots330$ K: $\varrho[\mu\Omega$ cm$] = 3.05\,T + 240$; paramagnetic with Curie-Weiss behavior: $C_m = 0.027$ cm^3 K mol^{-1}, $\Theta_p = -206$ K; i.e. $p_{eff} = 1.90\,\mu_B$/Fe^{III} [74B1]

Hall coefficient constant from 77 to 300 K, yielding $n = 1.3 \cdot 10^{21}$ electrons/cm^3 = 0.06 electrons/Fe atoms, corresponding to the number of Se atoms [74B1].

Fe$_{1-x}$Cr$_x$As$_2$

loellingite-type solid solutions up to x = 0.83 for samples quenched from 873 K, x = 0.81 for samples quenched from 1123 K [79K1].
decomposition temperature decreases to 1138 K at x = 0.8 [79K1].
For magnetic susceptibility in the range 80$\cdots$950 K, see Fig. 17.

Fe$_{1-x}$Co$_x$As$_2$

complete miscibility; monoclinic for x > 0.8 (Fig. 2) [63R, 68R1, 74K3].

Physical property	Numerical value	Experimental conditions	Experimental method remarks	Ref.

$Fe_{1-x}Ni_xAs_2$

homogeneity range: $x = 0 \cdots \approx 0.7$ [63R, 74K3, 79K1].
Fe and Ni keep their d-electron configuration regardless of composition [65P].
maximum of the ^{57}Fe Mössbauer chemical shift parameter at $x = 0.3$ [74K3].
$x = 0.5$: no magnetic order at 10 K [79K1].
$a(T)$, $b(T)$, $c(T)$ [77K3].

volume expansion coefficient:

β	$3.6 \cdot 10^{-5} K^{-1}$	$T = 300$ K	$T = 300 \cdots 600$ K, increases at higher temperatures	77K3

$FeSb_2$

Energy level scheme is similar to that of $FeAs_2$ (Fig. 8).
The observed Schottky anomaly (excess heat capacity, Fig. 26) can be interpreted by assuming that the [4]b band is in fact split, so that the highest occupied [2]b band is succeeded by a [2]$a_\parallel$ and a [4]σ^* band at E_1 and E_2, respectively, in Goodenoughs picture. $E_1/hc = 300$ cm^{-1} ($E_1 = 0.037$ eV), $E_2/hc = 500$ cm^{-1} ($E_2 = 0.062$ eV) [77G].

energy gap and activation energy (in eV, all values from $\log \varrho \propto E_g/2kT$):

$E_{g,th}$	≈ 0.17		polycrystalline, $T = 550 \cdots 800$ K	60D
E_A	≈ 0.05		sintered sample, $T = 500 \cdots 700$ K	59H
	≈ 0.03		sintered at 873 K, $T = 80 \cdots 160$ K	65J
	0.033		single crystal, $T = 50 \cdots 300$ K	72S, 72F1
	≈ 0.02		pressed pellets, $T = 80 \cdots 300$ K	80Y

Debye temperature:

Θ_D	327 K		from Mössbauer isomer shift $270 \cdots 560$ K	72S

electrical resistivity: Figs. 20$\cdots$22, anisotropy of the resistivity: Fig. 20, anisotropy of the Hall constant: Fig. 23.

thermoelectric power:

S	$+30$ μV K^{-1}	$T = 295$ K	polycrystal and single crystal of unknown orientation; for temperature dependence in the range $77 \cdots 400$ K, see Fig. 18	60D, 72F1

magnetic susceptibility (in 10^{-6} cm^3 mol^{-1}):

χ_m	522	$T = 300$ K	measured on sintered material, annealed at 873 K; χ in CGS-emu	80Y
	8	77 K		

For temperature dependence in the range $77 \cdots 550$ K, see Fig. 24.

peritectic temperature (in K):

T_{perit}	1005			58H
	1011			72M
	1021		from heat capacity	69S, 77G
	1018			77K3

The statement [72F1, 72R] that $FeSb_2$ decomposes under vacuum towards 600 K into $FeSb + Sb$ was refuted by [77G].
heat capacity: C_p in the range $100 \cdots 700$ K: Fig. 25, Schottky anomaly: Fig. 26.

Physical property	Numerical value	Experimental conditions	Experimental method remarks	Ref.
Brinell hardness:				
H_B	200 kg mm^{-2}			60K2

unit-cell dimensions $a(T)$, $b(T)$, $c(T)$ fairly linear in T [77K3].

volume expansion coefficient:				
β	$6.1 \cdot 10^{-5}$ K^{-1}	$T = 300$ K	from graphical representation; unit-cell volume linear in T up to ≈ 1000 K	77K3

FeSb$_{2-x}$Te$_x$

Sintered and annealed (873 K) samples (x = 0···2) show monoclinic arsenopyrite type for x ≈ 0.8···1.2 [80Y]; for resistivity in the range 77···300 K, see Fig. 22.

Fe$_{1-x}$Co$_x$Sb$_2$

complete miscibility for x = 0···1 [70B1, 74K3], loellingite-type for x < 0.8, arsenopyrite-type structure for x ≳ 0.8 [74K3]. See also Fig. 2.

Fe$_{1-x}$Ni$_x$Sb$_2$

homogeneity region: x = 0···≈ 0.50 and ≈ 0.97···1 [70B1]. Fe$_{0.5}$Ni$_{0.5}$Sb$_2$ decomposes at 915 K [77K3]. $a(T)$, $b(T)$ linear in T, $c(T)$ increases more rapidly [77K3].

volume expansion coefficient (in 10^{-5} K^{-1}):				
β	3.4	$T = 300$ K	$T = 300$···500 K, taken from graphic representation	77K3
	5.0	600 K	$T = 600$···900 K no magnetic order at 10 K [79K1]	

RuP$_2$

Energy level scheme is similar to that given for FeAs$_2$ (Fig. 8).

energy gap and activation energies (in eV):				
E_g	≈ 1	RT	from diffuse reflectance of sintered samples	63H1
	0.8	RT	optical absorption edge determined on single crystals containing 0.85 wt% tin from flux	77K1
E_A	0.01	$T < 160$ K	from log $\varrho \propto E_A/kT$ in the extrinsic range at 77···300 K	77K1
	0.012	< 300 K	from log $R_H \propto E_A/kT$ below room temperature	
resistivity:				
ϱ	0.09···0.26 Ω cm	RT	crystals with 0.85 wt% Sn For temperature dependence in the range 77···300 K; see Fig. 27	77K1
carrier concentration:				
n	$7.4 \cdot 10^{17}$ cm^{-3}	RT	from Hall coefficient ($n = 1/eR_H$)	77K1
thermoelectric power (in μV K^{-1}):				
S	−350	RT	sintered samples	63H1
	−170	RT	single crystals	77K1

Physical property	Numerical value	Experimental conditions	Experimental method remarks	Ref.
magnetic susceptibility (in 10^{-6} cm^3 mol; χ in CGS units):				
		T [K]		
χ_m	-63	60$\cdots$300	Sn-containing single crystals of random orientation	77K1
	-44	RT	sintered powder, Gouy method	63H1
	-33	85		
	-43	90, $B<0.8$ T	polycrystalline sample, Faraday method	68H1
	-46	295		
	-56	765		

far infrared absorption: for spectrum in the range 180$\cdots$500 cm^{-1}, see [77L].

RuPAs

energy gap:

E_g	≈ 0.8 eV	RT	from diffuse reflectance of sintered powder	63H1

thermoelectric power:

S	≈ -400 µV K^{-1}	RT	measured on a sintered sample	63H1

magnetic susceptibility (in 10^{-6} cm^3 mol^{-1}):

χ_m	-61	$T=295$ K	χ in CGS-emu; powder sample, Gouy method	63H1
	-46	85 K		

RuAs$_2$

Energy level scheme is similar to that given for FeAs$_2$ (Fig. 8).

energy gap:

E_g	≈ 0.8 eV	$T=295$ K	from diffuse reflectance of powder	63H1

thermoelectric power:

S	≈ -350 µV K^{-1}	$T=295$ K	sintered sample	63H1

magnetic susceptibility (in 10^{-6} cm^3 mol^{-1}, χ in CGS units):

χ_m	-73	$T=295$ K	powder sample	63H1
	-67	85 K		
	-42	90$\cdots$770 K, $B<0.8$ T	polycrystalline sample	68H1

far infrared absorption: for spectrum in the range 50$\cdots$400 cm^{-1}, see [83L].

RuSb$_2$

Energy level scheme is similar to that given for FeAs$_2$ (Fig. 8).

energy gap:

$E_{g,th}$	>0.3 eV		from $\log\varrho \propto E_g/2kT$ above 450 K, sintered sample	63H1

resistivity:

ϱ	≈ 1 Ω cm	RT	sintered sample	63H1

thermoelectric power:

S	≈ -250 µV K^{-1}		sintered sample	63H1

Physical property	Numerical value	Experimental conditions	Experimental method remarks	Ref.
magnetic susceptibility (in 10^{-6} cm^3 mol^{-1}, χ in CGS units):				
		T [K]		
χ_m	-82	295	powder sample	63H1
	-74	85		
	-78	90, $B<0.8$ T	linear up to 770 K, polycrystalline sample	68H1
	-80.4	770		

far infrared absorption: for spectrum in the range $50 \cdots 400$ cm^{-1}, see [83L].

melting temperature:

T_m	≈ 1600 K		estimated visually	63H1

Brinell hardness:

H_B	420 kg mm^{-2}		polycrystalline sample reacted at 1373 K	60K2

OsP$_2$

Energy level scheme is similar to that given for FeAs$_2$ (Fig. 8).

energy gap:

E_g	≈ 1.2 eV	RT	from diffuse reflectance of sintered samples	63H1

magnetic susceptibility (in 10^{-6} cm^3 mol^{-1}, χ in CGS units):

χ_m	-74	$T=295$ K	powder sample	63H1
	-50	85 K		
	-97	$90 \cdots 760$ K, $B<0.8$ T	polycrystalline sample	68H1

far infrared absorption: for spectrum in the range $150 \cdots 500$ cm^{-1}, see [77L].

OsAs$_2$

Energy level scheme is similar to that given for FeAs$_2$ (Fig. 8).

energy gap:

E_g	≈ 0.9 eV	RT	from diffuse reflectance	63H1

magnetic susceptibility (in 10^{-6} cm^3 mol^{-1}; χ in CGS units):

		T [K]		
χ_m	-78	85	powder sample, Gouy method	63H1
	-100	295		
	-110	90	$\chi(T)$ linear from 90 to 775 K, polycrystalline sample	68H1
	-115	775		

far infrared absorption: for spectrum in the range $50 \cdots 400$ cm^{-1}, see [83L].

OsSb$_2$

Energy level scheme is similar to that given for FeAs$_2$ (Fig. 8).

energy gap (in eV):

$E_{g,th}$	>0.3		from $\log \varrho \propto E_g/2kT$ above 600 K; sintered material	63H1
	>0.2		from $\log \varrho \propto E_g/2kT$, sintered material	65J

Physical property	Numerical value	Experimental conditions	Experimental method remarks	Ref.
resistivity:				
ϱ	$\approx 150\ \Omega$ cm	RT	sintered sample	63H1
thermoelectric power (in μV K^{-1}):				
S	-200	RT	sintered sample	63H1
	-115	RT	sintered material	61J, 65J
magnetic susceptibility (in 10^{-6} cm^3 mol^{-1}; χ in CGS-units):				
χ_m	-92	$T=85$ K	powder sample	63H1
	-94	295 K		
	-136	90$\cdots$800 K, $B<0.8$ T	polycrystalline sample	68H1

far infrared absorption: for spectrum in the range 50$\cdots$400 cm^{-1}, see [83L].

Brinell hardness:				
H_B	560 kg mm^{-2}		polycrystalline sample reacted at 1373 K	60K2

OsSb$_{1.95}$Te$_{0.05}$

activation energy:				
E_A	≈ 0.018 eV	$T<160$ K	from $\log \varrho \propto E_A/kT$ in the range 77$\cdots$160 K	65J

resistivity, temperature dependence in the range 77$\cdots$900 K: Fig. 21.
thermoelectric power, temperature dependence in the range 77$\cdots$800 K: Fig. 18.

CoP$_2$

Energy level scheme is similar to that given for CoAs$_2$ (Fig. 9).

energy gap:				
$E_{g,th}$	>0.02 eV		from $\varrho(T)$ below room temperature	72D2
electrical resistivity (in Ω cm):				
ϱ	0.0057	$T=298$ K	crystals grown from a Ge flux at $T=1100\cdots1500$ K, $p=65$ kbar	72D2
	0.22	4.2 K		

magnetic susceptibility: diamagnetic at room temperature [72D2].
far infrared absorption: for spectrum in the range 50$\cdots$400 cm^{-1}, see [83L].

CoAs$_2$

energy level scheme: Fig. 9.

energy gap (in eV):				
$E_{g,th}$	≈ 0.15		from $\log \varrho \propto E_g/2kT$, sintered sample	59H
	≈ 0.35		from $\log \varrho \propto E_g/2kT$, $T=450\cdots650$ K, single crystal	83S

Physical property	Numerical value	Experimental conditions	Experimental method remarks	Ref.
thermoelectric power:				
S	$\approx -200\ \mu V\ K^{-1}$	$T = 300$ K	measured on the sintered, weakly paramagnetic sample of [63H3]	80H
magnetic susceptibility (in 10^{-6} cm^3 mol^{-1}; χ in CGS units):		T [K]		
χ_m	-4	80···900	sample sintered at 1073 K ($\chi_g = 0.02 \cdot 10^{-6}$ cm^3 g^{-1})	71K1
	-16	300···400, $B = 0.5$ T	non-oriented single crystals grown by an iodine transport reaction, Faraday method	83S
	-10	600		
	0	720		
	$+23$	810	transition temperature	
	$+28$	920	linear increase above T_{tr}	

The weak paramagnetism of sintered samples [63H3, 66B] is obviously due to impurity phases.
lattice parameters $a(T)$, $b(T)$, $c(T)$, $\beta(T)$ at $T = 300···1270$ K [77K3].

Physical property	Numerical value	Experimental conditions	Experimental method remarks	Ref.
volume expansion coefficient:				
β	$4.2 \cdot 10^{-5}$ K^{-1}	$T = 300$ K	from graphic representation; unit-cell volume linear in T in the range 300···1300 K	77K3
structural transition (monoclinic → orthorhombic (Fig. 3).):				
T_{tr}	870 K		second-order transition to the loellingite structure	77K3
	810 K		from magnetic susceptibility	83S
	796···800 K		from DTA	83S

Co$_{1-x}$Ni$_x$As$_2$

homogeneity range $x = 0···1$, monoclinic for $x < \approx 0.25$ [63R].

CoSb$_2$

Energy level scheme is similar to that shown in Fig. 9 for CoAs$_2$.

Physical property	Numerical value	Experimental conditions	Experimental method remarks	Ref.
energy gap (in eV):				
$E_{g,th}$	0.2 (?)	polycrystal	from $\log \varrho \propto E_g/2kT$, $T = 550···800$ K, measured across the arsenopyrite → marcasite transition	56D, 57D
	0.17	single crystal	from $\log \varrho \propto E_g/2kT$, $T = 450···600$ K, semiconducting also in the marcasite modification (?)	83S
resistivity:				
ϱ	$2.5 \cdot 10^{-3}$ Ω cm	$T \approx 550$ K	polycrystalline material	56D
thermoelectric power (in $\mu V\ K^{-1}$):		T [K]		
S	≈ 40	300		80H
	≈ 54	400	polycrystalline material	56D
	≈ 75	600		
	≈ 60	700		
	≈ 70	800		

Physical property	Numerical value	Experimental conditions	Experimental method remarks	Ref.
magnetic susceptibility (in 10^{-6} cm^3 mol^{-1}, χ in CGS units):				
		T [K]		
χ_m	0.0	80$\cdots$900, $B=0.8$ T	samples sintered at 1073 K, Faraday method	71K1
	-4	300, 0.5 T	non-oriented single crystals grown by an iodine transport reaction; Faraday method	83S
	0	465		
	$+10$	585		
	$+22$	640	transition temperature (monoclinic$\rightarrow$orthorhombic)	
	$+36$	900	linear increase above T_{tr}	

lattice parameters $a(T)$, $b(T)$, $c(T)$, $\beta(T)$ at $T=300\cdots1200$ K: Fig. 4.

volume expansion coefficient (in 10^{-5} K^{-1}):

β	6.8	$T=300$ K	from graphical representation, $T=300\cdots400$ K	77K3
	4.2	500 K	unit-cell volume linear in T, $T=450\cdots1200$ K	

structural transition (monoclinic$\rightarrow$orthorhombic (Fig. 4)):

T_{tr}	650 K		second-order transition to the loellingite structure	77K3
	640 K		from magnetic susceptibility	83S
	643$\cdots$646 K		from DTA	83S

peritectic (decomposition) temperature (in K):

T_{perit}	1192			56D, 60Z, 78A1
	1204			77K3

Co$_{1-x}$Ni$_x$Sb$_2$

homogeneity ranges: $x=0\cdots\approx0.02$ and $\approx0.97\cdots1$ [70B1].

CoSb$_{2-x}$Te$_x$

marcasite structure for $x=0.2\cdots2$ [76Y]; peritectic temperature decreases with x [78A1].

RhP$_2$

Energy level scheme is similar to that given for CoAs$_2$ (Fig. 9).

energy gap:

E_g	≈ 1 eV	RT	from diffuse reflectance of powder samples	63H3

magnetic susceptibility (in 10^{-6} cm^3 mol^{-1}; χ in CGS-emu):

χ_m	-37	$T=295$ K	(temperature dependence probably due to paramagnetic impurities)	63H3
	-14	77 K		
	-77.5	80$\cdots$1000 K	($\chi_g=0.47\cdot10^{-6}$ cm^3 g^{-1})	71K1

far infrared absorption: for spectrum in the range 50$\cdots$400 cm^{-1}, see [83L].

RhAs$_2$

Energy level scheme similar to that given for CoAs$_2$ (Fig. 9).

energy gap:

E_g	1.15 eV	RT	from diffuse reflectance of powder samples	63H3

Physical property	Numerical value	Experimental conditions	Experimental method remarks	Ref.
magnetic susceptibility (in 10^{-6} cm^3 mol^{-1}; χ in CGS-emu):				
χ_m	-70	$T = 80$ K		66B
	-66	540 K		
	-104	295 K	(χ_g [cm^3 g^{-1}] $= -0.400 - 0.00004\,T$; $T = 80 \cdots 850$ K)	71K1

far infrared absorption: spectrum in the range $50 \cdots 400$ cm^{-1} [83L].
lattice parameters $a(T)$, $b(T)$, $c(T)$, $\beta(T)$ at $T = 300 \cdots 1320$ K: monoclinic→orthorhombic transformation above 1320 K, estimated at 1350 K [77K3].

volume expansion coefficient:				
β	$2.4 \cdot 10^{-5}$ K^{-1}	$T = 300$ K	from graphical representation, unit-cell volume linear in T in the range $300 \cdots 1300$ K	77K3

RhAsSb

energy gap:				
$E_{g,\,th}$	≈ 0.4 eV		from $\log \varrho \propto E_g/2kT$, $T = 300 \cdots 500$ K; sintered sample	63H3
resistivity:				
ϱ	$\approx 10\,\Omega$ cm	RT	polycrystalline sample	63H3
Seebeck coefficient:				
S	$\approx +100\,\mu$V K^{-1}	RT	polycrystalline sample	63H3
magnetic susceptibility:				
χ_m	$-75 \cdot 10^{-6}$ cm^3 mol^{-1}	$T = 77$ and 295 K	χ in CGS-emu, powder sample	63H3

RhSb$_2$

Energy level scheme is similar to that given for CoAs$_2$ (Fig. 9).
semiconductor [63H3].

resistivity:				
ϱ	$\approx 0.002\,\Omega$ cm	RT	polycrystalline sample, extrinsic conductivity range	65J
thermoelectric power:				
S	$+30\,\mu$V K^{-1}	RT	polycrystalline material	65J
magnetic susceptibility (in 10^{-6} cm^3 mol^{-1}; χ in CGS-emu):		T [K]		
χ_m	-66	77	powdered sample	63H3
	-70	295		
	-125	$80 \cdots 750$, $B = 0.8$ T	polycrystalline sample, sintered at 1073 K	71K1

far infrared absorption: for spectrum in the range $50 \cdots 400$ cm^{-1}, see [83L].
lattice parameters $a(T)$, $b(T)$, $c(T)$, $\beta(T)$ at $T = 300 \cdots 1300$ K: Fig. 5; unit-cell volume linear in T [77K3].

volume expansion coefficient:				
β	$3.7 \cdot 10^{-5}$ K^{-1}	$T = 300$ K	from graphic representation, unit-cell volume linear in T in the range $300 \cdots 1300$ K	77K3

Physical property	Numerical value	Experimental conditions	Experimental method remarks	Ref.
structural transformation (monoclinic→orthorhombic):				
T_{tr}	1070 K		second-order transition to the loellingite structure	77K3
peritectic temperature:				
T_{perit}	≈1373 K (?)		based on tentative phase diagram	60Z

RhBi$_2$(r) (α-RhBi$_2$)
semiconductor?

magnetic susceptibility (in 10^{-6} cm^3 mol^{-1}, χ in CGS-units):				
χ_m	−161	$T=295$ K	polycrystalline sample	71K1
	$-154-0.026\,T$	80···650 K		

lattice parameters $a(T)$, $b(T)$, $c(T)$, $\beta(T)$ in the range 300···690 K [77K3].

volume expansion coefficient (in 10^{-5} K^{-1}):				
β	2.3	$T=300$ K	from graphic representation; β increases continuously with T	77K3
	1.2	650 K		
transition temperature:				
T_{tr}	698 K		transition to another monoclinic (?) structure	77K3

The arsenopyrite→loellingite transition should take place near 700 K [77K3].

IrP$_2$

The energy level scheme is expected to be similar to that given for CoAs$_2$ (Fig. 9).

energy gap and activation energy (in eV):

		T [K]		
E_g	≈1	295	from diffuse reflectance, powder sample	63H3
	1.1	295	from optical absorption on single crystals containing 0.24 wt% Sn and 0.006 wt% Cu	77K1
E_A	0.009	77···160	from $\log\varrho \propto E_A/kT$	77K1
	0.022	160···300	from $\log\varrho \propto E_A/kT$	

resistivity, temperature dependence in the extrinsic range 77···300 K: Fig. 28.

carrier concentration:				
p	$6.9 \cdot 10^{18}$ cm^{-3}	RT	from Hall effect ($R_H > 0$) on single crystal grown from Sn (Cu) flux	77K1
thermoelectric power:				
S	+250 μV K^{-1}	RT	non-oriented single crystal	77K1

magnetic susceptibility (in 10^{-6} cm^3 mol^{-1}, χ in CGS units):

		T [K]		
χ_m	−61	295	non-oriented single crystals from Sn-flux	77K1
	−45	≈60		
	−106	295	polycrystalline sample	71K1
	$-100-0.02\,T$	80···900		

far infrared absorption: for spectrum in the range 50···400 cm^{-1}, see [83L].

Physical property	Numerical value	Experimental conditions	Experimental method remarks	Ref.

structural transformation: The arsenopyrite→loellingite transformation is expected to take place at 1620 K (provided that the peritectic decomposition occurs at a higher temperature) [77K3].

IrAs$_2$

Energy level scheme is similar to that given for CoAs$_2$ (Fig. 9).

energy gap:

E_g	≈1 eV	RT	from diffuse reflectance of powder samples	63H3

magnetic susceptibility (χ_m in 10^{-6} cm^3 mol^{-1}, χ in CGS units):

		T [K]		
χ_m	−88	77	powder sample	63H3
	−93	295		
	−94	80⋯540		66B
	−118	295	polycrystalline sample	71K1
	−108 − 0.034 T	80⋯800		

far infrared absorption: spectrum in the range 50⋯400 cm^{-1} [83L].
lattice parameters $a(T)$, $b(T)$, $c(T)$, $\beta(T)$ in the range 300⋯1340 K [77K3].

volume expansion coefficient:

β	$2.4 \cdot 10^{-5}$ K^{-1}	$T = 300$ K	from graphic representation, unit-cell volume linear in T	77K3

The arsenopyrite structure persists up to 1340 K. The temperature of the monoclinic→orthorhombic transition is estimated from the monoclinic angle $\beta(T)$: $T_{tr} \approx 1560$ K [77K3].

IrAsSb

energy gap:

$E_{g,th}$	≈0.4 eV		from log $\varrho \propto E_g/2kT$ up to 800 K	63H3

resistivity:

ϱ	≈20 Ω cm	RT	sintered sample	63H3

Seebeck coefficient:

S	≈ +100 μV K^{-1}	RT	sintered sample	63H3

IrSb$_2$

semiconductor [63H3].

magnetic susceptibility:

χ_m	$-179 \cdot 10^{-6}$ cm^3 mol^{-1}	$T = 80⋯1000$ K	χ in CGS-emu; sintered polycrystal	71K1

far infrared absorption: for spectrum in the range 50⋯400 cm^{-1}, see [83L].

peritectic temperature:

T_{perit}	>1373 K?		from tentative phase diagram	57K, 60Z

lattice parameters $a(T)$, $b(T)$, $c(T)$, $\beta(T)$ in the range 300⋯1340 K: Fig. 6 [77K3].

volume expansion coefficient (in 10^{-5} K^{-1}):

β	2.4	$T = 300$ K	unit-cell volume linear in T in the range 300⋯800 K	77K3
	0.9	1000 K	unit-cell volume linear in T in the range 900⋯1300 K	

arsenopyrite→loellingite transition estimated at $T_{tr} \approx 1425$ K [77K3].

Physical property	Numerical value	Experimental conditions	Experimental method remarks	Ref.
NiP_2 (monoclinic)				
energy gap (in eV):				
E_g	>0.3	$T=0$ K	from $\log \varrho \propto E_g/2kT$, $T=500\cdots600$ K, single crystal	78O
	0.5	0 K	from $\log \varrho \propto E_g/2kT$, polycrystalline sample	63H3
	0.73	RT	from optical absorption ($0.5\cdots3.5$ eV), single crystal	78O
activation energy of resistivity:				
E_A	0.002 eV		from $\log \varrho \propto E_A/kT$, $T=35\cdots100$ K, single crystal	78O
resistivity (in Ω cm):				
ϱ	0.29	$T=77$ K	crystal grown from a tin flux, extrinsic range	78O
	0.39	RT	van der Pauw method	
	$0.09\cdots0.42$	RT	crystals from different batches	
carrier concentration (in 10^{17} cm^{-3}):				
n	7.3	$T=77$ K	from the Hall coefficient (van der Pauw method)	78O
	8.7	290 K		
thermoelectric power (in μV K^{-1}):				
S	$+392$	RT	crystal from Sn flux	78O
	-100	RT	sintered sample	63H2
magnetic susceptibility:				
χ_m	$-19.0 \cdot 10^{-6}$ cm^3 mol^{-1}	$T=77\cdots300$ K, $B=1$ T	no field dependence; χ in CGS-emu; single crystals from tin flux	78O
$NiAs_2$ (r) (α-$NiAs_2$) (pararammelsbergite) certainly semiconducting.				
magnetic susceptibility (in 10^{-6} cm^3 mol^{-1}):				
χ_m	≈ -20	$T=80$ K	χ in CGS-emu, powder sample	66B
	≈ -10	530 K		
very narrow homogeneity range [79K2].				
structural transition temperature (in K):				
T_{tr}	853			77K3, 79K2
	863		at the Ni-rich side	61Y
	871		at the As-rich side	
$Ni_{0.94}Co_{0.06}As_{1.86}S_{0.14}$ pararammelsbergite type, diamagnetic [72F2].				
$NiAs_2$ (β-$NiAs_2$) (rammelsbergite, marcasite type)				
energy gap:				
$E_{g,th}$	≈ 0.05 eV		from $\log \varrho \propto E_g/2kT$, $T=400\cdots500$ K, sintered sample	59H
magnetic susceptibility: constant paramagnetism $373\cdots700$ K, weakly T-dependent below; paramagnetic due to impurities [66B].				

Physical property	Numerical value	Experimental conditions	Experimental method remarks	Ref.
melting point:				
T_m	>1313 K (?)		congruent, from tentative phase diagram (?)	61Y
peritectic temperature:				
T_{perit}	1125 K			77K3

range of homogeneity: $NiAs_{2-x}$, $x \lessapprox 0.01$ [63R, 68D].
lattice parameters $a(T)$, $b(T)$, $c(T)$: above 700 K increase of $a(T)$ and $b(T)$ steeper than linear in T, while $c(T)$ decreases [77K3].

volume expansion coefficient:				
β	$3.0 \cdot 10^{-5}$ K^{-1}	$T = 300$ K	from graphic representation; linear increase of unit-cell volume in the range 300···1100 K	77K3

The metastable marcasite structure persists down to 4 K [77K3].

$NiAs_{2-x}S_x$

maximum sulfur content of β-$NiAs_2$: 1.1 wt% at 973 K. Substitution of S for As lowers the $\beta \rightarrow \alpha$ inversion from 863 K to 773 K [62Y].

$Ni_{1-x}Co_xAs_2$

homogeneity range: $x = 0···1$ at ≈ 1100 K [63R], $x \leq 0.3$ at 573 K [79K2].
$\alpha \rightleftarrows \beta$ transformation temperature: Fig. 29.

$Ni_{1-x}Pd_xAs_2$

$x = 0.1$: $\alpha \rightleftarrows \beta$ transition between 923 and 1023 K [66B].
$x = 0.2$: $\alpha \rightleftarrows \beta$ transition between 1023 and 1048 K [66B].
$x > 0.43$: pyrite structure [66B].

$NiAs_{2-x}S_x$

homogeneity range: $x \leq 0.05$ at 1023 K [79K2], $x \leq 0.04$ at 573 K [79K2].
$\alpha \rightleftarrows \beta$ transformation temperature: Fig. 29.

$NiAs_{2-x}Se_x$

homogeneity range: $x \leq 0.07$ at 1023 K [79K2], $x \leq 0.05$ at 573 K [79K2].
$\alpha \rightleftarrows \beta$ transformation temperature: Fig. 29.

$NiSb_2$

appears to be metallic [57D, 59H].

Physical property	Numerical value	Experimental conditions	Experimental method remarks	Ref.
magnetic susceptibility (in 10^{-6} cm^3 mol^{-1}, χ in CGS units):				
χ_m	−30	$T = 77···295$ K	polycrystalline sample	53R
	−31	90···750 K	polycrystalline sample	68H1
peritectic (decomposition) temperature (in K):				
T_{perit}	889		from differential thermal analysis	57D, 68H2
	893			77K3
	894		polished-section examination of charges quenched from 891 to 897 K	70C

lattice parameters $a(T)$ and $b(T)$ roughly linear in T, $c(T)$ nearly constant above 550 K [77K3].

Physical property	Numerical value	Experimental conditions	Experimental method remarks	Ref.
volume expansion coefficient:				
β	$5.3 \cdot 10^{-5}\,\mathrm{K}^{-1}$	$T = 300\,\mathrm{K}$	from graphic representation; unit-cell volume linear in T up to $\approx 550\,\mathrm{K}$	77K3
microhardness:				
H_B	$\approx 550\,\mathrm{kg\,mm}^{-2}$	$T = 295\,\mathrm{K}$	wide spread $300\cdots760\,\mathrm{kg\,mm}^{-2}$ for 13 indentations; mineral nisbite $\approx 480\,\mathrm{kg\,mm}^{-2}$	70C

no transformation to the pararammelsbergite structure on long-term annealing at $\geqq 573\,\mathrm{K}$ [77K3, 79K2].

PdP$_2$

energy gap:				
$E_\mathrm{g,th}$	$0.6\cdots0.7\,\mathrm{eV}$		from $\log \varrho \propto E_\mathrm{g}/2kT$, sintered sample	63H2
thermoelectric power:				
S	$\approx -100\,\mathrm{\mu V\,K}^{-1}$	$T = 300\,\mathrm{K}$	sintered sample	63H2
magnetic susceptibility:				
χ_m	$-6 \cdot 10^{-6}\,\mathrm{cm}^3\,\mathrm{mol}^{-1}$	$T = 295\,\mathrm{K}$	χ in CGS-emu, powdered sample	63H2

no structural transition at pressures up to 65 kbar/1400 K [68D].

PdPAs

energy gap:				
$E_\mathrm{g,th}$	$0.45\,\mathrm{eV}$		from $\log \varrho \propto E_\mathrm{g}/2kT$ above 550 K, sintered sample	63H2
thermoelectric power:				
S	$-240\,\mathrm{\mu V\,K}^{-1}$	$T = 300\,\mathrm{K}$	sintered sample	63H2

PtP$_2$

energy gap:				
$E_\mathrm{g,th}$	$\geqq 0.6\,\mathrm{eV}$		from $\log \varrho \propto E_\mathrm{g}/2kT$ above 600 K, sintered sample	63H2

resistivity: temperature dependence in the extrinsic range $77\cdots500\,\mathrm{K}$: Fig. 30.

carrier concentration (in $10^{18}\,\mathrm{cm}^{-3}$):				
n	4.5	$T = 295\,\mathrm{K}$	from Hall effect in the extrinsic range, varying from sample to sample (crystals grown from tin flux containing 0.02% Sn)	74B2
	1.8	77 K		
thermoelectric power (in $\mathrm{\mu V\,K}^{-1}$):				
S	-90	RT	single crystal from tin flux	74B2
	$\approx +100$	RT	sintered sample	63H2
	$+150$	$T = 300\,\mathrm{K}$	hot-pressed at $\approx 1100\,\mathrm{K}$; $\varrho(\mathrm{RT}) = 0.24\,\Omega\,\mathrm{cm}$	65J
	$+180$	300 K	cold-pressed and annealed at $\approx 1100\,\mathrm{K}$; $\varrho(\mathrm{RT}) = 30\,\Omega\,\mathrm{cm}$	

For temperature dependence of S in the range $100\cdots900\,\mathrm{K}$ for a hot-pressed sample, see Fig. 31.

Physical property	Numerical value	Experimental conditions	Experimental method remarks	Ref.
magnetic susceptibility (in 10^{-6} cm^3 mol^{-1}):				
χ_m	−64.0	$T = 295$ K	χ in CGS-emu, single crystal from tin flux	74B2
	−63.5	77 K		
thermal conductivity:				
κ	0.26 W cm^{-1} K^{-1}	$T = 300$ K	hot-pressed sample	65J

far infrared absorption: for spectrum in the range 80⋯440 cm^{-1}, see [77L].

PtPAs

energy gap:				
$E_{g,th}$	≥ 0.4 eV		from $\log \varrho \propto E_g/2kT$, sintered sample	63H2
thermoelectric power:				
S	−10⋯+10 μV K^{-1}	$T = 300$ K	sintered samples	63H2
magnetic susceptibility (in 10^{-6} cm^3 mol^{-1}):				
χ_m	−62	$T = 295$ K	χ in CGS-emu, powdered sample	63H2
	−59	77 K		

PtAs$_2$

energy gap (in eV):				
$E_{g,th}$	≈ 0.5		from $\log \varrho \propto E_g/2kT$, sintered sample	63H2
	≈ 0.8		from $\log \varrho \propto E_g/2kT$, $T = 700⋯1000$ K, sintered sample	65J
	0.17		from $\log \varrho \propto E_g/2kT$, $T = 300⋯500$ K	66B

resistivity: temperature dependence: Fig. 30.

thermoelectric power (in μV K^{-1}):				
S	≈ −200	$T = 300$ K	sintered sample	63H2
	+310	295 K	sintered sample; for temperature dependence, see Fig. 31	65J
	< 0	> 815 K		
magnetic susceptibility (in 10^{-6} cm^3 mol^{-1}; χ in CGS units):				
		T [K]		
χ_m	−48	295	powder	63H2
	−45	77		
	≈ −62	80, $B > 0.8$ T	values extrapolated to infinite field, sintered sample	66B
	≈ −60	≈ 540		
melting temperature:				
T_m	> 1533 K		arsenic vapor pressure at this temperature > 1 atm.	65J

Se doping:

PtAs$_{1.99}$Se$_{0.01}$: $\varrho = 8 \cdot 10^{-4}$ Ω cm, $S = -54$ μV K^{-1} at RT [65J].

metal substitutions:

Ni$_x$Pt$_{1-x}$As$_2$: x ≈ 0 at 1023 K and 1373 K [66B].
Pd$_x$Pt$_{1-x}$As$_2$: x ≈ 0 at 1023 K, x ≈ 0.2 at 1373 K, x ≈ 0.5 at 1473 K [66B].

Physical property	Numerical value	Experimental conditions	Experimental method remarks	Ref.

PtSb$_2$

Electronic properties

The valence band maxima are six ellipsoids on [100] axes, with $K = m_\parallel/m_\perp = 0.87$ (low T)···0.67 (high T). The eight conduction band minima lie on [111] axes, with $m_\parallel/m_\perp = 1.43$ (low T)···1.36 (high T). Both the valence and conduction bands are derived from Pt d-orbitals with an additional higher-energy s-like conduction band [65E, 72D1, 83D].

energy gaps (in eV):

$E_{g,th}$	$\geqq 0.05$		from $\log \varrho \propto E_g/2kT$, polycrystalline sample	63H2
	0.06		from $\varrho(T)$ and $R_H(T)$, single crystal and polycrystal	65J
	0.075	$T = 0$ K	from $\log \varrho \approx 0.4 \, E_g/kT$, assuming $(\mu_n + \mu_p)(p \cdot n)^{1/2}$ temperature independent	65D
	0.11		from resistivity, Hall effect and Nernst-Ettingshausen effect on single crystals	73A
	0.112		from $\log \varrho \propto E_g/2kT$ in the intrinsic region, $T = 120$···300 K, single crystals	68E
	$0.110 - 0.00015\,T$		from resistivity and Hall effect on Te doped single crystals; ac method, 100 Hz, frequency independent $50\cdots10^4$ Hz, T in K	83D
$E_{g,ind}$	≈ 0.10		indirect gap from optical absorption	65D, 68R2, 70O, 72D1
	0.104···0.113	$T = 10$ K	band-to-band indirect absorption threshold	68R2, 70O
$E_{g,dir}$	> 0.4	4 K	from the energy dependence of the cyclotron mass $m_{\omega_c}(E) = m_{\omega_c}(0)/1 + (2E/E_{g,dir})$; m_{ω_c} changes by less than 5% when E increases by 0.11 eV	72D1

donor levels (in meV):

E_d	2.3		ionization energy of an isolated donor (Te doped sample)	76A
	3···4		from $\varrho(T)$ of Te-doped samples	68R2
	3···5.5			83D

effective masses (in units of m_0):

$m_{n,p} = n_{C,V}^{2/3}(m_1 m_2 m_3)^{1/3} = n_{C,V}^{2/3} m_\parallel^{1/3} m_\perp^{2/3} = n_{C,V}^{2/3}(m_N)_{C,V}$; n_C, n_V are the numbers of ellipsoids in conduction and valence band, respectively

		T [K]		
$m_{n\parallel}$	0.53···0.66	low T···high T	calculated from theoretical band model based on experimental data from [65D]	65E
$m_{n\perp}$	0.37···0.49	low T···high T		
$(m_N)_C$	0.42···0.54	low T···high T	density of states effective mass	65E
	≈ 0.5	low T		65D

(continued)

Physical property	Numerical value	Experimental conditions	Experimental method remarks	Ref.
effective masses (continued)				
m_n^c	1.6	360	conductivity effective mass of a single ellipsoid $3\left(\dfrac{1}{m_1}+\dfrac{1}{m_2}+\dfrac{1}{m_3}\right)^{-1}$ derived from the edge of the plasma reflection	73A
	0.35	50⋯100	conductivity effective mass for a single valley in the conduction band	83D
	0.34⋯0.57	77	conductivity effective mass, from optical absorption, assuming acoustic-phonon scattering or ionized-impurity scattering (upper limit)	68R2
$m_{p\parallel}$	0.52⋯0.55	low T⋯ high T	based on theoretical band model and data from [65D]	65E
	0.35	77	using the theoretical model of [65E] with $m_\parallel/m_\perp = 0.77$	73A
	0.66	310		
$m_{p\perp}$	0.60⋯0.82	low T⋯ high T		65E
	0.45	77		73A
	0.85	310		
$(m_N)_V$	0.57⋯0.72	low T⋯ high T	density of states mass of a single ellipsoid	65E
	0.22	≈4		65D, 72D1
	0.41	77	from $S(T)$ assuming acoustic-phonon scattering only ($m_p = 1.34\, m_o$)	73A
m_p^c	0.17⋯0.30	77	conductivity effective mass, from optical absorption from plasma reflection	68R2
	0.2	300		83D
	1.0	360	conductivity effective mass from the edge of the plasma reflection	73A
m_{ω_c}	0.168	1.3⋯4.2, $B \leq 6\,T$	least cyclotron mass $(m_1 m_2)^{1/2}$ from high-field magnetoresistance on p-type single crystals	72D1

anisotropy of valence band principal effective masses: $m_1:m_2:m_3 = 0.61:1:1.64$ ($= 0.134:0.22:0.361$) from Shubnikov-de Haas oscillations [72D1]. Group-theoretical arguments, however, require that in space group $Pa3$ of the pyrite structure the [100] ellipsoids should have two equal principal masses [83D].

Lattice properties

elastic moduli (in 10^{11} dyn cm^{-2}):

c_{11}	26.0	RT	calculated from sound velocity measurements made by the pulse-echo technique	65D
c_{12}	6.8	RT		
c_{44}	5.9	RT		

Transport properties (see also table page 106)

electrical resistivity (in Ω cm): (Figs. 33⋯36, see also Fig. 30 and 55b)

		T [K]		
ϱ_i	1	≈70	intrinsic value	68R2, 83D
	0.1	≈90		
	0.01	≈135		
	0.001	≈250		

Physical property	Numerical value	Experimental conditions	Experimental method remarks	Ref.

Hall coefficients (in cm^3 C^{-1}): (Figs. 36···39)

		T [K]		
R_H	$+4$	2···120, $B \leq 3$ T	single crystal, intrinsic near RT	73A
	$+3.0$	77···120, $B \leq 1.1$ T	Te doped single crystal with $\varrho = 0.0035$ Ω cm at 77 K	65D
	$+103$	77	single crystal with $\varrho = 0.0176$ Ω cm at 77 K	83D
	-110	77	single crystal with $\varrho = 0.58$ Ω cm at 77 K	68R2
	-600	62.5		

thermoelectric powers (in μV K^{-1}): (Figs. 40 and 40α)

		T [K]		
S	$+80$	300	polycrystalline sample	63H2
	$+180$	80	pure and Au doped (n-type PtSb$_2$ by doping with Te)	65J
	-200	≈4	Te doped single crystal with donor excess $n_d - n_a \approx 10^{17}$ cm^{-3} and $n_a/n_d \approx 0.8$; $\Delta T \leq 0.2$ K	83D
	-825	≈25		
	0	≈100		
	$+150$	≈125		

mobility of charge carriers (in cm^2 V^{-1} s^{-1}):

		T [K]		
μ_n	420	300	from $\varrho(T)$ and $R_H(T)$ in the mixed-conduction range; $\mu_n = 2.07 \cdot 10^6 \, T^{-3/2} \exp(12.6 \, \text{K}/T)$	68R2
	3700	77		
	2100	50	maximum of $R_H \sigma$	68E
	$3 \cdot 10^6 \, T^{-3/2}$	7···50	acoustic-phonon-limited mobility, from $\varrho(T)$ and $R_H(T)$	83D
μ_p	830	300	$\mu_p = 4.15 \cdot 10^6 \, T^{-3/2} \exp(12.6 \, \text{K}/T)$ $\approx 6.55 \cdot 10^6 \, T^{-1.57}$ above 100 K	68R2
	7300	77		
	$7.5 \cdot 10^6 \, T^{-3/2}$	7···50		83D
			temperature dependence: Figs. 41···43	

In the impurity-conduction range the mobility is temperature independent which suggests that a metallic rather than a hopping type of impurity conduction is dominant [68R2]. Between 10 and 40 K the mobilities are limited primarily by ionized-impurity scattering [68R2]. The occurrence of the near $T^{-3/2}$ dependence of μ_p and μ_n over a wide temperature range shows that acoustic-mode lattice scattering limits the free-carrier mobilities in sufficiently pure PtSb$_2$ [68R2, 68E]. Better agreement with experimental data by taking into account scattering by ionized impurities at low T and by optical phonons at higher T [83D]. Hole and electron mobility are independent of the electric field up to the impact ionization threshold, 1 and 3 kV cm^{-1}, respectively [80D].

mobility ratio: (Fig. 41)

		T [K]		
b	0.57	≈110	from $R_H^{max}/R_H^{exh} = (1-b)^2/4b$, taking R_H^{max} at 110 K and R_H^{exh} in the exhaustion region at 74 K	68E
	0.5	78···200	same procedure as above. Te doped single crystals with $n_d - n_a = 0.3 \cdots 3.2 \cdot 10^{16}$ cm^{-3}	76A

(continued)

Physical property	Numerical value	Experimental conditions	Experimental method remarks	Ref.
mobility ratio (continued)				
	0.47⋯0.52	80⋯300	from R_H in the acoustic phonon scattering region of different samples	68R2
	0.4	≈240	from $R_H(T)$ and $S(T)$	73A
	1	≈560	sign change of $R_H(T)$	
	0.4	10⋯100	acoustic phonon scattering only; total b increases with T	83D
intrinsic carrier concentration (in cm^{-3}):				
n_i	$7.7 \cdot 10^{18}$	$T = 300$ K	calculated from the experimental value of ϱ_i and the equations given above for μ_n and μ_p	68R2
	$2.2 \cdot 10^{15}$	77 K	calculated as $n_i = 2.72\, A \exp(-0.11\,\mathrm{eV}/2kT)$ with $A = 4.84 \cdot 10^{15}\, T^{3/2}$ [$K^{-3/2}\,cm^{-3}$]	
	$1.3 \cdot 10^{16}\, T^{3/2} \exp(-640/T)$	>100 K	from $\varrho(T)$ and $R_H(T)$ using a single parabolic conduction and valence band model	83D

For temperature dependence of $n_i/T^{3/2}$, see Figs. 44 and 45.

magnetoresistance in p-type crystals of low carrier concentration: Fig. 46.
oscillatory magnetoresistance in crystals of high carrier concentration: Figs. 47⋯49.
values for the coefficients of $\Delta\varrho/\varrho\, H^2$: see table page 106.
dependence of the resistivity on hydrostatic pressure: Fig. 50.
piezoresistance coefficients in the range 80⋯300 K: Fig. 51.

Optical properties

optical absorption at 10 and 77 K: Figs. 52⋯54.

dielectric constant:

$\varepsilon(\lambda)$	30	$\lambda = 16$ μm, $T = 10$ K	from refractive index (n^2), p-type crystal with ≈$2 \cdot 10^{17}$ holes cm^{-3} at 77 K	68R2
	31			83D
	32	IR, 360 K	from reflection coefficient 0.49 and low absorption (cited from inaccessible Russian paper)	79K3
	31.4	$f = 25.7$ GHz, 6⋯40 K	Te doped (≈$10^{17}\,cm^{-3}$) single crystal, $n_a/n_d \approx 0.7⋯1$. Permittivity due to the crystal lattice; contribution from free carriers negligible	79K3
refractive index:				
n	5.5	$\lambda = 16$ μm	from transmission interference fringes at 10 K	68R2

Further properties

magnetic susceptibility (in $10^{-6}\,cm^3\,mol^{-1}$):

χ_m	−56	$T = 295$ K	χ in CGS-emu; powdered sample, Gouy method, $B \leq 1$ T	63H2
	−52	77 K		
melting point:				
T_m	1499 K		congruent	58H, 65D

Fig. p. 411

9.15.1.3.2 References for 9.15.1.3

Physical property	Numerical value	Experimental conditions	Experimental method remarks	Ref.
Debye temperature (in K):		T [K]		
Θ_D	≈ 330	0	from heat capacity on 7 g single crystal which at 77 K showed $\varrho = 0.037\,\Omega\,\text{cm}$, $R_H = -2.89\,\text{cm}^3\,\text{C}^{-1}$, $n = 2.26 \cdot 10^{18}\,\text{cm}^{-3}$, close to degeneracy	82K
	≈ 260	30		
	≈ 280	100		
	≈ 325	300		
thermal conductivity (in W/K cm): (see Fig. 55)		T [K]		
κ	1.13	77	p-type single crystal with $p = 8 \cdot 10^{17}\,\text{cm}^{-3}$	73A
	0.39	300	same sample, intrinsic range	
	0.4	≈ 300	single crystal with $1\cdots5 \cdot 10^{17}\,\text{cm}^{-3}$ electrically active impurities (current carrier contribution is 25% at RT)	82K
	4	30	maximum of $\kappa(T)$	

doping, metal substitution: $\text{Pt}_{0.997}\text{Au}_{0.003}\text{Sb}_2$: $S = +180\,\mu\text{V}\,\text{K}^{-1}$ at 80 K, the same value as in the undoped sample [65J].

Very low solubility of the metallic PdSb_2 and AuSb_2 in PtSb_2 [63H2].

Values derived from different p-type samples for the coefficients of

$$\Delta\varrho/\varrho H^2 = b + c(\sum_i l_i \gamma_i)^2 + d\sum_i l_i^2 \gamma_i^2 + e\sum_{ijk} \varepsilon_{ijk} l_j^2 \gamma_k^2,$$

where l_i and γ_i are the direction cosines of the current and magnetic field referred to the crystal axes and ε_{ijk} is the unit antisymmetric tensor. b/μ_H^2, c/μ_H^2, ... are functions of $K_1 = m_3\tau_2/m_2\tau_3$ and $K_2 = m_1\tau_2/m_2\tau_1$; m_1, m_2, m_3 are the principal effective masses and τ_1, τ_2, τ_3 the principal relaxation times of the valence-band ellipsoids centered on the [100] axes [72D1].

T [K]	R_H [cm³ C⁻¹]	σ [Ω⁻¹ cm⁻¹]	μ_H [cm² V⁻¹ s⁻¹]	b/μ_H^2	c/μ_H^2	d/μ_H^2	e/μ_H^2	K_1	K_2	Ref.
77	1.89	913	1720	0.056	–	–	0.0065			65D
77	2.19	810	1770	0.058	–	$-(b+c)$	–			65D
77	1.57	978	1540	0.072	0.020	-0.092	0.013			65D
77	2.0	810	1620	0.076	0.015	-0.094	–			65D
77	1.54	1020	1570	0.164	–	$-(b+c)$	–			65D
77				0.080	≈ 0.03		0.014	1.4	0.7	72D1
4				0.12	≈ 0.05		0.022	1.6	0.7	72D1

9.15.1.3.2 References for 9.15.1.3

53R	Rosenqvist, T.: Acta Met. **1** (1953) 761.
56A	Abrikosov, N.Kh., Bankina, V.F.: Dokl. Akad. Nauk SSSR **108** (1956) 627.
56D	Dudkin, L.D., Abrikosov, N.Kh.: Zhur. Neorg. Khim. **1** (1956) 2096.
57A	Abrikosov, N.Kh.: Izv. Akad. Nauk SSSR, Ser. Fiz. **21** (1957) 141.
57D	Dudkin, L.D., Abrikosov, N.Kh.: Zhur. Neorg. Khim. **2** (1957) 212; translation: Russ. J. Inorg. Chem. **2** (1957) 325.
57K	Kuz'min, R.N., Zhdanov, G.S., Zhuravlev, N.N.: Kristallografiya **2** (1957) 48; translation: Sov. Phys. Cryst. **2** (1957) 42.
58H	Hansen, M., Anderko, K.: Constitution of Binary Alloys, McGraw Hill, New York, second ed. **1958**, 1138.
59H	Hulliger, F.: Helv. Phys. Acta **32** (1959) 615.

9.15.1.3.2 References for 9.15.1.3

60D	Dudkin, L.D., Vaidanich, V.I.: Fiz. Tverd. Tela **2** (1960) 1526; translation: Sov. Phys. Solid State **2** (1960) 1384.
60H	Heyding, R.D., Calvert, L.D.: Can. J. Chem. **38** (1960) 313.
60K1	Kjekshus, A.: Acta Chem. Scand. **14** (1960) 1450.
60K2	Kuz'min, R.N., Zhuravlev, N.N., Losievskaya, S.A.: Kristallografiya **4** (1960) 218; translation: Sov. Phys. Cryst. **4** (1960) 202.
60R	Rundqvist, S.: Nature (London) **185** (1960) 31.
60S	Swanson, H.E., Cook, M.I., Evans, E.H., de Groot, J.H.: NBS Circ. 539, **10** (1960) 42.
60Z	Zhuravlev, N.N., Zhdanov, G.S., Kuz'min, R.N.: Kristallografiya **5** (1960) 553; translation: Sov. Phys. Cryst. **5** (1961) 532.
61J	Johnston, W.D.: J. Inorg. Nucl. Chem. **22** (1961) 13.
61K	Kuz'min, R.N., Zhuravlev, N.N.: Kristallografiya **6** (1961) 269; translation: Sov. Phys. Cryst. **6** (1961) 209.
61R	Rundqvist, S.: Acta Chem. Scand. **15** (1961) 451.
61Y	Yund, R.A.: Econ. Geol. **56** (1961) 1273.
61Z	Zhdanov, G.S., Kuz'min, R.N.: Kristallografiya **6** (1961) 872; translation: Sov. Phys. Cryst. **6** (1962) 704.
62P	Pleass, C.M., Heyding, R.D.: Can. J. Chem. **40** (1962) 590.
62Q	Quesnel, J.C., Heyding, R.D.: Can. J. Chem. **40** (1962) 814.
62W	Wintenberger, M.: Bull. Soc. Fr. Minéral. Cristallogr. **85** (1962) 107.
62Y	Yund, R.A.: Amer. J. Sci. **260** (1962) 761.
63H1	Hulliger, F.: Nature (London) **198** (1963) 1081.
63H2	Hulliger, F.: Nature (London) **200** (1963) 1064.
63H3	Hulliger, F.: Phys. Lett. **4** (1963) 282.
63M	Matthias, B.T., Geballe, T.H., Compton, V.B.: Rev. Mod. Phys. **35** (1963) 1.
63R	Roseboom, E.H.: Amer. Mineral. **48** (1963) 271.
63Z	Zachariasen, W.H.: Acta Cryst. **16** (1963) 1253.
64L	Larsson, E.: Arkiv Kemi **23** (1964) 335.
65D	Damon, D.H., Miller, R.C., Sagar, A.: Phys. Rev. A **138** (1965) 636.
65E	Emtage, P.R.: Phys. Rev. A **138** (1965) 246.
65F	Furuseth, S., Selte, K., Kjekshus, A.: Acta Chem. Scand. **19** (1965) 735.
65H	Hulliger, F., Mooser, E.: Progr. Solid State Chem. (ed. H. Reiss) Vol. 2, Pergamon, New York **1965**, 330.
65J	Johnston, W.D., Miller, R.C., Damon, D.H.: J. Less-Common Met. **8** (1965) 272.
65M	Murray, J.J., Heyding, R.D.: Can. J. Chem. **43** (1965) 2675.
65P	Pearson, W.B.: Z. Krist. **121** (1965) 449.
65Z	Zhuravlev, N.N., Smirnova, E.M.: Kristallografiya **10** (1965) 828; translation: Sov. Phys. Cryst. **10** (1966) 694.
66B	Bennett, S.L., Heyding, R.D.: Can. J. Chem. **44** (1966) 3017.
66D	Darmon, R., Wintenberger, M.: Bull. Soc. Fr. Minéral. Cristallogr. **89** (1966) 213.
66H	Holseth, H., Kjekshus, A.: Acta Chem. Scand. **23** (1966) 3043.
66S	Swanson, H.E., Mc Murdie, H.F., Morris, M.C., Evans, E.H.: NBS Monogr. 25, Section 4 (1966) 10.
67P	Pearson, W.B.: A Handbook of Lattice Spacings and Structures of Metals and Alloys (Pergamon, Oxford) Vol. 2, **1967**.
68D	Donohue, P.C., Bither, T.A., Young, H.S.: Inorg. Chem. **7** (1968) 998.
68E	Elliott, C.T., Hiscocks, S.E.R.: J. Mater. Sci. **3** (1968) 174.
68H1	Holseth, H., Kjekshus, A.: J. Less-Common Met. **16** (1968) 472.
68H2	Holseth, H., Kjekshus, A.: Acta Chem. Scand. **22** (1968) 3273.
68H3	Holseth, H., Kjekshus, A.: Acta Chem. Scand. **22** (1968) 3284.
68H4	Hulliger, F.: Structure and Bonding, Vol 4 (eds. C.K. Jørgensen et al.) Springer Berlin **1968**, 83.
68M	Munson, R.A.: Inorg. Chem. **7** (1968) 389.
68R1	Radcliffe, D., Berry, L.G.: Amer. Mineral. **53** (1968) 1856.
68R2	Reynolds, R.A., Brau, M.J., Chapman, R.A.: J. Phys. Chem. Solids **29** (1968) 755.
68S	Stassen, W.N., Heyding, R.D.: Can. J. Chem. **46** (1968) 2159.
69A	Adachi, K., Sato, K., Matsuura, M.: J. Phys. Soc. Japan **26** (1969) 906.
69D	Dahl, E.: Acta Chem. Scand. **23** (1969) 2677.
69H	Holseth, H., Kjekshus, A.: Acta Chem. Scand. **23** (1969) 3043.

9.15.1.3.2 References for 9.15.1.3

69S	Shunk, F.A.: Constitution of Binary Alloys, Second Supplement. McGraw-Hill, New York **1969**.
70B1	Bjerkelund, E., Kjekshus, A.: Acta Chem. Scand. **24** (1970) 3317.
70B2	Brostigen, G., Kjekshus, A.: Acta Chem. Scand. **24** (1970) 2993.
70C	Cabri, L.J., Harris, D.C., Stewart, J.M.: Can. Mineralogist **10** (1970) 232.
70H	Holseth, H., Kjekshus, A., Andresen, A.F.: Acta Chem. Scand. **24** (1970) 3309.
70O	O'Shaughnessy, J., Smith, C.: Solid State Commun. **8** (1970) 481.
71B	Boda, G., Stenström, B., Sagredo, V., Beckman, O., Carlsson, B., Rundqvist, S.: Phys. Scripta **4** (1971) 132.
71K1	Kjekshus, A.: Acta Chem. Scand. **25** (1971) 411.
71K2	Kjekshus, A., Nicholson, D.G.: Acta Chem. Scand. **25** (1971) 866.
72D1	Damon, D.H., Miller, R.C., Emtage, P.R.: Phys. Rev. **B5** (1972) 2175.
72D2	Donohue, P.C.: Mat. Res. Bull. **7** (1972) 943.
72F1	Fan, A.K.L., Rosenthal, G.H., McKinzie, H.L., Wold, A.: J. Solid State Chem. **5** (1972) 136.
72F2	Fleet, M.E.: Amer. Mineral. **57** (1972) 1.
72G	Goodenough, J.B.: J. Solid State Chem. **5** (1972) 144.
72H	Hulliger, F., Mooser, E.: in "Chemical Bonds in Solids" Vol. 1 (ed. N.N. Sirota) Consultants Bureau, New York **1972**, 5.
72M	Maier, J., Wachtel, E.: Z. Metallkde. **63** (1972) 411.
72R	Rosenthal, G., Kershaw, R., Wold, A.: Mat. Res. Bull. **7** (1972) 479.
72S	Steger, J., Kostiner, E.: J. Solid State Chem. **5** (1972) 131.
73A	Abdullaev, A.A., Angelova, L.A., Kuznetsov, V.K., Ormont, A.B., Pashintsev, Yu. I.: Phys. Status Solidi (a) **18** (1973) 459.
74B1	Baghdadi, A., Wold, A.: J. Phys. Chem. Solids **35** (1974) 811.
74B2	Baghdadi, A., Finley, A., Russo, P., Arnott, R.J., Wold, A.: J. Less-Common Met. **34** (1974) 31.
74K1	Kjekshus, A., Rakke, T.: Structure and Bonding Vol. 19, (eds. Dunitz, J.D., et al.) Springer Berlin **1974**, 85.
74K2	Kjekshus, A., Rakke, T., Andresen, A.F.: Acta Chem. Scand. **A23** (1974) 996.
74K3	Kjekshus, A., Rakke, T.: Acta Chem. Scand. **A28** (1974) 1001.
75F	Fleet, M.E.: Z. Krist. **142** (1975) 332.
75J	Jeitschko, W., Donohue, P.C.: Acta Crystallogr. **B31** (1975) 574.
76A	Alekseeva, V.G., Kuznetsov, V.K., Morenkov, A.D.: Fiz. Tekh. Poluprovodn. **10** (1976) 458; translation: Sov. Phys. Semicond. **10** (1976) 274.
76Y	Yamaguchi, G., Shimada, M., Koizumi, M.: J. Solid State Chem. **19** (1976) 63.
77G	Grønvold, F., Highe, A.J., Westrum, Jr., E.F.: J. Chem. Thermodyn. **9** (1977) 773.
77K1	Kaner, R., Castro, C.A., Gruska, R.P., Wold, A.: Mat. Res. Bull. **12** (1977) 1143.
77K2	Kjekshus, A., Rakke, T., Andresen, A.F.: Acta Chem. Scand. **A31** (1977) 253.
77K3	Kjekshus, A., Rakke, T.: Acta Chem. Scand. **A31** (1977) 517.
77L	Lutz, H.D., Willich, P.: Z. Anorg. Allg. Chem. **428** (1977) 199.
78A1	Abrikosov, N.Kh., Petrova, L.I.: Izv. Akad. Nauk SSSR, Neorg. Mater. **14** (1978) 457; translation: Inorg. Mater. **14** (1978) 346.
78A2	Alles, A., Falk, B., Westrum, E.F., Grønvold, F.: J. Chem. Thermodyn. **10** (1978) 103.
78O	Odile, J.P., Soled, S., Castro, C.A., Wold, A.: Inorg. Chem. **17** (1978) 283.
79G	Guérin, R., Potel, M., Sergent, M.: Mat. Res. Bull. **14** (1979) 1335.
79K1	Kjekshus, A., Peterzens, P.G., Rakke, T., Andresen, A.F.: Acta Chem. Scand. **A33** (1979) 469.
79K2	Kjekshus, A., Rakke, T.: Acta Chem. Scand. **A33** (1979) 609.
79K3	Kundrotas, J.P., Dargys, A.J.: Litov. Fiz. Sb. **19** (1979) 549.
80B	Bhatt, Y.C., Schubert, K.: Z. Metallkde. **71** (1980) 581.
80D	Dargys, A., Kundrotas, J.: Phys. Status Solidi (b) **100** (1980) K9.
80G	Gonçalves da Silva, C.E.T.: Solid State Commun. **33** (1980) 63.
80H	Hulliger, F.: unpublished results **1980**.
80R	Rühl, R., Jeitschko, W.: Z. Anorg. Allg. Chem. **466** (1980) 171.
80Y	Yamaguchi, G., Shimada, M., Koizumi, M., Kanamaru, F.: J. Solid State Chem. **34** (1980) 241.
81S	Siegrist, T., Petter, W.: unpublished **1981**.
81T	Tossell, J.A., Vaughan, D.J., Burdett, J.K.: Phys. Chem. Minerals **7** (1981) 177.
82K	Kundrotas, J., Dargys, A.: Litov. Fiz. Sb. **22** (1982) 74.
83D	Dargys, A., Kundrotas, J.: J. Phys. Chem. Solids **44** (1983) 261.
83L	Lutz, H.D., Schneider, G., Kliche, G.: Phys. Chem. Minerals **9** (1983) 109.
83S	Siegrist, T., Hulliger, F.: unpublished.

9.15.1.4 T(V)$_3$ compounds

T = Co, Rh, Ir; V = P, As, Sb

9.15.1.4.0 Structure, chemical bond (including lattice properties)

The semiconducting transition-element tripnictides all crystallize in the cubic skutterudite structure (see table below), which is related to the pyrite type. In this polyanion structure the cations have again a distorted octahedral coordination (T–6X). Each octahedron shares corners with six neighboring octahedra. Layers of octahedra are similar to those in pyrite, however, the octahedra are tilted in a different way so that rectangular (nearly square) anion rings are formed. Each anion is bonded to two anions and two cations which form the corners of a strongly distorted tetrahedron. In skutterudite-type pnictides, semiconductivity is possible only with d^6 cations of an average oxidation number three. Arsenides appear to be slightly anion-deficient [62R]. Probably all skutterudite-type representatives form peritectically [60Z, 61K]. Peritectic temperatures are 859 °C for CoSb$_3$ [56D, 58D, 60Z] and ≈ 900 °C for RhSb$_3$ and IrSb$_3$ [60Z].

Doping of the skutterudite pnictides is possible with cations of the iron and nickel group. The anion can be partially substituted by group IV and group VI neighbors of the Periodic System.

No measurable homogeneity range was found for Co$_{1-x}$M$_x$Sb$_3$ and CoSb$_{3-y}$X$_y$ with M = Cu, Zn, Al, Ti and X = Si, Ge, Pb, Se [59D] and for Rh$_{1-x}$Ru$_x$As$_3$, Rh$_{1-x}$Ag$_x$As$_3$ and Ir$_{1-x}$Os$_x$As$_3$ at 750 °C [66B]. Isoelectronic substitutions, however, are possible. Thus, semiconducting ternary skutterudites (with rhombohedral superstructure due to anion ordering) are known: Co$_2$Ge$_3$S$_3$, Co$_2$Ge$_3$Se$_3$ [77K], Ir$_2$Ge$_3$S$_3$, Ir$_2$Ge$_3$Se$_3$, Ir$_2$Sn$_3$S$_3$, Rh$_2$Ge$_3$S$_3$ [78L].

Isoelectronic substitution is possible also for the cation. Whereas the antimonide Fe$_{0.5}$Ni$_{0.5}$Sb$_3$ [70B] is stoichiometric and thus truly isoelectronic, the corresponding arsenide is found to exist within a broad solubility range Fe$_{1-x}$Ni$_x$As$_3$, x = 0.25···0.46, but apparently not with the isoelectronic composition [62P].

The skutterudite structure contains two large voids (at the positions 2(a) of the cubic unit cell), which can be filled with additional cations. A large family of electron-deficient ternaries LnT$_4$X$_{12}$ exists, where Ln is one of the larger rare-earth elements La, Ce, …; T = Fe, Ru, Os, and X = P, As, Sb [77J, 80B1, 80B2]. Since the iron-group cation is divalent, semiconductivity could be expected only with tetravalent thorium or cerium, as met in ThFe$_4$P$_{12}$, CeFe$_4$P$_{12}$ and CeFe$_4$As$_{12}$ [80B1, 80B3]. However, an intermediate valence was claimed not only for Ce in CeFe$_4$P$_{12}$ but also for Th in ThRu$_4$P$_{12}$ [80B3], though finally CeFe$_4$P$_{12}$ was found to be semiconducting [82G]. CeFe$_4$Sb$_{12}$, on the other hand, was reported to behave as a metal [80B2].

Crystallographic data for semiconducting skutterudite-type compounds TX$_3$ (all data for RT)

Space group T_h^5 – Im3; Z = 8. T in 8(e), X in 24(g).

Compound	a [Å]	d_x [g cm^{-3}]	T–X [Å]	X–X [Å]		Ref.
				2z a	(1–2y) a	
CoP$_3$	7.7073	4.41	2.22	2.24	2.34	68R
	7.702					77A
CoAs$_3$	8.2055	6.82	2.34	2.48	2.56	74K
	8.208					77A
	8.204					60S
CoSb$_3$	9.0347	7.64	2.52	2.89	2.98	74K
	9.034					60Z, 77A
RhP$_3$	7.9951	5.05	2.34	2.23	2.32	68R
	7.991					78O
RhAs$_3$	8.4507	7.21	2.43	2.47	2.56	74K
RhSb$_3$	9.2322	7.90	2.62	2.80	2.92	74K
	9.229					60Z, 56Z
IrP$_3$	8.0151	7.36	2.35	2.23	2.34	68R
IrAs$_3$	8.4673	9.12	2.44	2.46	2.58	74K
	8.4691		2.45	2.47	2.54	61K

(continued)

Crystallographic data for semiconducting skutterudite-type compounds TX$_3$ (continued)

Compound	a [Å]	d_x [g cm^{-3}]	T–X [Å]	X–X [Å]		Ref.
				$2za$	$(1-2y)a$	
IrSb$_3$	9.2533	9.35	2.62	2.85	2.95	74K
	9.248					60Z, 57K, 56Z
Co$_{0.87}$Fe$_{0.11}$Ni$_{0.13}$As$_3$(?)	8.195	7.00	2.33	2.46	2.57	71M
Fe$_{0.5}$Ni$_{0.5}$Sb$_3$	9.0904	7.47	2.54	2.89	3.00	74K, 70B

Co$_{1-x-y}$Fe$_x$Ni$_y$As$_3$: a [Å] = 8.2060 − 0.0246x + 0.1240 y [62R].

9.15.1.4.1 Physical properties

Physical data for skutterudite-type semiconductors (see also Fig. 1···7).

Compound	E_g [eV]	ϱ [Ω cm]		$\varepsilon(\infty)$	$10^6 \cdot \chi_m$*) [cm^3 mol^{-1}] RT	S [μV K^{-1}] RT	Ref.
		RT	77 K				
CoP$_3$	0.45 (opt.)	5.6·10^{-4}	8.1·10^{-4}	g)	−14	≈ +50^b)	77A^a)
						≈ +150	61H
CoAs$_3$	<0.4 (opt.)	3.1·10^{-4}	1.6·10^{-4}		−30	≈ +1	77A^a)
	≈0.25					± ≈100	59H, 61H
	≈0.13				−44.8^e)	≈ +135^f)	62P
				6.47^g)			82L
CoSb$_3$	<0.4 (opt.)	1.4·10^{-4}	0.7·10^{-4}		−47	≈ +1	77A^a)
	^{c})					−400	61H
	0.49	0.12				−300	56D
		0.03···0.1				−200···+30	59D
				10.84^g)			82L
RhP$_3$	metallic^d)	14.5·10^{-5}	4.9·10^{-5}		−33.6	+32	78O
	^{c})				−55	+50	61H
RhAs$_3$	^{c})				−70	+70	61H
	>0.1				−71.8^e)		62P
RhSb$_3$	^{c})				−95	+60	61H
IrP$_3$						+200	61H
IrAs$_3$	^{c})				−105	+150	61H
					−108		61K
					−105.5^e)		62P
IrSb$_3$	^{c})				−145	+200	61H
					−134		61K

*) χ_m in CGS-emu.
^{a}) Measurements made on single crystals grown by a chlorine transport reaction.
^{b}) Hall coefficient $R_H > 0$ at room temperature [77A].
^{c}) Measurements on sintered samples. From log $\varrho \propto E_g/2kT$ values of 0.2···1.3 eV were derived for E_g.
^{d}) Single crystals grown from a tin flux. Carrier concentration: 3.7·10^{19} cm^{-3} at 77 K and 6·10^{19} cm^{-3} at 290 K [78O].
^{e}) χ measured between 100 and 700 K (Fig. 5). Slight decrease in diamagnetic susceptibilities at higher temperatures, consistent with the semiconducting behavior; see also Fig. 7.
^{f}) The Seebeck coefficient increases from 10 μV K^{-1} at 100 K to a maximum of 140 μV K^{-1} at 320 K, and then decreases to 85 μV K^{-1} at 600 K [62P].
^{g}) Far infrared spectra at various temperatures: [81L1, 81L2, 82L].

Crystallographic and physical data for semiconducting skutterudite-type pseudo-pnictides $TX_{1.5}Y_{1.5}$
(All data for RT except stated otherwise)
Space group $C_3^4 - R3$, $Z = 8$, verified for $CoGe_{1.5}S_{1.5}$ [77K].

Compound	a [Å]	α [°]	d_X [g cm^{-3}]	E_A [eV]	$10^6 \cdot \chi_m$ c) [cm^3 mol^{-1}] 77 K	RT	Ref.
$CoGe_{1.5}S_{1.5}$ a)	8.017	90	5.57		-50.6	-50.6	77K
$CoGe_{1.5}Se_{1.5}$ a)	8.299	90	6.65		-38.4	-38.4	77K
$RhGe_{1.5}S_{1.5}$	8.2746	90	6.09			<0	78L
	8.282	89.85					81L2
$RhGe_{1.5}Se_{1.5}$	8.546	89.86	7.03				81L2
$IrGe_{1.5}S_{1.5}$ a)	8.2970	90	8.12	0.11 b)		<0	78L
$IrGe_{1.5}Se_{1.5}$ a)	8.5591	90	8.89	0.076 b)		<0	78L
$IrSn_{1.5}S_{1.5}$	8.7059	90	8.42			<0	78L

a) Far infrared spectra: [81L2].
b) From $\log \varrho \propto E_A/kT$ above 220 K, sintered samples [78L].
c) χ in CGS-emu.

Doping and ternary phases

$Co_{1-x}Fe_xAs_3$

Solubility: $x \leq 0.1$ [62R], ≤ 0.16 [62P].
Metallic behavior results already at $x = 0.01$, although additional substitution increases the resistivity. At $x = 0.01$ the conduction is p-type and changes to n-type at higher Fe concentrations [62P] (Fig. 2).

$Co_{1-x}Fe_xSb_3$

Solubility: $x \leq 0.25$ [59D].
For $x = 0.01$ an exponential increase of $\sigma(T)$ is observed above 620 K; $E_A \approx 0.07$ eV, assuming $\log \sigma \propto -E_A/kT$ [59D]. Increasing Fe concentrations strongly reduce the resistivity and change the thermopower from large negative to small positive values [59D].

$Co_{1-x}Ni_xAs_3$

Solubility: $x \leq 0.6$ [62R], ≤ 0.65 [62P].
Nickel behaves as a donor impurity. Conduction becomes n-type with 1% substitution of Ni for Co. At $x \approx 0.1$ the system becomes metallic [62P] (Fig. 1).

$Co_{1-x}Ni_xSb_3$

Solubility: $x \leq 0.1$ [59D].
Ni is adopting a valence of 4 and thus acts as a donor, fully ionized at room temperature. Up to $x \approx 0.01$ the electron mobility and the microhardness increase (occasional purity effect?), then decrease. The thermoelectric power reaches a maximum of ≈ -350 µV K^{-1} at $x = 0.025 \cdots 0.05$ [59D]; Table below and Figs. (3), 4, 5 and 6.

Room-temperature values for the electron concentration n (from Hall effect), the electron mobility μ_n, the thermal conductivity κ, the microhardness H [57D], and the thermoelectric power S [59D].

x	n [cm^{-3}]	μ_n [cm^2 V^{-1} s^{-1}]	κ [W cm^{-1} K^{-1}]	H [kg mm^{-2}]	S [µV K^{-1}]*)
0	$2.2 \cdot 10^{17}$	290	0.052	370	-170
0.5	$3.0 \cdot 10^{18}$	610			-350
1	$2.4 \cdot 10^{18}$	510	0.049	390	-280
2	$1.3 \cdot 10^{19}$	110	0.048		-220
4	$1.2 \cdot 10^{20}$	22	0.043	382	-150
8			0.029	299	

*) Compare Fig. 5.

$CoSb_{3-x}Te_x$

Solubility: $x \geq 0.03$? [59D].

With increasing Te content the electrical conductivity rises drastically up to $x = 0.006$, while the thermal conductivity drops to half its value in pure $CoSb_3$. The Seebeck coefficient sharply increases on doping with Te, reaching a maximum value of $-300\,\mu V\,K^{-1}$ at $x = 0.0015$ [59D].

At $x = 0.015$, $E_{g,th} \approx 0.13$ eV above 650 K, assuming $\log \varrho \propto E_g/2kT$ [59D].

$CoSb_{3-x}Sn_x$

Solubility: $x \geq 0.015$? [59D].

Sn doping leads to hole conductivity. A sample with $x = 0.015$ shows extrinsic hole conduction up to ≈ 700 K. The Seebeck coefficient is constant $\approx +55\,\mu V\,K^{-1}$ up to 670 K, then decreases [59D].

$Rh_{1-x}Pd_xAs_3$

Solubility: $x \approx 0.07$ at 1023 K [66B].

$Ir_{1-x}Pt_xAs_3$

Solubility: $x < 0.1$ at 1023 K [66B].

Substitutions at constant valence-electron concentration:

$CoP_{3-x}As_x$

Solubility: $x = 0 \cdots 3$, Végard's law obeyed [81L2].
Far infrared spectra: [81L2].

$CoAs_{3-x}Sb_x$

Solubility: $x = 0 \cdots 0.4$ and $2.8 \cdots 3$; miscibility gap [81L2].
Far infrared spectra: [81L2].

$CoSb_{3-x}Bi_x$

Solubility: $x \geq 0.06$? [59D].

Small Bi substitutions do not much change the electrical conductivity and the thermopower [59D].
At $x = 0.06$: $E_{g,th} \approx 0.35$ eV above 500 K, assuming $\log \varrho \propto E_g/2kT$ [59D].

$Co_{1-x}Fe_{x/2}Ni_{x/2}As_3$

Solubility: $x = 0 \cdots 1$ [62R].

All compositions are expected to be nonmetallic. [62P] did not observe $Fe_{0.5}Ni_{0.5}As_3$: They reported a solubility in $Fe_yNi_{1-y}As_3$ of $y = 0.25 \cdots 0.46$. $Fe_{0.46}Ni_{0.54}As_3$ was found to be an n-type semiconductor. With higher nickel concentration the system becomes metallic [62P] (Fig. 2).

$Fe_xCo_{1-x}As_{3-x}S_x$ and $Fe_xCo_{1-x}As_{3-x}Se_x$

Solubility: x very small.

Strong increase of free-carrier concentration with x [82L].

9.15.1.4.2 References for 9.15.1.4

56D	Dudkin, L.D., Abrikosov, N.Kh.: Zhur. Neorg. Khim. **1** (1956) 2096.
56Z	Zhuravlev, N.N., Zhdanov, G.S.: Kristallografiya **1** (1956) 509 (translation: Sov. Phys. Cryst. **1** (1956) 3).
57D	Dudkin, L.D., Abrikosov, N.Kh.: Zhur. Neorg. Khim. **2** (1957) 212 (translation: Russ. J. Inorg. Chem. **2** (1957) 325).
57K	Kuz'min, R.N., Zhdanov, G.S., Zhuravlev, N.N.: Kristallografiya **2** (1957) 48 (translation: Sov. Phys. Cryst. **2** (1957) 42).
58D	Dudkin, L.D.: Zhur. Tekh. Fiz. **28** (1958) 240 (translation: Sov. Phys. Tech. Phys. **3** (1958) 216).
59D	Dudkin, L.D., Abrikosov, N.Kh.: Fiz. Tverd. Tela **1** (1959) 142 (translation: Sov. Phys. Solid State **1** (1959) 126).
59H	Hulliger, F.: Helv. Phys. Acta **32** (1959) 615.
59R	Rundqvist, S., Larsson, E.: Acta Chem. Scand. **13** (1959) 551.
60S	Swanson, H.E., Cook, M.I., Evans, E.H., de Groot, J.H.: NBS Circ. 539, Vol. **10** (1960) 21.

60Z	Zhuravlev, N.N., Zhdanov, G.S., Kuz'min, R.N.: Kristallografiya **5** (1960) 553 (translation: Sov. Phys. Cryst. **5** (1961) 532).
61H	Hulliger, F.: Helv. Phys. Acta **34** (1961) 782.
61K	Kjekshus, A., Pedersen, G.: Acta Crystallogr. **14** (1961) 1065.
62P	Pleass, C.M., Heyding, R.D.: Can. J. Chem. **40** (1962) 590.
62R	Roseboom, E.H.: Am. Mineral. **47** (1962) 310.
66B	Bennett, S.L., Heyding, R.D.: Can. J. Chem. **44** (1966) 3017.
68M	Munson, R.A., Kasper, J.S.: Inorg. Chem. **7** (1968) 390.
68R	Rundqvist, S., Ersson, N.O.: Arkiv Kemi **30** (1968) 103.
70B	Bjerkelund, E., Kjekshus, A.: Acta Chem. Scand. **24** (1970) 3317.
71M	Mandel, N., Donohue, J.: Acta Crystallogr. **B27** (1971) 2288.
74K	Kjekshus, A., Rakke, T.: Acta Chem. Scand. **A28** (1974) 99.
77A	Ackermann, J., Wold, A.: J. Phys. Chem. Solids **38** (1977) 1013.
77J	Jeitschko, W., Braun, D.J.: Acta Crystallogr. **B33** (1977) 3401.
77K	Korenstein, R., Soled, S., Wold, A., Collin, G.: Inorg. Chem. **16** (1977) 2344.
78L	Lyons, A., Gruska, R.P., Case, C., Subbarao, S.N., Wold, A.: Mat. Res. Bull. **13** (1978) 125.
78O	Odile, J.P., Soled, S., Castro, C.A., Wold, A.: Inorg. Chem. **17** (1978) 283.
80B1	Braun, D.J., Jeitschko, W.: J. Solid State Chem. **32** (1980) 357.
80B2	Braun, D.J., Jeitschko, W.: J. Less-Common Met. **72** (1980) 147.
80B3	Braun, D.J., Jeitschko, W.: J. Less-Common Met. **76** (1980) 33.
81L1	Lutz, H.D., Kliche, G.: Z. Anorg. Allg. Chem. **480** (1981) 105.
81L2	Lutz, H.D., Kliche, G.: J. Solid State Chem. **40** (1981) 64.
82G	Grandjean, F., Gérard, A., Braun, D.J., Jeitschko, W.: VII Int. Conf. Solid Compounds of Transition Elements, Grenoble **1982**, Abstr. III A15.
82L	Lutz, H.D., Kliche, G.: Phys. Status Solidi (b) **112** (1982) 549.

9.15.1.5 TP$_4$ compounds

MnP$_4$, TcP$_4$, ReP$_4$, FeP$_4$, RuP$_4$, OsP$_4$

9.15.1.5.0 Structure, chemical bond (including lattice properties)

Since the same cation d electron configurations as met in the transition-element tetraphosphides occur also in the TX$_2$ phases the corresponding structures bear similar features. In all these structures the metal atom is octahedrally coordinated by 6 P atoms. The phosphorus atoms form ten-membered rings which are condensed to two-dimensional nets. Thus, half the P atoms are connected to 3P+1T, while the remaining P atoms have 2P+2T neighbors, all being in roughly tetrahedral coordination. In order to fill the anion valence band two cation valence electrons are thus required for the cation-anion bonds. The various TP$_4$ structure types differ in the way of connecting the [TP$_6$] octahedra, which is dictated by the cation d electron configuration. Thus, in the orthorhombic modification of FeP$_4$, in CdP$_4$-type RuP$_4$(r) and OsP$_4$(r), the [TP$_6$] octahedra are linked via corners as they are in the low-spin d^6 pyrites, e.g. in PtP$_2$. Monoclinic FeP$_4$ and triclinic RuP$_4$ are reminiscent of the normal d^6 marcasites. In marcasites, however, the strings of edge-sharing octahedra have infinite length, while in FeP$_4$ and RuP$_4$ there are units of three and two octahedra, respectively. Based on the low-spin d^5 configuration of the cation the orthorhombic structure of ReP$_4$ is the analog to the arsenopyrite structure of CoP$_2$: Anion double octahedra containing the Re pairs are connected via corners. The same cation d electron configuration is met also in MnP$_4$, which occurs in three polymorphs. The structure of each of these MnP$_4$ modifications as well as that of CrP$_4$ and of orthorhombic FeP$_4$ can be generated by an appropriate stacking of identical puckered T$_n$P$_{4n}$ nets. The differences in the structures arise through different linking of the [TP$_6$] octahedra. In the monoclinic 8-layer modification of MnP$_4$ four octahedra share edges to form a linear array of two Mn pairs. These units are connected with each other via corners only. In the triclinic 6-layer modification such units are linked via edges. In the 2-layer modification the T$_n$P$_{4n}$ nets are stacked on top of each other. Distortions arise through the pairwise bonding of Mn atoms of adjacent nets. The near-neighbor environments in the three MnP$_4$ modifications are very similar. Half the phosphorus atoms are surrounded by 2Mn+2P. The remaining P atoms have 1Mn+3P neighbors. Thus the distorted tetrahedral coordination of the P "anions" is due to bonds to the cations based on electron back donating.

The cations in CrP$_4$ and in isotypical MoP$_4$ have a d^4 configuration as it is the case for the cations in the loellingites. Again the cations form infinite chains which, however, go zigzag instead of straight, and this may account for the (probably) metallic character of CrP$_4$ and MoP$_4$ [72J].

9.15.1.5 Transition-metal compounds TP$_4$

Physical property	Numerical value	Experimental conditions	Experimental method remarks	Ref.

Crystallographic data for the transition-element tetraphosphides (RT values).

Compound	Space group	Z	a [Å]	b [Å]	c [Å]	α [°]	β [°]	γ [°]	d_x [g cm^{-3}]	Ref.
MnP$_4$	C_{2h}^6–C2/c	16	10.513	5.0944	21.804		94.71		4.082	75J
	C_i^1–P$\bar{1}$	6	16.347	5.847	5.108	115.66	95.15	89.21	4.07	81R
	C_i^1–P$\bar{1}$	2	5.861	5.104	5.836	93.82	107.31	115.81	4.065	80J
			5.8622	5.1059	5.8360	93.803	107.338	115.830		80N
TcP$_4$(r)	D_{2h}^{15}–Pbca	8	6.238	9.215	10.837				4.75	82R
ReP$_4$	D_{2h}^{15}–Pbca	8	6.227	9.231	10.854				6.60	79J
FeP$_4$	C_{2h}^5–P2$_1$/c	6	4.619	13.670	7.002		101.48		4.13	78J
FeP$_4$(p)	D_2^5–C222$_1$	4	5.005	10.212	5.530				4.22	78S
RuP$_4$(r)	C_{2h}^5–P2$_1$/c	2	4.686	4.678	7.102		80.53		4.86	82F
RuP$_4$(h)	C_i^1–P$\bar{1}$	3	7.519	7.145	4.713	100.48	90.35	111.08	4.84	78B
OsP$_4$(r)	C_{2h}^5–P2$_1$/c	2	4.694	4.683	7.096		80.43		6.78	82F
OsP$_4$(h)	C_i^1–P$\bar{1}$	3	7.540	7.153	4.718	100.38	90.34	111.20	6.73	78B

Interatomic distances in transition-element tetraphosphides

Compound	T–6P [Å]	T–1T [Å]	P–2P [Å]	P–3P [Å]	Ref.
8L–MnP$_4$ (monocl.)	2.215···2.380 av. 2.282	2.941	2.177···2.278 av. 2.228	2.177···2.252 av. 2.224	75J
6L–MnP$_4$ (triclinic)	2.212···2.368 av. 2.275	2.963, 2.923 (2×)	2.172···2.241 av. 2.221	2.172···2.269 av. 2.232	81R
2L–MnP$_4$ (triclinic)	2.235···2.334 av. 2.273	2.936	2.173···2.230 av. 2.216	2.173···2.278 av. 2.239	80J
TcP$_4$	2.334···2.530 av. 2.398	3.002	2.207···2.253 av. 2.214	2.177···2.253 av. 2.197	82R
ReP$_4$	2.347···2.521 av. 2.402	3.012	2.179···2.251 av. 2.211	2.177···2.251 av. 2.196	79J
FeP$_4$ (monocl.)	2.187···2.328 av. 2.259		2.165···2.241 av. 2.209	2.165···2.275 av. 2.239	78J
FeP$_4$ (orthorh.)	2.175···2.308 av. 2.249		2.216 and 2.315 av. 2.266	2.183···2.315 av. 2.240	78S
RuP$_4$ (triclinic)	2.316···2.403 av. 2.367		2.158···2.240 av. 2.198	2.158···2.290 av. 2.228	78B

9.15.1.5.1 Physical properties

MnP$_4$

monoclinic modification (8L–MnP$_4$):

energy gap:

E_g	0.28 eV		given as $E_A = 0.14$ eV, from $\log \varrho \propto E_A/kT$	75J

resistivity:

ϱ	30 Ω cm	RT	four-probe technique, crystals prepared at 1500···1700 K and 30···35 kbar, unknown orientation	75J

Physical property	Numerical value	Experimental conditions	Experimental method remarks	Ref.
triclinic 6-layer modification (6L–MnP$_4$):				
energy gap:				
E_g	0.54 eV		given as $E_A = 0.27$ eV, from $\log \varrho \propto E_A/kT$ between 300 and 420 K, crystals from tin flux	80J
triclinic 2-layer modification (2L–MnP$_4$):				
energy gap:				
E_g	>0.08 eV		intrinsic conductivity range not reached at 420 K, crystals prepared by a transport reaction with iodine	80J

Crystals of all three MnP$_4$ modifications are shiny black, stable in air and in nonoxidizing acids; diamagnetic at room temperature [75J, 80J].

ReP$_4$
diamagnetic, shiny black crystals similar to MnP$_4$, probably semiconducting [79J].

FeP$_4$
monoclinic ambient-pressure modification:

energy gap and activation energy (in eV):				
E_A	0.0014	$T = 4 \cdots 70$ K	from $\log \varrho \propto E_A/kT$	80G
E_g	0.32	near 300 K	given as $E_A = 0.16$ eV, from $\log \varrho \propto E_A/kT$	80G
magnetic susceptibility:				
χ_m	$-1.14 \cdot 10^{-6}$ cm^3 mol^{-1}	RT	χ in CGS-emu	80G

black and lustrous crystals of acicular habit [80G].

orthorhombic high-pressure modification:

activation energy:				
E_A	0.052 eV		from $\log \varrho \propto E_A/kT$, $T = 77 \cdots 300$ K	78S
resistivity:				
ϱ	$3 \cdot 10^4$ Ω cm	RT		78S

weak magnetization 77$\cdots$300 K (second phase?) [78S].

RuP$_4$(r) (CdP$_4$-type modification)

energy gap:				
E_g	0.38 eV		given as $E_A = 0.19$ eV, from $\varrho = \varrho_0 \exp(E_A/kT)$, $T = 300 \cdots 500$ K	82F
electrical resistivity (in Ω cm):				
ϱ	≈ 800	RT	pressed pellets of microcrystalline powder from tin flux; two-probe method (ϱ estimated to be correct to within a factor of 3). Data from graphical representation.	82F
	≈ 125	$T = 400$ K		
	≈ 45	500 K		

9.15.1.5 Transition-metal compounds TP$_4$

Physical property	Numerical value	Experimental conditions	Experimental method remarks	Ref.
magnetic susceptibility:				
χ_m	$-95.5 \cdot 10^{-6}$ cm^3 mol^{-1}	RT	Faraday balance; no dependence on field strength; χ in CGS-emu.	82F

stability range:

$T < 900$ K, converts irreversibly into the high-temperature modification by annealing at 1073 K [82F].

RuP$_4$(h)

energy gap:

E_g	0.64 eV		given as $E_A = 0.32$ eV, from $\varrho = \varrho_0 \exp(E_A/kT)$, $T = 400 \cdots 550$ K	82F

electrical resistivity (in 10^5 Ω cm):

ϱ	≈ 5	RT	pressed powder from tin flux; two-probe method (ϱ estimated to be correct to within a factor of 3)	82F
	≈ 1	$T = 400$ K		

magnetic susceptibility:

χ_m	$-93.9 \cdot 10^{-6}$ cm^3 mol^{-1}	RT	Faraday balance, no field dependence; χ in CGS-emu	82F

OsP$_4$(r) (CdP$_4$-type modification)

energy gap:

E_g	0.40 eV		given as $E_A = 0.20$ eV, from $\varrho = \varrho_0 \exp(E_A/kT)$, $T = 300 \cdots 500$ K	82F

electrical resistivity (in Ω cm):

ϱ	≈ 500	RT	pressed powder from tin flux; two-probe method (correct to within a factor of 3). Data from graphical representation	82F
	≈ 80	$T = 400$ K		
	≈ 28	500 K		

magnetic susceptibility:

χ_m	$-87.6 \cdot 10^{-6}$ cm^3 mol^{-1}	RT	Faraday method; no field dependence; χ in CGS-emu	82F

OsP$_4$(h)

energy gap:

E_g	0.60 eV		given as $E_A = 0.30$ eV, from $\varrho = \varrho_0 \exp(E_A/kT)$, $T = 450 \cdots 500$ K	82F

electrical resistivity (in 10^4 Ω cm):

ϱ	≈ 40	$T = 400$ K	pressed powder from tin flux; two-probe method; data from graphical representation	82F
	≈ 5	500 K		

magnetic susceptibility:

χ_m	$-81.3 \cdot 10^{-6}$ cm^3 mol^{-1}	RT	Faraday method; no field dependence; χ in CGS-emu	82F

9.15.1.5.2 References for 9.15.1.5

72J	Jeitschko, W., Donohue, P.C.: Acta Crystallogr. **B28** (1972) 1893.
75J	Jeitschko, W., Donohue, P.C.: Acta Crystallogr. **B31** (1975) 574.
78B	Braun, D.J., Jeitschko, W.: Z. Anorg. Allg. Chem. **445** (1978) 157.
78J	Jeitschko, W., Braun, D.J.: Acta Crystallogr. **B34** (1978) 3196.
78S	Sugitani, M., Kimomura, N., Koizumi, M., Kume, S.: J. Solid State Chem. **26** (1978) 195.
79J	Jeitschko, W., Rühl, R.: Acta Crystallogr. **B35** (1979) 1953.
80G	Grandjean, F., Gérard, A., Krieger, U., Heiden, C., Braun, D.J., Jeitschko, W.: Solid State Commun. **33** (1980) 261.
80J	Jeitschko, W., Rühl, R., Krieger, U., Heiden, C.: Mat. Res. Bull. **15** (1980) 1755.
80N	Noläng, B.I., Tergenius, L.E.: Acta Chem. Scand. **A34** (1980) 311.
81R	Rühl, R., Jeitschko, W.: Acta Crystallogr. **B37** (1981) 39.
82F	Flörke, U., Jeitschko, W.: J. Less-Common Met. **86** (1982) 247.
82R	Rühl, R., Jeitschko, W., Schwochau, K.: J. Solid State Chem. **44** (1982) 134.

9.15.1.6 T–V–VI compounds

a) Ternary pyrites: T = Co, Rh, Ir; V = P, As, Sb, Bi; VI = S, Se, Te
b) Ternary marcasites: T = Co; V = P, As, Sb; VI = S, Se, Te
c) Ternary pararammelsbergites: T = Co; V = P, As, Sb; VI = S, Se
d) Arsenopyrites: T = Fe, Ru, Os; V = P, As, Sb, Bi; VI = S, Se, Te
e) PdPS-type compounds: T = Pd; V = P; VI = S, Se

9.15.1.6.0 Structure, chemical bond (including lattice properties)

Most of the semiconducting TXY pnigochalcogenides are ternary analogs of the corresponding binary phases with the same cation d electron configuration. Therefore, these phases may exist in a disordered high-temperature modification as well as in an ordered low-temperature modification. In the ternary pyrites, marcasites, pararammelsbergites and in the arsenopyrites the cations are in distorted octahedral coordination while the anions possess $3T+1(X,Y)$ neighbors at the apices of a distorted tetrahedron. In the ternary pyrites anion ordering is possible in at least four variants [73E], but only two of them were detected until now. One of these is the cubic ullmannite type, space group T^4-P2_13, met in the high-pressure modification of CoSbS. The other ordered variant is the cobaltite type of pseudocubic CoAsS and of orthorhombic PtGeSe, space group $C_{2v}^5-Pca2_1$. In the known ordered variant of marcasite, which is similar to arsenopyrite, one of the glide planes is absent, and this reduction of the symmetry transforms the space group from $D_{2h}^{12}-Pnnm$ to $C_{2v}^7-Pn2_1m$.

In pararammelsbergite α-NiAs$_2$ the anions occupy two sets of the general positions 8(c) of space group $D_{2h}^{15}-Pbca$. Thus, ordering is possible within the same space group. In this structure type layering of the cation coordination octahedra perpendicular to the c-axis is based on pyrite-type {100} planes alternating with marcasite-type {101} planes. In the pararammelsbergite-type modification of CoSbS (paracostibite) these alternating layers have every second "arsenopyrite" layer in reverse orientation. The pyrite-type layers contain the Sb atoms while the S atoms are located within the marcasite-type layers [75R].

In the ternary marcasites (as well as in the ternary pyrites and pararammelsbergites) with d^6 cations short cation-cation contacts are avoided. No ternary d^4 loellingite (like e.g. "MnAsSe", in analogy to FeAs$_2$) has been synthesized until now. In d^5 arsenopyrites nonmetallic properties require a splitting of the half-filled d-subband, and this is attained by displacing the cations within the chains along [101], which corresponds to the original marcasite c-axis. In these strings the [TX$_3$Y$_3$] octahedra share X···X edges alternating with Y···Y edges. The distortion of the structure on forming cation pairs leads to monoclinic symmetry and space group $C_{2h}^5-P2_1c$. All atoms occupy general positions 4(e), and ordering is again possible within the same space group.

A unique structure with no binary analog is exhibited by PdPS and isostructural PdPSe. In these compounds the metal atoms adopt a diamagnetic d^8 configuration with square-planar coordination as in PdS$_2$ and PdP$_2$. In the layer structure of PdPS structural elements of both binaries can be detected (PdS$_2$ crystallizes in a strongly elongated, layered pyrite structure). Instead of the infinite P chains of PdP$_2$, however, polyanions [S–P–P–S]$^{4-}$ are present in PdPS. Phosphorus is tetrahedrally coordinated by 1S + 1P + 2Pd. Sulfur has one P and two Pd neighbors. The lone electron pair which complements the tetrahedral coordination of S points towards the adjacent sandwich and is responsible for the Van der Waals bonding.

TXY. Crystallographic and physical data for pyrite-type derivatives (RT values). Seebeck coeff. from [62H].

Compound	a [Å]	d_x [g cm^{-3}]	Ref.	E_g [eV]	Ref.	$10^6 \chi_m$ [cm^3 mol^{-1}]	Ref.	S [μV K^{-1}]
CoPS (tetr.)	5.414 ($c = 5.430$)	5.08	74N	0.4	59H			
	5.422 (h ?)		63H1					
CoAsS (r) (C_{2v}^5 – Pca2$_1$)	5.582		65G	0.75	59H			−80
pseudocell*)	5.578	6.35	63H1	0.60	76A			
	5.572		65K					
CoSbS (p) (T^4 – P2$_1$3)	5.844	7.08	75H			≈ 0	75H	
CoPSe (p) (T_h^6 – Pa3)	5.63	6.29	75H					
CoAsSe (p) (T_h^6 – Pa3 or T^4 – P2$_1$3)	5.76**)		75H 63H1					
RhPS	5.640	6.14	63H1					+10
RhAsS	5.780	7.22	63H1					+50
RhSbS (ordered)	6.027	7.79	63H1					−300
RhBiS (T^4 – P2$_1$3)	6.138	9.88	63H1					−350
RhPSe (ordered)	5.795	7.26	63H1			−15	62H	
RhAsSe	5.934	8.16	63H1					+100
RhSbSe (ordered)	6.176	8.56	63H1					+50
RhBiSe (ordered)	6.283	10.47	63H1					−250
RhAsTe	6.165	8.66	63H1					−100
RhSbTe	6.392	8.96	63H1	>0.15	80H	−70	62H	−50
RhBiTe	6.504	10.61	63H1					−30
IrPS	5.650	9.40	63H1					
IrAsS	5.791	10.23	63H1			−66	62H	+10
IrSbS (ordered)	6.036	10.45	63H1					−1000
IrBiS (ordered)	6.143	12.41	63H1			−78	62H	−70
IrPSe (ordered)	5.798	10.30	63H1					
IrAsSe	5.940	10.97	63H1					+10
IrSbSe (ordered)	6.184	11.04	63H1					
IrBiSe (ordered)	6.290	12.82	63H1					−50
IrAsTe	6.164	11.19	63H1					+60
IrSbTe (ordered)	6.397	11.20	63H1					
IrBiTe	6.500	12.79	63H1					−50

*) (h): annealing at 850 °C completely disorders As and S [65G]. Most mineral specimens are partially ordered [65G, 82B].   **) Stabilized by impurities?

Crystallographic data for ternary marcasite-type semiconductors (RT values)

Compound	a [Å]	b [Å]	c [Å]	d_x [g cm^{-3}]	Ref.
"CoAsS"^{a})	4.661	5.603	3.411	5.95	76S
CoSbS*)^b)	4.873	5.852	3.608	6.87	75R
(r?)*)^b)	4.868	5.838	3.603		70C1
CoAsSe (h)	4.751	5.753	3.584	7.22	75H
**)	4.7562	5.7514	3.5628		79K
CoSbSe	5.056	6.031	3.686	7.68	75H
CoSbTe	≈ 5.24	≈ 6.26	≈ 3.84	8.13	78A
	5.257	6.248	3.844	8.11	76Y

*) Ordered, space group C_{2v}^7 – Pn2$_1$m.   **) Disordered, space group D_{2h}^{12} – Pnnm.
^{a}) Mineral alloclasite of composition Co$_{0.68}$Fe$_{0.19}$Ni$_{0.025}$As$_{1.01}$S$_{0.99}$, monoclinic, space group C_2^2 – P2$_1$, $\beta = 90.03°$.
^{b}) Mineral costibite Co$_{0.96}$Fe$_{0.02}$Ni$_{0.01}$As$_{0.01}$SbS.

Interatomic distances (in Å):

"CoAsS" (alloclasite):	Co	−3 As	2.282,	2.308,	2.309
		3 S	2.299,	2.309,	2.372
	As	− S	2.311		76S
CoSbS (costibite):	Co	−3 Sb	2.480,	2.502 (2×)	
		3 S	2.303,	2.354 (2×)	
	Sb	− S	2.521		75R
CoAsSe (h):	Co	−2 (As, Se)	2.352		
		4 (As, Se)	2.375		
	(As, Se)	− (As, Se)	2.485		79K

Crystallographic data for ternary pararammelsbergite-type compounds (RT values)
(space group D_{2h}^{15} – Pbca)

Compound	a [Å]	b [Å]	c [Å]	d_x [g cm^{-3}]	Ref.
CoSbS[a]	5.768	5.949	11.666	7.07	70C2
	5.840	5.958	11.673		59H, 70C2
	5.834	5.953	11.664		75H
[b]	5.842	5.951	11.666	6.97	75R
	5.8351	5.9600	11.6632		79K
CoPSe (r)	5.548	5.659	11.185		75H
	5.5475	5.6588	11.185	6.39	79K
CoAsSe (r)[b]	5.7285	5.7741	11.365	7.52	79K

[a] High-temperature modification (h)? [70C1].
[b] Ordered.

Interatomic distances (in Å):

CoSbS:	Co−3 Sb	2.521,	2.525,	2.551	
	3 S	2.291,	2.306	(2×)	
	Sb − S	2.510			75R
CoAsSe (r):	Co−3 As	2.331,	2.343,	2.349	
	3 S	2.358,	2.375,	2.378	
	As − S	2.482			79K

Crystallographic data for the arsenopyrite-type compounds (RT values)
(space group C_{2h}^{5} – P2$_1$/c)

Compound	a [Å]	b [Å]	c [Å]	β [°]	d_x [g cm^{-3}]	Ref.
FePS	5.63	5.54	5.66	112.5	4.84	65H
	5.617	5.534	5.637	112.2		83L
FeAsS	5.828	5.720	5.792	113.20	6.09	61M
	5.749	5.724	5.812	113.20	6.15	65K
	5.750	5.702	5.787	112.3		83L
FeSbS	6.02	5.93	6.02	112.13	6.95	39B
FePSe	5.76	5.74	5.81	112.3	6.20	65H
	5.768	5.707	5.832	112.9		83L
FeAsSe	5.95	5.89	5.95	112.7	7.24	65H
	5.877	5.871	5.974	113.4		83L
FeSbSe	6.24	6.18	6.27	113.0	7.65	65H
FeAsTe	6.19	6.10	6.21	113.6	7.98	65H
FeSbTe	6.56	6.45	6.58	114.1	7.97	65H
	6.543	6.439	6.578	113.8		83L
RuPS	5.77	5.77	5.81	111.2	6.04	63H2
	5.762	5.766	5.819	111.1		83L

(continued)

Crystallographic data for the arsenopyrite-type compounds (continued)

Compound	a [Å]	b [Å]	c [Å]	β [°]	d_x [g cm^{-3}]	Ref.
RuAsS	5.95	5.92	6.02	113.1	7.08	63H2
	5.945	5.920	6.025	112.92		72S
	5.934	5.913	6.016	112.9		83L
RuSbS	6.18	6.14	6.19	111.7	7.76	63H2
	6.183	6.144	6.198	111.7		83L
RuPSe	5.97	5.92	5.98	111.8	7.14	63H2
	5.969	5.919	5.989	111.6		83L
RuAsSe	6.08	6.06	6.14	112.7	8.11	63H2
	6.089	6.067	6.165	112.5		83L
RuSbSe	6.351	6.299	6.396	113.39	8.54	80H
	6.358	6.307	6.404	113.3		83L
RuAsTe	6.36	6.30	6.40	113.6	8.58	63H2
	6.361	6.302	6.410	113.5		83L
RuSbTe	6.603	6.538	6.687	114.57	8.87	80H
	6.543	6.601	6.651	114.1		83L
RuBiSe						66B
Ru$_{0.5}$Os$_{0.5}$AsS	5.992	5.944	6.044	113.08	8.47	72S
OsPS	5.78	5.78	5.83	110.9	9.24	63H2
	5.771	5.779	5.846	110.8		83L
OsAsS	5.94	5.92	6.01	111.9	10.07	63H2
	5.964	5.925	6.006	112.03	10.03	72S
	5.933	5.919	6.017	112.1		83L
OsSbS	6.21	6.15	6.22	111.7	10.35	63H2
	6.210	6.151	6.227	111.7		83L
OsPSe	5.98	5.94	6.00	111.4	10.05	63H2
	5.965	5.938	6.001	111.3		83L
OsAsSe	6.10	6.09	6.19	112.6	10.77	63H2
	6.099	6.082	6.199	112.5		83L
OsSbSe	6.38	6.34	6.42	113.4	10.89	66B
	6.391	6.333	6.437	112.3		83L
OsBiSe	6.52	6.46	6.57	113.8	12.54	66B
OsAsTe	6.37	6.31	6.42	113.1	10.99	63H2
	6.358	6.313	6.415	113.1		83L
OsSbTe	6.57	6.62	6.66	113.1	10.96	66B

Interatomic distances in FeSbS (in Å) [39B]:

Fe – 3 Sb   2.547, 2.563, 2.598
    3 S   2.221, 2.242, 2.244
Sb – S   2.613

Crystallographic data for the PdPS-type representatives (RT values)
(space group D_{2h}^{14} – Pbcn)

Compound	a [Å]	b [Å]	c [Å]	d_x [g cm^{-3}]	Ref.
PdPS	13.305	5.678	5.693		71B
	13.305	5.6777	5.6932	5.23	74J
PdPSe	13.569	5.824	5.856	6.21	71B, 74J

Interatomic distances in PdPS (in Å) [74J]:

Pd – 2 P   2.289, 2.292
    2 S   2.356, 2.364
    1 Pd   3.198

(continued)

Physical property	Numerical value	Experimental conditions	Experimental method remarks	Ref.

Interatomic distances in PdPS (continued)

P — 2 Pd   2.289, 2.292
     1 P   2.206
     1 S   2.111
S — 2 Pd   2.356, 2.364
     1 P   2.111

9.15.1.6.1 Physical properties

FePS

energy gap:

$E_{g,th}$	0.25 eV		from $\log \varrho \propto E_g/2kT$ above 500 K; sintered sample	59H

far infrared absorption: for spectrum in the range $60 \cdots 550$ cm^{-1}, see [83L].

FeAsS

energy gap (in eV):

$E_{g,th}$	$0.3 \cdots 0.5$		from $\log \varrho \propto E_g/2kT$ measured on minerals and sintered samples	59H
	0.06		from resistivity measurements at $180 \cdots 470$ K on a mineral sample	62W

far infrared absorption: for spectrum in the range $60 \cdots 550$ cm^{-1}, see [83L].

decomposition temperature (in K):

T_{dec}	≈ 970 (?)			62W
	$< 860 \cdots 970$			65K

$FeAs_{1+x}S_{1-x}$: According to published analyses the composition of natural arsenopyrites varies from $FeAs_{1.2}S_{0.8}$ to $FeAs_{0.9}S_{1.1}$, 2/3 of the minerals being sulfur-rich, corresponding to a lower growth temperature. In pure Fe arsenopyrites the As to S ratio, however, appears to be closer to 1 [65K].

$Fe_{1-x}Co_xAsS$: monoclinic structure up to $x = 0.25$ [65K].

$Fe_{1-x}Ni_xAsS$: monoclinic structure up to $x = 0.15$ [65K].

FeSbS

peritectic temperature very low [51L].

FePSe

far infrared absorption: for spectrum in the range $60 \cdots 550$ cm^{-1}, see [83L].

FeAsSe

energy gap:

$E_{g,th}$	0.6 eV		from $\log \varrho \propto E_g/2kT$, $T = 600 \cdots 900$ K; sintered sample	59H

thermoelectric power:

S	$+30\ \mu V\ K^{-1}$	$T = 300$ K	measured on the sample of [59H]	80H

far infrared absorption: for spectrum in the range $50 \cdots 500$ cm^{-1}, see [83L].

FeSbTe

activation energy:

E_A	0.04 eV		from $\log \varrho \propto E_A/kT$ ($E_g = 0.08$ eV?)	80Y

Physical property	Numerical value	Experimental conditions	Experimental method remarks	Ref.
magnetic susceptibility (in 10^{-6} cm^3 mol^{-1}): diamagnetic				
χ_m	-24.23	$T=300$ K	the original data were 100 times larger (misprints?); χ in CGS-emu	80Y
	-23.26	77 K		
far infrared absorption: for spectrum in the range $50\cdots500$ cm^{-1}, see [83L].				
FeSb$_{1.2}$Te$_{0.8}$ (arsenopyrite structure)				
magnetic susceptibility (in 10^{-6} cm^3 mol^{-1}):				
χ_m	-52.4	$T=300$ K	the original data were 100 times larger (misprints?); χ in CGS-emu	80Y
	-54.5	77 K		
FeSb$_{0.8}$Te$_{1.2}$ (arsenopyrite structure)				
magnetic susceptibility (in 10^{-6} cm^3 mol^{-1}):				
χ_m	-12.5	$T=300$ K	the original data were 100 times larger (misprints?); χ in CGS-emu	80Y
	-12.1	77 K		
RuPS				
energy gap:				
E_g	>1.4 eV	RT	from diffuse reflectance	63H2
thermoelectric power:				
S	-1000 μV K^{-1}	$T=300$ K	measured on sintered samples	63H2
magnetic susceptibility (in 10^{-6} cm^3 mol^{-1}):				
χ_m	-35	$T=295$ K	χ in CGS-emu, powdered sample; Gouy method, $B<1$ T	63H2
	-21	77 K		
far infrared absorption: for spectrum in the range $60\cdots550$ cm^{-1}, see [83L].				
RuAsS				
energy gap:				
E_g	≈1.2 eV	RT	from diffuse reflectance	63H2
thermoelectric power:				
S	$+250$ μV K^{-1}	$T=300$ K	sintered sample	63H2
magnetic susceptibility (in 10^{-6} cm^3 mol^{-1}):				
χ_m	-56	RT	χ in CGS-emu, powdered sample; Gouy method, $B<1$ T	63H2
	-52	$T=77$ K		
far infrared absorption: for spectrum in the range $60\cdots550$ cm^{-1}, see [83L].				
RuSbS				
thermoelectric power:				
S	$+200$ μV K^{-1}	$T=300$ K	measured on a sintered sample	63H2
magnetic susceptibility:				
χ_m	$-71\cdot10^{-6}$ cm^3 mol^{-1}	RT	χ in CGS-emu, powdered sample; Gouy method, $B<1$ T	63H2
far infrared absorption: for spectrum in the range $50\cdots500$ cm^{-1}, see [83L].				

Physical property	Numerical value	Experimental conditions	Experimental method remarks	Ref.
RuPSe				
thermoelectric power:				
S	$\approx -1000\ \mu V\ K^{-1}$	$T = 300\ K$	sintered sample	63H2
far infrared absorption: for spectrum in the range $60\cdots550\ cm^{-1}$, see [83L].				
RuAsSe				
thermoelectric power:				
S	$-250\ \mu V\ K^{-1}$	$T = 300\ K$	sintered sample	63H2
magnetic susceptibility (in $10^{-6}\ cm^3\ mol^{-1}$):				
χ_m	-77	$T = 295\ K$	χ in CGS-emu, powdered sample;	63H2
	-78	$77\ K$	Gouy method, $B < 1\ T$	
far infrared absorption: for spectrum in the range $50\cdots500\ cm^{-1}$, see [83L].				
RuSbSe				
energy gap (in eV):				
E_g	0.35	$T = 0\ K$	from $\log \varrho \propto E_g/2kT$, $T = 400\cdots700\ K$	63H2
	≈ 0.9	RT	from diffuse reflectance	
thermoelectric power:				
S	$-200\ \mu V\ K^{-1}$	$T = 300\ K$	sintered sample	63H2
magnetic susceptibility (in $10^{-6}\ cm^3\ mol^{-1}$):				
χ_m	-86	$T = 295\ K$	χ in CGS-emu, powdered sample;	63H2
	-78	$77\ K$	Gouy method, $B < 1\ T$	
far infrared absorption: for spectrum in the range $50\cdots500\ cm^{-1}$, see [83L].				
RuAsTe				
thermoelectric power:				
S	$-180\ \mu V\ K^{-1}$	$T = 300\ K$	sintered sample	63H2
far infrared absorption: for spectrum in the range $50\cdots500\ cm^{-1}$, see [83L].				
RuSbTe				
energy gap:				
$E_{g,th}$	$0.5\ eV$		from $\log \varrho \propto E_g/2kT$, above $500\ K$	63H2
thermoelectric power:				
S	$-150\ \mu V\ K^{-1}$	$T = 300\ K$	sintered sample	63H2
far infrared absorption: for spectrum in the range $50\cdots500\ cm^{-1}$, see [83L].				
OsPS				
energy gap:				
E_g	$> 1.4\ eV$	RT	from diffuse reflectance	63H2
thermoelectric power:				
S	$\approx -1000\ \mu V\ K^{-1}$	$T = 300\ K$	sintered sample	63H2
far infrared absorption: for spectrum in the range $60\cdots550\ cm^{-1}$, see [83L].				
OsAsS				
energy gap:				
E_g	$\approx 1.3\ eV$	RT	from diffuse reflectance	63H2

Physical property	Numerical value	Experimental conditions	Experimental method remarks	Ref.
thermoelectric power:				
S	$\approx -1000\ \mu V\ K^{-1}$	$T = 300$ K	sintered sample	63H2
magnetic susceptibility (in $10^{-6}\ cm^3\ mol^{-1}$):				
χ_m	-70	$T = 295$ K	χ in CGS-emu, powdered sample; Gouy method, $B < 1$ T	63H2
	-55	77 K		

far infrared absorption: for spectrum in the range $60 \cdots 550\ cm^{-1}$, see [83L].

OsSbS
energy gap:

$E_{g,th}$	1.2 eV		from $\log \varrho \propto E_g/2kT$, above 700 K	63H2
thermoelectric power:				
S	$+150\ \mu V\ K^{-1}$	$T = 300$ K	sintered sample	63H2
magnetic susceptibility:				
χ_m	$-92 \cdot 10^{-6}\ cm^3\ mol^{-1}$	$T = 295$ K	χ in CGS-emu; powdered sample; Gouy method, $B < 1$ T	63H2

far infrared absorption: for spectrum in the range $50 \cdots 500\ cm^{-1}$, see [83L].

OsPSe
energy gap:

E_g	≈ 1.4 eV	RT	from diffuse reflectance	63H2
thermoelectric power:				
S	$\approx -1000\ \mu V\ K^{-1}$	$T = 300$ K	sintered sample	63H2

far infrared absorption: for spectrum in the range $60 \cdots 550\ cm^{-1}$, see [83L].

OsAsSe
thermoelectric power:

S	$+250\ \mu V\ K^{-1}$	$T = 300$ K	sintered sample	63H2
magnetic susceptibility:				
χ_m	$-81 \cdot 10^{-6}\ cm^3\ mol^{-1}$	$T = 295$ K	χ in CGS-emu, powdered sample; Gouy method, $B < 1$ T	63H2

far infrared absorption: for spectrum in the range $50 \cdots 500\ cm^{-1}$, see [83L].

OsSbSe
energy gap (in eV):

E_g	$< 0.05\ (?)$	RT	60 μm thick samples were completely opaque to light with wavelengths from 0.5 to 25 μm	66B
	$> 0.1\ (?)$		estimated from ϱ of sample No. 7	
resistivity (in Ω cm):				
ϱ	1.59	RT	sample No. 7, prepared at $T \approx 1250$ K, $p = 45$ kbar	66B
	376	$T = 77$ K		
	0.12	RT	sample No. 9, prepared in a similar way	
	1.18	$T = 77$ K		
thermoelectric power (in $\mu V\ K^{-1}$):				
S	$+137$	RT	sample No. 7	66B
	$+157$	RT	sample No. 9	

Physical property	Numerical value	Experimental conditions	Experimental method remarks	Ref.
Knoop microhardness:				
H_K	672 kg mm^{-2}	RT		66B
decomposition temperature:				
T_{dec}	<1370 K		OsSbSe was converted to the elements when heated above 1370 K under a pressure of about 40 kbar	66B
OsBiSe				
decomposition temperature:				
T_{dec}	<1270 K		OsBiSe decomposed when heated above 1270 K under a pressure of about 40 kbar	66B
OsAsTe				
thermoelectric power:				
S	$+250\ \mu V\ K^{-1}$	$T = 300$ K	sintered sample	63H2
magnetic susceptibility:				
χ_m	$-114 \cdot 10^{-6}$ cm^3 mol^{-1}	$T = 295$ K	χ in CGS-emu, powdered sample; Gouy method, $B < 1$ T	63H2

far infrared absorption: for spectrum in the range 50···500 cm^{-1}, see [83L].

OsSbTe				
energy gap:				
E_g	<0.05 eV (?)	RT	60 μm thick samples were completely opaque in the wavelength range 0.5 to 25 μm	66B
resistivity (in $10^{-3}\ \Omega$ cm):				
ϱ	82.2	RT	sample No. 7 } both samples	66B
	459	$T = 77$ K	sample No. 7 } sintered at	
	38	RT	sample No. 10 } 950···1050 K	
	228	$T = 77$ K	sample No. 10 } and 42···47 kbar	
thermoelectric power (in $\mu V\ K^{-1}$):				
S	+138	$T = 300$ K	sample No. 7	66B
	+188	300 K	sample No. 10	
Knoop microhardness:				
H_K	196 kg mm^{-2}	$T = 300$ K		66B
decomposition temperature:				
T_{dec}	<1370 K		OsSbTe decomposes when heated up to 1370 K under a pressure of ≈ 40 kbar	66B

CoPS
see first table in 9.15.1.6.0
$CoP_{1-x}S_{1+x}$ completely miscible, tetragonal cell up tp $x \approx 0.4$; decomposition at temperatures higher than 1120 K [74N].

Physical property	Numerical value	Experimental conditions	Experimental method remarks	Ref.

CoAsS
see first table in 9.15.1.6.0
low-temperature modification ($C_{2v}^5 - Pca2_1$): pronounced optical anisotropy [65G, 65K, 69B].

transition temperature:

T_{tr}	823 K		sharp endothermal peak on DTA curve	65K

Samples quenched from ≈ 1100 K are disordered ($T_h^6 - Pa3$) and optically isotropic [65G].

$CoAs_{1-x}S_{1+x}$: $x \leq 0.58$ at 823 K [69B].
$x \leq 1$ at 1100 K [71M, 74N, 76A].

energy gap for $x = 0.04$ ($CoAs_{0.96}S_{1.04}$):

$E_{g,th}$	0.18 eV		from $\log \varrho \propto E_g / 2kT$	76A

semiconductor → metal transition at $x \approx 0.06$ ($CoAs_{0.94}S_{1.06}$) [76A].

$CoAs_{1+x}S_{1-x}$: $x < 0.05$ (range of existence) [74M].
$Co_{1-x}Fe_xAsS$: cubic up to $x = 0.47$ [65K].
$Co_{1-x}Ni_xAsS$: completely miscible, lattice parameter linear in x [65K].

CoSbS (pararammelsbergite-type modification)
energy gap:

$E_{g,th}$	0.5 eV		from $\log \varrho \propto E_g / 2kT$, above 600 K	59H

thermoelectric power:

S	≈ -500 μV K^{-1}	$T = 300$ K	measured on sintered sample of [59H]	80H

peritectic temperature:

T_{perit}	1149 K		from DTA measurement	70C2, 51L

A 20 min treatment at $T = 1420$ K, $p = 60$ kbar completely transformed the $D_{2h}^{15} - Pbca$ structure to the cubic ullmannite type [75H].

CoPSe(r) (pararammelbergite-type modification)
A 30 min treatment at $T = 1770$ K, $p = 60$ kbar completely transformed the orthorhombic $D_{2h}^{15} - Pbca$ structure to the disordered pyrite structure [75H].

CoAsSe(r) (pararammelsbergite-type modification)
energy gap:

$E_{g,th}$	0.2 eV		from $\log \varrho \propto E_g / 2kT$ up to 670 K (refers possibly to the marcasite-type modification)	59H

transition temperatures (in K):

T_{tr}	753			79K
	$\approx 753 - 8100\,x$	$x \leq 0.02$	for Se rich compositions $CoAs_{1-x}Se_{1+x}$	
	$\approx 753 - 8760\,y$	$y \leq 0.02$	for As rich compositions $CoAs_{1+y}Se_{1-y}$	

Physical property	Numerical value	Experimental conditions	Experimental method remarks	Ref.
CoAsSe (marcasite-type modification)				
thermoelectric power:				
S	$-50\ \mu V\ K^{-1}$	$T=300$ K	on sintered material (quenched ?)	80H

A 30 min treatment at $T=1870$ K, $p=60$ kbar partially transformed the marcasite phase to the cubic pyrite derivative [75H].

CoSbSe				
thermoelectric power:				
S	$-25\ \mu V\ K^{-1}$	$T=300$ K	on sintered sample	80H

A 30 min treatment at $T=1870$ K, $p=60$ kbar did not transform the marcasite structure [75H].

CoSbTe				
peritectic temperature:				
T_{perit}	≈ 1200 K			78A
microhardness:				
H	≈ 500 kg mm^{-2}		monotonous increase from CoTe$_2$ (300 kg mm^{-2}) up to CoSb$_{1.5}$Te$_{0.5}$; hardness not specified	78A
PdPS				
energy gap (in eV):				
E_g	1.4		given as $E_A=0.7$ eV, from $\log \varrho \propto E_A/kT$, $T=300\cdots450$ K, on single crystals	71B
	1.38	$T=298$ K	from optical transmission on a crystal platelet	
resistivity (in $10^6\ \Omega$ cm):				
ϱ	90	$T=298$ K	from four-probe resistivity measurements on single crystals	71B
	0.03	425 K		
refractive index:				
n	$3.1\cdots3.6$	$\lambda=0.2\cdots2\ \mu m$	from periodic variation of transmission with wavelength	71B

magnetic susceptibility: diamagnetic [71B].

The silvery blade-like crystals of PdPS are stable to about 1070 K, when heated in an argon atmosphere. PdPS is stable also at pressures up to at least 65 kbar. It has no detectable range of homogeneity [71B].

PdPSe				
activation energy:				
E_A	0.15 eV		from $\log \varrho \propto E_A/kT$, $T=300\cdots400$ K on single crystals ($E_g>0.3$ eV ?)	71B
resistivity (in Ω cm):				
ϱ	$4\cdot10^3$	$T=60$ K	from four-probe resistivity measurements on single crystals	71B
	30	300 K		
	1	400 K		

PdPS$_{1-x}$Se$_x$
complete miscibility $x=0\cdots1$ [71B].

9.15.1.6.2 References for 9.15.1.6

39B	Buerger, M.J.: Z. Kristallogr. Mineral. **101** (1939) 290.
51L	Lange, W., Schlegel, H.: Z. Metallkde. **42** (1951) 257.
59H	Hulliger, F.: Helv. Phys. Acta **32** (1959) 615.
61M	Morimoto, N., Clark, L.A.: Amer. Mineral. **46** (1961) 1448.
62H	Hulliger, F.: Helv. Phys. Acta **35** (1962) 535.
62W	Wintenberger, M.: Bull. Soc. Fr. Minéral. Cristallogr. **85** (1962) 107.
63H1	Hulliger, F.: Nature (London) **198** (1963) 382.
63H2	Hulliger, F.: Nature (London) **201** (1963) 381.
63H3	Hulliger, F.: Phys. Lett. **5** (1963) 226.
65G	Giese, Jr., R.F., Kerr, P.F.: Amer. Mineral. **50** (1965) 1002.
65H	Hahn, H., Klingen, W.: Naturwiss. **52** (1965) 494.
65J	Johnston, W.D., Miller, R.C., Damon, D.H.: J. Less-Common Met. **8** (1965) 272.
65K	Klemm, D.D.: Neues Jahrb. Mineral. Abhandl., **103** (1965) 205.
66B	Banus, M.D., Lavine, M.C.: Mat. Res. Bull. **1** (1966) 3.
69B	Bayliss, P.: Amer. Mineral. **54** (1969) 426.
70C1	Cabri, L.J., Harris, D.C., Stewart, J.M.: Amer. Mineral. **55** (1970) 10.
70C2	Cabri, L.J., Harris, D.C., Stewart, J.M.: Can. Mineralogist **10** (1970) 232.
71B	Bither, T.A., Donohue, P.C., Young, H.S.: J. Solid State Chem. **3** (1971) 300.
71M	Mikkelsen, J.C., Wold, A.: J. Solid State Chem. **3** (1971) 39.
72S	Snetsinger, K.G.: Amer. Mineral. **57** (1972) 1029.
73E	Entner, P., Parthé, E.: Acta Crystallogr. **B29** (1973) 1557.
74J	Jeitschko, W.: Acta Crystallogr. **B30** (1974) 2565.
74M	Maurel, C., Picot, P.: Bull. Soc. Fr. Minéral. Cristallogr. **97** (1974) 251.
74N	Nahigian, H., Steger, J., Arnott, R.J., Wold, A.: J. Phys. Chem. Solids **35** (1974) 1349.
75H	Henry, R., Steger, J., Nahigian, H., Wold, A.: Inorg. Chem **14** (1975) 2915.
75R	Rowland, J.F., Gabe, E.J., Hall, S.R.: Can. Mineralogist **13** (1975) 188.
76A	Adachi, K., Togawa, E., Kimura, F.: J. Phys. (Paris) **37** (1976) C4-29.
76S	Scott, J.D., Nowacki, W.: Can. Mineralogist **14** (1976) 561.
76Y	Yamaguchi, G., Shimada, M., Koizumi, M.: J. Solid State Chem. **19** (1976) 63.
78A	Abrikosov, N.Kh., Petrova, L.I.: Izv. Akad. Nauk SSSR Neorg. Mater. **14** (1978) 457 (translation: Inorg. Mater. **14** (1978) 346).
79K	Kjekshus, A., Rakke, T.: Acta Chem. Scand. **A33** (1979) 609.
80H	Hulliger, F.: unpublished **1980**.
80Y	Yamaguchi, G., Shimada, M., Koizumi, M., Kanamaru, F.: J. Solid State Chem. **34** (1980) 241.
82B	Bayliss, P.: Amer. Mineral. **67** (1982) 1048.
83L	Lutz, H.D., Schneider, G., Kliche, G.: Phys. Chem. Minerals **9** (1983) 109.

9.15.2 Binary transition-metal oxides

9.15.2.0 Introduction

9.15.2.0.1 Considerations unique to transition-metal compounds

Three types of outer (outside closed atomic shells) electrons may be distinguished in solids: localized, itinerant with strong correlations, and itinerant with weak correlations. The former are adequately described by crystal-field theory supplemented, where interatomic interactions between like atoms are present, by magnetic superexchange or double-exchange theories. The latter are well described by conventional band theory. An adequate description of the intermediate case is more difficult; in general it requires the introduction of both strong correlations and strong electron-phonon coupling into the first-order band theory.

Conventional semiconductors contain only broad bands that are well described by normal band theory. Transition-metal compounds contain, in addition, narrow d bands; and the d electrons may be localized, itinerant with strong correlations, or itinerant with weak correlations. The physical properties they impart to the solid vary with both the population and character of the d electrons present. Therefore interpretation of the physical properties of transition-metal compounds depends upon two essential features: (1) placement of the d-state energies relative to the edges of the broad bands due to outer s and p electrons and (2) the character of the d electrons.

9.15.2.0.2 Placement of d bands and formal valence

The binary oxides and chalcogenides contain a broad bonding band of primarily anion-p character separated by an energy gap from a broad antibonding band of primarily cation-s character. In oxides, the gap is large and the Fermi energy generally lies within it. Therefore we may use formal valence states, assuming O^{2-} ions, to obtain a count of the number of d electrons per transition-metal atom. Compounds may contain the same transition-metal atom in either a single or a mixed valence state, the latter occurring where the number of d electrons per transition-metal atom is non integral. In general, only two different valence states are present simultaneously, but exceptional cases are known where three quite different crystallographic sites stabilize three different valence states of the same atom in one structure. On the other hand, if two different valence states occupy crystallographically equivalent sites, valence transfer may be so rapid that available measuring techniques are unable to realise two distinguishable ions; in this case it is more meaningful to assign each atom an average, non integral valence state. Such is certainly the case in the itinerant-electron limit.

Although the situation is similar for the binary transition-metal chalcogenides of lower valence state, caution must be exercized in the interpretation of apparent higher valence states. Attempts to create higher valence states on the transition-metal ions of a chalcogenide may introduce holes into the broad, primarily anion-p bands instead of into the d states. These broad-band holes either render the compound metallic or become trapped in polyanion antibonding orbitals as in an $(S_2)^{2-}$ unit. Polyanion formation can generally be detected from the structure; it is classically illustrated in the pyrite and marcasite structures of FeS_2.

The location of the Fermi energy relative to the edge of the broad anion-p bands clearly reflects the position of the d bands relative to this band edge. The d bands are more stable the higher the formal valence state on the transition-metal atom and, for a given valence state, the further to the right in any long series.

9.15.2.0.3 Character of d electrons

The character of an atomic outer electron, once the atom has been incorporated into a solid, depends upon the relative magnitudes of three competitive energies: the bandwidth W_b arising from interatomic interactions between like atoms on equivalent lattice sites in the crystal, the correlation energy U arising from intraatomic electrostatic interations between electrons of the same manifold, and a stabilization of electronic energies Δ_{el} arising from electron-phonon interactions that produce a local or a cooperative distortion of the structure.

1. Bandwidth: Interatomic interactions between N like atoms on a periodic array broaden the energy of an atomic state into a band of energies containing $2N/v$ states, where the factor two is due to the spin degeneracy and v is the number of like atoms per primitive unit cell of the crystal structure. In the

9.15.2.0 Introduction

tight-binding approximation, the width of the band of $2N$ energies in a Bravais lattice ($v=1$) is given by

$$W_b \approx 2zb$$

where b is the transfer (resonance) integral for electron tunneling between like atoms on nearest-neighbour equivalent sites:

$$b_{ij} \equiv (\psi_i, \mathcal{H}' \psi_j) \simeq \varepsilon_{ij}(\psi_i, \psi_j).$$

$\mathcal{H}'$ is the perturbation of the potential at R_j due to an atom at R_i, ε_{ij} is a one-electron energy, and (ψ_i, ψ_j) is the wave function-overlap integral for degenerate, localized states at R_i and R_j. The wave functions ψ_i and ψ_j contain covalent admixtures of anion s and p wave functions, which makes M−X−M interactions competitive with M−M interactions.

2. Electrostatic correlation energies: The magnitude of the intraatomic electron-electron electrostatic energies is sensitive to the radial extension of the wave functions. On the free atom it is responsible for differences U in the successive ionization potentials associated with electrons from the same shell. For outer s and p electrons, the free-atom U is generally less than about 2 eV; for 4f electrons it is generally greater than 10 eV. The d electrons at free atoms have intermediate energies U. In a solid, covalent mixing with nearest-neighbour anions extends the wave functions, reducing U. Moreover, covalent mixing raises the d-state energies; therefore, since covalency increases with the formal valence on the cation, the energy differences between successive valence states is further decreased in a solid. Finally, interactions between like atoms further decrease U through a screening parameter $\xi(b)$ that increases with b: for example

$$U = (|\psi_m(1)|^2, V|\psi_m(2)|^2),$$
$$V = (e^2/r_{12}) \exp(-\xi r_{12})$$

for two electrons occupying the same atomic state with wave function ψ_m. In the absence of a bandwidth W_b, the energy U splits the energies of different formal valence states as indicated schematically in Fig. 1 for a single valence state corresponding to the configuration d^n. If this splitting is larger than the bandwidth ($U > W_b$), then a compound with a single formal valence state may be semiconducting whereas it is metallic if the bandwidth is larger ($U < W_b$). Atomically localized electrons have $W_b \ll U$, itinerant electrons have $W_b \gg U$, and "intermediate" electrons have $W_b \approx U$. Localized and intermediate electrons may impart a spontaneous atomic moment; itinerant electrons do not. Unless U is augmented by crystal-field splittings or separates a high-spin d^5 from a d^6 configuration, its magnitude for d electrons in solids appears to be not greater than about 3 eV, which is why it is possible to stabilize several valence states with the same anion. For example, VO, V_2O_3, VO_2, and V_2O_5 all have a Fermi energy between the broad V:4s and O^{2-}:2p^6 bands, which are not separated by more than about 6 eV. Thus an average correlation splitting $\bar{U} \lesssim 2$ eV is indicated for the vanadium oxides, and the intermediate condition $W_b \approx U$ is indeed encountered in V_2O_3 and VO_2.

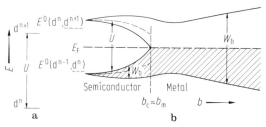

Fig. 1. a) Electrostatic energy U in a d-manifold, b) The effect of increasing b and hence the bandwidth W_b on the energies $E(d^n)$ and $E(d^{n+1})$. At a critical value of b the electron will become itinerant.

3. Crystallographic phase changes and semiconductor-metal transitions: The electron-phonon interactions, local or cooperative, that introduce an electronic stabilization Δ_{el} of occupied relative to empty states may either increase an existing energy gap between valence and conduction bands or it may introduce one. The former case represents a semiconductor-semiconductor transition; the latter a semiconductor-metal transition. In general, a semiconductor-metal transition is caused by either the edge crossing of filled and empty bands (or level and a band) or the collapse of a band splitting. It is instructive for interpretation of the data in the following tables to consider the variety of ways that Δ_{el} can influence the relative band (or level) energies.

9.15.2.0 Introduction

a) Mixed valence. The spinels hausmannite (Mn_3O_4) and magnetite (Fe_3O_4) represent two different classes of mixed-valent compounds. Hausmannite, $Mn^{2+}[Mn_2^{3+}]O_4$, is a normal spinel with Mn^{2+} ions in tetrahedral sites and Mn^{3+} ions in octahedral sites. The d electrons are localized ($W_b < U$), and the octahedral-site Mn^{2+} level lies discretely above the tetrahedral-site Mn^{2+} level as well as the octahedral-site Mn^{3+} level. Since valence transfer between ions on different sites is thus inhibited, the compound is an insulator. Magnetite, on the other hand, is an inverse spinel, $Fe^{3+}[Fe^{2+}Fe^{3+}]O_4$; and in the cubic phase, stable at room temperature, the mixed valence occurs on crystallographically equivalent sites. Under these conditions, valence exchange may readily occur via electron tunneling from one equivalent site to its neighbour, and Fe_3O_4 is a good conductor at room temperature. However, the tunneling time $\tau \approx h/zb$ may be long compared to a local optical-mode vibration period $\tau_R \approx 10^{-12} \cdots 10^{-13}$ s. In such a case ($\tau > \tau_R$), the electron (or hole) becomes trapped in a local deformation, and thermal energies must then overcome an activation-energy barrier ε_a before valence exchange via tunneling can occur. An electron (or hole) "dressed" in its local deformation is called a **small polaron,** and the mobility of a small polaron is described by diffusion theory as

$$\mu = (e D_0/kT) \exp(-\varepsilon_a/kT)$$

rather than by the scattering theory used for itinerant electrons, which gives $\mu = e \tau_s/m^*$ with τ_s the mean free time between collisions and m^* the effective mass of the charge carrier. An activated mobility makes the temperature dependence of the conductivity, $\sigma = ne\mu$, typical of a semiconductor even though the density n of charge carriers does not change with temperature. As the concentration of small polarons increases, interactions between them tend not only to reduce ε_a, but also to introduce a polaron ordering that creates distinguishable lattice positions for the two valence states. Long-range order is accompanied by a cooperative electron-phonon interaction, which replaces local and random deformations. In the case of Fe_3O_4, where the polaron concentration is a maximum, a complex long-range valency ordering occurs below the Verwey temperature $T_V = 119$ K to render the compound semiconducting. At temperatures just above T_V, strong electron-phonon interactions are evident and the conductivity σ continues to increase with increasing temperature as if the abrupt transition at T_V were a semiconductor-semiconductor transition. However, the temperature dependence of σ decreases with increasing temperature, becoming essentially temperature-independent by room temperature. These data are consistent with short-range ordering between small polarons persisting above T_V, but with ε_a vanishing as a result of polaron-polaron interactions in the totally disordered state.

A non integral number of itinerant electrons per transition-metal atom may give rise to the formation of charge-density waves, see Fig. 2 and associated discussion. Some argue for this alternative in Fe_3O_4.

b) Single valence. In crystalline compounds containing transition-metal compounds in a single valence state, three situations need to be distinguished: $W_b \lesssim \Delta_{el} < U$, $W_b \approx U \lesssim \Delta_{el}$, and $U < W_b \lesssim \Delta_{el}$.

(i) Localized-electron manifolds ($W_b \ll U$): Semiconductor-semiconductor transitions may be induced by localized-electron stabilization. Important examples include:

Low-spin to high-spin transitions are induced by the greater entropy associated with the higher multiplicity of the high-spin configurations. Since such a transition requires a transfer of antibonding electrons from π-bonding to σ-bonding states, the effective radius of the high-spin configuration is larger, causing an enhanced thermal expansion of the lattice. Such a transition may be smooth with an order parameter defined as the fraction of configurations that are low-spin.

Cooperative Jahn-Teller distortions remove an accidental orbital degeneracy. Although the enthalpy of the localized-electron manifold is increased linearly with the atomic displacements of the distortion, the sign of the distortion is only determined – as are the elastic restoring forces – to order square of these displacements. Nevertheless, a cooperative, static distortion is commonly stabilized below a finite temperature T_{tr}. If the orbital angular momentum of the manifold is not quenched in the high-temperature phase, the internal magnetic fields generated by a collinear-spin configuration in a magnetically ordered state will stabilize a distortion that preserves the spin-orbit coupling unless T_{tr} occurs above the magnetic-ordering temperature. Such distortions occurring below the antiferromagnetic Néel temperature T_N are illustrated by $Fe_{1-\delta}O$ and CoO. In Mn_3O_4, the high-spin $^5E_g(d^4)$ configuration at octahedral-site Mn^{3+} ions has its orbital angular momentum quenched by the crystalline fields, and T_{tr} for the tetragonal ($c/a > 1$) to cubic transition is quite independent of the ferrimagnetic Curie temperature T_C.

Exchange striction below a magnetic-ordering temperature reflects changes in the interatomic bonding between magnetic atoms as a result of the long-range magnetic order. The distortion of MnO from cubic to rhombohedral symmetry below T_N provides a classic illustration.

9.15.2.0 Introduction

(ii) "Intermediate" electrons ($W_b \approx U$): Semiconductor-metal transitions may be induced by a change from strong to weak correlations with increasing temperature. Such a transition is also marked by a change from a magnetically ordered state to a phase containing no spontaneously magnetic atoms. These transitions may be understood from Fig. 1 to represent a change from $W_b < U$ to $W_b > U$. Such a transition is first-order, the phase containing strongly correlated itinerant electrons ($W_b < U$) having the larger volume. Such a transition is illustrated by metastable NiS with B8 structure, which is stabilized at low temperatures by quenching. It is also illustrated in a more complex way by V_2O_3.

(iii) Itinerant electrons with weak correlations ($W_b > U$): Semiconductor-metal transitions may also be induced by other mechanisms. Among these are:

Band-edge crossing induced by axial-ratio changes: Ti_2O_3 exhibits a smooth semiconductor-metal transition with increasing temperature that is correlated with a remarkable increase in the c/a ratio. At low temperatures, electrons are trapped in molecular orbitals of c-axis Ti–Ti pairs; the a_1 bands of t_{2g} parentage directed parallel to the c-axis are split in two by the Ti–Ti pairing along this axis in the corundum structure. The e_π bands formed by Ti–Ti interactions in the basal plane lie discretely above the lower, filled a_1 band if c/a is small. For larger c/a ratios, the two bands overlap.

Band splitting by cation clustering: If partially filled bands are formed by M–M interactions, a change in translational periodicity of the metal atom M may split the band into filled and empty bands. Such changes are generally commensurate with the original crystalline array, as in the formation of cationic clusters or a commensurate charge-density wave with wave vector $q = n\,a$; but incommensurate charge-density waves with $q = \lambda\,a$, where λ is non-integral, have also been reported. Fig. 2 illustrates cation clustering or a charge-density wave with $q = 2a$ for a one-dimensional band. (Fig. 2 also illustrates band splitting via a disproportionation reaction.) In a two-dimensional band, the energy gap opened up at the new Brillouin-zone boundaries by a charge-density wave may not be large enough to make the crystal semiconducting; in this case a semimetal-metal transition is observed instead of a semiconductor-metal transition. Several of the layered chalcogenides exhibit the formation of two-dimensional charge-density waves at low temperatures.

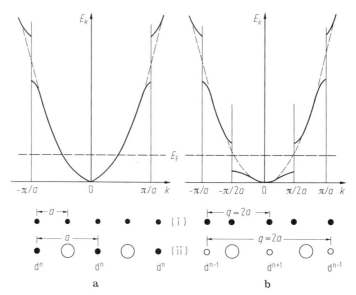

Fig. 2. Effect of cation pairing (i) or disproportionation (ii) on the band structure of a one-dimensional system (see text).

The B8 structure of FeS exhibits a triangular-cluster formation below the temperature T_α. This example is particularly interesting because the high-spin Fe^{2+} ions retain atomically localized majority-spin electrons at all temperatures; the single minority-spin electron per high-spin Fe^{2+} ion is itinerant for $T > T_\alpha$, but localized to triangular Fe_3 cluster below T_α. This observation emphasizes the fact that with magnetic atoms the majority-spin and minority-spin electrons see different potential energies. This difference is particularly important for high-spin d^6 ions in which a large intraatomic-exchange stabilization is present for the majority-spin electrons, but absent for the minority-spin electron.

The presence of a magnetic moment on the cation also makes the transfer integrals b_{ij} no longer spin-independent, which is the basis of the exchange striction mentioned above. The spin dependence of the transfer integrals also introduces a spin-disorder scattering that is responsible for the maximum in the resistivity at T_N in the "metallic" phase of FeS; in CrO_2 it produces a sharp cusp in resistivity versus temperature at the ferromagnetic Curie temperature T_C. Where there are mixed valencies, but with a small concentration of charge carriers, this effect may produce magnetic polarons just above T_C that give rise to a giant magnetoresistance.

On lowering the temperature of FeS, the orientation of the magnetic moments changes from within the basal planes to parallel to the c-axis at a spin-flip temperature $T_s \gtrsim T_\alpha$. With spins parallel to the c-axis, spin-orbit coupling orders the minority-spin electrons into the orbitally twofold degenerate $d_\perp$ (or e_π) minority-spin bands of t_{2g} parentage within the basal planes; these bands are then half-filled because the spin degeneracy is removed. Triangular-cluster formation below T_α splits these bands in two to give a semiconductor-metal transition.

In FeS, spin reorientation and spin-orbit coupling removed a band overlap to make possible the formation of a semiconducting phase by a cluster formation (a commensurate charge-density wave) that splits a narrow, partially filled d band. In VO_2, ferroelectric-type displacements perform the same function. A cation displacement from the center of symmetry of an octahedral interstice destabilizes two t_{2g} orbitals relative to the third if it is towards a terminal anion; it destabilizes all t_{2g} orbitals equally if toward an octahedral face. If the d shell is empty, as at a Ti^{4+} ion, displacements may occur to a terminal oxygen, an octahedral edge, or an octahedral face, as is illustrated by the three ferroelectric phases of $BaTiO_3$. Such displacements are common where the gain in energy of the binding states is larger than the elastic restoring force (both vary as the square of the atomic displacements, and static distortions are anticipated by a "softening" of specific optical-vibration modes). For the V^{4+} ion, with one outer d electron, displacement is toward a terminal oxygen to form a vanadyl complex ion $(VO)^{2+}$. If z is parallel to the axis of the complex, the single d electron is stabilized in the d_{xy} orbital. Above $T_{tr} \approx 67\,°C$, VO_2 has the tetragonal rutile structure; $V-V$ interactions form a $d_\parallel$ band within chains of edge-shared octahedra aligned parallel to the c-axis and $V-O-V$ interactions between chains create an overlapping π^* band, so the metallic conductivity is nearly isotropic. Below T_{tr}, V^{4+}-ion displacements produce a semiconducting, monoclinic phase. The displacements have two components: a $\langle 110 \rangle$ displacement within the basal planes toward a terminal oxygen, which stabilizes the $d_\parallel$ band relative to the π^* bands, and a $V-V$ pairing within the c-axis chains to split the now half-filled $d_\parallel$ band in two. Small substitutions of Cr for V reveal the two-component character of the displacements; for a limited Cr concentration half the c-axis chains show only $M-M$ pairing and half show only the zig-zag antiferroelectric displacements due to vanadyl-complex formation. Where pairing occurs, the V^{4+} ions are non-magnetic; where only vanadyl formation occurs, the V^{4+} ions carry a magnetic moment. Thus VO_2 represents a situation where $U \approx W_b \lesssim \Delta_{el}$.

Although $W_b > U$ clearly holds in the Magnéli shear phase Ti_4O_7, the situation is analogous. In this structure, TiO_2 slabs $4c$ thick are connected by "shear" planes across which Ti atoms of one slab are in interstitial positions relative to those in the other slab. Ferroelectric-type displacements allow for a strong relaxation that stabilizes condensation of O^{2-}-ion vacancies into the shear planes; but these displacements destabilize the d states at Ti atoms of a slab face relative to those in the bulk of the slabs. Thus within the bulk of the slabs the situation is analogous to VO_2, and at low temperatures Ti–Ti pairing within c-axis chains of the slab traps out the electrons to create a semiconducting phase as in low-temperature VO_2. However, just above T_{tr} in this case the Ti–Ti pairing becomes disordered and mobile. Such mobile homopolar bands have been named **bipolarons** as they represent two-electron species localized and dressed by a local lattice deformation; they migrate as a small polaron with an activated mobility.

9.15.2.1 Titanium oxides

9.15.2.1.0 Introduction

The titanium-oxygen phase diagram (Fig. 1) contains one commercially important semiconducting phase, TiO_2 [66W2, 67G]. The Magnéli phases Ti_nO_{2n-1} ($4 < n < 28$) as well as Ti_3O_5 and the sesquioxide Ti_2O_3 exhibit semiconductor-metal transitions with increasing temperature; these transitions have been of considerable theoretical interest. TiO (high- and low-temperature phases), Ti_2O and α-Ti containing interstitial oxygen are metallic at all temperatures.

9.15.2.1.1 Titanium oxide (TiO_2)

A) Crystal structure

TiO_2 crystallizes in three polymorphic forms: anatase, brookite and rutile. Anatase and brookite transform to rutile if heated to 700 °C and 900 °C, respectively.

| Physical property | Numerical value | Experimental conditions | Experimental method, remarks | Ref. |

thermodynamic parameters of phase transformation (ΔH_{tr}, ΔG_{tr} in kcal mol^{-1}):
anatase-rutile:

ΔH_{tr}°	+2.79			66V
	−0.78 (9)			79M2
	−1.57 (19)			67N
	−0.10 (5)			61R
	+0.1			68R
ΔG_{tr}°	−0.66			79M2

brookite-rutile

ΔH_{tr}°	−0.17 (9)			79M2

Each phase consists of TiO$_6$ octahedra sharing four edges (anatase), three edges and a corner (brookite) or two edges and two corners (rutile).

a) Anatase

structure: tetragonal, space group D_{4h}^{19}–I4$_1$/amd, $Z=4$ [70R, 72H2]. The structure may be derived from the B1 structure of NaCl by ordering Ti^{4+} ions and Ti-vacancies as shown in Fig. 2; the bridging oxygens are displaced towards one another to produce shorter O−O distances.

lattice parameters (in Å) at various temperatures:

		T [°C]		
a	3.7845	28		70R
c	9.5143			
a	3.7855	84		
c	9.5185			
a	3.7866	161		
c	9.5248			
a	3.7875	210		
c	9.5294			
a	3.7884	258		
c	9.5342			
a	3.7894	306		
c	9.5374			
a	3.7907	354		
c	9.5432			
a	3.7923	449		
c	9.5548			
a	3.7939	497		
c	9.5595			
a	3.7948	534		
c	9.5669			
a	3.7962	571		
c	9.5754			
a	3.7970	608		
c	9.5794			
a	3.7989	645		
c	9.5872			
a	3.7998	679		
c	9.5933			
a	3.8009	712		
c	9.5975			

density:

d_{calc}	3.894 g cm^{-3}	$T=298$ K		70R, 72H2

9.15.2.1.1 Titanium oxide (TiO$_2$)

Physical property	Numerical value	Experimental conditions	Experimental method remarks	Ref.

interatomic distances (in Å) and angles (in °) (from [72H2]):
(see also Fig. 3)

	$T=25$ °C	$T=300$ °C	$T=600$ °C	$T=900$ °C	
d(Ti–O)	1.9338 (5)	1.9365 (5)	1.9410 (5)	1.9452 (7)	(s) shared edge
d(Ti–O∥[001])	1.9797 (23)	1.9831 (22)	1.9909 (23)	1.9957 (31)	
d(O–O) (s)	2.4658 (29)	2.4681 (29)	2.4745 (30)	2.4777 (39)	
d(O–O)	2.7924 (13)	2.7968 (13)	2.8037 (14)	2.8108 (18)	
≮O–Ti–O	101.90 (7)	101.96 (7)	102.01 (7)	102.10 (9)	

coefficient of linear thermal expansion (in K^{-1}):

$\alpha_\parallel$	$7.380\cdot 10^{-6} + 6.620\cdot 10^{-9}(T-273) + 1.771\cdot 10^{-11}(T-273)^2$	see Fig. 4	70R
$\alpha_\perp$	$3.533\cdot 10^{-6} + 5.610\cdot 10^{-9}(T-273) + 4.315\cdot 10^{-12}(T-273)^2$		

b) Rutile

structure: tetragonal, space group D_{4h}^{14}–P4$_2$/mnm, $Z=2$ [79M1]. The structure may be derived from the hexagonal B8 structure of NiAs by ordering Ti^{4+} ions into alternate sites of the c-axis chains of face-shared octahedra. Within a hexagonal plane, the Ti^{4+} ions are ordered into alternate [100] strings. Reduction of the bridging O–O band produces, in this case with $c_{hex}=\sqrt{3}\,a$, a tetragonal unit cell having as c the Ti–Ti distance in the strings of edge-shared TiO$_6$ octahedra (Fig. 5). Each oxygen is coplanar with three nearest-neighbour Ti^{4+} ions and has a p$_\pi$ orbital directed perpendicular to this plane toward a cation vacancy on one side, a pair of cation vacancies on the other.

lattice parameters (in Å):

		T [°C]	
a	4.5941	30	70R
c	2.9589		
a	4.5963	84	
c	2.9601		
a	4.5981	161	
c	2.9619		
a	4.6007	210	
c	2.9639		
a	4.6025	258	
c	2.9654		
a	4.6035	306	
c	2.9668		
a	4.6059	354	
c	2.9684		
a	4.6074	401	
c	2.9699		
a	4.6091	449	
c	2.9710		
a	4.6105	497	
c	2.9726		
a	4.6123	534	
c	2.9741		
a	4.6141	571	
c	2.9757		
a	4.6163	608	
c	2.9772		
a	4.6176	645	
c	2.9788		

Figs. p. 414f.   9.15.2.1.1 Titanium oxide (TiO$_2$)   [Ref. p. 148

Physical property	Numerical value	Experimental conditions	Experimental method remarks	Ref.

density:

d_{calc}	4.250 g cm^{-3}	$T = 298$ K		70R

interatomic distances (in Å) and angles (in °) (from [79M1]):

	$T = 25$ °C	$T = 300$ °C	$T = 600$ °C	$T = 900$ °C	
$d(Ti-O_{Ia})$ (4)	1.947 (2)	1.954 (2)	1.961 (2)	1.962 (2)	Number behind d in parentheses:
$d(Ti-O_{Ib})$ (2)	1.982 (3)	1.983 (3)	1.988 (3)	1.995 (3)	multiplicity of interatomic
$d(Ti-O)_{av}$	1.959	1.964	1.970	1.973	distance
$d(O_{Ia}-O_{Ia})$ (s) (2)	2.532 (5)	2.544 (6)	2.552 (6)	2.547 (5)	(s) shared edge,
$d(O_{Ib}-O_{Ia})$ (8)	2.778 (2)	2.784 (2)	2.792 (2)	2.799 (2)	(u) unshared edge
$d(O_{Ia}-O_{Ia})$ (u) (2)	2.959 (2)	2.966 (2)	2.977 (2)	2.986 (2)	
$d(O-O)_{av}$	2.767	2.774	2.783	2.788	
$d(Ti-Ti)$	2.959 (2)	2.966 (2)	2.977 (2)	2.986 (2)	
$\sphericalangle(O_{Ia}-Ti-O_{Ia})$ (s)	81.12 (12)	81.23 (14)	81.22 (14)	80.93 (12)	
$\sphericalangle(Ti-O_{Ia}-Ti)$	130.56 (6)	130.62 (7)	130.61 (7)	130.47 (6)	

coefficient of linear thermal expansion (in K^{-1}):

$\alpha_\parallel$	$8.816 \cdot 10^{-6} + 3.653 \cdot 10^{-9}(T-273) + 6.329 \cdot 10^{-12}(T-273)^2$	see Fig. 4	70R
$\alpha_\perp$	$7.249 \cdot 10^{-6} + 2.198 \cdot 10^{-9}(T-273) + 1.298 \cdot 10^{-12}(T-273)^2$		

c) Brookite

structure: orthorhombic, space group D_{2h}^{15}–Pbca, $Z = 2$ [79M1]. The bridging oxygens again have shorter O–O bonds (Fig. 6). Antiferroelectric displacements of the Ti^{4+} ions parallel to the b-axis (opposite to the arrows in the figure) produce three shorter Ti–Ti bonds. At RT, the Ti–Ti distances in zig-zag chains parallel to the c-axis (3.062 (2) Å) are greater than the Ti–Ti distance (2.951 (2) Å) across the third shared edge [79M2] of shared octahedral-site edges but the octahedral volume also decreases (Fig. 7).

lattice parameters (in Å):

		T [°C]	
a	9.174 (2)	25	79M1
b	5.449 (2)		
c	5.138 (2)		
a	9.175 (4)	280	
b	5.459 (4)		
c	5.149 (2)		
a	9.191 (4)	425	
b	5.463 (4)		
c	5.157 (4)		
a	9.211 (4)	625	
b	5.472 (4)		
c	5.171 (4)		

density:

d_{calc}	4.133 g cm^{-3}	$T = 25$ °C	79M1

interatomic distances (in Å) and angles (in °) (from [79M1]):

	$T = 25$ °C	$T = 280$ °C	$T = 425$ °C	$T = 625$ °C
$d(Ti-O_{Ia})$	1.990 (3)	1.989 (3)	1.985 (3)	1.991 (5)
$d(Ti-O_{Ib})$	1.863 (3)	1.870 (4)	1.877 (4)	1.881 (5)
$d(Ti-O_{Ic})$	1.999 (3)	2.000 (3)	2.003 (3)	2.005 (5)
$d(Ti-O_{IIa})$	2.052 (3)	2.043 (3)	2.045 (3)	2.050 (5)
$d(Ti-O_{IIb})$	1.930 (3)	1.938 (3)	1.944 (4)	1.945 (5)
$d(Ti-O_{IIc})$	1.923 (3)	1.928 (3)	1.930 (3)	1.937 (4)
$d(Ti-O)_{av}$	1.960	1.961	1.964	1.968

(continued)

Physical property	Numerical value	Experimental conditions	Experimental method remarks	Ref.

interatomic distances and angles (continued)
(* angles opposite shared edges, ** shared edge)

	$T=25\,°C$	$T=280\,°C$	$T=425\,°C$	$T=625\,°C$
$d(O_{Ia}-O_{IIa})$	2.734 (4)	2.733 (5)	2.736 (5)	2.752 (7)
$d(O_{IIa}-O_{Ic})**$	2.534 (5)	2.533 (5)	2.536 (5)	2.548 (7)
$d(O_{Ic}-O_{IIb})$	2.869 (5)	2.874 (5)	2.876 (5)	2.872 (7)
$d(O_{IIb}-O_{Ib})$	2.799 (5)	2.800 (5)	2.809 (5)	2.806 (7)
$d(O_{Ib}-O_{IIc})$	2.993 (4)	3.000 (5)	3.003 (5)	3.016 (7)
$d(O_{IIc}-O_{IIb})$	2.746 (4)	2.751 (2)	2.753 (2)	2.756 (2)
$d(O_{IIb}-O_{IIc})$	2.860 (3)	2.870 (3)	2.873 (3)	2.879 (4)
$d(O_{IIc}-O_{IIa})$	2.982 (3)	2.984 (3)	2.989 (4)	2.996 (5)
$d(O_{Ia}-O_{Ic})$	2.794 (2)	2.801 (3)	2.805 (3)	2.814 (4)
$d(O_{Ic}-O_{Ib})**$	2.495 (6)	2.496 (6)	2.505 (6)	2.509 (8)
$d(O_{Ib}-O_{Ia})$	2.814 (2)	2.820 (2)	2.821 (2)	2.827 (3)
$d(O_{Ia}-O_{IIc})**$	2.534 (5)	2.533 (5)	2.536 (5)	2.548 (7)
$d(O-O)_{av}$	2.763	2.766	2.770	2.777
$d(Ti-Ti)$	2.951 (2)	2.960 (2)	2.965 (2)	2.976 (3)
$d(Ti-Ti)$	3.062 (2)	3.060 (2)	3.060 (2)	3.063 (2)
$\angle(O_{Ia}-Ti-O_{IIa})$	85.14 (13)	85.36 (14)	85.50 (14)	85.87 (19)
$\angle(O_{IIa}-Ti-O_{Ic})*$	77.43 (12)	77.57 (13)	77.58 (13)	77.88 (18)
$\angle(O_{Ic}-Ti-O_{IIb})$	93.76 (13)	93.71 (14)	93.52 (14)	93.27 (20)
$\angle(O_{IIb}-Ti-O_{Ib})$	95.07 (13)	94.60 (15)	94.61 (15)	94.30 (21)
$\angle(O_{Ib}-Ti-O_{IIc})$	104.49 (13)	104.33 (14)	104.14 (14)	104.33 (20)
$\angle(O_{IIc}-Ti-O_{Ia})*$	80.70 (12)	80.56 (13)	80.72 (14)	80.87 (19)
$\angle(O_{IIa}-Ti-O_{IIb})$	87.13 (8)	87.37 (9)	87.26 (9)	87.19 (12)
$\angle(O_{IIb}-Ti-O_{IIc})$	95.85 (7)	95.85 (8)	95.72 (9)	95.71 (12)
$\angle(O_{IIc}-Ti-O_{IIa})$	97.20 (13)	97.38 (14)	97.50 (14)	97.38 (20)
$\angle(O_{Ia}-Ti-O_{Ic})$	88.94 (12)	89.21 (14)	89.39 (15)	89.54 (20)
$\angle(O_{Ic}-Ti-O_{Ib})*$	80.41 (13)	80.29 (13)	80.38 (13)	80.08 (18)
$\angle(O_{Ib}-Ti-O_{Ia})$	93.78 (5)	93.78 (7)	93.82 (7)	93.77 (8)
$\angle(Ti-O_{Ib}-Ti)$	99.59 (13)	99.71 (13)	99.62 (13)	99.92 (18)
$\angle(Ti-O_{Ic}-Ti)$	100.26 (16)	100.21 (16)	100.20 (17)	100.08 (23)
$\angle(Ti-O_{Ia}-Ti)$	159.20 (16)	159.07 (17)	159.17 (17)	158.97 (23)
$\angle(Ti-O_{IIa}-Ti)$	123.92 (15)	124.16 (15)	124.11 (15)	124.10 (21)
$\angle(Ti-O_{IIb}-Ti)$	134.80 (16)	134.50 (16)	134.67 (17)	134.83 (23)
$\angle(Ti-O_{IIc}-Ti)$	100.72 (14)	100.76 (15)	100.63 (16)	100.35 (22)

B) Physical properties of rutile

Electronic properties (including optical spectra)

Band structure: Fig. 8 (tight binding calculation with next Ti–O, Ti–Ti, O–O neighbour interaction only and two empirical fitting parameters), Brillouin zone: Fig. 9. Alternate calculation [77P, 78M] using tight binding for valence bands, pseudopotential for conduction band gives broader bands than observed and conduction band minimum at Γ with a close minimum at X, but not M.

band structure energies (in eV):

theoretical:

$W(\pi^*)$	1.4	π^*-bandwidth	77V1
$W(\sigma^*)$	2.7	σ^*-bandwidth	
$W(2p)$	5.5	2p-bandwidth	
$E(\pi^*)$	−5.6	centre of π^*-band	
$E(\sigma^*)$	−3.3	centre of σ^*-band	

band edges: valence band: Γ_3^+, conduction band: $M_{1,2}$ with R_1^+ only 0.12 eV higher (continued)

Physical property	Numerical value	Experimental conditions	Experimental method remarks	Ref.

band structure energies (continued)

experimental:

		assignment:		
E_{peak}	$-5.6 \brace -7.7$ (4.7)	π / σ	peak positions in XPE spectra; see Fig. 10 for low energy part;	77R
	-22.4 (3.4)	O:2s	total spectrum: [72F];	
	-37.5 (2.4)	Ti:3p	values in brackets: half-max. widths	

Ti L_{III} emission:

$E(F)$	-1.6	$1t_{2g} \to 2p_{3/2}$	peaks in X-ray band spectra;	72F
$E(A)$	-5.6	$2e_g \to 2p_{3/2}$	Fig. 11 shows the construction	
$E(G)$	-7.9	$2a_{1g} \to 2p_{3/2}$	of an empirical band scheme	
$E(X)$	...	extended tailing	on which the assignment is made. Data of Ti K, O K, Ti L_{III} X-ray	
$E(C)$	-20.4	$1e_g \to 2p_{3/2}$	band spectra, optical DOS,	
$E(D)$	-22.9	$1a_{1g} \to 2p_{3/2}$	absorption and photoconductivity have been used. Component X	

Ti L_{III} absorption:

$E(f)$	-2	exciton (?)	artificially introduced to match lang-wavelength tail. Components
$E(b)$	$+1.5$	$2p_{3/2} \to 2t_{2g}$	B, b, c primarily due to Ti; F, A,
$E(c)$	$+3.6$	$2p_{3/2} \to 3e_g$	G, C, D primarily due to O. All energies relative to E_F.

Ti K emission:

$E(K\beta_5)$	-4.4	$3t_{1u} \to 1s$
	-6.1	$2t_{1u} \to 1s$
$E(K\beta'')$	-20.6	$1t_{1u} \to 1s$

Ti K absorption:

$E(b)$	$+1.5$	$1s \to 2t_{2g}$
$E(c)$	$+3.6$	$1s \to 3e_g$
$E(d)$	$+9.9$	$1s \to 3a_{1g}$
$E(e)$	$+17.3$	$1s \to 4t_{1u}$

O K emission:

$E(B)$	-3.1	$t_{2u}, t_{1g} \to 1s$
$E(A)$	-5.3	$2e_g \to 1s$
$E(C)$	-6.4	$2t_{1u} \to 1s$

O K absorption:

$E(f)$	-1	exciton (?)	
$E(b)$	$+1.5$	$1s \to 2t_{2g}$	
$E(c)$	$+3.6$	$1s \to 3e_g$	
$W(2p)$	6.3		bandwidths and differences in
$E(\sigma)-E(\pi)$	4.0		mean band energies deduced
$W(2s)$	2.5		from data above
$E(\sigma)-E(2s)$	14.5		
$E(\sigma^*)-E(\pi^*)$	2.1		

energy gaps (in eV):

There is considerable controversy over the assignment of the fundamental optical absorption edge in rutile. Assignments based on two different band-structure calculations [77V2, 77D] have been reported, and the most important difference is that [78P] conclude, on the basis of [77D], that the edge is indirect for $E \parallel c$ but there is a lower-energy, direct forbidden edge for $E \perp c$, whereas [77V2] concludes that the edge is indirect on both polarizations.

(continued)

Physical property	Numerical value	Experimental conditions	Experimental method remarks	Ref.
energy gaps (continued)				
		T [K]		
$E_{g,dir}$	3.033	1.6; $E \perp c$	direct forbidden edge preceded by a 1s exciton at 3.031 eV (the transition is indirect for $E \parallel c$)	78P
	3.038	80		
	3.062	298		
$E_{g,ind}$	3.049	1.6; $E \parallel c$, $E \perp c$	indirect edge; edges at 3.060 and 3.064 due to emission of TA and O_1 phonons, respectively. Identical edges found for $E \perp c$, see Fig. 13.	78P
	3.057	80; $E \parallel c$, $E \perp c$	Transition edges at 3.042 and 3.046 eV due to absorption of O_1 and TA phonons and at 3.068 eV due to TA phonon emission	
	3.101	298; $E \parallel c$, $E \perp c$		

Support for the above assignment comes from optical measurements on stressed crystals; they suggest that the two phonons, O_1 at 15 meV and TA at 11 meV, are associated with the point 0.3 $(2\pi/a)$ [100] in the Δ direction, and the direct transition with $\Gamma_{3v} \to \Gamma_{1c}$ [78M].

Rather different conclusions have been drawn from the electroreflectance data of [77V2].

$E_{g,ind}(\Gamma_{3v}^+ \to M_{1,2c})$	3.005	80; $E \perp c$	edge associated with a M_9^+ phonon at 3.037 eV	77V2
$(\Gamma_{5v}^+ \to M_{1,2c})$	3.013	80; $E \parallel c$	edge associated with a M_9^- phonon at 3.069 eV	
further interband transitions:				
$E(\Gamma_{3v}^+ \to R_{1c}^+)$	3.06 $\cdots$ 3.11	$T = 80$ K; $E \perp c$, $E \parallel c$	associated with R_1^- phonon at 3.124 eV	77V2
$E_{g,dir}(\Gamma_{5v}^+ \to \Gamma_{1,4c}^+)$	3.57	$T = 80$ K; $E \perp c$	Pronounced structure above this edge in Fig. 12 due to transitions between energy-band regions near the Σ symmetry line of BZ. Further direct transitions due to $\Gamma_{1v}^- \to \Gamma_{1c}^+$ and $\Gamma_{5v}^+ \to \Gamma_{5c}^-$ are found above 3.6 eV in $E \parallel c$	

fundamental edge dichroism (ascribed to $\Gamma_{3v}^+ - M_{1,2c}$ transitions for $E \perp c$ and $\Gamma_{5v}^+ - M_{1,2c}$ transitions for $E \parallel c$, in meV):

		T [K]		
$D(E_{g,ind}(E \parallel c))$	11.5	298		61S1
$D(E_{g,ind}(E \perp c))$	12	77		75A
	15	77		66A1
	33	77		74V
	33	4.2		75A

temperature and pressure dependence of energy gaps:

$dE_{g,ind}/dp$	$1.19 \cdot 10^{-6}$ eV bar^{-1}	RT	wavelength-modulated absorption	77P, 78M
$dE_{g,ind}/dT$	$1.98 \cdot 10^{-4}$ eV K^{-1}	$T > 60$ K		78P
$dE_{g,dir}/dp$	$1.17 \cdot 10^{-6}$ eV bar^{-1}	RT		78M
$dE_{g,dir}/dT$	$1.77 \cdot 10^{-4}$ eV K^{-1}	$T > 60$ K		78P

(continued)

Physical property	Numerical value	Experimental conditions	Experimental method remarks	Ref.

temperature and pressure dependence of energy gaps (continued)

$(E_{g,ind})_T = (E_{g,ind})_0 + \dfrac{2.363 \cdot 10^{-4} T^2}{T + (\Theta_D/T)}$ [eV] Θ_D: Debye temperature taken as 758 K, $(E_{g,ind})_0 = 3.0489$ eV, $(E_{g,dir})_0 = 3.0329$ eV   80G3

$(E_{g,dir})_T = (E_{g,dir})_0 + \dfrac{1.245 \cdot 10^{-4} T^2}{T + (\Theta_D/10)}$ [eV]

exciton binding energy:

E_b	4 meV		from E_b(eV) $= 13.6\,(\mu/m_0)(\varepsilon(0)_{\|a}\,\varepsilon(0)_{\|c})^{-1}$ and using $\varepsilon(0)_{\|c} = 257$, $\varepsilon(0)_{\|a} = 111$, an effective mass of 8.4 m_0 is obtained	78P
	6 meV		effective mass $10 \cdots 12\,m_0$	80A

absorption spectra: Figs. 13, 14, electron-energy loss spectra: Fig. 15.

Lattice properties (including further optical properties)

symmetry of optical modes: Fig. 16, phonon dispersion curves: Figs. 17, 18.

frequencies of optical lattice modes at 4 K (in 10^{12} Hz):

$\nu(\Gamma_4^+, B_{2g})$	24.78	R	$T = 4$ K	R: Raman data from [67P]	
	24.72 (25)	N		I: IR data from [64E]	
$\nu(\Gamma_5^+, E_u)_{LO}$	24.18	I		N: neutron data from [71T]	
	25.24 (34)	N		Symmetries of modes are given in	
$\nu(\Gamma_1^-, A_{2u})_{LO}$	24.33	I		Koster notation (as in Fig. 17)	
	24.30	N		and Mulliken notation.	
$\nu(\Gamma_1^+, A_{1g})$	18.36	R		For the correspondence of the	
	18.30	N		data between Γ, X, M and Z	
$\nu(\Gamma_5^+, E_u)_{TO}$	15.00	I		points, see Figs. 17, 18	
	14.81 (15)	N			
$\nu(\Gamma_5^-, E_g)$	13.41	R			
	13.339 (181)	N			
$\nu(\Gamma_5^+, E_u)_{LO}$	13.74	I			
	12.853 (129)	N			
$\nu(\Gamma_4^-, B_{1u}^2)$	12.182 (122)	N			
$\nu(\Gamma_5^+, E_u)_{TO}$	11.64	I			
$\nu(\Gamma_5^+, E_u)_{LO}$	11.19	I			
	11.232 (112)	N			
$\nu(\Gamma_5^+, E_u)_{TO}$	5.49	I			
	5.661 (75)	N			
$\nu(\Gamma_1^-, A_{2u})_{TO}$	5.01	I			
	5.177 (52)	N			
$\nu(\Gamma_3^+, B_{1g})$	4.29	R			
	4.246 (94)	N			
$\nu(\Gamma_4^-, B_{1u})$	3.389 (57)	N			
				corresponding to:	
$\nu(X_1)$	25.38 (25)			(Γ_4^+)	71T
$\nu(X_1)$	24.38 (25)			(Γ_5^+)	
$\nu(X_2)$	13.528 (135)			(Γ_5^-)	
$\nu(X_1)$	11.756 (118)			(Γ_5^+)	
$\nu(X_1)$	9.130 (125)			(Γ_5^+)	
$\nu(X_2)$	9.033 (90)			(Γ_4^-) and (Γ_1^-)	

(continued)

Physical property	Numerical value	Experimental conditions	Experimental method remarks	Ref.
frequencies of optical lattice modes at 4 K (continued)				
$\nu(X_1)$	8.071 (81)		(Γ_5^+) and (Γ_3^+)	71T
$\nu(X_1)$	5.821 (58)		acoustic LA and TA branches (Γ_5^+)	
$\nu(X_2)$	3.084 (31)		(Γ_4^-) and acoustic TA branch (Γ_1^-)	
$\nu(M_{1,2})$	23.26 (23)		(Γ_4^+) and (Γ_5^+)	
$\nu(M_9^+)$	13.522 (135)		(Γ_5^-)	
$\nu(M_9^+)$	9.442 (94)		(Γ_5^+)	
$\nu(M_{5,6})$	9.398 (94)		(Γ_4^-) and (Γ_1^-)	
$\nu(M_9^+)$	7.811 (78)		(Γ_5^+) and (Γ_3^+)	
$\nu(M_{5,6})$	3.058 (56)		(Γ_4^-) and acoustic TA branch (Γ_1^-)	
$\nu(M_9^+)$	2.936 (70)		acoustic LA and TA branches (Γ_5^+)	
$\nu(Z_1)$	12.397 (124)		(Γ_4^-) and acoustic LA-branch (Γ_1^-)	
$\nu(Z_2)$	12.312 (123)		(Γ_5^-)	
$\nu(Z_4)$	9.720 (97)		(Γ_5^+)	
$\nu(Z_3)$	9.040 (90)		(Γ_3^+)	
$\nu(Z_2)$	3.697 (69)		acoustic TA branch (Γ_5^+)	

temperature dependence of vibrational modes: A_{2u} and E_u(TO) modes, with displacements along [001] and [110], have an anomalous increase in intensity and decrease in frequency with decreasing temperature (Fig. 19) characteristic for mode softening on the approach to a ferroelectric catastrophe. These modes are responsible for the high dielectric constant.

pressure dependence of Raman modes: E_g (417 cm^{-1}) and A_{1g} (612 cm^{-1}) peaks shift with rates of $+0.43$ and $+0.41$ cm^{-1} kbar^{-1}, but the B_{1g} (145 cm^{-1}) mode softens by -0.36 cm^{-1} kbar^{-1} [71N]. Above 26 kbar a transition to a new spectrum is observed [71N, 79H], which has been interpreted as a transition from an anharmonic to a harmonic material at high pressure [79H] and as a transition to a new phase of PbO$_2$ structure [71N, 80M2] or CaCl$_2$ structure [78N]. Uniaxial stress measurements [80M1] show that all modes displace linearly with pressure, but the B_{1g} mode shows non-classical behaviour. At very high pressures (≥ 300 kbar) another phase is observed in Raman spectrum whose structure is unknown [81M2].

paraelectric Curie temperature:

Θ_p	-540 K		from $\nu[A_{2u}]$ vs. T (100 K $< T <$ 300 K)	71T, 61P
	-510 K		from $\nu[A_{2u}]$ vs. T (90 $< T <$ 500 K)	73K2
	-700 K		from $\nu[E_u]$ vs. T	

elastic moduli (in 10^{12} dyn cm^{-2}):

c_{11}	2.660 (66)			62W
	2.674 (2)			76G
c_{12}	1.733 (71)			62W
	1.808 (37)			76G
c_{13}	1.362 (81)			62W
	1.466 (29)			76G
c_{33}	4.699 (81)			62W
	4.790 (2)			76G
c_{44}	1.239 (57)		for temperature dependence of c_{44}, see [79R]	62W
	1.238 (8)			76G
c_{66}	1.886 (50)			62W
	1.894 (15)			76G

Debye temperature: Fig. 20.

Physical property	Numerical value	Experimental conditions	Experimental method remarks	Ref.
dielectric constants:				
$\varepsilon(0)$	257	$\|c$, $T\to 0$	for temperature dependence, see Fig. 21	61P
	170	$\|c$, RT		
	111	$\|a$, $T\to 0$		
	86	$\|a$, RT		
$\varepsilon(\infty)$	8.427	$\|c$, RT		51C
	6.843	$\|a$, RT		

Defects in non-stoichiometric TiO_{2-x}

nature of defects – direct evidence: The most important distinction is between ordered shear planes [72B] and point defects [72K]. Estimates [72H1, 76B, 77B1] put x as $\simeq 2\cdot 10^{-4}$ and $8\cdot 10^{-3}$ as the upper limit of the point defect region. In this region the main argument has lain between which process of the two shown is dominant:

(I) $TiO_2 \to TiO_{2(1-x)} + xO_2 + 2xV_O^{\cdot\cdot} + 4xe^-$

(II) $TiO_2 \to (1-x)TiO_2 + xO_2 + xTi_i^{3\cdot} + 3xe^-$, $xTi_i^{3\cdot} \to xTi_i^{4\cdot} + xe^-$. (For $V_O^{\cdot\cdot}$, Ti_i; etc., see below.)

Although theoretical calculations [77J] suggest that (I) should dominate, the experimental situation is confused. No ESR study on single-crystal TiO_{2-x} has identified $V_O^{\cdot\cdot}$, though indirect methods have strongly implied that $V_O^{\cdot\cdot}$ may dominate for very small x [72K, 67A2]. Larger values of x and higher reduction temperatures have been suggested as favouring (II) [67K, 69B1], but oxygen diffusion experiments [72I] have been interpreted as implying that oxygen vacancies also form even at considerable deviations from stoichiometry.

native defects in pure n-type TiO_{2-x} [72H1, 73K1]:

$V_O^{\cdot\cdot}$: doubly ionized oxygen vacancy.
Strong relaxation of Ti^{4+} ions away from neighbouring $V_O^{\cdot\cdot}$ appears to raise the energy of $V_O^{\cdot\cdot}$ above bottom of conduction band; so no direct observation on the defect by ESR is possible.

$Ti_i^{3\cdot}$: triply ionized Ti interstitial; shares two octahedral faces, 4 octahedral edges. A- or C-center in ESR, see below.

$Ti_i^{4\cdot}$: quadruply ionized Ti interstitial; shares two octahedral faces, 4 octahedral edges.

$(Ti_i^{3\cdot})_2$: di-interstitial = pair of triply ionized Ti in nearest-neighbour interstitial octahedra. Internal friction peak for vibrational stress $\|[100]$, no peak for stress $\|[110]$ or $[001]$. X center in ESR, see below.

$(Ti^{3+}-Ti^{3+})$ APB: Ti pairs of face-shared octahedra intersected by antiphase boundary across which normal, interstitial sites become inverted; each Ti octahedron also shares three edges. Ti-pair displacement $\|\frac{1}{2}[0, -0.90, 0.90]$ instead of "ideal" $\frac{1}{2}[0, \bar{1}, 1]$ because of electrostatic repulsion force.

$(Ti^{3+}-Ti^{4+})$ APB: as above. W center in ESR, see below.

$(Ti^{4+}-Ti^{4+})$ APB: as above.

$(Ti^{3+}-Ti^{3+})_{bulk}$: displacements parallel to the c-axis to create c-axis dimer trapping spin paired electrons in homopolar Ti–Ti bond across a shared octahedral-site edge. Cooperative dimerization and bipolaron formation are both found in homologous series Ti_nO_{2n-1}. See section on Ti_4O_7.

$(Ti^{4+}-Ti^{4+})$ CS: Ti pairs of face-shared octahedra intersected by an [hkl] crystallographic shear plane (Fig. 22). Ti-pair same as for APB. Theory suggests CS occurs only in crystals with high static dielectric constant [77J]! Strong cooperative relaxation of Ti^{4+} ions away from CS produces large Ti–Ti separation of face-shared pairs (Magnéli), raising energy of homopolar bond formation above conduction band edge.

Long-range (elastic) attractive force and short-range repulsive force between [312] CS planes (see Fig. 23 for interaction potential) produces clustering of planes into aggregates containing from two to about thirty planes before member of homologous series Ti_nO_{2n-1} is identified in electron microscope. Within an aggregate, the CS spacing D_{sp} varies systematically: $D_{sp} \simeq 8.7$ nm for a "lone pair" to $D_{sp} \simeq 3.8$ nm at center of largest aggregate, increasing to about 6.3 nm at aggregate edge [72B].

ESR parameters of native defects:

	e_{CB}^-	S	principal axes	g-value	
					e_{CB}^-: number of conduction band electrons per defect
					S: spin quantum number
$V_O^{\cdot\cdot}$	2	0			
$Ti_i^{3\cdot}$	3	$\frac{1}{2}$	[001]	1.941 $(g_\parallel)$	At lowest x, ESR signal is broader and has principal axes in basal plane [100]+26°, [010]+26° with g=1.974 and 1.977 and 1.941 respectively. At higher x, a new ESR signal C is seen which may be of the same origin as A [72H1] but motionally narrowed by electron hopping from centre to centre. Superhyperfine field shows electron spread over near-neighbour Ti. [73O, 74S] have identified low temperature A-signal as due to interstitial protons, but C-centre has been reaffirmed [77Y2] as derived from isolated Ti_i^{3+}.
				1.976 (2) $(g_\perp)$	
$Ti_i^{4\cdot}$	4	0			
$(Ti_i^{3\cdot})_2$	6	1	[001]	1.951 (2)	X-centre
			[110]	1.9803 (5)	
			[$\bar{1}$10]	1.9747 (5)	
$(Ti^{3+}-Ti^{3+})$ APB	0	0			trapped e^- spin-paired
$(Ti^{3+}-Ti^{4+})$ APB	1	$\frac{1}{2}$	[001]	1.791 (3)	W-centre. Whether (Ti–Ti) APB traps one or two electrons depends on Ti–Ti separation; only one trapped electron gives ESR signal as two trapped electrons are spin-paired in Ti–Ti bond.
			[100]+4.5°	2.053	
			[010]+4.5°	1.835	
$(Ti^{4+}-Ti^{4+})$ APB	2	0			
$(Ti^{3+}-Ti^{3+})_{bulk}$	2	0			
$(Ti^{4+}-Ti^{4+})$ CS	2	0			

Fig. 24 shows the variation of ESR signal intensity with oxygen deficit. The low-temperature A-signal is not intrinsic [74S] but associated with interstitial hydrogen, and the so-called vacuum reduction of TiO_2 can only be effected in vacuum system with residual impurities; TiO_2 cannot be reduced by heating in a clean vacuum [73O]. The weight of ESR evidence is nevertheless in favour of interstitial Ti rather than $V_O^{\cdot\cdot}$ at moderate ($x \simeq 10^{-4}$) deviations from stoichiometry and possible interstitial sites are shown in Fig. 25. Other evidence for interstitials derives from proton channeling experiments [77Y1] on "vacuum reduced" TiO_2.

Ordered defects exist on shear planes [69B3, 77Y2]. At low defect concentrations the main defect line is along [312] (Fig. 22) and isolated planar {312} defects have been found for $x \gtrsim 0.001$. The (312) CS planes show a net attraction (Fig. 23) and at higher deviation from stoichiometry they order to produce higher members of the homologous series Ti_nO_{2n-1} ($15 \leq n \leq 37$) [69B3, 71B1, 71B2]. As n decreases the shear planes are observed to swing around to (121) via a very large number of intermediate orientations (hkl) that can be resolved into $(hkl) = p(121) + q(011) = (p, 2p+q, p+q)$ (p, q integers). This "swinging process" has little effect on the CS separation, and its onset appears to be at $x \simeq 0.05$ though work on $(Ti,Cr)O_{2-x}$ suggests that CS swinging commences at much smaller values of x ($\simeq 0.02$) in this system.

The type of defect formed as well as the level of ordering depends critically on the purity and previous history of the sample, and reproducible results are only obtained for samples annealed under controlled conditions for lengthy periods. The ubiquity of interstitial hydrogen revealed by EPR may also be of considerable significance. Apparently reduction by CO/CO_2 mixtures rather than by "vacuum" can arrest the formation of shear planes [80B] though this surprising result lacks confirmation.

point defect thermodynamics: Theoretical reviews: [72K, 73D]. High-temperature studies of stoichiometry: Figs. 26, 27. Lines in the figures are derived from a model [67K] that assumes the concentration of conduction band electrons n arising from point defects only:

$$n = 3[\text{Ti}_i^{3\cdot}] + 4[\text{Ti}_i^{4\cdot}] + 2[V_0^{\cdot\cdot}] \quad \text{with} \quad [V_0^{\cdot\cdot}]n^2 = k_1 p_{O_2}^{-1/2}, \quad [\text{Ti}_i^{3\cdot}]n^3 = k_2 p_{O_2}^{-1}, \quad [\text{Ti}_i^{4\cdot}]n = k_i[\text{Ti}_i^{3\cdot}]$$

with

k_1 $4 \cdot 10^2 \exp(-105(\text{kcal/mol})/RT)$ [atm$^{1/2}$]
k_2 $9.3 \cdot 10^9 \exp(-210(\text{kcal/mol})/RT)$ [atm]
k_i $3 \cdot 10^2 \exp(-35(\text{kcal/mol})/RT)$

where the defect concentration in number of moles per mol of oxide is

$$x = (2([\text{Ti}_i^{3\cdot}] + [\text{Ti}_i^{4\cdot}]) + [V_0^{\cdot\cdot}])/(1 + [\text{Ti}_i^{3\cdot}] + [\text{Ti}_i^{4\cdot}]).$$

If oxygen vacancies are the dominant defect, $n \propto 2[V_0^{\cdot\cdot}]$ and $n \propto p_{O_2}^{-1/6}$. If $\text{Ti}_i^{4\cdot}$ is the dominant defect, $[\text{Ti}_i^{3\cdot}] \propto n[\text{Ti}_i^{4\cdot}] \propto n^2$, so that $n \propto p_{O_2}^{-1/5}$. If $\text{Ti}_i^{3\cdot}$ is dominant, $[\text{Ti}_i^{3\cdot}] \propto n$ and $n \propto p_{O_2}^{-1/4}$. For the application of this model to transport data, see below. For impurities in doped TiO_2, see the last subsection of this section.

Transport properties

a) stoichiometric TiO_2:

preparation: Single crystal 1mm thick requires annealing in oxygen at about 800 °C for $t > 60$ h. Resistivity $\varrho > 10^{13} \,\Omega$ cm in the a and c-direction [72H1, 69G]. For $T < 1000$ K, conductivity is mainly ionic [79P] with an activation energy of 0.55···0.75 eV along c-axis and ≈ 2 eV along a-axis.

photocurrent: Photoconductivity is dominated by traps [69G, 73G]. Early reports of significant anisotropy [62H] may have been caused by contact problems [63A]. For highly oxidized samples, the spectra had a maximum at 410 nm [73W] and showed very little intrinsic band gap response [69G]. For more reduced samples, the intrinsic response increases substantially (Fig. 28). Similar results have been reported from photoelectrolytic investigations [80G1]. Temperature dependence of hole and electron mobilities derived from photoconduction: Fig. 29. $E_A(\mu)$ is negative for $T > 100$ K and numerically equal to the Debye temperature, suggesting optical-mode scattering. The decrease in the mobility below 100 K is ascribed to shallow trapping effects. Earlier work suggests field assisted hopping with $E_A^* = E_A - \beta E^{1/2}$, $\beta = (e^3/\pi \varepsilon_0 \varepsilon)^{1/2}$, $E =$ electric field, $E_A \simeq 0.08$ eV [72T].

surface photocurrent: Results quite different from bulk photocurrent, Fig. 30.

b) non-stoichiometric TiO_{2-x}:

conductivity and defect distribution: Results of calculations using the model discussed above are compared with experimental conductivity data in Fig. 31. Additional data: Fig. 32. In the compositional range $1 \cdot 10^{19}$ cm$^{-3} < O_d < 5 \cdot 10^{19}$ cm^{-3} (O_d: oxygen deficiency), the equilibrium low-temperature defect concentration decreases with increasing O_d as $\text{Ti}_i^{3\cdot}$ centres first form pairs $(\text{Ti}_i^{3\cdot})_2$ and then clusters of pairs at an APB prior to establishment of (hkl) CS planes [72H1]. Formation of (hkl) CS aggregates increases with O_d (D_{sp} within aggregate corresponds to $n \simeq 35$ to 40 of a $\text{Ti}_n\text{O}_{2n-1}$ phase) prior to condensation into a definite phase of the homologous series. From table above, n increases sharply with O_d in range of point defects and $(\text{Ti}_i^{3\cdot})_2$ pairs, but decreases with increasing O_d in compositional range where point defects are suppressed by formation of planar defects. Fig. 24 shows evolution of defect populations as observed by direct methods. The associated decrease in n with increasing O_d in the range $1 \cdot 10^{19}$ cm$^{-3} < O_d < 5 \cdot 10^{19}$ cm^{-3} is manifest in the low temperature conductivity (Figs. 33, 34) as a minimum in ϱ vs. O_d. Note that this behaviour is to be distinguished from the impurity-band formation with increasing impurity concentration found in conventional broad-band semiconductors.

resistivity vs. T^{-1} above 300 K (Fig. 35) for samples with oxygen deficiencies in the critical range $1.8 \cdot 10^{19}$ cm$^{-3} < O_d < 2 \cdot 10^{20}$ cm^{-3} exhibits three distinct temperature regions: Region "a" having an O_d-independent $E_A^a = 0.028$ eV, region "b" having an O_d-dependent $E_A^b = E_0 + \beta O_d^{1/3}$ with $E_0 \simeq 0.0080$ eV and $\beta \simeq 8.8 \cdot 10^{-9}$ eV cm, and an exhaustion region "c" [74I]. Typical data for such samples, obtained by measurements in the [110] direction, are given in the following table (cf. Fig. 35) [74I]:

Specimen	1	2	3	4	5	6
Reduction (temperature [K] · time [h])	1073 · 368	1073 · 368 + 1173 · 164	1173 · 160	1173 · 160 + 1273 · 45	1073 · 368 + 1173 · 164 + 1273 · 230	1173 · 160 + 1273 · 45 + 1373 · 40
Exhaustion temperature [K]	≈900	≈1000	≈1000	≈1100	≈1100	≈1200
ϱ [Ω cm] in exhaustion range	2.76	1.60	1.06	0.91	0.71	0.50
μ [cm^{2}V^{-1}s^{-1}] in exhaustion range	$6.42 \cdot 10^{-2}$	$4.93 \cdot 10^{-2}$	$4.93 \cdot 10^{-2}$	$3.89 \cdot 10^{-2}$	$3.89 \cdot 10^{-2}$	$3.12 \cdot 10^{-2}$
Concentration of electrons, n [cm^{-3}]	$3.5 \cdot 10^{19}$	$7.9 \cdot 10^{19}$	$1.2 \cdot 10^{20}$	$1.8 \cdot 10^{20}$	$2.3 \cdot 10^{20}$	$4.00 \cdot 10^{20}$
Concentration of deficient oxygens, O_d [cm^{-3}]	$1.8 \cdot 10^{19}$	$4.0 \cdot 10^{19}$	$6.0 \cdot 10^{19}$	$8.8 \cdot 10^{19}$	$1.1 \cdot 10^{20}$	$2.00 \cdot 10^{20}$

The peculiar increase in E_A with increasing O_d appears to reflect a clustering of donor centres, i.e. the condensation of $(Ti_i^{3+})_2$ into $(Ti^{3+}-Ti^{3+})$ APB's of increasing area. Hall measurements on single crystal samples cut perpendicular to the c-axis also showed a change from simple to complex behaviour with increasing O_d suggestive of defect clustering at the higher concentrations [64A1].

At high oxygen partial pressures, it is necessary to take into account the influence of impurities (see below).

It seems reasonably well established [76B, 77B1, 2], that at elevated temperatures, the onset of diphasic behaviour occurs at a somewhat larger value of x than suggested by low-temperature EPR and electron microscopy data. Thus the apparent success of point defect models for TiO$_2$ at $T > 1300$ K [67K] for values of x as high as 0.01 can be understood on this basis.

electron mobility: Measurements of the drift mobility are plagued by uncertainties in the estimate of n. For $O_d < 1 \cdot 10^{19}$ cm^{-3}, comparison of point-defect theory for n [72K] with measured (nearly isotropic) conductivity gives $\mu_{dr} \simeq 0.2$ cm^{2}V^{-1}s^{-1} at elevated temperatures ($T \simeq 1200$°C).

The IR reflectivity data could be fitted to a model in which the phonon modes for free carriers in n-type TiO$_2$ are coupled to LO phonon modes [77B2]. The mobility was found unactivated, and "dressed" effective masses of $8 \cdots 10 \, m_0$ were obtained.

Thermal conductivity and thermoelectric power of single crystals show maxima at 15 K (see Fig. 36). Whereas κ is anisotropic ($\kappa_c/\kappa_a = 1.5 \, (1)$) for $T > 25$ K, S is not within $\pm 0.5\%$ (Fig. 36) [65T]. Sample 2 in Fig. 36 had a thermal gradient along the c-axis, sample 1 along the a-axis. The steep increase in S with decreasing $T < 50$ K is attributed to phonon drag [55H].

From Seebeck data at 1260 K a mobility $\mu_a = 0.17$ cm^{2}V^{-1}s^{-1} was obtained by [75O, 77B3], in agreement with [72K]. Similar point defect model analysis by [81M1] gives average μ-values of 0.06 cm^{2}V^{-1}s^{-1} at 1100°C up to ≈ 0.1 cm^{2}V^{-1}s^{-1} at 800°C. Extrapolation to RT gives $\mu \approx 0.3$ cm^{2}V^{-1}s^{-1} [81M1].

At low p_{O_2}, S varies as $p_{O_2}^{-1/5}$, in agreement with a point defect model involving interstitial Ti. At higher p_{O_2}, S rises rapidly and may even change sign (Fig. 37) implying a p-type conducting regime.

In contrast to [65T] a slight anisotropy in S ($5 \cdots 7\%$) is reported by [67B1, 2; 75B] which is very enhanced at higher donor concentrations. RT mobilities deduced from Seebeck data are $\mu_a = 0.16$ cm^2/Vs and $\mu_c = 0.56$ cm^2/Vs. μ_a becomes activated above 300 K, μ_c above 500 K. Very high effective masses were reported ($100 \cdots 150 \, m_0$) [68B] leading to the suggestion of small polaron behaviour [69A].

The Hall coefficients for fields parallel or perpendicular to the c-axis are different. With the assumption of a single conduction band, the effective mass can be calculated from R_H and $S = -(k/e)(A_n + (E_F/kT))$, where A_n depends on the scattering mechanism. For a RT resistivity of $1 \cdots 10 \, \Omega$ cm: $m_n = 30 \cdots 35 \, m_0$ at 300 K with $A_n = 2.0$, $m_n = 20 \, m_0$ with $A_n = 2.5$.

In the temperature interval $100 \, K \leq T \leq 300$ K the mobile electrons can be considered to have a lower effective mass ($m_n = 3 \cdots 5 \, m_0$) [64A2] provided electron-phonon coupling gives a large-polaron mobility [62F]. The temperature dependence of μ_H is given in Fig. 38 below room temperature; Fig. 39 shows some representative results for temperatures above RT.

Defect configuration	Experimental conditions	Remarks	Ref.

Fig. 40 shows dispersion of conductivity and dielectric constant at 300 K for a vacuum-reduced sample ($O_d \simeq 1.8 \cdot 10^{17}$ cm^{-3}). The magnitude of the dispersion of conductivity is essentially temperature-independent below 300 K, which is compatible with strong electron-coupling to optical-mode phonons [68G].

The assumption of a single conduction band may need refinement. At the centre of the Brillouin zone, the Ti:3d orbitals of t_{2g} parentage are split by the tetragonal symmetry of the octahedral interstice; distinguishing the two orbitals π-bonding with the oxygen nearest neighbours from the orbital that σ-bonds with like nearest-neighbour cation orbitals along the c-axis gives the nomenclature π* and $d_\parallel$ in Fig. 8 for bands of t_{2g} parentage. Interpretation of the Hall coefficient anisotropy (Fig. 41) has been made on a model consisting of two conduction bands having their bottoms separated by about 0.05(1) eV [65B]. The Hall mobilities $\mu_{H\parallel c}(R_H(\parallel a)/\varrho_{\parallel c})$ and $\mu_{H\perp c}(R_H(\parallel c)/\varrho_{\parallel a})$ are essentially independent of carrier concentration over a range of 10^6, indicating they represent intrinsic properties. At 300 K, the Hall mobilities fall in the ranges $0.77 \leq \mu_{H\parallel c} \leq 1.25$ cm^2/Vs and $0.15 \leq \mu_{H\perp c} \leq 0.35$ cm^2/Vs [65B]. This model has been refined [66A2] to a single ellipsoidal multivalley conduction band with longitudinal masses of ca. 50 m_o and transverse mass of ca. m_o. If an isotropic model is used, the same data [66A2] gives an average mass of 5···13 m_o at 300 K. Alternatively, anisotropy in the phonon modes, apparent in the anisotropic thermal conductivity, may introduce anisotropy into a polaron mobility. This possibility has not been explored.

Impurity-doped rutile

trivalent cations: Johnson Mathey Specpure TiO$_2$ powder contains cationic impurities; the number and concentration of impurities generally increases with sample treatment.

A typical starting powder contained Al^{3+} (32 ppm), Fe^{3+} (16 ppm), and Cr^{3+} (10 ppm) [77B3].

trivalent impurities;
(n, p: electron, hole concentration)
defect configuration

Defect configuration	Experimental conditions	Remarks	Ref.
$2 M'_{Ti} + V_O^{\cdot\cdot}$		analysis for M = Al at 1000···1500 K [63Y]; $M'_{Ti} = M^{3+}$ on Ti^{4+} lattice site	63Y
$(M_{Ti} - V_O)^{\cdot} = M'_{Ti} - V_O^{\cdot\cdot}$ pair	resonance peak in dielectric relaxation (0···200 °C)	activation energy for mobility of $V_O^{\cdot\cdot}$ about M'_{Ti}: 0.13 eV	63Y, 58K
$(M_{Ti} - V_O) = M'_{Ti} - V_O^{\cdot}$ pair	resonance peak in dielectric relaxation ($T < 20$ K)	activation energy for mobility of $V_O^{\cdot}$ about M'_{Ti}: $1 \cdot 10^{-3}$ eV	67D1
$3 M'_{Ti} + M_i^{3\cdot}$	solubility Al$_2$O$_3$ in TiO$_2$ is 1···2 wt%	$M_i^{3\cdot}$ = triply ionized impurity interstitial	69S2
$M_i^{3\cdot}$	resonance peak in internal friction (50 °C)	activation energy for $M_i^{3\cdot}$ mobility: 0.66 eV	66W
$(M_i^{3\cdot} - Ti_i^{3\cdot})$	resonance peak in internal friction (200 °C)	activation energy for $M_i^{3\cdot} - Ti_i^{3\cdot}$ rotation: 0.93 eV	66W
$(M_{Ti} - Ti_i - M_{Ti})^{\cdot}$	resonance peak in dielectric relaxation (16 K)	Ti$_i^{3\cdot}$ linearly coordinated by 2 M'_{Ti} activation energy $2 \cdot 10^{-2}$ eV	67D2
$(M_{Ti} - Ti_i)^{\cdot\cdot} = M'_{Ti} - Ti_i^{3\cdot}$ pair	ESR ($S = 1/2$)		72H1

| Physical property | Numerical value | Experimental conditions | Experimental method remarks | Ref. |

At high oxygen partial pressures, where oxygen vacancies are suppressed the principal defects at high temperatures appear to be quadruply ionized Ti-interstitials ($Ti_i^{4\cdot}$), substitutional M^{3+}-ions (M'_{Ti}), and Ti^{4+}-ion vacancies ($V_{Ti}^{4'}$) (see also below). Hence the general electroneutrality equation becomes $4[Ti_i^{4\cdot}] + p = n + [M'_{Ti}] + 4[V_{Ti}^{4'}]$, and the equilibrium constants are $K_1 = [Ti_i^{4\cdot}] n^4 p_{O_2}$, $K_2 = [V_{Ti}^{4'}] p^4 p_{O_2}^{-1}$, $K_i = np = K_i^0 \exp(-E_g/kT)$. With the assumption that $[V_{Ti}^{4'}] \simeq 0$, the Seebeck data (Fig. 42) gives $E_g(T) = E_g(0) - \beta T$ where $E_g(0) = 3.2$ eV and $\beta = 6.6 \cdot 10^{-4}$ eV K^{-1} [77B3]. An independent value of β from photoconductivity measurements is $8.9 \cdot 10^{-4}$ eV K^{-1} [66B3]. Analysis also gives K_1 as $2.4 \cdot 10^{84}$ at 1273 K, $5.3 \cdot 10^{89}$ at 1473 K and $7.5 \cdot 10^{92}$ atm cm^{-15} at 1623 K. Similarly, $K_2 K_i^{-4}$ takes the values $3.5 \cdot 10^{-58}$, $9.4 \cdot 10^{-61}$ and $4.1 \cdot 10^{-62}$ cm^9 atm^{-1} at the same temperatures [77B3]. Fig. 43 shows the conductivity of Al-doped TiO_2 as a function of oxygen partial pressure; the conductivity changes from n to p type with increasing p_{O_2} at the minimum conductivity. The critical p_{O_2} varies with T.

Nb-doped TiO_2: Rutile dissolves 4 to 8 at% Nb^{5+} at 1330 K under oxidizing conditions before $TiNb_2O_7$ appears [70E]. EPR evidence shows that the Nb^{5+} enters substitutionally [61C, 73Z] and for low doping $4[Ti_i^{4\cdot}] + p + [Nb_{Ti}^{\cdot}] = n + [M'_{Ti}] + 4[V_{Ti}^{4'}]$, since ionization is complete for $T > 30$ K [73Z]. At high p_{O_2} where point-defect model applies, Ti-vacancies may compensate Nb-interstitials, even reaching $4[V_{Ti}^{4'}] = [Nb_{Ti}^{\cdot}]$, to give $n = ([Nb_{Ti}^{\cdot}]/4K_2)^{1/4} K_i p_{O_2}^{-1/4}$ [77B3]. The resultant defect situation is shown schematically in Fig. 44 and the conductivity vs. p_{O_2} in Fig. 45, vs. Nb-content in Fig. 46. The Seebeck coefficients for three concentrations of Nb are shown in Fig. 47. For Seebeck coefficient and calculated effective masses, see Fig. 48. Similar behaviour is found for Ta-doped TiO_2 [80T].

Cr-doped TiO_2: For doping with M^{3+} ions such as Cr one has the equilibria

$$2M_2O_3 + Ti_{Ti} \rightleftarrows 4M'_{Ti} + Ti_i^{4\cdot} + 6O_O$$

or under high oxygen pressure

$$M_2O_3 + \tfrac{1}{2}O_2(g) \rightleftarrows 2M'_{Ti} + 4O_O + 2e^+.$$

The local charge balance gives $4[Ti_i^{4\cdot}] + p = n + [M'_{Ti}]$ and this gives rise to the defect situation of Fig. 49 [80T]. Conductivity vs. p_{O_2} at 1273 K (Fig. 49a) shows an n−p transition. The crystal chemistry of the Cr_2O_3−TiO_2 system is complex; shear phases are found close to pure TiO_2 and extensive CS phases in the intermediate region [70F, 72B, 74P, 78S, 79I]. At low Cr_2O_3 concentrations disordered CS planes lying on (253) planes of the rutile structure are found. As Cr_2O_3 increases, these CS planes become ordered in a series of oxides down to $(Ti, Cr)O_{1.93}$. Between $(Ti, Cr)O_{1.93}$ and $(Ti, Cr)O_{1.90}$ the planes swing round towards (121) and increasing the Cr_2O_3 content further gives a homologous series of oxides down to $(Ti, Cr)_6O_{11}$ at high temperatures and $Ti_6Cr_2O_{15}$ at low temperatures. The dc resistivity is given vs. T in Fig. 50 and the ac resistivity in Fig. 51.

other dopants: see [72M, 79M]. Some results are summarized in Fig. 52.

C) Physical properties of anatase
Electronic and optical properties

energy gap:

E_g	3.23 eV			81G

Real and imaginary parts of the refractive index: Fig. 53.
Optical absorption spectrum: Fig. 14.

Defects

Very little systematic work has been reported on defects in anatase; evidence has however been adduced to support the argument that defects similar to the "A" centre in rutile can be formed on irradiation [77M].

Lattice properties

Under the space group D_{4h}^{19} the local site vibrational fundamentals transform as $a_{1g} + a_{2u} + 2b_{1g} + b_{2u} + 3e_g + 2e_u$, of which a_{1g}, b_{1g}, and e_g are Raman active.

Figs. p. 431

Physical property	Numerical value	Experimental conditions	Experimental method remarks	Ref.
phonon wavenumbers (in cm^{-1}, RT values):				
$\bar{\nu}$	144 vvs	Mode: e_g	w: weak, m: medium,	67B3
	144		s: strong, vs: very strong	78O
	197 w	e_g	Weak second order peaks also	67B3
	197		seen at 316, 696 and 796 cm^{-1}	78O
	400 m	b_{1g}	[67B3]. The e_g band at 197 cm^{-1}	67B3
	399		is considerably softened from	78O
	515 mw	b_{1g}	the value reported from a	67B3
	513		rigid-ion calculation [79K];	78O
	519 mw	a_{1g}	for temperature dependence,	67B3
	519		see Fig. 54; for pressure	78O
	640 m	e_g	dependence, see Fig. 55	67B3
	639			78O

Above 25.6 kbar a transition to a new form of TiO$_2$ is observed which may have the PbO$_2$ or CaCl$_2$-structure [79O]. A similar transition has been reported for rutile (see above).

Transport properties

Resistivities of as grown crystals very variable: $10^4 \cdots 10^{12}$ Ω cm [78V] at RT. Some typical data are shown in Fig. 56. For the more conducting samples two regions of behaviour are found: for $T > $ RT, $\log \varrho \propto E_A/kT$, where $E_A \approx 0.38$ eV. For $T < $ RT, random hopping behaviour is found, with $\log \varrho \propto (T_0/T)^{1/4}$ and $T_0 \approx 22$ K [78V].

References for 9.15.2.1.0 and 9.15.2.1.1

51C	Cronemeyer, D.C.: MIT Laboratory for Insulation Research Rept. 46, **1951**.
55H	Herring, C.: Bell Syst. Tech. J. **34** (1955) 237.
58B	Bevan, H., Dawes, S.W., Ford, R.A.: Spectrochim. Acta. **13** (1958) 43.
58K	Van Keymeulen, J.: Naturwissenschaften **45** (1958) 56.
59G	Grant, F.A.: Rev. Mod. Phys. **31** (1959) 646.
59W	Weyl, R.: Z. Kristallogr. **111** (1959) 401.
60M	Moch, P., Balkanski, M., Aigrain, P.: C.R. Acad. Sci. **251** (1960) 1373.
61C	Chester, P.F.: J. Appl. Phys. **32** (1961) 866.
61P	Parker, R.A.: Phys. Rev. **124** (1961) 1719.
61R	Rao, C.N.R.: Can. J. Chem. **39** (1961) 498.
61S1	Soffer, B.: J. Chem. Phys. **35** (1961) 940.
61S2	Straumanis, M.E., Ejima, T., James, W.J.: Acta Crystallogr. **14** (1961) 493.
62F	Feynmann, R.P., Hellworth, R.W., Iddings, C.K., Platzmann, P.M.: Phys. Rev. **127** (1962) 1004.
62H	von Hippel, A., Kalnejs, J., Westphal, W.B.: J. Phys. Chem. Solids **23** (1962) 779.
62K	Kofstad, P.: J. Phys. Chem. Solids **23** (1962) 1579.
62S	Spitzer, W.G., Miller, R.C., Kleinman, D.A., Howarth, L.E.: Phys. Rev. **126** (1962) 1710.
62W	Wachtman, J.B., Tefft, W.E., Lam, D.G.: J. Res. Nat. Bur. Stand. **A66** (1962) 465.
63A	Acket, G.A., Volger, J.: Physica **29** (1963) 225.
63B	Becker, J.H., Hosler, W.R.: J. Phys. Soc. Jpn. **23 II** (1963) 152.
63Y	Yahia, J.: Phys. Rev. **130** (1963) 1711.
64A1	Acket, G.A., Volger, J.: Phys. Lett. **8** (1964) 244.
64A2	Acket, G.A., Volger, J.: Physica **30** (1964) 1667.
64E	Eagles, D.M.: J. Phys. Chem. Solids **25** (1964) 1243.
64F	Førland, K.S.: Acta Chem. Scand. **18** (1964) 1267.
65B	Becker, J.H., Hosler, W.R.: Phys. Rev. **A137** (1965) 1872.
65M	Moser, J.B., Blumenthal, R.N., Whitmore, D.H.: J. Am. Ceram. Soc. **48** (1965) 384.
65P	Pauley, H.N.: Phys. Status Solidi **11** (1965) 743.
65T	Thurber, W.R., Maute, A.J.H.: Phys. Rev. **A139** (1965) 1655.
66A1	Arntz, F., Yacoby, Y.: Phys. Rev. Lett. **17** (1966) 857.
66A2	Acket, G.A., Volger, J.: Physica **32** (1966) 1680.
66B1	Blumenthal, R.N., Coburn, J., Baukus, J., Hirthe, W.M.: J. Phys. Chem. Solids **27** (1966) 643.
66B2	Blumenthal, R.N., Kirk, J.C., Hirthe, W.M.: J. Phys. Chem. Solids **28** (1966) 1077.

References for 9.15.2.1.0 and 9.15.2.1.1

66B3	Bube, R.H.: Photoconductivity of Solids, New York: J. Wiley **1966**, p. 237.
66V	Vahldiek, F.W.: J. Less-Common Met. **11** (1966) 99.
66W1	Wachtman, J.B., Spinner, S., Brower, W.S., Fridinger, T., Dickson, R.W.: Phys. Rev. **148** (1966) 811.
66W2	Wahlbeck, P.G., Gilles, P.W.: J. Am. Ceram. Soc. **49** (1966) 180.
67A1	Alcock, C.B., Zadov, S., Steele, B.C.H.: Proc. Br. Ceram. Soc. **8** (1967) 231.
67A2	Anderson, J.S., Hyde, B.G.: J. Phys. Chem. Solids **28** (1967) 1393.
67B1	Bogomolov, V.N., Zhuze, U.P.: Fiz. Tverd. Tela **8** (1967) 1904.
67B2	Bogomolov, V.N., Kudinov, E.K., Firsov, Yu.A.: Fiz. Tverd. Tela **9** (1967) 3175.
67B3	Beattie, I.R., Gillson, T.R.: Proc. R. Soc. (London) Ser. A **307** (1967) 407.
67D1	Dominik, L.A.K., MacCrone, R.K.: Phys. Rev. **156** (1967) 910.
67D2	Dominik, L.A.K., MacCrone, R.K.: Phys. Rev. **163** (1967) 756.
67G	Gilles, P.W., Carlson, K.D., Fransen, H.F., Wahlbeck, P.G.: J. Chem. Phys. **46** (1967) 2463.
67K	Kofstad, P.: J. Less-Common Met. **13** (1967) 635.
67N	Navrotsky, A., Kleppa, O.J.: J. Am. Ceram. Soc. **50** (1967) 626.
67P	Porto, S.P.S., Fleury, P.A., Damen, T.C.: Phys. Rev. **154** (1967) 522.
68B	Bogomolov, V.N., Kudinov, E.K., Firsov, Yu.A.: Sov. Phys. Solid State (English Transl.) **9** (1968) 2502.
68G	Goto, T., Okada, T.: J. Phys. Soc. Jpn. **25** (1968) 289.
68R	Robie, R.A., Waldbaum, D.R.: U.S. Geol. Surv. Bull. **1259** (1968) 146.
69A	Austin, T.G., Mott, N.F.: Adv. Phys. **18** (1969) 41.
69B1	Barbanel', V.I., Bogomolov, V.N., Borodin, S.A., Budarina, S.I.: Fiz. Tverd. Tela **11** (1969) 534.
69B2	Bransky, I., Tannhauser, D.S.: Solid State Commun. **7** (1969) 245.
69B3	Bursill, L.A., Hyde, B.G., Teresaki, O., Watanabe, D.: Philos. Mag. **20** (1969) 347.
69G	Ghosh, A.K., Wakim, F.G., Addiss, R.R.: Phys. Rev. **184** (1969) 979.
69S1	Sandin, T.R., Keesom, P.H.: Phys. Rev. **177** (1969) 1370.
69S2	Slepetys, R., Vaughan, P.A.: J. Phys. Chem. **73** (1969) 2157.
70E	Eror, N.G., Smyth, D.M.: The Chemistry of Extended Defects in Non-Metallic Solids, L. Eyring and M. O'Keeffe, eds., N. Holland, **1970**.
70F	Flörke, O.W., Lee, C.W.: J. Solid State Chem. **1** (1970) 144.
70H	Hasiguti, R.R., Yagi, E., Aono, M.: Radiat. Eff. **3–4** (1970) 137.
70R	Rao, K.V.K., Nagender Naidu, S.V., Iyengar, L.: J. Am. Ceram. Soc. **53** (1970) 124.
71A	Abraham, S.C., Bernstein, J.L.: J. Chem. Phys. **55** (1971) 3206.
71B1	Bursill, L.A., Hyde, B.G.: Acta Crystallogr. **827** (1971) 210.
71B2	Bursill, L.A., Hyde, B.G.: Philos. Mag. **23** (1971) 3.
71N	Nicol, M., Fong, M.Y.: J. Chem. Phys. **54** (1971) 3167.
71T	Traylor, J.G., Smith, H.G., Nicklow, R.M., Wilkinson, M.K.: Phys. Rev. **B3** (1971) 3457.
72B	Bursill, L.A., Hyde, B.G.: Prog. Solid State Chem. **7** (1972) 177.
72F	Fischer, D.W.: Phys. Rev. **B5** (1972) 4219.
72H1	Hasiguti, R.: Adv. Mater. Sci. **2** (1972) 69.
72H2	Horn, M., Schwertfeger, C.F., Meagher, E.P.: Z. Kristallogr. **136** (1972) 273.
72I	Iguchi, E., Yajima, K.: J. Phys. Soc. Jpn. **32** (1972) 1415.
72K	Kofstad, P.: Nonstoichiometry, Diffusion and Electrical Conductivity in Binary Metal Oxides, New York: Wiley, **1972**.
72M	Mizushima, K., Tanaka, M., Iida, S.: J. Phys. Soc. Jpn. **32** (1972) 1519.
72T	Tsuchiya, Y., Segawa, H., Tagaki, H., Kawakubo, T.: J. Phys. Soc. Jpn. **33** (1972) 859.
73D	DeFord, J.W., Johnson, O.W.: J. Appl. Phys. **44** (1973) 3001.
73G	Ghosh, A.K., Lauer, R.B., Addiss, R.R.: Phys. Rev. **B8** (1973) 4842.
73K1	Kersson, J., Volger, J.: Physica **69** (1973) 535.
73K2	Knyazev, A.S., Zakharov, V.P., Mityureva, I.A., Poplavko, Yu.M.: Fiz. Tverd. Tela **15** (1973) 2371.
73O	Ohlsen, W.D., Johnson, O.W.: J. Appl. Phys. **44** (1973) 1927.
73W	Wakim, F.G.: J. Appl. Phys. **44** (1973) 496.
73Z	Zimmermann, P.H.: Phys. Rev. **B8** (1973) 3917.
74G1	Gervais, F., Pirim, B.: J. Phys. **C7** (1974) 2374.
74G2	Gervais, J., Pirim, B.: Phys. Rev. **B10** (1974) 1642.
74I	Iguchi, E., Yagima, K., Asahima, T., Kanamori, T.: J. Phys. Chem. Solids **35** (1974) 597.
74P	Philp, D.K., Bursill, L.A.: Acta Crystallogr. **A30** (1974) 265.
74S	Shen, L.N., Johnson, O.W., Ohlsen, W.D., DeFord, J.W.: Phys. Rev. **B10** (1974) 1823.

References for 9.15.2.1.0 and 9.15.2.1.1

74T	Tsuchiya, Y., Segawa, H.: J. Phys. Soc. Jpn. **36** (1974) 1566.
74V	Vos, K., Krusemeyer, H.J.: Solid State Commun. **15** (1974) 949.
75A	Agekyan, V.T., Stepanov, Yu.A.: Fiz. Tverd. Tela **17** (1975) 3676.
75B	Barbanel', V.I., Bogomolov, V.N.: Fiz. Tverd. Tela **17** (1975) 1782.
75O	Odier, P., Baumard, J.F., Panis, D., Anthony, A.M.: J. Solid State Chem. **12** (1975) 324.
75W	Wells, A.F.: Structural Inorganic Chemistry O.U.P. **1975**.
76B	Baumard, J.F.: Solid State Commun. **20** (1976) 859.
76G	Grimsditch, M.H., Ramdas, A.K.: Proc. Int. Conf. Raman Spectrosc. 5th. (1976) 670; E.D. Schmid, J. Bradmueller and W. Kiefer (eds.) Freiburg: H.F. Schultz Verlag.
77B1	Baumard, J.F., Panis, D., Anthony, A.M.: J. Solid State Chem. **20** (1977) 43.
77B2	Baumard, J.F., Gervais, E.: Phys. Rev. **B15** (1977) 2316.
77B3	Baumard, J.F., Tani, E.: Phys. Status Solidi (a) **39** (1977) 373.
77B4	Baumard, J.F., Tani, E.: J. Chem. Phys. **67** (1977) 3018.
77B5	Balabanova, L.A., Stepin, E.V.: Fiz. Tverd. Tela **19** (1977) 3018.
77B6	Blondeau, G., Froelicher, M., Froment, M., Hugot-le-Goff, A.: Thin Solid Films **42** (1977) 147.
77D	Daude, N., Gout, C., Jouanin, C.: Phys. Rev. **B15** (1977) 3229.
77J	James, R., Catlow, C.R.A.: J. Phys. Colloq. **38** (1977) C7–32.
77M	MacKenzie, K.J.D., Whitehead, N.E.: Trans. Br. Ceram. Soc. **76** (1977) 36.
77P	Pascual, J., Camassel, J., Mathieu, H.: Phys. Rev. Lett. **39** (1977) 1490.
77R	Riga, J., Teuret-Noel, C., Pireaux, J.J., Caudano, P., Verbist, J.J., Gobillon, Y.: Physica Scr. **16** (1977) 351.
77V1	Vos, K.: J. Phys. **C10** (1977) 3917.
77V2	Vos, K., Krusemeyer, H.J.: J. Phys. **C10** (1977) 3893.
77Y1	Yagi, E., Koyawa, A., Sakairi, H., Hasiguti, R.R.: J. Phys. Soc. Jpn. **42** (1977) 939.
77Y2	Yagi, E., Hasiguti, R.R.: J. Phys. Soc. Jpn. **43** (1977) 1998.
78C	Cristea, V., Babes, V.: Phys. Status Solidi (a) **48** (1978) 617.
78M	Mathieu, H., Pascual, J., Camassel, J.: Phys. Rev. **B18** (1978) 6920.
78N	Nicol, M., Hara, Y.: Proc. Int. Conf. Raman Spectrosc. 6th **1978**, **2** 320; E.D. Schmid, R.S. Krishnan and W. Kiefer (eds.) London: Heyden.
78O	Ohsaka, T., Izumi, F., Fujiki, Y.: J. Raman Spectrosc. **7** (1978) 321.
78P	Pascual, J., Camassel, J., Mathieu, H.: Phys. Rev. **B18** (1978) 5606.
78S	Somiya, S., Hirano, S., Kamiya, S.: J. Solid State Chem. **25** (1978) 273.
78V	Vorotilova, L.S., Ioffe, V.A., Razumeenko, M.V.: Fiz. Tekh. Poluprovodn. **12** (1978) 36.
79C	Camassel, J., Pascual, J., Mathieu, H.: Phys. Rev. **B20** (1979) 5292.
79H	Hara, Y., Nicol, M.: Phys. Status Solidi (b) **94** (1979) 317.
79I	Inoue, A., Iguchi, E.: J. Phys. **C12** (1979) 5157.
79J	Jourdan, J.L., Gout, C., Albert, J.P.: Solid State Commun. **31** (1979) 1023.
79K	Krishnamurthy, N., Haridisan, T.M.: Indian J. Pure Appl. Phys. **17** (1979) 67.
79M1	Meagher, E.P., Lager, G.A.: Can. Mineral. **17** (1979) 77.
79M2	Mitsuhashi, T., Kleppa, O.J.: J. Am. Ceram. Soc. **62** (1979) 356.
79M3	Mizushima, K., Tanaka, M., Asai, A., Iida, S., Goodenough, J.B.: J. Phys. Chem. Solids **40** (1979) 1129.
79O	Ohsaka, T., Yamaooki, S., Shinomura, O.: Solid State Commun. **30** (1979) 345.
79P	Popov, V.P., Shvaiko-Shvaikovskii, V.E., Andreev, A.A.: Fiz. Tverd. Tela **21** (1979) 383.
79R	Rimai, D.S., Dunn, M.A., Jamieson, J.C., Manghnani, M.H.: Phys. Rev. **B19** (1979) 3215.
80A	Agekyou, V.T., Berezlinaya, A.A., Lutsenko, V.V., Stepanov, Yu.A.: Fiz. Tver. Tela **22** (1980) 12.
80B	Blanchin, M.G., Faisiant, P., Picard, C., Ezzo, M., Fontaine, G.: Phys. Status Solidi (a) **60** (1980) 357.
80G1	Gautron, J., Lemasson, P., Marucco, J.-F.: Faraday Discuss. Chem. Soc. **70** (1980) 81.
80G2	Goodenough, J.B.: Adv. Chem. Ser. **186** (1980) 113.
80G3	Gupta, V.P., Ravindra, N.M.: J. Phys. Chem. Solids **41** (1980) 591.
80M1	Merle, P., Pascual, J., Camassel, J., Mathieu, H.: Phys. Rev. **B21** (1980) 1617.
80M2	Mammone, J.F., Sharma, S.K., Nicol, M.: Solid State Commun. **34** (1980) 799.
80O	Ohsaka, T.: J. Phys. Soc. Jpn. **48** (1980) 1661.
80T	Tani, E., Baumard, J.F.: J. Solid State Chem. **32** (1980) 105.
81G	Gusev, V.B., Lenev, L.M., Kalinichenko I.I.: Zh. Prikl. Spectrosk. **34** (1981) 939.
81M1	Marucco, J.-F., Gautron, J., Lemasson, P.: J. Phys. Chem. Solids **42** (1981) 363.
81M2	Mammone, J.F., Nicol, M., Sharma, S.K.: J. Phys. Chem. Solids **42** (1981) 379.

9.15.2.1.2 Titanium oxide (Ti_2O_3)

Crystal structure

structure: same as α-Al_2O_3, trigonal space group D_{3d}^6 – $R\bar{3}c$, $Z=2$ (rhombohedral system), $Z=6$ (hexagonal system). Projection of the structure on the {110} plane: Fig. 1. If the temperature is raised, the unit cell dimensions alter in a very anomalous manner so as to increase the c/a ratio dramatically [77R1, 76R]. This change is reflected in the considerable expansion in the Ti–Ti separation across a shared face (Fig. 2).

lattice parameters, interatomic distances and angles (distances in Å, angles in °) (from [77R1]):

Parameter	$T=23$°C	117°C	177°C	217°C	292°C	348°C	440°C	595°C
a_{hex}	5.1570 (4)	5.1571 (4)	5.1417 (3)	5.1348 (5)	5.1286 (3)	5.1260 (4)	5.1257 (4)	5.1251 (4)
c_{hex}	13.610 (1)	13.631 (1)	13.731 (1)	13.783 (2)	13.839 (2)	13.878 (2)	13.914 (2)	13.957 (2)
V(Å^3)	313.46	313.95	314.38	314.71	315.24	315.81	316.59	317.48
c/a	2.639	2.643	2.670	2.684	2.698	2.707	2.715	2.723
d(Ti(1)–Ti(2))	2.582 (2)	2.590 (2)	2.622 (2)	2.645 (1)	2.670 (2)	2.685 (1)	2.699 (1)	2.722 (2)
	2.592	2.601	2.636	2.659	2.686	2.701	2.717	2.743
d(Ti(1)–Ti(3))	2.994 (1)	2.994 (1)	2.987 (1)	2.984 (1)	2.983 (1)	2.983 (1)	2.984 (1)	2.985 (1)
	3.002	3.003	2.997	2.996	2.995	2.995	2.998	3.003
d(Ti(1)–O(1))	2.068 (2)	2.070 (2)	2.074 (2)	2.077 (2)	2.081 (2)	2.083 (2)	2.088 (2)	2.093 (2)
	2.081	2.086	2.093	2.096	2.102	2.104	2.113	2.123
d(Ti(1)–O(5))	2.024 (1)	2.024 (1)	2.021 (1)	2.020 (1)	2.018 (1)	2.018 (1)	2.017 (2)	2.017 (2)
	2.037	2.040	2.040	2.040	2.040	2.040	2.042	2.048
d(O(1)–O(2))	2.798 (4)	2.798 (4)	2.784 (4)	2.773 (4)	2.764 (4)	2.759 (4)	2.762 (4)	2.753 (5)
	2.808	2.811	2.801	2.790	2.781	2.777	2.782	2.778
d(O(1)–O(4))	2.791 (1)	2.793 (1)	2.803 (1)	2.807 (1)	2.812 (1)	2.816 (1)	2.821 (1)	2.824 (1)
	2.802	2.807	2.819	2.825	2.830	2.834	2.842	2.850
d(O(1)–O(5))	2.879 (1)	2.882 (1)	2.893 (1)	2.900 (1)	2.907 (1)	2.912 (1)	2.916 (1)	2.923 (1)
	2.889	2.894	2.908	2.916	2.923	2.928	2.934	2.946
d(O(4)–O(5))	3.071 (2)	3.071 (2)	3.065 (2)	3.065 (2)	3.064 (2)	3.065 (2)	3.063 (2)	3.067 (2)
	3.080	3.082	3.079	3.080	3.080	3.080	3.080	3.090
∢O(1)–Ti(1)–O(2)	85.16 (7)	85.03 (7)	84.30 (7)	83.77 (6)	83.23 (7)	82.93 (7)	82.80 (8)	82.34 (8)
∢O(1)–Ti(1)–O(4)	85.98 (2)	86.02 (2)	86.35 (2)	86.46 (2)	86.60 (3)	86.69 (2)	86.81 (3)	86.85 (3)
∢O(1)–Ti(1)–O(5)	89.41 (5)	89.45 (5)	89.86 (5)	90.12 (5)	90.33 (6)	90.47 (6)	90.48 (6)	90.64 (6)
∢O(1)–Ti(1)–O(6)	169.96 (8)	169.42 (8)	169.42 (8)	168.98 (7)	168.54 (8)	168.32 (8)	168.24 (9)	167.78 (9)
∢O(4)–Ti(1)–O(5)	98.64 (3)	98.66 (3)	98.59 (3)	98.68 (3)	98.78 (4)	98.80 (3)	98.80 (4)	98.98 (4)
∢Ti(1)–O(1)–Ti(2)	77.24 (9)	77.42 (9)	78.41 (9)	79.13 (9)	79.85 (10)	80.25 (10)	80.44 (10)	81.06 (11)
∢Ti(1)–O(2)–Ti(3)	94.02 (2)	93.98 (2)	93.65 (2)	93.54 (2)	93.40 (3)	93.30 (3)	93.19 (3)	93.15 (3)
∢Ti(2)–O(2)–Ti(3)	132.29 (5)	132.30 (5)	132.45 (5)	132.56 (4)	132.61 (5)	132.65 (5)	132.59 (5)	132.76 (5)

density:

d	4.590 (1) g cm^{-3}	RT		77R1

Earlier workers [73E, 68R] reported similar data, [75B] presents thermal expansion data below room temperature.

Electronic properties (including optical spectra)

Band structure: Fig. 3, Brillouin zone: Fig. 4. Calculations reported in [69N, 71N, 75A]. Phenomenological band scheme: Fig. 5.

The symmetry at each Ti is D_3 and the degeneracy of the $t_{2\pi}$ orbitals lifted to form $a_1 + e_\pi$. The a_1 orbitals on the two face-sharing Ti interact strongly to form a_1 and a_1^* bands that are broadened considerably to form two bands $\simeq 1$ eV wide.

The e_π orbitals form a narrow band essentially confined to the basal planes and ca. 1/3 eV wide. As the temperature is raised, the Ti–Ti separation across a shared face increases whereas the basal-plane separation remains constant. The effect is to stabilize the e_π band with respect to the a_1 band, and the two bands intersect at 460 K (Fig. 6). This behaviour can be fitted quantitatively to the elastic-constant variation [73C], resistivity and Seebeck experiments [73S1] and heat capacity data [61N, 72P, 73B] (Fig. 7). Further support to this model comes from optical data (see below) and from the photoelectron spectrum [78V] which shows a shift of 0.6 eV to lower binding energy of the Ti 3d peak.

energy gap (in eV):

E_g	0.13	$T=300$ K	estimates using the model	73B
	0.027···0.050		discussed above. [68H, 59M, 61Y]	68H
	0.056···0.079		are determined from activation	61Y
	0.079		energies of conductivity	59M
	0.14		(100···300 K) and Hall data	66A
	0.10···0.23		(100···300 K), respectively	73S1

optical spectra: near infrared reflectance: Fig. 8, visible and UV reflectance: Fig. 9, XPE spectrum: Fig. 10.

IR reflectance shows a feature extending from 1200 to 3000 cm^{-1} whose polarization dependence suggests that it is the a_1-e_π transition. ε_2 exhibits considerable structure at ca. 1 eV which may contain both a_1-e_π and $e_\pi-e_\pi^*$ transitions, the latter playing a role due to a_1/e_π mixing for $k \neq 0$. At 3.0 eV further structure is found and assigned to $a_1-a_1^*$. Peaks $C^\perp$ and $C^\parallel$ are assigned to a_1-e_σ though both O−Ti charge transfer and Ti d ··· s transitions should also be considered. The XPE spectrum reveals at $\simeq 300$ K a d-band of width 1···1.5 eV, in agreement with the above analysis, showing that Ti$_2$O$_3$ is clearly a semiconductor at 300 K. However, at 580 K there is a clear metal-like Fermi edge.

Fig. 11 shows the temperature dependence of the a_1-e_π and mixed a_1-e_π, $e_\pi-e_\pi^*$ transitions. Thermoreflectance [79B] and resonant Raman spectra [75S] show the presence of an additional small feature at 2.3···2.4 eV assigned variously to $a_1-e_\pi^*$ and $e_\pi-e_\pi^*$, and [79B] confirms the assignment of the structure near 3.0 eV to $a_{1g}···a_{1g}^*$. The thermoreflectance [79B] also assign structure at 1.8 eV to plasmon resonance.

Lattice properties

wavenumbers of lattice modes (in cm^{-1}):

IR active modes at RT (2A$_{2u}$ and 4E$_u$):

$\bar{\nu}_{TO}(E_u)$	280	oscillator 0.15	from IR reflection spectrum	77L
	376	strength 5.15		
	451	8.88		
	511	0.43		
$\bar{\nu}_{LO}(E_u)$	281			
	391			
	502			
	537			
$\bar{\nu}_{TO}(A_{2u})$	343	oscillator 3.0		
	448	strength 13.4		
$\bar{\nu}_{LO}(A_{2u})$	351			
	552			

Raman active modes at RT (2A$_{1g}$ and 5E$_g$):

$\bar{\nu}_R(E_g)$	279	274	the low-frequency E$_g$ and A$_{1g}$	
	308	302	modes show pronounced softening	
	350	347	(Fig. 12) as the temperature is	
	465	452	raised; first column [74S],	
	567	564	second column [71M]	
$\bar{\nu}_R(A_{1g})$	238	228		
	513	530		

dielectric constants:

$\varepsilon(0)$	44.1	$E \parallel c$	classical oscillator fit	77L
	45.8	$E \perp c$		
$\varepsilon(\infty)$	27.7	$E \parallel c$		
	31.2	$E \perp c$		

Physical property	Numerical value	Experimental conditions	Experimental method remarks	Ref.

elastic moduli:

temperature dependence: Fig. 13. c_{14} and c_{44} can be fitted quantitatively to a simple anharmonic model. c_{11}, c_{12} and c_{33} can only be explained by invoking electronic effects.

Debye temperature:

Θ_D	672 (12) K	$T = 3 \cdots 15$ K	low-temperature heat capacity	73S2

Data calculated from elastic constants are shown in Fig. 14.

Transport properties

Figures: resistivity, Seebeck coefficient and Hall coefficient at low temperatures: Figs. 15···17, resistivity and Seebeck coefficient at high temperatures: Figs. 18, 19; magnetoresistance: Fig. 20, conductivity: Fig. 21, piezoresistivity: Fig. 22. The very small band gap implies that intrinsic excitation must play an important role, and all transport parameters at low temperatures have been interpreted in terms of a two-band model. At high temperatures anomalous behaviour of ρ and S has been interpreted as band-crossing phenomenon.

transport parameters at 300 K:

σ_n	$23 \cdots 48 \, \Omega^{-1} \text{cm}^{-1}$	$\parallel a$		73S1
	$7.8 \cdots 16 \, \Omega^{-1} \text{cm}^{-1}$	$\parallel c$		
σ_p	$139 \cdots 163 \, \Omega^{-1} \text{cm}^{-1}$	$\parallel a$		
	$46 \cdots 54 \, \Omega^{-1} \text{cm}^{-1}$	$\parallel c$		
μ_n	$2.5 \cdots 7.5 \, \text{cm}^2/\text{Vs}$	$\parallel a$		
	$1 \cdots 2.5 \, \text{cm}^2/\text{Vs}$	$\parallel c$		
μ_p	$16 \cdots 20 \, \text{cm}^2/\text{Vs}$	$\parallel a$		
	$5 \cdots 8 \, \text{cm}^2/\text{Vs}$	$\parallel c$		
σ_p/σ_n	$3 \cdots 7$			
n	$3.9 \cdots 5.9 \cdot 10^{19} \, \text{cm}^{-3}$			

The conductivity may also be calculated from the dielectric function by fitting the free-carrier contribution to ε_2 by the formula.

$\varepsilon_{2, \text{free carrier}} = -\bar{\nu}_p^2/\bar{\nu}(\bar{\nu} + iG_p)$, where $\bar{\nu}_p$ is the plasma wavenumber and G_p proportional to the reciprocal relaxation time [78L1]. For these parameters:

$\bar{\nu}_p$	$2.81 \cdot 10^3 \, \text{cm}^{-1}$	$E \perp c$		78L1
G_p	$766 \, \text{cm}^{-1}$			
$n/(m_n/m_0)$	$8.81 \cdot 10^{19} \, \text{cm}^{-3}$		derived from plasma wavenumber	
$\bar{\mu}(m_n/m_0)$	$12.2 \, \text{cm}^2/\text{Vs}$		derived from relaxation time	
$\bar{\nu}_p$	$1.52 \cdot 10^3 \, \text{cm}^{-1}$	$E \parallel c$		
G_p	$507 \, \text{cm}^{-1}$			
$n/(m_n/m_0)$	$2.5 \cdot 10^{19} \, \text{cm}^{-3}$		mean conductivity in a and c	
$\bar{\mu}(m_n/m_0)$	$17.4 \, \text{cm}^2/\text{Vs}$		directions calculated from these data is shown in Fig. 21	

Magnetic properties

Early work [63A] found evidence for antiferromagnetic ordering, but later studies [68K] showed that the apparent superlattice peaks were due to multiple Bragg scattering. There is essentially no localized moment on the Ti^{3+} ions, a result confirmed by neutron polarization studies [69M].

Doped Ti_2O_3

Ti_2O_3:V: Addition of small amounts of V appears to mimic the temperature increase. Studies of lattice parameters: [74R, 76C1, 77R2, 77R3], axial and equatorial interatomic distances: Fig. 23. Lattice properties show marked changes [75B, 74S]. Raman spectrum shows the softening of the lowest A_{1g} und E_g modes (Fig. 24) to follow closely the behaviour of the pure material on heating (Fig. 12). The temperature variation of the A_{1g} and E_g modes is essentially suppressed for $x \gtrsim 0.12$ in $(Ti_{1-x}V_x)_2O_3$ (Fig. 25). The heat capacity anomaly (see below) at ca. 470 K is also suppressed for $x \gtrsim 0.04$ [73B].

The heat capacity of V-doped Ti_2O_3 shows a marked anomaly at low temperature [73S2, 79M]. Initially this was interpreted in terms of a one dimensional conductivity model [72S]; but this idea was shown to be untenable [73Z1, 73Z2], and an electronic model invoking a very narrow V a_1 band was proposed. Subsequent magnetic measurements [75D1, 75D2, 75E, 77D, 78D, 79D1] have established that, for $x<0.02$, localized magnetic moments exist on V, but decrease with increasing x. Above $x=0.007$ a spin-glass forms at very low temperature, and it is this transition from spin-glass to a paramagnetic metal with no magnetic moment on the V that gives rise to the heat capacity anomaly [79D1].

Conductivity in Ti_2O_3:V: Fig. 26. The transition onset temperature remains constant up to the point where it disappears, and the apparent activation energy falls smoothly from 0.0254 eV in pure Ti_2O_3 to zero at 2.6 at % V [70C]. Similar Seebeck data: Fig. 27. Phase diagram: Fig. 28 and [80D].

Ti_2O_3 : Sc: Structural changes reported in [77R4], electrical properties in [74C]. ϱ first decreases with x and then increases. Seebeck coefficient shows much reduced phonon effect at low temperatures, perhaps because impurity scattering dominates.

References for 9.15.2.1.2

59M	Morin, F.J.: Phys. Rev. Lett. **3** (1959) 34.
61N	Nomura, S., Kawakubo, T., Yanayi, T.: J. Phys. Soc. Jpn. **16** (1961) 706.
61Y	Yahia, J., Frederikse, H.P.R.: Phys. Rev. **123** (1961) 1257.
63A	Abrahams, S.C.: Phys. Rev. **130** (1963) 2230.
66A	Adler, D.: Phys. Rev. Lett. **17** (1966) 139.
67R	Reed, T.B., Fahey, R.E., Honig, J.M.: Mater. Res. Bull. **2** (1967) 561.
68H	Honig, J.M., Reed, T.B.: Phys. Rev. **174** (1968) 1020.
68K	Kendrick, H., Arnott, A., Werner, S.A.: J. Appl. Phys. **39** (1968) 585.
68R	Rao, C.N.R., Loehmann, R.E., Honig, J.M.: Phys. Lett. **A27** (1968) 271.
68Z	van Zandt, L.L., Honig, J.M., Goodenough, J.B.: J. Appl. Phys. **39** (1968) 594.
69M	Moon, R.M., Riste, J., Koehler, W.C., Abrahams, S.C.: J. Appl. Phys. **40** (1969) 1445.
69N	Nebenzahl, I., Weger, M.: Phys. Rev. **184** (1969) 936.
70C	Chandrashekhar, G.V., Won Choi, Q., Moyo, J., Honig, J.M.: Mater. Res. Bull. **5** (1970) 999.
71M	Mooradian, A., Raccah, P.M.: Phys. Rev. **B3** (1971) 4253.
71N	Nebenzahl, I., Weger, M.: Philos. Mag. **24** (1971) 1119.
72P	Paukov, I.E., Berezovskii, G.A.: Zh. Fiz. Khim. **46** (1972) 2683.
72S	Sjostrand, M.E., Keesom, P.H.: Phys. Lett. **A39** (1972) 147.
73B	Barros, H.L., Chandrashekhar, G.V., Chi, T.C., Honig, J.M., Sladek, R.J.: Phys. Rev. **B7** (1973) 5147.
73C	Chi, T.C., Sladek, R.J.: Phys. Rev. **B7** (1973) 5080.
73E	Eckert, L.J., Bradt, R.C.: J. Appl. Phys. **44** (1973) 3470.
73S1	Shin, S.H., Chandrashekhar, G.V., Loehman, R.E., Honig, J.M.: Phys. Rev. **B8** (1973) 1364.
73S2	Sjostrand, M.E., Keesom, P.H.: Phys. Rev. **B7** (1973) 3558.
73Z1	van Zandt, L.L., Eklund, P.C.: Phys. Rev. **B7** (1973) 1454.
73Z2	van Zandt, L.L.: Phys. Rev. Lett. **31** (1973) 598.
74B	Bennett, J.G., Sladek, R.J.: IEEE Ultrasonics Symposium Proceedings **1974**, 517.
74C	Chandrashekhar, G.V., van Zandt, L.L., Honig, J.M., Jayaraman, A.: Phys. Rev. **B10** (1974) 5063.
74R	Robinson, W.R.: J. Solid State Chem. **9** (1974) 255.
74S	Shin, S.H., Aggarwal, R.L., Lax, B., Honig, J.M.: Phys. Rev. **B9** (1974) 583.
75A	Ashkenazi, J., Chuchem, T.: Philos. Mag. **32** (1975) 763.
75B	Bennett, J.G., Sladek, R.J.: J. Solid State Chem. **12** (1975) 370.
75D1	Dumas, J., Schlenker, C., Natoli, P.C.: Solid State Commun. **16** (1975) 493.
75D2	Dumas, J., Schlenker, C., Tholence, J.L., Tournier, R.: Solid State Commun. **17** (1975) 1215.
75E	Eagen, C.F., Koon, N.C., van Zandt, L.L.: AIP Conf. Proc. **29** (1975) 590.
75H	Honig, J.M., van Zandt, L.L.: Annu. Rev. Mater. Sci. **5** (1975) 225.
75S	Shin, S.H., Pollak, F.H., Halpern, T., Raccah, P.M.: Solid State Commun. **16** (1975) 687.
76B	Bennett, J.G., Sladek, R.I.: Solid State Commun. **18** (1976) 1055.
76C1	Caponi, J.J., Marezio, M., Dumas, J., Schlenker, C.: Solid State Commun. **20** (1976) 893.
76C2	Chen, H.L.S., Sladek, R.J.: Bull. Am. Phys. Soc. **11** (1976) 1355.
76R	Rice, C.E., Robinson, W.R.: Mater. Res. Bull. **11** (1976) 1355.
76S	Sinha, A.P.B., Honig, J.M.: J. Solid State Chem. **19** (1976) 391.

77D	Dumas, J., Schlenker, C., Tholence, J.L., Tournier, R.: Physica **86–88b** (1977) 867.
77L	Lucovsky, G., Sladek, R.J., Allen, J.W.: Phys. Rev. **B16** (1977) 5452.
77R1	Rice, C.E., Robinson, W.R.: Acta Crystallogr. **B33** (1977) 1342.
77R2	Rice, C.E., Robinson, W.R.: Mater. Res. Bull. **12** (1977) 421.
77R3	Rice, C.E., Robinson, W.R.: J. Solid State Chem. **21** (1977) 145.
77R4	Rice, C.E., Robinson, W.R.: J. Solid State Chem. **21** (1977) 155.
78C1	Chen, H.L.S., Sladek, R.J.: Phys. Rev. **B18** (1978) 6824.
78C2	Courbin, P., Bonnet, J.-P., Chaminade, J.-P., Le Fleur, G.: Rev. Chim. Minér. **15** (1978) 508.
78D	Dumas, J., Schlenker, C.: J. Mag. Magn. Mater. **7** (1978) 252.
78L1	Lucovsky, G., Allen, J.W., Allen, R.: Phys. Semicond. – Proc. Int. Conf. 14th **1978**.
78L2	Lu, S.S.M., Pollak, F.H., Raccah, P.M.: Phys. Rev. **B17** (1978) 1970.
78V	Vaandevan, S., Hedge, M.S., Rao, C.W.R.: Solid State Commun. **27** (1978) 131.
79B	Bianconi, A., Stizza, S., Bernardini, R., Nannarone, S.: Phys. Status Solidi **B93** (1979) 5767.
79D1	Dumas, J., Schlenker, C., Tholence, J.L., Tournier, R.: Phys. Rev. **B20** (1979) 3913.
79D2	Dumas, J., Schlenker, C.: J. Phys. **C12** (1979) 2381.
79M	Miyako, Y., Sato, T., Kimishima, Y., Yuochumas, Y.G.: J. Phys. Soc. Jpn. **46** (1979) 1379.
80D	Dumas, J.: Phys. Rev. **B22** (1980) 5085.

Physical property	Numerical value	Experimental conditions	Experimental method, remarks	Ref.

9.15.2.1.3 Phases between Ti_2O_3 and TiO_2

Phase diagram of the Ti–O system: see Fig. 1 in section 9.15.2.1.0.

Crystal structure

Ti_3O_5

Structurally, Ti_3O_5 differs from the other Magnéli phases; disagreement exists on details of its structure. Two crystal structures have been reported [57A1, 57A2, 57A3, 59A, 69I].

a) low-temperature phase ($T < T_{tr}$):

structure: Fig. 1b. Characteristic groups of six edge-sharing octahedra arranged in "steps"; some relationship to anatase (Fig. 2), to which it is converted on oxidation [69I], though reduction of anatase at 1250 °C for 3 h with H_2 gives a metastable form of the high-temperature phase [69I], which may be stabilized by quite small quantities of e.g. Fe^{3+} [59A]. Space group $C_{2h}^3 - C2/m$, $Z = 4$.

atomic position parameters [59A]:

In C2/m unit cell, all atoms occupy point position 4 (i); $(0, 0, 0; \frac{1}{2}, \frac{1}{2}, 0); x, 0, z; \bar{x}, 0, \bar{z}$.

	x	z
Ti(1)	0.1280	0.0440
Ti(2)	0.7786	0.2669
Ti(3)	0.0538	0.3659
O(1)	0.676	0.060
O(2)	0.241	0.245
O(3)	0.588	0.345
O(4)	0.953	0.158
O(5)	0.866	0.441

lattice parameters (distances in Å):

a	9.752	$T = 20$ °C	59A
b	3.802		
c	9.442		
β	91.55°		
a	9.76	RT	69I
b	3.80		
c	9.43		
β	91°35′		

Physical property	Numerical value	Experimental conditions	Experimental method remarks	Ref.
density:				
$d_{calc} = 4.16$ g cm^{-3}		RT		69I
interatomic distances (in Å):				
d(Ti(1) − 2·Ti(1))	3.174 (10)	$T = 20\,°\mathrm{C}$	partial bond-strength analysis suggests that Ti(1) is Ti^{3+} but the charges on Ti(2) and Ti(3) are $3\tfrac{1}{3}$ and $3\tfrac{2}{3}$, respectively	59A
− Ti(1)	2.610 (10)			
−2·Ti(2)	3.168 (7)			
− Ti(2)	3.097 (7)			
− Ti(3)	3.143 (7)			
−2·O(1)	1.96 (4)			
− O(1)	2.17 (4)			
− O(2)	2.17 (3)			
− O(4)	2.04 (3)			
− O(4)	2.04 (3)			
d(Ti(2) − 2·Ti(3))	3.067 (7)			
− Ti(3)	2.818 (7)			
− O(1)	2.18 (4)			
−2·O(2)	1.95 (3)			
− O(3)	2.01 (3)			
− O(4)	2.01 (3)			
− O(5)	1.83 (2)			
d(Ti(3) − Ti(3))	2.767 (10)			
− O(2)	2.18 (3)			
−2·O(3)	1.94 (3)			
− O(4)	2.17 (3)			
− O(5)	1.99 (2)			
− O(5)	1.97 (2)			
d(O(1) − 2·O(1))	2.66 (7)			
−2·O(2)	2.65 (5)			
− O(2)	3.01 (5)			
− O(3)	2.84 (4)			
− O(4)	2.83 (4)			
−2·O(4)	3.05 (4)			
−2·O(4)	3.04 (4)			
d(O(2) − 2·O(3))	2.61 (4)			
−2·O(4)	2.94 (4)			
− O(4)	2.91 (4)			
−2·O(5)	2.90 (4)			
− O(5)	3.17 (4)			
d(O(3) − 2·O(4))	2.89 (4)			
− O(5)	2.83 (4)			
−2·O(5)	3.04 (4)			
−2·O(5)	2.81 (4)			
d(O(4) − O(4))	3.14 (5)			
− O(5)	2.83 (3)			
d(O(5) − 2·O(5))	3.17 (4)			
− O(5)	2.82 (4)			

b) high-temperature phase ($T > T_{tr}$):

structure: monoclinic, space group $C_{2h}^3 - C2/m$, $Z = 4$. Similar to anasovite or monoclinically deformed pseudobrookite in which all Ti atoms are equivalent [59A]. Structurally related to rutile [69I].

Physical property	Numerical value	Experimental conditions	Experimental method remarks	Ref.
lattice parameters (distances in Å):				
a	9.82	$T = 393$ K		59A
b	3.78			
c	9.97			
β	91.0°		$\beta \to 90°$ with increasing temperature	
a	9.828	RT	apparently a RT metastable form	69I
b	3.776		of the high-temperature	
c	9.898		modification reported by [59A]	
β	91°19′			
density:				
d_{calc}	4.11 g cm^{-3}	RT		69I
transition temperature:				
T_{tr}	393 K			59A
	450 K		from DTA and XRD measurements ($\Delta H_{tr} = 1.5\,(4)$ kcal mol^{-1})	71R, 77R
	460 K		from discontinuities in electric and magnetic properties	71R

Unit cell volume and c-parameter change abruptly at 450 K, see Fig. 3.

Ti$_4$O$_7$

There are two crystallographic transitions at ≈ 150 K and ≈ 130 K on cooling. The lattice parameters only show a first-order transition at 150 K (Fig. 4). Data see below.

structure: room-temperature phase: space group $C_i^1 - A\bar{1}$, $Z = 4$. Rutile-like slabs of TiO$_6$ octahedra extending infinitely in the a- and b-directions and 4 octahedra thick in the c-direction (Fig. 5).

Two inequivalent strings of Ti ions labelled $3-1-1-3$ and $4-2-2-4$ can be distinguished at 298 K, though the average Ti–O bond length for the four octahedra are very similar. Below 125 K, a major rearrangement occurs (Fig. 6), characterized by Ti^{3+}–Ti^{3+} homopolar bonding (dimer formation) within the $3-1-1-3$ strings. In the intermediate phase, the average interatomic distances differ only slightly from the room-temperature phase, but the thermal factors are anomalously high, consistent with disordering of the dimers (mobile dimers = bipolarons) [80S]. The relationship of the rutile lattice to Ti$_4$O$_7$ is shown in Fig. 7.

transition parameters:				
T_{tr}	154 K (3 K wide)	upper transition	from heat capacity and DTA	74S
ΔH_{tr}	468 (5) cal mol^{-1}			
ΔS_{tr}	3.40 (5) cal K^{-1} mol^{-1}			
T_{tr}	142 K↑ (10 K wide)	lower transition		
	130 K↓			
ΔH_{tr}	95 (5) cal mol^{-1}			
ΔS_{tr}	0.70 (5) cal K^{-1} mol^{-1}			
lattice parameters (distances in Å, angles in °):				
a	5.593 (1)	$T = 298$ K		73M
b	7.125 (1)			
c	12.456 (3)			
α	95.02 (1)			
β	95.21 (1)			
γ	108.73 (1)			
a	5.590 (1)	140 K		
b	7.128 (1)			
c	12.483 (2)			
α	95.03 (1)			
β	95.34 (1)			
γ	108.89 (1)		(continued)	

9.15.2.1.3 Phases between Ti_2O_3 and TiO_2

Physical property	Numerical value	Experimental conditions	Experimental method, remarks	Ref.
lattice parameters (continued)				
a	5.591 (1)	120 K		73M
b	7.131 (1)			
c	12.487 (2)			
α	95.00 (1)			
β	95.33 (1)			
γ	108.88 (1)			
density:				
d_{calc}	4.32 g cm^{-3}	$T = 298$ K		73M

interatomic distances in TiO_6-octahedra (in Å) (from [73M]):

	$T = 298$ K	140 K	120 K	
Ti(1)–O(1)	1.940 (2)	1.949 (4)	2.057 (4)	
–O(1)	1.977 (2)	2.004 (4)	2.058 (4)	
–O(2)	1.935 (2)	1.936 (4)	1.992 (4)	
–O(4)	2.041 (2)	2.030 (4)	2.030 (4)	
–O(5)	2.070 (2)	2.088 (4)	2.037 (4)	
–O(6)	2.071 (2)	2.058 (4)	2.086 (4)	
Average	2.006	2.011	2.043	
O(5)–O(6) $e^{1/3}$	2.694 (2)	2.686 (6)	2.696 (5)	$e^{1/3}$ or $f^{5/8}$ next to an O–O distance
O(5)–O(4) $e^{1/3}$	2.693 (2)	2.745 (6)	2.853 (6)	indicate that the distance is either
O(5)–O(2)	2.883 (3)	2.904 (5)	3.002 (5)	a shared edge or part of a shared
O(5)–O(1)	2.918 (2)	2.919 (5)	2.920 (5)	face, between Ti(1)–Ti(3) and
O(1)–O(6)	2.834 (3)	2.839 (6)	2.854 (5)	Ti(5)–Ti(8), respectively.
O(1)–O(4)	3.048 (2)	3.038 (5)	3.059 (5)	
O(1)–O(2)	2.896 (3)	2.893 (6)	2.952 (6)	
O(1)–O(1) $e^{1/1}$	2.639 (4)	2.659 (9)	2.667 (8)	
O(2)–O(4)	2.987 (3)	3.009 (6)	3.158 (6)	
O(2)–O(1)	2.786 (2)	2.784 (5)	2.813 (5)	
O(5)–O(4)	2.811 (2)	2.804 (6)	2.808 (5)	
O(5)–O(1)	2.734 (3)	2.723 (6)	2.733 (6)	
Ti(2)–O(1)	1.931 (2)	1.902 (4)	1.776 (4)	
–O(2)	1.997 (2)	2.001 (4)	1.964 (4)	
–O(3)	2.010 (2)	2.011 (4)	1.995 (4)	
–O(3)	1.976 (2)	1.972 (4)	1.958 (4)	
–O(4)	2.064 (2)	2.068 (5)	2.148 (4)	
–O(7)	2.061 (2)	2.044 (4)	1.995 (4)	
Average	2.006	2.000	1.973	
O(7)–O(1)	2.866 (2)	2.864 (5)	2.845 (5)	
O(7)–O(2) $e^{2/4}$	2.749 (2)	2.765 (6)	2.700 (5)	
O(7)–O(3)	2.972 (2)	2.953 (5)	2.936 (5)	
O(7)–O(4) $e^{2/4}$	2.706 (2)	2.654 (6)	2.566 (6)	
O(3)–O(1)	2.806 (3)	2.806 (6)	2.772 (6)	
O(3)–O(2)	2.948 (2)	2.924 (5)	2.921 (5)	
O(3)–O(3) $e^{2/2}$	2.691 (3)	2.691 (8)	2.547 (7)	
O(3)–O(4)	2.965 (2)	2.949 (6)	2.897 (5)	
O(3)–O(1)	2.774 (2)	2.760 (5)	2.725 (5)	
O(3)–O(4)	2.841 (2)	2.825 (6)	2.770 (6)	
O(2)–O(1)	2.896 (2)	2.887 (6)	2.840 (6)	
O(2)–O(4)	2.798 (2)	2.810 (5)	2.791 (5)	

(continued)

interatomic distances in TiO$_6$-octahedra (continued)

	$T = 298$ K	140 K	120 K
Ti(3)–O(5)	2.156(2)	2.146(4)	2.138(4)
–O(6)	1.935(2)	1.952(4)	2.049(4)
–O(7)	2.022(2)	2.039(4)	2.109(4)
–O(5)	1.994(2)	2.018(4)	2.029(4)
–O(4)	2.064(2)	2.070(4)	2.037(4)
–O(3)	1.855(2)	1.862(4)	1.903(4)
Average	2.004	2.015	2.044
O(6)–O(5) $e^{3/1}$	2.694(2)	2.686(6)	2.696(5)
O(6)–O(7)	2.829(3)	2.845(7)	2.866(7)
O(6)–O(5)	2.911(2)	2.898(5)	2.896(5)
O(6)–O(3)	2.872(2)	2.900(5)	2.962(5)
O(4)–O(5) $f^{3/4}$	2.695(2)	2.670(5)	2.613(5)
O(4)–O(7) $f^{3/4}$	2.695(2)	2.716(5)	2.752(5)
O(4)–O(5) $e^{1/3}$	2.693(3)	2.745(6)	2.853(6)
O(4)–O(3)	3.008(2)	3.033(6)	3.138(6)
O(5)–O(7) $f^{3/4}$	2.649(2)	2.657(5)	2.644(5)
O(5)–O(5) $e^{3/3}$	2.601(3)	2.594(8)	2.588(7)
O(3)–O(7)	3.075(2)	3.103(6)	3.184(6)
O(3)–O(5)	2.926(2)	2.950(5)	3.042(5)
Ti(4)–O(4)	2.118(2)	2.117(4)	2.023(4)
–O(2)	1.939(2)	1.934(4)	1.945(4)
–O(7)	1.995(2)	1.985(4)	1.950(4)
–O(7)	2.176(2)	2.190(4)	2.314(4)
–O(6)	1.876(2)	1.882(4)	1.787(4)
–O(5)	2.001(2)	1.967(4)	1.958(4)
Average	2.018	2.012	1.996
O(4)–O(7) $e^{2/4}$	2.706(2)	2.654(6)	2.566(5)
O(4)–O(7) $f^{3/4}$	2.695(2)	2.716(5)	2.752(5)
O(4)–O(6)	3.001(2)	2.993(6)	2.918(5)
O(4)–O(5) $f^{3/4}$	2.695(2)	2.670(5)	2.613(5)
O(2)–O(7)	2.897(3)	2.909(5)	2.875(5)
O(2)–O(7) $e^{2/4}$	2.749(2)	2.765(6)	2.700(5)
O(2)–O(6)	2.954(2)	2.947(5)	2.936(5)
O(2)–O(5)	2.868(3)	2.866(7)	2.775(6)
O(7)–O(7) $e^{4/4}$	2.584(3)	2.571(8)	2.614(8)
O(7)–O(5) $f^{3/4}$	2.649(2)	2.657(5)	2.644(5)
O(6)–O(7)	2.948(2)	2.923(5)	2.884(5)
O(6)–O(5)	3.072(2)	3.062(6)	3.024(6)
Ti(1)–Ti(1)[a] edge	2.895(1)	2.926(2)	3.133(2)
–Ti(3)[a]	3.020(1)	2.990(2)	2.802(2)
–Ti(2)[a] corner	3.604(1)	3.609(1)	3.550(2)
–Ti(2)[a]	3.502(1)	3.491(1)	3.461(1)
–Ti(2)[a]	3.572(1)	3.574(3)	3.465(3)
–Ti(2)[a]	3.553(1)	3.554(3)	3.671(3)
–Ti(4)[a]	3.793(1)	3.783(2)	3.784(2)
–Ti(4)[a]	3.530(1)	3.526(2)	3.534(2)
–Ti(4)[a]	3.425(1)	3.434(2)	3.498(2)
–Ti(3)[b] edge	3.111(1)	3.124(2)	3.159(2)
–Ti(4)[b] corner	3.806(1)	3.795(2)	3.699(2)
Ti(2)–Ti(2)[a] edge	2.942(1)	2.937(2)	3.023(2)
–Ti(4)[a]	3.019(1)	3.000(2)	3.083(2)
–Ti(1)[a] corner	3.604(1)	3.609(1)	3.550(2)
–Ti(1)[a]	3.502(1)	3.481(1)	3.461(1)
–Ti(1)[a]	3.572(1)	3.574(3)	3.465(3)

[a] Distances inside the rutile blocks.
[b] Distances between rutile blocks.

(continued)

interatomic distances in TiO_6-octahedra (continued)

	$T=298$ K	140 K	120 K	
—Ti(1)[a]	3.553 (1)	3.554 (3)	3.671 (3)	[a] Distances inside the rutile blocks.
—Ti(3)[a]	3.751 (1)	3.746 (2)	3.778 (1)	[b] Distances between rutile blocks.
—Ti(3)[a]	3.538 (1)	3.539 (2)	3.557 (1)	
—Ti(3)[a]	3.464 (1)	3.473 (2)	3.439 (2)	
—Ti(4)[b] edge	3.067 (1)	3.101 (2)	3.104 (1)	
—Ti(3)[b] corner	3.788 (1)	3.773 (2)	3.765 (2)	
Ti(3)—Ti(1)[a] edge	3.020 (1)	2.990 (2)	2.802 (2)	
—Ti(2)[a] corner	3.751 (1)	3.746 (2)	3.778 (1)	
—Ti(2)[a]	3.538 (1)	3.339 (2)	3.557 (1)	
—Ti(2)[a]	3.464 (1)	3.473 (2)	3.493 (2)	
—Ti(4)[a]	3.556 (1)	3.573 (2)	3.576 (2)	
—Ti(4)[a]	3.569 (1)	3.555 (2)	3.568 (2)	
—Ti(4)[b] face	2.811 (1)	2.806 (1)	2.838 (1)	
—Ti(4)[b] corner	3.417 (1)	3.434 (1)	3.414 (1)	
—Ti(3)[b] edge	3.237 (1)	3.261 (2)	3.267 (1)	
—Ti(1)[b]	3.111 (1)	3.124 (2)	3.159 (2)	
—Ti(2)[b] corner	3.788 (1)	3.773 (2)	3.765 (2)	
Ti(4)—Ti(2)[a] edge	3.019 (1)	3.000 (2)	3.083 (2)	
—Ti(1)[a] corner	3.793 (1)	3.783 (2)	3.784 (2)	
—Ti(1)[a]	3.530 (1)	3.526 (2)	3.534 (2)	
—Ti(1)[a]	3.425 (1)	3.434 (2)	3.498 (2)	
—Ti(3)[a]	3.556 (1)	3.573 (2)	3.576 (2)	
—Ti(3)[a]	3.569 (1)	3.555 (2)	3.568 (2)	
—Ti(3)[b] face	3.811 (1)	2.806 (1)	2.838 (1)	
—Ti(3)[b] corner	3.417 (1)	3.434 (1)	3.414 (1)	
—Ti(4)[b] edge	3.280 (1)	3.295 (2)	3.389 (2)	
—Ti(2)[b]	3.067 (1)	3.101 (1)	3.104 (2)	
—Ti(1)[b] corner	3.806 (1)	3.795 (2)	3.699 (2)	

Ti_5O_9

structure: triclinic, space group $C_i^1 - P\bar{1}$, $Z=2$. As for Ti_4O_7 there are two distinguishable strings of octahedra, labelled 5—3—1—3—5 and 6—4—2—4—6 (Fig. 8), and there are two phase transitions marked by discontinuities in the lattice parameters (Fig. 9).

transition parameters:

T_{tr}	142 K (6 K wide)	upper transition	77M1
ΔH_{tr}	135 (5) cal mol^{-1}		
ΔS_{tr}	0.97 cal K^{-1} mol^{-1}		
T_{tr}	131 K (4 K wide)	lower transition	
ΔH_{tr}	110 (5) cal mol^{-1}		
ΔS_{tr}	0.85 cal K^{-1} mol^{-1}		

Report on additional crystallographic transition at ≈ 127 K corresponding to a change in superstructure without appreciable change in subcell parameters. This transition is also reported in conductivity data [83I].

lattice parameters (distances in Å, angles in °):

a	5.569	5.577 (2)	RT	first column [60A], second
b	7.120	7.127 (2)		column [63A], [82L]
c	8.865	8.872 (3)		
α	97.55	97.561 (7)		
β	112.34	112.356 (9)		
γ	108.50	108.524 (7)		

9.15.2.1.3 Phases between Ti_2O_3 and TiO_2

interatomic distances (in Å, at RT) (from [77M1]):
The symbols c, e, and f refer to Ti–Ti distances across a shared octahedral corner, edge, or face, respectively. The symbol b indicates Ti–Ti distances between rutile blocks.

Interatomic distances	Numerical value	Interatomic distances	Numerical value
Ti(1)–O(2) (2×)	1.9543 (9)	Ti(4)–O(1)	1.9406 (9)
–O(3) (2×)	1.9666 (9)	–O(3)	1.9060 (11)
–O(4) (2×)	2.0164 (11)	–O(4)	1.9953 (11)
Average	1.979	–O(5)	1.9887 (12)
O(2)–O(3) (2×)	2.598 (2)	–O(6)	2.0937 (11)
O(2)–O(3) (2×)	2.937 (1)	–O(9)	2.0893 (8)
O(2)–O(4) (2×)	2.797 (1)	Average	2.002
O(2)–O(4) (2×)	2.819 (1)	O(1)–O(3)	2.830 (1)
O(3)–O(4) (2×)	2.780 (1)	O(1)–O(4)	2.979 (1)
O(3)–O(4) (2×)	2.853 (2)	O(1)–O(5)	2.690 (2)
Average	2.787	O(1)–O(6)	2.959 (1)
Ti(2)–O(1) (2×)	2.0090 (11)	O(9)–O(3)	2.843 (1)
–O(2) (2×)	1.9733 (11)	O(9)–O(4)	2.712 (2)
–O(5) (2×)	2.0044 (8)	O(9)–O(5)	2.920 (1)
Average	1.996	O(9)–O(6)	2.696 (1)
O(1)–O(2) (2×)	2.765 (2)	O(4)–O(3)	2.907 (2)
O(1)–O(2) (2×)	2.865 (2)	O(4)–O(6)	2.810 (1)
O(1)–O(5) (2×)	2.690 (2)	O(5)–O(3)	2.763 (1)
O(1)–O(5) (2×)	2.979 (1)	O(5)–O(6)	2.780 (2)
O(2)–O(5) (2×)	2.802 (1)	Average	2.824
O(2)–O(5) (2×)	2.824 (1)	Ti(5)–O(5)	1.8425 (12)
Average	2.820	–O(6)	2.0517 (10)
Ti(3)–O(1)	1.9073 (12)	–O(7)	1.9915 (9)
–O(2)	1.9264 (10)	–O(8)	1.9394 (9)
–O(3)	1.9756 (9)	–O(9)	2.0384 (10)
–O(6)	2.0342 (9)	Average	2.003
–O(7)	2.0673 (9)	O(9)–O(5)	3.087 (2)
–O(8)	2.0866 (11)	O(9)–O(6)	2.698 (1)
Average	2.000	O(9)–O(7)	2.648 (1)
O(3)–O(1)	2.769 (2)	O(9)–O(8)	2.838 (2)
O(3)–O(2)	2.598 (2)	O(7)–O(5)	2.942 (2)
O(3)–O(7)	2.928 (1)	O(7)–O(6)	2.681 (1)
O(3)–O(8)	2.705 (1)	O(7)–O(7)	2.573 (1)
O(6)–O(1)	2.986 (2)	O(7)–O(8)	2.895 (1)
O(6)–O(7)	2.693 (2)	O(7)–O(8)	2.697 (2)
O(6)–O(8)	2.824 (1)	O(7)–O(6)	2.693 (2)
O(1)–O(2)	2.866 (2)	O(5)–O(8)	2.833 (1)
O(1)–O(7)	2.852 (1)	O(5)–O(6)	3.027 (2)
O(8)–O(2)	2.849 (1)	Average	2.801
O(8)–O(7)	2.697 (2)	Ti(6)–O(4)	1.8803 (8)
Average	2.817	–O(6)	2.1314 (10)
		–O(7)	2.0027 (11)
		–O(8)	1.8563 (11)
		–O(9)	1.9799 (11)
		–O(9)	2.1878 (11)
		Average	2.006

(continued)

9.15.2.1.3 Phases between Ti_2O_3 and TiO_2

Interatomic distances	Numerical value	Interatomic distances	Numerical value
interatomic distances (continued)			
O(4)–O(7)	2.809 (1)	Ti(4)–Ti(3) c	3.4660 (4)
O(4)–O(8)	2.908 (1)	–Ti(3) c	3.5527 (5)
O(4)–O(9)	2.712 (2)	–Ti(3) c	3.5653 (5)
O(4)–O(9)	2.878 (1)	–Ti(5) c	3.5540 (5)
O(6)–O(7)	2.681 (1)	–Ti(5) c	3.7704 (5)
O(6)–O(8)	3.000 (2)	–Ti(1) c	3.5225 (4)
O(6)–O(9)	2.696 (1)	–Ti(1) c	3.6078 (4)
O(6)–O(9)	2.698 (1)	–Ti(2) e	2.9191 (4)
O(8)–O(7)	3.039 (2)	–Ti(6) e	3.0362 (4)
O(8)–O(9)	2.947 (2)	–Ti(5) cb	3.8210 (5)
O(9)–O(7)	2.648 (1)	–Ti(6) eb	3.1114 (4)
O(9)–O(9)	2.581 (1)	Ti(5)–Ti(6) c	3.5561 (4)
Average	2.800	–Ti(6) c	3.5622 (4)
Ti(1)–Ti(4) c (2×)	3.5225 (4)	–Ti(3) e	2.9988 (4)
–Ti(4) c (2×)	3.6078 (4)	–Ti(4) c	3.5540 (5)
–Ti(6) c (2×)	3.4774 (4)	–Ti(4) c	3.7704 (5)
–Ti(2) c (2×)	3.5590 (2)	–Ti(2) c	3.4668 (3)
–Ti(3) e (2×)	2.9234 (4)	–Ti(6) fb	2.8323 (5)
Ti(2)–Ti(3) c (2×)	3.5324 (4)	–Ti(5) eb	3.2586 (6)
–Ti(3) c (2×)	3.6109 (4)	–Ti(6) cb	3.3900 (5)
–Ti(1) c (2×)	3.5590 (2)	–Ti(4) cb	3.8210 (5)
–Ti(5) c (2×)	3.4668 (3)	–Ti(3) eb	3.1210 (4)
–Ti(4) e (2×)	2.9191 (4)	Ti(6)–Ti(5) c	3.5561 (4)
Ti(3)–Ti(6) c	3.5349 (5)	–Ti(5) c	3.5622 (4)
–Ti(6) c	3.7894 (5)	–Ti(4) e	3.0362 (4)
–Ti(4) c	3.4660 (4)	–Ti(3) c	3.5349 (5)
–Ti(4) c	3.5527 (5)	–Ti(3) c	3.7894 (5)
–Ti(4) c	3.5653 (5)	–Ti(1) c	3.4774 (4)
–Ti(1) e	2.9234 (4)	–Ti(5) fb	2.8323 (5)
–Ti(2) c	3.5324 (4)	–Ti(6) eb	3.2788 (6)
–Ti(2) c	3.6109 (4)	–Ti(5) cb	3.3900 (5)
–Ti(5) e	2.9988 (4)	–Ti(3) cb	3.8036 (5)
–Ti(6) cb	3.8036 (5)	–Ti(4) eb	3.1114 (4)
–Ti(5) eb	3.1210 (4)		

electrostatic charges at the Ti-ions (in units of e) (from [77M1]):

e^*/e of:	$T = 298$ K	135 K	115 K
Ti(1)	3.85	3.91	3.92
Ti(2)	3.65	3.60	3.65
Ti(3)	3.60	3.60	3.56
Ti(4)	3.57	3.60	3.59
Ti(5)	3.56	3.52	3.50
Ti(6)	3.52	3.52	3.52
Average	3.60	3.60	3.59

calculated from Ti–O bond lengths. There is no clearcut evidence of low-temperature electron condensation into bipolarons as in Ti_4O_7, since there is no marked change in the partial cation charges with temperature

Physical property	Numerical value	Experimental conditions	Experimental method, remarks	Ref.
Ti_nO_{2n-1}				
Later members have not been refined.				
lattice parameters (distances in Å, angles in °, RT values):				
a	5.566 (2)	Ti_6O_{11}	space group $A\bar{1}$	63A, 71T, 82L
b	7.144 (2)			
c	24.066 (7)			
α	98.473 (4)			
β	120.802 (8)			
γ	108.520 (6)			
a	5.547 (2)	Ti_7O_{13}	space group $C_i^1 - P\bar{1}$	
b	7.140 (3)			
c	15.370 (7)			
α	98.905 (6)			
β	125.457 (9)			
γ	108.517 (6)			
a	5.534 (2)	Ti_8O_{15}	space group $C_i^1 - A\bar{1}$	
b	7.144 (2)			
c	37.613 (12)			
α	97.167 (3)			
β	128.384 (6)			
γ	108.470 (4)			
a	5.527 (2)	Ti_9O_{17}	space group $C_i^1 - P\bar{1}$	
b	7.141 (3)			
c	22.278 (8)			
α	97.264 (4)			
β	131.338 (7)			
γ	108.499 (4)			

Estimated atomic positions [82L] show that in these shear phases the pseudo-rutile chains deviate considerably from the rutile axis, which allows the face-sharing Ti ions to move considerably further apart and reduce electron repulsion.

EPR spectra and defects [75H, 71R, 76H, 71H, 74H1, 72H, 74H2].

Ti_3O_5: Spectra on single crystals [72S], on powder samples [74H3]. Twinned single crystals gave 2 EPR signals with $g_x = 1.960$, $g_y = 1.973$, $g_z = 1.978$ (1). The x and y axes are in the $a-c$ plane and the x-axis is 22.5° to the a-axis for one signal and $-22.5°$ for the other. Spin concentrations are estimated to be $\simeq 10^{-4} n(Ti^{3+})$. Powder EPR data show hyperfine coupling; they were interpreted on the basis of the crystal structure, and Fig. 10 shows a detail of the structure in the $a-c$ plane. Strong Ti(1)–Ti(1) and weaker Ti(3)–Ti(3) bonds exist at low temperatures. [76H] estimates that the Ti(1)–Ti(1) pair has a trap energy of 0.24 eV, and Ti(3)–Ti(3) of $\simeq 0.067$ eV. EPR signal is then due to excitation of Ti(3)–Ti(3) pairs to the conduction band. The observed EPR signal intensity at 77 K can be rationalized on this approach. The effective symmetry at Ti(3) is D_{2h} and $e_g \to a_g + b_{3g}$; $t_{2g} \to a_g + b_{2g} + b_{1g}$. Analysis gives $\bar{v}(b_{2g}) - \bar{v}(a_g) = 3100$ cm^{-1}, $\bar{v}(b_{1g}) - \bar{v}(a_g) = 6300$ cm^{-1}, $\bar{v}(b_{3g}) - \bar{v}(a_g) = 26000$ cm^{-1}. The EPR signals in Ti_3O_5 and the Magnéli phases have also been ascribed to $(Ti^{3+} - V_O - Ti^{4+})$ defects; axial g-values have been listed for all oxides in [74H]. From optical data [72P] a trigonal distortion of 5000 cm^{-1} was estimated, and spin-orbit coupling constants 79.3···93.0 cm^{-1} have been calculated (cf. free-ion value of 135 cm^{-1}).

Ti_4O_7: Spectra interpreted in terms of an isolated substitutional Ti^{3+} ion associated with a cation vacancy, see [76L].

Transport properties

conductivity: for Ti_3O_5: Fig. 11; Ti_4O_7: Fig. 12, 12α; Ti_5O_9: Fig. 13, 13α; Ti_6O_{11}: Fig. 14, 14α; Ti_7O_{13}: Fig. 15α; Ti_8O_{15}: Fig. 15β; Ti_9O_{17}: Fig. 15γ.

The following table [75H] presents a comparison of calculated and measured values of the conductivity and the carrier mobility, the hopping frequencies f_h and mobilities μ_h and concentrations as well as effective masses of localized (n_{loc}) and delocalized (n_{del}) electrons.

9.15.2.1.3 Phases between Ti_2O_3 and TiO_2

Physical property	Numerical value	Experimental conditions	Experimental method, remarks	Ref.
transport parameters:				
σ_{exp} [Ω^{-1}cm^{-1}]	$5.5 \cdot 10^{-3}$	Ti_3O_5	low-temperature phase (see also Figs. 11, 12). $T = 120$ K. σ measured along b-axis of single crystal sample; [77M1] gives σ two orders of magnitude lower for Ti_5O_9	69B
	$3.10 \cdot 10^{-3}$	Ti_4O_7		
	6.3	Ti_5O_9		
	$7.0 \cdot 10^{-2}$	Ti_6O_{11}		
	$2.5 \cdot 10^{-1}$	Ti_8O_{15}	no phase transition	
σ_{exp} [Ω^{-1}cm^{-1}]	$3.5 \cdot 10^1$	Ti_3O_5	high-temperature phase; $T = 450$ K for Ti_3O_5 and 300 K for the others; [77M1] again reports significantly lower conductivity for Ti_5O_9 at 300 K ($\approx 0.6\,\Omega^{-1}$cm^{-1})	69B
	$1.5 \cdot 10^3$	Ti_4O_7		
	$5.5 \cdot 10^2$	Ti_5O_9		
	$1.1 \cdot 10^2$	Ti_6O_{11}		
μ_{calc} [cm^2/Vs]	0.21	Ti_3O_5	$T = 450$ K	75H
	0.6	Ti_4O_7	300 K	
	4.1		160 K	
	0.76	Ti_5O_9	300 K	
	0.9	Ti_6O_{11}	300 K	
μ_{exp} [cm^2/Vs]	1.0	Ti_4O_7	$T = 300$ K	75H
	4.0		160 K	
f_h [Hz]	$4.9 \cdot 10^{11}$	Ti_3O_5	$T = 120$ K	75H
	$1.6 \cdot 10^{12}$	Ti_4O_7		
	$9.3 \cdot 10^{12}$	Ti_5O_9		
	$1.4 \cdot 10^{13}$	Ti_6O_{11}		
	$1.5 \cdot 10^{12}$	Ti_8O_{15}		
μ_h [cm^2/Vs]	$7.0 \cdot 10^{-5}$	Ti_3O_5	$T = 120$ K	75H
	$4.0 \cdot 10^{-6}$	Ti_4O_7		
	$4.2 \cdot 10^{-4}$	Ti_5O_9		
	$2.7 \cdot 10^{-7}$	Ti_6O_{11}		
	$2.1 \cdot 10^{-4}$	Ti_8O_{15}		
n_{loc} [cm^{-3}]	0.00	Ti_3O_5	high-temperature region, calculated from magnetic susceptibility ($T = 450$ K for Ti_3O_5 and 300 K for the others)	75H
	$1.71 \cdot 10^{21}$	Ti_4O_7		
	$1.10 \cdot 10^{21}$	Ti_5O_9		
	$3.7 \cdot 10^{21}$	Ti_6O_{11}		
	$3.85 \cdot 10^{21}$	Ti_8O_{15}		
n_{del} [cm^{-3}]	$2.29 \cdot 10^{22}$	Ti_3O_5	high-temperature region, calculated from magnetic susceptibility (T as above)	75H
	$1.54 \cdot 10^{22}$	Ti_4O_7		
	$1.24 \cdot 10^{22}$	Ti_5O_9		
	$7.4 \cdot 10^{21}$	Ti_6O_{11}		
	$4.6 \cdot 10^{21}$	Ti_8O_{15}		
n_{del} [cm^{-3}]	$9.1 \cdot 10^{20}$	Ti_3O_5	low-temperature region, from EPR data; $T = 120$ K $< T_{tr}$, except for Ti_8O_{15}, where no T_{tr} were observed	75H
	$2.4 \cdot 10^{21}$	Ti_4O_7		
	$1.7 \cdot 10^{21}$	Ti_5O_9		
	$3.8 \cdot 10^{21}$	Ti_6O_{11}		
	$5.3 \cdot 10^{21}$	Ti_8O_{15}		
m_n/m_0	22.4 (6.7)	Ti_3O_5	high-temperature region, in brackets: low temperature-region defined above	75H
	15.4 (5.6)	Ti_4O_7		
	19.5 (6.4)	Ti_5O_9		
	15.4 (5.4)	Ti_6O_{11}		
	13.2 (12.5)	Ti_8O_{15}		

further transport data and remarks:

[76L] finds that σ increases in Ti_4O_7 between 160 and 300 K in a similar way to that reported by [69B] for Ti_5O_9. [83I] reports both types of behaviour depending on the crystal orientation. For one free electron

Physical property	Numerical value	Experimental conditions	Experimental method, remarks	Ref.

per Ti^{3+} calculated mobilities are $0.55 \text{ cm}^2/\text{Vs}$ [69B], $0.05 \text{ cm}^2/\text{Vs}$ [76L] at 300 K. Below T_{tr} no activation energies can be obtained for Ti_3O_5, Ti_4O_7, Ti_5O_9, Ti_6O_{11}.

However, Ti_8O_{15} shows no transition and an activation energy of 0.087 eV. In Ti_3O_5, the transition involves a structural rearrangement in which $Ti^{3+}-Ti^{3+}$ bonds are broken at high temperatures and the electrons become quasi-itinerant. The two transitions seen in Ti_4O_7 and Ti_5O_9 have been ascribed to solid-liquid-gas. The central unit remains a $Ti^{3+}-Ti^{3+}$ bond in the intermediate phase, but this "bipolaron" becomes mobile. The Seebeck coefficient suggests n-type conduction [79S], and in the low-temperature region variable-range hopping occurs [79S]. [83I] reports fine structure in the hysteresis range of Ti_4O_7 (Fig. 12α) and three transitions, at 127(2), 132(1) and 139(1) K in the conductivity of Ti_5O_9 (Fig. 13α). Unlike [69B], who reports a transition at 130 K in Ti_6O_{11}, [83I] reports two transitions, at 119(1) and 147(1) K (Fig. 14α). Interestingly, Ti_7O_{13} appears to show no metallic region (Fig. 15α), but a transition at 115 K↓, 123 K↑ between two semiconducting regions. Below T_{tr}, for n = 5, 6 or 7, ln σ varies linearly with T, a result ascribed to temperature-dependent incoherent tunneling [83I].

Optical properties

Diffuse reflectance spectra of Ti_nO_{2n-1}: Fig. 16. A very broad peak that shifts steadily to the red with increasing n is apparent, and a linear relationship between log n and $\bar{v}_{max}$ is found (Fig. 17). In the low-temperature phase of Ti_4O_7, there is an absorption onset at $\lambda \simeq 5 \mu m$, suggesting a bandgap of 0.25 eV [77K]. Above the first transition, with rising temperature, at 143 K the transmission at 5.6 μm falls dramatically. The photoelectron spectrum of Ti_3O_5 shows a distinct d-band split into a doublet below T_{tr} (≈ 0.8 eV); this doublet merges above the transition temperature [78V].

Magnetic properties

susceptibility: Figs. 18 (Ti_nO_{2n-1}), 19 (Ti_5O_9) [66K, 69M, 72D, 65V, 75H, 77M1]. Noteworthy is the 30 K hysteresis in Ti_3O_5, $T_{tr} = 462$ K↑, 432 K↓. Transitions in Ti_4O_7, Ti_5O_9 and Ti_6O_{11} were observed at 150 K, 130 K and 122 K, respectively, with no hysteresis [75H].

Above the transition temperatures, the susceptibilities of the flux-grown (not oriented) single crystals were found to obey $\chi_m = C_m/(T+\Theta_p) + \alpha$, where χ_m is the susceptibility per mol Ti atoms.

Curie-Weiss parameters:

α	$246.5 \cdot 10^{-6}$ cm^3 mol^{-1}	Ti_3O_5	p_{eff} [μ_B] = 2.858 $C_m^{1/2}$ (the numerical factor is of the dimension (K cm^3 mol^{-1})$^{-1/2}$ in CGS units); C_m: molar Curie constant; Θ_p: paramagnetic Curie temperature. Ti_3O_5 shows only T.I.P. above T_{tr} (426 K). Temperature range for measurements $T_{tr}\cdots$500 K. [77M1] reports only temperature independent paramagnetism above 138 K	72D
C_m	0.01270 K cm^3 mol^{-1}	Ti_4O_7		
Θ_p	28.84 K			
p_{eff}	0.319 μ_B			
α	$148.4 \cdot 10^{-6}$ cm^3 mol^{-1}			
C_m	0.00756 K cm^3 mol^{-1}	Ti_5O_9		
Θ_p	−46.86 K			
p_{eff}	0.245 μ_B			
α	$184.1 \cdot 10^{-6}$ cm^3 mol^{-1}			
C_m	0.02900 K cm^3 mol^{-1}	Ti_6O_{11}		
Θ_p	18.34 K			
p_{eff}	0.481 μ_B			
α	$120.1 \cdot 10^{-6}$ cm^3 mol^{-1}			
C_m	0.03058 K cm^3 mol^{-1}	Ti_8O_{15}		
Θ_p	55.22 K			
p_{eff}	0.494 μ_B			
α	$77.0 \cdot 10^{-6}$ cm^3 mol^{-1}			
C_m	0.06809 K cm^3 mol^{-1}	$Ti_{10}O_{19}$		
Θ_p	210.10 K			
p_{eff}	0.738 μ_B			
α	$41.6 \cdot 10^{-6}$ cm^3 mol^{-1}			

V-doped Ti_4O_7

This system has been extensively explored [79H, 79S, 77M2, 76L, 78A, 79S].

Phase diagram: Fig. 20. The lower transition temperature decreases rapidly with increasing x in $(Ti_{1-x}V_x)_4O_7$ and vanishes for $x \gtrsim 0.0035$. For $x \simeq 0.0025$ two crystallographic transitions are seen (Fig. 21). The entropy changes at the two transition temperatures (Fig. 22). Magnetic susceptibility shows Curie behaviour below the upper transition temperature, $x > 0$ (Fig. 23) with $p_{eff} \simeq 2.0\,\mu_B$ per V-atom. Electrical resistivity: Fig. 24. The activation energy in the intermediate phase drops sharply with x. In the low-temperature phase, the resistivity varies with T as $\varrho \propto A \exp(-(T_0/T)^{1/4})$ (Fig. 25) characteristic of variable-range hopping, with $T_0 \approx 10^8$ K [79S].

References for 9.15.2.1.3

57A1	Andersson, S., Collén, B., Kuylestrina, U., Magnéli, A.: Acta Chem. Scand. **11** (1957) 1641.
57A2	Andersson, S., Collén, B., Kruuse, G., Kuylestrina, U., Magnéli, A., Pestinalis, H., Åsbrink, S.: Acta Chem. Scand. **11** (1957) 1653.
57A3	Åsbrink, S., Magnéli, A.: Acta Chem. Scand. **11** (1957) 1606.
59A	Åsbrink, S., Magnéli, A.: Acta Crystallogr. **12** (1959) 575.
60A	Andersson, S.: Acta Chem. Scand. **14** (1960) 1161.
63A	Andersson, S., Jahnberg, L.: Arkiv Kemi **21** (1963) 413.
65V	Vasil'ev, Ya.V., Ariya, S.M.: Izv. Akad. Nauk SSSR, Neorg. Mater. **1** (1965) 347.
66K	Keys, L.K., Mulay, L.N.: Appl. Phys. Lett. **9** (1966) 248.
67K	Keys, L.K., Mulay, L.N.: Phys. Rev. **154** (1967) 453.
69B	Bartholomew, R.F., Frankl, D.R.: Phys. Rev. **187** (1969) 828.
69I	Iwasaki, H., Bright, N.F.H., Rowland, J.F.: J. Less-Common Met. **17** (1969) 99.
69M	Mulay, L.N., Danley, M.J.: J. Appl. Phys. **41** (1969) 877.
70M	Marezio, M., McWhan, D.B., Dernier, P.D., Remeika, J.P.: Mater. Res. Bull. **5** (1970) 1015.
71H	Houlihan, J.F., Mulay, L.N.: Mater. Res. Bull. **6** (1971) 737.
71M	Marezio, M., Dernier, P.D.: J. Solid State Chem. **3** (1971) 340.
71R	Rao, C.N.R., Randes, S., Loehman, R.E., Honig, J.M.: J. Solid State Chem. **3** (1971) 83.
71T	Teresaki, O., Watanabe, D.: Jpn. J. Appl. Phys. **10** (1971) 292.
72D	Danley, W.J., Mulay, L.N.: Mater. Res. Bull. **7** (1972) 739.
72M	Marezio, M., McWhan, D.B., Dernier, P.D., Remeika, J.P.: Phys. Rev. Lett. **28** (1972) 1390.
72P	Porter, V.R., White, W.B., Roy, R.: J. Solid State Chem. **4** (1972) 250.
72S	Schlenker, C., Buder, R., Schlenker, C., Houlihan, J.F., Mulay, L.N.: Phys. Status Solidi (b) **51** (1972) 247.
73M	Marezio, M., McWhan, D.B., Dernier, P.D., Remeika, J.P.: J. Solid State Chem. **6** (1973) 213.
74H1	Houlihan, J.F., Mulay, L.N.: Inorg. Chem. **13** (1974) 745.
74H2	Houlihan, J.F., Mulay, L.N.: Phys. Status Solidi (b) **61** (1974) 647.
74H3	Houlihan, J.F., Mulay, L.N.: Phys. Status Solidi (b) **65** (1974) 513.
74S	Schlenker, C., Lakkis, S., Coey, J.M.D., Marezio, M.: Phys. Rev. Lett. **32** (1974) 1318.
75H	Houlihan, J.F., Danley, W.J., Mulay, L.N.: J. Solid State Chem. **12** (1975) 265.
76H	Houlihan, J.F., Madassi, D.P., Mulay, L.N.: Mater. Res. Bull. **11** (1976) 307.
76L	Lakkis, S., Schlenker, C., Chakraverty, B.K., Buder, R., Marezio, M.: Phys. Rev. **B14** (1976) 1429.
77K	Kaplan, D., Schlenker, C., Since, J.J.: Philos. Mag. **36** (1977) 1275.
77M1	Marezio, M., Tranquin, D., Lakkis, S., Schlenker, C.: Phys. Rev. **B16** (1977) 2811.
77M2	Miller, V.I., Perelyaev, V.A., Shveikin, G.P., Alyanovskii, S.I.: Izv. Akad. Nauk SSSR, Neorg. Mater. **13** (1977) 566.
77R	Rao, C.N.R., Rao, G.R.: Phys. Lett. **A61** (1977) 247.
77S	Schlenker, C., Lakkis, S., Ahmed, S., Hodeau, J.L., Marezio, M.: J. Phys. **C10** (1977) L151.
78A	Ahmed, S., Schlenker, C., Buder, R.: J. Mag. Magn. Mater. **7** (1978) 338.
78V	Vaandevan, S., Hedge, M.S., Rao, C.N.R.: Solid State Commun. **27** (1978) 131.
79H	Hodeau, J.L., Marezio, M.: J. Solid State Chem. **29** (1979) 47.
79S	Schlenker, C., Ahmed, S., Buder, R., Gourmala, M.: J. Phys. **C12** (1979) 3503.
80S	Schlenker, C., Marezio, H.: Philos. Mag. **42** (1980) 453.
82L	Le Page, Y., Strobel, P.: J. Solid State Chem. **43** (1982) 314.
83I	Inglis, A.D., LePage, Y., Strobel, P., Hurd, C.M.: J. Phys. **C16** (1983) 317.

Physical property	Numerical value	Experimental conditions	Experimental method, remarks	Ref.

9.15.2.2 Vanadium oxides

9.15.2.2.0 Introduction

The vanadium-oxygen phase diagram is very complex (Fig. 1 and 2, [63K, 67K, 73K]). In the following sections we shall consider first the shear phases V_nO_{2n-1} (VO_x with $x=1.5\cdots<2.0$), than the phases V_nO_{2n+1} (VO_x with $x=2.0\cdots2.5$) and than the compounds V_2O_3 ($VO_{1.5}$), VO_2 and V_2O_5 ($VO_{2.5}$) separately.

9.15.2.2.1 The shear phases V_nO_{2n-1}

Crystal structure

V_3O_5

The crystal structure is controversial. Space group: C_{2h}^4-P2/c [80A] or $C_{2h}^5-P2_1/c$ [76H] at RT and C_{2h}^6-C2/c above a crystallographic transition temperature $T_{tr}=154.7°C\uparrow$, $153.7°C\downarrow$ [79A]. For thermal properties at T_{tr}, see tables p. 171, 172.

lattice parameters (distances in Å, angles in °):

a	10.004	RT		76H
b	5.040			
c	9.854			
β	137.9			
a	9.859	RT		80A
b	5.0416			
c	6.991			
β	109.478			
a	9.870	$T \lesssim T_{tr}$	$Z=4$; for variation with	78C
b	5.052		temperature, see Fig. 1	
c	7.012			
β	109.1			
V	330.2 Å^3			
a	9.850	$T \gtrsim T_{tr}$	high-temperature phase	78C
b	5.042			
c	7.022			
β	109.3			
V	329.2 Å^3			
Δa	-0.13%	at $T=T_{tr}$		
Δb	-0.18%			
Δc	-0.13%			

density:

d_{calc}	4.720 g cm^{-3}	RT		78C

interatomic distances in the low-temperature phase [80A]:

In the table below, e or f next to an O−O distance means that the distance is a shared edge or part of a shared face, respectively. e or f next to a V−V distance means that the distance is across a shared edge or across a shared face, respectively. Averages in parentheses are corrected for the irregularity of the actual octahedron. Further numbers in parentheses are e.s.d.'s in the least significant digits. Distances are in Å units. RT values.

(continued)

interatomic distances in the low-temperature phase (continued)

V(11) octahedron		V(12) octahedron	
V(11)–O(11)	1.7240 (8)	V(12)–O(12)	1.8912 (8)
–O(12)	1.9240 (9)	–O(22)	1.9756 (9)
–O(21)	1.9638 (9)	–O(22)	1.9806 (9)
–O(21)	1.9650 (9)	–O(11)	2.0270 (10)
–O(31)	2.0336 (9)	–O(32)	2.0913 (10)
–O(21)	2.1734 (8)	–O(22)	2.1316 (8)
Average	1.9640	Average	2.0162
	(1.9337)		(2.0060)
O(21)–O(21) e	2.578 (2)	O(22)–O(22) e	2.579 (2)
O(21)–O(21) f	2.585 (2)	O(22)–O(22) f	2.566 (2)
O(21)–O(31) f (2×)	2.607 (2)	O(22)–O(32) f (2×)	2.711 (2)
O(21)–O(31) e	2.670 (1)	O(22)–O(32) e	2.670 (1)
O(21)–O(12) e	2.686 (1)	O(22)–O(11) e	2.789 (1)
O(21)–O(12)	2.725 (2)	O(22)–O11)	2.833 (1)
O(11)–O(12)	2.797 (2)	O(12)–O(11)	2.908 (2)
O(21)–O(11)	2.815 (1)	O(22)–O(12)	2.897 (1)
O(21)–O(12)	2.895 (2)	O(22)–O(11)	3.016 (2)
O(11)–O(31)	2.930 (1)	O(12)–O(32)	3.062 (1)
O(21)–O(11)	2.957 (1)	O(22)–O(12)	3.137 (1)
Average	2.738	Average	2.823

V(21) octahedron		V(22) octahedron	
V(21)–O(12) (2×)	1.9609 (8)	V(22)–O(32) (2×)	1.9815 (7)
–O(31) (2×)	2.0249 (7)	–O(22) (2×)	1.9954 (9)
–O(21) (2×)	2.0392 (9)	–O(11) (2×)	2.0414 (8)
Average	2.0083	Average	2.0061
	(2.0064)		(2.0051)
O(21)–O(31) e (2×)	2.670 (1)	O(22)–O(32) e (2×)	2.670 (1)
O(21)–O(12) e (2×)	2.686 (1)	O(22)–O(11) e (2×)	2.789 (1)
O(12)–O(31) (2×)	2.793 (1)	O(11)–O(32) (2×)	2.842 (1)
O(12)–O(31) (2×)	2.844 (1)	O(11)–O(32) (2×)	2.848 (1)
O(12)–O(21) (2×)	2.965 (2)	O(11)–O(22) (2×)	2.919 (2)
O(21)–O(31) (2×)	3.064 (2)	O(22)–O(32) (2×)	2.948 (2)
Average	2.837	Average	2.836

V–V distances			
V(11)–V(11) f	2.8171 (4)	V(12)–V(12) f	2.7632 (4)
V(11)–V(21) e	2.9708 (3)	V(12)–V(22) e	2.9730 (3)
V(11)–V(21) e	3.0239 (3)	V(12)–V(22) e	3.0020 (3)
V(11)–V(11) e	3.2438 (4)	V(12)–V(12) e	3.2067 (4)

The structure is based on a shear plane (see Fig. 2). Four quite different octahedra are found at low temperatures as seen in the table above. The vanadium octahedra are shown in Fig. 3; partial charges calculated are (in units of e) V(11) 3.86, V(12) 3.02, V(21) 3.06, V(22) 3.07.

At the transition, the different octahedral coordinations of V^{III} inside the shear planes become equivalent, and the vanadiums at the shear-phase edges also become equivalent. Thus, if subscript s denotes an atom at the shear plane, low-temperature V_3O_5 can be formulated as $(V^{IV})_s(V^{III})_s(V^{III}_{1/2}, V^{III}_{1/2})O_5$ and the high-temperature form as $(V^{IV}, V^{III})_s (V^{III})O_5$.

Physical property	Numerical value	Experimental conditions	Experimental method, remarks	Ref.

V_4O_7

Space group: $C_i^1 - A\bar{1}$, $Z=4$. For transition temperature T_{tr} and thermal properties, see tables p. 170, 172.

lattice parameters (distances in Å, angles in °):

a	5.509 (1)	RT	variation with temperature: see	78H,
b	7.008 (2)		Fig. 4 (transition at ≈ 240 K);	73M1,
c	12.256 (2)		positional parameters are given	72H
α	95.10 (2)		for three temperatures in [73K]	
β	95.17 (1)			
γ	109.25 (2)			
V	441.33 Å^3			

The structure is a shear phase closely related to Ti_4O_7 (q.v.) and there are two independent strings of vanadium atoms V(4)–V(2)–V(2)–V(4) and V(3)–V(1)–V(1)–V(3). At room temperature, the partial charges are (in units of e) 3.48–3.35–3.34–3.48 and 3.64–3.53–3.53–3.64 with a much shorter V(1)–V(1) distance than V(2)–V(2), V(3)–V(1) or V(2)–V(4). At ca. 240 K a transition is seen, in which charge differentiation takes place to give charge distributions of 3.11–3.22–3.22–3.11 and 3.84–3.83–3.83–3.84 at 200 K and 3.09–3.24–3.24–3.09 and 3.84–3.83–3.83–3.84 at 120 K. The main structural alteration is the development of two short V(2)–V(4) distances below the transition temperature.

V_5O_9

Space group $C_i^1 - A\bar{1}$, $Z=2$. For transition temperature and thermal properties at T_{tr}, see tables p. 170, 172.

lattice parameters (distances in Å, angles in °):

a	5.472 (1)	RT	variation with temperature: see	74M
b	7.003 (1)		Fig. 5; a clear transition is	
c	8.727 (1)		seen at ca. 130 K	
α	97.49			
β	112.40			
γ	109.01			

A view of the RT structure looking down the triclinic a-axis is seen in Fig. 6. There are two independent strings V(5)–V(3)–V(1)–V(3)–V(5) and V(6)–V(4)–V(2)–V(4)–V(6). Below the transition, charge ordering occurs, but is not complete, and there is no crystallographic evidence for any particular pairing scheme. The estimated change in partial charges at the transition is (in units of e): V(1) 3.8–4.1, V(2) 3.7–3.5, V(3) 3.6–3.8, V(4) 3.5–3.2, V(5) 3.6–3.8, V(6) 3.5–3.3.

V_nO_{2n-1} with $n>5$

The higher members have not been refined though structural aspects of the metal-insulator transition (MIT) in V_6O_{11} have been investigated [74D]. They are all triclinic, space group $C_i^1 - P\bar{1}$. Thermal properties may be found in the table p. 172.

lattice parameters (distances in Å, angles in °):

a	5.448	for V_6O_{11},	$Z=2$, $T_{tr}=170$ K	76H
b	6.998	RT		
c	30.063			
α	41.0			
β	72.5			
γ	108.9			
V	338.58 Å^3			
a	5.439	for V_7O_{13},	$Z=2$	76H
b	7.005	RT		
c	35.516			
α	40.9			
β	72.6			
γ	109.0			
V	398.32 Å^3			

(continued)

Physical property	Numerical value	Experimental conditions	Experimental method, remarks	Ref.
lattice parameters (continued)				
a	5.42	for V_8O_{15},	$Z=2$, $T_{tr}=68$ K	71H
b	7.02	RT		
c	40.65			
α	40.8			
β	73.2			
γ	109.5			
V	456.0 Å^3			
a	5.418	for V_9O_{17},	$Z=2$	81K
b	7.009	RT		
c	45.213			
α	39.3			
β	74.5			
γ	108.9			

Electronic and lattice properties, optical and magnetic data

No band-structure calculations appear to be available.

Optical spectra: Fig. 7, optical bandgap for V_3O_5: 0.62 eV at RT [78C].

magnetic data for several V_nO_{2n-1} compounds (see also Figs. 8, 9. Data from [73K, 79N1, 79N2]):

V_nO_{2n-1}	$T_{tr}^{a)}$ K	$T_N^{b)}$ K	$T > T_{tr}$						$T_{tr} > T > T_N$	
			C_{obs} cm^{3}K mol^{-1}	$C_{calc}^{c)}$ cm^{3}K mol^{-1}	p_{eff} μ_B	Θ_p K	$\chi_m^{d)} \cdot 10^4$ cm^{3}mol^{-1}	$g(E_F)$ $^{e)}$	C_{obs} cm^{3}K mol^{-1}	p_{eff} μ_B
V_2O_3	155	155	0.82	0.98	2.96	-600	320	9.8	–	–
V_3O_5	430	76	0.78	0.77	2.51	-97	85	–	–	–
V_4O_7	238	33.3	0.57	0.67	2.14	$-20(10)$	275	8.6	0.057	0.699
V_5O_9	135	28.8	0.57	0.61	2.13	-18	290	9.0	0.056	0.674
V_6O_{11}	170	24.0	0.53	0.58	2.06	$-30(10)$	287	8.8	0.24	1.364
V_7O_{13}	–	43.0	0.52	0.55	2.04	-38	162	5.0	–	–
V_8O_{15}	68	6.1	–	–	–	–	–	–	–	–
VO_2	340	–	0.68	0.37	2.33	-629	61	1.9	–	–

^{a}) Transition temperature.
^{b}) Néel temperature.
^{c}) C_{calc} is the Curie constant calculated assuming the localized V^{3+} and V^{4+} ions.
^{d}) For $T > T_{tr}$, $\chi_m \approx \chi_{m_0} + \dfrac{C_m}{T - \Theta_p}$.
^{e}) State density of electron (electrons per eV per V atom) at Fermi level E_F; calculated assuming Pauli paramagnetism for χ_m.

V_3O_5 is apparently a low-dimensional magnet [80G, 77U]. Although the susceptibility peaks at 120 K, the Néel temperature is known to be 75 K from NMR data. The magnetic susceptibility shows a discontinuity at 430 K (Fig. 10).

IR phonon wavenumbers (in cm^{-1}):

$\bar{\nu}$	710 (696)	V_3O_5, RT	figure in brackets refers to	80R
	587 (585)		stretching frequency in	
	537 (528)		20 at % ^{18}O doped material	
	492 (487)			
	462 (459)			

Transport properties

Electrical properties of several V_nO_{2n-1} compounds are collected in the table below. Further data are:

Physical property	Numerical value	Experimental conditions	Experimental method, remarks	Ref.

V_3O_5

Resistivity and thermoelectric power: Fig. 11. At 430 K a low-order transition is seen at which the conductivity changes abruptly by a factor of 10···20.

activation energy for conductivity:

E_A	0.3 eV	below T_{tr}	non-linear nature of $\ln \varrho$ vs. $1/T$	78C
	0.29 eV		plot ascribed in [77T] to a strongly	76K
	0.13 eV	above T_{tr}	temperature dependent band gap	78C

Hall mobility:

μ_H	0.43 cm^2 V^{-1} s^{-1}	below T_{tr}	orientation of sample not specified	78C

lowering of T_{tr} by doping:

ΔT_{tr}	0.5 K/at%	by Nb, Zn, Mg	T_{tr} is increased by +1.1 K on	76T,
	1.2 K/at%	by Mo	doping with 20 at% ^{18}O	78T
	2 K/at%	by Mn, W		

V_4O_7

Resistivity and thermoelectric power: Figs. 12, 13.
Immediately below T_{tr} (250 K) $E_A \approx 0.8$ eV but decreases rapidly as T is lowered [70O1]. Resistivity varies with pressure as shown in Fig. 14.
T_{tr} increases by +3.5 K on doping with 17.9 at% O [80R].

V_5O_7

Resistivity and thermoelectric power: Figs. 15, 16.

V_6O_{11}, V_7O_{13}, V_8O_{15}, V_9O_{17}

Resistivity and thermoelectric power: Figs. 17···20. V_7O_{13} is metallic, $m_n \approx 32 m_0$ [73M2].

summary of electrical properties of V_nO_{2n-1} compounds (from [73K, 81N, 78C]):

V_nO_{2n-1}	Characteristic of electrical conduction	T_{tr} K cooling	T_{tr} K heating	ϱ (metallic phase) Ω cm	S (metallic phase) μV/K	E_A (semiconducting phase) eV	E_F eV
V_3O_5	(semiconductive, n-type)	430	430	10^{-2}···10^{-3}	-10	0.3···0.4	0.1
V_4O_7	Metal (semiconductive, n-type)	244	250	10^{-2}···10^{-3}	-10	0.08···0.1^a)	0.15
V_5O_9	Metal (semiconductive, n-type)	129	135	10^{-2}···10^{-3}	-20	0.1···0.2	0.2
V_6O_{11}	Metal (semiconductive n-type)	174	177	10^{-2}···10^{-3}	-10	0.12	–
V_7O_{13}	Metal	–	–	10^{-3}	0···-1	–	–
V_8O_{15}	Metal (semiconductive, n-type)	–	70	10^{-3}	-5···-20	0.13	–
V_9O_{17}	Metal (semiconductive, n-type)	–	79	10^{-3}···10^{-4}	–	–	–

^{a}) Mean value; $\log \sigma$ vs. $1/T$ non-linear.

Fig. p. 454

Thermal properties at T_{tr}

heat and entropy changes of transition (from [73K, 78C, 79N2]):

V_nO_{2n-1}	ΔH_{tr} cal/mol [73K]	T_{tr} K [73K]	[79N2]	ΔS_{tr} cal mol^{-1} K^{-1} ($\Delta H_{tr}/T_{tr}$)
V_3O_5	215	430^a)	430	0.50
V_4O_7	142	250	238	0.57
V_5O_9	215	135	135	1.59
V_6O_{11}	222	170	170	1.31
V_7O_{13}	–	–	–	–
VO_2	1012	340		3.0
V_8O_{15}			68	

^{a}) From [78C].

Heat capacity at low temperature: Fig. 21.

some further data:

V_3O_5: $\Delta H_{tr} = 650\,(30)$ cal mol^{-1} according to [77K], $dT_{tr}/dp = 2.1$ K kbar^{-1}, $\Delta V = 0.3\%$ [78C].

V_4O_7: $\Theta_D = 729$ K, $dT_{tr}/dp = -0.65$ K kbar^{-1} [79T], $= -0.20$ K kbar^{-1} [73M].

V_5O_9: $dT_{tr}/dp = -0.8$ K kbar^{-1} [79T].

V_6O_{11}: $dT_{tr}/dp = -2.4$ K kbar^{-1} [79T].

References for 9.15.2.2.0 and 9.15.2.2.1

63K	Killingbeck, S.: Ph.D. Thesis, University of Kansas **1963**.
67K	Kosuge, K.: J. Phys. Chem. Solids **28** (1967) 1617.
70M	Marezio, M., McWhan, D.B., Dernier, P.D., Remeika, J.P.: Mater. Res. Bull. **5** (1970) 1015.
70O1	Okinaka, H., Nagasawa, K., Kosuge, K., Bando, Y., Kachi, S., Takada, T.: J. Phys. Soc. Jpn. **28** (1970) 798.
70O2	Okinaka, H., Nagasawa, K., Kosuge, K., Bando, Y., Kachi, S., Takada, T.: J. Phys. Soc. Jpn. **29** (1970) 245.
70O3	Okinaka, H., Kosuge, K., Kachi, S., Nagasawa, K., Bando, Y., Takada, T.: Phys. Lett. **A33** (1970) 370.
70O4	Okinaka, H., Nagasawa, K., Kosuge, K., Bando, Y., Kachi, S., Takada, T.: J. Phys. Soc. Jpn. **28** (1970) 803.
71H	Horiuchi, H., Tokonami, M., Nagasawa, K., Morimato, N., Bando, Y., Takada, T.: Mater. Res. Bull. **6** (1971) 833.
72H	Horiuchi, H., Tokonami, M., Morimoto, N., Nagasawa, K.: Acta Crystallogr. **B28** (1972) 1404.
72K	Kachi, S.: AIP Conf. Proc. **10** (1972) 714.
72P	Porter, V.R., White, W.B., Roy, R.: J. Solid State Chem. **4** (1972) 250.
73K	Kachi, S., Kosuge, K., Okinaka, H.: J. Solid State Chem. **6** (1973) 258.
73M1	Marezio, M., McWhan, D.B., Dernier, P.D., Remeika, J.P.: J. Solid State Chem. **6** (1973) 419.
73M2	McWhan, D.B., Remeika, J.P., Maita, J.P., Okinaka, H., Kosuge, K., Kachi, S.: Phys. Rev. **B7** (1973) 326.
74D	Dernier, P.D.: Mater. Res. Bull. **9** (1974) 955.
74M	Marezio, M., Dernier, P.D., McWhan, D.B., Kachi, S.: J. Solid State Chem. **11** (1974) 301.
76H	Horiuchi, H., Morimato, N., Tokonami, M.: J. Solid State Chem. **17** (1976) 407.
76K1	Kartenko, N.F., Terukov, E.I., Chudnovskii, F.A.: Fiz. Tverd. Tela **18** (1976) 1874.
76K2	Khoi, N.N., Simon, T.R., Eastwood, H.K.: Mater. Res. Bull. **11** (1976) 873.
76T	Terukov, E.I., Chudnovsky, F.A., Brückner, W., Reichelt, W., Brückner, H.P., Moldenhauer, W., Oppermann, H.: Phys. Status Solidi (a) **38** (1976) K23.
77K	Keen, H.V., Honig, J.M.: Mater. Res. Bull. **12** (1977) 277.
77U	Ueda, Y., Kosuge, K., Kachi, S.: Mater. Res. Bull. **12** (1977) 763.
77T	Terukov, E.I., Thomskii, D.I., Chudnovskii, F.A.: Zh. Eksp. Teor. Fiz. **73** (1977) 2217.
78C	Chudnovskii, F.A., Terukov, E.I.: Solid State Commun. **25** (1978) 573.
78H	Hodeau, J.-L., Marezio, M.: J. Solid State Chem. **23** (1978) 253.

78K	Khattak, G.D., Keesom, P.H., Faile, S.P.: Phys. Rev. **B 18** (1978) 6181.
78T	Terukov, E.I., Reichelt, W., Oppermann, H.: Phys. Status Solidi (a) **45** (1978) K77.
79A	Åsbrink, S., Hong, S.-H.: Nature (London) **279** (1979) 624.
79N1	Nagata, S., Griffing, B.F., Khattak, G.D., Keesom, P.H.: J. Appl. Phys. **50** (1979) 7575.
79N2	Nagata, S., Keesom, P.H., Faile, S.P.: Phys. Rev. **B 20** (1979) 2886.
79T	Terukov, E.I., Saufirov, Yu.Z., Zyuzui, A.Yu.: Fiz. Tverd. Tela **21** (1979) 1563.
80A	Åsbrink, S.: Acta Crystallogr. **B 36** (1980) 1332.
80G	Griffing, B.F., Faile, S.P., Honig, J.M.: Phys. Rev. **B 21** (1980) 154.
80R	Reichelt, W., Oppermann, H., Wagner, H., Terukov, E.I., Wolf, E.: Z. Anorg. Allg. Chem. **463** (1980) 193.
81K	Kuwamoto, H., Otsuka, N., Sato, H.: J. Solid State Chem. **36** (1981) 133.
81N	Nagata, S., Keesom, P.H., Kuwamoto, H., Otsuka, N., Sato, H.: Phys. Rev. **B 23** (1981) 411.

Physical property	Numerical value	Experimental conditions	Experimental method, remarks	Ref.

9.15.2.2.2 The phases V_nO_{2n+1}

Only three oxides have been reported with stoichiometry V_3O_7, V_4O_9 and V_6O_{13}. Of these, V_4O_9 is uncertain and no transport data have been reported for V_3O_7.

Crystal structure

V_6O_{13}

High-temperature phase: monoclinic, space group $C_{2h}^3 - C2/m$, $Z = 2$ [73S1].
Low-temperature phase: monoclinic, space group $C_{2h}^5 - P2_1/a$ [73S2, 78K].
Crystallographic transition temperature $T_{tr} = 151$ K↑, 147 K↓ [73S1].

lattice parameters (distances in Å, angles in °):

a	11.921	high-tempera-	variation with temperature: Fig. 1	74D
b	3.6811	ture phase	(refinements given by	
c	10.147	at RT	[74D, 78K, 71W])	
β	100.88		unit cell: Fig. 2	
a	11.963	low-tempera-		74D
b	3.707	ture phase		
c	10.064	at 120 K		
β	100.96			

interatomic distances (in Å) and angles (in °) (at 20 °C; from [71W]):

Two types of structural elements can be distinguished: a single zig-zag string (Fig. 2I) and a double zig-zag ribbon (Fig. 2II) with infinite extension perpendicular to the plane of the paper.

(a) *Structure element I*, the single zig-zag string

Metal oxygen distances		Bond angles		Bond angles at V(11)	Distances between the oxygen atoms
V(11)—O(41)	1.766 (1)	O(41)—V(11)—O(11)		179.6 (2)	3.830 (4)
—O(14) (2×)	1.876 (1)	O(41)	O(14) (2×)	101.2 (2)	2.815 (3)
—O(51)	1.964 (5)	O(41)	O(51)	89.0 (2)	2.618 (4)
—O(61)	1.993 (4)	O(41)	O(61)	89.4 (2)	2.649 (4)
—O(11)	2.064 (4)	O(11)	O(14) (2×)	78.8 (2)	2.504 (5)
		O(11)	O(51)	90.7 (2)	2.865 (6)
		O(11)	O(61)	91.0 (2)	2.894 (6)
		O(14)	O(14)	157.6 (3)	3.680 (1)
		O(14)	O(51) (2×)	90.2 (2)	2.721 (4)
		O(14)	O(61) (2×)	90.1 (2)	2.740 (4)
		O(51)	O(61)	178.4 (2)	3.958 (6)
		Metal-metal separations (<3.50 Å)			
		V(11)—V(14)	3.047 (2)		(continued)

Physical property	Numerical value	Experimental conditions	Experimental method, remarks		Ref.

interatomic distances and angles (continued)

(b) *Structure element II*, the double zig-zag ribbon

Metal-oxygen distances		Bond angles		Bond angles at V(21) and V(31)	Distances between the oxygen atoms
V(21)–O(51)	1.655(5)	O(51)–V(21)–O(32)		102.3(2)	2.925(6)
–O(71)	1.761(4)	O(51)	O(23)(2×)	97.4(2)	2.677(4)
–O(23)(2×)	1.902(1)	O(51)	O(71)	104.9(3)	2.708(6)
–O(32)	2.084(4)	O(51)	O(72)	176.5(2)	3.930(6)
–O(72)	2.277(5)	O(32)	O(23)(2×)	76.1(2)	2.462(4)
V(31)–O(61)	1.641(4)	O(32)	O(71)	152.8(2)	3.738(6)
–O(33)(2×)	1.919(1)	O(32)	O(72)	74.1(2)	2.633(6)
–O(72)	1.928(4)	O(23)	O(23)	150.7(3)	3.680(1)
–O(22)	1.981(4)	O(23)	O(71)(2×)	100.2(2)	2.810(4)
–O(32)	2.261(4)	O(23)	O(72)(2×)	81.8(2)	2.751(4)
		O(71)	O(72)	78.6(3)	2.589(9)
		O(61)–V(31)–O(32)		176.8(2)	3.901(6)
		O(61)	O(33)(2×)	103.4(2)	2.800(4)

Valence balance gives V(1): $+4.16e$, V(2): $+4.60e$, V(3): $+4.34e$ [74D]. The only crystallographic change below 147 K is that V(1) shifts by 0.14 Å along the *b*-axis, destroying the mirror plane [78K].

V_3O_7

structure: monoclinic, space group C_{2h}^6–C2/c, unit cell: Fig. 3 [74W].

structural elements:

(1) single chains of V(1)O$_6$ octahedra which are corner-linked; mean V–O distance 1.947 Å, typical of V^{4+}, (2) double chains of V(2)O$_6$ octahedra edge-shared in the *a*–*c* plane and corner-shared along the *b*-axis; mean V–O distance 1.938 Å, typical of V^{4+}, (3) zig-zag strings of trigonal bipyramids, V(3)O$_5$, linked by edges; mean V–O distance 1.83 Å, typical of V^{5+}, (4) zig-zag strings of V(4) and V(5) polyhedra linked by edges; mean V–O distances: V(4)–O 1.813 Å, V(5)–O 1.826 Å, both typical of V^{5+}. The structural data suggest that the compound may best be formulated $V_2^{5+}V^{4+}O_7$ [72C].

lattice parameters (distances in Å, angles in °):

a	21.921	$T=298$ K	$Z=12$	74B
b	3.679			
c	18.341			
β	95.61			
a	21.91	4 K		74B
b	3.68			
c	18.29			
β	95.7			

V_4O_9

Different formulations:

lattice parameters (in Å):

a	17.926(4)	RT	orthorhombic, space group D_{2h}^7–Pmna, $Z=4$	70W
b	3.631(1)			
c	9.396(2)			
a	8.235	RT	orthorhombic, $Z=8$	77G
b	10.32			
c	16.47			
a=b	8.215	RT	$Z=4$	69T
c	10.32			

Physical properties

V_6O_{13}

Magnetic susceptibility: Figs. 4, 5. $T_N \approx 55$ K. The transition at 150 K has been reported as semiconductor-semiconductor [73S1] and semiconductor-metal [74K]. Considerable anisotropy of conductivity has been reported (Fig. 6). The Seebeck coefficient shows the material is p-type below 150 K (Fig. 7) [74K]. IR bands at 895 and 865 cm^{-1} shift to 890 and 862 cm^{-1} on 22.76 at % ^{18}O doping; T_{tr} increases by +1.0 K on same doping [80R].

V_3O_7

Magnetic susceptibility: Fig. 8. For high temperatures $\chi_m \propto C_m/(T-\Theta_p)$ with $C_m = 0.40$ cm^3 K mol^{-1} and $\Theta_p = 10(5)$ K according to [74B], $C_m = 0.446$ cm^3 K mol^{-1} and $\Theta_p = 4$ K, $p_{eff} = 1.89 \mu_B$ according to [79N]. At low temperatures the material orders antiferromagnetically; $T_N \approx 18$ K [74B], <5 K [79N]. It appears to be insulating but no detailed data are available.

References for 9.15.2.2.2

69T	Théobald, F., Cabala, R., Barnard, J.: C. R. Acad. Sci. Ser. C **269** (1969) 1209.
70W	Wilhelmi, K.A., Waltersson, K.: Acta Chem. Scand. **24** (1970) 3409.
71W	Wilhelmi, K.A., Waltersson, K., Kihlborg, L.: Acta Chem. Scand. **25** (1971) 2671.
72C	Casalot, A.: Mater. Res. Bull. **7** (1972) 903.
73S1	Saeki, M., Kimizuka, N., Ishii, M., Kawada, I., Ichinose, A., Nakahira, M.: J. Cryst. Growth **18** (1973) 101.
73S2	Saeki, M., Kimizuka, N., Ishii, M., Kawada, I., Nakahira, M.: J. Less Common Met. **32** (1973) 171.
74B	Bayard, M., Grenier, J.C., Pouchard, M., Hagenmuller, P.: Mater. Res. Bull. **9** (1974) 1137.
74D	Dernier, P.D.: Mater. Res. Bull. **9** (1974) 955.
74G	Gossard, A.C., DiSalvo, F.J., Erich, L.C., Remeika, J.P., Yasuoka, H., Kosuge, K., Kachi, S.: Phys. Rev. **B10** (1974) 4178.
74K	Kawashima, K., Ueda, Y., Kosuge, K., Kachi, S.: J. Cryst. Growth **26** (1974) 321.
74W	Waltersson, K., Forslund, B., Wilhelmi, K.A., Andersson, S., Galy, J.: Acta Crystallogr. **B30** (1974) 2644.
76U	Ueda, Y., Kosuge, K., Kachi, S.: Mater. Res. Bull. **11** (1976) 293.
76W	Waltersson, K.: Chem. Commun. Univ. Stockholm **1976**, No. 7.
77G	Grymonprez, G., Fiermans, L., Vennik, J.: Acta Crystallogr. **A33** (1977) 834.
78K	Kiwizuka, N., Nahano-Onoda, M., Kato, K.: Acta Crystallogr. **B34** (1978) 1037.
79N	Nishihara, H., Ueda, Y., Kosuge, K., Yasuoka, H., Kachi, S.: J. Phys. Soc. Jpn. **47** (1979) 790.
80R	Reichelt, W., Oppermann, H., Wagner, H., Terukov, E.I., Wolf, E.: Z. Anorg. Allg. Chem. **463** (1980) 193.

Physical property	Numerical value	Experimental conditions	Experimental method, remarks	Ref.

9.15.2.2.3 Vanadium oxide (V_2O_3)

(See also data on V_nO_{2n-1})

Crystal structure
Pure material:

a) room-temperature phase:

Corundum structure, trigonal space group $D_{3d}^6 - R\bar{3}c$, $Z=6$. This phase is normally V-deficient, phase diagram: Fig. 1. Detailed structure: Fig. 2.

lattice parameters (distances in Å):

a	4.9517	$T = 298$ K	lattice parameters as a function of x in V_2O_{3+x} and as a function of pressure: Fig. 3	78B1, 78B2
c	14.005			
c/a	2.8283			

(continued)

9.15.2.2.3 Vanadium oxide (V_2O_3)

Physical property	Numerical value	Experimental conditions	Experimental method, remarks	Ref.
lattice parameters (continued)				
a	$4.954 + 1.619 \cdot 10^{-4} T$ $- 6.905 \cdot 10^{-8} T^2$	298···1273 K	anomalous variation of lattice parameters above room temperature: Fig. 4	73E
c	$13.963 - 1.68 \cdot 10^{-4} T$ $+ 1.577 \cdot 10^{-7} T^2$			
a	4.9492		variation of c/a ratio with temperature: Fig. 5	75R
c	13.998			
c/a	2.828			
V	296.94 Å^3			
a	4.951			73P
c	14.007			
a_{rh}	5.474	300 K	rhombohedral cell parameters	78B
α_{rh}	53.80°			

interatomic distances (in Å) and angles (in °) (from [75R]): (see also Fig. 6)
(Two values for each distance correspond to the lower and upper limits.)

	$T=$ 23°C	115°C	212°C	300°C	400°C	600°C
V(1)–V(2)	2.697 (1)	2.705 (1)	2.718 (1)	2.728 (1)	2.734 (1)	2.738 (1)
	2.705	2.714	2.730	2.741	2.748	2.755
V(1)–V(3)	2.880 (1)	2.888 (1)	2.900 (1)	2.910 (1)	2.916 (1)	2.924 (1)
	2.885	2.895	2.909	2.920	2.927	2.939
V(1)–O(1)	2.051 (1)	2.053 (1)	2.057 (2)	2.061 (1)	2.064 (1)	2.066 (1)
	2.061	2.065	2.072	2.078	2.083	2.090
V(1)–O(5)	1.968 (1)	1.970 (1)	1.973 (1)	1.975 (2)	1.977 (2)	1.981 (1)
	1.978	1.983	1.989	1.993	1.997	2.006
O(1)–O(2)	2.676 (3)	2.675 (3)	2.675 (4)	2.676 (3)	2.680 (3)	2.681 (3)
	2.685	2.685	2.687	2.690	2.696	2.701
O(1)–O(4)	2.804 (1)	2.802 (1)	2.800 (1)	2.799 (2)	2.799 (2)	2.800 (1)
	2.813	2.813	2.814	2.814	2.815	2.821
O(1)–O(5)	2.889 (1)	2.891 (2)	2.893 (1)	2.895 (2)	2.897 (1)	2.901 (1)
	2.894	2.898	2.903	2.906	2.909	2.916
O(4)–O(5)	2.952 (1)	2.962 (2)	2.978 (1)	2.990 (1)	2.997 (2)	3.007 (1)
	2.958	2.973	2.989	3.002	3.010	3.023
O(1)–V(1)–O(2)	81.45 (7)	81.30 (7)	81.11 (7)	80.95 (7)	80.93 (7)	80.89 (7)
O(1)–V(1)–O(4)	88.46 (2)	88.27 (2)	87.99 (2)	87.77 (2)	87.67 (2)	87.52 (2)
O(1)–V(1)–O(5)	91.90 (5)	91.85 (5)	91.74 (5)	91.65 (5)	91.55 (5)	91.52 (5)
O(1)–V(1)–O(6)	168.62 (8)	168.26 (8)	167.74 (8)	167.33 (8)	167.17 (8)	167.00 (7)
O(4)–V(1)–O(5)	97.17 (3)	97.49 (3)	97.98 (3)	98.36 (3)	98.56 (3)	98.73 (3)
V(1)–O(1)–V(2)	82.23 (9)	82.44 (9)	82.69 (9)	82.90 (9)	82.92 (9)	82.98 (9)
V(1)–O(2)–V(3)	91.54 (3)	91.73 (2)	92.01 (2)	92.23 (2)	92.35 (2)	92.48 (2)
V(2)–O(2)–V(3)	133.21 (5)	133.18 (4)	133.11 (4)	133.04 (4)	132.98 (4)	132.98 (4)

Only single-phase behaviour is seen in this temperature range [75M1, 74C1, 76R, 78B], and the anomalous variations in lattice parameters appear to correspond to a broad higher-order transition centred at 533 K.

b) low-temperature phase:

At 170 K↑ [80U, 70N], 155 K↓ [70D1, 80U, 70N] a transition from or to a low-temperature phase occurs. Structure: monoclinic, space group $C_{2h}^6 - I2/a$, $Z = 4$ (Fig. 3) [81W].

lattice parameters (distances in Å, angle in °):

a	7.255	$T = 148$ K		81W
b	5.002			
c	5.548			
β	96.75			

9.15.2.2.3 Vanadium oxide (V_2O_3)

Physical property	Numerical value	Experimental conditions	Experimental method, remarks	Ref.

changes in interatomic distances (in Å) (from [70D1]): (see also Fig. 6)

	Monoclinic V_2O_3 (T = 148 K)	Corundum*) (trigonal) V_2O_3 (T = 148 K)
V(1)–V(2)	2.745 (2)	2.700 (1)
V(1)–V(3) (a–c plane)	2.987 (4)	2.872 (1)
V(1)–V(3) (adjacent layer)	2.861 (9)	2.872 (1)
V(1)–V(3) (adjacent layer)	2.876 (9)	2.872 (1)
V(1)–O(1)	2.028 (8)	2.046 (1)
V(1)–O(5)	2.065 (14)	2.046 (1)
V(1)–O(3)	2.108 (12)	2.046 (1)
V(1)–O(2)	1.989 (10)	1.967 (1)
V(1)–O(4)	1.954 (11)	1.967 (1)
V(1)–O(6)	1.963 (11)	1.967 (1)
O(1)–O(2)	2.84 (1)	2.803 (1)
O(1)–O(3)	2.67 (3)	2.664 (2)
O(1)–O(5)	2.68 (2)	2.664 (2)
O(3)–O(5)	2.68 (2)	2.664 (2)
O(1)–O(6)	2.83 (1)	2.890 (1)
O(2)–O(6)	2.99 (3)	2.947 (1)
O(2)–O(4)	2.99 (3)	2.947 (1)
O(4)–O(6)	2.97 (1)	2.947 (1)

*) Corundum data obtained from low-temperature lattice parameters and RT positional parameters.

thermodynamic data at the phase transition:

ΔH_{tr}	1.49 (1) kJ mol^{-1}	heat capacity anomaly at 150···170 K	72K
	2.0 kJ mol^{-1}		76K1
	2.95 kJ mol^{-1}		36A
ΔS_{tr}	9.0 (1) J mol^{-1} K^{-1}		72K
	13 J mol^{-1} K^{-1}		76K1
	20 J mol^{-1} K^{-1}		36A
dT_{tr}/dp	$-4.1 \cdot 10^{-3}$ K bar^{-1}		76K1
	$-3.1 \cdot 10^{-3}$ K bar^{-1}		64M

Chromium and aluminium doped material:

Doping with Cr has a marked effect on the lattice parameters [70D2]. Two forms of V_2O_3:Cr have been detected [74H, 75R] labelled α and β. The α-form converts smoothly into the β-form over a wide temperature range. For anomalies in Cr and Al doped samples, see p. 178.

interatomic distances (in Å) and **angles** (in °) (from [75R]): (see also Fig. 6)
(two values correspond to upper and lower limits on crystallographic distances)

	α Cr–V_2O_3	β Cr–V_2O_3		
	T = 23 °C	T = 23 °C	T = 113 °C	T = 310 °C
M(1)–M(2)	2.700 (1)	2.747 (1)	2.746 (1)	2.739 (1)
	2.706	2.754	2.754	2.750
M(1)–M(3)	2.884 (1)	2.917 (1)	2.919 (1)	2.920 (1)
	2.889	2.922	2.925	2.929

(continued)

9.15.2.2.3 Vanadium oxide (V_2O_3)

Physical property	Numerical value	Experimental conditions	Experimental method, remarks	Ref.

interatomic distances and angles (continued)

	α Cr–V_2O_3	β Cr–V_2O_3		
	$T=23\,°C$	$T=23\,°C$	$T=113\,°C$	$T=310\,°C$
M(1)–O(1)	2.050 (1)	2.061 (1)	2.061 (1)	2.062 (2)
	2.060	2.070	2.072	2.078
M(1)–O(5)	1.970 (1)	1.976 (1)	1.977 (1)	1.979 (1)
	1.981	1.986	1.989	1.996
O(1)–O(2)	2.673 (3)	2.661 (2)	2.663 (2)	2.670 (2)
	2.682	2.669	2.673	2.683
O(1)–O(4)	2.802 (1)	2.792 (1)	2.793 (1)	2.796 (1)
	2.813	2.801	2.803	2.810
O(1)–O(5)	2.890 (1)	2.897 (1)	2.898 (1)	2.899 (1)
	2.899	2.904	2.905	2.910
O(4)–O(5)	2.958 (2)	3.004 (1)	3.006 (1)	3.006 (1)
	2.966	3.011	3.015	3.016
O(1)–M(1)–O(2)	81.36 (5)	80.42 (4)	80.47 (4)	80.70 (4)
O(1)–M(1)–O(4)	88.36 (2)	87.51 (1)	87.47 (1)	87.51 (1)
O(1)–M(1)–O(5)	91.91 (4)	91.71 (3)	91.68 (3)	91.63 (3)
O(1)–M(1)–O(6)	168.43 (6)	166.50 (5)	166.52 (5)	166.81 (5)
O(4)–M(1)–O(5)	97.32 (2)	98.93 (2)	98.95 (2)	98.79 (2)
M(1)–O(1)–M(2)	82.36 (6)	83.60 (5)	83.54 (6)	83.24 (6)
M(1)–O(2)–M(3)	91.64 (2)	92.49 (1)	92.53 (2)	92.47 (1)
M(2)–O(2)–M(3)	133.23 (3)	133.05 (2)	133.04 (3)	133.04 (3)

lattice parameters of Cr doped V_2O_3 (distances in Å):

a	4.9540	α-form with		74R
c	13.9906	1 at % Cr, RT		
c/a	2.828			
a	4.9974	β-form with		
c	13.9260	1 at % Cr, RT		
c/a	2.787			
a	4.998	3 at % Cr, RT	only one crystallographic form reported	74C2
c	13.92			
c/a	2.785			
$\dfrac{1}{a}\dfrac{da}{dT}$	$1.0 \cdot 10^{-5}\,K^{-1}$			
$\dfrac{1}{c}\dfrac{dc}{dT}$	$0.4 \cdot 10^{-5}\,K^{-1}$			

Crystallographic effects of Ti-substitution: Fig. 3.

anomalies in Cr and Al doped material (ΔH_{tr} in J mol^{-1}):

ΔH_{tr}	2000	$T=176\,K\uparrow$, 165 K$\downarrow$	1 at % Cr doped	76K1
	160	335 K$\uparrow$		
	330	266 K$\downarrow$		
	2400	186 K$\uparrow$, 174 K$\downarrow$	14 at % Cr doped	76K1
	150	270 K$\uparrow$, 210 K$\downarrow$		
	2000	174 K$\uparrow$, 164 K$\downarrow$	1 at % Al doped	76K1
	160···180	345···365 K$\uparrow$		
	210···510	310 K$\downarrow$		

Physical property	Numerical value	Experimental conditions	Experimental method, remarks	Ref.

extended Cr- and Al-data:

Cr-data (first column: mean transition temperature (in K); second column: actual span of transition (in K); third column: ΔH_{tr} (in J mol^{-1}), precision ± 80 J mol^{-1}; fourth column: mol % Cr_2O_3):

173	18	1995···2025	0	hysteresis of ca. 3···15 K is	77K1
181	25	2115···2160	1	observed when measurements are	
186	30	2355	2	performed while cooling	
187	30	2355	3.3		
183	20	1730	4		
176	20	1630	6	only the 1% Cr-sample showed an	
164	22	1470	9	additional anomaly at 325···360 K	
151	25	1130	12		

Al-data (first column: mean transition temperature (in K); second column: ΔH_{tr} (in J mol^{-1}), precision ± 105 J mol^{-1}; third column: mol % Al_2O_3):

171	2386	0.33	only the 1% Al-sample showed a	76K2
182	2721	1	clear high temperature anomaly	
187	2345	3.3	(see above)	
179	1863	5		
166	1172	8		
148	1047	10		

Electronic properties (see also optical properties below)

Brillouin zone – see section on Ti_2O_3. Two major calculations have been reported [76A2, 78C] based on earlier work [71N, 71W, 73A2, 75A]. The major controversy is over the ordering of the e_π and a_1 components of the trigonally split t_{2g} band. [76A2] finds, on inclusion of correlation effects, that e_π lies below a_1 (Figs. 7···9), but [78C] assumes a configuration $a_1^1 e^1$ for V_2O_3. This latter is inconsistent with the antiferromagnetic coupling observed in the low-temperature form [70G]. In monoclinic V_2O_3, the e_π band is further split into e_1 and e_2 components (Figs. 10···12).

calculated orbital occupancies:

e_π	1.927	metallic phase	76A2
a_1	0.073	($T > T_{tr}$)	
$e_1 (\uparrow)$	0.900	semiconducting	
$e_1 (\downarrow)$	0.025	phase	
$e_2 (\uparrow)$	0.958	($T < T_{tr}$)	
$e_2 (\downarrow)$	0.039		
$a_1 (\uparrow)$	0.070		
$a_1 (\downarrow)$	0.070		

Several empirical band schemes have been suggested: Fig. 13 and [70G, 72H, 76S].

Lattice properties

wavenumbers of phonons (in cm^{-1}):

$\bar{v}$	203	$T = 300$ K	assignment: E_g	77K2
	209			77M
	231		A_{1g}	77K2
	238			77M
	290		E_g	77K2
	290			77M
	500		A_{1g}	77K2
	500			77M
	590		E_g	77K2

Raman spectrum in the temperature range 4···300 K: Fig. 14. A peak at 450 cm^{-1} in the antiferromagnetic phase disappears at T_{tr} and is ascribed to a magnon [77K2].

Physical property	Numerical value	Experimental conditions	Experimental method, remarks	Ref.
elastic moduli (in dyn cm^{-2}):				
c_{11}	$25.6(6) \cdot 10^{11} (1 - 7.1(3) \cdot 10^{-4} (T-273))$	$T = 150 \cdots 273$ K		76A1
c_{12}	$6.7(8) \cdot 10^{11} (1 - 2.2(1) \cdot 10^{-4} (T-273))$			
c_{14}	$1.7(3) \cdot 10^{11} (1 - 13.6(4) \cdot 10^{-4} (T-273))$			
c_{13}	$15.0(2) \cdot 10^{11} (1 - 8.8(3) \cdot 10^{-4} (T-273))$			
c_{33}	$33.6(3) \cdot 10^{11} (1 - 1.9(2) \cdot 10^{-4} (T-273))$			
c_{44}	$8.0(3) \cdot 10^{11} (1 + 5.3(2) \cdot 10^{-4} (T-273))$			
c_{11}	$26.5 \cdot 10^{11}$	RT		79Y
c_{12}	$7.15 \cdot 10^{11}$			
c_{33}	$31.6 \cdot 10^{11}$			
c_{44}	$8.55 \cdot 10^{11}$			
Debye temperature (in K):				
Θ_D	643		from elastic constants, $T = 150 \cdots 273$ K	76A1
	610		from heat capacity, $T = 150 \cdots 273$ K	66G
	585		from heat capacity, $T = 0.4 \cdots 30$ K	75W
	575		from heat capacity, $T = 0.3 \cdots 30$ K under pressure	73M2
Transport properties				
a) high-temperature phase, $T > T_{tr}$:				
resistivity:				
ϱ	$6.3(6) \cdot 10^{-4}$ Ω cm	$\| a$, $T = 273$ K	ϱ almost isotropic, $\varrho_c/\varrho_a \approx 0.9$ at 273 K and increases to ≈ 1.1 at 625 K	67F
	$5.6(6) \cdot 10^{-4}$ Ω cm	$\| c$, 273 K		
$\varrho(T)/\varrho(0°C)$	$0.36 + 2.3 \cdot 10^{-3} T$	$T = 160 \cdots 340$ K		
$\dfrac{d \ln \varrho}{dT}$	$1.8 \cdot 10^{-3}$ K^{-1}	$T = 298$ K		69M1, 69M2
$\dfrac{d \ln \varrho}{dp}$	$-1.0 \cdot 10^{-2}$ kbar^{-1}	$T = 298$ K, $p = 25$ kbar	conductivity vs. hydrostatic pressure: Fig. 15. Some anisotropy reported	69M1, 69M2
	$-1.4 \cdot 10^{-2}$ kbar^{-1}	$\varrho \| a$, $T = 298$ K		67F
	$-7.6 \cdot 10^{-4}$ kbar^{-1}	$\varrho \| c$, 298 K		
change of transition temperature with pressure (see Fig. 15):				
dT_{tr}/dp	$-3.78 \cdot 10^{-3}$ K bar^{-1}		hydrostatic pressure $1 \cdots 6000$ bar	67F
$\dfrac{d \ln T_{tr}}{dp}$	$-2.6 \cdot 10^{-5}$ bar^{-1}		T_{tr} increases by 2.3 K on doping with 11 at % ^{18}O [80R]	
effective mass:				
m_n	$50 \, m_o$			67F
	$31 \, m_o$		from heat capacity near 0 K at 20 kbar	73M2
	$52 \, m_o$		$V_{1.97}O_3$, near 0 K	73M2
carrier mobility:				
μ	0.2 cm^2 V^{-1} s^{-1}	RT	orientation not specified	67F
μ_H	$0.2 \cdots 0.6$ cm^2 V^{-1} s^{-1}	RT	current in basal plane ($\| a$), $B \| c$	69A
Hall coefficient:				
R_H	$2.3(1) \cdot 10^{-4}$ cm^3 C^{-1}	$T = 300 \cdots 800$ K	n-type material, Fig. 16; $B \| c$; I in basal plane	69A

Physical property	Numerical value	Experimental conditions	Experimental method, remarks	Ref.
carrier concentration:				
n	$2.7 \cdot 10^{22}$ cm^{-3}		0.6 carriers per V-atom	69A
Seebeck coefficient:				
S	$11 \cdots 13$ μV K^{-1}	along c-axis, $T = 170 \cdots 800$ K	positive(!), Fig. 17	69A
b) low-temperature phase, $T < T_{tr}$:				
resistivity:				
ϱ	$0.5 \cdot 10^4$ Ω cm	$\|c$, $T = 132$ K	activation energy 0.12 eV	67F
	$1.3 \cdot 10^4$ Ω cm	$\|b$, 132 K	activation energy 0.18 eV	

For stress X the resistivity varies as $\varrho(X,T) = A(0)\exp((E + aX)/kT)$ with $a = -1.5 \cdot 10^{-6}$ eV bar^{-1} for uniaxial stress along b-axis, $= -1.1 \cdot 10^{-6}$ eV bar^{-1} for hydrostatic pressure [67F].

variation of the transition temperature with uniaxial stress:

dT_{tr}/dX	$-6.8(5) \cdot 10^{-3}$ K bar^{-1}	$X\|a$		67F
	$-4.1(5) \cdot 10^{-3}$ K bar^{-1}	$X\|b$		
	$-0.5(5) \cdot 10^{-3}$ K bar^{-1}	$X\|c$		

Seebeck coefficient: Fig. 18. S is large and negative.

Hall data: No data reported.

transport properties for V_2O_{3+x}: Fig. 19. No semiconducting phase formed for $x \gtrsim 0.04$ (see also Fig. 1). Discontinuity in resistivity $\Delta\varrho$ is independent of x; $\Delta\varrho \simeq 10^7$ Ω cm at T_{tr} [80U]. T_{tr} decreases with x (Fig. 19).

high-temperature transition: $\varrho(T)/\varrho(0°C)$ rises anomalously rapidly in the region of 525 K (Fig. 16).

Cr doped material: Conductivity: Fig. 21, Seebeck coefficient: Fig. 22, conductivity of $(V_{1-x}Cr_x)_2O_{3+y}$: Fig. 23.

Al doped material: Conductivity: Fig. 24.

Ti doped material: Conductivity: Fig. 25.

Optical properties

XPE spectrum: Fig. 26, reflectivity and optical conductivity: Fig. 27, electroreflectance: Fig. 28. Further PE data, showing development of a clear Fermi edge above T_{tr} in [80B].

Absorption spectra: measurements for different crystalline directions at 93 K [73A1] show some anisotropy, with a peak found at 0.2 eV ($E\|c$), of width 0.14 eV, assigned to an optical exciton. Otherwise absorption above 0.1 eV is assigned to a plasmon [71Z] with $m_{\omega_p} = 11 m_o$ (calculated from $n = 3.85 \cdot 10^{22}$ cm^{-3} [68Z, 69A]) or $m_{\omega_p} = 4 m_o$ [76S]. An absorption peak at 0.32 eV ($E \perp c$) at 93 K is ascribed to impurities, possibly associated with V-defects [73A1].

characteristic peak energies in optical spectra (in eV, RT values):

E	0.48	thermo-reflectance		71S
	0.95	thermo-reflectance	assignment acc. to [76S]: plasmon	76S
	1.45	reflectance	$e_\pi \to a_1$	71V
	1.4	thermoreflectance, $E \perp c$		76S
	1.4	electro-reflectance		71V
	1.1	SXS	SXS: soft X-ray spectrum	69F2
	1.9	reflectance	$e_\pi \to e_\pi^*$	71V
	1.9	thermoreflectance, $E\|c$		76S

(continued)

Physical property	Numerical value	Experimental conditions	Experimental method, remarks	Ref.
characteristic peak energies in optical spectra (continued)				
	2.0	electro-reflectance	shoulder	71V
	1.9	SXS		69F2
	2.3···2.7	reflectance	$e_\pi \to a_1^*$	71V
	2.2	thermoreflectance, $E \perp c$		76S
	2.98	thermo-reflectance	$e_\pi \to e_\sigma$	71S
	2.9	electro-reflectance		71V
	2.9	SXS		69F2
	3.4	thermo-reflectance, unpolarized	$O2p \to 3d_\pi$	76S
	4.0	electro-reflectance		71V
	3.5	SXS		69F2
	4.6	reflectance	$O2p \to a_1, a_1^*$	71V
	4.9	thermo-reflectance, unpolarized		76S
	4.5	electro-reflectance		71V
	5.6	reflectance	$O2p \to a_1, a_1^*$	71V
	5.5	thermo-reflectance, unpolarized		76S
	5.5	electro-reflectance	shoulder	71V
	5.5 (1.0)	EELS	assignment: $O2p \to V3d$	74S
	10.4 (1)	EELS	surface plasmon	
	22.1 (1)	EELS	volume plasmon	
	32.3 (2)	EELS	volume and surface plasmon	
	40.2 (5)	EELS	$VM_{2,3}$ (V: valence orbitals, M_i: 3s, 3p core levels)	
	49.8 (1)	EELS	localized interband transition	
	68.5 (5)	EELS	VM_1	

Magnetic properties

Magnetic susceptibility for V_2O_{3+x}: Fig. 29. V_2O_3 is antiferromagnetically ordered below 155 K, with spin structure shown in Fig. 2 [81W]. The spins are ordered in (010) planes of monoclinic cell or (110) hexagonal plates with moments canted at 71° to the c-axis [75M].

exchange constants and anisotropy (in meV):

J_α	−27.2	81W
J_β	45.9	
J_γ	−23.2	
J_δ	2.81	
J_ε	−4.76	
A_u	9.51	

Physical property	Numerical value	Experimental conditions	Experimental method, remarks	Ref.

Magnon dispersion: Fig. 30. Effective magnetic moment in the monoclinic phase is $1.25\,\mu_B$ for the V^{3+} ion [70H] or $1.2\,\mu_B$ [70M2]. The magnetic susceptibility in the semiconducting phase ($T<T_{tr}$) is highly anisotropic [71G1]. For p_{eff}, χ_m and C_m, see also table above (magnetic data for V_nO_{2n-1} compounds). For $T>T_N$ ($=T_{tr}$) the susceptibility is given by $\chi \approx \chi_0 + C/(T-\Theta_p)$ with $\chi_0 = 2.8 \cdot 10^{-6}\,\text{cm}^3\,\text{g}^{-1}$ and:

C	$0.00926\,\text{cm}^3\,\text{K}\,\text{g}^{-1}$	$T<400\,\text{K}$, $\chi_\perp$		68A
Θ_p	$-649\,\text{K}$			
p_{eff}	$2.352\,\mu_B$			
C	$0.00922\,\text{cm}^3\,\text{K}\,\text{g}^{-1}$	$T<400\,\text{K}$, $\chi_\parallel$	a small magnetic anomaly is	
Θ_p	$-660\,\text{K}$		reported in $\chi_\parallel$ at 518 K	
p_{eff}	$2.347\,\mu_B$		[77P, 78K]	
C	$0.01134\,\text{cm}^3\,\text{K}\,\text{g}^{-1}$	$T>500\,\text{K}$, $\chi_\perp$		
Θ_p	$-743\,\text{K}$			
p_{eff}	$2.603\,\mu_B$			
C	$0.01183\,\text{cm}^3\,\text{K}\,\text{g}^{-1}$	$T>500\,\text{K}$, $\chi_\parallel$		
Θ_p	$-740\,\text{K}$			
p_{eff}	$2.659\,\mu_B$			

magnetic phase diagrams: for V_2O_{3+x}, see Fig. 1; combined magnetic phase pressure/composition diagram for V_2O_3 : Cr and V_2O_3 : Ti, see Fig. 31; for $V_2O_3 - Cr_2O_3$ system, see Fig. 32; magnetic data, see [71G2].

References for 9.15.2.2.3

36A	Anderson, C.T.: J. Am. Chem. Soc. **58** (1936) 564.
64M	Minomura, S., Nagasaki, H.: J. Phys. Soc. Jpn. **19** (1964) 131.
66G	Gopal, E.S.R.: "Specific Heats of Low Temperature"; London: Heywood **1966**.
67F	Feinleib, J., Paul, W.: Phys. Rev. **155** (1967) 841.
68A	Arnold, D.J., Mires, R.W.: J. Chem. Phys. **48** (1968) 2231.
68Z	Zhuze, V.P., Andreev, A.A., Shelykh, A.I.: Fiz. Tverd. Tela **10** (1968) 3674.
69A	Austin, I.G., Turner, C.E.: Philos. Mag. **19** (1969) 939.
69F1	Feinleib, J.: "Electronic Structure in Solids"; New York: Plenum Press, **1969**.
69F2	Fisher, D.W.: J. Appl. Phys. **40** (1969) 4151.
69M1	McWhan, D.B., Rice, T.M.: Phys. Rev. Lett. **22** (1969) 887.
69M2	McWhan, D.B., Rice, T.M., Remeika, J.P.: Phys. Rev. Lett. **23** (1969) 1384.
70B	Barker, A.S., Remeika, J.P.: Solid State Commun. **8** (1970) 1521.
70D1	Dernier, P.D., Marezio, M.: Phys. Rev. **B2** (1970) 3771.
70D2	Dernier, P.D.: J. Phys. Chem. Solids **31** (1970) 2569.
70G	Goodenough, J.B.: Proc. 10th Int. Conf. Phys. Semicond. **1970**, 304.
70H	Heidemann, A.: Z. Phys. **238** (1970) 208.
70M1	McWhan, D.B., Remeika, J.P.: Phys. Rev. **B2** (1970) 3734.
70M2	Moon, R.M.: Phys. Rev. Lett. **25** (1970) 527.
70N	Nakatina, M., Horuichi, S., Ooshina, H.: J. Appl. Phys. **41** (1970) 836.
71G1	Greenwood, M., Mires, R.W., Smith, A.R.: J. Chem. Phys. **54** (1971) 1417.
71G2	Gossard, A.C., Menth, A., Warren, W.W., Remeika, J.P.: Phys. Rev. **B3** (1971) 3993.
71N	Nebenzahl, I., Weger, M.: Philos. Mag. **24** (1971) 1119.
71S	Shtorch, P., Yacoby, Y.: Phys. Lett. **A36** (1971) 89.
71V	Valiev, K.A., Kopaev, Yu.V., Mokerov, M.G., Rakov, A.V., Solov'ev, S.G.: Zh. Eksp. Teor. Fiz. **60** (1971) 2175.
71W	Weger, M.: Philos. Mag. **24** (1971) 1095.
71Z	Zhuze, V.P., Lukurskii, D.P., Startsev, G.P.: Fiz. Tverd. Tela **13** (1971) 317.
72H	Honig, J.M., Van Zandt, L.L., Board, R.D., Weaver, H.E.: Phys. Rev. **B6** (1972) 1323.
72K	Khlyustrov, V.G., Borukhovich, A.D., Perelyaev, V.A.: Fiz. Tverd. Tela **14** (1972) 2142.
72R	Reid, A.F., Sabine, T.M., Wheeler, D.A.: J. Solid State Chem. **4** (1972) 400.
73A1	Andrianov, G.O., Aronov, A.G., Smirnova, T.V., Chudnovskii, F.A.: Phys. Status Solidi (b) **60** (1973) 79.
73A2	Ashkenazi, J., Weger, M.: Adv. Phys. **22** (1973) 207.

References for 9.15.2.2.3

73E	Eckert, L.J., Bradt, R.C.: J. Appl. Phys. **44** (1973) 3470.
73M1	McWhan, D.B., Menth, A., Remeika, J.P., Brinkmann, W.F., Rice, T.M.: Phys. Rev. **B7** (1973) 1920.
73M2	McWhan, D.B., Remeika, J.P., Bader, S.D., Triplett, B.B., Phillips, W.E.: Phys. Rev. **B7** (1973) 3079.
73M3	McWhan, DB., Remeika, J.P., Maita, J.P., Okinaka, H., Kosuge, K., Kachi, S.: Phys. Rev. **B7** (1973) 326.
73P	Pouchard, M., Launay, J.C.: Mater. Res. Bull. **8** (1973) 95.
74C1	Chandrashekhar, G.V., Sinha, A.P.B., Honig, J.M.: Phys. Lett. **A47** (1974) 185.
74C2	Chandrashekhar, G.V., Sinha, A.P.B.: Mater. Res. Bull. **9** (1974) 787.
74H	Honig, J.M., Chandrashekhar, G.V., Sinha, A.P.B.: Phys. Rev. Lett. **32** (1974) 13.
74R	Robinson, W.R.: Mater. Res. Bull. **9** (1974) 1091.
74S	Szalkowski, F.J., Bertrand, P.A., Samorjai, G.A.: Phys. Rev. **B9** (1974) 3369.
75A	Ashkenazi, J., Chuchem, T.: Philos. Mag. **32** (1975) 763.
75K	Keem, J.E., Honig, J.M.: Phys. Status Solidi (a) **28** (1975) 335.
75M1	McWhan, D.B., Jayaraman, A., Remeika, J.P., Rice, T.M.: Phys. Rev. Lett. **34** (1975) 547.
75M2	Motoye, K., Yasuoka, H., Nakamura, Y., Kosuge, K., Kachi, S.: J. Phys. Soc. Jpn. **39** (1975) 1137.
75R	Robinson, W.R.: Acta Crystallogr. **B31** (1975) 1153.
75S	Sinha, A.P.B., Chandrashekhar, G.V., Honig, J.M.: J. Solid State Chem. **12** (1975) 402.
75W	Wenger, L.E., Keesom, P.H.: Phys. Rev. **B12** (1975) 5288.
76A1	Andrianov, G.O., Drichke, I.L.: Fiz. Tverd. Tela **18** (1976) 1392.
76A2	Ashkenazi, J., Weger, M.: J. Phys. **37** (1976) C4–189.
76K1	Keer, H.V., Dickerson, D.L., Kuwamoto, H., Barros, H.L.C., Honig, J.M.: J. Solid State Chem. **19** (1976) 95.
76K2	Kuwamoto, H., Dickerson, D.L., Keer, H.V., Honig, J.M.: Mater. Res. Bull. **11** (1976) 1301.
76K3	Kuwamoto, H., Keer, H.V., Keem, J.E., Shivashenkar, S., Van Zandt, L.L., Honig, J.M.: J. Phys. **37** (1976) C4–35.
76R	Rice, C.E., Robinson, W.R.: Phys. Rev. **B13** (1976) 3655.
76S	Shuker, P., Yacoby, Y.: Phys. Rev. **B14** (1976) 2211.
77K1	Keer, H.V., Barros, H.L.C., Dickerson, D.L., Barfknecht, A.T., Honig, J.M.: Mater. Res. Bull. **12** (1977) 137.
77K2	Kuroda, N., Fan, H.Y.: Phys. Rev. **B16** (1977) 5003.
77M	Mirlin, D.M., Reshina, I.I.: Fiz. Tverd. Tela **19** (1977) 201.
77P	Parette, G., Rao, L.M.: Solid State Commun. **23** (1977) 179.
78B1	Belbéoch, R., Kleinberger, R., Roulliay, M.: Solid State Commun. **25** (1978) 1043.
78B2	Belbéoch, R., Kleinberger, R., Roulliay, M.: J. Solid State Chem. **39** (1978) 1007.
78C	Castellani, C., Natoli, C.R., Ranninger, J.: Phys. Rev. **B18** (1978) 4945, 4967, 5001.
78K	Kajzer, F., Parette, G.: Solid State Commun. **25** (1978) 535.
79Y	Yelon, W.B.: Solid State Commun. **29** (1979) 775.
80B	Beatham, N., Fragala, I.L., Orchard, A.F., Thornton, G.: J. Chem. Soc. Faraday Trans. 2 **76** (1980) 929.
80K1	Kuwamoto, H., Honig, J.M., Appel, J.: Phys. Rev. **B22** (1980) 2626.
80K2	Kuwamoto, H., Honig, J.M.: J. Solid State Chem. **32** (1980) 335.
80R	Reichelt, W., Oppermann, H., Wagner, H., Terukov, E.I., Wolf, E.: Z. Anorg. Allg. Chem. **463** (1980) 193.
80U	Ueda, Y., Kosuge, K., Kachi, S.: J. Solid State Chem. **31** (1980) 171.
81W	Word, R.E., Werner, S.A., Yelon, W.B., Honig, J.M., Shivashenkar, S.: Phys. Rev. **B23** (1981) 3533.

9.15.2.2.4 Vanadium oxide (VO$_2$)

Crystal structure

This oxide exists in a narrow stoichiometry range VO$_{2-\delta}$ ($\delta < 0.006$) [75B2]. Below this stoichiometry, (121) CS planes have been identified [78G]. Intergrowths of V$_8$O$_{15}$, or, more rarely, V$_9$O$_{17}$, have been observed in slightly reduced material; no phase V$_n$O$_{2n-1}$ with n > 10 has been found [78G]. Values of $\delta < 0$ have also been reported [70K, 74K1, 79M].

At ca. 340 K, VO$_2$ exhibits a metal-semiconductor transition, which is accompanied by a discontinuous transition in other physical properties. Detailed structure: Fig. 1.

low-temperature phase:

structure: monoclinic, space group $C_{2h}^5 - P2_1/c$, $Z = 4$

lattice parameters (distances in Å, angles in °) (sometimes subscript m = monoclinic is used with parameters a, b, c):

		T [K]		
a	5.743	298		56A
b	4.517			
c	5.375			
β	121.61			
a	5.7517$_{30}$	298		70L
b	4.5378$_{25}$			
c	5.3825$_{25}$			
β	122.646$_{96}$			
a	5.75173	298		79K
b	4.52596			
c	5.38326			
β	122.6148			

high-temperature phase:

structure: tetragonal, space group D_{4h}^{14}-P4$_2$/mnm, $Z = 2$ (subscript R for rutile form)

lattice parameters (in Å):

		T [K]		
a_R	4.5546$_3$	360		74M
c_R	2.8514$_2$			
a_R	4.55396	360	temperature dependence: Fig. 2	79K
c_R	2.85028			
a_R	4.5555$_4$	413		74M
c_R	2.8552$_3$			
a_R	4.5561$_5$	473		
c_R	2.8598$_3$			
a_R	4.5573$_5$	523		
c_R	2.8626$_3$			
a_R	4.5580$_4$	573		
c_R	2.8663$_2$			

The chains of edge-shared VO$_6$ octahedra along the c-axis in the high-temperature (metallic) phase (Fig. 1) have V−V distances of 2.851 Å across the edge at 350 K and 2.869 Å at 470 K. For V−V across corner, V−V distance is 3.522 Å at 360 K and 3.525 Å at 470 K. Below 340 K, the vanadium ions pair up and rock off the c_R-axis (Fig. 1). Short V−V distance is 2.61914 Å [70L]. The coordination of V becomes very irregular: V−O distances (in Å, $\pm 12 \cdot 10^{-4}$ Å): 1.7628, 1.8633, 1.8935, 2.0149, 2.0241, 2.0637 [70L]. Longer V−V distance: 3.12 Å [56A].

Physical property	Numerical value	Experimental conditions	Experimental method, remarks	Ref.
coefficient of linear thermal expansion (in 10^{-6} K^{-1}):		T [K]		
α_{11}	12.11	298⋯334	1, 2 and 3 are the three monoclinic axes $\alpha_{ij} = \frac{1}{2\Delta t}\left(\frac{\partial u_i}{\partial x_j} + \frac{\partial u_j}{\partial x_i}\right)$ where u_i is the strain in the ith direction	79K
α_{22}	1.22			
α_{33}	2.57			
α_{13}	-5.43			
α_{av}	5.70			
α_{11}	4.87	339⋯360	for $T > T_{tr}$; 1, 3 refer to the a and c-axes of the rutile structure	79K
α_{33}	30.48			
α_{av}	13.35			
α_{av}	6.4	290⋯335		64K
α_{av}	17.1	459⋯470		64K
α_{11}	5.22⋯3.95	373⋯753		67R
α_{33}	26.96⋯22.37			
α_{av}	12.97			
α_{11}	3.5	360⋯573		74M
α_{33}	29.3			
α_{av}	10.4			
α_{11}	$5.828 \cdot 10^{-6} - 7.091 \cdot 10^{-9}(T-273) + 6.946 \cdot 10^{-12}(T-273)^2$ [K^{-1}]			67R
α_{33}	$29.683 \cdot 10^{-6} - 2.930 \cdot 10^{-8}(T-273) + 2.576 \cdot 10^{-11}(T-273)^2$ [K^{-1}]			
	for the high-temperature phase (T in K)			
parameters of the phase transition:				
$\Delta V/V$	0.001 (1)	at T_{tr}		70M
	0.00044			79K
$\Delta a_m/a_m$	$9.85 \cdot 10^{-3}$	at T_{tr}	change occurs over a temperature range of 0.1 K at 338 K (m: monoclinic)	79K
$\Delta b_m/b_m$	$6.12 \cdot 10^{-3}$	at T_{tr}		
$\Delta c_m/c_m$	$2.20 \cdot 10^{-3}$	at T_{tr}		
$\Delta\beta$	0.5967°	at T_{tr}		
ΔH_{tr}	4.27 kJ mol^{-1}		see also table in 9.15.2.2.1	69B1
	3.4 kJ mol^{-1}			64K
	4.68 kJ mol^{-1}			73C3
	4.33 kJ mol^{-1}			47C
change in transition temperature with pressure and uniaxial stress:				
dT_{tr}/dX	$-1.2 \cdot 10^{-3}$ K bar^{-1}	$X \| c$		69L
dT_{tr}/dp	$6 \cdot 10^{-5}$ K bar^{-1}			
dT_{tr}/dp	$0.082 \cdot 10^{-8}$ K Pa^{-1}		variation of T_{tr} with oxygen content: Fig. 3	69B2
influence of ^{18}O substitution on T_{tr}:				
ΔT_{tr}	0.8 °C	for 21 at % ^{18}O		78T

effect of dopant ions:

Two different types of aliovalent dopant have marked effects: M^{3+} and M^{5+}.

phase diagrams: $Ga_xV_{1-x}O_2$: Fig. 4, $Al_xV_{1-x}O_2$: Fig. 5, $0 \leq x \leq 0.012$ M$_1$-phase, $0.012 \leq x \leq 0.030$ T-phase, $0.030 \leq x \leq 0.045$ M$_2$-phase, $x > 0.045$ AlVO$_4$ [74D], $Fe_xV_{1-x}O_2$: Fig. 6, extended phase diagram [76K], $0.05 \leq x \leq 0.11$ M$_4$-phase, $0.11 \leq x \leq 0.13$ orthorhombic phase, $Cr_xV_{1-x}O_2$: Fig. 7, $Ti_xV_{1-x}O_2$: Fig. 8, extended phase diagram [76H1], $0.14 \leq x \leq 0.24$ M$_4$-phase, $x > 0.24$ tetragonal semiconducting rutile R$_s$-phase, $Nb_xV_{1-x}O_2$: Fig. 9.

For the M^{3+}-ion substituents, the following phases are formed:

M$_1$: monoclinic; isostructural with low temperature VO$_2$; schematic (110)$_R$ section: Fig. 10a,
M$_2$: a new monoclinic phase; schematic plan of (110)$_R$ section: Fig. 10b. Half the V c-axis pairs tilt but do not pair, the others pair but do not tilt, though the V^{4+} ions are localized [76P],

T: (also referred to in earlier literature as M_3): a new triclinic phase intermediate between M_1 and M_2,
M_4: a monoclinic phase closely related to M_2 is found for larger values of x. Some long range order has disappeared compared to M_2.

$Al_{0.015}V_{0.985}O_2$:

Detailed evolution of lattice parameters: Fig. 11, refined interatomic distances and effective charges: (Table from [77G1])

(Phases labelled as in Fig. 5. The V(1) chains are paired along the c_R ($=a_m$) axis in M_2 and the V(2) chains are tilted: Fig. 10b. In the T-phase this distinction is less clearcut.)

Phases:	R		M_2	T_{298K}	T_{173K}
Effective charge (in units of e)	3.88	V(1)	3.90	3.96	3.78
		V(2)	3.96	4.00	4.17
V–O average distances (in Å)	1.926	V(1)	1.945	1.934	1.946
		V(2)	1.946	1.935	1.921
V–V distances across the sheared face (in Å)	2.853	V(1)	2.540	2.547	2.545
			3.261	3.226	3.200
		V(2)	2.935	2.819	2.747
			2.935	3.005	3.051

$Cr_{0.024}V_{0.976}O_2$:

Detailed evaluation of lattice parameters: Fig. 12, temperature-pressure diagram: Fig. 13.

$Nb_xV_{1-x}O_2$:

Detailed structural investigations apparently not reported. Phase diagram (Fig. 9) shows M_1 structure which transforms into M' by loss of long-range order in $(110)_R$ and/or $(1\bar{1}0)_R$ planes of M_1 structure. The R' phase may similarly be derived from the rutile R phase [74C, 76P].

$W_xV_{1-x}O_2$:

Monoclinic phases and tetragonal phases (for $x>0.017$) also found [69N, 72H].

An increase in pressure brings about a reduction in the $M_2 \rightarrow R$ transition of $M_x^{3+}V_{1-x}O_2$ and an increase in the $T \rightarrow M_2$ transition. Above a critical pressure p_{crit} that depends on x the M_2 phase disappears and the $T \rightarrow R$ transition temperature above this pressure is independent of pressure [80T].

Compound	x	p_{crit} [kbar]	dT_{tr}/dp (T→M_2) [K kbar^{-1}]	dT_{tr}/dp (M_2→R) [K kbar^{-1}]
$Fe_xV_{1-x}O_2$	0.00011	7.5	−1.21	+3.97
	0.00015	10.5	−1.15	+3.75
	0.0002	14.4	−0.90	+3.47
$Ga_xV_{1-x}O_2$	0.000073	6.3	−0.79	+4.29
$Al_xV_{1-x}O_2$	0.000075	6.7	−0.75	+4.33
	0.00024	26.0	−0.88	+1.88

(Table from [80T])

Electronic properties

Calculations of the electronic structure: cluster type [76L, 75S], tight binding type [71M, 73M2, 77A1], APW type [72M, 73C1, 73C2], band structure types [77G2], calculation of Hubbard-U: [73H1, 78S], semi-empirical band structures [69B1, 73G]. Photoelectron spectrum [78V] shows a shift of the V3d band by 0.3 eV to lower binding energies at T_{tr}. Other PE data show the development of a clear Fermi edge above T_{tr} [80B].

low-temperature phase (semiconducting):

Brillouin zone: Fig. 14, energy bands: Fig. 15, density of states: Fig. 16. For E_g-values, see under "Optical properties" (absorption coefficient).

9.15.2.2.4 Vanadium oxide (VO$_2$)

Physical property	Numerical value	Experimental conditions	Experimental method, remarks	Ref.

high-temperature phase (metallic):
Brillouin zone: Fig. 17, energy bands: Fig. 18, density of states: Fig. 19, semi-empirical band-structure: Fig. 20.

Hubbard self-energy ("Hubbard-U") (in eV):

U	0.5 (5)···0.7 (5)		theoretical estimates	79S
	1.22			78S
	2			73H1
	1.1		experimental upper limit	79S
	0.92		photoconductivity	80K

Lattice properties

wavenumbers of phonon modes (in cm^{-1}):
infrared-active modes:

$\bar{\nu}$	189	$E \perp a_m$	low-temperature phase	66B
	270		(no IR modes seen in high-	
	310		temperature phase)	
	340		Vibrational absorption peaks at	
	505		435, 535 and 680 cm^{-1} reported	
	600		by [80R]; on 18.7 at% ^{18}O doping	
	710		these decrease to 428, 530 and	
	227.5	$E \| a_m$	667 cm^{-1}, respectively [80R].	
	285			
	324			
	355			
	392.5			
	478			
	530			
	700			

Raman-active modes:

$\bar{\nu}_R$	149	low-	assignment: —	71S,
	200	temperature	A_g	77A2
	226	phase	A_g	
	259		A_g	
	313		A_g	
	340		A_g	
	392		A_g	
	501		A_g	
	594		B_g	
	620		A_g	
	550	high-	broad band	
	300	temperature	weak band	
		phase	at $T \to T_{tr}$ the bands are broadened	
			considerably (Fig. 21). This	
			suggests a considerable role for the	
			lattice in the transition	

Debye temperature:

Θ_D	750 (20) K		heat capacity data 150···300 K	73C3

heat capacity:

C_p	$0.068(4) \cdot 10^{-4} T^3$ cal K^{-1} mol^{-1}	$T \lesssim 25$ K	for VO$_2$	73M1
	$80(5) T + (0.127(10) \cdot 10^{-4}) T^3$ cal K^{-1} mol^{-1}	$T \lesssim 25$ K	for W$_{0.14}$V$_{0.86}$O$_2$	73M1

Transport properties

resistivity:

The first order phase transition at 340 K is accompanied by a discontinuity in ϱ of ca. $10^4 \cdots 10^5 \, \Omega$ cm [69L]. High quality crystals show a transition in σ from ca. $10^{-1} \, \Omega^{-1}$ cm^{-1} to $10^4 \, \Omega^{-1}$ cm^{-1} at 339(1) K over about a 0.1 K interval with hysteresis of $0.1 \cdots 1$ K [69L].

$\dfrac{d \ln \varrho}{dX}$	$-4 \cdot 10^{-5}$ bar^{-1}	$\varrho \| c$	temperature region just below T_{tr}	69L
$\dfrac{d \ln \varrho}{dp}$	$-1.5 \cdot 10^{-5}$ bar^{-1}	$\varrho \| c$, RT		

The conductivity is slightly anisotropic (Fig. 22) with $\varrho_\perp / \varrho_\| \simeq 0.5$ on both sides of the transition [65B].

Effects of stoichiometry: Figs. 23, 24, pressure effects in both phases: Figs. 25, 26.

No clear activation energy found in the semiconducting phase [74S, 75B1] but $T^{-1/4}$ plot found to be linear (Fig. 27). Alternatively the band gap may show marked temperature dependence [80K]. Additional evidence for this comes from photoconductivity measurements [74S, 70P, 80K]. With 18.7 at % ^{18}O substitution, T_{tr} increases by +0.6 K [80R].

ac conductivity:

follows a power law for $T < T_{tr}$ [78M]: $\sigma(\omega) \propto \omega^s$ ($0.7 \leq s \leq 1$). Variation with temperature: Fig. 28. Some workers report only a slight dispersion at 300 K save at very high frequencies [67K] though dispersion does not increase at lower temperatures [72K1]. Dispersion can be accounted for by invoking a distribution of relaxation times [78M]. Detailed investigation of transport measurements suggest at low temperatures, conductivity is essentially due to impurity states in the bandgap [76R2] which might arise from so-called "phasons" and that just below T_{tr} intrinsic excitation across the bandgap has become significant, giving rise to band-type conduction. These impurity states may also be extrinsic, and are a strong function of stoichiometry. Their concentration has been estimated at $10^{18} \cdots 10^{20}$ cm^{-3} [76R2, 69B1, 74S].

photoconductivity:

Response broadens considerably at high temperature (Fig. 29). Variation with energy at 150 K: Fig. 29 α. Onset ≈ 0.5 eV corresponds to band gap transition $d_{x^2-y^2} \to d_{xz}$ [80K, 75S]. Peak at 0.72 eV, $d_{x^2-y^2} \to d_{yz}$ and peak at 0.92 eV transition across Mott-Hubbard gap [73S].

Seebeck coefficient (in µV K^{-1}):

low-temperature phase:

		T [K]		
S	$-30 \cdots -400$	298	no measurable anisotropy in LT phase.	69B1
	$-30 \cdots -150$	303, VO$_{1.994}$	S increases in magnitude with	75B2
	$-880 \cdots -1000$	303, VO$_{2.000}$	increasing activation energy;	75B2
	$-780 \cdots -910$	293	the low values are apparently	66K2
	$-700 \cdots -800$	333	associated with large concentrations of interband states which pin the Fermi level.	66K2
	-1000	295		65B

high-temperature phase:

-23.1 (2)	$\| c$		69B1
-21.1 (2)	$\perp c$		69B1
-23	353	measurement on film	68H
-21	373		75B2
$+30$	$430 \cdots 480$	no anisotropy in S measured	65B

There is a strong discontinuity in S at the transition temperature. Detailed behaviour in the semiconducting phase: Fig. 30. Effect of stoichiometry on S: Fig. 31. The fact that σ and $\exp(-eS/k)$ track parallel suggests an activationless mobility in the semiconducting phase. If an effective electron mass of $30 \, m_0$ is assumed, a drift mobility of 0.14 cm^2/V s can be calculated from S [66K2] for $T < T_{tr}$ in the range $293 \cdots 333$ K. For data on Nb-doped VO$_2$, see [75Z2].

Physical property	Numerical value	Experimental conditions	Experimental method, remarks	Ref.
Hall mobility (in cm^2/Vs):				
		T [K]		
μ_H	0.5	$\|c_R, T \lesssim T_{tr}$	little change when passing through T_{tr}, almost constant with T, decreasing by 0.6% per K [73R]	73R
	0.35	353		
	0.08···0.13	335···273 (ceramic sample)		66K1
	0.07···0.14	300	measurements on films	68H, 72K2
	0.5···0.7	$\|c_R, 295$		66B
	0.1···1.0	$\|c_R, T < T_{tr}$	some anisotropy reported; $\mu_{\|c_R} \approx 2\mu_{\perp c_R}$; carrier concentration $n = 10^{18}···10^{19}\,cm^{-3}\,(T < T_{tr})$	69B1
	16···19	$\|c_R, T > T_{tr}$	values of $1···10\,cm^2\,V^{-1}\,s^{-1}$ calculated by [69B1]. Carrier concentration estimated at $3 \cdot 10^{21}\,cm^{-3}$ [66B]. $T > T_{tr}$	66B

For Hall measurements (mobility and carrier concentration), see Figs. 32, 33.

effective electron masses (in units of m_0):				
		T [K]		
m_n	3.5	353	optical data, $T > T_{tr}$ from thermopower on films	73R
	7.1	353		68H
	65 (10)	$T \lesssim T_{tr}$	from thermopower data on W doped material, lower values of m_n: $28···60\,m_0$ reported for lightly doped samples	76R1
	60	$T \lesssim T_{tr}$	magnetic susceptibility data	75Z2
	36	4···26	heat capacity data, $T < T_{tr}$	73M1
plasmon data:				
m_{ω_p}	$1···4\,m_0$	$T < T_{tr}$	effective plasmon mass	66B
α	2		Fröhlich coupling constant	69B1
m_{ω_p}	$0.5\,m_0$	$T > T_{tr}$		66B
	$0.81\,m_0$			70B
μ_{opt}	$2\,cm^2/Vs$			66B

thermal conductivity: Disagreement in the literature; [69B1] reports little change on passing through the transition, Fig. 34(a), whereas [78A] report a marked change, Fig. 34(b).

transport in doped material:

$Ga_xV_{1-x}O_2$: conductivity: Fig. 35, thermoelectric power: Fig. 36.
$Al_xV_{1-x}O_2$: conductivity: Fig. 37, thermoelectric power: Fig. 38.
$Cr_xV_{1-x}O_2$: conductivity: Fig. 39, thermoelectric power: Fig. 40.
$Fe_xV_{1-x}O_2$: conductivity: Fig. 41, thermoelectric power: Fig. 42.
$Ti_xV_{1-x}O_2$: conductivity: Fig. 43 (also [74K3, 73C4]).
$Nb_xV_{1-x}O_2$: behaviour very complex. For low values of x ($x \lesssim 0.10$), the conductivity of the semiconducting state rises, but as x increases, the conductivity falls, and the metal-semiconductor transition is suppressed: Figs. 44···46 (also [76M2, 76T]).
$W_xV_{1-x}O_2$: conductivity: Fig. 47, thermoelectric power: Fig. 48 (also [74K2]).
$Mo_xV_{1-x}O_2$: conductivity: Fig. 49 (also [73P]).
$Re_xV_{1-x}O_2$: conductivity: Fig. 50.
$VO_{2-x}F_x$: conductivity: Fig. 51 (also [71C]).

Physical property	Numerical value	Experimental conditions	Experimental method, remarks	Ref.

Optical properties

Photoelectron spectrum: Fig. 52, real and imaginary parts of the refractive index for $T < T_{tr}$: Fig. 53, UV data reported in [72G, 68V, 74V1, 76M3]).

absorption coefficient:

For $T < T_{tr}$: Fig. 54, comparison of absorption coefficients: Fig. 55. Structure at $E < 1.8$ eV assigned to d–d transitions with threshold at 0.6 eV [72G]. At 1.82 eV, threshold for O2p–V3d transitions with peaks at 2.64 eV and 3.56 eV. Optical absorption in VO_2: Cr shows no distinct edge, but for $\hbar\omega < 0.6$ eV the absorption tail can be fitted to $K \propto A \exp(E/E_o)$ where $E_o = 0.088$ eV. Hence $E_g \simeq 0.60(5)$ eV and absorption below 0.6 eV is excitonic in origin. Thermoreflectance spectra are not observed either above or below T_{tr} save in a region very close to T_{tr}, and the electroreflectance signal also behaved in an anomalous fashion near T_{tr} [80M].

dielectric constants:

$\varepsilon(0)$	40.6	$E \perp a_m$, $T < T_{tr}$	from infrared lattice modes	66B
	25.9	$E \| a_m$		
	24	$E \| a_m$, $T < T_{tr}$		72K1
	43	$T = 77$ K	$f = 10$ Hz; true static ε estimated as $70 \cdots 100$	76R2
	ca. 100	300 K		78M
	18.3	$E \| a_m$, RT		75Z1
	39	$E \perp a_m$		
$\varepsilon(\infty)$	10.0	$E \perp a_m$, $T < T_{tr}$	from infrared lattice modes	66B
	9.7	$E \| a_m$		
	9	$E \| c_R$, $T > T_{tr}$ (353 K)		66B

carrier contribution to dielectric constant:

$\varepsilon_{carrier} = -\omega_p^2/(\omega^2 + i\omega\omega_c)$ with $\omega_p = 8000$ cm^{-1} and $\omega_c = 10000$ cm^{-1} for high-temperature phase [66B].

Magnetic properties

(see also table p. 170)

susceptibility:

For pure VO_2, see Fig. 56; for VO_2 doped with Ga, Al, Fe, Cr, Ti, Nb, W, Mo, Re see Figs. 57···65.

In the semiconducting phase, electrons are paired and no magnetic effects are found. However, the material is borderline and the Mössbauer spectra of ^{119}Sn doped in VO_2 shows hyperfine splitting, which the authors attribute to regions of antiferromagnetically coupled V located around the impurity atom [74F].

References for 9.15.2.2.4

47C	Cook, O.A.: J. Am. Chem. Soc. **69** (1947) 331.
56A	Anderson, G.: Acta Chem. Scand. **10** (1956) 623.
64K	Kawakubo, T., Nagakawa, T.: J. Phys. Soc. Jpn. **19** (1964) 517.
65B	Bongers, P.F.: Solid State Commun. **3** (1965) 275.
66B	Barker, A.S., Verleur, H.W., Guggenheim, H.J.: Phys. Rev. Lett. **17** (1966) 1286.
66K1	Kitahoro, I., Ohashi, T., Watanabe, A.: J. Phys. Soc. Jpn. **21** (1966) 2422.
66K2	Kitahiro, I., Watanabe, A.: J. Phys. Soc. Jpn. **21** (1966) 2423.
67K	Kabashima, S., Tshuchiya, Y., Kawakuba, T.: J. Phys. Soc. Jpn. **22** (1967) 932.
67R	Rao, K.V.K., Naidu, S.V.N., Iyengar, L.: J. Phys. Soc. Jpn. **23** (1967) 1380.
68H	Hensler, D.H.: J. Appl. Phys. **39** (1968) 2360.
68V	Verleur, H.W., Barker, A.S., Berglund, C.N.: Phys. Rev. **172** (1968) 788.
69B1	Berglund, C.N., Guggenheim, H.J.: Phys. Rev. **185** (1969) 1022.
69B2	Berglund, C.N., Jayaraman, A.: Phys. Rev. **185** (1969) 1034.
69L	Ladd, L.A., Paul, W.: Solid State Commun. **7** (1969) 425.

References for 9.15.2.2.4

69M	MacChesney, J.B., Guggenheim, H.J.: J. Phys. Chem. Solids **30** (1969) 225.
69N	Nygren, M., Israelsson, M.: Mater. Res. Bull. **4** (1969) 881.
70B	Borisov, B.S., Koretskaya, S.T., Mokerov, V.G., Rakov, A.V., Solov'ev, S.G.: Fiz. Tverd. Tela **12** (1970) 2209.
70G	Guntersdorfer, M.: Solid State Electron. **13** (1970) 355.
70K	Kimizuka, N., Saeki, M., Nakahira, M.: Mater. Res. Bull. **5** (1970) 69.
70L	Longo, J.M., Kierkegaard, P.: Acta Chem. Scand. **24** (1970) 420.
70M	Maurezio, M., Dernier, P.D., McWhan, D.B., Remeika, J.P.: Mater. Res. Bull. **5** (1970) 1015.
70P	Paul, W.: Mater. Res. Bull. **5** (1970) 691.
71B	Bayard, M.L.F., Reynolds, T.G., Vlasse, M., McKinzie, H.L., Arnott, R.J., Wold, A.: J. Solid State Chem. **3** (1971) 484.
71C	Chamberland, B.L.: Mater. Res. Bull. **6** (1971) 425.
71M	Mitra, T.K., Chatterjee, S., Hyland, G.J.: Phys. Lett. **A37** (1971) 221.
71S	Srivastava, R., Chase, L.L.: Phys. Rev. Lett. **27** (1971) 727.
71V	Villeneuve, G., Bordet, A., Casalot, A., Hagenmuller, P.: Mater. Res. Bull. **6** (1971) 119.
72G	Gavini, A., Kwan, C.C.Y.: Phys. Rev. **B5** (1972) 3138.
72H	Horlin, T., Niklewski, T., Nygren, M.: Mater. Res. Bull. **7** (1972) 1515.
72K1	Kabashima, S., Goto, T., Nishimura, K., Kanwakubo, T.: J. Phys. Soc. Jpn. **32** (1972) 158.
72K2	Kwan, C.C.Y., Griffith, C.H., Eastwood, H.K.: Appl. Phys. Lett. **20** (1972) 93.
72M	Marezio, M., McWhan, D.B., Remeika, J.P., Dernier, P.D.: Phys. Rev. **B5** (1972) 2541.
72P	Pouget, J.P., Lederer, P., Schreiber, D.S., Launois, H., Wohlleben, D., Casalot, A., Villeneuve, G.: J. Phys. Chem. Solids **33** (1972) 1961.
72V	Villeneuve, G., Bordet, A., Casalot, A., Pouget, J.P., Launois, H., Lederer, P.: J. Phys. Chem. Solids **33** (1972) 1953.
73C1	Caruthers, E., Kleinman, L., Zhang, H.I.: Phys. Rev. **B7** (1973) 3753.
73C2	Caruthers, E., Kleinman, L.: Phys. Rev. **B7** (1973) 3760.
73C3	Chandrashekhar, G.V., Barros, H.L.C., Honig, J.M.: Mater. Res. Bull. **8** (1973) 369.
73C4	Chase, L.L.: Phys. Lett. **A46** (1973) 215.
73G	Goodenough, J.B., Hong, H.Y.-P.: Phys. Rev. **B8** (1973) 1323.
73H1	Hearn, C.J., Hyland, G.J.: Phys. Lett. **A43** (1973) 87.
73H2	Horlin, T., Niklewski, T., Nygren, M.: Mater. Res. Bull. **8** (1973) 179.
73M1	McWhan, D.B., Remeika, J.P., Maita, J.P., Okinaka, H., Kosuge, K., Kachi, S.: Phys. Rev. **B7** (1973) 326.
73M2	Mitra, T.K., Chatterjee, S., Hyland, G.J.: Can. J. Phys. **51** (1973) 352.
73P	Prodan, A., Marinkovič, V., Prošek, M.: Mater. Res. Bull. **8** (1973) 551.
73R	Rosevear, W.H., Paul, W.: Phys. Rev. **B7** (1973) 2109.
73V	Villeneuve, G., Drillon, M., Hagenmuller, P.: Mater. Res. Bull. **8** (1973) 1111.
74D	Drillon, M., Villeneuve, G.: Mater. Res. Bull. **9** (1974) 1199.
74F	Fabritchinyi, P.B., Bayard, M., Pouchard, M., Hagenmuller, P.: Solid State Commun. **14** (1974) 603.
74G	Gomes, P., Félix, P., Lambert, M., Villeneuve, G.: Acta Crystallogr. **30** (1974) 55.
74K1	Kimizuka, N., Ishii, M., Kawada, I., Saeki, M., Nakahira, M.: J. Solid State Chem. **9** (1974) 69.
74K2	Kleinschmidt, P.: Phys. Lett. **A47** (1974) 205.
74K3	Kristensen, I.K.: Mater. Res. Bull. **9** (1974) 1677.
74M	McWhan, D.B., Marezio, M., Remeika, J.P., Dernier, P.D.: Phys. Rev. **B10** (1974) 490.
74P1	Pouget, J.P., Launois, H., Rice, T.M., Dernier, P.D., Gossard, A., Villeneuve, G., Hagenmuller, P.: Phys. Rev. **B10** (1974) 1801.
74P2	Palmier, J.F., Bellini, Y., Merenda, P.: Solid State Commun. **14** (1974) 575.
74S	von Schulthess, G., Wachter, P.: Solid State Commun. **15** (1974) 645.
74V1	Valiev, K.A., Mokerov, V.G., Galiev, G.B.: Fiz. Tverd. Tela **16** (1974) 2361.
74V2	Villeneuve, G., Launay, J.-C., Hagenmuller, P.: Solid State Commun. **15** (1974) 1683.
75B1	Blaauw, C., Leenhouts, F., van der Woude, F., Sawatzky, G.A.: J. Phys. **C8** (1975) 459.
75B2	Brückner, W., Moldenhauer, W., Wich, H., Wolf, E., Oppermann, H., Gerlach, U., Reichelt, W.: Phys. Status Solidi (a) **29** (1975) 63.
75S	Sommers, C., de Groot, R., Kaplan, D., Zylbersteijn, A.: J. Phys. (Paris) Lett. **36** (1975) L-157.
75V	Villeneuve, G., Launay, J.-C., Marquestant, E., Hagenmuller, P.: Solid State Commun. **17** (1975) 657.

References for 9.15.2.2.4

75Z1	Zylberstejn, A., Pannetier, B., Merenda, P.: Phys. Lett. **A54** (1975) 145.
75Z2	Zylberstejn, A., Mott, N.F.: Phys. Rev. **B11** (1975) 4383.
76B1	Brückner, W., Gerlach, U., Moldenhauer, W., Brückner, H.-P., Thuss, B., Oppermann, H., Wolf, E., Storbeck, I.: J. Phys. (Paris) Colloq. 37 (1976) C4-63.
76B2	Brückner, W., Gerlach, U., Moldenhauer, W., Brückner, H.-P., Mattern, N., Oppermann, H., Wolf, E.: Phys. Status Solidi (a) **38** (1976) 93.
76B3	Brückner, W., Moldenhauer, W.: Phys. Status Solidi (a) **38** (1976) K17.
76H1	Hörlin, T., Niklewski, T., Nygren, M.: J. Phys. (Paris) Colloq. 37 (1976) C4-69.
76H2	Hörlin, T., Niklewski, T., Nygren, M.: Acta Chem. Scand. **A30** (1976) 619.
76K	Kosuge, K., Kachi, S.: Mater. Res. Bull. **11** (1976) 255.
76L	Lazukova, N.I., Gubanov, V.A.: Solid State Commun. **20** (1976) 649.
76M1	Magarino, J., Tuchendler, J., D'Haenens, J.P.: Proc. Semicond. 13th Int. Conf. **1976**, 318.
76M2	Merenda, P., Kaplan, D., Sommers, C.: J. Phys. (Paris) 37 (1976) C4-59.
76M3	Mokerov, V.G., Saraikin, V.V.: Fiz. Tverd. Tela **18** (1976) 1801.
76P	Louget, J.P., Launois, H.: J. Phys. (Paris) Colloq. 37 (1976) C4-39.
76R1	Reyes, J.M., Sayer, M., Chen, R.: Can. J. Phys. **54** (1976) 408.
76R2	Reyes, J.M., Sayer, M., Mansingh, A., Chen, R.: Can. J. Phys. **54** (1976) 413.
76T	Tuchendler, J., Magarino, T., D'Haenens, J.P.: Proc. Semicond. 13th Int. Conf. **1976**, 318.
77A1	Altanham, T., Hyland, G.J.: Phys. Lett. **A61** (1977) 426.
77A2	Aronov, A.G., Mirlin, D.N., Reshina, I.I., Chudnovskii, F.A.: Fiz. Tverd. Tela **19** (1977) 193.
77G1	Ghedira, M., Vincent, H., Marezio, M., Launay, J.C.: J. Solid State Chem. **22** (1977) 423.
77G2	Gupta, M., Freeman, A.J., Ellis, D.: Phys. Rev. **B16** (1977) 3338.
77S	Savberg, Ö., Nygren, M.: Phys. Status Solidi (a) **43** (1977) 645.
77V	Villeneuve, G., Drillon, M., Hagenmuller, P., Nygren, M., Pouget, J.P., Carmona, F., Delhaes, P.: J. Phys. **C10** (1977) 3621.
78A	Andreev, V.N., Chudnovskii, F.A., Pettov, A.V., Terukov, E.I.: Phys. Status Solidi (a) **48** (1978) K153.
78G	Gannon, J.R., Tilley, R.J.D.: J. Solid State Chem. **25** (1978) 301.
78M	Mansingh, A., Singh, R., Sayer, M.: Phys. Status Solidi (a) **49** (1978) 773.
78P	Pintchovskii, F., Glausinger, W.S., Navrotsky, A.: J. Phys. Chem. Solids 39 (1978) 941.
78S	Sommers, C., Doniach, S.: Solid State Commun. **28** (1978) 133.
78T	Terukov, E.T., Reichelt, W., Wolf, M., Henschik, H., Oppermann, H.: Phys. Status Solidi (a) **48** (1978) 377.
78V	Vorotilova, L.S., Ioffe, V.A., Razumeenko, M.V.: Fiz. Tekh. Poluprovodn. **12** (1978) 36.
79K	Kucharezyk, D., Niklewski, T.: J. Appl. Crystallogr. **12** (1979) 370.
79M	Mokerov, V.G., Begishov, A.B., Ignat'ev, A.S.: Fiz. Tverd. Tela **21** (1979) 1482.
79S	Sawatsky, G.A., Post, D.: Phys. Rev. **B20** (1979) 1546.
80B	Beatham, N., Fragala, I.L., Orchard, A.F., Thornton, G.: J. C. S. Faraday Trans. II **76** (1980) 929.
80K	Kaminskii, V.V., Terukov, E.I., Shelykh, A.I.: Fiz. Tverd. Tela **22** (1980) 1686.
80M	Mokerov, V.G., Begishov, A.B.: Fiz. Tverd. Tela **22** (1980) 1079.
80R	Reichelt, W., Oppermann, H., Wagner, H., Terukov, E.I., Wolf, E.: Z. Anorg. Allg. Chem. **463** (1980) 193.
80T	Terukov, E.I., Orchinnikov, S.G., Zyuzui, A.Yu., Gerlach, U., Oppermann, H., Reichelt, W.: Fiz. Tverd. Tela **22** (1980) 1374.

9.15.2.2.5 Vanadium oxide (V_2O_5)

Crystal structure

structure: orthorhombic, layer-type, space group D_{2h}^{13}-Pmm, $Z=2$.

Projection of the structure along [001] is shown in Fig. 1, and along [010] in Fig. 2. Perspective view: Fig. 3. Monovalent dopants occupy interstitial positions in the lattice as shown in Figs. 1 and 2.

lattice parameters (in Å):

a	11.510_8	$T=293$ K	61B, 50B
b	4.369_5		
c	3.563_3		

density:

d_{calc}	3.371 g cm^{-3}	$T=293$ K	61B, 50B

interatomic distances (in Å) **and angles** (in °) at RT:

Values in parentheses give the estimated standard deviation [61B].
Atoms are labelled as in Fig. 3.

V–O$_1$(1)	1.585 (4)
V–O$_1$(2)	2.785 (5)
V–O$_2$(1)	2.021 (3)
V–O$_2$(2)	1.878 (2)
V–O$_2$(3)	1.878 (2)
V–O$_3$	1.780 (2)
O$_1$(1)–O$_2$(1)	2.873 (5)
O$_1$(1)–O$_2$(2)	2.744 (4)
O$_1$(1)–O$_2$(3)	2.744 (4)
O$_1$(1)–O$_3$	2.668 (5)
O$_2$(1)–O$_2$(2)	2.388 (3)
O$_2$(1)–O$_2$(3)	2.388 (3)
O$_2$(1)–O$_3$	3.673 (4)
O$_1$(2)–O$_2$(3)	3.563 (3)
O$_1$(2)–O$_3$	2.740 (3)
O$_1$(3)–O$_3$	2.740 (3)
O$_1$(2)–O$_2$(1)	3.042 (5)
O$_1$(2)–O$_2$(2)	2.954 (4)
O$_1$(2)–O$_2$(3)	2.954 (4)
O$_1$(2)–O$_3$	2.846 (5)
O$_1$(1)–V–O$_2$(1)	105.0 (2)
O$_1$(1)–V–O$_2$(2)	104.5 (3)
O$_1$(1)–V–O$_2$(3)	104.5 (3)
O$_1$(1)–V–O$_3$	104.8 (3)
O$_2$(1)–V–O$_2$(2)	75.5 (2)
O$_2$(1)–V–O$_2$(3)	75.5 (2)
O$_2$(1)–V–O$_3$	150.1 (2)
O$_2$(2)–V–O$_2$(3)	143.1 (2)
O$_2$(2)–V–O$_3$	97.0 (2)
O$_2$(3)–V–O$_3$	97.0 (2)

coefficients of linear thermal expansion (in 10^{-6} K^{-1}):

α_a	2.0	$T=25\cdots 600$ °C	67K
α_b	55.4		
α_c	8.0		

9.15.2.2.5 Vanadium oxide (V_2O_5)

Physical property	Numerical value	Experimental conditions	Experimental method, remarks	Ref.

Electronic properties

No band structure has been reported. Calculations using the CNDO/2 method have been carried out on clusters $(VO_n)^{5-2n}$, n = 4, 5, 6, where – with reference to Fig. 3 – the n = 6 cluster is shown. The n = 5 cluster is derived from this by removal of $O_1(2)$ to give an approximate square pyramid and in the n = 4 cluster the $O_2(1)$ atom is also removed to leave an approximately tetrahedral VO_4^{3-} ion.

bonds, populations and energies E_B in clusters n = 4, 5, 6 (from [77L]):

Bond	n = 6		n = 5		n = 4	
	population	E_B [eV]	population	E_B [eV]	population	E_B [eV]
$V-O_1(1)$	0.438	−1.319	0.490	−1.244	0.518	−1.122
$V-O_1(2)$	0.121	−0.766	−	−	−	−
$V-O_2(1)$	0.289	−0.975	0.296	−0.915	−	−
$V-O_2(2)$	0.350	−1.118	0.351	−1.035	0.370	−0.936
$V-O_2(3)$	0.350	−1.103	0.351	−1.035	0.370	−0.936
$V-O_3$	0.377	−1.175	0.381	−1.119	0.420	−0.963

energy gap (in eV):

E_g	2.17	$E \| c$	see also the spectra shown in	76V
	2.34	$E \| c$	the section on optical	66K
	2.25	$E \| c$	properties	73M
	2.19	$E \| b$		76V
	2.363	$E \| b$		66K
	2.23	$E \| a$		73M

According to [66K, 67B] the edge is direct and forbidden. Diffuse reflectance spectra [71K] give $E_g = 2.31$ eV at RT but band edge determined to be direct and allowed.

temperature dependence of energy gap:

dE_g/dT	$-6.1 \cdot 10^{-4}$ eV K^{-1}	$E \| c$	$E_g(T \to 0) = 2.49$ eV	67B
	$-7.3 \cdot 10^{-4}$ eV K^{-1}	$E \perp c$	$E_g(T \to 0) = 2.54$ eV	

Lattice properties

wavenumbers of lattice vibration modes (in cm^{-1}):

The following data are taken from: A [72R], B [73G], C [73B], D [76C2], Kramers-Kronig analysis, E [76C2], classical oscillator dispersion, F [76C2], 4-parameter oscillator dispersion, G [76C2] single crystal transmission, H [76C2] powder transmission. First column: $\bar{\nu}_{TO}$, second column: $\bar{\nu}_{LO}$.

$\bar{\nu}_1$	74		B	$E \| a$,	the lattice vibrations of crystals
	73.5	78.5	C	b_{3u}-type	with space group D_{2h}^{13} – Pmnm can
	71.9	78.1	D		be classified by
	72.0	77.0	E		
	72.4	76.2	F		$\Gamma = 7a_g + 7b_{1g} + 3b_{2g} + 4b_{3g} + 3a_u$
$\bar{\nu}_2$	255.3	266	A		$+ 3b_{1u} + 6b_{2u} + 6b_{3u}$
	265		B		for E and F the paper cited [76C2]
	261	269	C		also contains data on other
	259.1	267.0	D		fitting parameters
	260.0	267.8	E		
	261.0	265.5	F		
	262		H		
$\bar{\nu}_3$	308.8	389	A		
	386		B		
	305	392	C		
	301.6	389.4	D		
	303.0	387.9	E		
	303.0	390.5	F		
	370		H		

(continued)

9.15.2.2.5 Vanadium oxide (V$_2$O$_5$)

Physical property	Numerical value		Experimental conditions	Experimental method, remarks	Ref.
wavenumbers of lattice vibration modes (continued)					
$\bar{\nu}_4$	408.5	582	A		
	520		B		
	412	602	C		
	410.5	585.0	D		
	410.5	582.7	E		
	411.0	586.0	F		
	472		H		
$\bar{\nu}_5$	778.0	957	A		
	880		B		
	796	975	C		
	765.5	965.0	D		
	765.0	952.4	E		
	767.5	959.0	F		
	813		H		
$\bar{\nu}_6$	995		B		
	980.7	987.0	D		
	981.0	986.4	E		
	980.5	982.0	F		
	982		G		
	982		H		
$\bar{\nu}_1$	144		B	$E \| b$, b_{2u}-type	
$\bar{\nu}_2$	481		B		
	476.5	494	C		
	472.2	492.0	D		
	473.5	494.2	E		
	473.0	490.0	F		
	1038		G		
$\bar{\nu}_3$	1035		B		
	974.1	1039.5	D		
	975.0	1037.7	E		
	975.5	1038.0	F		
	1023		H		
$\bar{\nu}_1$	213.4	225	A	$E \| c$, b_{1u}-type	
	220		B		
	214	226	C		
	212.6	226.2	D		
	212.6	226.0	E		
	212.0	225.0	F		
	217		H		
$\bar{\nu}_2$	284.2	312	A		
	308		B		
	287	315	C		
	285.4	314.3	D		
	283.5	314.4	E		
	284.0	312.5	F		
	294		H		
$\bar{\nu}_3$	515.0	850	A		
	700		B		
	532	865	C		
	506.0	844.0	D		
	505.7	842.4	E		
	506.5	842.5	F		
	605		H		

Physical property	Numerical value	Experimental conditions	Experimental method, remarks	Ref.
wavenumbers of Raman active lattice vibrations (in cm^{-1}):				
$\bar{\nu}_R$	104, 198, 306, 406, 482, 530, 995		a_g-type	73G
	201, 310, 355, 502?, 954?, 995		b_{1g}-type	
	144, 284, 701		b_{2g}-type	
	147, 230, 287, 701		b_{3g}-type	

wavenumbers of additional far IR peaks (in cm^{-1}): (All values for $T = 4 \cdots 78$ K, except $\bar{\nu} = 14$ cm^{-1}.)
Such peaks have been found for doping with monovalent ions.

$\bar{\nu}$	29.5		doping with Li	77B
	56.5			
	14	$T \simeq 4$ K		
	94			
	29.5		doping with Na	
	53.0			
	28.5		doping with Cu	
	53.0			

The EPR evidence suggests that the e^- from these ions is delocalized over four vanadium ions in the b-planes. The 4-centre MO is split into components by the local field, and the FIR absorption has been ascribed to transitions between these components [77B].

Transport properties

V_2O_5 shows three quite distinct conductivity regions (Fig. 4): a low-temperature region ($T \lesssim 140$ K) of very low activation energy, an intermediate region (140 K $\lesssim T \lesssim$ 350 K) where good linearity in the ln ϱ vs. T^{-1} curve is obtained, and a final high-temperature region (350 K $\lesssim T \lesssim$ 600 K) where substantial non-linearity is found [70I, 73H, 71P].

a) low-temperature range:

Disagreements exist about the conductivity mechanism: [71P] report an activated mobility with $E_A = 0.0781$ eV at $T < 170$ K, [73H] find evidence for a random range hopping between defects with a $T^{-1/4}$-law (Fig. 5). [73H] reports activated conductivity in the b direction to much lower temperatures than the a and c directions (Fig. 6).

No Seebeck data have been reported in this temperature region. ac data supports the hopping model (Fig. 7). EPR data on $Na_{0.01}V_2O_5$ [74S] suggest that direct hopping between defects occurs at low temperature with $E_A \approx 0.07$ (2) eV.

b) intermediate temperature range (140 $\cdots$ 400 K):

activation energy for conductivity (in eV):

E_A	0.27			70I
	0.26			73H
	0.170 $\cdots$ 0.189			71P
	0.21 $\cdots$ 0.29			65P
	0.20 $\cdots$ 0.24			72V
	0.28			71S
	0.22 $\cdots$ 0.32			69V
	0.19 $\cdots$ 0.21			70N

The conductivity remains anisotropic (Fig. 6). Seebeck coefficient shows anomalous behaviour (Fig. 8) which has been interpreted in [71P] in terms of a two level system. Analysis suggests for the activation energy two contributions: a hopping energy (polaron energy) of 0.15 $\cdots$ 0.17 eV and an energy required to free an electron from an impurity trap of 0.10 $\cdots$ 0.12 eV [70I].

Hall mobility (in cm^2/V s):

μ_H	$3 \cdot 10^{-2}$	$T = 298$ K, $\parallel c$	shows smaller temperature dependence than conductivity, a decreasing temperature	70I
	0.2	RT, $\parallel c$		71P

(continued)

Physical property	Numerical value	Experimental conditions	Experimental method, remarks	Ref.
Hall mobility (continued)				
μ_H	$3.5 \cdot 10^{-2}$	$T = 328$ K, $\|c$	dependence is shown in Fig. 9	70I
	$3.8 \cdots 9.7 \cdot 10^{-2}$	293 K	$\mu_H \propto T^{-3}$; orientation not specified	72V
drift mobility:				
μ_{dr}	$\mu_0 \exp(-0.26 \text{ eV}/kT)$	along b-axis	see Fig. 10; in the a-direction the mobility decreases with temperature (T in K)	69V
	$\propto \mu_0 \, T^{0.32}$	along c-axis		
	$\propto \mu_0 \, T^{-1.25}$	along a-axis		
	$0.1 \cdots 0.8$ cm^2/V s	$T = 293$ K	orientation not specified	72V

c) high-temperature range:

A plot of the form $\sigma = AT^{-3/2} \exp(-0.16 \text{ eV}/kT)$ seems to fit the data along the c-axis (Fig. 11) [70I] (T in Kelvin).

Drift mobility $\mu = \mu_0 \exp(-0.06 \text{ eV}/kT)$, T in K, along all three axes; μ_0 is axis dependent [69V]; resistance vs. pressure: Fig. 12.

ac conductivity: Oxygen deficient V_2O_5 samples are inhomogenous containing, in layers parallel to (010) plane, planar inclusions (probably of lower oxides) whose conductivity was much higher than the native V_2O_5 matrix [81I]. The ac conductivity and permittivity are shown in Fig. 12 α in the temperature range $70 \cdots 250$ K. An analysis shows $\sigma \propto \omega^s$, $s \approx 0.4 \cdots 0.5$, activation energy for the V_2O_5 matrix is 0.2 eV and for inclusions 0.08 eV. The break at 140 K in Fig. 12 α corresponds to a transition from σ dominated by the inclusions at low temperature to σ dominated by the matrix above 140 K [81I].

transport properties for doped material:

V_2O_5 may be doped interstitially with monovalent ions. Conductivity and Seebeck coefficients for Li- and Na-doped samples: Figs. 13, 14, conductivity of Cu-doped material: Fig. 15, effect of substitutional fluorine doping: Fig. 16 [76C1, 71P, 71B].

electrical conductivity and charge carrier concentration (at 25°C) for polycrystalline Li- and Na-doped samples (A obtained from alkali metal concentration, B from electrical conductivity using $\mu = 0.03$ cm^2/V s) (Table from [76C1]):

	σ Ω^{-1}cm$^{-1} \cdot 10^{-3}$	n [cm^{-3}]		E_A [eV]	
		A	B	low temperature	high temperature
V_2O_5	0.05		$1.04 \cdot 10^{16}$	0.21 (300···595 K)	0.37
Li$_{0.002}$V$_2$O$_5$	0.115	$1.43 \cdot 10^{19}$	$2.4 \cdot 10^{16}$	0.20 (300···562 K)	0.36
Li$_{0.006}$V$_2$O$_5$	0.38	$4.29 \cdot 10^{19}$	$7.9 \cdot 10^{16}$	0.19 (300···515 K)	0.26
Li$_{0.02}$V$_2$O$_5$	0.525	$1.43 \cdot 10^{20}$	$1.09 \cdot 10^{17}$	0.18 (300···610 K)	0.10
Li$_{0.06}$V$_2$O$_5$	1.51	$4.29 \cdot 10^{20}$	$3.14 \cdot 10^{17}$	0.16 (300···582 K)	0.10
Na$_{0.002}$V$_2$O$_5$	0.24	$1.43 \cdot 10^{19}$	$5.0 \cdot 10^{16}$	0.19 (300···549 K)	0.26
Na$_{0.006}$V$_2$O$_5$	1.05	$4.29 \cdot 10^{19}$	$2.19 \cdot 10^{17}$	0.17	0.17
Na$_{0.02}$V$_2$O$_5$	3.8	$1.43 \cdot 10^{20}$	$7.92 \cdot 10^{17}$	0.15	0.15
Na$_{0.06}$V$_2$O$_5$	19.5	$4.29 \cdot 10^{20}$	$4.06 \cdot 10^{18}$	0.13 (300···544 K)	0.10

activation energies and conductivity (at RT along the c-axis) for Cu-doped samples (from [71P]):

	E_A [eV] ($T > 170$ K)	E_A [eV] ($T < 170$ K)	$\sigma_{\|c}$ [Ω^{-1}cm^{-1}] ($T = 298$ K)
V_2O_5	0.170		$8.3 \cdot 10^{-4}$
V_2O_5	0.189	0.0781	$8.8 \cdot 10^{-4}$
Cu$_{0.0043}$V$_2$O$_5$	0.193	0.0758	$8.6 \cdot 10^{-3}$
Cu$_{0.0043}$V$_2$O$_5$	0.208	0.0752	$8.4 \cdot 10^{-3}$
Cu$_{0.0062}$V$_2$O$_5$	0.235		$9.1 \cdot 10^{-3}$
Cu$_{0.0083}$V$_2$O$_5$	0.222	0.0759	$14.4 \cdot 10^{-3}$
Cu$_{0.02}$V$_2$O$_5$	0.237	0.0790	$12.9 \cdot 10^{-3}$

Physical property	Numerical value	Experimental conditions	Experimental method, remarks	Ref.

activation energies, conductivity (at RT) and Seebeck coefficient for W-doped polycrystalline samples (from [79P]):

$x = W/V_2O_5$	S [μV K^{-1}]	σ [Ω^{-1} cm^{-1}]	E_A [eV]
0	−200	$1.5 \cdot 10^{-5}$	0.20
0.005	−50	$2.86 \cdot 10^{-2}$	0.24
0.01	−48	$3.13 \cdot 10^{-2}$	0.24
0.02	−40	$3.63 \cdot 10^{-2}$	0.23
0.03	−30	$4.17 \cdot 10^{-2}$	0.22

conductivity: Fig. 16 α (note discrepancy to values in table above); Seebeck coefficient: Fig. 16 β. Evidence suggests that W^{6+} ions are interstitial and main contribution to activation energy is from small polaron hopping [79P].

Optical properties

optical spectra: electroreflectance and ε_2 spectra: Figs. 17, 18, reflectivity in the far UV: Fig. 19 (other reflectivity data: [73V, 72M, 69M]), absorption coefficient near the band edge: Fig. 20.

absorption coefficient:

In the sub-bandgap region K shows a strong Urbach tail $K = K_0 \exp(\beta \hbar\omega/kT)$ with $\beta = 0.55$ for $E \| a$ and 0.52 for $E \| c$ [66K], $\beta = 0.50$ ($E \perp c$) and 0.47 ($E \| c$) [67B]. Strong absorption is found for $E \| a$ at 1.24 eV with a shoulder at 1.50 eV which derive from defect centres. Careful studies [74C2, 78V] have shown that this F1 absorption is due to electrons trapped at or near an oxygen vacancy. The EPR principal axes are coplanar with the layers in V_2O_5, but the x and z axes lie at 25° to the c and a axes of the crystal and a fit to a two-electron centre is found with electrons located on 2 or 3 nearest neighbour vanadium atoms.

energies of experimental absorption maxima (in eV):

a, b, c indicate the polarization, the three columns to the right give the corresponding energies and (in brackets) oscillator strengths (10^{-2}) for the cluster models mentioned in the section on electronic properties for n = 4, 5 and 6, respectively. (Table from [77L].)

			n = 4		n = 5		n = 6		
E	2.17	(a)	2.92 (0.12)	(a)	2.79 (2.34)	(a)	2.18	(0.084)	(a)
								(1.33)	(b)
	2.2	(a)	3.03 (0.26)	(a)	2.00 (0.68)	(a)	2.55	(0.36)	(a)
								(3.93)	(b)
	2.5	(a)	3.14 (0.55)	(c)	3.30 (0.77)	(a)	2.71	(0.74)	(a)
	2.54	(c)	3.31 (0.15)	(a)	3.50 (1.52)	(c)	2.89	(3.68)	(a)
	2.9	(a)	3.31 (0.53)	(c)	3.69 (1.98)	(a)	3.05	(4.91)	(a, b)
	3.0	(c)	3.49 (0.43)	(c, b)	4.39 (1.29)	(a)	3.17	(0.71)	(c)
	3.25	(c)	3.68 (0.42)	(a, b)	4.51 (2.63)	(c)	3.25	(0.39)	(c)
	3.38	(a)	3.88 (0.52)	(a, b)	4.66 (2.05)	(a, b)	3.45	(1.41)	(b, c)
	3.7	(c)	3.89 (0.47)	(c)	4.68 (1.63)	(c)	3.75	(3.04)	(a)
	3.87⋯3.9	(a)	4.18 (1.86)	(c)	4.85 (1.07)	(a, b)	4.26	(1.89)	(a)
	4.15	(c)	4.41 (2.25)	(a)	5.00 (0.47)	(a, b)	4.59	(1.58)	(a)
	4.4	(a)	4.76 (1.71)	(a)	5.16 (0.39)	(a)	4.72	(1.23)	(a, b)
	4.9	(c)	5.12 (0.36)	(a)	5.82 (1.42)	(a)	4.80	(1.40)	(c)
	5.3	(a)	5.60 (1.35)	(a, b)			5.06	(1.48)	(c)
	6.5	(a, c)					5.13	(2.49)	(a)

refractive index:

$n(\infty)$	2.07	$E \| a$	parameters of the equation	66K
A	0.284 μm^2		$n = n(\infty) + A \lambda^{-2}$ (λ in μm)	
$n(\infty)$	1.97	$E \| b$		
A	0.045 μm^2			
$n(\infty)$	2.12	$E \| c$		
A	0.149 μm^2			

Physical property	Numerical value	Experimental conditions	Experimental method, remarks	Ref.
dielectric constants:				
$\varepsilon(0)$	37.2	$E\|a$	dielectric function in the	76C2
	20.1	$E\|b$	infrared: Fig. 21	
ε	4.28	$E\|b$, $\lambda = 0.671$ μm		76C2
	4.41	0.589 μm		
	6.00	$E\|c$, $\lambda = 0.671$ μm		
	6.50	0.589 μm		
	7.29	$E\|a$, $\lambda = 0.671$ μm		
$\varepsilon(\infty)$	4.28	$E\|a$	from refractive index	66K
	3.88	$E\|b$		
	4.49	$E\|c$		

The dielectric behaviour at lower frequencies is very anomalous. At very low frequencies relaxation phenomena are especially marked along the c-axis (Fig. 22, see also Fig. 12 α). This has been explained as relaxation of electron hopping about structural defects located in the $\langle 010 \rangle$ planes. For 100···200 K the activation energy is 0.045 eV and for $T > 200$ K 0.19 eV [74C1].

ESR data on $W_xV_2O_5$ show two defects. One is intrinsic to V_2O_5 and derives from oxygen vacancies whereas the other is derived from $W^{6+} - O - V^{4+}$ units [81G].

References for 9.15.2.2.5

50B	Bystrom, A., Wilhelmi, K.A., Brotzan, O.: Acta Chem. Scand. **4** (1950) 1119.
55K	King, B.W., Suber, L.L.: J. Am. Ceram. Soc. **38** (1955) 306.
61B	Bachmann, H.G., Ahmed, F.R., Barnes, W.H.: Z. Kristallogr. **115** (1961) 110.
63M	Minomura, S., Drickamer, H.G.: J. Appl. Phys. **34** (1963) 3043.
65P	Patrina, I.B., Ioffe, V.A.: Sov. Phys. Solid State (English Transl.) **6** (1965) 2425, 2581.
66K	Kenay, N., Kannewurf, O.R., Whitmore, D.H.: J. Phys. Chem. Solids **27** (1966) 1237.
67B	Bodó, Z., Hevesi, I.: Phys. Status Solidi **20** (1967) K45.
67K	Kennedy, T.N., Hakim, R., McKenzie, J.D.: Mater. Res. Bull. **2** (1967) 193.
69M	Mokerov, V.G., Rakov, A.V.: Fiz. Tverd. Tela **11** (1969) 197.
69V	Volzhenskii, D.S., Pashkovskii, M.V.: Sov. Phys. Solid State (English Transl.) **11** (1969) 950.
70I	Ioffe, V.A., Patrina, I.B.: Phys. Status Solidi **40** (1970) 389.
70N	Nagels, P., Denayer, M.: Proc. 10th Int. Conf. Semiconductors **1970**, 321.
71B	Bayard, M.L.F., Reynolds, T.G., Vlasse, M., McKinzie, H.L., Arnott, R.J., Wold, A.: J. Solid State Chem. **3** (1971) 484.
71K	Karvaly, B., Hevesi, I.: Z. Naturforsch. Teil A **26** (1971) 245.
71P	Perlstein, J.H.: J. Solid State Chem. **3** (1971) 217.
71S	Scott, A.B., McCulloch, J.C., Mar, K.M.: Proc. 2nd Int. Conf. Low Mobility Materials **1971**.
72M	Mokerov, V.G., Sigelov, B.L.: Fiz. Tverd. Tela **14** (1972) 3405.
72R	Reshina, I.I.: Sov. Phys. Solid State (English Transl.) **14** (1972) 287.
72V	Vinogradov, A.A., Shelykh, A.I.: Fiz. Tverd. Tela **13** (1972) 3310.
73B	Bootz, B., Finkenrath, H., Franz, G., Uhle, N.: Solid State Commun. **13** (1973) 1477.
73G	Gilson, T.R., Bizri, O.F., Cheetham, N.: J. Chem. Soc. **A1973**, 291.
73H	Haemers, J., Baetens, E., Vennik, J.: Phys. Status Solidi (a) **20** (1973) 381.
73M	Mokerov, V.G.: Fiz. Tverd. Tela **15** (1973) 2393.
73V	Valiev, K.A., Mokerov, V.G., Rakov, A.V.: Fiz. Tverd. Tela **15** (1973) 361.
74C1	Chernenko, I.M., Ivon, A.I.: Fiz. Tverd. Tela **16** (1974) 2130.
74C2	Clauws, P., Vennik, J.: Phys. Status Solidi (b) **66** (1974) 553, 953.
74S	Sperlich, G., Zimmermann, P.: Solid State Commun. **14** (1974) 897.
76C1	Chakrabarty, D.K., Guha, D., Biswas, A.B.: J. Mater. Sci. **11** (1976) 1347.
76C2	Clauws, P., Vennik, J.: Phys. Status Solidi (b) **76** (1976) 707.
76V	Volzhenski, D.S., Grin', V.A., Savitskii, V.G.: Kristallografiya **21** (1976) 1238.
77B	Broeskx, J., Clauws, P., Vennik, J.: Solid State Commun. **22** (1977) 577.
77D	Dziembaj, R., Piwowarczyk, J.: J. Solid State Chem. **21** (1977) 387.
77L	Lazukova, N.I., Gubanov, V.A., Mokerov, V.G.: Int. J. Quantum Chem. **12** (1977) 915.

78D	Dziembaj, R.: J. Solid State Chem. **26** (1978) 159.
78I	Ivon, A.I.: Izv. Vyss. Ucheb. Zaved. Fiz. **21** (1978) 130.
78V	Vanhaelet, M., Clauws, P.: Phys. Status Solidi (b) **87** (1978) 719.
79P	Palanna, O.G.: Indian J. Chem. **17A** (1979) 442.
81G	Gaillard, B., Blanchard, C., Deville, A., Livage, J.: Phys. Status Solidi (b) **104** (1981) 627.
81I	Ivon, A.I., Chernenko, M.: Fiz. Tekh. Poluprovodn. **15** (1981) 263.

Physical property	Numerical value	Experimental conditions	Experimental method, remarks	Ref.

9.15.2.3 Monoxides

The oxides MnO, $Fe_{1-x}O$, CoO and NiO are all semiconducting type II antiferromagnets, in which the next-nearest neighbour superexchange parameter J_2 (see below) is large and negative. Controversy still surrounds the electronic structure, assignment of optical and photoelectron spectra and transport data.

9.15.2.3.1 Manganese oxide (MnO)

Phase diagram: $MnO - Mn_3O_4$ boundary: Figs. 1···3 and [59D, 67S1, 67H1, 70F].

Crystal structure

structure: rocksalt structure at room temperature, rhombohedral structure at 4.2 K ($T_{tr} \simeq 117$ K).

lattice parameters:

a	4.4475 (2) Å	RT	space group $O_h^5 - Fm3m$	70M
a	4.4316 (3) Å	$T = 4.2$ K	from neutron diffraction	58R1
α	90.624 (8)°			

Temperature dependence of lattice parameters: Figs. 4,5. The distortion below $T_N = 117$ K is apparently an exchange striction which compresses the crystal along $\langle 111 \rangle$, increasing the nearest neighbour and decreasing the next-nearest neighbour distances (Fig. 6). Under high pressure a transformation to tetragonal or lower symmetry is observed. Variation of $a/a(0)$ with pressure: Fig. 7.

melting point:

T_m	1815 °C	in purified He		62S

standard free energy:

$\Delta F°$	$-54.56 + 0.0277(2) T$	$T < 1433$ K	for reaction:	70F
	$-49.41 + 0.0241(2) T$	$T > 1433$ K	$(3/(1-4x)) Mn_{1-x}O + \tfrac{1}{2} O_2$	
			$\rightarrow ((1-x)/(4-x)) Mn_3O_4$	

Second-order transformations in $Mn_{1-x}O$ have been reported [70F]. The extent of non-stoichiometry in $Mn_{1-x}O$ is $0 < x < 0.12$ at 1773 K as $\log_{10} p_{O_2}$ changes from -14.7 to -1.4 [74N].

Electronic properties

Brillouin zone: Fig. 8, partial APW band structure: Fig. 9. This calculation overestimates the O2p–Mn3d energy separation [78K]. For higher transition energies from reflectance spectra, see under "Optical properties".

energy gaps (in eV):

$E_{g,th}$	2.43 (8)		from transport data	67H2
	2.6			68A
	2.6 (2)			71G
	$1.90 - 6 \cdot 10^{-4} T$		T in K	76K1
	2.7···3.0		from photoconductivity, Fig. 10	77U
E_g	3.6	$T = 300$ K	from UV absorption; shift of 500 cm^{-1} in E_g below T_N	74C

For higher transition energies, see section on "Optical properties".

Defects: see "Transport properties" and "Lattice properties".

Physical property	Numerical value	Experimental conditions	Experimental method, remarks	Ref.

Lattice properties

phonon dispersion curves: Fig. 11, neutron scattering experiments, see [69H1, 71H, 77K].

phonon wavenumbers (in cm^{-1}):

$\bar{\nu}_{TO}$	262	RT	from IR reflectance	69P
	268	RT	from IR reflectance	69K
	262	$T = 4.2$ K	from UV absorption at 4.2 K	77Y
$\bar{\nu}_{LO}$	552	RT		69P
	550	RT		69K
	485	$T = 4.2$ K		77Y

elastic moduli (in 10^{12} dyn cm^{-2}):

		T [K]		
c_{11}	2.23	298	At $T \approx T_N$, c_{11} decreases by 19.2%	69O
	2.22	296	over 0.3 °C as T is lowered,	72U
	1.768 (6)	above T_N	whereas c_{44} decreases strongly	70C1
c_{12}	1.2	298	at 10 °C above T_N, falling by 30%	69O
	1.099	296	at just above T_N [70C1]. The slow	72U
c_{44}	0.79	298	shear mode, $[\frac{1}{2}(c_{11}-c_{12})]$ shows	69O
	0.783	296	complex behaviour at T_N and a	72U
	0.68 (1)	above $T_N + 20$ K	further anomaly at 42 K [80P].	70C1

Debye temperature:

Θ_D	533 K	$T = 298$ K	from elastic constants	78S
	525.7 K	298 K	from elastic constants	81F

Transport properties

defect equilibria:

the predominant defects are Mn vacancies in pure MnO, for which the following equilibria hold:

$$\tfrac{1}{2}O_2 \rightleftharpoons V_{Mn} + O_0; \quad [V_{Mn}]/p_{O_2}^{\frac{1}{2}} = K_x = K_{x0}\exp(-\Delta g_v/kT)$$

$$V_{Mn} \rightleftharpoons V'_{Mn} + e^+; \quad p \cdot [V'_{Mn}]/[V_{Mn}] = K_1 = K_{10}\exp(-E_1/kT)$$

$$V'_{Mn} \rightleftharpoons V''_{Mn} + e^+; \quad p \cdot [V''_{Mn}]/[V'_{Mn}] = K_2 = K_{20}\exp(-E_2/kT)$$

67H1, 67H2, 70O

where Δg_v is the Gibbs energy of formation of a neutral cation vacancy and E_1 and E_2 are the ionization energies of the vacancy. For the defects, standard Kröger-Vink notation is used throughout sect. 9.15.2.

Because the mobility of the conduction band electrons is so much greater than that of the holes, intrinsic conduction at high temperatures gives an additional significant contribution to the conductivity. As shown in Fig. 12, three regions can be distinguished:

region A: $[V''_{Mn}] \gg [V'_{Mn}], [V_{Mn}]$; $\mu_n n \gg \mu_p p$

region B: $[V''_{Mn}] \gg [V'_{Mn}], [V_{Mn}]$; $\mu_p p \gg \mu_n n$

region C: $[V'_{Mn}] \gg [V''_{Mn}]$ $\mu_p p \gg \mu_n n$.

MnO is unusual in that it can be doped both p-type and n-type [81P, 71K, 70O], the latter by addition of Ti, Sb, Cr.

defect formation energy (in eV):

$\Delta g_v + E_1 + E_2$	1.41	$T > 1375$ K		67H2
$d(\ln x)/d(kT^{-1})$	0.94		x is deviation from stoichiometry	67H2
	1.26 (5)		in Mn$_{1-x}$O	72B

In lithium-doped samples, intrinsic effects are very much reduced. The main equilibrium is

$$\tfrac{1}{2}O_2 + 2MnO + Li_2O \rightleftharpoons 2Li'_{Mn} + 4O_0 + 2Mn^{\cdot}_{Mn}$$

and a dielectric process corresponding to the holes on $Mn^{\cdot}_{Mn}$ hopping around Li'_{Mn} has been identified [68C] with a relaxation energy of 0.29 eV (Fig. 13) [70C2].

Physical property	Numerical value	Experimental conditions	Experimental method, remarks	Ref.

The trapping energy of the $Li'_{Mn} - Mn^{\cdot}_{Mn}$ site is 0.4 eV [70C2] on the basis of an activated mobility or 0.64 eV [68A] if the mobility is unactivated. The extent of compensation in Li-doped MnO is controversial, being regarded as slight [70C2, 68A] or extensive [76K1].

Equilibrium constants for the reactions:

$$2MnO + \tfrac{1}{2}O_2 \rightleftharpoons 2Mn^{\cdot}_{Mn} + V''_{Mn} + 3O_0 : K_{ox}$$
$$Cr_2O_3 \rightleftharpoons 2Cr^{\cdot}_{Mn} + V''_{Mn} + 3O_0 : K_{Cr}$$

are shown in Fig. 19.

transport data:

at **low and intermediate** temperatures (< 500 K), single crystal data on pure MnO are collected in Figs. 14a, 14b. Material is p-type in this region with characteristic properties:

activation energy for conductivity:

E_A	0.7···0.8 eV			81P, 80J, 76K1, 71K

hole mobility:

μ_p	$1.8 \cdot 10^{-5}$ ···$1.7 \cdot 10^{-2}$ cm^2/V s	$T = 230$···500 K	from thermopower and conductivity data	81P
	10^{-3}···10^{-2} cm^2/V s	420···870 K	from Seebeck data; in this study [74A] the activation energy associated with the Seebeck coefficient decreased from 0.65···0.45 eV with stoichiometry; the more nearly stoichiometric the sample the lower the energy	74A

activation energy for hole mobility:

E_A	0.45 eV	$T = 250$···500 K	thermopower measurement	81P
	0.22 eV	420···820 K	thermopower measurement	74A

electron mobility:

μ_n	5···18 cm^2/V s	$T = 350$···900 K	Hall mobility of an n-type sample; the mobility showed a maximum near 500 K; see also Fig. 15	71K

At **higher** temperatures, activation energy for conduction in MnO increases substantially.

activation energy for conductivity (in eV, $T = 1000$···1300 °C):

E_A	1.41			67H2
	1.33			71B
	1.39		results from a variety of Cr-doped and pure samples	70O
	1.8			71K

hole mobility (in cm^2 V^{-1} s^{-1}):

		T [°C]		
μ_p	0.15	1200	$\mu_n/\mu_p \geq 10$; see also Fig. 14c	67H2
	0.17	1000		71B
	$6.6 \cdot 10^{-3}$	1000	$\mu_n/\mu_p = 1500$. Actually the mobility ratio has been measured and μ_p calculated assuming $\mu_n = 10$ cm^2/V s.	70O
	$9.3 \cdot 10^{-3}$	1100	$= 1070$	
	$1.2 \cdot 10^{-2}$	1200	$= 820$	
	$5 \cdot 10^{-3}$	730···1230		71K
	0.1	943···1210	from Hall measurement	74A

Physical property	Numerical value	Experimental conditions	Experimental method, remarks	Ref.
activation energies for hole mobility (in eV):				
E_A	0.42(5)	$T < 1375\,°C$	activation energy for hopping	67H2
	0.32(7)	$T > 1375\,°C$		
	0.57		corrected for assumed $T^{-3/2}$ pre-exponential factor	
	0.35	$T = 943 \cdots 1210\,°C$	Hall measurement	74A
electron mobility (in $cm^2\,V^{-1}\,s^{-1}$):				
μ_n	11.3	$T = 1000\,°C$	Hall mobility	67W
	9.3	$1200\,°C$		
	10.0	$943\,°C$	Hall mobility	68G

In addition to these data, photoconductivity experiments at and below RT give electron mobilities of $10 \cdots 20\,cm^2/V\,s$ [77U]; see also Fig. 15.

dependence of transport parameters on stoichiometry: conductivity: Fig. 16, Seebeck coefficient: Fig. 18. (For ϱ and S of MnO, see also Figs. 20, 23.)

effect of n-type dopants: Fig. 17.

transport in Li-doped MnO: electrical resistivity of Li-doped samples: Fig. 20, 21, Hall coefficient: Fig. 21, Hall mobility: Fig. 22, Seebeck data: Fig. 23 (see also Fig. 13 of 9.15.2.5.2 (Fe_2O_3)).

The Hall mobility of Fig. 22 is essentially identical to that reported in [68A], though the risks of comparing data on ceramic Li-doped samples with data on pure single crystals have been emphasized [81P]. Above 500 K the Hall mobility is constant suggesting that all the observed activation energy in the conductivity is due to freeing the holes from the $Mn_{Mn}^{\cdot} - Li'_{Mn}$ sites. This interpretation is not supported by the thermopower data of Fig. 23. The slope of S vs. $1/T$ differs from $\ln \varrho$ vs. $1/T$ suggesting an **activated** mobility of activation energy $0.25 \cdots 0.30\,eV$ [70C2] which lies between the values quoted above [67H2, 71B, 81P]. The trapping energy of the $Li'_{Mn} - Mn_{Mn}^{\cdot}$ trap is $0.4\,eV$ [70C2] in the range $RT \cdots 1000\,K$ if the mobility is activated, and the mobility of holes derived from these traps is $0.01\,cm^2/V\,s$ at 1000 K [70C2] and $3 \cdot 10^{-5}\,cm^2/V\,s$ at 300 K. This compares with a Hall mobility of $3 \cdots 8 \cdot 10^{-2}\,cm^2/V\,s$ in the T-range $350 \cdots 750\,K$ [70C2]. The apparently larger Hall mobility may be a consequence of the small-polaron nature of these holes.

Optical properties

higher energy peaks in the optical spectra:

Photoemission spectrum, decomposed into Mn 3d and O 2p components: Fig. 24. The spectrum can be understood in terms of ionization of local Mn^{2+} electrons to give ^5E and $^5T_2$ states separated by 1.56 eV and broadened by Franck-Condon factors.

peak wavenumbers in optical spectra (in cm^{-1}, at 4.2 K):

For spectra, see Fig. 25 and [69H1, 69H2, 59P]

$\bar{v}_{peak}$	15532		band 1, transition $^6A_{1g} - {}^4T_{1g}$ (I)	77Y
	ca. 19000		band 2, transition $^6A_{1g} - {}^4T_{2g}$ (I)	
	23322		band 3, transition $^6A_{1g} - {}^4A_{1g}$	
	23389		transition $^6A_{1g} - {}^4E_g$ (I)	
	25659		band 4, transition $^6A_{1g} - {}^4T_{2g}$ (II)	
			fine structure associated with these peaks was assigned to a magnon side-band at $112\,cm^{-1}$ and optical phonons.	
	29000		interband absorption edge, see Fig. 26; assigned to Mn 3d – Mn 4s	74C

Physical property	Numerical value	Experimental conditions	Experimental method, remarks	Ref.
peak energies in reflectance spectra (in eV):				
		T [K]		
E_{peak}	4	RF, 300	onset of reflectance; RF: reflectance spectrum, TF: thermoreflectance assignment: Mn e_g – Mn 4s	76K2, 72M3
	4.6	RF, 22		
	4.7	RF, 300		
	4.9	TF, TI	TI: temperature independent; thermoreflectance spectrum: Fig. 27; derived empirical band diagram: Fig. 28	
	5.2	RF, 22		
	5.4	TF, TI	Mn e_g – Mn t_{2g}	
	5.5	RF, 22	Mn t_{2g} – Mn 4s	
	5.7	TF, 85	shifted to 5.55 eV at 200 K O 2p – Mn 4s	
	6.3	RF, 22	Mn e_g – Mn e_g; Mn t_{2g} – Mn t_{2g}	
	6.4	TF, TI		
	6.6	RF, 300	peak at 22 K, shoulder at 300 K also TF, O 2p – Mn 4s	
	6.9	RF, 300		
	7.0	TF, 200	shifts to 6.7 eV at 310 K	
	7.2	RF, 22	Mn t_{2g} – Mn e_g	
		TF, TI		
	4···7.5	300	Mn e_g – Mn t_{2g}	76K2
	8···10.6	300	Mn t_{2g} – Mn 4s	
	11···19	300	O 2s – Mn 4s	
	16(1)	RF, 22		72M3
	5.3	RT	EELS data O 2p – Mn t_{2g}	78A
	8.2		O 2p – Mn e_g	
	9.8		O 2p – Mn 4s	
dielectric constants:				
$\varepsilon(0)$	22.5	RT	from IR reflectance	69P
	18.8	RT	from IR reflectance	69K
$\varepsilon(\infty)$	4.95	RT		69P
	4.47	RT	from $n=2.11_5$ in the near IR	69K
ε_{AF}	18.6	RT	AF: audio-frequency	69K
	16.5	$T=4.2$ K		69K

refractive index, real and imaginary parts: Fig. 29.

Magnetic properties

Néel temperature (in K):

T_N	118.5		heat capacity	81K
	117.9		magnetic susceptibility	81J
	116.9		Moessbauer data	67S2
	116		magnetic susceptibility	61B
	117		magnetic susceptibility	65L
	117.8		heat capacity	51T
	117.7		neutron data	66L
	119.7		neutron data	74P
	117.5		critical neutron scattering	78B

Physical property	Numerical value	Experimental conditions	Experimental method, remarks	Ref.

susceptibility:

high temperatures:
Curie-Weiss law satisfied: $\chi = C/(T + \Theta_p)$ with $\Theta_p = 540$ K [65L] $= 461$ K [61B]. Above $T = 5T_N$ data is conflicting: according to [65L] Curie-Weiss law not obeyed, but according to [61B] a good fit to the expression was found for $T \lesssim 1200$ K with $C_m = 3.78$ K cm^3/mol, a value considerably lower than the expected spin-only value.

lower temperatures:
susceptibility maximum at 122.5(5) K [66L], 117.9 K [81J] and marked field dependence (Fig. 30) and stress dependence [81J]. Neutron scattering data [51S, 58R1, 58R2, 66B] and Moessbauer [67S2] show that the spins order into (111) sheets that are coupled antiferromagnetically. The spins are confined within the (111) sheets and point in the [1$\bar{1}$0] direction or in one of the two directions at 120° to this. There is little anisotropy within the plane [66B, 79B]. Long-range spin-ordering: Fig. 31. Low temperature magnetic moment is 2.34 μ_B/Mn^{2+} at 4.2 K, reduced from the expected 2.5 by covalency effects [73J].

spin-wave dispersion: Fig. 32.

spin-wave Hamiltonian parameters:

For $T > T_N$ the data can be analysed in terms of a simple Hamiltonian of the form

$$\mathcal{H} = -\sum_{nn} J_1 \, s_i \cdot s_j - \sum_{nnn} J_2 \, s_i \cdot s_j$$

where the first summation is over all the nearest neighbours and the second sum is over all the next-nearest neighbours. For $T < T_N$, the rhombohedral distortion leads to two distinct nearest neighbour coupling constants J_1^+ and J_1^- corresponding to the nearest neighbours with spins parallel and antiparallel, respectively. The total Hamiltonian also contains an anisotropy term and hence has the form

$$\mathcal{H} = -\sum_{nnp} J_1^+ \, s_i \cdot s_j - \sum_{nna} J_1^- \, s_i \cdot s_j - \sum_{nnn} J'_{2i} \, s_i \cdot s_j + \sum_i D_1 (s_i^x)^2 + \sum_i D_2 (s_i^y)^2. \quad [74P]$$

For $T > T_N$ ($J_{1,2}$ in meV):

J_1	−0.366	nearest neighbour coupling	74P
	−0.383	constant	74K
	−0.861	calculated value	72B
J_2	−0.414	next nearest neighbour coupling	74P
	−0.458	constant	74K
	−0.947	calculated	72B
ε_1	25.8	$\varepsilon_1 = -\partial \ln J_1 / \partial \ln r$, where r is	74P
	23.0	the nearest-neighbour distance,	74K
	23.0	similar definition for ε_2	72B
	27(5)		73S
ε_2	14.0		74P
	11.8	calculated value	72B

for $T < T_N$ ($J_{1,2}$; $D_{1,2}$ in meV):

J_1^+	−0.428	$T = 4.2$ K		74P
J_1^-	−0.324			
J'_2	−0.420			
D_1	0.059			
D_2	0.0015			

References for 9.15.2.3.1

51S	Shu, C.G., Strauser, W.A., Wollan, E.O.: Phys. Rev. **83** (1951) 333.
51T	Todd, S.S., Bonnickson, K.R.: J. Am. Ceram. Soc. **73** (1951) 3894.
58R1	Roth, W.L.: Phys. Rev. **110** (1958) 1333.
58R2	Roth, W.L.: Phys. Rev. **111** (1958) 772.
59D	Davies, M.W., Richardson, F.D.: Trans. Faraday Soc. **55** (1959) 604.
59P	Pratt, G.W., Coelho, R.: Phys. Rev. **116** (1959) 281.
61B	Benewicz, J., Heidelberg, R.F., Luxem, A.H.: J. Phys. Chem. **65** (1961) 615.
62S	Singleton, E.L., Carpenter, L., Lundquist, R.V.: U.S. Bur. Mines, Rep. Invest. **1962**, No. 5938.
65L	Lines, M.E., Jones, E.D.: Phys. Rev. **A139** (1965) 1313.
65R	Rodbell, D.S., Osika, L.M., Lawrence, P.E.: J. Appl. Phys. **35** (1965) 666.
66B	Blech, I.A., Averbach, B.L.: Phys. Rev. **142** (1966) 237.
66C	Clendenen, R.L., Drickamer, H.G.: J. Chem. Phys. **44** (1966) 4223.
66L	Lindsay, R., Michelsohn, F.: Bull. Am. Phys. Soc. **11** (1966) 108.
67B	Bloch, D., Feron, J.L., Georges, R., Jacobs, I.S.: J. Appl. Phys. **38** (1967) 1474.
67H1	Hed, A.Z., Tannhauser, D.S.: J. Electrochem. Soc. **114** (1967) 314.
67H2	Hed, A.Z., Tannhauser, D.S.: J. Chem. Phys. **47** (1967) 2090.
67N	Nagels, P., Denayer, M.: Solid State Commun. **5** (1967) 193.
67S1	Schwertfeger, K., Muan, A.: Trans. AIME **239** (1967) 1115.
67S2	Siegworth, J.D.: Phys. Rev. **155** (1967) 285.
67W	de Wit, H.J., Crevecoeur, C.: Phys. Lett. **A25** (1967) 393.
68A	Ali, M., Fridman, M., Denayer, M., Nagels, P.: Phys. Status Solidi **28** (1968) 193.
68B	Bloch, D., Cherbit, P., Georges, R.: Compt. Rend. **266B** (1968) 430.
68C	Crevecoeur, C., de Wit, H.J.: Solid State Commun. **6** (1968) 135.
68G	Gvishi, M., Tallon, N.M., Tannhauser, D.S.: Solid State Commun. **6** (1968) 135.
68N	Nagels, P., Denayer, M., de Wit, H.J., Crevecoeur, C.: Solid State Commun. **6** (1968) 695.
69H1	Haywood, B.C.G., Collins, M.F.: J. Phys. **C2** (1969) 46.
69H2	Huffmann, D.R., Wild, R.L., Shinmei, M.: J. Chem. Phys. **50** (1969) 4092.
69K	Kinney, T.B., O'Keefe, M.: Solid State Commun. **7** (1969) 977.
69O	Oliver, D.W.: J. Appl. Phys. **40** (1969) 893.
69P	Plendl, J.N., Mansur, L.C., Mitra, S.S., Chang, I.F.: Solid State Commun. **7** (1969) 107.
70C1	Cracknell, M.F., Evans, R.G.: Solid State Commun. **8** (1970) 359.
70C2	Crevecoeur, C., de Wit, H.J.: J. Phys. Chem. Solids **31** (1970) 783.
70F	Fender, B.E.F., Riley, F.D.: "Chemistry of Extended Defects in Non-metallic Solids", Eyring, L., O'Keeffe, M. (eds.), Amsterdam: North-Holland Publ. Comp., **1970**, p. 54.
70M	Morosin, B.: Phys. Rev. **B1** (1970) 236.
70O	O'Keeffe, M., Valigi, M.: J. Phys. Chem. Solids **31** (1970) 947.
71B	Bransky, I., Tallan, N.M.: "Conduction in Low-mobility Materials", Klein, N., Tannhauser, D.S., Pollak, M. (eds.), London: Publ. Taylor, **1971**, p. 31.
71G	Gvishi, M.: Solid State Commun. **9** (1971) 2089.
71H	Haywood, B.C.G., Collins, M.F.: J. Phys. **C4** (1971) 1299.
71K	Ksendzov, Y.M., Makarov, V.V.: Fiz. Tverd. Tela **12** (1971) 3166.
72B	Bartel, C.L.: Solid State Commun. **11** (1972) 55.
72M1	Mattheiss, L.F.: Phys. Rev. **B5** (1972) 290.
72M2	Mattheiss, L.F.: Phys. Rev. **B5** (1972) 306.
72M3	Messick, L., Walker, W.C., Glosser, R.: Phys. Rev. **B6** (1972) 3941.
72U	Uchiba, N., Saito, S.: J. Acoust. Soc. Am. **51** (1972) 1602.
73J	Jacobsen, A.J., Tofield, B.C., Fender, B.E.F.: J. Phys. **C6** (1973) 1615.
73M	Messick, L., Walker, W.C., Glosser, R.: Surf. Sci. **37** (1973) 267.
73S	Seino, D., Miyahara, S., Naro, Y.: Phys. Lett. **A44** (1973) 35.
74A	Albella, J.M., Pajares, J.A., Soria, J.A.: Z. Phys. Chem. (NF) **92** (1974) 101.
74B	Bloch, D., Many, R., Vater, C., Yelon, W.B.: Phys. Lett. **A49** (1974) 354.
74C	Chou, H.-H., Fan, H.Y.: Phys. Rev. **B10** (1974) 901.
74K	Kohgi, M., Ishikawa, Y., Hawada, I., Motuzuki, K.: J. Phys. Soc. Jpn. **36** (1974) 112.
74N	Navrotsky, A.: MTP International Review of Science, Inorganic Chemistry, Series 2, volume 5, D.W.A. Sharp (ed.) (Butterworths, 1974, U.K.).
74P	Pepy, G.: J. Phys. Chem. Solids **35** (1974) 433.

75E	Eastman, D.E., Freeouf, J.L.: Phys. Rev. Lett. **34** (1975) 395.
76K1	Kleinpenning, T.G.M.: J. Phys. Chem. Solids **37** (1976) 925.
76K2	Ksendzov, Y.M., Korobova, I.L., Sidorin, K.K., Startsev, G.P.: Fiz. Tverd. Tela **18** (1976) 173.
76W	Wagner, V., Reichardt, W., Kress, W.: Proc. Int. Conf. Neutron Scatt. Moon, R.M. (ed.) **1976** 175.
77K	Kress, M.: Proc. Int. Conf. Lattice Dynamics, Balkanski, M. (ed.), **1977**, p. 77.
77U	Usani, T., Masumi, T.: Physica **86–88** B+C (1977) 985.
77Y	Yokogawa, M., Taniguchi, K., Hamaguchi, C.: J. Phys. Soc. Jpn. **42** (1977) 591.
78A	Akimoto, K., Sakisaki, Y., Nishijama, M., Onchi, M.: J. Phys. **C11** (1978) 2535.
78B	Betsuyaki, H.: Nip. Gaish. Keuk., Ann. Rep. Neutron Scatt. Stud. JAERI-M-8089 (1978).
78K	Kunz, A.B., Surratt, G.T.: Solid State Commun. **25** (1978) 9.
78S	Subhadra, K.G., Sirdesmukh, D.B.: Indian J. Pure Appl. Phys. **16** (1978) 693.
79A	Agarwal, S.: Solid. State Commun. **29** (1979) 197.
79B	Bidoux, R., Coute, R., Nasser, J.A.: J. Appl. Phys. **50** (1979) 1683.
80J	Joshi, G.M.: Ph.D. Thesis. University of Purdue, 1980.
80P	Palmer, S.B., Waintal, A.: Solid State Commun. **34** (1980) 663.
81F	Freer, R.: J. Mater. Sci. **16** (1981) 3225.
81J	Jagadesh, M.S., Seehra, M.S.: Phys. Rev. **B23** (1981) 1185.
81K	Kleinclauss, J., Mainard, R., Fousse, H., Ciret, N., Bour, D., Pointon, A.J.: J. Phys. **C14** (1981) 1163.
81P	Pai, M., Honig, J.M.: J. Solid State Chem. **40** (1981) 59.

Physical property	Numerical value	Experimental conditions	Experimental method, remarks	Ref.

9.15.2.3.2 Iron oxide (FeO and $Fe_{1-x}O$)
(See also Landolt-Börnstein, Vol. III/12b.)

Phase diagram: Fig. 1. Above 560 °C a non-stoichiometric cubic phase $Fe_{1-x}O$ is stable; below this temperature $Fe_{1-x}O$ is metastable with respect to disproportionation. Smallest x value in equilibrium is 0.05, though stoichiometric FeO can be made under pressure. There is no complete solid solution range between $Fe_{1-x}O$ and Fe_3O_4, even at the highest temperature. Thermogravimetric [64V, 65V1, 2] and EMF measurements [69F] have suggested that several order-disorder or other higher order transitions are found in the $Fe_{1-x}O$ domain, but X-ray diffraction studies at high temperatures [72H] have failed to reveal any structural evidence for these phases.

Stoichiometric FeO can be synthesized for $T \geq 770$ °C, $p \geq 36$ kbar [67K2].

Crystal structure

structure ($Fe_{1-x}O$): cubic, space group $O_h^5 - Fm3m$, $Z = 4$, above $T_N \simeq 200$ K [63W].

For $T < T_N$ the structure shows a rhombohedral distortion ($\alpha < 60°$). At constant T this distortion increases as $x \rightarrow 0$ (Fig. 2). Under pressure at 300 K the value of a alters as in Fig. 3.

lattice parameters in the cubic phase (in Å):

a	4.3390	$T = 300$ K	extrapolated to x = 0 from the	33J
	4.3370		da/dx data below	66L
	4.3297			68F1
	4.3453			60R
	4.3202			65V1
	4.323		measured on stoichiometric FeO	67K2
	4.332			70H
da/dx	0.5643	300 K	x-dependence of a: Figs. 4, 5	33J
	0.5286		T-dependence of a: Fig. 6	66L
	0.4256			68F1
	0.625			60R
	0.3025			65V1

density

d	5.87 g cm^{-3}		X-ray density	70H

Electronic properties

Band structure calculations: [72M, 78K2]. Results compared to recent photoemission data [75E]: Fig. 7. Substantial overlap between the d-orbital energies and the O 2p band is evident. A local cluster calculation on FeO_6^{10-} has also been reported [76T] (see Fig. 8). The qualitative conclusions are similar to the later band calculation [78K2] with the $t_{2g}\uparrow(\Gamma'_{25})$ lying below the O 2p valence band edge.

Analysis of photoemission and absorption spectra: see also under "Optical properties".

Defects

Review: [80B]. It is established that the predominant type of defect arises from the migration of iron as Fe^{3+} to an interstitial tetrahedral site. To minimize cation repulsion, Fe^{2+} vacancies then cluster in a tetrahedral arrangement around Fe_i^{3+}. The elementary defect is thus $Fe_i^{3+}\ (V''_{Fe})_4$. Such a defect is strongly negatively charged and will attract octahedral site Fe^{3+} ions [75C, 77C]. Some suggested defect morphologies are shown in Fig. 9. Calculations favour the 6:2 and 8:3 clusters and neutron diffraction data [71C, 79B, 79G] suggest that the ratio $[V''_{Fe}]/[Fe_i^{3+}]$ does fall from 4 towards 3 with increasing x. X-ray diffraction [69K, 73H] and electron diffraction [74I, 77A] on annealed samples show ordered arrangements of 13:4, 10:3 or 16:6 clusters in a low symmetry structure. The present interpretation of the available experimental evidence is against the formation of corner-shared spinel-like aggregates (Fig. 10) [79G, 77C].

Deviation from stoichiometry vs. oxygen partial pressure: Fig. 11. No simple defect model can fit the data. [68K] suggests the complex defect $Fe_c = Fe_i^{3+}\ (V''_{Fe})\ (V'_{Fe})$ with the following equilibrium reactions:

$$Fe_{Fe} + V_i + \tfrac{1}{2}O_2 \rightleftharpoons Fe_c + O_0, \qquad K_1 = 1.75\cdot10^{-9}\exp(79400\,[\text{cal mol}^{-1}]/RT)$$

$$Fe_c + Fe_{Fe} \rightleftharpoons Fe'_c + Fe_{Fe}\cdot e^+, \qquad K_i = 0.13\cdot10^2 \exp(-17900\,[\text{cal mol}^{-1}]/RT)$$

Fe_c and Fe'_c are neutral and singly ionized complex defects, respectively. V_i represents an interstitial site available for occupancy by the complex defect, Fe_{Fe} is a divalent Fe atom on a normal site, $Fe_{Fe}\cdot e^+$ atom is a Fe atom on an octahedral site with one trapped hole (a trivalent Fe ion) and O_0 is an oxygen atom on an oxygen lattice site. K_1: equilibrium constant for the formation of neutral complex species, K_i: equilibrium constant for the first ionization.

Lattice properties

phonon dispersion spectrum: Fig. 12, density of states: Fig. 13.

wavenumbers of optical phonons (in cm^{-1}):

$\bar{\nu}_{LO}$	493.5	RT	from IR spectrum	77P
	526.6 (133)	RT	from phonon dispersion spectrum	77K
$\bar{\nu}_{TO}$	337.7	RT		77P
	320.0 (133)	RT		77K

other lattice parameters in fitting of the dielectric function (RT values):

$$\varepsilon(\omega) = \varepsilon(\infty) + 4\pi\varrho_0\,\bar{\nu}_0^2/(\bar{\nu}_0^2 - \bar{\nu}^2 - i\gamma_0\bar{\nu}) - \varepsilon(\infty)\,\bar{\nu}_p^2/(\bar{\nu}^2 + i\bar{\nu}/\tau)$$

$\bar{\nu}_0$	337.28 cm^{-1}	for the dielectric constant, see	77P
$\bar{\nu}_p$	265.94 cm^{-1}	section on "Optical properties"	
γ_0	96.80 cm^{-1}		
$4\pi\varrho_0$	12.94		
τ^{-1}	$143\cdot10^{12}$ Hz		

elastic moduli (in 10^{12} dyn cm^{-2}):

c_{11}	3.02	RT	77K
c_{12}	1.21		
c_{44}	0.70		

Debye temperature:

Θ_D	494 K	$T=298$ K	from velocity of sound	81F

Physical property	Numerical value	Experimental conditions	Experimental method, remarks	Ref.

Transport properties

conductivity:

dependence on p_{O_2}: Fig. 14, temperature dependence in $Fe_{0.91}O$: Fig. 15, at high pressures: Fig. 16, resistivity near T_N: Fig. 17.

At high temperatures σ varies in a manner similar to the stoichiometry with p_{O_2} but there are differences in detail. At low temperatures there is disagreement in the literature (Fig. 15). Above 170 K $\sigma = 300 [\Omega^{-1} cm^{-1}] \exp(-0.07 (eV)/kT)$. The optical conductivity at 300 K can be calculated from the plasmon frequency using the formula $\sigma_{opt} [\Omega^{-1} cm^{-1}] = \pi e v_p^2 \varepsilon(\infty) \tau/30$ and has the value 17.3 $(\Omega cm)^{-1}$ [77P]. Below this, the data of [62T] show a fall off not reflected in the data of [74B1]. The latter show a sharp kink at 120 K, the activation energy doubling below this temperature to 0.14 eV. Below 100 K, the authors find a $T^{-1/4}$ law, suggesting random-range hopping.

conductivity in nearly stoichiometric FeO: Quite different behaviour has been observed in nearly stoichiometric samples of FeO [74B2, 74B3]. A resistivity anomaly is found near T_N (Fig. 17); in this region, the effective mass $m_n = 4 m_0$ and the mobility is $\approx 10 cm^2/V s$, suggesting large polaron coupling instead of the hopping conduction apparently found in more defect samples.

Seebeck coefficient: Fig. 18. Unusual switch from positive to negative values as the deviation from stoichiometry increases [70S] which could be understood within a complex defect model of the form $Fe_c = Fe_i^{3+} (V_{Fe})_2$ as described above. If $\beta = [Fe_c']$, then the formula $S = k/e \{\ln[(1-4\beta)/\beta^2] + A\}$ can be derived for S [70S]. However, a best fit to the data of Fig. 18 gives $A = -0.89$ ($\approx 1000 \cdots 1300 °C$). This is ascribed to "negative phonon drag". Given that the change from p- to n-type behaviour coincides with a change in the Moessbauer spectrum [69J] it is more probable that clustering of defects is responsible. The p–n transition can also be suppressed by doping with magnesium [67H, 67B1] (Fig. 19).

hole mobility (in $cm^2/V s$):

μ_p	$\mu_o \beta^{1/2} (1-4\beta)$ $\exp(-0.01 [eV]/kT[K])$		formula used to fit data in Fig. 14; $\beta = [Fe_c']$ as defined above	70S
	≈ 1	$T = 1000$ $\cdots 1300 °C$		70S
	0.137 (7)	1000 °C		67B1
	$\simeq AT^{-1} \exp(-0.2 [eV]/kT[K])$		for higher temperatures, see Fig. 20	67B1
	10	near T_N in "stoichiometric" FeO	analysed from Fig. 17 assuming large-polaron behaviour	74B2

Optical properties

Photoemission spectrum: Fig. 21, absorption spectra: Figs. 22, 23.

Detailed interpretation of the PE spectrum is complicated by final-state effects [75E, 77B, 79V, 80G] and a detailed calculation [77B, 80G] and comparison with XPS suggests that the Fe 3d emission is far broader than had earlier been suggested (Fig. 21); this result is essentially traceable to extensive excited state configuration interaction.

The rather ill-defined structure in the absorption spectrum is assigned to Fe 3d – Fe 4s transitions [78B], but earlier work [74B1] has shown that absorption in this region was strongly dependent on the degree of non-stoichiometry (Fig. 22). Above 2.4 eV, the absorption rises strongly, and this edge has also been ascribed to a d–s transition.

dielectric constants:

$\varepsilon(0)$	24	RT	from IR reflectivity	77P
	31.4	RT	from dispersion spectrum	77K
$\varepsilon(\infty)$	11.13	RT		77P
	9.24	RT		77K

Physical property	Numerical value	Experimental conditions	Experimental method, remarks	Ref.

Magnetic properties

Below T_N, the spins of Fe^{2+} lie in ferromagnetic (111) sheets which are antiferromagnetically coupled along [111]. The spin direction is thought to vary depending on the distance from a defect cluster; near a cluster the spins tend to lie close to the (111) plane and in the defect free region they align along [111] [79B]. At high temperatures (970···1500 K) a Curie-Weiss law is obeyed: $\chi = C/(T - \Theta_p)$.

parameters of the Curie-Weiss law for $Fe_{1-x}O$:

C_m	3.30 (3.30)	$x = 0$	values in brackets: Curie constant	70M
	3.24 (3.27)	0.01	C_{Fe} referred to gram-atom of Fe;	
	3.23 (3.30)	0.02	C_m in K cm³/mol, Θ_p in K.	
	3.06 (3.28)	0.04	Mean value of C_{Fe}: $3.25 \frac{\text{K cm}^3}{\text{g-atom}}$,	
	2.93 (3.25)	0.06		
	2.96 (3.26)	0.09	corresponding to $p_{eff} \simeq 5.1 \mu_B$	
Θ_p	−10 (10)	0		
	−40 (10)	0.01		
	−40 (10)	0.02		
	−60 (10)	0.04		
	−90 (10)	0.06		
	−100 (10)	0.09		

magnetic susceptibility: temperature dependence near T_N shown in Fig. 25.

Néel temperature:

T_N	200 K	$x = 0.02$	varies in a rather complex way	68F2
	195 K	0.068	with x	67K1
	203 K	0.104		

spin wave spectrum: Fig. 26. Analysis of neutron scattering data [79B] shows that the spin of the interstitial Fe_i^{3+} may be oriented in the (111) plane, pulling neighbouring octahedral Fe^{2+} and Fe^{3+} spins also into the (111) plane (Fig. 27).

parameters of the spin wave Hamiltonian:

$\mathcal{H} = -\sum_{ij} J_{ij} s_i \cdot s_j + \sum_i D(s_i^z)^2$

J_1^+	$+3.70$ cm^{-1}	$T = 4.2$ K	various coupling constants defined	78K1
J_1^-	$+2.06$ cm^{-1}	$(T < T_N)$	in MnO section 9.15.2.3.1	
J_2'	-6.52 cm^{-1}			
D_1	-0.68 cm^{-1}			
ε	65		ε is defined as $d \ln J_1 / d \ln r$ with r = internuclear distance	

References for 9.15.2.3.2

33J	Jette, E.R., Foote, F.: J. Chem. Phys. **1** (1933) 33.
45D	Darken, L.S., Gurry, R.W.: J. Am. Chem. Soc. **67** (1945) 1398.
46D	Darken, L.S., Gurry, R.W.: J. Am. Chem. Soc. **68** (1946) 798.
51C	Cirilli, V., Brisi, C.: Ann. Chim. **41** (1951) 508.
53W	Willis, B.T.M., Rooksby, H.P.: Acta Crystallogr. **6** (1953) 827.
56F	Foster, P.K., Welch, A.J.E.: Trans. Faraday Soc. **52** (1956) 1626.
58R	Roth, W.L.: Phys. Rev. **110** (1958) 1333.
60R	Roth, W.L.: Acta Crystallogr. **13** (1960) 140.
62T	Tannhauser, D.S.: J. Phys. Chem. Solids **23** (1962) 25.
63W	Wyckoff, R.W.G.: Crystal structures, New York: John Wiley, Vol. 1, 2nd ed. **1963**.
64C	Carel, C., Weigel, D., Vallet, P.: Compt. Rend. **258** (1964) 6126.
64V	Vallet, P., Raccah, P.: Compt. Rend. **258** (1964) 3679.
65V1	Vallet, P., Raccah, P.: Compt. Rend. **260** (1965) 4325.
65V2	Vallet, P., Raccah, P.: Mem. Sci. Rev. Metall. **62** (1965) 1.
66C	Clendenen, R.L., Drickamer, H.G.: J. Chem. Phys. **44** (1966) 4223.

References for 9.15.2.3.2

66G	Geiger, G.H., Levin, R.L., Wagner, J.B.: J. Phys. Chem. Solids **27** (1966) 947.
66L	Levin, R.L., Wagner, J.B.: Trans. AIME **236** (1966) 516.
67B1	Bransky, I., Tannhauser, D.S.: Physica **37** (1967) 547.
67B2	Bransky, I., Tannhauser, D.S.: Trans. AIME **239** (1967) 75.
67H	Hillegas, W.J., Wagner, J.B.: Phys. Lett. **A25** (1967) 742.
67K1	Koch, F.B., Fine, M.E.: J. Appl. Phys. **38** (1967) 1470.
67K2	Katsura, T., Iwasaki, B., Kimura, S., Akimoto, S.: J. Chem. Phys. **47** (1967) 4559.
68F1	Fujii, C.T., Meussner, R.A.: Trans. AIME **242** (1968) 1259.
68F2	Fine, M.E., Koch, F.B.: J. Appl. Phys. **39** (1968) 2478.
68K	Kofstad, P., Hed, A.Z.: J. Electrochem. Soc. **115** (1968) 102.
69F	Fender, B.E.F., Riley, F.D.: J. Phys. Chem. Solids **30** (1969) 793.
69J	Johnson, D.P.: Solid State Commun. **7** (1969) 1785.
69K	Koch, F., Cohen, J.B.: Acta Crystallogr. **B25** (1969) 275.
70H	Hentschel, B.: Z. Naturforsch. **A25** (1970) 1997.
70M	Michel, A., Poix, P., Bernier, J.-C.: Ann. Chim. **5** (1970) 261.
70S	Seltzer, M.S., Hed, A.Z.: J. Electrochem. Soc. **117** (1970) 815.
71C	Cheetham, A.K., Fender, B.E.F., Taylor, R.I.: J. Phys. **C4** (1971) 2160.
72H	Hayakawa, M., Cohen, J.B., Reed, T.B.: J. Am. Ceram. Soc. **55** (1972) 160.
72M	Mattheiss, L.F.: Phys. Rev. **B5** (1972) 290.
72S	Slater, J.C., Johnson, K.H.: Phys. Rev. **B5** (1972) 844.
73H	Hoang, T.K., Romanov, A.D., Shayovitch, Y.L., Zvinchuk, R.A.: Vestn. Leningr. Univ., Fiz. Khim. **4** (1973) 144.
74B1	Bowen, H.K., Adler, D., Auker, B.H.: J. Solid State Chem. **12** (1975) 355.
74B2	Balberg, I.: Phys. Semicond. Proc. Int. Conf. **1974**, 920.
74B3	Balberg, I., Alexander, S., Helman, J.S.: Phys. Rev. Lett. **33** (1974) 836.
74I	Iijima, S.: Proc. Electron Microsc. Soc. Am. **32** (1974) 352.
74K	Kawai, N., Nishiyawa, A.: Proc. 4th Int. Conf. High Press. Kyoto **1974**, 324.
74T	Thouzelin, B.: Rev. Int. Haut. Temp. Refract. **2** (1974) 219.
75C	Catlow, C.R.A., Fender, B.E.F.: J. Phys. **C8** (1975) 3287.
75E	Eastman, D.E., Freeouf, J.L.: Phys. Rev. Lett. **34** (1975) 395.
76A	Alvarado, S.F., Erbudak, M., Munz, P.: Phys. Rev. **B14** (1976) 2740.
76T	Tossell, J.A.: J. Electron Spectrosc. Relat. Phenom. **8** (1976) 1.
77A	Anderson, B., Sletner, J.O.: Acta Crystallogr. **A33** (1977) 268.
77B	Bagus, P.S., Brundle, C.R., Chuang, T.J., Wandelt, K.: Phys. Rev. Lett. **39** (1977) 1229.
77C	Catlow, C.R.A., Fender, B.E.F., Muxworthy, D.G.: J. Phys. (Paris) Colloq. **38** (1977) C7-67.
77K	Kugel, G., Carabatos, C., Hennion, B., Prévot, B., Reviolevschi, A., Tocchetti, D.: Phys. Rev. **B16** (1977) 378.
77P	Prévot, B., Briellman, J., Meftah, M.P., Sieskind, M.: Phys. Status Solidi **A40** (1977) 503.
78B	Balberg, I., Pinch, H.L.: J. Magn. Magn. Mater. **7** (1978) 12.
78K1	Kugel, G.E., Hennion, B., Carabatos, C.: Phys. Rev. **B18** (1978) 1317.
78K2	Kunz, A.B., Surratt, G.T.: Solid State Commun. **25** (1978) 9.
79B	Battle, P., Cheetham, A.K.: J. Phys. **C12** (1979) 337.
79G	Gavarri, J.-R., Cavel, C., Weigel, D.: J. Solid State Chem. **29** (1979) 81.
79V	Vasudevan, S., Hegde, M.S., Rao, C.N.R.: J. Solid State Chem. **29** (1979) 253.
80B	Bauer, E., Pianelli, A.: Mater. Res. Bull. **15** (1980) 177.
80G	Grenet, G., Jugnet, Y., Duc, T.M., Kilder, M.: J. Chem. Phys. **72** (1980) 218.
81F	Freer, R.: J. Mater. Sci. **16** (1981) 3225.

Physical property	Numerical value	Experimental conditions	Experimental method, remarks	Ref.

9.15.2.3.3 Cobalt oxide (CoO)

Phase diagram: Fig. 1. The spinel Co_3O_4 is unstable at high temperatures and the monoxide shows a wide range of stoichiometry [64F].

phase boundaries:

$\log_{10} a_{O_2}$	16.5 − 20300/T[K] for $Co_{1-\delta}O/Co_3O_4$ 7.2 − 24100/T[K] for $Co/Co_{1-\delta}O$		a_{O_2} is the oxygen activity	77D

Crystal structure

$T > T_N$: NaCl cubic structure, space group $O_h^5 - Fm3m$, $Z = 4$ [53G].

$T < T_N$: Jahn-Teller distortion leading to an approximately tetragonal structure that preserves orbital angular momentum (Fig. 2) [53G], space group $C_{2h}^3 - C2/m$, $Z = 2$ [66S1].

lattice parameters:

a	4.2581 (5) Å 4.258 Å	high-temperature phase, $T = 295$ K	pressure dependence: Fig. 3	53G 66C1
a b c β	5.183 Å 3.015 Å 3.017 Å 125°33′	low-temperature phase, $T = 77$ K	within the NaCl setting this can be described as $a = b = 4.265$ Å, $c = 4.217$ Å, $c/a = 0.9887$, $\alpha = \beta = \gamma = 89°58′$	66S1

Electronic properties

band structures: Fig. 4 (see also [72M1, 78K]).

energy gaps:

E_g	2.6 eV	RT	absorption: Fig. 5; fundamental edge shifts to higher energy at lower temperature. Below 290 K, sample showed optical anisotropy.	59P, 70P
$E_{g, th}$	3.6 (5) eV		from Hall data in Ti doped CoO	72G

width of bands:

W (O 2p band)	$\simeq 4$ eV	RT	XPE spectra, (core region, see Fig. 6)	79J, 75K
W (Co 3d band)	≤ 3 eV	RT		

higher transition energies from optical spectra (in eV):

For a discussion of the spectra, see below under "Optical properties".

E_{peak}	5.5 (5.7) 7.5 12.6 17.5	RT (shoulder at 8.5 eV found at 120 K)	absorption (Fig. 7). Value in parentheses from [72M2].	70P
			thermoreflectance (Fig. 8), assignment of final state (cf. Fig. 9)	72M2
	2.8	RT	I: $d^6(t_\alpha^3 t_\beta e_\alpha^2) + 4s$	
	3.2	RT	II: $d^6(t_\alpha^3 t_\beta e_\alpha^2) + d^8(t_\alpha^3 t_\beta^3 e_\alpha^2)$	
	3.8	RT	III: $d^6(t_\alpha^3 t_\beta^2 e_\alpha) + 4s$	
	4.2	$T = 80$ K	IV: $d^6(t_\alpha^3 t_\beta e_\alpha^2) + d^8(t_\alpha^3 t_\beta^2 e_\alpha^2 e_\beta)$ and $d^6(t_\alpha^3 t_\beta^2 e_\alpha) + d^8(t_\alpha^3 t_\beta^3 e_\alpha^2)$	
	4.75 5.0	RT 80 K	V: $d^6(t_\alpha^2 t_\beta^2 e_\alpha^2) + 4s$	
	5.35	80 K	VI: $d^6(t_\alpha^3 t_\beta^2 e_\alpha) + d^8(t_\alpha^3 t_\beta^2 e_\alpha^2 e_\beta)$	
	5.75 6.0	RT 80 K	O 2p − Co 4s (onset)	

(continued)

Physical property	Numerical value	Experimental conditions	Experimental method, remarks	Ref.
higher transition energies from optical spectra (continued)				
E_{peak}	6.25	80 K	VII: $d^6(t_\alpha^2 t_\beta^2 e_\alpha^2) + d^8(t_\alpha^3 t_\beta^2 e_\alpha^2 e_\beta)$	72M2
	7.2	80 K ⎫	O 2p – Co 4s	
	7.0	RT ⎭		
			empirical energy level scheme: Fig. 9. For the meaning of the t, e, α, β, see caption of Fig. 28 of 9.15.2.3.1 (MnO)	
		RT	EELS, radically different assignment:	76A
	2.2		not assigned	
	3.3 ⎫		O 2p – Co t_β	
	4.6 ⎭			
	5.6 ⎫		O 2p – Co e_β	
	6.8 ⎭			
	7.6 ⎫		O 2p – Co 4s	
	9.7 ⎭			

Lattice properties

phonon wavenumbers (in cm^{-1}):

$\bar{\nu}_{TO}$	262	RT	IR reflectance	69P
	347···350	RT	IR reflectance	65G
	347		neutron scattering, Figs. 10, 11	68S1
$\bar{\nu}_{LO}$	552	RT		69P
	544···547	RT		65G
	520			68S1

elastic moduli (in 10^{11} dyn cm^{-2}):

		T [K]		
c_{11}	27.7	425		68S1
	30.7	110		68S1
	25.56 (11)	296		72U
	26.2	295	c_{11} shows a sharp decrease at $T = T_N$ to $\approx 18 \cdot 10^{11}$ for $T < T_N$	74D
c_{12}	18.0	425		68S1
	18.3	110		68S1
	14.36 (15)	296		72U
c_{44}	9.1	425		68S1
	9.0	110		68S1
	8.03 (05)	296		72U

Debye temperature:

Θ_D	513 K	$T = 298$ K	from elastic constants	78S
	517 K	298 K	from elastic constants	81F

Transport properties

Deviation from stoichiometry in $Co_{1-\delta}O$: Fig. 12.

$T-x$ diagram: Fig. 13. For temperature dependence of relaxation time for local hole-hopping, see Fig. 13 of 9.15.2.3.1 (MnO).

Physical property	Numerical value	Experimental conditions	Experimental method, remarks	Ref.

dominant equilibria in solid $Co_{1-\delta}O$ [77D]:

$$Co_{Co} + \tfrac{1}{2}O_2 \rightleftharpoons V_{Co} + CoO \quad K_1 = 1.6 \cdot 10^{-2} \exp(-26 [\text{kJ mol}^{-1}]/RT)$$
$$V_{Co} \rightleftharpoons V'_{Co} + e^+ \quad K_2 = 2.4 \exp(-51 [\text{kJ mol}^{-1}]/RT) \text{ in the range } 1000 \cdots 1400\,°C$$
$$V'_{Co} \rightleftharpoons V''_{Co} + e^+ \quad K_3 = 0.17 \exp(-72 [\text{kJ mol}^{-1}]/RT)$$

77D

ΔH for the formation of singly ionized vacancies 94 kJ mol^{-1} [72B], 54 kJ mol^{-1} [66F], 55$\cdots$60 kJ mol^{-1} [68N]. Second ionization energy E_2 is 63 kJ mol^{-1} [66F], 74.3 kJ mol^{-1} [72B].

Three regions noted by [66F]: region A_1 has unionized vacancies dominant, region A_2 has singly ionized vacancies dominant and region B has doubly ionized vacancies dominant.

activation energies for conductivity (in eV):

		T [K]		
E_A	0.22$\cdots$0.57	220$\cdots$290	quenched samples in which δ of $Co_{1-\delta}O$ varies from 0.012$\cdots$0.001	66F
	0.24$\cdots$0.5	200$\cdots$300	quenched samples: $\delta = 0.012\cdots0.002$	67B
	0.29$\cdots$0.43	150$\cdots$300	Li doped samples; smaller energies correspond to larger lithium content	69B
	0.41$\cdots$0.49	290$\cdots$420	Li doped single crystal	67A
	0.58	1270$\cdots$1670	region A	66F
	0.70	1270$\cdots$1670	region B	66F
	0.48	1270$\cdots$1470		68E
	0.742	300$\cdots$500	single crystal	80J

Conductivity dependence on oxygen partial pressure: Fig. 14, on Li addition: Figs. 18, 19, 21; see also Fig. 13 of 9.15.2.5.2 (Fe_2O_3). The deviation from linearity at low values of Li dopant is due to Co vacancies dominating the conductivity. This is borne out by the Seebeck data (Fig. 20).

activation energy for Seebeck coefficient (in eV):

		T [K]		
E_A	0.54	1270$\cdots$1670	region A	66F
	0.94	1270$\cdots$1670	region B	66F
	0.27$\cdots$0.42	290$\cdots$420	Li-doped single crystal	67A
	0.306	300$\cdots$1000	single crystal	80J

Positive sign of Seebeck coefficient (Fig. 15, 15α) supports conduction by holes. Sign inversion at high temperatures near Co/CoO boundary [70G, 72G]. For lithium doped samples the similarity of the slopes for the thermopower and conductivity suggests that conduction is of the band or large-polaron type [71B, 67A, 67B, 65Z] (see also Fig. 16), but results on single crystals that have been annealed at high temperature suggest the possibility of activated conductivity [77D, 66F, 80J]. For Seebeck coefficient of Li doped samples, see also Fig. 20. For reduced Seebeck coefficient, see Fig. 13 of 9.15.2.5.2 (Fe_2O_3).

hole drift mobilities (in cm^2/V s):

		T [K]		
μ_p	0.4	1620	activated; $E_A = 0.3$ eV in the range 1000$\cdots$1300 °C, extrapolated to RT gives 10^{-5} cm^2/V s	66F
	$2.5 \cdot 10^{-4}$	555	single crystal data, activated with energy 0.436 eV	80J
	$5.0 \cdot 10^{-7}$	333		
	0.31	1470	activated with energy 0.089 eV	77D
	0.45	1470	unactivated	72B
	0.4$\cdots$0.5	295$\cdots$425	Li doped, unactivated	67A
	0.3 (1)	800$\cdots$1500	undoped and Li doped, unactivated, Li content 0.02$\cdots$3.0 at%	67B
	0.25	1200	Li doped, unactivated	69B
	0.3	1500	unactivated	65S
	0.06	1480$\cdots$1550		72G

Physical property	Numerical value	Experimental conditions	Experimental method, remarks	Ref.

Seebeck coefficient varies with oxygen partial pressure [66F, 67B].
For mobility of Li doped material, see also Fig. 22.

hole Hall mobilities (in cm^2/V s):

		T [K]		
μ_p	0.05···0.085	295···425	Li doped single crystal, Hall mobility unactivated	67A
	0.05···0.1	RT···1500	mobility varies only slightly with temperature	67B
	0.04···0.07	300	increases slightly with T	65Z
	0.06	1480···1550	Hall mobility at low oxygen pressures: Fig. 17	72G

electron Hall mobilities:

μ_n	0.3···0.6 cm^2/V s	T = 980···1140 °C	unactivated; pure and Ti doped material	70G, 72G

See also Figs. 16, 21, 22 and 24 for dependence of mobility on stoichiometry dopant and temperature. For energy level scheme for Cr and Ti doped CoO obtained from transport data, see Fig. 23.

Optical properties

optical spectra: photoemission: Figs. 6, 25, 26, absorption: Figs. 5, 7, thermoreflectance: Fig. 8.

For peak energies in these spectra, see above under "Electronic properties". The photoemission spectrum has been interpreted in terms of a localized model involving a large number of final states. High-resolution XPS data [72W, 73H, 75K, 76H, 79J] show d-state photoemission to occur over $\simeq 12$ eV, a result supported by calculation (Figs. 25, 26). The long d-energy spread in Fig. 6 is attributed to extensive ground-state configuration mixing. An upper limit of $\simeq 3$ eV for the Co 3d band width can be assigned and an upper limit of ≈ 3 eV to the process $2Co^{2+}(3d^7) \rightarrow Co^{3+}(3d^6) + Co^+(3d^8)$. Satellite structure in the core region of the XPS spectrum: 6 eV wide structure at 9.8 eV ascribed to O 2p (e_g) → Co 3d (e_g) charge transfer as are peaks $\simeq 6$ eV below the main Co 2s, 2p, 3d and (?) 3s, 3p signals [75K]. Structures in the absorption spectrum: see above. Below the band threshold of 470 nm (2.64 eV) transitions are excitonic d−d in nature; detailed assignments [59P].

dielectric constants:

		T [K]		
$\varepsilon(0)$	22.50	295	IR reflectance fit	69P
	13	295	IR reflectance fit	65G
	10.70	425	neutron scattering data	68S1
	12.73	110		
	12.19	296		
$\varepsilon(\infty)$	4.95	295		69P
	5.76	295		65R
	5.3	295		65G
	5.4	295		59N
	5.19	425		68S1
	5.28	110		
	5.3	296		

For real and imaginary parts of the dielectric function vs. photon energy, see Fig. 7.

Magnetic properties

magnetic susceptibility: Fig. 27.

effective magnetic moment:

p_{eff}	5.25 μ_B		from Curie-Weiss law, for $T > 400$ K	56S
	4.96 μ_B			51H

Physical property	Numerical value	Experimental conditions	Experimental method, remarks	Ref.
paramagnetic Curie temperature:				
Θ_p	-330 K			56S
	-280 K			51H
Néel temperature (in K):				
T_N	292.2		susceptibility	51H
	292.2		dilatation	48F
	292		susceptibility	51T
	295		neutron diffraction	68K
	290		thermal conductivity	69Z
	290		heat capacity	69Z
	289		heat capacity	74S
	291		Moessbauer effect	66C2
	287		heat capacity	81K
$\dfrac{d \ln T_N}{d \ln V}$	$\simeq -3$		(V = unit cell volume)	66C2, 66B

The spin structure has been described in two different ways:
1. spins coupled antiferromagnetically between adjacent (111) planes, spins lying in [$\bar{1}\bar{1}7$] direction at an angle of 11°30′ with (001) [58R],
2. spins lie in (1$\bar{1}$0) plane making an angle of 27.4(5)° with (001) and ca. 8° with the (111) plane [65L, 78H].

The second description is now preferred; canting of the spins from the c-axis is due to dipole-dipole forces that partially decouple spin-orbit coupling.

spin-wave spectrum: Figs. 10, 11.

coupling constants of the Hamiltonian (in meV):

see preceding section for definition

		T [K]		
J_1	-0.08	110	J_1, J_2: averaged values	68S1
	-0.47	4.2···135		72R1
	-0.26			72R2
J_2	-1.42	110		68S1
	-2.36	4.2···135		72R1
	-1.52			72R2
	-1.45 or -1.8	9.3		76W

Short-range magnetic order above T_N also reproted [72R2].

References for 9.15.2.3.3

36W	Wagner, C., Koch, E.: Z. Phys. Chem. **B32** (1936) 439.
48F	Foëx, M.C.: Compt. Rend. **227** (1948) 193.
51H	Henry La Blanchetais, C.: J. Phys. Radium **12** (1951) 765.
51T	Trombe, M.F.: J. Phys. Radium **12** (1951) 170.
53G	Greenwald, S.: Acta Crystallogr. **6** (1953) 396.
54C	Carter, R.E., Richardson, F.D.: Trans. AIME **200** (1954) 1244.
56S	Singer, J.R.: Phys. Rev. **104** (1956) 929.
58R	Roth, W.L.: Phys. Rev. **110** (1958) 1333.
59N	Newman, R., Chrenko, R.M.: Phys. Rev. **114** (1959) 1507.
59P	Pratt, G.W., Coelho, R.: Phys. Rev. **116** (1959) 281.
64F	Fisher, B., Tannhauser, D.S.: J. Electrochem. Soc. **111** (1964) 1194.
64R	Roilos, M., Nagels, P.: Solid State Commun. **2** (1964) 285.
65G	Gielisse, P.J., Plendl, J.N., Mansur, L.C., Marshall, R., Mitra, S.S., Mykolojewicz, R., Smakula, A.: J. Appl. Phys. **36** (1965) 2446.

References for 9.15.2.3.3

65L	von Laar, B.: Phys. Rev. **A138** (1965) 584.
65R	Rao, K.V., Smakula, A.: J. Appl. Phys. **36** (1965) 2031.
65S	Shelykh, A.I., Artenov, K.S., Shvaiko-Shvaikovskii, V.E.: Fiz. Tverd. Tela **8** (1965) 1287.
65Z	Zhuze, V.P., Shelykh, A.I.: Fiz. Tverd. Tela **8** (1965) 629.
66B	Bloch, D., Chaissé, F., Pauthenot, R.: J. Appl. Phys. **37** (1966) 1401.
66C1	Clendenen, R.L., Drickamer, H.G.: J. Chem. Phys. **44** (1966) 4223.
66C2	Creston, C.J., Ingalls, R., Drickamer, H.G.: J. Appl. Phys. **37** (1966) 1400.
66F	Fisher, B., Tannhauser, D.S.: J. Chem. Phys. **44** (1966) 1663.
66S1	Saito, S., Nakahiyashi, K., Shinomura, Y.: J. Phys. Soc. Jpn. **21** (1966) 850.
66S2	Shelykh, A.I., Artenov, K.S., Shvaiko-Shvaikowskii, V.E.: Fiz. Tverd. Tela **8** (1966) 883.
67A	Austin, I.G., Springthorpe, A.J., Smith, B.A., Turner, C.E.: Proc. Phys. Soc. (London) **90** (1967) 157.
67B	Bruck, A., Tannhauser, D.S.: J. Appl. Phys. **38** (1967) 2520.
67D	van Daal, H.J., Bosman, A.J.: Phys. Rev. **158** (1967) 736.
67N	Nagels, P., Denayer, M.: Solid State Commun. **5** (1967) 193.
68E	Eror, N.G., Wagner, J.B.: J. Phys. Chem. Solids **29** (1968) 1597.
68K	Khan, D.C., Erickson, R.A.: J. Phys. Chem. Solids **29** (1968) 2087.
68S1	Sakurai, J., Buyers, W.J.L., Cowley, R.A., Dolling, G.: Phys. Rev. **167** (1968) 510.
68S2	Sockel, H.-G., Schmalzried, H.: Ber. Bunsenges. Phys. Chem. **72** (1968) 745.
69B	Bosman, A.J., Crevecoeur, C.: J. Phys. Chem. Solids **30** (1969) 1151.
69P	Plendl, J.N., Mansur, L.C., Mitra, S.S., Chang, I.F.: Solid State Commun. **7** (1969) 109.
69Z	Zhuze, V.P., Novuzov, O.N., Shelykh, A.I.: Fiz. Tverd. Tela **11** (1969) 1287.
70G	Griski, M., Tannhauser, D.S.: Solid State Commun. **8** (1970) 485.
70P	Powell, R.J., Spicer, W.E.: Phys. Rev. **B2** (1970) 2182.
71B	Bransky, I., Tallan, N.M.: Cond. Low-mobility Mat.; (ed.) Klein, N.; London: Taylor, **1971**, 31.
72B	Bransky, I., Wimmer, J.M.: J. Phys. Chem. Solids **33** (1972) 801.
72G	Gvishi, M., Tannhauser, D.S.: J. Phys. Chem. Solids **33** (1972) 893.
72M1	Mattheiss, L.F.: Phys. Rev. **B5** (1972) 290.
72M2	Messick, L., Walker, W.C., Glosser, R., Phys. Rev. **B6** (1972) 3941.
72R1	Rechtin, M.D., Averbach, B.L.: Phys. Rev. **B6** (1972) 4294.
72R2	Rechtin, M.D., Averbach, B.L.: Phys. Rev. **B5** (1972) 2693.
72U	Uchiba, N., Saito, S.: J. Acoust. Soc. Am. **51** (1972) 1602.
72W	Wertheim, G.K., Hüfner, S.: Phys. Rev. Lett **28** (1972) 1028.
73H	Hüfner, S., Wertheim, G.K.: Phys. Rev. **B8** (1973) 4857.
73M	Messick, L., Walker, W.C., Glosser, R.: Surf. Sci. **37** (1973) 267.
74C	Chou, H.-H., Fan, H.Y.: Phys. Rev. **B10** (1974) 901.
74D	Dentschuk, P., Palmer, S.B.: Phys. Lett. **A47** (1974) 343.
74S	Salomon, M.B., Garnier, P.R., Gidding, B., Buchler, E.: J. Phys. Chem. Solids **35** (1974) 851.
75E	Eastman, D.E., Freeouf, J.L.: Phys. Rev. Lett. **34** (1975) 395.
75K	Kim, K.S.: Phys. Rev. **B11** (1975) 2177.
75M	Morin, F.: Can. Metall. Q. **14** (1975) 105.
76A	Akimoto, K., Sakisaka, Y., Nishijima, M., Onchi, M.: J. Phys. **C11** (1976) 2535.
76F	Fryt, E.: Oxid. Met. **10** (1976) 311.
76H	Huang, T.J., Brundle, C.R., Rice, D.W.: Surf. Sci. **59** (1976) 413.
76W	Wagner, V., Drexel, W.: J. Magnetism. Magn. Mater. **2** (1976) 106.
77D	Dieckmann, R.: Z. Phys. Chem. (N.F.) **107** (1977) 189.
78H	Hermann-Rouzaud, D., Burlet, P., Rossat-Mignod, J.: J. Phys. **C11** (1978) 2123.
78K	Kunz, A.B., Surratt, G.T.: Solid State Commun. **25** (1978) 9.
79J	Jugnet, Y., Duc, T.M.: J. Phys. Chem. Solids **40** (1979) 29.
79W	Wahlgren, U., Johansen, H.: Int. J. Quantum Chem. **15** (1979) 403.
78S	Subhadra, K.G., Sirdesmukh, D.B.: Indian J. Pure Appl. Phys. **16** (1978) 693.
79S	Seehra, M.S., Silinsky, P.: Solid State Commun. **31** (1979) 183.
80J	Joshi, G.M., Pai, M., Harrison, H.R., Sandberg, C.J., Aragón, R., Honig, J.M.: Mater. Res. Bull. **15** (1980) 1575.
81F	Freer, R.: J. Mater. Sci. **16** (1981) 3225.
81K	Kleinclauss, J., Mainard, R., Fousse, H., Ciret, N., Bour, D., Pointon, A.J.: J. Phys. **C14** (1981) 1163.

Physical property	Numerical value	Experimental conditions	Experimental method, remarks	Ref.

9.15.2.3.4 Nickel oxide (NiO)

Crystal structure

Only the monoxide NiO has been well characterized though other oxides, e.g. Ni_3O_4, Ni_2O_3 and NiO_2 have been suggested on the basis of electrochemical evidence [72B1]. However, no higher phases have been prepared using conventional solid-state techniques, even under very high oxygen partial pressures [71D2, 72S1]. As normally synthesized, NiO is cation deficient, the colour varying from green near the stiochiometric limit to black as δ increases in $Ni_{1-\delta}O$.

structure: NaCl-type, cubic, space group $O_h^5 - Fm3m$, $Z = 4$.

Below the Néel temperature ($T_N \approx 522$ K) there is a small rhombohedral distortion derived from exchange striction.

lattice parameter (in Å):

		T [K]		
a	4.1705	$T \to 0$	variation with T: Fig. 1, variation	71B1
	4.1759	298	with pressure: Fig. 2	71B1
	4.1758	298		54S
	4.1752	273		63P
	4.1767	300		
	4.1796	354		
	4.1886	515		
	4.1991	688		
	4.2070	821		
	4.2101	890		
	4.2183	1023		

		p_{O_2} [atm]		
	4.1738	298, 10^{-6}	samples quenched from 1673 K after	69B
	4.1725	10^{-4}	equilibration at the pressures	
	4.1719	10^{-2}	of O_2 shown	
	4.1713	10^{-1}		
	4.1701	$10^{-0.7}$		
	4.1696	10^0		

coefficient of linear thermal expansion (in K^{-1}):

α	$3.2 \cdot 10^{-5}$	RT		66C
	$7.93 \cdot 10^{-6}$	at high temperatures		71B1
	$1.39 \cdot 10^{-5}$	$T = 160 \cdots 1300\,°C$		63G

rhombohedral distortion (at $T = 300$ K) (see Fig. 3):

Δ	3.5′			71B1
	3.5′			48R
	4.2′			60S
	3.8′			54S
	3.8′			70V1
	4.2′			63G
	4.5′			71T

NiO can be doped with Li, with marked effect on many physical properties. Structurally, doping with Li causes a contraction in the lattice consistent with the presence of Ni^{3+}.

lattice parameter in Li doped NiO (in Å):

		Li content [at %]		
a	4.1764	0	room temperature data for	76D
	4.1758	1.00	$p_{O_2} \leq 10^{-6}$ atm and incorporation	
	4.1753	1.30	of Li_2O at 1073 K;	
	4.1746	1.70	at all concentrations $\alpha = 90.06°$	

At $p_{O_2} = 10^{-6}$ atm the maximum Li content before seggregation of a new phase is 1.9 at%. At higher p_{O_2} higher Li contents are possible [76D].

(continued)

Physical property	Numerical value	Experimental conditions	Experimental method, remarks	Ref.
lattice parameter in Li doped NiO (continued)				
		x δ		
a	4.170	0 0	third column gives x and fourth δ in $Li_xNi_{1-x}O_{1+\delta}$ at RT	66S
	4.167	0.005 0.057		
	4.164	0.025 0.063		
	4.153	0.075 0.078		
	4.140	0.135 0.097		

Electronic properties

There is no literature consensus on the detailed electronic structure of NiO or the interpretation of the UV absorption spectra. The results of three different calculations of dispersion in the ΓX direction are shown in Fig. 4. A complete band structure calculation at the non-self-consistent Hartree-Fock level is shown in Fig. 5. At the other extreme, the results of a local SCF–Xα cluster calculation are given in Fig. 6. A large number of phenomenological electronic structures have been proposed and are shown in Figs. 7···9. These differ primarily in the widths and positions of the d^8 and d^9 Ni bands. The extremes are represented by [70A], who places a very narrow d^9 band some 15 eV above the d^8 band which in turn is placed below the $O\,2p^6$ band, and [81D1] who places a much broader d^9 band 3.6 eV above the $O\,2p$ and 2.2 cV above the Ni d^8 level. In this model the extinction coefficient for charge exchange between Ni^{2+} ions is assumed to be extremely small near threshold.

For a discussion of optical spectra in terms of band transitions, see the section on "Optical properties".

Lattice properties

phonon dispersion curves: Fig. 10, calculations: [75R, 76C, 79A], density of states: Fig. 11.

wavenumbers of infrared active phonons (in cm^{-1}, at RT):

$\bar{\nu}_{TO}$	400	green NiO		65G
	440	black NiO		
	387 (5)		neutron study, $\nu_{TO}(\Gamma)$	75R
$\bar{\nu}_{LO}$	580	green NiO		65G
	560	black NiO		
	577 (7)		neutron study, $\nu_{LO}(\Gamma)$	75R

wavenumbers of Raman active phonons (in cm^{-1}, at RT):

$\bar{\nu}_R$	400	black NiO		71D1
	500			
	1100		assigned to $2\bar{\nu}_{LO}$	
	1560		assigned to two-magnon peak with cut-off at 1880 cm^{-1}	

reststrahlen frequency:

ν	$12.2 \cdot 10^{12}$ Hz			65G
	$11.45 \cdot 10^{12}$ Hz			75R

elastic moduli (in 10^{12} dyn cm^{-2}):

c_{11}	2.70 (13)	$T = 296$ K		72U
	2.24			71P
c_{12}	1.25 (28)	296 K		72U
	0.97			71P
c_{44}	1.05 (21)	296 K		72U
	1.10			71P

heat capacity: Fig. 12

Debye temperature:

Θ_D	595 (20) K	$T \rightarrow 0$ K	see Fig. 13	74W
	317.4 K	$T = 298$ K	from elastic constants	81F
	404 K	298 K	from elastic constants	40S

| Physical property | Numerical value | Experimental conditions | Experimental method, remarks | Ref. |

Transport properties

a) pure NiO:

defects: NiO is approximately stoichiometric at the Ni/NiO phase boundary, but at higher oxygen partial pressures, it is cation deficient, the dominant vacancies being variously-charged Ni vacancies. The deviation from stoichiometry as a function of temperature is shown in Fig. 14, and averaged values of x in $Ni_{1-x}O$ from the literature [65F, 70V2, 61M, 68S, 70T, 71O] are given by $x = 0.1093 \exp(-18525 \,[\text{cal mol}^{-1}]/RT)$ (i.e. an activation energy of 0.804 eV). The deviation from stoichiometry in $Ni_{1-x}O$ as calculated from diffusion data is similar to the data given in Fig. 14, varying for representative samples between $x = 0.2 \cdot 10^{-4} \cdots 1.49 \cdot 10^{-4}$ at 900 K to $1.92 \cdot 10^{-4} \cdots 11.70 \cdot 10^{-4}$ at 1300 K [73D]. No agreement exists over the variation of $[V_{Ni}]$ with p_{O_2} and approximations to two different power laws have been reported for $T \geq 1000$ K: $[V_{Ni}] \propto p_{O_2}^{1/6}$ [68S, 69T] and $[V_{Ni}] \propto p_{O_2}^{1/5}$ [70T]. The former would correspond to doubly ionized Ni vacancies as the predominant defects whereas the latter will correspond to a mixture of defects. It can be shown [71O, 73D] that the power law and the parameters derived from a point defect model will be extremely sensitive to the presence of small quantities of compensating donors.

conductivity:

Ionic conductivity remains very small; at 1000 °C in air, $t_{Ni^{2+}} \approx 2 \cdot 10^{-7}$ [79D].

At high temperatures ($T \geq 1200$ K) σ varies with p_{O_2} (Fig. 15). No precise power law has been derived. Widely different exponent values have been published, though a $p_{O_2}^{1/5}$ law at high temperatures seems to be favoured:

parameter n in a power law $\sigma \propto p_{O_2}^{1/n}$:

		T [°C]	oxygen pressure region (p in atm):	
n	5.57	1100	$1 \cdots 10^{-4}$	71O
	5.74	1200		
	5.76	1300		
	5.65	1400		
	3.97	950	$1 \cdots 10^{-4}$	69E
	4.08	1000		
	4.17	1050		
	4.26	1100		
	4.27	1150		
	4.44	1200		
	4.8	950	$1 \cdots 10^{-2}$	73D
	5	1000		
	4.8	1050		
	4.8	1100		
	5.1	1150		
	4.9	1200		
	4.6	1050	$1 \cdots 10^{-5}$	70V2
	4.8	1205		
	5	1245		
	5.3	1377		
	6	1300	$1 \cdots 10^{-6}$	61M
	4.6	1500	$1 \cdots 10^{-6}$	68B
	4.2	1400		
	4.1	1300		
	4.2	1200		
	4.05	1100		
	4	1000		
	4.28 $\cdots$ 4.30	1000	$1 \cdots 1.89 \cdot 10^{-4}$	78F
	4.27 $\cdots$ 5.18	1200		
	4.40 $\cdots$ 5.88	1400		
	4.6	1050	$1 \cdots 10^{-3}$	74I
	4.5	1000		

(continued)

Physical property	Numerical value	Experimental conditions	Experimental method, remarks	Ref.
parameter n in a power law $\sigma \propto p_{O_2}^{1/n}$ (continued)				
n	4	900	$1 \cdots 10^{-6}$	79D
	4.7	1000		
	4.8	1100		
	5.1	1200		
	5.4	1250		

activation energies for conductivity (in eV):

high-temperature region, data at constant partial pressure of oxygen (normally in air)

		T [°C], p [atm]		
E_A	1.07 (1)	$1182\cdots1762$, $1\cdots5\cdot10^{-7}$	$\sigma = 65\,(11) \cdot 10^6\,[\Omega^{-1}\,cm^{-1}]\exp(-E_1/RT)$, $E_1 = 36.4\,(6)$ kcal mol^{-1} in pure CO_2; $\sigma = 7.5 \cdot 10^6\,[\Omega^{-1}\,cm^{-1}]\exp(-E_A/kT)$ in pure O_2; single crystal	70V2
	1.00	$1000\cdots1250$, $1\cdots10^{-3}$	single crystal, ac data	67U
	0.92 (2)	$1000\cdots1600$, $1\cdots10^{-6}$		68B
	$0.94\cdots1.10$	$900\cdots1200$, $1\cdots10^{-4}$	single crystal	69E
	0.81	$900\cdots1400$, $1\cdots10^{-4}$	single crystal and ceramic	71O
	0.86	$900\cdots1300$, $1\cdots10^{-4}$	single crystal and ceramic	73D
	1.01		see Fig. 16 for further values	79D
	0.3	$T_N\cdots730$	single crystal, air	78K1
	≈ 0.3	$T_N\cdots690$	polycrystalline, air	73N
	$0.4\cdots0.8$	$T > T_N$	air	66V

lower temperature region

E_A	0.69	$-20\cdots130$	flame-fusion-grown crystal, see Fig. 17, dry nitrogen	68A1
	$0.5\cdots0.7$	$20\cdots130$	air	66V
	$0.6\cdots0.9$	$130 < T < T_N$	air	
	≈ 0.6	$T < T_N$	single crystal, air	78K1
	≈ 0.6	$T < T_N$	polycrystalline material, air	73N

The similarity of the high-temperature activation energies for conduction and defect formation suggests that μ is not activated in this temperature region [71O]. This conclusion is supported by the Seebeck data, which approximately tracks the conductivity (Fig. 19). Combined thermogravimetric and conductivity data give a drift mobility of 0.53 cm^2/V s in the temperature range $1000\cdots1400$°C [69T, 70T, 71O]. The constancy of this result is not supported by other work; [61M] suggests an activation energy for mobility of 5.5 kcal/mol in this region, close to that expected from a $T^{3/2}$ impurity scattering law. Work on very high purity crystals [78K1] suggests that the Seebeck and conductivity activation energies do not track at all in the range RT$\cdots$1000 K but the discrepancy becomes markedly worse below T_N with the activation energy rising to 0.6 eV compared to 0.2 eV for the Seebeck coefficient (which does not change at T_N). This conclusion is suggested also by [68B] who find an activation energy for σ of 0.92 eV and for the thermopower of 0.66 eV at $1000\cdots1600$°C.

Physical property	Numerical value	Experimental conditions	Experimental method, remarks	Ref.

conduction mechanism at lower temperatures:

ac conductivity shows a marked dispersion at low temperatures (Fig. 19) suggesting the hopping of bound carriers (hopping energy: 0.7 eV) [68A1].

mobility of carriers:

Seebeck coefficient: Fig. 21. Comparison of Hall and drift mobilities: Fig. 22. The calculated drift mobility was an order of magnitude larger than the Hall mobility for Li doped samples ($\mu_{dr}/\mu_H \approx 10$ at 300 K) [67A], leading to $m^{**} \approx 12 \cdots 36\, m_0$, polaron coupling constant $\alpha \approx 6$. High temperature pure single crystal data [68G] give an activated mobility of the form $\mu_p = 15959/T\,[\text{K}]\exp(-0.37\,\text{eV}/kT)\,[\text{cm}^2/\text{V s}]$.

Mobility data of Figs. 21, 22 are representative, but considerably higher values ($20 \cdots 50\,\text{cm}^2/\text{V s}$ at RT) have been reported using transient techniques: Fig. 23. From these data, $m^{**} = 1.5\,m_0$ and $\alpha = 1.6$. Far smaller values of μ_{dr} have also been reported [78K1] ranging from $\leq 10^{-2}\,\text{cm}^2/\text{V s}$ at 1000 K to as little as $10^{-5}\,\text{cm}^2/\text{V s}$ at 300K and a marked kink at T_N [78K1].

These values could only be consistent with a hopping type of mobility, and would make the observation of photoconductivity most unlikely. However, disagreement in the literature exists about the appearance [65K2, 67M] or non-appearance [70P, 78K2, 75K] of photocurrent at the optical threshold.

At intermediate temperatures, the Hall mobility changes sign, usually at a temperature near T_N, though this is strongly dependent on sample stoichiometry (Fig. 24). A quantitative treatment led to a thermal bandgap of only 1.7 eV, substantially lower than that observed optically [75F].

b) Li doped NiO:

Lithium enters NiO substitutionally, compensation being by the formation of $Ni^{3+}-Li^+$ dipoles that are only slightly dissociated at 300 K. In addition some compensation by V_O formation is found, with profound effects on semiconductor statistics [70A].

Figures: effect of lithiation on T_N: Fig. 25, on resistivity: Fig. 26 (change in activation energy closely following T_N), resistivity in the exhaustion region: Fig. 27, Seebeck coefficient: Fig. 28, and comparison with resistivity: Fig. 29. (For temperature dependence of resistivity and reduced Seebeck coefficient, see also Fig. 13 of 9.15.2.5.2 (Fe_2O_3).) The activation energy of 0.3 eV observed is entirely due to the variation in free-hole concentration above 140 K. Below this temperature impurity conduction apparently dominates.

Hall mobilities: Figs. 20, 30; calculated from Seebeck coefficient: Figs. 21, 31. Sign inversion of R_H occurs at T_N as in pure NiO. Further data: Fig. 22 and [67K, 65K2, 64R]. It has been established that the activation energy for conduction falls with increasing Li content both above and below the break near T_N (Fig. 32).

μ_H	$0.1 \cdots 0.4\,\text{cm}^2/\text{V s}$	$T = 300\,\text{K}$	Li content: $0.005 \cdots 0.088$ at % Li	70B
μ_{dr}	$0.5 \cdots 5.0\,\text{cm}^2/\text{V s}$			
	$0.4\,\text{cm}^2/\text{V s}$	higher temperatures, exhaustion region		
μ_H	$\exp(0.08\,[\text{eV}]/kT)$	$T = 200 \cdots 400\,\text{K}$	data support large-polaron model ($m^{**} \simeq 6\,m_0$) [70B]	70B, 67K

dielectric relaxation: At low temperatures and low frequencies, the change $\Delta\varepsilon$ in dielectric constant on doping with Li has the form $\Delta\varepsilon = 4\pi N p^2/1.5\,kT$ (CGS units), where N is the concentration of Li^+-Ni^{3+} dipoles and p the associated dipole moment. At higher frequencies dispersion and deviations from the formula appear. Very similar dielectric loss peaks are found for Na doped NiO [70B] (Fig. 33). Relaxation processes due to Ni^{3+} holes hopping around Li^+ have been identified [68A2]. These appear to behave as small polarons with a hopping energy of $0.20 \cdots 0.25$ eV.

Corresponding data on Na doped NiO showed significantly higher resistivities (by some four orders of magnitude) and activation energies than the corresponding Li doped samples. The thermoelectric power of the Na doped samples is also much higher [81D2].

Physical property	Numerical value	Experimental conditions	Experimental method, remarks	Ref.

Optical properties

optical spectra: photoemission: Fig. 34; XPS: Fig. 36; absorption: Figs. 37···40; electroreflectance: Fig. 41; thermoreflectance: Fig. 42; real and imaginary parts of the refractive index: Fig. 43, of the dielectric constant: Fig. 33.

For the photoemission spectra (Fig. 34) two interpretations in terms of local cluster-type calculations are given (Fig. 35). Interpretation in terms of a band structure calculation: Fig. 4. The O 2p band is ≈ 5 eV wide but the width of the 3d signals may not represent the true bandwidth. A finite signal width would be expected even in the zero-overlap limit arising from vibronic effects. The core region XPE spectrum (Fig. 36) shows two satellites on Ni $2p_{3/2}$ which have been assigned to Ni d−d or multiplet splitting (1.8 eV) and O 2p → Ni 3d (6.0 eV) [74K]. From [73H] the value of the Hubbard self-energy U has been estimated as 1···3 eV. The 6.0 eV peak does not correspond to the minimum O 2p → 3d energy since selection rules will favour the O $2p(e_g^\sigma) \to$ Ni $3d(e_g^{\sigma*})$ transition [74K].

Electroreflectance spectra (Fig. 41) show pronounced structure at the optical threshold (3.5 eV). The assignment of the optical edge at 3.5 eV is controversial, with $d^8 \to d^7 s$ [70P, 72M2], O 2p → Ni d^9 [78M, 81D1] and 2Ni $d^8 \to$ Ni d^9 + Ni d^7 [76A] all suggested. Problems with all these assignments have been noted. Assignments of the EELS spectrum is also difficult, though the weight of evidence would suggest O 2p − Ni 3d at 4.6 and 6.9 eV and O 2p − Ni 4s at 9.5 and 11.7 eV [77S1].

refractive index (at RT):

n	2.26		in the visible, see also Fig. 43	58D
	2.33			70P
	2.38			65G

dielectric constants (at RT):

$\varepsilon(0)$	11.75			65G
	13.0			75R
$\varepsilon(\infty)$	5.70			65G
	5.7			75R

Magnetic properties

magnetic susceptibility (dependence on stress): Fig. 44.

$T > T_N$: Curie-Weiss law obeyed with $\Theta_p = -2000$ K [56S].

$T < T_N$: magnetic structure consists of ferromagnetic (111) planes antiferromagnetically coupled [58R1, 58R2, 60R]. The spins lie in the (111) planes along the $[\bar{1}\bar{1}2]$ direction [67S].

Néel temperature (in K):

T_N	524.5			67S
	523 (1)			73N1
	523.0			78D
	523.7			74G
	525			73L
dT_N/dp	1.58 K kbar^{-1}			78D

sublattice magnetization near T_N:

M	$\propto C(T_N - T)^{2\beta}$	$\beta = 0.35$	for powders	73N1
	$\propto > C(1 - T/T_N)^\beta$	$\beta = 0.37$ (3)	for single crystals	78D

spin-wave dispersion: shows an initial slope of 250 meV Å^{-1} and a maximum energy of 117 meV; density of spin-wave states: Fig. 45.

References for 9.15.2.3.4

Physical property	Numerical value	Experimental conditions	Experimental method, remarks	Ref.

parameters of spin wave Hamiltonian:

$\mathcal{H} = -\sum_j \sum_i J_{ij} s_i \cdot s_j + \sum_i D_1 S_i^{x^2} + \sum_i D_2 S_i^{y^2}$ ($x \perp (111)$ plane, z along spin direction, D_1 represents out-of-plane anisotropy, D_2 in-plane anisotropy)

		T [K]		
J_1/k	$+16$ K	78		71H
J_2/k	-221 K	78		71H
	-213 K	1.4		71D1
	-201 K		mean value from powder susceptibility measurements $0 \cdots 1200$ K	73S2
$\varepsilon_1 = -\dfrac{d \ln J_1}{d \ln r}$	27 (5)		$r =$ nearest neighbour distance	73S1
D_1/k	1.13 K	78		71H
D_2/k	0.06 K	78		71H
J_1^+/k	$+15.7$ K	78	splitting into values for spin aligned parallel and anti-parallel by the small rhombohedral distortion	72H
J_1^-/k	$+16.1$ K			
$(J_1^+ - J_1^-)/k$	-0.5 (2) K		equal to $J_1 \varepsilon_1 \Delta$ with the trigonal distortion $\Delta = 1.3 \cdot 10^{-3}$ [71B1]	72B2

References for 9.15.2.3.4

40S	Seltz, H., de Witt, B.J., McDonald, H.J.: J. Amer. Chem. Soc. **62** (1940) 88.
48R	Rooksby, H.P.: Acta Crystallogr. **1** (1948) 226.
54S	Shinomura, Y., Tsubokawa, I., Kojuma, M.: J. Phys. Soc. Jpn. **9** (1954) 521.
56S	Singer, J.R.: Phys. Rev. **104** (1956) 929.
58D	Doyle, W.P., Lonergan, G.A.: Discuss. Faraday Soc. **26** (1958) 27.
58R1	Roth, W.L.: Phys. Rev. **110** (1958) 1333.
58R2	Roth, W.L.: Phys. Rev. **111** (1958) 772.
59N	Newman, R., Chrenko, R.M.: Phys. Rev. **114** (1959) 1507.
60R	Roth, W.L., Slack, G.A.: J. Appl. Phys. **31** (1960) 352S.
60S	Slack, G.A.: J. Appl. Phys. **31** (1960) 1571.
61M	Mitoff, S.P.: J. Chem. Phys. **35** (1961) 882.
63G	Gillam, E., Holden, J.P.: J. Am. Ceram. Soc. **46** (1963) 601.
63P	Pistorius, C.W.F.T.: Neues Jahrb. Mineral., Monatsh. **1963**, 30.
64R	Roilos, M., Nagels, P.: Solid State Commun. **2** (1964) 285.
65F	Fueki, K., Wagner, J.: J. Electrochem. Soc. **112** (1965) 384.
65G	Gielisse, P.J., Plendl, J.N., Mansur, L.C., Marshall, R., Mitra, S.S., Mykolojewicz, R., Smakula, A.: J. Appl. Phys. **36** (1965) 2446.
65K1	Koide, S.: J. Phys. Soc. Jpn. **20** (1965) 123.
65K2	Ksendzov, Ya.M., Drabkin, I.A.: Fiz. Tverd. Tela **7** (1965) 1886.
66B	Bosman, A.J., Crevecoeur, C.: Phys. Rev. **144** (1966) 763.
66C	Clendenen, R.L., Drickamer, H.G.: J. Chem. Phys. **44** (1966) 4223.
66S	Sastry, R.L., Mehrotra, P.N., Rao, C.N.R.: J. Inorg. Nucl. Chem. **28** (1966) 2167.
66V	Vernon, M.W., Loval, M.C.: J. Phys. Chem. Solids **27** (1966) 1125.
67A	Austin, I.G., Springthorpe, A.J., Smith, B.A., Turner, C.E.: Proc. Phys. Soc. **90** (1967) 157.
67D	van Daal, H.J., Bosman, A.J.: Phys. Rev. **158** (1967) 736.
67K	Ksendszov, Ya.M., Avdeenko, B.K., Makarov, V.V.: Fiz. Tverd. Tela **9** (1967) 1058.
67M	Makaraov, V.V., Ksendszov, Ya.M., Kruglov, V.I.: Fiz. Tverd. Tela **9** (1967) 663.
67S	Siegworth, J.D.: Phys. Rev. **155** (1967) 285.
67U	Uno, R.: J. Phys. Soc. Jpn. **22** (1966) 1502.
68A1	Aiken, J.G., Jordan, A.G.: J. Phys. Chem. Solids **29** (1968) 2153.
68A2	Austin, I.G., Clay, B.D., Turner, C.E.: J. Phys. **C1** (1968) 1418.
68B	Bransky, I., Tallan, N.M.: J. Chem. Phys. **49** (1968) 1243.

68K	Kobashima, S., Kawakubo, T.: J. Phys. Soc. Jpn. **24** (1968) 493.
68S	Sockel, H.G., Schmalzried, H.: Ber. Bunsenges. Phys. Chem. **72** (1968) 745.
69B	Bhatt, S.J., Merchant, H.D.: J. Am. Ceram. Soc. **52** (1969) 452.
69E	Eror, N.G., Wagner, J.B.: Phys. Status Solidi **35** (1969) 641.
69M	McNatt, J.L.: Phys. Rev. Lett. **23** (1969) 915.
69R	Rossi, C.E., Paul, W.: J. Phys. Chem. Solids **30** (1969) 2295.
69T	Tretyakov, Y.D., Raff, R.A.: Trans. AIME **245** (1969) 1235.
70A	Adler, D., Feinleib, J.: Phys. Rev. **B2** (1970) 3112.
70B	Bosman, A.J., van Daal, H.J.: Adv. Phys. **19** (1970) 1.
70P	Powell, R.J., Spicer, W.E.: Phys. Rev. **B2** (1970) 2182.
70T	Tripp, W.C., Tallan, N.M.: J. Am. Ceram. Soc. **53** (1970) 531.
70V1	Vernon, M.W.: Phys. Status Solidi **37** (1970) K1.
70V2	Volpe, M.L., Reddy, J.: J. Chem. Phys. **53** (1970) 1117.
71B1	Bertel, L.C., Morosin, B.: Phys. Rev. **B3** (1971) 1039.
71B2	Bransky, I., Tallan, N.M.: Conductivity in Low-mobility Material, ed. Klein, N.; London: Taylor **1971**.
71D1	Dietz, R.E., Parisot, G.I., Meisner, A.E.: Phys. Rev. **B4** (1971) 2302.
71D2	Drakeford, R.W., Quinn, C.M.: J. Mater. Sci. **6** (1971) 175.
71G	Glosser, R., Walker, W.C.: Solid State Commun. **9** (1971) 1599.
71H	Hutchings, M.I., Samuelson, S.J.: Solid State Commun. **9** (1971) 1011.
71O	Osburn, C.M., Vest, R.W.: J. Phys. Chem. Solids **32** (1971) 1331, 1341.
71P	du Plessis, P. de V., van Tonder, S.J., Alberts, L.: J. Phys. **C4** (1971) 1983.
71T	Toussaint, C.J.: J. Appl. Crystallogr. **4** (1971) 293.
72B1	Brusič, V.: Oxides Oxide Films **1** (1972) 2.
72B2	Bartel, L.C.: Solid State Commun. **11** (1972) 55.
72H	Hutchings, M.I., Samuelson, S.J.: Phys. Rev. **B6** (1972) 3447.
72K	Kolber, M.A., MacCrone, R.K.: Phys. Rev. Lett. **29** (1972) 1457.
72M1	Mattheiss, L.F.: Phys. Rev. **B5** (1972) 290, 306.
72M2	Messick, L., Walker, W.C., Glosser, R.: Phys. Rev. **B6** (1972) 3941.
72S1	Sachse, H.: J. Mater. Sci. **7** (1972) 255.
72S2	Slater, J.C., Johnson, K.H.: Phys. Rev. **B5** (1972) 844.
72U	Uchiba, N., Saito, S.: J. Acoust. Soc. Am. **51** (1972) 1602.
72W	Wertheim, G.K., Hüfner, S.: Phys. Rev. Lett. **28** (1972) 1028.
73D	Deren, J., Mrowec, S.: J. Mater. Sci. **8** (1973) 545.
73H	Hüfner, S., Wertheim, G.K.: Phys. Rev. **B8** (1973) 4857.
73J	Johnson, K.H., Messmer, R.P., Lonnolly, J.W.D.: Solid State Commun. **12** (1973) 313.
73L	Lewis, F.B., Saunders, N.H.: J. Phys. **C6** (1973) 2525.
73N1	Negetovič, I., Konstantinovič, J.: Solid State Commun. **13** (1973) 219.
73N2	Notis, M.R., Spriggs, R.M., Hahn, W.C.: J. Appl. Phys. **44** (1973) 4165.
73S1	Seino, D., Miyahira, S., Naro, Y.: Phys. Lett. **A44** (1973) 35.
73S2	Shanker, R., Singh, R.A.: Phys. Rev. **B7** (1973) 5000.
73S3	Spear, W.E., Tannhauser, D.S.: Phys. Rev. **B7** (1973) 831.
74G	Germann, K.H., Maier, K., Strauss, E.: Solid State Commun. **14** (1974) 1309.
74I	Ikeda, Y., Nii, K., Beranger, G., Lacombe, P.: Trans. Jpn. Inst. Met. **15** (1974) 441.
74K	Kim, K.S.: J. Electron Spectrosc. Rel. Phenom. **3** (1974) 217.
74W	White, H.W.: J. Chem. Phys. **61** (1974) 4907.
75C	Collins, T.C., Kunz, A.B., Ivey, J.L.: Int. J. Quantum Chem. Symp. **9** (1975) 519.
75E	Eastman, D.E., Freeouf, J.L.: Phys. Rev. Lett. **34** (1975) 395.
75F	Friedman, F., Weichman, F.L., Tannhauser, D.S.: Phys. Status Solidi **A27** (1975) 273.
75K	Kharif, Ya.L., Galaktionov, S.S., Dergacheva, N.M., Chashchim, V.A.: Fiz. Tverd. Tela **17** (1975) 987.
75R	Reichardt, W., Wagner, V., Kress, W.: J. Phys. **C8** (1975) 3955.
76A	Allen, G.C., Dyke, J.M.: Chem. Phys. Lett. **37** (1976) 391.
76C	Coy, R.A., Thompson, C.W., Girman, E.: Solid State Commun. **18** (1976) 845.
76D	Degraix, H., Gravelle, P.C., Teichner, S.T., Turlier, P.: J. Solid State Chem. **18** (1976) 69, 79.
77S1	Sakisaki, Y., Akimoto, K., Nichijima, M., Onchi, M.: Solid State Commun. **24** (1977) 105.
77S2	Subhadra, K.G., Sirdeshmukh, D.G.: Indian J. Pure Appl. Phys. **16** (1977) 693.
78D	von Doorn, C.F., du Plessis, P. de V.: Phys. Lett **A66** (1978) 141.

78F	Farki, R., Petot-Ervas, G.: J. Phys. Chem. Solids **39** (1978) 1169, 1175.
78K1	Keem, J.E., Honig, J.M., van Zandt, L.L.: Philos. Mag. **B 37** (1978) 537.
78K2	Keem, J.E., Wittenauer, M.A.: Solid State Commun. **26** (1978) 213.
78K3	Kunz, A.B., Surratt, G.T.: Solid State Commun. **25** (1978) 9.
78M	Merlin, R., Martin, T.P., Polian, A., Cardona, M., Audlauer, B., Tannhauser, D.: J. Magnetism Magn. Mater. **9** (1978) 83.
78P	Propach, V., Reinen, D., Drenkhahn, H., Müller-Buschbaum, H.: Z. Naturforsch. **33 B** (1978) 619.
78S	Subhadra, K., Sirdesmuth, D.B.: Indian J. Pure Appl. Phys. **16** (1978) 693.
79A	Agarwal, S.: Solid State Commun. **29** (1979) 197.
79D	Duclot, M., Deportes, C.: J. Solid State Chem. **30** (1979) 231.
81D1	Dare-Edwards, M.P., Goodenough, J.B., Hamnett, A., Nicholson, N.D.: Faraday Trans. Chem. Soc. II **77** (1981) 643.
81D2	Dutt, M.B., Banerjee, R., Barua, A.K.: Phys. Status Solidi **a 65** (1981) 365.
81F	Freer, R.: J. Mater. Sci. **16** (1981) 3225.

Physical property	Numerical value	Experimental conditions	Experimental method, remarks	Ref.

9.15.2.3.5 Palladium oxide (PdO)

Crystal structure

structure: tetragonal, space group, $D_{4h}^9 - P4_4/mmc$, $Z = 2$ [71R].

Each Pd is surrounded by a slightly distorted square of oxygen atoms at 2.02369 Å.

The four oxygens are actually in the form of a rectangle with sides 3.0434 and 2.682 Å. Coordinates in the unit cell: 2 Pd atoms at 0, 0, 0; $\frac{1}{2}, \frac{1}{2}, \frac{1}{2}$; 2 O atoms at $\frac{1}{2}, 0, \frac{1}{4}$; $\frac{1}{2}, 0, \frac{3}{4}$ [71R].

lattice parameters (in Å):

a	3.03		53W
	3.0434		68M
	3.0434(2)		71R
c	5.33		53W
	5.337		68M
	5.3363(4)		71R

variation with temperature:

		T [°C]		
a	3.0475	20		74B
c	5.3430			
a	3.0525	220		
c	5.3516			
a	3.0557	420		
c	5.3582			
a	3.0593	620		
c	5.3662			
a	3.0625	750		
c	5.3731			

linear thermal expansion coefficients:

α_a	$6.7(5) \cdot 10^{-6}$ K^{-1}	$T = 20 \cdots 750$ °C	in a-direction	74B
α_c	$7.7(7) \cdot 10^{-6}$ K^{-1}	$20 \cdots 750$ °C	in c-direction	

PdO dissociates at high temperature.

thermodynamical parameters:

ΔH_f^0	-25.8 kcal mol^{-1}	reaction: Pd(s) + $\frac{1}{2}$O$_2$ → PdO(s);	66B
ΔS_f^0	-22.5 cal K^{-1} mol^{-1}	see [66B, 78R]	
ΔH_f^0	-26.8 kcal mol^{-1}		
ΔS_f^0	-23.6 cal K^{-1} mol^{-1}		
ΔH_f^0	-27.37 kcal mol^{-1}		
ΔS_f^0	-24.00 cal K^{-1} mol^{-1}		74B

Figs. p. 529f.

Physical property	Numerical value	Experimental conditions	Experimental method, remarks	Ref.
oxygen dissociation pressure:				
p	-11220 K		$\log_{10} p_{O_2} = p/T + q$, T in K	74B
q	12.63		(p_{O_2} in atm)	
p	-13257 K			76T
q	12.11			

Electronic properties

Photoelectron spectra: HeI, HeII, and AlK$_\alpha$: Fig. 1.

energy gaps (in eV) (RT values quoted in the literature):

E_g	0.6		diffuse reflectance	65H
	0.8		Kramers-Kromig analysis	79N
	2.13		extrapolated from optical density	78R
	2.67		extrapolated from photoconductivity	78R
$E_{g,th}$	1.5		estimated from onset of intrinsic conduction, Fig. 2	67O

Two possibilities have been proposed: (1) The main band gap at ca. 2 eV is the Pd d−d transition with a very long Urbach-tail [78R], (2) the main bandgap is at ca. 0.8 eV and is the Pd d−d transition; it is highly indirect, with the first direct transition at ca. 2 eV [79N].

Transport properties

PdO is reported as a p-type semiconductor [65H, 67O, 71R]. Temperature dependence of conductivity: Fig. 2.

transport parameters:

		T[K]		
S	$+250\,\mu\text{V K}^{-1}$	300		65H
E_A	$0.04 \cdots 0.1$ eV	$4.2 \cdots 300$	activation energy for conductivity of single crystals	71R
μ	17 cm^2/Vs	300	polycrystalline films	67O
R_H	5 cm^3 C^{-1}	300	polycrystalline films	67O
ϱ	$10 \cdots 1000\,\Omega$ cm	300	typical values	71R
σ	$9 \cdot 10^{-2} \cdots 30\,\Omega^{-1}$ cm^{-1}	$77 \cdots 560$	polycrystalline films	67O

Optical properties

Optical density and photoconductivity: Fig. 3, real and imaginary parts of the refractive index: Fig. 4, optical conductivity: Fig. 5.

Refractive index in the visible: $n = 2.8$, static dielectric constant $\varepsilon(0) \approx 8$ [79N].

Magnetic properties

magnetic susceptibility:

χ_m	$-4 \cdot 10^{-6}$ cm^3 mol^{-1}	$T = 295$ K	χ_m in CGS-emu	65H
	$-2 \cdot 10^{-6}$ cm^3 mol^{-1}	80 K		

References for 9.15.2.3.5

53W	Waser, J., Levy, H.A., Peterson, S.W.: Acta Crystallogr. **6** (1953) 661.
65H	Hulliger, F.: J. Phys. Chem. Solids **26** (1965) 639.
66B	Bell, W.E., Inyard, R.E., Tagami, M.: J. Phys. Chem. **70** (1966) 3735.
67O	Okamoto, H., Aso, T.: Jpn. J. Appl. Phys. **6** (1967) 779.
68M	Macdaniel, C.L., Schneider, S.J.: J. Res. Nat. Bur. Stand **72A** (1968) 27.
71R	Rogers, D.B., Shannon, R.D., Gillson, J.L.: J. Solid State Chem. **3** (1971) 314.

74B	Bayer G., Wiedmann, H.G.: Thermal Anal. Proc. Int. Conf. 4th. **1** I. Buzas (ed.) London: Heiden, **1975**, p. 763.			
76T	Tagirov, V.K., Chizhikov, D.M., Kazanas, E.K., Shubochkin, L.K.: Zh. Neorg. Khim. **21** (1976) 2565.			
78R	Rey, E., Kamal, M.R., Miles, R.B., Joyce, B.S.H.: J. Mater. Sci. **13** (1978) 812.			
79H	Holl, Y., Krill, G., Amamou, A., Legare, P., Hilaire, L., Maire, G.: Solid State Commun. **32** (1979) 1189.			
79N	Nilsson, P.D., Shvaraman, M.S.: J. Phys. C **12** (1979) 1423.			

Physical property	Numerical value	Experimental conditions	Experimental method, remarks	Ref.

9.15.2.4 Spinel oxides

(See also Landolt-Börnstein, Vols. III/4b and 12b).

9.15.2.4.1 Magnetite (Fe$_3$O$_4$)

Magnetite shows a highly unusual transition at 119 K – the Verwey transition. Above this temperature, the material shows the properties of a poor metal, but below T_V a distortion to a semiconducting phase of much lower symmetry is found.

Crystal structure

a) high-temperature phase

structure at room temperature: cubic, inverse spinel type, space group O_h^7 – Fd3m, $Z=8$ [75S1, 58H].

positions of atoms in the unit cell:

O atoms at 32(e) sites $u u u$; $u \bar{u} \bar{u}$; $\frac{1}{4}-u, \frac{1}{4}-u, \frac{1}{4}-u$; $\frac{1}{4}-u, u+\frac{1}{4}, u+\frac{1}{4}$;
$\bar{u}\bar{u}u$; $\bar{u}u\bar{u}$; $u+\frac{1}{4}, \frac{1}{4}-u, u+\frac{1}{4}$; $u+\frac{1}{4}, u+\frac{1}{4}, \frac{1}{4}-u$ + f.c. translations
Fe atoms at 16(d) sites $\frac{5}{8}\frac{5}{8}\frac{5}{8}$; $\frac{5}{8}\frac{7}{8}\frac{7}{8}$; $\frac{7}{8}\frac{5}{8}\frac{7}{8}$; $\frac{7}{8}\frac{7}{8}\frac{5}{8}$ + f.c. translations.
Fe atoms at 8(a) sites $0\,0\,0$; $\frac{1}{4}\frac{1}{4}\frac{1}{4}$ + f.c. translations.
The 16(d) sites are octahedral and the 8(a) sites are tetrahedral.

lattice parameters (in Å):

		T [K]		
a	8.3940(5)	295		51T
	8.3905	165		
	8.3940	298		82I
	8.386	130		76S2
	8.3963	298	see also Fig. 1	65G
	8.4135	473		
	8.4374	673		
	8.4556	798		
	8.4625	823		
	8.4658	848		
	8.4724	873		
	8.4731	888		
	8.4746	898		
	8.4783	923		
	8.4804	948		
	8.4821	973		
	8.4956	1073		
	8.5336	1273		
		Fe$_2$O$_3$/FeO =		
	8.379	1.013	variation with stoichiometry; RT values	41V
	8.381	1.021		
	8.381	1.042		
	8.379	1.049		
	8.375	1.020	(continued)	

Physical property	Numerical value	Experimental conditions	Experimental method, remarks	Ref.
lattice parameters (continued)				
$d(Fe-O)$	2.0590 (16)		Fe on d-sites	
	1.8871 (29)		Fe on a-sites	
$d(O-O)$	3.0817 (46)		for the three	58H
	2.9689 (3)		inequivalent	
	2.8588 (46)		pairs of sites	
value of u:				
u	0.379			25C
	0.3798			58H
lattice constant after equilibration under various oxygen pressures:				
			equilibration at 1400°C; p_{O_2} [atm]	69B
a	8.3960	$T=298$ K	10^{-6}	
	8.3953		10^{-4}	
	8.3943		10^{-2}	
	8.3938		$10^{-1.5}$	
	8.3933		10^{-1}	
	8.3925		$10^{-0.7}$	
pressure dependence of a:				
da/dp	$-1.5 \cdot 10^{-3}$ Å kbar^{-1}			69S
linear thermal expansion coefficient (in 10^{-6} K^{-1}):				
		T [°C]		
α	10.41	25		65G
	11.38	100		
	12.68	200		
	13.97	300		
	15.26	400		
	16.54	500		
	19.05	700		
	20.35	800		
	21.61	900		
	22.85	1000		
density:				
d	5.238 g cm^{-3}		X-ray density	55W

b) low-temperature phase:

structure: monoclinic, space group C_s^4-Cc, very close to $D_{2h}^{11}-Pmca$ or $C_{2v}^2-Pmc2_1$ [77Y, 82I].

At 119 K, Fe_3O_4 undergoes a first-order crystallographic transition to a structure of lower symmetry. The unit cell parametes vary in an unusual fashion (Fig. 2). The low-temperature phase has been studied by X-ray diffraction [77Y, 77I], NMR [78M1, 78M2, 81Y], neutron diffraction [75S2, 75I, 75F, 76S2], Moessbauer spectroscopy [70H, 71R] and magnetoelectric effect [79S1, 79K1, 78M3].

Initially, Verwey [41V, 47V] predicted an orthorhombic LT symmetry as a result of ordering of Fe^{2+} and Fe^{3+} on the B-sites, (Fig. 3). Early neutron work [58H] supported this model, but later studies [68S2, 75S2] revealed the true symmetry to be much lower.

lattice parameters (distances in Å, β in °):

		T [K]		
a	11.888	84		77Y
	11.868 (2)	10		82I
b	11.847	84		77Y
	11.851 (2)	10		82I
				(continued)

Physical property	Numerical value	Experimental conditions	Experimental method, remarks	Ref.
lattice parameters (continued)				
c	16.773	84		77Y
	16.752	10		82I
β	89.76	84		77Y
	90.20 (3)	10		82I

Such a low symmetry suggests an alternation of Fe^{3+} and Fe^{2+} ions in the [110] direction as in the ab model of Fig. 3. Neutron diffration data have been used to suggest that the transition involves the condensation of a particular optical mode $\Delta_5^{(1)}$ (Fig. 4) [79L, 75I]. Even more complex structures have been proposed as a result of NMR experiments [78M1]. Three-dimensional neutron scattering data [82I] have been refined in the two primitive orthorhombic space-groups Pmca and Pmc2$_1$. The resulting ionic displacements and Fe–O distances are shown in Fig. 5. The charge-density distribution in the LT phase has not been finally determined; discrepancies exist in the interpretations of Moessbauer and NMR data [78M1] relative to the neutron data [82T], probably as a result of the different time scales of the measurements.

Electronic properties

Only empirical energy level schemes are available [71B2, 72C, 78T]. Quantitative calculations [73C, 75L, 79L] have been restricted to the one $t_{2g}\downarrow$ electron, an important feature of these calculations is that fractional charge density is permitted for $T<T_V$.

Empirical schemes from [72C] and [78T] are given in Fig. 6. Both predict O 2p–Fe 4s separations of $\approx 6\cdots 7$ eV at RT. The main difference is in the relative position of the O 2p level to the Fe 3d orbitals. UPS-spectra (Figs. 7$\cdots$9) clearly show that the initial absorption is due to the $t_{2g}\downarrow$ electron, then $e_g\uparrow$ B-site ionization before $t_2\downarrow$ A-site ionization, in agreement with [78T]. High temperature transport data [81W] indicate a random distribution of mobile electrons over both A and B sites above 1450 °C.

For E_g-values, see Transport and Optical properties below.

Lattice properties

Phonon dispersion spectrum: Fig. 10, Brillouin zone: Fig. 11.

In the cubic phase the symmetries of lattice modes at Γ are given by:

$$\Gamma = a_{1g} + e_g + t_{1g} + 3 t_{2g} + 2 a_{2u} + 2 e_u + 4 t_{1u} + 2 t_{2u}.$$

wavenumbers of IR and Raman active modes (in cm^{-1}):

IR-active modes (see Figs. 12, 13):

$\bar{\nu}_1(t_{1u})$	565	$T>T_V$	Fe–O stretch at both (tetrahedral) A and (octahedral) B sites	72I
	570			72G
	570			77K
	615, 585	$T<T_V$		77K
$\bar{\nu}_2(t_{1u})$	360	$T>T_V$		72I
	390			72G
	380			77K
	420, 405, 375	$T<T_V$		77K
$\bar{\nu}_3(t_{1u})$	268	$T>T_V$	motion of Fe$_{tetr.}$ vs. Fe$_{oct.}$	72G
$\bar{\nu}_4(t_{1u})$	178	$T>T_V$	O–Fe–O bending mode	72G
Raman active modes (see Fig. 14):				
$\bar{\nu}_R(a_{1g})$	680	$T=77$ K	see also Fig. 12. Five Raman active modes are expected in the cubic phase and 15 in the orthorhombic phase ($T<119$ K) but only five are observed in both phases. No dramatic changes are noted in passing through the Verwey transitions. All lines are anomalously broad (30 cm^{-1}) even at low temperatures.	74V
$\bar{\nu}_R(t_{2g})$	560			
	320			
	300			
$\bar{\nu}_R(e_g)$	420			

No soft mode observed at T_V.

Physical property	Numerical value	Experimental conditions	Experimental method, remarks	Ref.

heat capacity: shows two transitions near T_V [69W, 75E], but this bifurcation appears to be due to unannealed stress [77M2].

Debye temperature:

Θ_D	548 K	RT	IR absorption	72G
	550 K	$T = 5 \cdots 100$ K	heat capacity, above 100 K, Θ_D increases slowly to 571 K at 140 K.	69W

Transport properties

Fe_3O_4 is commonly cation deficient and the deviation from stoichiometry can be very large (Figs. 15(a) and 15(b)). Tracer diffusion experiments [77D1] show that $(d \log D^*_{Fe}/d \log p_{O_2})$ is $\approx 2/3$ at high oxygen pressures and $-2/3$ at low pressures. The equilibria involved are

$$2/3 O_2 + 3 Fe^{3+}_{oct} = V_{Fe_{oct}} + 2 Fe^{3+}_{oct} + 1/3 Fe_3O_4; \quad [V_{Fe_{oct}}] \propto p_{O_2}^{2/3}; \quad K_2$$
$$Fe^{2+}_{oct} = V_{Fe_{oct}} + Fe^{2+}_i; \quad [Fe^{2+}_i] \propto p_{O_2}^{-2/3}; \quad K_3.$$

At low oxygen pressures, diffusion is therefore by interstitials, and at high oxygen pressures by cation vacancies. A more detailed analysis at high temperature (1300 °C) [77D2] indicates that a better fit to the data can be obtained if the following equilibria hold:

$$Fe^{3+}_{tetr} + Fe^{2+}_{oct} = Fe^{3+}_{oct} + Fe^{2+}_{tetr}; \quad K \approx 1$$
$$V_{Fe_{oct}} + Fe^{3+}_{tetr} = V_{Fe_{tetr}} + Fe^{3+}_{oct}; \quad K \approx 1.$$

From an analysis of the data of [77D2, 68S3, 35G, 46D, 57S, 41S] values of $K_2 = 24$ and $K_3 = 10^{-8}$ at 1200 °C are obtained; plot of $\ln K_2$ vs. $1/T$: Fig. 15(b). This random distribution of Fe^{2+} on octahedral and tetrahedral sites is also supported by thermopower measurements [70G] and by structural investigations [69K, 71W]. Originally, only vacancies on the 16(d) sites of cation deficient Fe_3O_4 were considered, since heating Fe_3O_4 in air at 400 °C for long periods led to γ-Fe_2O_3 (maghemite), which contains only such vacancies ordered in a rather complex way [58F, 61U]. More recently, by heating for very long periods [69K], or by heating in sulfur [69K, 71W] **tetrahedral** vacancies were also found, cation vacancies being randomly distributed between tetrahedral and octahedral sites according to Moessbauer evidence [71W].

conductivity: Figs. 16 ··· 18, 21. The Verwey transition is observed as a discontinuity in σ at 120 (2) K [50D, 68S1] and σ also shows a maximum at ≈ 300 K [57M], 360 K [50D], and a further minimum at ≈ 780 K [70G], 800 K [76P, 57M]. For the low-temperature phase some anisotropy in the conductivity is found, with the properties showing further strong dependence on magnetic field direction.

The conductivity minimum at high temperatures is correlated with the Curie temperature; it apparently represents a maximum in the spin-disorder scattering. The conductivity maximum at 300 K is associated with the onset of strong, dynamic electron-lattice interactions that correlate the mobile electrons [81S].

The Verwey transition decreases with increasing iron deficit, Fig. 17, a result supported by Moessbauer measurements [71W]. Below T_V, Fe_3O_4 behaves as a semiconductor, with an activation energy that decreases continuously with decreasing T. Alternatively, [71D] suggests that $\ln \sigma$ is linear with $T^{-1/4}$ from the Verwey temperature down to ≈ 10 K. Below this, the conductivity reaches a minimum of 10^{-15} $(\Omega \text{ cm})^{-1}$ at 5.3 K and then increases [71D].

activation energy for conductivity:

		T [K]		
E_A	0.10	78 ··· 90		54C
	0.10 (1) eV	100 ··· T_V		68S1
	0.10	110		57M
	0.09	56 ··· 77		54C
	0.06 eV	40 ··· 52		54C
	0.03 eV	40		57M

stress and pressure dependence of transport parameters:

stress dependence of conductivity: Fig. 19

$d \ln T_V/dp$	$-4.0 (2) \cdot 10^{-3}$ kbar^{-1}		68S1
$d \ln E_A/dp$	$-3.0 (1.0)$ $\cdot 10^{-3}$ kbar^{-1}	$T \lesssim T_V$	68S1
$d \ln \varrho/dp$	$-4.9 (2) \cdot 10^{-2}$ kbar^{-1}	$T = 77$ K	68S1

Seebeck coefficient: Figs. 20, 21. At lower temperatures the Seebeck coefficient has been interpreted in terms of a cation deficient model $Fe^{3+}_{2+2\gamma}Fe^{2+}_{1-3\gamma}(V_{Fe})_\gamma O_4$ with $E_A = 0.10 \cdots 0.11$ eV and both n and p carriers, $(E_A)_p - (E_A)_n \approx 0.01$ eV, where E_A is the activation energy for drift mobility. Values of $(E_A)_n$ of ≈ 0.11 eV at $T = T_V$ and 0.04 eV at 80 K have been reported [76K]. The data of [76K, 71C, 79K2] all show that at low temperatures, S is positive suggesting that below T_V, Fe_3O_4 is a p-type semiconductor.

At high temperatures ($T > T_C$) the Seebeck coefficient is well described by a diffusional (electron-hopping) model as $S = -(k/e) \ln[\beta(1-c)/c]$, where $\beta = 2$ is the spin-degeneracy and the concentration of mobile carriers, c, on the B sites is temperature dependent due to thermal excitations to the A sites [81W]. On lowering the temperature from $T > T_C$ to 300 K, the spin-degeneracy parameter decreases from $\beta = 2$ to $\beta = 1$ and correlated electron hopping changes the Seebeck coefficient to $S = -(k/e) \ln(1/c)$ [81S].

magnetoresistance: The resistivity in [110] direction decreases by 0.1% on application of a magnetic field in [110] or [111] direction but increases by 0.1% if the field is parallel to [100] [50D].

Hall effect: Hall coefficient and Hall mobility: Fig. 22. The Hall mobility is apparently unactivated for $T < T_V$ and the Hall coefficient is positive.

substituted Fe_3O_4:

^{18}O substitution shifts T_V to larger values; $\delta T_V = 6.1(5)$ K for 43% ^{18}O substitution [79T].

F substitution causes T_V to decrease and the Verwey transition eventually to be suppressed [77W] (see Figs. 23, 24 for conductivity and Seebeck coefficient). Qualitatively similar but less marked effects are reported by [77C].

Ti substitution: Figs. 25···27. Energy gap in low temperature phase found to be 0.12 eV. Verwey transition is associated with an increase in the number of carriers rather than a sharp change in carrier mobility. The mobility has been found to be activated both below and above the transition temperature (Fig. 27).

Other transition metal substituents: Fig. 28, 29.

Optical properties

Optical spectra: UPS-spectra; Figs. 7···9 (for XPS spectra, see Fig. 17 of 9.15.2.5.2 (Fe_2O_3)); absorption coefficient: Figs. 30, 31; real and imaginary parts of the dielectric constant: Fig. 32, 33; dielectric constant in the microwave region: Fig. 34. See also [77K, 79S2, 74M, 72B, 71B1].

The maximum in the absorption coefficient at 0.72(2) eV has been interpreted in terms of small polaron theory [74M], band-to-band transitions [71B1, 72B, 75B] or d−s charge transfer [79S2]. Only slight changes in absorption coefficient occur at the Verwey transition (Fig. 33). Most important is the apparent formation of a small optical gap of $E_g \approx 0.12$ eV [72B].

Magnetic properties

Fe_3O_4 is ferrimagnetic. Real and imaginary parts of the permeability are given in Fig. 35.

Curie temperature (in K):

T_C	848.5	74G
	850	69S
	848.41	56S
	848	78R
$\frac{1}{T_C}\frac{dT_C}{dp}$	$2.42 \cdot 10^{-3}$ kbar^{-1}	69S

At 300 K, the magnetic structure [48N, 51S] has a-site Fe^{3+}_{tetr} spins along [001] coupled antiferromagnetically to spins on d-site Fe^{3+}_{oct} (spins along [00$\bar{1}$]). The d-site Fe^{2+}_{oct} ions are coupled ferromagnetically to the d-site Fe^{3+}_{oct} ions and their spins hence also point along [00$\bar{1}$].

Spin wave spectrum in the acoustic branch: Fig. 36.

References for 9.15.2.4.1

Physical property	Numerical value	Experimental conditions	Experimental method, remarks	Ref.
spin-spin coupling constants (in meV):				
J_{AB}	−0.44	$T = 1.8 \cdots 4.2$ K	heat capacity	56K
	−1.10	$1.8 \cdots 4.2$ K		65D
	−2.4	$T > T_V$	neutron scattering	63G, 62W
	−2.5	$T > T_V$		66M
	−1.0	$T < T_V$		66M
	−2.55	$T > T_V$		62W
	−2.35	$T \gtrsim T_V$		67A
	−2.32	$T \gtrsim T_V$		67T
J_{BB}	0.25	$T > T_V$		66M
	0.68	$T \gtrsim T_V$		67T
magnetic moments (in units of μ_B, at RT(?)):				
p_m	3.47	single crystal powder	mean magnetic moment per formula unit	78R, 77R, 77M1
$p_A(A)$	3.82	single crystal	A-site moment; 1.2% Ti	77R, 78R
	4.0	powder		
$p_A(B)$	3.6	single crystal	B-site moment	
	3.7	powder		

References for 9.15.2.4.1

25C	Claasen, A.A.: Proc. Phys. Soc. **38** (1925) 482.
35G	Greig, W., Posnjak, E., Merwin, H.E., Sosman, R.B.: Am. J. Sci. **30** (1935) 239.
41S	Schmahl, N.G.: Z. Elektrochem. **47** (1941) 821.
41V	Verwey, E.J.W., Haayman, P.W.: Physica **8** (1941) 979.
46D	Darken, L.S., Gurry, R.W.: J. Am. Chem. Soc. **68** (1946) 798.
47V	Verwey, E.J.W., Haayman, P.W., Romeijn, F.C.: J. Chem. Phys. **15** (1947) 181.
48N	Néel, L.: Ann. Phys. **3** (1948) 137.
50D	Domenciali, C.A.: Phys. Rev. **78** (1950) 458.
51S	Shull, C.G., Wollan, E.O., Koehler, W.C.: Phys. Rev. **84** (1951) 912.
51T	Tombs, N.C., Rooksby, H.P.: Acta Crystallogr. **4** (1951) 474.
52S	Smith, D.O.: Prog. Rep. Lab. Ins. Res. MIT **11** (1952) 53.
54C	Calhoun, B.A.: Phys. Rev. **94** (1954) 1577.
55W	Waldron, R.D.: Phys. Rev. **99** (1955) 1727.
56K	Kouvel, J.S.: Phys. Rev. **102** (1956) 1489.
56S	Smith, D.O.: Phys. Rev. **102** (1956) 959.
57M	Miles, P.A., Westphal, W.B., von Hippel, A.: Rev. Mod. Phys. **29** (1957) 279.
57S	Smiltens, J.: J. Am. Chem. Soc. **79** (1957) 4877.
58F	Ferguson, G.A., Haas, M.: Phys. Rev. **112** (1971) 1130.
58H	Hamilton, W.C.: Phys. Rev. **110** (1958) 1050.
61U	Ueda, R., Hasegawa, K.: J. Phys. Soc. Jpn. **17** (1961) Suppl. BII 391.
62W	Watanabe, H., Brockhouse, B.N.: Phys. Lett. **1** (1962) 189.
63G	Glasser, M.L., Milford, F.J.: Phys. Rev. **130** (1963) 1783.
65D	Dixon, M., Hoare, F.E., Holden, T.M.: Phys. Lett. **14** (1965) 184.
65G	Gorton, A.J., Bitsianes, G., Joseph, T.L.: Trans. AIME **233** (1965) 1519.
66M	Mills, R.E., Kenan, R.P., Milford, F.J.: Phys. Rev. **145** (1966) 704.
67A	Alperin, H.A., Steinsvoll, O., Nathaus, R., Shirane, G.: Phys. Rev. **154** (1967) 508.
67T	Torrie, B.H.: Solid State Commun. **5** (1967) 715.
68S1	Samara, G.: Phys. Rev. Lett. **21** (1968) 795.
68S2	Samuelson, E.J., Blecker, E.J., Dobrzynski, L., Riste, T.: J. Appl. Phys. **39** (1968) 1114.
68S3	Sockel, H.-G., Schmalzried, H.: Ber. Bunsenges. **72** (1968) 745.
69B	Bhatt, S.J., Merchant, H.D.: J. Am. Ceram. Soc. **52** (1969) 452.

References for 9.15.2.4.1

69K	Kullerud, G., Donnay, G., Donnay, J.D.H.: Z. Kristallogr. **128** (1969) 1.
69S	Samara, G.A., Giardini, A.A.: Phys. Rev. **186** (1969) 577.
69W	Westrum, E.F., Grønveld, F.: J. Chem. Thermodyn. **1** (1969) 543.
70G	Griffith, B.A., Elwell, D., Parker, R.: Phil. Mag. **22** (1970) 163.
70H	Hargrave, R.S., Kündig, W.: Solid State Commun. **8** (1970) 303.
70S	Siemans, W.J.: IBM J. Res. Develop. **14** (1970) 245.
71B1	Balberg, I., Pankove, J.I.: Phys. Rev. Lett. **27** (1971) 596.
71B2	Balberg, I., Pankove, J.I.: Phys. Rev. Lett. **27** (1971) 1371.
71C	Constantin, C., Rosenberg, M.: Solid State Commun. **9** (1971) 675.
71D	Drabble, J.R., Whyte, T.D., Hooper, R.M.: Solid State Commun. **9** (1971) 275.
71H	Haudek, H.: Z. Angew. Phys. **32** (1971) 149.
71R	Rubinstein, M., Forester, D.W.: Solid State Commun. **9** (1971) 1675.
71W	Weber, H.-P., Hafner, S.S.: Z. Kristallogr. **133** (1971) 327.
72B	Buchenau, U., Müller, I.: Solid State Commun. **11** (1972) 1291.
72C	Camhausen, D.L., Coey, J.M.D., Chakraverty, B.K.: Phys. Rev. Lett. **29** (1972) 657.
72G	Grimes, N.W.: Philos. Mag. **26** (1972) 1217.
72I	Ishii, M., Nakahiri, M.: Solid State Commun. **11** (1972) 209.
72K	Kofstad, P.: Non-Stoichiometry, Diffusion and Transport in Transition Metal Oxides, New York: J. Wiley, **1972**.
73C	Cullen, J.R., Callen, E.R.: Phys. Rev. **B7** (1973) 400.
74G	Grønveld, F., Sveen, A.: J. Chem. Thermodyn. **6** (1974) 859.
74M	Muret, P.: Solid State Commun. **14** (1974) 1119.
74S	Samuelson, E.J., Steinsvoll, O.: Phys. Status Solidi **B61** (1974) 615.
74V	Verble, J.L.: Phys. Rev. **B9** (1974) 5236.
75A	Alvarado, S.F., Eib, W., Meier, F., Pierce, D.T., Sattler, K., Siegmann, H.C., Remeika, J.P.: Phys. Rev. Lett. **34** (1975) 319.
75B	Buchenau, U., Müller, I.: Physica **80B** (1975) 75.
75E	Evans, B.J.: AIP Conf. Proc. **24** (1975) 73.
75F	Fujii, Y., Shirane, G., Yamada, Y.: Phys. Rev. **B11** (1975) 2036.
75I	Izumi, M., Shirane, G.: Solid State Commun. **17** (1975) 433.
75L	Lorenz, B., Ihle, D.: Phys. Status. Solidi **B69** (1975) 451.
75S1	Samuelson, E.J.: J. Phys. **C7** (1975) L115.
75S2	Shirane, G., Chikazumi, S., Akimitsu, J., Chiba, K., Matsui, M., Fujii, Y.: J. Phys. Soc. Jpn. **39** (1975) 949.
76A	Alvarado, S.F., Erbudak, M., Munz, P.: Phys. Rev. **B14** (1976) 2740.
76K	Kuipers, A.J.M., Brabers, V.A.M.: Phys. Rev. **B14** (1976) 1401.
76P	Pan, L.S., Evans, B.J.: AIP Conf. Proc. **34** (1976) 181.
76S1	Samuelson, E.J.: Phys. Rev. **B14** (1976) 200.
76S2	Shapiro, S.M., Izumi, M., Shirane, G.: Phys. Rev. **B14** (1976) 200.
77C	Claverie, J., Portier, T., Hagenmuller, P.: J. Phys. **38** (1977) C1–169.
77D1	Dieckmann, R., Schmalzried, H.: Ber. Bunsenges. **81** (1977) 344.
77D2	Dieckmann, R., Schmalzried, H.: Ber. Bunsenges. **81** (1977) 414.
77I	Iida, S., Mizushima, K., Mizoguchi, M., Mada, J., Umemura, S., Yoshida, J., Nakao, K.: Physica **86–88B** (1977) 957.
77K	Kuipers, A.J.M., Brabers, V.A.M.: Phys. Rev. Lett. **39** (1977) 488.
77M1	Murthy, N.S.S.: J. Phys. **38** (1977) C1–79.
77M2	Matsui, M., Todo, S., Chikazumi, S.: J. Phys. Soc. Jpn. **42** (1977) 1517.
77R	Rakheda, V.C., Chakraverty, R., Murthy, N.S.S.: J. Phys. **38** (1977) C1–107.
77W	Whall, T.E., Rigo, M.O., Jones, M.R.B., Pointon, A.J.: J. Phys. **38** (1977) C1–229.
77Y	Yoshida, J., Iida, S.: J. Phys. Soc. Jpn. **42** (1977) 230.
78M1	Mizoguchi, M.: J. Phys. Soc. Jpn. **44** (1978) 1501.
78M2	Mizoguchi, M.: J. Phys. Soc. Jpn. **44** (1978) 1512.
78M3	Matsumoto, S., Goto, T., Syono, Y., Nakagawa, Y.: J. Phys. Soc. Jpn. **44** (1978) 162.
78M4	Mizushima, K., Nakao, K., Tanaka, S., Iida, S.: J. Phys. Soc. Jpn. **44** (1978) 1831.
78R	Rakheda, V.C., Murthy, N.S.S.: J. Phys. **C11** (1978) 4389.
78T	Tossell, J.A.: Phys. Rev. **B17** (1978) 484.
79K1	Kita, E., Siratori, K., Kohn, K., Tasaki, A., Kimura, S., Shindo, I.: J. Phys. Soc. Jpn. **47** (1979) 1788.

79K2	Kuipers, A.J.M., Brabers, V.A.M.: Phys. Rev. **B20** (1979) 594.			
79L	Lorenz, B., Ihle, D.: Phys. Status Solidi **B96** (1979) 659.			
79S1	Siratori, K., Kila, E., Kaji, G., Tasaki, A., Kimura, S., Shindo, I., Kohn, K.: J. Phys. Soc. Jpn. **47** (1979) 1779.			
79S2	Schlegel, A., Alvarado, S.F., Wachter, P.: J. Phys. **C12** (1979) 1157.			
79T	Terukov, E.I., Reichelt, W., Ihle, D., Oppermann, H.: Phys. Status Solidi **B95** (1979) 491.			
80I	Iwaucji, K., Kita, Y., Koizumi, N.: J. Phys. Soc. Jpn. **49** (1980) 328.			
81S	Srinavasan, G., Srivastava, C.M.: Phys. Status Solidi **103** (1981) 665.			
81W	Wu, C., Mason, T.O.: J. Amer. Ceram. Soc. **64** (1981) 520.			
81Y	Yamai, K., Mizoguchi, M., Iida, S.: J. Phys. Soc. Jpn. **50** (1981) 65.			
82I	Iizumi, M., Koetzle, T.F., Shirane, G., Chikazumi, S., Matsui, M., Todo, S.: Acta Crystallogr. **B38** (1982) 2121.			

Physical property	Numerical value	Experimental conditions	Experimental method, remarks	Ref.
		9.15.2.4.2 Cobalt oxide (Co_3O_4)		
Crystal structure				
structure: spinel, space group $O_h^7 - Fd3m$. The tetragonal 8(a) sites are occupied by Co^{2+}, and the octahedral 16(d) sites by Co^{3+}. The $32O^{2-}$ ions occupy 32(e) sites.				
position parameter of the (e)-sites:				
u	0.3881			64R
	0.392			73S
	0.3887			68K
lattice parameter (in Å):				
a	8.084		lattice parameters of Ni- and Zn-doped samples: Figs. 1, 2	73S
	8.065			64R
	8.0835			68K
	8.080			78A, 79G
interatomic distances (in Å):				
$d(Co^{2+}-O)$	1.93		tetrahedral site	64R
	1.99			73S
	1.946			68K
$d(Co^{3+}-O)$	1.92		octahedral site	64R
	1.89			73S
	1.915			68K

low-temperature crystal structure:

At low temperatures, magnetic ordering gives space group $T_d^2 - F\bar{4}3m$ [64R]. This structure has $4Co^{2+}$ on the 4(a) sites (f.c. +000), $4Co^{2+}$ on the 4(c) sites (f.c. $+\frac{1}{4}\frac{1}{4}\frac{1}{4}$), $16Co^{3+}$ on 16(e) sites ($x=\frac{5}{8}$), $16O^{2-}$ on 16(e) sites ($x=u$) and $16O^{2-}$ on 16(e) sites ($x=\frac{1}{4}-u$).

Electronic and optical properties

No band structure or cluster calculation appears to be available. XPE and UPE spectra have been reported (Figs. 3, 4). The assignment of the Co 3d region (Fig. 4) shows that very little energy apparently separates the Co^{2+} and Co^{3+} ionizations (ca. 0.5···1.0 eV). The O 2p band is ≈4 eV wide and is most clearly shown in the He II excitation (Fig. 3). Satellites in the core region have been explored for various Co-spinels, and tetrahedral Co^{2+} in $CoAl_2O_4$ and $CoCr_2O_4$ is characterized by Co 2p satellites at ca. 5···6 eV relative to the main $2p_{\frac{3}{2},\frac{1}{2}}$ core levels whereas octahedral Co^{3+} is characterized by Co 2p satellites at ca. 9 eV [76O].

EELS peaks: 8.3(8) eV, 35.8(10) eV, 62.2(5) eV [77B].

References for 9.15.2.4.2   Figs. p. 544 ff.

Physical property	Numerical value	Experimental conditions	Experimental method, remarks	Ref.

Transport properties

conductivity: Fig. 5. p-type semiconductor with room temperature resistivity of $\simeq 10^4$ Ω cm.

activation energy for conductivity:

E_A	≈ 0.4 eV	$T = 400 \cdots 600$ K		79G
	0.68 eV	$450 \cdots 500$ K	$Co_3O_{4.02}$	38H

Seebeck coefficient:

S	600 μV K^{-1}	$T = 300$ K		78A
	1050 μV K^{-1}	300 K		79G

Co_3O_4 can be synthesized grossly cation deficient [78B]. X-ray measurements indicate that the cation vacancies order to provide solid solutions between Co_3O_4 and the hypothetical defect spinel γ-Co_2O_3. The formulation $Co(Co^{3+}_{2-x}Co^{4+}_{x/4} \square_{3x/4})O_4$ has been suggested [78B] and conductivity and Seebeck data for films of non-stoichiometric oxide are shown in Fig. 6.

On doping with Zn the material remains p-type and semiconducting (Fig. 7). Doping with Ni leads to a semiconductor-semimetal transition at $x \simeq 0.5$ (Fig. 8). Doping with Li has a marked effect on conductivity, σ increasing from $\simeq 10^{-5}$ (Ω cm)$^{-1}$ to $\simeq 1$ (Ω cm)$^{-1}$ at 1 at% Li and RT [77V].

Magnetic and further properties

Co_3O_4 has antiferromagnetically coupled spins on the Co^{2+} tetrahedral sites at low temperature, with $T_N = 33.0\,(10)$ K [69K], 46K [63B], 40K [64R]. The basic magnetic structure at low temperatures is shown in Fig. 1. Moessbauer spectroscopy has suggested that the Co^{2+} spins actually cant away from the z-axis [69K]. The main spin coupling mechanism is apparently via the diamagnetic, low-spin Co^{3+}.

In the ordered magnetic state the magnetic moment on Co^{2+} sites is $p_A(A) = 3.26\,\mu_B$ at 4.2 K. No moment is found on B-sites. At high temperatures, the magnetic susceptibility in the paramagnetic phase follows the Curie-Weiss law $\chi = \alpha + C/(T - \Theta_p)$ (Fig. 10) [64R, 63B].

magnetic parameters:

		T [K]		
α	$0.74 \cdot 10^{-3}$ cm^3 mol^{-1}	$100 \cdots 300$		64R
	$0.71 \cdot 10^{-3}$ cm^3 mol^{-1}	$300 \cdots 1000$		58C
Θ_p	-53 K			64R
	-110 K			64P
p_{eff}	$4.14\,\mu_B$	$300 \cdots 1000$		58C
$p_A(A)$	$3.25\,\mu_B$	$100 \cdots 300$		64R
J_1/k	$-7.27\,(25)$ K	155	exchange coupling constant, from neutron scattering	75S
	-5.3 K	$100 \cdots 300$	from Curie-Weiss law	75S
	-4.0 K	$100 \cdots 300$		64R
	-7.08 K	$90 \cdots 1200$	from magnetic susceptibility data of [63B]	75S

Debye temperature:

Θ_D	525 K	RT	from neutron scattering temperature factor	64R

References for 9.15.2.4.2

38H	Hochberg, B., Sominski, M.S.: Phys. Z. Sowjetunion **13** (1938) 198.
58C	Cossee, P.: J. Inorg. Nucl. Chem. **8** (1958) 483.
63B	Blasse, G.: Philips Res. Rep. **18** (1963) 383.
64P	Perthel, R., Jahn, H.: Phys. Status Solidi **5** (1964) 563.
64R	Roth, W.L.: J. Phys. Chem. Solids **25** (1964) 1.
68K	Knop, O., Reid, K.I.G., Sutarno, S., Nakagawa, Y.: Can. J. Chem. **46** (1968) 3463.
69K	Kündig, W., Kobalt, M., Appel, H., Constabaris, G., Lundqvist, R.H.: J. Phys. Chem. Solids **30** (1969) 819.

73S	Smith, W.L., Hobson, A.D.: Acta Crystallogr. **B29** (1975) 362.
75S	Scheerlinck, D., Hauteder, S., Wegener, W.: Phys. Status Solidi **B68** (1975) 535.
76O	Oku, M., Hirokawa, K.: J. Electron Spectrosc. Relat. Phenom. **8** (1976) 475.
77B	Balabanova, L.A., Stepin, E.V.: Fiz. Tverd. Tela **19** (1977) 3018.
77V	Voloskin, A.G., Kamarenko, N.I., Kolesnikova, I.P., Korolenko, S.D.: Elektrokhimiya **13** (1977) 1724.
78A	Appandairajan, N.K., Gopalakrishnan, J.: Proc. Indian Acad. Sci. Sect. A **87** (1978) 115.
78B	Belova, I.D., Shalaginov, V.V., Galyanov, B.Sh., Roginskaya, Yu.E., Shub, D.M.: Zh. Neorg. Khim. **23** (1978) 286.
79G	Gopalakrishnan, J., Appandairajan, N.K., Viswanathan, B.: Proc. Indian Acad. Sci. Sect. A **88** (1979) 217.
79J	Jugnet, Y., Duc, T.M.: J. Phys. Chem. Solids **40** (1979) 29.

Physical property	Numerical value	Experimental conditions	Experimental method, remarks	Ref.

9.15.2.4.3 Manganese oxide (Mn_3O_4)

Phase diagram: Fig. 2 (see also Fig. 2 of 9.15.2.3.1 (MnO)). In the temperature range 860···975°C equilibrium with Mn_2O_3 has the quantitative form: $6Mn_2O_3 \rightleftharpoons 4Mn_3O_4(\alpha) + O_2$, whereas at 1248···1562°C the equilibrium with MnO is $2Mn_3O_4(\beta) \rightleftharpoons 6MnO + O_2$. For phase diagrams, see further below (Mn_2O_3).

phase boundaries:

$\log p_{O_2} = a - b/T$ (p_{O_2} in atm)

a	7.190	$T = 860···975$°C	for Mn_3O_4/Mn_2O_3 boundary	64O
b	9004 K			
a	8.05		see also [76P]	60H
b	10100 K			
a	8.2682			67S
b	10240 K			
a	13.31	$T = 1248$ ···1562°C	for Mn_3O_4/MnO boundary	60H
b	26000 K			

Crystal structure

$T < 1170$°C: tetragonal distorted spinel, hausmannite (α-Mn_3O_4), space group $D_{4h}^{19} - I4_1/amd$, $Z = 4$, unit cell body centred tetragonal (Fig. 1)

$T > 1170$°C: cubic spinel (β-Mn_3O_4), space group $O_h^5 - Fm3m$.

lattice parameters of the low-temperature phase (distances in Å):

		T [K]		
a	5.753	RT	for relationship between tetragonal and cubic spinel unit cells, see [74N3]	58H
c	9.422			
a	5.747	RT		66R
c	9.433			
a	5.71	RT		74J
c	9.35			
a	5.763	RT		71B2
c	9.456			
a	5.76	RT		61S
c	9.46			
a	5.756	4.2		71B2
c	9.439			
a	5.756	77		
c	9.439			
a	5.768	300		
c	9.456			

(continued)

9.15.2.4.3 Manganese oxide (Mn$_3$O$_4$)

Physical property	Numerical value	Experimental conditions	Experimental method, remarks	Ref.
lattice parameters of the low-temperature phase (continued)				
a	5.76	523		58H
c	9.45			
a	5.76	723		
c	9.47			
a	5.77	923		
c	9.49			
a	5.78	1073		
c	9.50			
a	5.80	1273		
c	9.49			
a	5.81	1433		
c	9.46			
d(Mn—O)	1.88	in c-plane	local symmetry at Mn$_{oct}$ is tetragonal	61S
	2.28	$\parallel c$-axis		
	2.02	mean value of both values		
	2.07	in regular tetrahedral sites		
δ	0.0083 (15)		for definition, see Fig. 1	61S
ε	0.0320 (27)		for definition, see Fig. 1	
lattice parameter of the high-temperature phase (in Å):		T [°C]		
a	8.55	1170		58H
	8.56	1200		
	8.57	1270		
	8.68	1270		67D
	8.64	1270		50M
phase transition temperature (in °C):				
T_{tr}	1170↑		differential thermal analysis	48M
	1100↓			
	1167↑		differential thermal analysis	58H
	1130↓			
	1150↑		X-ray diffraction	67D
	1125↓			
	1162↑			58H
	1144↓			
	1172			42S
	1157			81M
thermodynamic parameters:				
ΔH_{tr}	+4.96 kcal mol^{-1}		enthalpy and entropy change at the transition temperature	74N1
	+4.5 kcal mol^{-1}			42S
	+5.9 (7) kcal mol^{-1}			70F
ΔS_{tr}	4.2 (5) cal K^{-1} mol^{-1}			70F
	3.93 cal K^{-1} mol^{-1}			74N1

Transport properties

conduction below phase transition temperature:

Conductivity very low at RT ($\varrho_n \simeq 1.6 \cdots 6.4 \cdot 10^8$ Ω cm [66R], $\simeq 10^{15}$ Ω cm extrapolated [75L]). Experiments are complicated by the transition to Mn$_2$O$_3$ at lower temperature (Fig. 3). The clear kink in the Mn$_3$O$_4$ data at higher p_{O_2} is ascribed to intergranular Mn$_2$O$_3$ formation. $E_A = 1.3$ eV in this region [75L]. Seebeck coefficient positive [75L] suggesting p-type conduction.

9.15.2.4.3 Manganese oxide (Mn_3O_4)

Physical property	Numerical value	Experimental conditions	Experimental method, remarks	Ref.

change in transport properties at T_{tr}:

A substantial resistivity anomaly is observed at T_{tr} (1430 K, 25 K hysteresis) with a marked change in slope, from 1.35 eV in the low-temperature phase to 0.65 eV in the high-temperature phase for sintered disks. The Seebeck coefficient has an associated activation energy of 1.2 eV in the tetragonal phase and 0.3 eV in the cubic phase. The melt grown crystals have an associated activation energy from Seebeck measurements of 1.04 eV in the tetragonal phase [81M].

further transport data (Fig. 4):

$E_A(\mu)$	0.28⋯0.38 eV	$T < T_{tr}$	activation energy for drift mobility	81M
	0.52 eV	$T > T_{tr}$		
E_a	1.04⋯1.17 eV	$T < T_{tr}$	binding energy of holes to Mn vacancies	81M
	0.26 eV	$T > T_{tr}$		
μ	5.3⋯8.1·10^{-3} cm²/V s	LT phase at T_{tr}	drift mobility	81M
	1.3·10^{-2} cm²/V s	HT phase at T_{tr}		

Optical properties

The IR spectrum shows two bands at 18.2 and 21.1 μm [61H].

Magnetic properties

Mn_3O_4 is ferrimagnetic below T_C. Easy directions of magnetization are [100] and [010]. Hard axis is [001] [60D, 74J].

Curie temperature (in K):

T_C	41.8 (1)	T_C is depressed by the addition of Zn	60D
	41		74J
	43		71B1
	54.4		75K

susceptibility: Fig. 5

Two different magnetic structures have been proposed at 4.2 K [71B2, 74J] with the magnetic unit cell being twice the size of the chemical cell. The magnetic ordering of [74J] is shown in Fig. 6.

magnetic moments per ion and direction cosines (α_1, α_2, α_3) **referred to** (a, b, c) ($|b|=2a$) (from [74J]): For ion numbering, see Fig. 6.

Type of site	Ion no.	p_A μ_B	$T=4.7$ K		
			α_1	α_2	α_3
A	1, 2, 3, 4, 5, 6, 7, 8	4.34 (10)	0.00 (15)	1.00	0.00 (15)
B	9, 10, 13, 14,	3.64 (8)	−0.09 (5)	−0.38 (2)	−0.92 (1)
	11, 12, 15, 16,	3.64 (8)	0.09 (5)	−0.38 (2)	0.92 (1)
B'	17, 20, 21, 24,	3.25 (6)	0.00 (6)	−0.33 (2)	−0.94 (1)
	18, 19, 22, 23	3.25 (6)	0.00 (6)	−0.33 (2)	0.94 (1)

Type of site	Ion no.	p_A μ_B	$T=29$ K		
			α_1	α_2	α_3
A	1, 2, 3, 4, 5, 6, 7, 8	3.78 (12)	0.00 (12)	1.00	0.00 (12)
B	9, 10, 13, 14,	3.05 (8)	−0.12 (7)	−0.40 (3)	−0.91 (1)
	11, 12, 15, 16,	3.05 (8)	0.12 (7)	−0.40 (3)	0.91 (1)
B'	17, 20, 21, 24,	3.17 (8)	0.00 (4)	−0.25 (3)	−0.97 (1)
	18, 19, 22, 23	3.17 (8)	0.00 (4)	−0.25 (3)	0.97 (1)

Physical property	Numerical value	Experimental conditions	Experimental method, remarks	Ref.

At 32.8 K [74J], 35 K [71B2] a phase transition to a new magnetic structure occurs. The transition appears to be a first order modulation of the cell doubling; evolution of the magnetic structure continues until at 39.4 K the magnetic and b.c. tetragonal chemical cells coincide.

Magnetic anisotropy persists above T_C up to $\simeq 100$ K, above which it is isotropic. The magnetic data at high temperature are consistent with the formula $Mn^{2+}_{tetr}(Mn^{3+}_{oct})_2O_4$.

At higher temperatures a Curie-Weiss law is obeyed approximately (Fig. 5) with $\Theta_p = -564$ K and $p_{eff} = 5.27\ \mu_B$ [71B2]. More accurate fit obtained with a Néel function:
$$\frac{1}{\chi} = \frac{T}{C} + \frac{1}{\chi_0} - \frac{\sigma}{T-\Theta}:$$

C_m	10.71 (7) cm^3 K mol^{-1}		74N2
	6.4 cm^3 K mol^{-1}		75K
$1/\chi_0$	53 (2) mol cm^{-3}		74N2
	36 mol cm^{-3}		75K
σ	1250 (50) mol K cm^{-3}		74N2
	600 mol K cm^{-3}		75K
Θ	40 (5) K		74N2
	44 K		75K

References for 9.15.2.4.3

42S	Southerd, J.C., Moore, G.E.: J. Am. Ceram. Soc. **64** (1942) 1769.
48M	McMurdie, H.F., Golovato, E.: J. Res. Nat. Bur. Stand. **41** (1948) 589.
50M	McMurdie, H.F., Sullivan, B.M., Mauer, F.A.: J. Res. Nat. Bur. Stand. **45** (1950) 35.
58H	van Hook, H.J., Keith, M.L.: Am. Mineral. **43** (1958) 69.
60D	Dwight, K., Menyuk, N.: Phys. Rev. **119** (1960) 1470.
60H	Hahn, W.C., Muan, A.: Am. J. Sci. **258** (1960) 66.
61H	Hafner, S.: Z. Kristallogr. **115** (1961) 331.
61S	Satomi, K.: J. Phys. Soc. Jpn. **16** (1961) 258.
64O	Otto, E.M.: J. Electrochem. Soc. **111** (1964) 88.
66R	Rozhdestvenskaya, M.V., Mokievskii, V.A., Stogova, V.A.: Kristallografiya **11** (1966) 903.
67D	Driessens, F.C.M.: Inorg. Chim. Acta **1** (1967) 193.
67S	Shenouda, F., Aziz, S.: J. Appl. Chem. **17** (1967) 258.
70F	Fender, B.E.F., Riley, F.D.: The chemistry of extended defects in non-metallic solids. L. Eyring, M. O'Keeffe (eds.), North-Holland, **1970** p. 54.
71B1	Boucher, B., Buhl, R., Perrin, M.: J. Appl. Phys. **42** (1971) 1615.
71B2	Boucher, B., Buhl, R., Perrin, M.: J. Phys. Chem. Solids **32** (1971) 2429.
74J	Jensen, G.B., Nielsen, O.V.: J. Phys. **C7** (1974) 409.
74N1	Navrotsky, A.: MTP International Review of Science; Inorganic Chemistry Series Two, 5, D.W.A. Sharp (ed.). London: Butterworths, **1974** p. 29.
74N2	Nogues, M., Poix, P.: Solid State Commun. **15** (1974) 463.
74N3	Nogues, M., Poix, P.: J. Solid State Chem. **9** (1974) 330.
75K	Kheer, H.V., Boda, M.G., Bhaduri, A., Biswas, A.B.: J. Inorg. Nucl. Chem. (USSR) **37** (1975) 1605.
75L	Logothetis, E.M., Park, K.: Solid State Commun. **16** (1975) 909.
76P	Pompe, R.: Acta Chem. Scand. **A30** (1976) 370.
81M	Metselaar, R., van Tol, R.E.J., Piercy, P.: J. Solid State Chem. **38** (1981) 335.

9.15.2.5 Sesquioxides and related oxides
9.15.2.5.1 Oxides of chromium

Phase diagram: see Fig. 1.

At oxygen pressures below $\simeq 900$ bar we have:

$$CrO_3 \xrightarrow{220°C} \beta\text{-oxide }(CrO_{2.65}) \xrightarrow{230°C} \gamma\text{-oxide }(CrO_{2.44}) \xrightarrow{250°C} (CrO_2) \xrightarrow{325°C} Cr_2O_3.$$

Above 1 kbar oxygen pressure we have:

$$CrO_3 \xrightarrow{200°C} (Cr_6O_{15}) \xrightarrow{220°C} (Cr_5O_{12}) \xrightarrow{250°C} (CrO_2) \xrightarrow{350°C} Cr_2O_3.$$

Little data is available on the higher oxides of Cr:

CrO_3 is reported to be an n-type semiconductor, activation energy for conductivity $E_A = 1.98$ eV ($T < 500$ K) [68C]. It has an unusual chain structure; orthorhombic unit cell, C_{2v}^{16}–Ama, $a = 5.743$ Å, $b = 8.557$ Å, $c = 4.789$ Å [60H]. $d = 2.70$ g cm^{-3} [83H].

Cr_5O_{12} is reported to be an n-type semiconductor, activation energy for conductivity $E_A = 0.82$ (11) eV ($T < 600$ K) [68C]. It has an orthorhombic structure, D_{2h}^{14}–Pbcn, $a = 12.04$ Å, $b = 8.21$ Å, $c = 8.18$ Å [65W]; $a = 12.044$ Å, $b = 8.212$ Å, $c = 8.177$ Å, $Z = 4$ [68W].

Cr_6O_{15} has been characterized as orthorhombic; $a = 8.47$ Å, $b = 12.90$ Å, $c = 10.08$ Å, $Z = 4$, space group D_{2h}^{17}–Cmcm [68W].

CrO_2 is a ferromagnetic metal with the rutile structure [77C1]; $a = 4.423$ Å, $c = 2.917$ Å [68W]; $a = 4.421$ Å, $c = 2.917$ Å [71B]; $a = 4.421$ Å, $c = 2.916$ Å [57T].

Chromium sesquioxide (Cr_2O_3)

structure: corundum, space group D_{3d}^6–R$\bar{3}$c, $Z = 6$. Rhombohedral unit cell: Fig. 2. Connection with hexagonal unit cell: $a_{hex} = 2a_{rh} \sin(\alpha_{rh}/2)$, $c_{hex} = 3a_{rh} \cos\gamma$, where $\sin\gamma = (2/\sqrt{3})\sin(\alpha_{rh}/2)$. In corundum, Cr atoms occupy the 12(c) sites at $\pm(0\,0\,z;\ 0\,0\,z+\tfrac{1}{2})$ and O atoms occupy the 18(e) sites at $\pm(x\,0\,\tfrac{1}{4};\ 0\,x\,\tfrac{1}{4};\ \bar{x}\,\bar{x}\,\tfrac{1}{4})$ where $z = 0.3475(3)$, $x = 0.306(4)$ [62N]. There is a lattice anomaly near the Néel temperature ($T_N \approx 33$°C) (Fig. 3).

lattice parameters (hexagonal unit cell, distances in Å):

		T [°C]	
a	4.9607	RT	62N
c	13.599		
a	4.95762 (8)	RT	67S
c	13.5874 (10)		
a	4.9590	25	72K
c	13.589		
a	4.9999	1070	
c	13.680		
a	5.0191	1470	
c	13.731		
a	5.0324	1670	
c	13.765		
d(Cr–Cr)	2.65	across a shared face	66L
	2.89	across a shared edge	
d(Cr–O$_1$)	2.016	$p \approx 1$ bar	66L
	1.965	28.6 kbar	
d(Cr–O$_2$)	1.965	$p \approx 1$ bar	66L
	1.906	28.6 kbar	

Physical property	Numerical value	Experimental conditions	Experimental method, remarks	Ref.

coefficients of linear thermal expansion (in 10^{-6} K^{-1}):

		T [°C]		
α_a	7.1	$T > T_N$		76A
α_c	5.1			
α_a	5.86	26···1000		64K
α_c	8.12			
α_a	7.85	25···1070		72K
α_c	6.38			
α_a	9.75	1070···1470		
α_c	9.30			
α_a	13.30	1470···1670		
α_c	12.40			

pressure dependence of a_{hex}, c_{hex}, a_{rh}, α_{rh}, **and** V/V_0:

From [66L]:

a_{hex} [Å]	c_{hex} [Å]	a_{rh} [Å]	α_{rh} [deg]	V/V_0	p [kbar]
4.961	13.600	5.362	55.11	1.000	10^{-3}
4.940	13.579	5.350	54.99	0.990	20
4.918	13.558	5.338	54.87	0.980	50
4.897	13.537	5.325	54.74	0.970	85
4.876	13.516	5.312	54.62	0.960	122
4.854	13.496	5.299	54.49	0.950	161
4.832	13.475	5.286	54.36	0.940	203
4.811	13.454	5.273	54.23	0.930	244
4.790	13.433	5.260	54.10	0.920	286
4.779	13.422	5.253	54.04	0.915	308

From [80F] at RT:

a_{hex} [Å]	c_{hex} [Å]	c/a	p [kbar]
4.9511	13.5656	2.740	0.001
4.9371	13.530	2.740	18.6
4.9233	13.4953	2.741	40.6
4.9128	13.4689	2.742	56.8

density:

d	5.21 g cm^{-3}			83H

melting point:

T_m	2266 (25) °C			83H

Electronic and optical properties

Cluster calculation: Fig. 4. The O 2p−Cr 3d separation is given as 3 eV. HeII UPS spectrum: Fig. 5. In first order, only one ionization is expected for Cr^{3+} corresponding to d^3 $^4A_{2g} \rightarrow$ d^2 $^3T_{1g}$. This is shown deconvoluted in Fig. 5, and the O 2p−Cr 3d separation experimentally is far smaller than 3 eV. The skewed O 2p band is composed of two separate transitions, the lower energy one corresponding to the simple O 2p^{-1} process and the higher energy one to an O 2p→Cr 3d charge transfer shake up process [80H].

The visible spectrum is dominated by ligand-field transitions (Fig. 6). The UV-spectrum from Kramers-Kronig analysis is shown in Fig. 7. The wavelength modulated reflectance spectrum shows structure at 2.1 eV ($^4A_{2g} \rightarrow ^4T_{2g}$), 2.4···2.5 eV ($^4A_{2g} \rightarrow ^2T_{2g}$) and 2.6 eV ($^4A_{2g} \rightarrow ^4T_{1g}$) [72B]. At higher energies, both spectra show the energy gap to be at 3.3 eV. The transition has been assigned to a charge transfer process: O 2p−Cr 3d. This assignment again suggests that the cluster calculation of Fig. 4 greatly overestimates the 3d−2p energy separation.

For thermal band gap, see "Transport properties" below.

dielectric constants (see also Fig. 8):

$\varepsilon(0)$	10.33	RT, $E \perp c$	from ir active E_u-modes	65R,
$\varepsilon(\infty)$	5.73			77L
$\varepsilon(0)$	10.93	RT, $E \parallel c$	from ir active A_{2u}-modes	
$\varepsilon(\infty)$	5.97			
$\varepsilon(0)$	13.0	$T = 303$ K	from ac measurement at $\nu \gtrsim 10^6$ Hz on single crystals with electric field along the a axis	77R

Physical property	Numerical value	Experimental conditions	Experimental method, remarks	Ref.

Lattice properties

wavenumbers of lattice modes (in cm^{-1}):

IR active modes:

$\bar{\nu}_{TO}(E_u)$	305	RT, $E \perp c$	for the space group $D_{3d}^6 - R\bar{3}c$	65R,
	440		seven Raman active and six	77L
	538		IR active modes exist	
	609			
$\bar{\nu}_{TO}(A_{2u})$	402	RT, $E \parallel c$		
	533			

Raman active modes:

$\bar{\nu}_R(E_g)$	235	$T \approx 330$ K		71H
	290			
	352			
	528			
	617			
$\bar{\nu}_R(A_{2g})$	266			
	547			

For $T < T_N$ two new features appear; broad second-order scattering from $500 \cdots 900$ cm^{-1} with peak at 685 cm^{-1} and a sharp feature at 396 cm^{-1}. Latter is ascribed to zone-centre optical magnon [71H].

elastic moduli (in 10^{12} dyn cm^{-2}):

c_{11}	3.74	RT	trigonal system	76A
c_{12}	1.48			
c_{13}	1.75			
c_{33}	3.62			
c_{44}	1.59			
c_{66}	−0.19			

Transport properties

Cr_2O_3 is a p-type semiconductor, but equilibrium with oxygen is only sluggishly established at temperatures below 1500 K, so thermogravimetric analysis has been unable to reveal the nature of the defect. Early work [51H] suggested: $Cr_{Cr} \rightarrow V'''_{Cr} + Cr_i^{2\cdot} + e^+$ with substantial mobility for the interstitial $Cr_i^{2\cdot}$ only found at high temperatures [61H]. More complex defect structures have been proposed to account for the oxygen pressure dependence [72M]. Ionic transport numbers remain less than 0.1% even at 1800 K [61H]; ionic diffusion coefficient of vacancies is $\approx 2 \cdot 10^{-7}$ $cm^2 s^{-1}$ at 1500 K with an activation energy of 0.25 eV [72M].

conductivity:

At **high temperatures** ($1150 \cdots 1400$ °C), the conductivity becomes intrinsic and independent of p_{O_2} [64C, 72M] (Fig. 9). In this region $\sigma \propto \sigma_0 \exp(-U/kT)$ with

U	1.63 (1) eV	unoriented sample		64C
σ_0	1.48 (10) $\cdot 10^4$ Ω^{-1} cm^{-1}			
U	1.68 (1) eV	a-direction		
σ_0	1.96 (16) $\cdot 10^4$ Ω^{-1} cm^{-1}			
U	1.59 (2) eV	c-direction		
σ_0	1.37 (15) $\cdot 10^4$ Ω^{-1} cm^{-1}			

thermal energy gap:

$E_{g,th}$	$3.2 \cdots 3.4$ eV	$T = 1000 \cdots 1400$ °C	suggested from data above	64C, 51H

Physical property	Numerical value	Experimental conditions	Experimental method, remarks	Ref.

At **low temperatures** ($T<1100°C$), an extrinsic region is found with

conductivity activation energies (in eV):

		T [K]		
E_A	0.26	670···1370	see Fig. 10; hot pressed sample, $\sigma \propto CT^{-3/2} \exp(-0.26 [\text{eV}]/kT)$	72M
	0.6	400···1000	ceramic, $\sigma = 3.8 \cdot 10^6 T^{-1} \cdot 2x(1-2x)\exp(-0.6[\text{eV}]/kT)$ $[\Omega^{-1} \text{cm}^{-1}]$ for Cr_2O_{3+x}	69C
	0.41	470···970	ceramic, from ac conductivity at 1kHz	71R
	0.4	RT		67V
	0.519	≈370	polycrystalline sample	74C

In the low T range the defects created at high temperatures, presumably $[V'''_{Cr}]$, are "frozen in" and the thermopower is constant [72M], or varies only slightly [69C] with T and p_{O_2}. The variation of σ with p_{O_2} is too small to be explained in terms of any simple defect equilibria. ac conductivity has an activation energy of 0.9 eV for $T=300°C$, $f=100···10^6$ Hz [77R].

influence of non-stoichiometry: Fig. 11 (conductivity), Fig. 12 (Seebeck coefficient).

hole mobilities (in cm^2/Vs):

μ_p	3···5·10^{-5}	$T=550$ K	from Seebeck data on cold-pressed ceramics	69C
	1···5·10^{-8}	350 K		
	0.76	873 K	ceramic sample	51H

pressure dependence of transport parameters:

E_A	0.4 eV	$p=1$ atm, low T	activation energy for conductivity	67V
dE_A/dp	$-0.8·10^{-6}$ eV bar^{-1}			
dST/dp	$-0.2·10^{-6}$ V bar^{-1}			

At very high pressures, collapse to a metallic state occurs (Fig. 13).

effect of dopants (polycrystalline samples):

ϱ [Ω cm] at 100°C	E_A [eV]	impurity	concentration [at %]		
3.0·10^6	0.519	undoped	–		74C
>1.0·10^{11}	1.466	Ti	1.00	E_A in the range	
1.3·10^8	0.558		0.01	$-50°C···400°$	
2.8·10^6	0.519	Nb	1.00		
2.9·10^6	0.521		0.01		
9.3·10^6	0.627	Fe	1.00		
1.3·10^6	0.511		0.01		
1.8·10^2	0.364	Mg	1.00		
2.7·10^4	0.444		0.01		
5.5·10^2	0.460	Ni	1.00		
1.8·10^5	0.531		0.01		

Li doping: Fig. 14, Mg doping: Fig. 15, Ni doping: Fig. 16. Much larger mobilities have been found for the Mg and Ni doped samples [77C2]; furthermore, the mobilities are apparently not activated.

Magnetic properties

Cr_2O_3 is antiferromagnetic. Spin direction along [111] axis of rhombohedral unit cell. Spins are paired antiparallel across shared octahedral face [65C].

Magnetic susceptibility: Fig. 17, magnetic phase diagram: Fig. 18, spin-wave spectrum: Fig. 19, sublattice magnetization: Fig. 20.

At high temperatures Curie-Weiss law is obeyed [39F].

References for 9.15.2.5.1

Physical property	Numerical value	Experimental conditions	Experimental method, remarks	Ref.
Néel temperature (in K):				
T_N	308		from powder susceptibility	65C
	307		from powder susceptibility	78N
	307.3		from magnetic phase diagram	77Y
	306.99		from heat capacity	77B
Curie-Weiss temperature:				
Θ_p	550 (50) K		from susceptibility	39F
	515 (88) K		from spin-wave spectrum	70S
further parameters:		T [K]		
$p_{A,s}$	2.76 (3) μ_B	$T \to 0$	saturation magnetic moment per Cr atom	65C
J_1	-85.7 cm^{-1}	78	exchange parameters J_i as defined in [70S]	70S
J_2	-38.8 cm^{-1}	78		
J_3	-2.7 cm^{-1}	78		
dT_N/dX	0.50 (5) K kbar^{-1}	$\parallel c$		73G
	0.30 (5) K kbar^{-1}	$\parallel a$		
dT_N/dp	1.50 (5) K kbar^{-1}			73G

References for 9.15.2.5.1

39F	Foëx, M., Graff, M.: C.R. Acad. Sci. **209** (1939) 160.
51H	Hauffe, K., Block, J.: Z. Phys. Chem. **198** (1951) 232.
56M	McGuire, T.R., Scott, E.J., Graunis, F.H.: Phys. Rev. **102** (1956) 1000.
57T	Thamer, B.J., Douglass, R.M., Staritzky, E.: J. Am. Chem. Soc. **79** (1957) 547.
60H	Hanic, F., Štempelová, D.: Chem. Zvesti **14** (1960) 165.
61H	Hagel, W.C., Seybolt, A.U.: J. Electrochem. Soc. **108** (1961) 1146.
62N	Newnham, R.E., de Haan, Y.M.: Z. Kristallogr. **117** (1962) 235.
63F	Foner, S.: Phys. Rev. **130** (1963) 183.
64C	Crawford, J.A., Vest, R.W.: J. Appl. Phys. **35** (1964) 2413.
64K	Kirchner, H.P.: Prog. Solid State Chem. **1** (1964) 1.
65C	Corliss, L.M., Hastings, J.M., Nathans, R., Shirane, G.: J. Appl. Phys. **36** (1965) 1099.
65H	Hagel, W.C.: J. Appl. Phys. **36** (1965) 2586.
65R	Renneke, D.R., Lynch, D.W.: Phys. Rev. **138** (1965) A530.
65W	Wilhelmi, K.-A.: Acta Chem. Scand. **19** (1965) 165.
66L	Lewis, G.K., Drickamer, H.G.,: J. Chem. Phys. **45** (1966) 224.
67L	Lal, H.B., Srivastava, R., Srivastava, K.G.: Phys. Rev. **154** (1967) 505.
67S	von Steinwehr, H.E.: Z. Kristallogr. **125** (1967) 377.
67V	Rimas Vaišnys, J.: J. Appl. Phys. **38** (1967) 2153.
68C	Cojocaru, L.N., Costea, T., Nagoescu, I.: Z. Phys. Chem. (Frankfurt am Main) **60** (1968) 152.
68W	Wilhelmi, K.-A.: Acta Chem. Scand. **22** (1968) 2565.
69C	Cojocaru, L.N.: Z. Phys. Chem. (Frankfurt am Main) **64** (1969) 255.
70S	Samuelson, E.J., Hutchings, M.I., Shirane, G.: Physica **48** (1970) 13.
71B	Baur, W.H., Khan, A.A.: Acta Crystallogr. **B27** (1971) 2133.
71H	Hart, T.R., Aggarwal, R.L., Lax, B.: Light Scattering in solids, Balkanski, M. (ed.), Paris: Flammarion Press **1971**.
71K	Kawai, N., Mochizuki, S.: Phys. Lett. **36A** (1971) 54.
71R	Rao, G.V.S., Wanklyn, B.M., Rao, C.N.R.: J. Phys. Chem. Solids **32** (1971) 345.
72B	Blazey, K.W.: Solid State Commun. **11** (1972) 371.
72K	Kudielka, H.: Monatsh. Chem. **103** (1972) 72.
72M	Meadowcroft, D.B., Hicks, F.G.: Proc. Br. Ceram. Soc. **23** (1972) 33.
73G	Goredetsky, G., Hornreich, R.M., Shtrikman, S.: Phys. Rev. Lett. **31** (1973) 938.
74C	de Cogan, D., Lonergan, G.A.: Solid State Commun. **15** (1974) 1517.
74G	Grefer, J., Reinan, D.: Z. Anorg. Allgem. Chem. **404** (1974) 167.

75E	Eastman, D.E., Freeouf, J.L.: Phys. Rev. Lett. **34** (1975) 395.	
76A	Alberts, H.L., Boeyens, J.C.A.: J. Mag. Magn. Mater. **2** (1976) 327.	
76T	Tossell, J.A.: J. Electron Spectrosc. Relat. Phenom. **8** (1976) 1.	
77A	Allos, T.I.Y., Birss, R.R., Parker, M.R., Ellis, E., Johnson, D.W.: Solid State Commun. **24** (1977) 129.	
77B	Bonie, R.H., Cahnell, D.S.: Phys. Rev. **B15** (1977) 4451.	
77C1	Chamberland, B.L.: CRC Critical Reviews in Solid State and Materials Science **1977** 1.	
77C2	de Cogan, D. Lonergan, G.A.: J. Phys. Chem. Solids **38** (1977) 333.	
77L	Lucovsky, G., Sladek, R.J., Allen, J.W.: Phys. Rev. **B16** (1977) 4716.	
77R	Rao, K.V., Subrahmanyan, A.: Ind. J. Pure Appl. Phys. **15** (1977) 359.	
77Y	Yacovitch, R.D., Shapira, Y.: Physica **86-88** B+C (1977) 1126.	
78N	Napijalo, M.L., Sreckovic, A., Novakovic, L.: Fizika (Zagreb) **10** (Suppl. 2) (1978) 169.	
80F	Finger, L.W., Hazen, R.M.: J. Appl. Phys. **51** (1980) 5362.	
80H	Howng, W.-Y., Thorn, R.J.: J. Phys. Chem. Solids **41** (1980) 75.	
83H	Handbook of Chemistry and Physics, Weast, R.C., (ed.), 64th edition **1983**, CRC Press. Inc.	

Physical property	Numerical value	Experimental conditions	Experimental method, remarks	Ref.

9.15.2.5.2 Hematite (α-Fe$_2$O$_3$)

(See also Landolt Börnstein, Vol. III/12b)

Fe$_2$O$_3$ is a line phase n-type semiconductor which, unusually, can also be doped p-type. Heating at 1 kbar oxygen and 500···700 °C for long periods gave no evidence of higher oxide phases [71D, 72S].

Crystal structure

structure: rhombohedral (hexagonal), space group $D_{3d}^6 - R\bar{3}c$, $Z = 6$. See Figs. 1, 2.

lattice parameters (distances in Å, angles in °):

a_{rh}	5.427	RT	temperature dependence: Fig. 3	62N
α_{rh}	55.271		relationship between trigonal	
a_{hex}	5.0345		and hexagonal unit cells: see	
c_{hex}	13.749		Cr$_2$O$_3$.	
a_{hex}	5.03490 (9)	$T = 298$ K		67S
c_{hex}	13.7524 (18)			
z(Fe)	0.355	RT	defined as in Cr$_2$O$_3$ section	62N
x(O)	0.300		defined as in Cr$_2$O$_3$ section	
d(Fe$_1$–O$_1$)	2.09 (2.116)	RT	for the Fe$_i$ and O$_i$ see Fig. 1;	62N
d(Fe$_1$–O$_5$)	1.96 (1.945)		values in parentheses: [69P]	
d(Fe$_1$–Fe$_2$)	2.89 (2.900)			
d(Fe$_1$–Fe$_3$)	2.97 (2.971)			
d(Fe$_1$–Fe$_4$)	3.37 (3.364)			
d(Fe$_1$–Fe$_5$)	3.70 (3.706)			
d(O$_1$–O$_2$)	2.62 (2.669)			
d(O$_1$–O$_3$)	3.06 (3.035)			
d(O$_1$–O$_4$)	2.76 (2.775)			
d(O$_1$–O$_5$)	2.89 (2.888)			
$\angle$(Fe$_1$–O$_1$–Fe$_2$)	87.4 (86.6)			
$\angle$(O$_1$–O$_2$–Fe$_3$)	94.1 (94.0)			
$\angle$(Fe$_1$–O$_5$–Fe$_4$)	118.2 (119.7)			
$\angle$(Fe$_1$–O$_1$–Fe$_5$)	132.1 (131.7)			
$\angle$(O$_1$–Fe$_1$–O$_2$)	77.5 (78.2)			
$\angle$(O$_1$–Fe$_1$–O$_5$)	85.9 (86.1)			
$\angle$(O$_5$–Fe$_1$–O$_6$)	102.7 (102.6)			
$\angle$(O$_1$–Fe$_1$–O$_6$)	161.5 (162.4)			

The cell constants show an anomaly in the region of the Néel temperature ($\simeq 690$ °C) (Fig. 3). Dilatometry also shows anomalies at 684 and 726 °C [62R], at 687 °C only [64I]. Anomalies in differential thermal analysis: 675 and 725 °C [62A].

For thermal expansion and elastic moduli, see Landolt Börnstein, Vol. III/12b, p. 8.

9.15.2.5.2 Hematite (α-Fe$_2$O$_3$)

Physical property	Numerical value	Experimental conditions	Experimental method, remarks	Ref.

pressure dependence of a_{hex}, c_{hex}, a_{rh}, α_{rh}, V/V_0:

From [66L]:

a_{hex} [Å]	c_{hex} [Å]	a_{rh} [Å]	α_{rh} [°]	V/V_0	p [kbar]
5.035	13.749	5.427	55.27	1.000	10^{-3}
5.017	13.705	5.409	55.26$^-$	0.990	48
5.000	13.661	5.392	55.24	0.980	79
4.982	13.617	5.374	55.22	0.970	106
4.964	13.574	5.356	55.20	0.960	130
4.946	13.530	5.338	55.18	0.950	154
4.928	13.487	5.320	55.17	0.940	176
4.910	13.444	5.302	55.15$^+$	0.930	198
4.893	13.400	5.284	55.14	0.920	220
4.884	13.375	5.275	55.13	0.915	230

From [80F] at RT:

a_{hex} [Å]	c_{hex} [Å]	c/a	p [kbar]
5.0346	13.7473	2.731	0.001
5.0250	13.7163	2.729	15.4
5.0143	13.6733	2.727	31.4
5.0080	13.6467	2.725	41.6
5.0067	13.6111	2.725	43.9
5.0020	13.6202	2.723	52.4

melting point: decomposes in air at atmospheric pressure [59S]; $T_m = 1565$ °C (nat. hematite) [83H].

density:

d	5.277 g cm^{-3}	X-ray	59S
	5.24 g cm^{-3}		83H

Electronic properties

No extended band structure calculation reported. Xα-cluster calculation: Fig. 4 (see [74T, 73T, 76T]). From this, the topmost O 2p level is $1t_{1g}\downarrow$, and Fe $d-t_{2g}\uparrow$ levels lie well below the O 2p edge. The highest occupied level is the Fe $d-e_g\uparrow$.

For thermal band gap, see "Transport properties" below.

peak energies in optical spectra (in eV) (see also section on "Optical properties"):

E_{peak}	2.27···2.30	specular	assignment: d−d	79G
	2.83	reflectance	d−d + "2-exciton"	
	3.10	from		
	3.97	hexagonal		
	4.71	plate at RT		
	5.39		charge-transfer	
	11.07		interband transitions	
	16.75			
	1.46	absorption		63G
	2.38	spectrum		
	3.16	of poly-		
	3.93	crystalline		
	4.84	film at RT		
	5.75			
	0.7	absorption	strongly enhanced by Ti doping	54M
	1.5	of poly- crystalline films at RT		
	1.49	absorption	d−d (broad reflectivity peaks	60B
	1.98	of (111) crystal sections at RT	d−d found at 24500 and 26000 cm^{-1})	
	1.44	peaks in diffuse	d−d: $^6A_{1g}-^4T_{1g}$	70T
	2.67	reflectance spectrum at RT	d−d: $^6A_{1g}-^4A_{1g}$	

(continued)

Physical property	Numerical value	Experimental conditions	Experimental method, remarks	Ref.
peak energies in optical spectra (continued)				
E_{peak}	1.44 (3)	absorption of single crystal platelets, light incident along c-axis (RT values)	d–d: $^6A_{1g}$ – $^4T_{1g}$	79M
	2.16		d–d: $^6A_{1g}$ – ^4E_g	
	2.92	wavelength modulated spectrum from unoriented single crystal at RT	d–d: $^6A_{1g}$ – $^4T_{2g}$ charge transfer	74B
	3.37			
	3.97			
	4.38			
	5.01			
	1.52	absorption	d–d: $^6A_{1g}$ – $^4T_{1g}$	78M
	2.024		d–d: $^6A_{1g}$ – $^4T_{2g}$ superimposed on a rising background	
	2.61	diffuse reflectance at RT	assignment: d–d	74T
	2.98		d–d	
	3.14		t_{1u} – t_{2g}, charge transfer	
	3.64		t_{2u} – t_{2g}	
	4.43		t_{1g} – e_g	
	5.29		t_{1u} – e_g	
	2.67	diffuse reflectance at RT	d–d	70T
	3.57		charge transfer	
	4.27		charge transfer	
	5.37		charge transfer	
	7.5	EEL	assignment: d–s	75B,
	24.0 (5)		3d–4p [75D] or plasma peak [75B]	75D

Lattice properties

wavenumbers of ir and Raman active lattice modes (in cm^{-1}, RT values):

symmetries: $\Gamma = 2A_{1g} + 2A_{1u} + 3A_{2g} + 3A_{2u} + 5E_g + 5E_u$. ir active: $2A_{2u} + 4E_u$, Raman active: $2A_{1g} + 5E_g$.

Physical property	Numerical value	Experimental conditions	Experimental method, remarks	Ref.
$\bar{\nu}_{TO}(A_{2u})$	299	RT	from ir reflectance	77O
	526			
$\bar{\nu}_{LO}(A_{2u})$	414			
	660			
$\bar{\nu}_{TO}(E_u)$	227			
	286			
	437			
	524			
$\bar{\nu}_{LO}(E_u)$	230			
	368			
	494			
	662			
$\bar{\nu}_{R}(A_{1g})$	225	RT		75H
	498			
$\bar{\nu}_{R}(E_g)$	247			
	293			
	299			
	412			
	613			

Physical property	Numerical value	Experimental conditions	Experimental method, remarks	Ref.

Transport properties
a) pure Fe_2O_3
defects:

The most important extrinsic defect is an oxygen vacancy. From thermogravimetric and thermodynamic studies [61S] we have $[V_O] = 3.27 \cdot 10^{26} p_{O_2}^{-1/4} \exp(-2.03 [eV]/kT)$ [cm^{-3}]; p_{O_2} in atm, T in K. Reasonably rapid equilibrium with the surrounding atmosphere occurs for $T > 1000\,°C$ [63G]. Single crystal studies are complicated by impurities and show very anomalous low-temperature effects unless crystals are annealed at 1300 °C and quenched [64T].

conductivity:

Three distinct regions for air-fired samples are observed (Fig. 5, see also Fig. 15).

(A) Region of variable activation energy, $T \lesssim 450\,°C$. Conductivity dominated by grain boundary resistance in polycrystalline samples, with $E_A \simeq 0.7$ eV. This region can be minimized by cooling under a reducing atmosphere. Grain boundary effects are clearly demonstrated by ac conductivity [51M].

(B) Region of low activation energy, $450\,°C \lesssim T \lesssim 800\,°C$; $E_A \simeq 0.1$ eV. Ascribed to extrinsic conduction with "frozen-in" equilibrium.

parameters in the formula $\sigma = n_d e A/T \cdot \exp(-\alpha/kT)$:

n_d	$4.5 \cdot 10^{18}$ cm^{-3}	firing	n_d: quenched donor concentration,	63G
A	2270 cm^2 K/Vs	temperature	A: constant, α: activation	
α	0.31 eV	$T = 1000\,°C$	energy for electron mobility.	
n_d	$1.7 \cdot 10^{19}$ cm^{-3}	1100 °C	Samples are ceramics.	
A	207 cm^2 K/Vs			
α	0.22 eV			
n_d	$5.5 \cdot 10^{19}$ cm^{-3}	1200 °C		
A	44 cm^2 K/Vs			
α	0.13 eV			
n_d	$1.5 \cdot 10^{20}$ cm^{-3}	1300 °C		
A	42 cm^2 K/Vs			
α	0.12 eV			
n_d	$4 \cdot 10^{20}$ cm^{-3}	Fe_2O_3		
A	232 cm^2 K/Vs	+1.0 at% Ti		
α	0.10 eV			

For unfired samples, containing no Fe^{2+}, the region of low activation energy is far smaller, and the intrinsic region extends to much lower temperatures (Fig. 6) [63G].

Higher activation energies have been reported for this region: 0.39 eV [65G], 0.38 eV [48B]. Activation energy decreases strongly with Fe^{2+} concentration [63G].

(C) Intrinsic region, $T > 800\,°C$.

activation energies in region (C) (in eV):

$E_A (= E_{g,\,th}/2)$	1.15	all samples are ceramics	65G
	1.0		63G
	1.17		51M
	1.15		71R
	1.06		48B

Resistance varies strongly with applied hydrostatic pressure; there is a collapse to a low-resistivity phase at $\approx 440\cdots 520$ kbar [79K2]. For dependence of resistance on high pressure, see Fig. 13 of 9.15.2.5.1 (Cr_2O_3).

Seebeck coefficient: S also shows three regions (Fig. 7). The sharp rise to positive values above 1000 K suggests $\mu_p > \mu_n$ in the intrinsic region. Values of $290\cdots 330$ µV K^{-1} have also been reported in this region [59L]. See also Fig. 16.

Physical property	Numerical value	Experimental conditions	Experimental method, remarks	Ref.
carrier mobilities:				
μ_n	$(232/T)\exp(-0.1\,[eV]/kT)\,[cm^2/Vs]$	T in K; ceramic samples		63G
	$0.073\,cm^2/Vs$	$T=1273$ K		
μ_p	$(1.73\cdot 10^5)\exp(-0.69\,[eV]/kT)\,[cm^2/Vs]$	T in K; ceramic samples		63G
	$0.253\,cm^2/Vs$	$T=1273$ K		

The much larger value of μ_p indicates that at temperatures higher than 1273 K, the conductivity should increase with increasing oxygen pressure owing to the dominance of intrinsic effects.

intrinsic carrier concentration:

n_i	$3.95\cdot 10^{22}\exp(-1.8\,[eV]/2kT)\,[cm^{-3}]$	T in K		63G

The value of 1.8 eV is close to the reported indirect optical threshold (see below) [63G, 79K1]. The origin of this transition would in principle be $Fe^{3+}+Fe^{3+}\to Fe^{4+}+Fe^{2+}$ or $Fe^{3+}+O^{2-}\to Fe^{2+}+O^-$. Cluster calculations support the first assignment [74T] whereas the optical data discussed below favours the second.

Hall effect: It shows highly anomalous dependence on the field strength and appears to vanish above the Néel temperature [51M].

b) doped Fe_2O_3

resistivity (first column) and activation energy (second column) for doped sinters at RT:

ϱ [Ω cm]:	E_A [eV]:			Ref.
$6.5\cdot 10^5$	0.728	undoped sample		74C
$1.0\cdot 10^0$	0.164	Ti doped,	1 at %	
$5.1\cdot 10^0$	0.191		0.1	
$2.7\cdot 10^3$	0.558		0.01	
$4.0\cdot 10^1$	0.207	Nb doped,	1 at %	
$1.2\cdot 10^2$	0.451		0.1	
$1.0\cdot 10^3$	0.602		0.01	
$2.7\cdot 10^7$	0.814	Cr doped,	1 at %	
$5.9\cdot 10^7$	0.814		0.1	
$20.0\cdot 10^7$	0.814		0.01	
$1.2\cdot 10^6$	0.722	Mn doped,	1 at %	
$6.3\cdot 10^6$	0.722		0.1	
$28.0\cdot 10^6$	0.722		0.01	
$3.9\cdot 10^7$	0.809	V doped,	1 at %	
$8.1\cdot 10^7$	0.809		0.1	
$25.0\cdot 10^7$	0.809		0.01	

Ti doping: The Ti concentration and the carrier concentration determined from Seebeck data assuming the density of states in the valence band $g_v=[Fe]$ and the kinetic term $=0$ are very similar [54M]. Mobility found to be activated [54M] with a temperature dependent activation energy; at high T, $E_A=0.1$ eV. Activation energy for conductivity is also a function of Ti concentration [65G]: Figs. 8, 9. Seebeck effect: Figs. 10; also [67D]. Electron drift mobility for ceramic sample 0.10 (2) cm^2/Vs at 1000 K. Hall coefficient: quite anomalous with $R_H>0$, for $T\gtrsim T_N$. In the region 1000 K $<T<$ 1100 K, $\mu_H\approx 0.025\,cm^2/Vs$ (ceramic sample) [67D].

Zr and Nb doping: Figs. 11, 12. From the exhaustion region, the drift mobility at 1200 K is 0.1 cm^2/Vs for ceramic sample. Comparison of Seebeck and conductivity data (Fig. 13) shows that S and $\log\varrho$ do not track, a result interpreted in terms of impurity conduction [70B]. Hall coefficient data have been interpreted in terms of spin-disorder scattering in a narrow band, with $m_n\simeq 9.5\,m_0$.

Ca doping: conductivity: Fig. 14, Seebeck coefficient: Fig. 10, p-type behaviour.

Mg doping: conductivity: Fig. 15, Seebeck coefficient: Fig. 16, p-type behaviour. In the temperature range 450···900 °C, for Mg doped ceramic samples [66G] containing 0.2 at% Mg: $\mu_p\simeq(262/T)\exp(-0.1\,[eV]/kT)\,[cm^2/Vs]$ (T in K), a result completely at variance with the data for pure material. The Mg content cannot be increased beyond this level owing to spinel formation, $MgFe_2O_4$.

Optical properties

optical spectra: He II and XPE: Fig. 17, absorption near edge: Fig. 18, far UV and visible spectrum: Fig. 19. For data on peak energies in these spectra, see above under "Electronic properties".

He II, XPE: straight forward interpretation not possible; configuration interaction gives rise to a very broad valence band due to Fe 3d ionization [77B].

Resonant Raman two-magnon scattering in Fe_2O_3 shows a strong maximum as the exciting laser line is scanned through 2.2 eV, supporting the $^6A_{1g} \rightarrow {}^4T_{2g}$ assignment for this transition; however, the overall intensity of absorption at 2.2 eV is too strong for a spin-forbidden transition, and a broad indirect absorption maximum with a threshold at 1.88 eV [79K1] has been suggested – this process has been assigned to charge transfer [78M].

dielectric constants (see also Fig. 19):

$\varepsilon(0)$	20.6	RT, $E\|c$	from A_{2u} modes	77O
	24.1	$E \perp c$	from E_u modes	
$\varepsilon(\infty)$	6.7	RT, $E\|c$	from A_{2u} modes	77O
	7.0	$E \perp c$	from E_u modes	

Magnetic properties

Low temperature magnetic transition: $T_M \approx 260$ K (Morin temperature).

$T \lesssim 260$ K: The magnetic arrangement has Fe^{3+} spins directed along the [111] axis and paired across the shared octahedral face.

$T \gtrsim 260$ K: The spins become essentially localized in (111) sheets directed towards the three nearest neighbours. However, the spins have canted slightly out of the plane, giving rise to a weak ferromagnetic moment along the [111] axis (Fig. 20).

spin-wave spectrum: Fig. 21.

exchange parameters (see preceding section) (J_i given as J/k in K):

$J_1 (J_{12})$	6.0	$J_i, J_{ij'}$ are defined in [70S]	70S
	5.86		78K
$J_2 (J_{23'})$	1.6		70S
	0.99		78K
$J_3 (J_{34'})$	−29.7		70S
	−37.3		78K
$J_4 (J_{24'})$	−23.2		70S
	−23.2		78K

Néel temperature (in K):

T_N	947···969		78K
	955	heat capacity anomaly over 3 K	75G
	963	Moessbauer spectrum	77N
	960	Moessbauer spectrum	64I
	950 (10)		65K
	968		69S

The two-magnon spectrum gives a peak at 1380 cm^{-1} [75H].

Magnetic ordering shows a very long short-range-order tail (Fig. 22) [75G].

Long-range order also shown in Fig. 22, see also [66W].

paramagnetic Curie temperature: $\Theta_p = -2940$ K at temperatures well above T_N [51G].

entropy of transition at T_N: 32.4 J K^{-1} mol^{-1} [75G]; close to value expected for randomistion of spin 5/2.

References for 9.15.2.5.2

48B	Bevan, D., Shelton, J., Anderson, J.: J. Chem. Soc. **1948**, 1729.
51G	Guillaud, C.: J. Phys. Radium **12** (1951) 490.
51M	Morin, F.J.: Phys. Rev. **83** (1951) 1005.
52W	Willis, B.T.M., Rooksby, H.P.: Proc. Phys. Soc. **65B** (1952) 950.
54M	Morin, F.J.: Phys. Rev. **93** (1954) 1195.
59L	Lessoff, H., Kersey, Y., Horne, R.A.: J. Chem. Phys. **31** (1959) 1141.
59S	Shirane, G., Pickart, S.J., Nathans, R., Istikawa, Y.: J. Phys. Chem. Solids **10** (1959) 37.
60B	Bailey, P.C.: J. Appl. Phys. **31** (1960) 39S.
61S	Salmon, O.N.: J. Phys. Chem. **65** (1961) 550.
62A	Aharoni, A., Frei, E.H., Schieber, M.: Phys. Rev. **127** (1962) 439.
62N	Newnham, R.E., de Haan, Y.M.: Z. Kristallogr. **117** (1962) 235.
62R	Robbrecht, G.G., Dodo, R.J.: Phys. Lett. **3** (1962) 85.
62T	Tannhauser, D.S.: J. Phys. Chem. Solids **23** (1962) 25.
63G	Gardner, R.F.G., Swett, F., Tanner, D.W.: J. Phys. Chem. Solids **24** (1963) 1183.
63T	Tasaki, A., Iida, S.: J. Phys. Soc. Jpn. **18** (1963) 1148.
64I	Iserentaut, C.M., Robbrecht, G.G., Dodo, P.T.: Phys. Lett. **11** (1964) 14.
64T	Tanner, D.W., Swett, F., Gardner, R.F.G.: Br. J. Appl. Phys. **15** (1964) 1041.
65G	Geiger, G.H., Wagner, J.B.: Trans. AIME **233** (1965) 2092.
65K	Kren, E., Szabo, P., Konzos, G.: Phys. Lett. **19** (1965) 103.
66B	Blake, R.L., Hassevick, R.E., Zoltai, T., Finger, L.W.: Am. Mineral. **51** (1966) 123.
66G	Gardner, R.F.G., Moss, R.L., Tanner, D.W.: Br. J. Appl. Phys. **17** (1966) 55.
66L	Lewis, G.K., Drickamer, H.G.: J. Chem. Phys. **45** (1966) 224.
66W	van der Woude, F.: Phys. Status Solidi **17** (1966) 419.
67D	van Daal, H.J., Bosman, A.J.: Phys. Rev. **158** (1967) 740.
67S	von Steinwehr, H.E.: Z. Kristallogr. **125** (1967) 377.
69P	Prewitt, C.J., Shannon, R.D., Rogers, D.B., Sleight, A.W.: Inorg. Chem. **8** (1969) 1985.
69S	Scharenberg, W.: Ber. Kernforschungsanlage Juelich, Jul-611-RX (1969) 1.
70B	Bosman, A.J., van Daal, H.J.: Adv. Phys. **19** (1970) 1.
70S	Samuelson, E.J., Shirane, G.: Phys. Status Solidi **42** (1970) 241.
70T	Tandon, S.P., Gupta, J.P.: Spectrosc. Lett. **3** (1970) 297.
71D	Drakeford, R.W., Quinn, C.M.: J. Mater. Sci. **6** (1971) 175.
71R	Rao, G.V.S., Wanklyn, B.M., Rao, C.N.R.: J. Phys. Chem. Solids **32** (1971) 345.
72S	Sachse, H.B.: J. Mater. Sci. **7** (1972) 255.
73T	Tossell, J.A., Vaughan, D.J., Johnson, K.H.: Nature (London) **244** (1973) 42.
74B	Blazey, K.W.: J. Appl. Phys. **45** (1974) 2273.
74C	de Logan, D., Lonergan, G.A.: Solid State Commun. **15** (1974) 577.
74T	Tossell, J.A., Vaughan, D.J., Johnson, K.H.: Am. Mineral. **59** (1974) 319.
75B	Bakulin, E.A., Bredow, M.M., Shcherlanina, V.V.: Fiz. Tverd Tela **17** (1975) 2653.
75D	Ditchfield, R.W.: Solid State Commun. **17** (1975) 1367.
75G	Grønvold, F., Samuelsen, E.J.: J. Phys. Chem. Solids **36** (1975) 249.
75H	Hart, T.R., Adams, S.B., Temkin, H.: Proc. Int. Conf. on "Light Scattering in Solids", ed. **1975**, 259.
76T	Tossell, J.A.: J. Electron Spectrosc. Relat. Phenom. **8** (1976) 1.
77B	Brundle, C.R., Chuang, T.J., Wandelt, K.: Surf. Sci. **68** (1977) 459.
77N	Neskovic, N.B., Babic, R., Konstantinovic, J.: Phys. Status Solidi (a) **41** (1977) K133.
77O	Onari, S., Arai, T., Kudo, K.: Phys. Rev. **B16** (1977) 1717.
78B	Balberg, I., Pinch, H.L.: J. Mag. Magn. Mater. **7** (1978) 12.
78K	Kowalska, A., Stoniowska, B.: Acta Physiol. Pol. **A54** (1978) 679.
78M	Merlin, R., Martin, T.P., Polian, A., Cardona, M., Audlemer, B., Tannhauser, D.: J. Mag. Magn. Mater. **9** (1978) 83.
79G	Galuza, A.I., Eremenko, V.V., Kirichenko, A.P.: Fiz. Tverd Tela **21** (1979) 1125.
79K1	Koffyberg, F.P., Dwight, K., Wold, A.: Solid State Commun. **30** (1979) 433.
79K2	Kondo, K., Sunakawa, T., Sawaoke, A.: Jpn J. Appl. Phys. **18** (1979) 851.
79M	Merchant, P., Collins, R., Kershaw, R., Dwight, K., Wold, A.: J. Solid State Chem. **27** (1979) 307.
79V	Vasudevan, S., Hedge, M.S., Rao, C.N.R.: J. Solid State Chem. **29** (1979) 253.
80F	Finger, L.W., Hazen, R.M.: J. Appl. Phys. **51** (1980) 5362.
83H	Handbook of Chemistry and Physics, Weast, R.C., (ed.), 64th edition **1983**, CRC Press, Inc.

Physical property	Numerical value	Experimental conditions	Experimental method, remarks	Ref.

9.15.2.5.3 Oxides of rhodium

Phase diagram: Fig. 1. At low temperatures the stable phase is metallic RhO_2 which has rutile structure ($a=4.489$ Å, $c=3.090$ Å, $Z=2$ [68M]). At higher temperatures the sesquioxide Rh_2O_3 is found and at yet higher temperatures decomposition to Rh and O_2 occurs.

Rh_2O_3

Crystal structure

Rh_2O_3 (I):

Low-temperature form, stable for $T<900\,°C$ [63W]. Irreversibly transformed to low-pressure high-temperature form $750\cdots1000\,°C$ [81P].

structure: corundum, space group $D_{3d}^6-R\bar{3}c$, $Z=6$. Structure shown in Fig. 2a.

lattice parameters at RT (distances in Å, angles in °):

a	5.108		63W
c	13.81		
a	5.127	for the connection between	70C
c	13.853	trigonal and hexagonal settings, see Cr_2O_3	
a_{rh}	5.485	lattice parameters in a	70C
α_{rh}	55.73	rhombohedral setting	
$d(Rh-O_1)$	2.03	each Rh has three O_1 and three	
$d(Rh-O_2)$	2.07	O_2. In the hexagonal setting [70C], the Rh occupy the 12 (c) sites ($z=0.348$) and the O the 18 (f) sites ($x=0.295$)	
$d(Rh-Rh)$	2.72	along c-axis	
	3.03	basal plane pairs	

density:

d_{calc}	8.02 g cm^{-3}	63W

temperature dependence of a_{hex} and c_{hex} and coefficient of linear thermal expansion:

a_{hex}	$5.1266+2.7427\cdot10^{-5}\,T+3.3579\cdot10^{-9}\,T^2$ [Å]	T in K	73E
c_{hex}	$13.8512+7.2677\cdot10^{-5}\,T+4.425\cdot10^{-8}\,T^2$ [Å]		
α_a	$5.350\cdot10^{-6}+1.281\cdot10^{-9}\,T-1.133\cdot10^{-14}\,T^2$ [K^{-1}]		
α_c	$5.246\cdot10^{-6}+6.369\cdot10^{-9}\,T-7.480\cdot10^{-14}\,T^2$ [K^{-1}]		

Rh_2O_3 (II):

High-temperature, high-pressure form, synthezized at 65 kbar, 1200 °C [70S]. This form can also be obtained as black crystals by chemical vapour transport [81P].

structure: orthorhombic, space group $D_{2h}^{14}-Pbna$, $Z=4$. Structure is given in Fig. 2b.

lattice parameters (in Å):

a	5.1686 (3)	RT		70S
	5.159 (1)	(meta-		81P
b	5.3814 (4)	stable)		70S
	5.381 (1)			81P
c	7.2426 (4)			70S
	7.241 (1)			81P

Physical property	Numerical value	Experimental conditions	Experimental method, remarks	Ref.
atomic parameters and interatomic distances (in Å):				
Rh: x	0.7498 (1)		Rh atoms in 8 (d) sites,	70S
y	0.0312 (2)		O(1) in 8 (d) sites and	
z	0.1058 (1)		O(2) in 4 (c) sites	
O(1) x	0.6037 (11)			
y	0.1161 (10)			
z	0.8494 (8)			
O(2) x	0.0505 (16)			
y	1/4			
z	0			
$d(Rh-O(1,1))$	2.055	$00\bar{1}$	Second number in the brackets	
$d(Rh-O(1,3))$	2.076	$0\bar{1}0$	after O indicates the symmetry	
$d(Rh-O(1,4))$	2.068	000	operation applied to the oxygen	
$d(Rh-O(1,5))$	2.017	$0\bar{1}0$	atom referred to by the first	
$d(Rh-O(2,1))$	2.094	100	number; transformations are:	
$d(Rh-O(2,5))$	1.985	$0\bar{1}0$	$xyz; \frac{1}{2}-x, \frac{1}{2}+y, \frac{1}{2}-z; \frac{1}{2}-x, \bar{y},$	
$d(Rh-O)$(av.)	2.050		$\frac{1}{2}-z; x, \frac{1}{2}-y, \bar{z}; \bar{x}\bar{y}\bar{z}; -\frac{1}{2}+x,$	
$d(Rh-O(1,5))$	3.449	$1\bar{1}0$	$-\frac{1}{2}-y, -\frac{1}{2}+z; -\frac{1}{2}+x, y,$	
$d(Rh-O(1,6))$	3.171	100	$-\frac{1}{2}-z; \bar{x}, -\frac{1}{2}+y, z$ in that	
$d(Rh-O(1,8))$	3.431	$0\bar{1}\bar{1}$	numerical order. The number	
$d(Rh-O(2,5))$	3.255	$0\bar{1}0$	pqr in the third column is	
			the translation applied to	
			the transformed coordinates	
density:				
d_{calc}	8.36 g cm^{-3}			70S

Rh$_2$O$_3$ (III):

High-temperature, low-pressure form (1 atm), stable for $T > 900$ °C [63W, 73B].

structure: orthorhombic, space group D_{2h}^{15}–Pbca, $Z = 8$, see Fig. 2c.

Physical property	Numerical value	Experimental conditions	Experimental method, remarks	Ref.
lattice parameters (in Å):				
a	5.149		all atoms in 8 (c) sets	63W
b	5.436		(data at RT on metastable	
c	14.688		samples)	
a	5.1477			73B
b	5.4425			
c	14.6977			
some short interatomic distances (in Å):				
$d(Rh(1)-Rh(2))$	2.69		figures in brackets refer to	73B
$d(Rh(1)-O(3))$	1.82		the atoms numbered in Fig. 2c;	
$d(Rh(1)-O(4))$	2.24		primes refer to corresponding	
$d(Rh(1)-O(5))$	1.99		atoms on neighbouring units	
$d(Rh(2)-O(3))$	1.82		in the figure.	
$d(Rh(2)-O(4))$	1.88		Positions of atoms in 8 (c)	
$d(Rh(2)-O(5))$	2.08		sets of Pbca:	
$d(O(3)-O(4))$	2.51		Rh(1) (0.000 (5), 0.220 (5), 0.081 (2))	
$d(O(3)-O(5))$	2.35		Rh(2) (0.000 (5), −0.190 (5), 0.183 (2))	
$d(O(4)-O(5))$	2.62		O(3) (0.13 (2), 0.12 (2), 0.190 (5))	
			O(4) (−0.30 (2), −0.04 (2), 0.135 (5))	
			O(5) (0.15 (2), −0.11 (2), 0.055 (5))	

(continued)

Physical property	Numerical value	Experimental conditions	Experimental method, remarks	Ref.

some short interatomic distances (continued)

$d(Rh(1)-O(4'))$	1.84			73B
$d(Rh(1)-O(5'))$	2.22			
$d(Rh(1)-O(5''))$	2.06			
$d(Rh(2)-O(3'))$	2.24			
$d(Rh(2)-O(3''))$	2.17			
$d(Rh(2)-O(4'))$	2.28			
$d(O(3)-O(4'))$	2.64			
$d(O(3)-O(4''))$	2.74			
$d(O(3)-O(5'))$	2.72			
$d(O(4)-O(4'))$	2.77			
$d(O(5)-O(5'))$	2.54			

density:

d_{calc}	8.19 g cm^{-3}			63W

The three structures are closely related; Rh$_2$O$_3$ (II) can be thought of as formed from layers of corundum cut parallel to $(10\bar{1}1)$ whereas Rh$_2$O$_3$ (III) can similarly be thought of as stacked layers of corundum cut parallel to $(0\bar{1}12)$ [70S].

Physical properties

UPE spectra: Figs. 3 and 4. The Rh 4d-band appears to be ca. 2 eV wide.

Conductivity data have been reported for Rh$_2$O$_3$ (II). This is p-type, presumably by virtue of Rh vacancies: $\varrho = \varrho_0 \exp(E_A/kT)$ [Ω cm] with ϱ (300 K) = 130 Ω cm and $E_A = 0.16$ eV [70S].

References for 9.15.2.5.3

63W	Wold, A., Arnott, R.J., Croft, W.J.: Inorg. Chem. **2** (1963) 972.
68M	Muller, O., Roy, R.: J. Less-Commun. Metals **16** (1968) 129.
70C	Coey, J.M.D.: Acta Crystallogr. B **26** (1970) 1876.
70S	Shannon, R.D., Prewitt, C.J.: J. Solid State Chem. **2** (1970) 134.
73B	Biesterbos, J.W.M., Hornstra, J.: J. Less-Commun. Metals **30** (1973) 121.
73E	Eckert, L.J., Bradt, R.C.: Mater. Res. Bull. **8** (1973) 375.
78B	Beatham, N.: D. Phil. Thesis, Oxford **1978**.
81P	Poeppelmeier, K.R., Ansell, G.B.: J. Cryst. Growth **51** (1981) 587.

9.15.2.5.4 Oxides of manganese

Phase diagram: see Fig. 1.

In the following we present data on Mn$_2$O$_3$ and MnO$_2$. The compound Mn$_3$O$_4$ has already dealt with in section 9.15.2.4.3.

Mn$_2$O$_3$

Crystal structure

a) low-temperature phase ($T \leq 302$ K):

structure: orthorhombically distorted bixbyite structure, space group D_{2h}^{15}–Pcab, $Z=16$ [71G, 67N]. Bixbyite structure: Fig. 2. Mn$_2$O$_3$ is the only sesquioxide having this structure, the origin is thought to lie in the expected substantial Jahn-Teller distortion associated with the d^4 Mn^{3+} ion.

lattice parameters (in Å):

a	9.4137	$T = 298$ K	temperature dependence, see Fig. 3	71G
b	9.4233			
c	9.4047			
a	9.4118 (8)	298 K		67N
b	9.4177 (7)			
c	9.4233 (7)			

In the undistorted bixbyite structure there are two distinguishable Mn sites at the 8(a) and 24(d) positions. In the setting shown in Fig. 2, the 8(a) sites are at $\frac{1}{4}\frac{1}{4}\frac{1}{4}; \frac{1}{4}\frac{3}{4}\frac{3}{4}; \frac{3}{4}\frac{1}{4}\frac{3}{4}; \frac{3}{4}\frac{3}{4}\frac{1}{4}$+b.c. and the 24(d) sites are at $\pm(u0\frac{1}{4}; \frac{1}{4}u0; 0\frac{1}{4}u; \bar{u}\frac{1}{2}\frac{1}{4}; \frac{1}{4}\bar{u}\frac{1}{2}; \frac{1}{2}\frac{1}{4}\bar{u})$+b.c. The oxygen atoms occupy the 48(e) sites described by the parameters $\pm(xyz; x\bar{y}\frac{1}{2}-z; \frac{1}{2}-xy\bar{z}; \bar{x}\frac{1}{2}-yz; zxy; \frac{1}{2}-zx\bar{y}; \bar{z}\frac{1}{2}-xy; z\bar{x}\frac{1}{2}-y; yzx; \bar{y}\frac{1}{2}-zx; y\bar{z}\frac{1}{2}-x; \frac{1}{2}-yz\bar{x})$+b.c. In the mineral bixbyite itself, $(Mn, Fe)_2O_3$, $u = -0.035$, $x = 0.384$, $y = 0.147$ and $z = 0.382$ [71G], for the composition $(Mn_{0.983}Fe_{0.017})_2O_3$. The high-temperature form of Mn_2O_3 itself is discussed below. For the low-temperature form, there are *five* distinguishable Mn sites, two derived from the 8(a) sites and three from the 24(d) sites. The sets derived from 8(a) are the 4(a) at $(\frac{1}{4}, \frac{1}{4}, \frac{1}{4})\cdots$ and the 4(b) which arise by loss of b.c. and are at $(\frac{1}{4}, \frac{1}{4}, \frac{3}{4})\cdots$. The 24(d) sites split into three sets of 8(c).

average Mn−O distances in the low-temperature form (in Å):

$d(Mn(1)-O)$	2.003 (87)	numbers of the Mn atoms as in Fig. 2	71G
$d(Mn(2)-O)$	2.002 (87)		
$d(Mn(3)-O)$	2.043 (408)		
$d(Mn(4)-O)$	2.043 (380)		
$d(Mn(5)-O)$	2.042 (384)		

b) high-temperature phase ($T \gtrsim 302$ K):

structure: undistorted cubic bixbyite structure, space group $T_h^7 - Ia3$. For this structure, at 302 K, $u = -0.035$, $x = 0.379$, $y = 0.147$, $z = 0.377$. The 8(a) sites are *regular* trigonal antiprisms and the 24(d) sites have *three* sets of two oxide ions.

Mn−O distances (in Å):

$d(Mn-O)$	2.003	Mn in 8(a) sites	71G
$d(Mn-O)$	1.987	Mn in 24(d) sites	
	1.898		
	2.242		

lattice parameter (in Å):

		T [K]	
a	9.414	314	68G
	9.418	375	
	9.426	482	
	9.443	702	

Doped Mn_2O_3

lattice parameter for the cubic phase in $(Mn_{1-x}Fe_x)_2O_3$ (in Å):

a	9.4146 (1)	$x = 0.017$	$T = 23$°C	71G
	9.4156 (2)	0.097		
	9.4158 (2)	0.49		
	9.4126 (3)	0.63		

interionic distances in $(Mn_{1-x}Fe_x)_2O_3$ (in Å) (RT values):

(Mn, Fe) in 8(a) sites:

$\bar{d}(Mn-O)$	2.003	$x = 0.017$		71G
	2.009	0.63		

(Mn, Fe) in 24(d) sites:

$\bar{d}(Mn-O)$	2.242	$x = 0.017$	$u = -0.0351$	
	1.898			
	1.987			
	2.147	0.63	$u = -0.0347$	
	1.930			
	2.033			

Substitution of Sc^{3+}, Ga^{3+}, and Cr^{3+} in the cubic phase, see Fig. 4 (T_{tr} and a).

Physical property	Numerical value	Experimental conditions	Experimental method, remarks	Ref.

Transport properties

Mn_2O_3 is a cation deficient p-type semiconductor. For data on samples derived from Mn_3O_4, see section 9.15.2.4.3.

conductivity: Fig. 5.

activation energy for conductivity (cubic phase):

E_A	0.54 eV	$T = 500 \cdots 800$ K	sample annealed at 624 °C; in the range $300 \cdots 500$ K this sample had an activation energy of 0.30 eV	70K
	0.64 eV	$700 \cdots 1200$ K	samples metastable with respect to Mn_3O_4 at higher temperatures	75L

Seebeck coefficient:

S	$+0.37$ mV K^{-1}	$T = 300 \cdots 400$ K	cubic phase	70K

Magnetic properties

Susceptibility of pure and doped samples: Figs. $6 \cdots 9$.

Two magnetic transitions are found; one, at $79 \cdots 80$ K (T_{N_1}) corresponds to incipient antiferromagnetic ordering and the other, at 25 K (T_{N_2}), is accompanied by a first-order change in χ by a factor of two. For Fe doped material T_{N_2} decreases and vanishes for $x \gtrsim 0.028$ (Fig. 10) [68G].

Hypothesized spin structure for the cubic bixbyite phase: Fig. 11.

MnO_2

Crystal structure

structure: rutile (β-MnO_2) (Fig. 12), tetragonal, space group D_{4h}^{14}–$P4_2/mnm$; $Z = 2$. In the rutile structure the metal ions occupy 2 (a) points at (0, 0, 0); ($\frac{1}{2}, \frac{1}{2}, \frac{1}{2}$) and the oxygen ions are at 4 (f) points $\pm(x, x, 0)$; ($x + \frac{1}{2}, \frac{1}{2} - x, \frac{1}{2}$), $x = 0.30293$ (6) for MnO_2 [76B].

lattice parameters at RT (distances in Å, angles in °):

a	4.3983		see also Fig. 12	76B
c	2.8730			
a	4.3980			69R
c	2.8738			
a	4.3977			73C
c	2.8778			
V	55.59 Å^3			76B
c/a	0.65321			
$d(Mn-O)_1$	1.8795		equatorial ⎱ mean value 1.8868 Å	76B
$d(Mn-O)_2$	1.8981		axial ⎰	
$d(Mn-Mn)$	2.86		closest distance	69R
$\angle(O-Mn-O)$	80.31		angle on a shared edge	76B
d log a/dT	$6.69 \cdot 10^{-6}$ K^{-1}			62B,
d log c/dT	$6.92 \cdot 10^{-6}$ K^{-1}			69K, 70H

density:

d_{calc}	5.026 g cm^{-3}			76B

heat capacity:

anomaly at 92.12 K [43K], 91.9 K [71O] associated with magnetic ordering.

thermodynamic data:

ΔH_f°	-124.64 kcal mol^{-1}			43K
ΔS_f°	-43.9 cal K^{-1} mol^{-1}			
ΔG_f°	-110.54 kcal mol^{-1}			

Physical property	Numerical value	Experimental conditions	Experimental method, remarks	Ref.
For the equilibrium $4\,MnO_2 \rightleftharpoons 2\,Mn_2O_3 + O_2$: $-4.5756 \log p_{O_2} = 37870/T - 51.28$		($T = 441 \cdots 642\,°C$)	p in atm	65O

β-MnO_2 usually made by controlled pyrolysis of $Mn(NO_3)_2$ [64W].

Transport properties

Single crystal conductivity data: Fig. 13. The marked anomaly at 92···95 K is associated with magnetic ordering (see also above under heat capacity), with the major loss mode in the 50···100 K range being magnon scattering.

Resistance vs. pressure: Fig. 14. Studies of powdered samples are bedevilled by the formation of Mn_2O_3 at grain boundaries [70K]. At $-50 \cdots +125\,°C$, activation energy ≈ 0.04 eV found [70K].

Seebeck effect: for low resistivity samples $S \simeq -0.60$ mV/K [70K] for polycrystalline samples.

Hall effect: $\mu_H \simeq 1 \cdots 10$ cm^2/Vs [58D, 70K] for polycrystalline samples.

Magnetic properties

MnO_2 is antiferromagnetic below T_N (92K [59Y], 92···95 K [69 R]) with a spin structure of the proper screw type derived from neutron diffraction studies [59Y]. The spin structure is such that the spins all lie in the (001) plane, but screw along the c axis with a pitch of $\frac{7}{2}c$ (Fig. 15). The exchange integrals, defined in Fig. 12, have the values

exchange integrals (J_i/k in K):

J_1	-8.9		71O
J_2	-5.5		
J_3	1.3		

Powder susceptibility: Fig. 16. The samples all showed Curie-Weiss behaviour above T_N, but Θ_p values are very scattered, 850···1300 K [71O].

References for 9.15.2.5.4

43K	Kelley, K.K., Moore, G.E.: J. Am. Ceram. Soc. **65** (1943) 782.
58D	Das, J.N.: Z. Phys. **151** (1958) 345.
59Y	Yoshinori, A.: J. Phys. Soc. Jpn. **14** (1959) 807.
62B	Bradt, R.C., Wiley, J.S.: J. Electrochem. Soc. **109** (1962) 651.
63M	Minomura, S., Drickamer, H.G.: J. Appl. Phys. **34** (1963) 3043.
64W	Wiley, J.S., Knight, H.T.: J. Electrochem. Soc. **111** (1964) 656.
65O	Otto, E.M.: J. Electrochem. Soc. **112** (1965) 367.
67N	Norrestam, R.: Acta Chem. Scand. **21** (1967) 2871.
67S	Shenouda, F., Aziz, S.: J. Appl. Chem. **17** (1967) 258.
68G	Grant, R.W., Geller, S., Cape, J.A., Espinosa, G.P.: Phys. Rev. **175** (1968) 686.
69K	Kirchner, H.P.: J. Am. Ceram. Soc. **52** (1969) 379.
69R	Rogers, D.B., Shannon, R.D., Sleight, A.W., Gillson, J.L.: Inorg. Chem. **8** (1969) 841.
70G	Geller, S., Espinosa, G.P.: Phys. Rev. **B1** (1970) 3763.
70H	Hazony, Y., Parkins, H.K.: J. Appl. Phys. **41** (1970) 5130.
70K	Klose, P.H.: J. Electrochem. Soc. **117** (1970) 854.
71G	Geller, S.: Acta Crystallogr. **27** (1971) 821.
71O	Ohama, N., Hamaguchi, Y.: J. Phys. Soc. Jpn. **30** (1971) 1311.
73C	Chamberland, B.L., Cloud, W.H., Frederick, C.G.: J. Solid State Chem. **8** (1973) 238.
75L	Logothetis, E.M., Park, K.: Solid State Commun. **16** (1975) 909.
75W	Wells, A.F.: Structural Inorganic Chemistry, OUP **1975**.
76B	Baur, W.H.: Acta Crystallogr. **B32** (1976) 2200.

9.15.2.6 Oxides of group V elements (Nb, Ta)

9.15.2.6.1 Oxides of niobium

Phase diagram: Fig. 1, detailed diagram for the NbO$_x$ system ($2.42 \leq x \leq 2.50$): Fig. 2. In addition to the phases Nb$_2$O$_5$ and NbO$_2$, oxides of composition Nb$_{32}$O$_{79}$, Nb$_{12}$O$_{29}$, Nb$_{25}$O$_{62}$ and Nb$_{22}$O$_{54}$ have been reported, as well as a number of intergrowth phases. We present in detail the two phases Nb$_2$O$_5$ and NbO$_2$. The structural principles of the other oxides mentioned above are discussed in the Nb$_2$O$_5$ section.

Nb$_2$O$_5$

Crystal structure

Structurally the chemistry of Nb$_2$O$_5$ is more complex than any other binary transition metal oxide.

reported modifications of the Nb$_2$O$_5$ structure:

(first column: crystal form according to [66S], second column: alternative notations and references)

TT	"pseudohexagonal" [55F], monoclinic [72T], δ or γ'-form [57H, 63T]
T	γ-form [57H, 63T], T-form [41B]
B	ζ-form [57H, 63T]
M	M-form [41B], β or α'-form [57H, 63T]
H	H-form [41B], α-form [57H, 63T]
N	–
P	η-form [64L2]
ε	ε-form [59R]
I-high	–
II	–
OXI···OXVI	–
–	R-form [66G]

All save the TT, ε, I, II and OXI to OXVI forms have been studied crystallographically. With the notable exception of the B- and T-forms, the essential structural principle is the sheared ReO$_3$ or NbO$_2$F structure [70W]; where idealized projections are shown, as in Figs. 4, 6, 9 the pseudo-cubic axes of the parent structures are used.

R−Nb$_2$O$_5$

structure: monoclinic, space group $C_{2h}^3 - C2/m$, $Z = 2$.

The structure is essentially identical to that of idealized V$_2$O$_5$ [66G] (Fig. 3). The unit cell contains 4 Nb and 4 O at the 4 (i) positions $x, 0, z$; $\bar{x}, 0, \bar{z}$; $x, \frac{1}{2}, \frac{1}{2}+z$; $\bar{x}, \frac{1}{2}, \frac{1}{2}-z$ where, for Nb, $x = 0.07$ (1), $z = 0.146$ (2), for O, $x = 0.00$ (4), $z = 0.68$ (1). There are two further O at the 2(a) positions 0, 0, 0; 0, $\frac{1}{2}, \frac{1}{2}$. The R-form is metastable and converts to the P-form on slow heating and to the N-form on rapid heating [75P].

lattice parameters (distances in Å, angle in °):

a	12.79	RT	66G, 70W
b	3.826		
c	3.983		
β	90.75		

P−Nb$_2$O$_5$

structure: tetragonal, space group $D_4^{10} - I4_122$, $Z = 4$.

Structure related to anatase (Figs. 4, 5). It can be obtained in crystalline form using Cl$_2$/H$_2$O as transporting gas, with $T_1 = 600$ °C and $T_2 = 800$ °C.

lattice parameters (in Å):

a	3.876	RT	65P
c	25.43		

Physical property	Numerical value	Experimental conditions	Experimental method, remarks	Ref.

$M-Nb_2O_5$

structure: tetragonal, space group $D_{4h}^{17}-I4/mmm$; $Z=16$. The idealized M-form is given in Fig. 6 and a projection of the unit cell in Fig. 7. $M-Nb_2O_5$ can be grown by chemical vapour transport provided a SnO_2 catalyst is present [68E].

lattice parameters (in Å):

a	20.01	RT		70W
c	3.84			
a	20.44			70M
c	3.832			

density:

d	4.4 g cm^{-3}		X-ray density	66S
	4.3···4.4 g cm^{-3}		pycnometric density	66S

atomic position parameters [70M]:

Atom	Position	x	y	z
Nb(1)	8 (i)	0.1294 (6)	0	0
Nb(2)	8 (i)	0.4047 (9)	0	0
Nb(3)	16 (l)	0.2738 (5)	0.1344 (5)	0
O(1)	8 (j)	0.3479 (74)	$\frac{1}{2}$	0
O(2)	8 (j)	0.0731 (64)	$\frac{1}{2}$	0
O(3)	16 (l)	0.2095 (55)	0.3402 (55)	0
O(4)	8 (h)	0.0678 (42)	$=x$	0
O(5)	8 (h)	0.2116 (56)	$=x$	0
O(6)	16 (l)	0.2095 (34)	0.0727 (33)	0
O(7)	16 (l)	0.3555 (45)	0.0630 (44)	0

Nb–O distances (in Å):

$d(Nb(1)-O(1))(2\times)$	1.971	RT		70M
$d(Nb(1)-O(4))(2\times)$	1.874			
$d(Nb(1)-O(6))(2\times)$	2.211			
$d(Nb(2)-O(2))(2\times)$	2.455			
$d(Nb(2)-O(2))(2\times)$	1.969			
$d(Nb(2)-O(7))(2\times)$	1.636			
$d(Nb(3)-O(3))$	2.048			
$d(Nb(3)-O(3))(2\times)$	2.014			
$d(Nb(3)-O(5))$	2.026			
$d(Nb(3)-O(6))$	1.823			
$d(Nb(3)-O(7))$	2.219			

$N-Nb_2O_5$

structure: monoclinic, space group C_{2h}^3-C2/m. Closely related to $M-Nb_2O_5$.

Unit cell: Fig. 8. $N-Nb_2O_5$ is made by chemical vapour transport and favourable growth conditions are discussed in [66S].

lattice parameters (distances in Å, angle in °):

a	28.51	RT		67A
b	3.830			
c	17.48			
β	120.80			

atomic position parameters [67A]:

Atom	Position	x	y	z
Nb(1)	4(i)	0.0615(3)	0	0.0990(5)
Nb(2)	4(i)	0.1577(3)	1/2	0.0995(5)
Nb(3)	4(i)	0.2903(3)	1/2	0.0958(5)
Nb(4)	4(i)	0.4265(3)	1/2	0.0968(5)
Nb(5)	4(i)	0.0194(3)	0	0.3716(5)
Nb(6)	4(i)	0.1553(3)	0	0.3694(5)
Nb(7)	4(i)	0.2511(3)	1/2	0.3653(5)
Nb(8)	4(i)	0.3885(3)	1/2	0.3670(5)
O(1)	4(i)	0.070	1/2	0.085
O(2)	4(i)	0.129	0	0.075
O(3)	4(i)	0.273	0	0.055
O(4)	4(i)	0.409	0	0.050
O(5)	4(i)	0.021	1/2	0.353
O(6)	4(i)	0.175	1/2	0.366
O(7)	4(i)	0.234	0	0.361
O(8)	4(i)	0.376	0	0.362
O(9)	4(i)	0.216	1/2	0.085
O(10)	4(i)	0.348	1/2	0.069
O(11)	4(i)	0.480	1/2	0.053
O(12)	4(i)	0.103	0	0.223
O(13)	4(i)	0.199	1/2	0.221
O(14)	4(i)	0.335	1/2	0.221
O(15)	4(i)	0.484	1/2	0.229
O(16)	4(i)	0.085	0	0.352
O(17)	4(i)	0.300	1/2	0.336
O(18)	4(i)	0.460	1/2	0.379
O(19)	4(i)	0.064	0	0.492
O(20)	4(i)	0.210	0	0.497

interatomic distances (in Å) [67A] (RT values):

Distance d		Mean value	Distance d		Mean value
2 Nb(1)–O(1)	1.96		2 Nb(5)–O(5)	1.95	
Nb(1)–O(2)	2.19		Nb(5)–O(15)	2.09	
Nb(1)–O(11)	1.98	2.01	Nb(5)–O(16)	2.08	1.93
Nb(1)–O(12)	2.21		Nb(5)–O(18)	1.77	
Nb(1)–O(12)	1.78		Nb(5)–O(19)	1.73	
2 Nb(2)–O(2)	2.03		2 Nb(6)–O(6)	2.01	
Nb(2)–O(1)	2.36		Nb(6)–O(7)	2.33	
Nb(2)–O(4)	2.18	2.03	Nb(6)–O(12)	2.10	2.03
Nb(2)–O(9)	1.82		Nb(6)–O(16)	1.85	
Nb(2)–O(13)	1.74		Nb(6)–O(20)	1.85	
2 Nb(3)–O(3)	2.00		2 Nb(7)–O(7)	1.97	
Nb(3)–O(3)	2.18		Nb(7)–O(6)	2.18	
Nb(3)–O(9)	2.02	1.99	Nb(7)–O(13)	2.07	1.99
Nb(3)–O(10)	1.95		Nb(7)–O(17)	1.74	
Nb(3)–O(14)	1.81		Nb(7)–O(20)	1.99	
2 Nb(4)–O(4)	2.03		2 Nb(8)–O(8)	1.94	
Nb(4)–O(2)	2.47		Nb(8)–O(14)	2.08	
Nb(4)–O(10)	2.00	2.08	Nb(8)–O(17)	2.26	2.03
Nb(4)–O(11)	2.06		Nb(8)–O(18)	1.93	
Nb(4)–O(15)	1.92		Nb(8)–O(19)	2.02	

Physical property	Numerical value	Experimental conditions	Experimental method, remarks	Ref.

H−Nb$_2$O$_5$

structure: monoclinic, space group C_2^1−P2 [64G], C_{2h}^1−P2/m [76K1], $Z=14$.

This is the stable form at high temperatures ($T>1100\,°C$). The idealized structure is shown in Fig. 9. The major difference from the preceding structure is the presence of tetrahedrally coordinated Nb. These tetrahedral sites lie in tunnels and $\frac{1}{2}$ of the available sites are occupied in H−Nb$_2$O$_5$. Detailed unit cell projection: Fig. 10.

lattice parameters (distances in Å, angles in °):

a	21.16	RT		64G
b	3.822			
c	19.35			
β	119.8			
a	21.153			76K1
b	3.8233			
c	19.356			
β	119.80			
a	21.1848 · $(1+0.56328 \cdot 10^{-5}\,T)$	$T=273\cdots 1273$ K	from thermal expansion (T in K)	72M, 73D
b	3.8236 (not dependent on T)			
c	19.3680 · $(1+0.58856 \cdot 10^{-5}\,T)$			
β	119.63 · $(1+0.26456 \cdot 10^{-5}\,T)$			
V	1363.4 · $(1+0.880152 \cdot 10^{-5}\,T)$ (Å^3)			

density:

$d_{calc}=4.55$ g cm^{-3}				64G

interatomic distances (in Å) [76K1]:

Nb atom	Surrounding oxygens and distances to them				Mean Nb−O distance
Nb(1)	O(19^{iv})	1.838(5)	O(19^{ii})	1.838(5)	1.826
	O(21)	1.813(4)	O(21^{viii})	1.813(4)	
Nb(2)	O(2)	1.9116(2)	O(2^{ii})	1.9116(2)	1.968
	O(9)	2.005(5)	O(9^v)	2.005(5)	
	O(11)	1.986(5)	O(11^v)	1.986(5)	
Nb(3)	O(5)	1.775(4)	O(7)	1.973(5)	2.006
	O(18^{iv})	2.327(3)	O(19^{iv})	1.989(6)	
	O(34^{iii})	1.986(1)	O(34^{iv})	1.986(1)	
Nb(4)	O(3)	2.267(4)	O(5)	2.112(4)	2.006
	O(6)	1.872(4)	O(8)	1.801(5)	
	O(24)	1.991(1)	O(24^{ii})	1.991(1)	
Nb(5)	O(4)	2.269(4)	O(6)	1.954(4)	1.999
	O(9)	1.807(5)	O(15^v)	2.001(3)	
	O(25)	1.982(1)	O(25^{ii})	1.982(1)	
Nb(6)	O(7)	1.866(5)	O(10)	1.753(4)	2.051
	O(12)	2.194(4)	O(16^{iv})	2.479(3)	
	O(26)	2.007(1)	O(26^{ii})	2.007(1)	
Nb(7)	O(8)	2.023(5)	O(10)	2.226(4)	1.987
	O(11)	1.824(5)	O(13)	1.991(5)	
	O(27)	1.930(1)	O(27^{ii})	1.930(1)	

(continued)

Physical property	Numerical value		Experimental conditions	Experimental method, remarks	Ref.
interatomic distances (continued)					
Nb(8)	O(12)	1.865 (4)	O(12^{iv})	2.340 (3)	2.034
	O(14)	1.769 (4)	O(16)	2.258 (4)	
	O(28)	1.986 (1)	O(28^{ii})	1.986 (1)	
Nb(9)	O(13)	1.808 (5)	O(14)	2.154 (4)	2.003
	O(15)	1.830 (4)	O(17)	2.243 (4)	
	O(29)	1.992 (1)	O(29^{ii})	1.992 (1)	
Nb(10)	O(3)	1.983 (1)	O(3^{i})	1.983 (1)	1.992
	O(21)	2.034 (5)	O(23)	1.868 (4)	
	O(24)	2.242 (4)	O(36)	1.843 (5)	
Nb(11)	O(4)	1.982 (1)	O(4^{i})	1.982 (1)	1.997
	O(22)	1.801 (5)	O(23)	1.949 (4)	
	O(25)	2.269 (4)	O(31^{v})	2.000 (5)	
Nb(12)	O(16)	2.013 (1)	O(16^{i})	2.013 (1)	2.018
	O(26^{iv})	2.276 (4)	O(28)	2.230 (4)	
	O(30)	1.774 (4)	O(32)	1.801 (4)	
Nb(13)	O(17)	1.987 (1)	O(17^{i})	1.987 (1)	1.999
	O(29)	2.225 (4)	O(30)	2.130 (4)	
	O(31)	1.836 (4)	O(33)	1.831 (5)	
Nb(14)	O(18)	1.974 (1)	O(18^{i})	1.974 (1)	2.000
	O(32)	2.034 (4)	O(34)	2.281 (4)	
	O(35)	1.769 (4)	O(36^{vi})	1.966 (5)	
Nb(15)	O(1^{v})	1.887 (1)	O(20)	1.920 (1)	1.972
	O(20^{i})	1.920 (1)	O(22^{vi})	2.027 (5)	
	O(33)	1.974 (5)	O(35)	2.103 (4)	

Symmetry code

(i)	x,	$1+y$,	z;	(ii)	x,	$-1+y$,	z;
(iii)	$1-x$,	$-1+y$,	$-z$;	(iv)	$1-x$,	y,	$-z$;
(v)	$1-x$,	y,	$1-z$;	(vi)	$1+x$,	y,	z;
(vii)	$-1+x$,	y,	z;	(viii)	$-x$,	y,	$-z$.

The interconversion of B, T and H-forms has been investigated under pressure in the temperature range RT⋯1300°C [72T]:

H—T transition:

ΔV_m	-5.1 cm^3 mol^{-1}	$p \approx 1$ bar	phase diagram showing stability	72T
p_{tr}	$(0.02\,T(°C)-15)$ kbar	$T=800$	regimes for T, B, P and H-forms,	
ΔH_{tr}	-3 kcal mol^{-1}	⋯1300°C	see [72K, 72T]	
ΔS_{tr}	-2 cal K^{-1} mol^{-1}			

T—B transition:

ΔV_m	-3.1 cm^3 mol^{-1}	$p \approx 1$ bar		72T
p_{tr}	$(0.18\,T(°C)-120)$ kbar	$T=800$		
ΔH_{tr}	-10 kcal mol^{-1}	⋯1000°C		
ΔS_{tr}	-10 cal K^{-1} mol^{-1}			

B—Nb$_2$O$_5$

structure: monoclinic, space group $C_{2h}^6 - C2/c$, $Z=4$.

The structure can be described as hexagonal close packed with 2/5 of the octahedral holes filled in such a way as to lead to hexagonal layers parallel to (2$\bar{1}$0) with the a-axis in (001) (Fig. 11).

lattice parameters (distances in Å, angle in °):

a	12.73	RT	64L2,
b	4.88		70W
c	5.56		
β	105.1		

Physical property	Numerical value	Experimental conditions	Experimental method, remarks	Ref.
density:				
d	5.29 g cm^{-3}		X-ray density	64L2, 70W
atomic position parameters:				
	$-x$ $\quad$ $-y$ $\quad$ $-z$			
8 Nb in (f) sites	0.140, $\quad$ 0.249, $\quad$ 0.238			64L2, 70W
4 O(1) in (c) sites	0, $\quad$ 0.25, $\quad$ 0.099			
8 O(2) in (f) sites	0.389, $\quad$ 0.031, $\quad$ 0.054			
8 O(3) in (f) sites	0.295, $\quad$ 0.375, $\quad$ 0.426			
Nb—O distances in each of the octahedra (in Å):				
d(Nb—O)	1.91	RT		64L2
	1.81			
	2.06			
	2.12			
	2.19			
	1.94			

T—Nb$_2$O$_5$

structure: orthorhombic, space group D_{2h}^9—Pbam, $Z=8.4$

The T-form exhibits a further type of structure containing 7-coordinate Nb [75K1]. Unit cell projection: Fig. 12. There are 42 O ions in the unit cell and the Nb atoms occupy 16.8 positions. This structure is closely related to that of the phase 45 Ta$_2$O$_5 \cdot$ Al$_2$O$_3 \cdot$ 4 WO$_3$, save that the missing O positions in the latter structure are filled in T—Nb$_2$O$_5$. The 16 Nb atoms lie in a sheet parallel to (001) and are surrounded by 6 or 7 O-atoms forming pentagonal bipyramids or distorted octahedra. These polyhedra are joined by edge or corner sharing within the (001) plane and exclusively by corner sharing along the [001] axis. The "spare" Nb occupy nine-coordinate additional sites at random between the sheets.

lattice parameters (distances in Å):

a	6.170	RT		72N1
b	28.25			
c	3.928			
a	6.175			75K1
b	29.175			
c	3.930			

density:

d_{calc}	5.236 g cm^{-3}			75K1

interatomic distances (in Å) [75K1]:

	Nb(1)		Nb(2)		Nb(3)		Nb(4)	
d(Nb—O)								
	O(11)	2.027(3)	O(5)	2.07(2)	O(6^{iii})	1.98(1)	O(9^{ii})	1.91(2)
	O(5)	2.04(2)	O(6)	2.07(1)	O(6)	1.99(1)	O(7)	1.96(2)
	O(8^{iv})	2.06(1)	O(9)	2.09(1)	O(10^{ii})	2.01(2)	O(8)	1.96(2)
	O(8^{iii})	2.09(1)	O(10)	2.09(2)	O(7)	2.08(2)	O(5^{iii})	2.02(2)
	O(9)	2.27(1)	O(7^{ii})	2.13(1)				
	O(1^{i})	1.792(5)	O(2^{i})	1.779(2)	O(3^{i})	1.782(5)	O(4^{i})	1.770(3)
	O(1)	2.142(5)	O(2)	2.154(3)	O(3)	2.151(6)	O(4)	2.193(3)
d(Nb—Nb)								
Min.		3.282(3)		3.286(3)		3.375(4)		3.282(3)
Max.		4.039(6)		4.057(4)		4.057(4)		3.841(3)
Av.		3.622		3.608		3.620		3.623

(continued)

Physical property	Numerical value	Experimental conditions	Experimental method, remarks	Ref.

interatomic distances (continued)

	Nb(1)	Nb(2)	Nb(3)	Nb(4)
$d[O(5-11)-O(5-11)]$				
Min.	2.32(3)	2.34(2)	2.34(2)	2.41(2)
Max.	2.59(2)	2.64(2)	3.55(1)	3.21(2)
Av.	2.45	2.45	2.77	2.74
$d[O(1-4)-O(1-4)]$				
Min.	2.78(1)	2.80(2)	2.76(2)	2.69(1)
Max.	2.93(1)	2.91(2)	2.82(2)	2.88(2)
Av.	2.87	2.87	2.80	2.79

	Nb(5)	Nb(6)	Nb(7)
$d(Nb-O)$			
Min.	2.11(6)	2.11(1)	1.94
Max.	2.66(4)	2.85(2)	3.00
Av.	2.46	2.45	2.47

symmetric code:

| (i) | x | y | $1+z$ | (iii) | $\frac{1}{2}+x$ | $\frac{1}{2}-y$ | z |
| (ii) | $-\frac{1}{2}+x$ | $\frac{1}{2}-y$ | z | (iv) | $\frac{1}{2}-x$ | $-\frac{1}{2}+y$ | z |

TT$-$Nb$_2$O$_5$

structure: Little seems to be known about this form save that it is apparently stabilized by impurities. Two different structures have been proposed:

"pseudohexagonal": $a=3.607$ Å, $c=3.925$ Å, $Z=0.5$, density $d_{calc}=4.99$ g cm^{-3} (RT values) [66S].
monoclinic: $a=7.23$ Å, $b=15.7$ Å, $c=7.18$ Å, $\beta=119°5'$, $V=711$ Å^3, $Z=8$ (RT values) [72Z].

Tentative phase diagram for the Nb(v) oxides, Fig. 12α [72K].

Electronic and optical properties

Only few data available.

energy gap:

| E_g | 3.9 eV | | from photoconductivity spectra (amorphous film) | 78D |

XPE spectrum: Fig. 13, IR spectrum: Fig. 14, Raman spectrum: Fig. 15.
He I and He II photoelectron spectra, Fig. 34.

refractive index:

		λ [nm]		
n	2.44	1050	thin films (amorphous)	78S
	2.66	574		
	2.85	546		
	3.40	488		
	3.68	397		
	4.15	355		
	2.64(11)	632.8	B$-$Nb$_2$O$_5$	68E
	2.5(2)	632.8	P$-$Nb$_2$O$_5$	
	2.6(3)	632.8	H$-$Nb$_2$O$_5$	

dielectric constants:

B$-$Nb$_2$O$_5$:

$\varepsilon(0)$	50(3)	a-axis	frequency range 70 Hz$\cdots$100 MHz, data at RT	68E
	29(3)	b-axis		
	35(3)	c-axis		

(continued)

Physical property	Numerical value	Experimental conditions	Experimental method, remarks	Ref.
dielectric constants (continued)				
$P-Nb_2O_5$:				
$\varepsilon(0)$	30.6	c-axis		68E
	29 (3)···33 (2)	a-axis	crystal twinned	
$T-Nb_2O_5$:				
$\varepsilon(0)$	180 (20)	1-axis	axes taken as along the needles (1), through the largest reproducible natural face (2), and orthogonal to these (3)	
	95 (10)	2-axis		
	73 (10)	3-axis		
$H-Nb_2O_5$:				
$\varepsilon(0)$	120 (20)	b-axis		68F
	30 (5)	$\perp$ b-axis	average value	
$M-Nb_2O_5$:				
$\varepsilon(0)$	200 (100)	$\parallel$ needle axis		68E
	75 (15)	$\perp$ needle axis		
Lattice properties				
wavenumbers of lattice modes (in cm^{-1}, RT data):				
$H-Nb_2O_5$:				
$\bar{\nu}$	995 (s)	$\bar{\nu}_1$ (LO) mode region	Nb$-$O stretch in edge-shared octahedron	76M2
	883 (m)	$\bar{\nu}_1$ (LO) mode region	Nb$-$O stretch in corner-shared octahedron	
	850 (w)	$\bar{\nu}_1$ (A$_1$) mode	NbO$_4$ tetrahedra	
	670 (doublet, m)	$\bar{\nu}_2$ (TO) region	(s: strong, m: medium, w: weak)	
	625 (doublet, s)	$\bar{\nu}_2$ (TO) region		
	550 (w)	$\bar{\nu}_5$ (T$_{2g}$) region		
	470 (w)			
	350 (w)			
	260 (s)	$\bar{\nu}_6$ (T$_{2u}$) region		
	232 (s)			
	204 (m)			
$N-Nb_2O_5$:				
$\bar{\nu}$	995 (s)	$\bar{\nu}_1$ (LO) mode		76M2
	885 (m)			
	670 (m)	$\bar{\nu}_2$ (TO) mode		
	618 (s)			
	550 (w)	$\bar{\nu}_5$ (T$_{2g}$) region		
	485 (w)			
	350 (w)			
	260 (s)	$\bar{\nu}_6$ (T$_{2u}$) region		
	200 (m)			
$B-Nb_2O_5$:				
$\bar{\nu}$	935 (w)	$\bar{\nu}_1$ (LO) mode		76M2
	760 (s)			
	625 (m)	$\bar{\nu}_2$ (TO) mode		
	605 (m)			
	550 (s)	$\bar{\nu}_5$ (T$_{2g}$) mode		
	475 (m)			
	450 (m)			
	370 (s)			
	260 (s)	$\bar{\nu}_6$ (T$_{2u}$) region		
	200 (s)			

Physical property	Numerical value	Experimental conditions	Experimental method, remarks	Ref.

Defect and transport properties

The departure of $H-Nb_2O_{5-x}$ from stoichiometry as a function of temperature and pressure has been extensively studied [61K, 68K, 65B, 72N2, 69S3, 68A, 73K, 74M, 76K2, 79M]. The earlier results at lower temperatures suggested that the defect concentrations varies as $p_{O_2}^{-1/4}$ near stoichiometry and at $p_{O_2}^{-1/6}$ at lower oxygen pressures. This can easily be explained with a *point defect model* as: $O_0 \rightleftharpoons V_0 + \frac{1}{2}O_2$ (H_v), $V_0 \rightleftharpoons V_0^{\cdot} + e^-$ (ε_1), $V_0^{\cdot} \rightleftharpoons V_0^{\cdot\cdot} + e^-$ (ε_2). If the dominant defects are $V_0^{\cdot}$, we expect $[V_0^{\cdot}] \propto p_{O_2}^{-1/4}$ and if the dominant defects are $V_0^{\cdot\cdot}$ we expect a $p_{O_2}^{-1/6}$-law. Analysis of the temperature variation of the stoichiometry from thermal measurements gives

$H_v + \varepsilon_1 + \varepsilon_2$	105.8 kcal mol^{-1}			74M
	103 kcal mol^{-1}			62K

At 1100°C, a discontinuity in resistance was found [79M] corresponding to the formation of a new phase (see below).

This simple view of point defects extending over a substantial stoichiometric range has been strongly challenged [72N2, 73K]. Typical stoichiometry data are shown in Figs. 16 and 17 and at very high temperature a series of breaks in the curve [73K] strongly support the idea that a significant number of quite distinct phases can form. Combined structural and thermal experiments at 1000°C [72N2] reveal the following domains for NbO_x:

(1) $2.500 \geq x \geq 2.495$. Only the single phase $H-Nb_2O_{5-x}$ found without any local rearrangement or extended Wadsley defects. It is probable that in this range point defect theory may be used.
(2) $2.495 > x \geq 2.480$. Two phases found corresponding to $H-Nb_2O_5$ and $Nb_{25}O_{62}$.
(3) $2.480 > x \geq 2.417$. Two phases found corresponding to $Nb_{25}O_{62}$ and $Nb_{12}O_{29}$ (monoclinic).

Thus, at 1000°C, the reduction traverses the route $H-Nb_{28}O_{70} \rightarrow Nb_{25}O_{62} \rightarrow Nb_{12}O_{29}$. The structural relationship between these three phases is shown in Fig. 18. Thus, although Fig. 16 apparently shows typical bivariant behaviour for the 900°C and 1000°C results, the material present for $x \leq 2.495$ is inhomogeneous; only at $x = 2.480$ did there appear to be a single phase: the $Nb_{25}O_{62}$ structure. The apparently bivariant behaviour must thus be traced to extremely slow kinetics [72N2].

At higher temperatures, a quite different type of behaviour is observed. In addition to the definite phases $Nb_{28}O_{70}$ (i.e. $H-Nb_2O_5$) and $Nb_{25}O_{62}$, a phase formed as an intergrowth, $Nb_{53}O_{132}$ can form. The process of intergrowth is shown schematically in Fig. 19, which shows the regular intergrowth phase $Nb_{53}O_{132}$ as a natural intermediate. The important point about these intergrowth phases is that, unlike the nucleation (presumably at isolated Wadsley defects) and subsequent growth of $Nb_{25}O_{62}$ from $H-Nb_2O_5$, considerable activation energy is needed to form the (thermodynamically more stable) $Nb_{53}O_{132}$. A similar intergrowth, $Nb_{47}O_{116}$, forms between $Nb_{25}O_{62}$ and $Nb_{12}O_{29}$ (Fig. 19). The reduction sequence at high temperatures is then $Nb_{28}O_{70} \rightarrow Nb_{53}O_{132} \rightarrow Nb_{25}O_{62} \rightarrow Nb_{47}O_{116} \rightarrow Nb_{23}O_{54} \rightarrow Nb_{12}O_{29}$. A considerable trivariant regime therefore still remains at high temperature, and materials formulated as $Nb_{22}O_{54-x}$ and $Nb_{25}O_{62-x}$ show quite wide ranges for x (Figs. 20, 21) as well as $Nb_{28}O_{70-x}$ itself. It can be seen that in the bivariant region $[V_0]_{tot} \propto p_{O_2}^{-1/2}$, where $[V_0]_{tot} = x/54$ for $Nb_{22}O_{54-x}$ and $x/62$ for $Nb_{25}O_{62-x}$.

The *nature of the point defects* in Nb_2O_{5-x} has been controversial even over the area where a single definite phase is seen and $[V_0]_{tot} \propto p_{O_2}^{-1/4}$. In addition to singly charged oxygen vacancies, interstitial Nb ions have been proposed in both $N-Nb_2O_5$ [72H] and $H-Nb_2O_5$ [73A]. For the latter, the interstitial tetrahedral Nb represents an attractive structural option; thermodynamically $Nb_{Nb} + \frac{5}{2}O_0 \rightleftharpoons Nb_i^{n\cdot} + ne^- + \frac{5}{2}O_2$. Hence, for Nb_2O_{5-x}, $x \propto p_{O_2}^{-5/4(n+1)}$ and the observed $p_{O_2}^{-1/4}$ dependence can be accounted for with n = 4. However, the change to a slope of $-1/6$ cannot be explained on a point defect interstitial Nb model.

In the $p_{O_2}^{-1/4}$ region, *the activation energy* for deviation from stoichiometry is found to be 3.10 eV in ceramic samples of $H-Nb_2O_5$ [73B] whereas the *resistance* varies as in Fig. 22 with an activation energy of 1.65 eV [62K, 73B]. This suggests a slightly activated *mobility*, $E_A(\mu) \approx 0.10$ eV and $\mu(1100°C) \simeq 0.10$ cm^2 V^{-1} s^{-1} [73B]. The Seebeck coefficient also suggests a low activation energy [73B].

In the $p_{O_2}^{-1/6}$ region, the thermogravimetric studies show an activation energy of 4.46 eV, for the formation of doubly ionized oxygen vacancies in $H-Nb_2O_5$ [62K]. The actual *conductivity* in this region shows a lower apparent activation energy (Fig. 23), which can be satisfactorily explained by allowing the *mobility* $\mu \propto T^{-3}$ (Fig. 24), a result consistent with normal optical-mode scattering [59S].

Physical property	Numerical value	Experimental conditions	Experimental method, remarks	Ref.

There are a number of puzzling features about the interpretation in terms of singly-ionized oxygen vacancies [62K], and we would tentatively suggest that the transition from the $p_{O_2}^{-1/4}$ to $p_{O_2}^{-1/6}$ regions may correspond to a transition from interstitial $Nb_i^{4\cdot}$ to doubly ionized oxygen vacancies ($V_O^{\cdot\cdot}$).

At higher temperatures, the conductivity shows marked discontinuities (Fig. 25) corresponding to the formation of new phases. The resistivity continues to show a $p_{O_2}^{-1/6}$ behaviour in this region, suggesting the presence of doubly ionized oxygen vacancies [74M].

NbO_2
Crystal structure
a) low-temperature form ($T \lesssim 800\,°C$):

structure: tetragonal, space group $C_{4h}^6 - I4_1/a$, $Z = 32$.

A projection of the structure along the c axis is shown in Fig. 26, the essential structural element in Fig. 27. The structure is related closely to rutile with NbO_6 octahedra linked to form strings. Along the strings the Nb–Nb distances alternate (at RT) between 2.80 Å [62M], 2.74 Å [76C] and 3.20 Å [62M], 3.26 Å [76C], forming Nb–Nb doublets that further tilt along the $\langle 110 \rangle$ or $\langle 1\bar{1}0 \rangle$ directions of the rutile subcell. The relationship between the tetragonal supercell and the rutile cell is $a_{tetr} = 2\sqrt{2}\,a_R$ and $c_{tetr} = 2\,c_R$.

lattice parameters (tetragonal supercell) (in Å):

a	13.70	RT	62M
c	5.987		
a	13.699		77G
c	5.982		
a	13.696		76C
c	5.981		

density:

d_{calc}	5.90 g cm^{-3}		62M

atomic position parameters:

Atom	x	y	z		
Nb(1)	0.116 (1)	0.123 (2)	0.488 (3)	all ions occupy 16(f) subsets of $I4_1/a$	76C
Nb(2)	0.133 (1)	0.124 (2)	0.031 (2)		
O(1)	0.987 (1)	0.133 (2)	−0.005 (3)		
O(2)	0.976 (1)	0.126 (2)	0.485 (3)		
O(3)	0.274 (1)	0.119 (2)	0.987 (3)		
O(4)	0.265 (1)	0.126 (2)	0.509 (3)		

interatomic distances (in Å):

$d(Nb(1)-Nb(2))$	2.74 (2)	76C
$d(Nb(1)-Nb(2'))$	3.26 (2)	
$d(Nb(1)-O(1'))$	2.10 (2)	
$d(Nb(1)-O(2))$	1.93 (2)	
$d(Nb(1)-O(2'))$	2.18 (2)	
$d(Nb(1)-O(3'))$	2.02 (2)	
$d(Nb(1)-O(4))$	2.04 (2)	
$d(Nb(1)-O(4'))$	2.01 (2)	
$d(Nb(2)-O(1))$	2.02 (2)	
$d(Nb(2)-O(1'))$	2.06 (2)	
$d(Nb(2)-O(2'))$	2.11 (2)	
$d(Nb(2)-O(3))$	1.93 (2)	
$d(Nb(2)-O(3')$	2.23 (2)	
$d(Nb(2)-O(4'))$	2.03 (2)	

Physical property	Numerical value	Experimental conditions	Experimental method, remarks	Ref.

b) high-temperature form ($T \gtrsim 800\,°C$):

structure: true rutile.

Differential thermal analysis reveals a marked endothermic anomaly at 795(5)°C [69S1]. For rutile or pseudo-rutile unit cell parameters as a function of temperature: Fig. 28.

linear thermal expansion coefficients (in $10^{-6}\,K^{-1}$):

α_a	2.5	$T=100\,°C$	a and c are the pseudorutile	69S1
α_c	9.3		unit cell parameters for the	
α_a	4.3	900°C	LT phase	
α_c	24.5			

Diffuse superstructure lines may persist even up to 850°C [67S] suggesting that short-range structural fluctuations occur even well above the transition temperature [78P]. Furthermore, the superstructure lines of the low-temperature phase decrease continuously to zero, rather than discontinuously, suggesting a higher than first-order transition [74S]. Neutron diffraction studies on the lines lead to a crystallographic transition temperature of 808.5(1)°C [78P]. Very detailed studies of the transition [76P, 78P] have shown that the principal atomic displacements associated with the transformation may be represented as linear combinations of normal modes of the sort shown in Fig. 29 [76P, 78P]. However, all attempts to find a "soft mode" in the phonon spectrum have been unsuccessful and there remain some unresolved questions about the role of the lattice in the NbO_2 transition.

Electronic properties (including optical spectra)

No calculation of low-temperature phase band structure has been reported. The high-temperature rutile phase has been calculated using an APW technique [79P]. Reduced Brillouin zone: Fig. 30, band structure: Fig. 31, density of states: Fig. 32.

The optical spectrum has been reported in preliminary form [76L] and a comparison of the theoretical joint density of states for the rutile form with experimental data is shown in Fig. 33 for both the low-temperature and the rutile form stabilized to room temperature by addition of 20% Ti. The transitions below 3.4 eV are assigned to d–d transitions on Nb, and O 2p – Nb 4d charge transfer only occurs at energies above this.

The He I and XPE spectra are shown in Figs. 34 and 35. Comparison with RuO_2 and MoO_2 is also facilitated by the empirical band structure diagram for the LT form shown in Fig. 36. The d bands are quite narrow, that in NbO_2 being ca. 1 eV.

Lattice properties

wavenumbers of phonon modes:

The following data are best fit oscillator parameters for IR reflectance (Fig. 37) at room temperature. When the mode is very weak and the TO–LO splitting very small only the TO wavenumber is indicated. First column: $\bar{\nu}_{TO,LO}$ (in cm^{-1}), second column: contribution of this mode to $\varepsilon(0)-\varepsilon(\infty)$, third column: damping parameter γ of the oscillator equation (in cm^{-1}).

$\bar{\nu}_{TO}$ ($E \parallel c$)	255	0.78	20	with these parameters the	79G
$\bar{\nu}_{LO}$	257		20	dielectric constants become:	
$\bar{\nu}_{TO}$	318	8.23	35	$\varepsilon(\infty)=7.4$, $\varepsilon(0)=17.5$	
$\bar{\nu}_{LO}$	366		8		
$\bar{\nu}_{TO}$	368	0.18	8		
$\bar{\nu}_{LO}$	415		16		
$\bar{\nu}_{TO}$	434	0.38	16		
$\bar{\nu}_{LO}$	480		11		
$\bar{\nu}_{(TO,O)}$	550	0.02	(20)		
$\bar{\nu}_{TO}$	613	0.16	20		
$\bar{\nu}_{LO}$	621		20		
$\bar{\nu}_{TO}$	689	0.42	25		
$\bar{\nu}_{LO}$	720		25		

(continued)

Physical property	Numerical value			Experimental conditions	Experimental method, remarks	Ref.
wavenumbers of phonon modes (continued)						
$\bar{\nu}_{TO}\,(E\perp c)$	164	2.1	2.4		with these parameters the dielectric constants become: $\varepsilon(\infty)=7.2,\ \varepsilon(0)=28$	79G
$\bar{\nu}_{LO}$	168		2.4			
$\bar{\nu}_{TO}$	222	10.9	8			
$\bar{\nu}_{LO}$	243		12			
$\bar{\nu}_{TO}$	245	0.73	12			
$\bar{\nu}_{LO}$	264		5			
$\bar{\nu}_{TO}$	275.5	1.37	5			
$\bar{\nu}_{(TO,LO)}$	291	0.13	(12)			
$\bar{\nu}_{LO}$	302.5		6			
$\bar{\nu}_{TO}$	306	0.19	6			
$\bar{\nu}_{LO}$	329		7			
$\bar{\nu}_{(TO,LO)}$	350	0.03	(10)			
$\bar{\nu}_{(TO,LO)}$	370	0.02	(10)			
$\bar{\nu}_{TO}$	397	0.15	18			
$\bar{\nu}_{LO}$	400		18			
$\bar{\nu}_{TO}$	464	0.22	30			
$\bar{\nu}_{LO}$	467		30			
$\bar{\nu}_{TO}$	499	0.25	30			
$\bar{\nu}_{LO}$	502		30			
$\bar{\nu}_{TO}$	572	4.24	29			
$\bar{\nu}_{LO}$	677		16			
$\bar{\nu}_{TO}$	696	0.31	15			
$\bar{\nu}_{(TO,LO)}$	770	0.01	(30)			
$\bar{\nu}_{LO}$	800.5		20			
elastic moduli (in $10^{12}\,\mathrm{dyn\,cm^{-2}}$):						
c_{11}	4.33			low T	low temperature variation: Fig. 38	76B
c_{12}	0.93					
c_{13}	1.71					
c_{16}	± 0.01 (8)					
c_{33}	3.88					
c_{44}	0.94					
c_{66}	0.57					
Debye temperature:						
Θ_D	596 (7) K				from elastic moduli (about 5% lower from heat capacity data of Fig. 38)	76B

Transport properties and stoichiometry

Considerable disagreement over the stoichiometric range has been reported:

$NbO_{1.9975} - NbO_{2.0030}$ at 1373 K			66J
$NbO_{2.000} - NbO_{2.024}$ at 1575 K			69S3
$NbO_{1.9} - NbO_{2.1}$		arc melted single crystals	75S
$NbO_{1.990} - NbO_{2.0336}$			77G
$NbO_{2.000} - NbO_{2+x}$		$x=0.0032$ at 1273 K, $=0.0046$ at 1323 K, $=0.0060$ at 1373 K	76M1

Isotherms are shown in Fig. 39. Calculated curves were obtained for a model which involves niobium vacancies.

For the processes $2O_0 + V_{Nb} \rightarrow 2O_\infty\ (E_V)$ and $2O_0 + V'_{Nb} + Nb\dot{}_{Nb} \rightarrow 2O_\infty + Nb_{Nb}\ (E_{V'})$: $E_V = 254\,\mathrm{kcal\,mol^{-1}}$, $E_{V'} = 285\,\mathrm{kcal\,mol^{-1}}$ [76M1], where O_∞ is an oxygen atom at infinity and the temperature range studied was 900···1100 °C.

Physical property	Numerical value	Experimental conditions	Experimental method, remarks	Ref.

Crystals can be grown by vapour transport and by Czochralski techniques [72S, 75K2, 74B]. Vapour transport crystals show conductivity and susceptibility data (Fig. 40) similar to results obtained earlier [69S2].

More detailed studies have shown that at both low and high temperatures, the conductivity and associated parameters are very anisotropic [74B]. For resistivity in the a and c directions, see Fig. 41. In the range 300···450 K, the Hall mobility (Fig. 42) and drift mobility (Fig. 43) are activated (save for μ_H along the c axis).

transport parameters [74B]:

(A): current parallel to the a axis, (B) current in an arbitrary direction perpendicular to the c axis, (C) current parallel to the c axis. Activation energies E_A in eV.

Property	Numerical value	T [K]	Remarks
ϱ [Ω cm]	3509	300 (A)	$E_A(\sigma) = 0.44$
	22.4	435	
	6315	300 (B)	0.45
	32.5	435	
	824	300 (C)	0.37
	11.6	435	
R_H [cm^3 C^{-1}]	1377	300 (A)	$E_A(R_H^{-1}) = 0.40$
	13.6	435	
	2543	300 (B)	0.43
	18.2	435	
	517	300 (C)	0.38
	6.7	435	
μ_H [cm^2 V^{-1} s^{-1}]	0.40	300 (A)	$E_A(\mu_H) = 0.035$
	0.60	435	
	0.40	300 (B)	0.029
	0.56	435	
	0.63	300 (C)	−0.007
	0.58	435	
μ_{dr} [cm^2 V^{-1} s^{-1}]	0.016	300 (A)	$E_A(\mu_{dr}) = 0.17$
	0.114	435	
	0.009	300 (B)	0.19
	0.079	435	
	0.068	300 (C)	0.102
	0.22	435	
$\mu_H T^{3/2}$ [K$^{3/2}$ cm^2 V^{-1} s^{-1}]	2045	300 (A)	$E_A(\mu_H T^{3/2}) = 0.083$
	5365	435	
	2084	300 (B)	0.076
	5027	435	
	3270	300 (C)	0.040
	5181	435	
$n_{polaron}$ [cm^{-3}]	$1.11 \cdot 10^{17}$	300 (A) (B) (C)	$E_A(n) = 0.27$
	$2.44 \cdot 10^{18}$	435	

Thermoelectric power over a wide temperature range is shown in Fig. 44, the crystal used was apparently n-type. From the transport data the conductivity transition is apparently semiconductor-semiconductor in the a direction and semiconductor-metal in the c direction.

Magnetic properties

Magnetic susceptibility: Fig. 45, see also Fig. 40.

References for 9.15.2.6.1

41B	Brauer, G.: Z. Anorg. Allgem. Chem. **248** (1941) 1.
55F	Frevel, L.K., Rinn, H.W.: Anal. Chem. **27** (1955) 1329.
57H	Holtzberg, F., Reisman, A., Berry, M., Berkenblit, M.: J. Am. Ceram. Soc. **79** (1957) 2039.
59R	Reisman, A., Holtzberg, F.: J. Am. Ceram. Soc. **81** (1959) 3182.
59S	Scanlon, W.W.: Solid State Phys. **9** (1959) 83.
61K	Kofstad, P., Anderson, P.B.: J. Phys. Chem. Solids **21** (1961) 280.
62K	Kofstad, P.: J. Phys. Chem. Solids **23** (1962) 1571.
62M	Marinder, B.-O.: Ark. Kemi **19** (1962) 435.
63T	Terao, N.: Jpn. J. Appl. Phys. **2** (1963) 156.
64G	Gatehouse, B.M., Wadsley, A.D.: Acta Crystallogr. **17** (1964) 1545.
64L1	Laves, F., Moser, R., Petter, W.: Naturwiss. **51** (1964) 356.
64L2	Laves, F., Petter, W., Wulf, H.: Naturwiss. **51** (1964) 633.
65B	Blumenthal, R.N., Moser, J.B., Whitmore, D.H.: J. Am. Ceram. Soc. **48** (1965) 617.
65P	Petter, W., Laves, F.: Naturwissenschaften **52** (1965) 617.
66G	Gruehn, R.: J. Less-Common Met. **11** (1966) 119.
66J	Janninck, R.F., Whitmore, D.H.: J. Phys. Chem. Solids **30** (1966) 1183.
66S	Schäfer, H., Gruehn, R., Schultz, F.: Angew. Chem. Int. Ed. Engl. **5** (1966) 40.
67A	Anderson, S.: Z. Anorg. Allgem. Chem. **351** (1967) 106.
67S	Sakata, T., Sakata, K., Nishida, I.: Phys. Status Solidi **20** (1967) K155.
68A	Abbatista, F., Chientaretto, G., Tkachenko, E.V.: Atti Accad. Sci. Torino Cl. Sci. Fis. Mat. Nat. **102** (1968) 866.
68E	Emmenegger, F.P., Robinson, M.L.A.: J. Phys. Chem. Solids **29** (1968) 1673.
68K	Kofstad, P.: J. Less-Comm. Met. **14** (1968) 153.
68R	Robinson, M.L.A., Roetschi, H.: J. Phys. Chem. Solids **29** (1968) 1503.
69C	Chang, L.L.Y., Phillips, B.: J. Am. Ceram. Soc. **52** (1969) 527.
69S1	Sakata, K.: J. Phys. Soc. Jpn. **26** (1969) 582.
69S2	Sakata, K.: J. Phys. Soc. Jpn. **26** (1969) 867.
69S3	Schäfer, H., Bergner, D., Gruehn, R.: Z. Anorg. Allgem. Chem. **365** (1969) 31.
70M	Mertin, W., Anderson, S., Gruehn, R.: J. Solid State Chem. **1** (1970) 419.
70W	Wadsley, A.D., Anderson, S.: Perspectives in Structural Chemistry, Dunitz, J.D., Ibers, J.A. (eds.) New York: Academic Press **3** (1970) 1.
71E	Emin, D.: Ann. Phys. (Paris) **64** (1971) 336.
72H	Hutchinson, J.L., Anderson, J.S.: Phys. Status Solidi (a) **9** (1972) 207.
72K	Kodoma, H., Kikuchi, T., Goto, M.: J. Less-Commun. Met. **29** (1972) 415.
72M	Manning, W.R., Hunter, O., Calderwood, F.W., Stacey, D.W.: J. Am. Ceram. Soc. **55** (1972) 342.
72N1	Nolander, B., Norin, R.: Acta Chem. Scand. **26** (1972) 3814.
72N2	Nimmo, K.M., Anderson, J.S.: J. Chem. Soc. Dalton Trans. **1972**, 2328,
72S	Sakata, T., Sakata, K., Höfer, G., Horiuchi, T.: J. Cryst. Growth **12** (1972) 88.
72T	Tamura, S.: J. Mater. Sci. **7** (1972) 298.
73A	Anderson, J.S., Browne, J.M., Cheetham, A.K., von Dreele, R., Hutchinson, J.L., Lincoln, F.J., Bevan, D.J.M., Straehle, J.: Nature **243** (1973) 81.
73B	Le Brusq, H., Delmaire, J.P.: Rev. Int. Hautes Temp. Refract. **10** (1973) 15.
73D	Diwedi, G.L., Subbarao, E.C.: J. Am. Ceram. Soc. **56** (1973) 443.
73K	Kimura, S.: J. Solid State Chem. **6** (1973) 438.
74B	Bélanger, G., Destry, J., Perluzzo, G., Raccah, P.M.: Can. J. Phys. **52** (1974) 2272.
74M	Marucco, J.F.: J. Solid State Chem. **10** (1974) 211.
74S	Shapiro, S.M., Axe, J.D., Shirane, G., Raccah, P.M.: Solid State Commun. **15** (1974) 377.
74T	Tamura, S., Kato, K., Goto, M.: Z. Anorg. Allgem. Chem. **410** (1974) 313.
75K1	Kato, K., Tamura, S.: Acta Crystallogr. **B31** (1975) 673.
75K2	Kodama, H., Goto, M.: J. Cryst. Growth **29** (1975) 222.
75P	Plies, Y., Gruehn, R.: J. Less-Common Met. **42** (1975) 77.
75S	Shin, S.H., Halpern, T.H., Raccah, P.M.: Mater. Res. Bull. **10** (1975) 1061.
76B	Boyle, W.F., Bennett, J.G., Shin, S.H., Sladek, R.J.: Phys. Rev. **B14** (1976) 526.
76C	Cheetham, A.K., Rao, C.N.R.: Acta Crystallogr. **B32** (1976) 1579.
76K1	Kato, K.: Acta Crystallogr. **B32** (1976) 764.

76K2	Kikuchi, T., Goto, M.: J. Solid State Chem. **16** (1976) 363.
76L	Lu, S.S.M., Shin, S.H., Pollak, F.H., Raccah, R.M.: Proc. 13th Int. Conf. Semicond. Rome **1976**, p. 330.
76M1	Marucco, J.F., Tetot, R., Gerdanian, P., Picard, C.: J. Solid State Chem. **18** (1976) 97.
76M2	McConnell, A.A., Anderson, J.S., Rao, C.N.R.: Spectrochim. Acta **A32** (1976) 1067.
76P	Pynn, R., Axe, J.D., Thomas, R.: Phys. Rev. **B13** (1976) 2965.
77G	Gannon, J.R., Tilley, R.J.D.: J. Solid State Chem. **20** (1977) 331.
78B1	Beatham, N.: D. Phil. Thesis, Oxford **1978**.
78B2	Bennett, J.G., Sladek, R.J.: Solid State Commun. **25** (1978) 1035.
78D	D'yakonov, M.N., Zamyslovskii, M.G., Koslov, D.V., Muzhdaba, V.M., Khanin, S.D., Shelekhin, Ya.L.: Fiz. Tverd. Tela **20** (1978) 2801.
78P	Pynn, R., Axe, J.D., Raccah, P.M.: Phys. Rev. **B17** (1978) 2196.
78S	Sayaz, G.I., Septier, A.: Thin Solid Films **55** (1978) 191.
79G	Gervais, F., Baumard, J.F.: J. Phys. **C12** (1979) 1977.
79M	Marucco, J.F.: J. Chem. Phys. **70** (1979) 649.
79P	Posternak, M., Freeman, A.J., Ellis, D.E.: Phys. Rev. **B19** (1979) 6555.
81G	Gervais, F.: Phys. Rev. **B23** (1981) 6580.

9.15.2.6.2 Oxides of tantalum (Ta_2O_5)

The Ta−O phase diagram appears to show only one thermodynamically stable phase, Ta_2O_5.

Ta_2O_5

Ta_2O_5 exists in two forms, L- or β-Ta_2O_5, stable below 1320°C, and H- or α-Ta_2O_5, which is the stable form at high temperature. Kinetics of transformation are slow and H-Ta_2O_5 is therefore the form available as crystals at 300 K and for which the data reported below pertain. H-Ta_2O_5 melts at 1872°C [56R].

L−Ta_2O_5
Crystal structure

Indexing of the diffraction pattern of this oxide proved very difficult. The most intense lines correspond to an orthorhombic lattice, $a' = 6.20$ Å, $b' = 3.66$ Å, $c' = 3.89$ Å [70R1], but numerous weak lines appear contingent on b' multiplied by a factor m which is dependent on impurity content and thermal history. A single phase L-form can be equilibrated by heating Ta_2O_5 to 1350°C for a long period, just below the upwards transition temperature [70R2]. This has $m = 11$, $a = 6.198$ Å, $b = 40.290$ Å, $c = 3.888$ Å, $V = 970.9$ Å^3, $Z = 11$, $d_{calc} = 8.31$ g cm^{-3} [71S1].

Electron microscopy also shows lattice imaging with $m = 11$. Equilibration of L−Ta_2O_5 at 1000°C for long periods gives a structure with $m = 14$, and at intermediate temperatures an apparently infinite number of ordered, or partially ordered intergrowths with $m = 11$ to 14 are possible. The L-form is also stabilized by small amounts of cation impurity, provided the ionic radius of the impurity is small; examples are W, Ge, B and Al [70R2].

detailed structure: Fig. 1. There is some relationship to T−Nb_2O_5. The structure shown in Fig. 1 in (001) projection would contain 22 Ta and 58 O, so three O atoms have to be eliminated, which the structure achieves by distortion in the planes $d_1 \cdots d_4$ of Fig. 1. Low temperature amorphous films appear to consist of edge-sharing pentagonal bipyramids, and it is the fusion of these, and concomitant elimination of OH and H_2O that leads to the structure of Fig. 1.

For a detailed table of interatomic distances, see [71S1].

H−Ta_2O_5
Crystal structure

structure: monoclinic, space group C2, $Z = 6$, retains the distorted bipyramids of the L-form [71S2], (010) projection of the structure: Fig. 2.

Physical property	Numerical value	Experimental conditions	Experimental method, remarks	Ref.
lattice parameters (distances in Å):				
a	35.966			71S2
b	3.810			
c	3.810			
β	96°7′			

For interatomic distances, see [71S2].

Other structures have been reported for Ta_2O_5. An orthorhombic form has been found in films grown below 700 °C with $a = 5.47$ Å, $b = 7.65$ Å, $c = 26.10$ Å. Its structure has been likened to the perovskite tantalates [72K1].

density:				
d	8.2 g cm^{-3}			83H

Electronic properties

energy gaps (in eV):				
E_g	4.20	RT	direct edge, absorption, Fig. 3	73K
	4.5···4.6			69H, 52A
$E_{g,\mathrm{th}}$	3.58	$T = 900$ ···1400 °C	from transport data, some evidence also in absorption, see Fig. 3	62K, 72K2, 74S

For peak energies in optical spectra, see below. No assignments have been made to band-band transitions.

Transport properties and non-stoichiometry

Ta_2O_5 is almost a line phase, and direct measurements of non-stoichiometry have only proved possible at the lowest oxygen partial pressures (Fig. 4).

Transport parameters vs. oxygen partial pressure: conductivity: Fig. 5a, transport number for ionic conductance, t_i: Fig. 5b, Seebeck coefficient: Fig. 6, for resistance vs. temperature, see Fig. 7.

The Seebeck coefficient shows a change from n- to p-type behaviour with increasing p_{O_2}.

The logarithm of the oxygen pressure at the p-n transition varies linearly with reciprocal temperature [62K]. At higher oxygen pressures t_i goes through a maximum and σ through a minimum. Activation energies vary with p_{O_2} (in atm) at $\log p_{O_2} = -16$, $E_A \approx 1.9$ eV ($T = 900$···1400 °C) whereas at higher p_{O_2}, E_A is lower (Fig. 7).

The resistivity has a marked maximum with p_{O_2} at the p-n transition.

Interpretation is bedevilled by the significant ionic contribution. Two regions may be distinguished: $n \gg 1$, p ($\sigma \propto p_{O_2}^{-1/6}$) and $p, n \ll 1$ ($\sigma \propto p_{O_2}^{-1/4}$). Further analysis showed [74S]:

(a) the ratio of ionic and electron mobilities is activated with $E_A = 2.0$ eV ($T = 900$···1400 °C),
(b) the ionic conductivity is activated with $E_A = 1.8$ eV ($T = 900$···1400 °C),
(c) $\mu_n \simeq 0.05$ cm^2/Vs, $\mu_{\mathrm{ion}} \simeq 2 \cdot 10^{-3}$ cm^2/Vs at 1100 °C for a ceramic sample,
(d) Seebeck data suggest motion by electrons in a very narrow conduction band,
(e) if the minimum in conductivity corresponds to $n = p$, then the activation energy for the electron conductivity is 1.8 eV, assuming unactivated mobility and $E_g \simeq 3.6$ eV.

Anodic thin amorphous film measurements at 300 K have led to very small values of $\mu_n \simeq 10^{-12}$ cm^2/Vs [73A, 74J] with the Fermi level apparently pinned 0.44 eV below the conduction band. The interpretation of these results is controversial [76Y, 78G].

ac conductivity results (Fig. 8) were analyzed in terms of a power law (real part of admittance proportional to ω^n). At low temperatures (near RT) n = 0.95, but for higher temperatures and lower frequencies, n decreases [77J]; at 380 K and $\nu < 1$ Hz, n = 0.3 [79S1]. The value of n is consistent with random range hopping models with $E_A \simeq 0.3$ eV.

Photoconductivity (Fig. 9) shows a threshold at $\simeq 1.5$ eV and a strong peak at 2.1 eV [73T, 74T]. The results have been interpreted in terms of an impurity level situated ca. 2.1 eV below the conduction band edge.

9.15.2.6.2 Oxides of tantalum (Ta$_2$O$_5$)

Physical property	Numerical value	Experimental conditions	Experimental method, remarks	Ref.
Optical properties				
peaks in optical spectra:				
infrared absorption results (wavenumbers in cm^{-1}):				
$\bar{\nu}$	220			73K
	325	shoulder	RT, 5000 Å film of poly-	
	380	weak	crystalline β-Ta$_2$O$_5$ on Si	
	425	shoulder		
	525	strong		
	630	shoulder		
	710	shoulder		
	840			
	900	shoulder		
	1040	shoulder		
	315		anodic film	64M
	455	shoulder		
	575	strong		
infrared reflection results (wavenumbers in cm^{-1}):				
$\bar{\nu}$	270	medium	RT, sample as above	73K
	425	broad		
	550	shoulder		
	630	medium		
	980	strong		
	1140	weak		
	810	weak to strong	intensity of these peaks very	70K
	950	weak to strong	dependent on voltage of anodization	
	1100	weak		
refractive index:				
n	2.21	$\lambda = 632.8$ nm		74L
n^2	$A + B\bar{\nu}^2$	visible range		79S1
A	4.2446		evaporated film as grown	
B	0.13158 μm^2			
A	4.2454		baked evaporated film	
B	0.06677 μm^2			
n	$1.85 + 1671(\lambda - 903)^{-1}$	$\lambda = 275.0$ ⋯ 1400 nm	anodic thin film	75A
	2.20 (3)	546.1 nm		73K
	2.12⋯2.20		for the principal optical axes	56K
	2.32⋯2.37		of β-Ta$_2$O$_5$ at 671.0 nm	
	2.22, 2.30, 2.325		for the principal optical axes of α-Ta$_2$O$_5$	56K

dielectric function:

Results on β-form ceramic (Fig. 10) and α-form single crystal (Fig. 11) quite different, with the α-form showing strong anisotropy and dispersion at high temperatures. There is also a strong rise in ε above 300 K for the α-form, a result also reported for thin films [73K]. The observed dielectric relaxation has an activation energy of 0.64 eV, close to that observed for film growth on tantalum [58V] and for conductivity in ceramics at low temperatures and high oxygen pressures [62K]. Capacitance data on H−Ta$_2$O$_5$−p−Si sandwiches also showed a dielectric relaxation process with an activation energy of 0.71 eV [74A].

References for 9.15.2.6.2

52A	Apker, L., Taft, E.A.: Phys. Rev. **88** (1952) 58.
56K	King, B.W., Schutz, T., Durbin, E.A., Duckworth, E.H.: Battelle Mem. Inst. Rep. **BMI-1106** (1956).
56R	Reisman, A., Holtzberg, F., Berkenblit, M., Berry, M.: J. Am. Chem. Soc. **78** (1956) 4514.
58V	Vermilyea, D.A.: Acta Metall. **6** (1958) 167.
62K	Kofstad, P.: J. Electrochem. Soc. **109** (1962) 778.
64M	McDevilt, N.T., Braun, W.L.: J. Am. Ceram. Soc. **47** (1964) 622.
64P	Pavlovic, A.S.: J. Chem. Phys. **40** (1964) 951.
67A	Alyamorskii, S.I., Shveikin, G.P., Gel'd, P.V.: Russ. J. Inorg. Chem. **12** (1967) 915.
68K	Kofstad, P.: Corrosion (Houston) **24** (1968) 379.
69H	Hortl, P., Schwartz, W.: Z. Naturforsch. **24A** (1969) 296.
70K	Kihara-Morishita, H., Takamura, T., Takeda, R.: Thin Solid Films **6** (1970) R29.
70R1	Roth, R.S., Waring, J.L., Parker, H.S.: J. Solid State Chem. **2** (1970) 445.
70R2	Roth, R.S., Waring, J.L., Brower, W.: J. Res. Nat. Bur. Stand. Sect. A **74** (1970) 485.
71S1	Stephenson, N.C., Roth, R.S.: Acta Crystallogr. **B27** (1971) 1037.
71S2	Stephenson, N.C., Roth, R.S.: J. Solid State Chem. **3** (1971) 145.
72K1	Khikova, V.I., Klechkovskaya, V.V., Pinsker, Z.G.: Kristallografiya **17** (1972) 506.
72K2	Kofstad, P.: "Nonstiochiometry, Diffusion and Electrical Conductivity in Binary Metal Oxides", New York: J. Wiley and Sons **1972**.
73A	Aris, F.C., Lewis, T.J.: J. Phys. **D 6** (1973) 1067.
73K	Knausenberger, W., Tauber, R.N.: J. Electrochem. Soc. **120** (1973) 927.
73T	Thomas, J.H.: Appl. Phys. Lett. **22** (1973) 406.
74A	El Arabi, A., Seve, G., Lassabatore, L.: Thin Solid Films **21** (1974) 11.
74J	Jones, M.W., Hughes, D.M.: J. Phys. **D 7** (1974) 11.
74L	Leslie, J.D., Knorr, K.: J. Electrochem. Soc. **121** (1974) 263.
74S	Stroud, J.E., Tripp, W.C., Wimmer, J.M.: J. Am. Ceram. Soc. **57** (1974) 172.
74T	Thomas, J.H.: J. Appl. Phys. **45** (1974) 835.
75A	Albella, J.M., Martinez-Duart, J.M., Rueda, F.: Opt. Acta **22** (1975) 973.
76Y	Young, P.L.: J. Appl. Phys. **47** (1976) 235.
77J	Jonscher, A.K.: Phys. Status Solidi **(b) 84** (1977) 159.
78G	Gubanski, S.M., Hughes, D.M.: Thin Solid Films **52** (1978) 119.
79S1	Smith, D., Baumeister, P.: Appl. Opt. **18** (1979) 111.
79S2	Savinova, N.A.: Fiz. Tverd. Tela **21** (1979) 2889.
83H	Handbook of Chemistry and Physics, Weast, R.C., (ed.), 64th edition **1983**, CRC Press, Inc.

Physical property	Numerical value	Experimental conditions	Experimental method, remarks	Ref.

9.15.2.7 Oxides of group VI elements (Mo, W)

9.15.2.7.1 Molybdenum oxides

A number of oxides have been reported in the range $MoO_2 - MoO_3$ but little seems to be known about their electronic properties save that they are apparently metallic. The only well characterized semiconducting phase is MoO_3. Phase diagram of the Mo−O system: Fig. 1 [69C].

MoO_3

Crystal structure

Layer structure, ideally the layers can be envisaged as octahedra fused together as shown in Fig. 2 [63K]. Space group: D_{2h}^{16}–Pbnm, $Z=4$. Projection of the structure on the (100), (010) and (001) planes: Fig. 3. The coordination about the Mo atoms is better thought of a distorted tetrahedron (Fig. 4 [63K]) and the fourfold coordination is built up into layers involving two further more distant neighbours about each Mo (Fig. 5 [63K]). Crystals of MoO_3 can be grown by a variety of techniques [68D, 79F].

lattice parameters (in Å):

a	3.9628	RT	63K
b	13.855		
c	3.6964		

Physical property	Numerical value	Experimental conditions	Experimental method, remarks	Ref.
interatomic distances and angles at RT (distances in Å):				
Mo–Mo 1–1'	3.4379	$\sigma = 0.0006$ Å	angle O–Mo–O, $\sigma \approx 0.3°$	63K
1–1	3.9628	‖ [100]	2–1–2' 73.5°	
1–1	3.6964	‖ [001]	2–1–3 91.5°	
Mo–O 1–2	2.332	$\sigma \approx 0.008$ Å	2–1–3' 76.3°	
1–2'	2 · 1.948		2–1–4 164.9°	
1–3	1.734		2'–1–2 143.1°	
1–3'	2.251		2'–1–3 2 · 98.3°	
1–4	1.671		2'–1–3' 2 · 78.4°	
O–O 2–2'	2 · 2.578	$\sigma \approx 0.011$ Å	2'–1–4 2 · 103.8°	
2–3	2.831		3–1–3' 167.8°	
2–3'	2.942		3–1–4 103.6°	
2–3''	2 · 2.664		3'–1–4 88.6°	
2–3'''	2 · 2.790		structurally the O-ions can be	
2–4	2 · 2.853		divided into three categories:	
3–2	2.942		O [1]: singly coordinated O at	
3–2'	2.831		1.67 Å from Mo atom (position 4)	
3–2''	2 · 2.664		O [2]: doubly coordinated O at	
3–2'''	2 · 2.790		1.73 Å and 2.25 Å	
3–3'	2 · 3.033		(positions 3, 3')	
3–4	2.677		O [3]: triply coordinated O at	
3–4'	2.770		1.95 Å and 2.33 Å	
3–4''	2 · 3.240		(positions 2, 2')	
4–2	2 · 2.853			
4–3	2.677			
4–3'	2.770			
4–3''	2 · 3.240			
4–4'	4 · 2.823			
linear thermal expansion coefficient (in K^{-1}):				
α_a	$0.9(1) \cdot 10^{-5}$	RT		64K
α_b	$18(2) \cdot 10^{-5}$			
α_c	≈ 0			
density:				
d	4.709 g cm^{-3}		calculated from lattice parameters	63K

Electronic properties

No band calculation has been reported. An Xα-calculation on an ideal octahedral MoO$_6^{6-}$ cluster gives the following energy levels (given in Ry):

E	-0.117		orbital: $4a_{1g}$ (Mo 5s)	although the 78B2
	-0.181		$3e_g$ (Mo 4d)	qualitative
	-0.308		$2t_{2g}$ (Mo 4d)	ordering of
	-0.374		$1t_{1g}$ (O 2p)	these levels
	-0.381		$4t_{1u}$ (O 2p)	is very
	-0.396		$1t_{2u}$ (O 2p)	reasonable
	-0.432		$3t_{1u}$ (O 2p)	the predicted
	-0.475		$3a_{1g}$ (O 2p)	band gap
	-0.488		$1t_{2g}$ (O 2p)	(O (2p) – Mo (4d))
	-0.495		$2e_g$ (O 2p)	is far too low

Physical property	Numerical value	Experimental conditions	Experimental method, remarks	Ref.

energy gap:
an analysis of the absorption coefficient – found to follow Urbach's rule in the threshold region ($10^2 < K < 10^4$ cm^{-1}) – gives the following data:

β	0.21	$E \perp c$	parameters of the Urbach formula	68D
	0.24	$E \parallel c$	$K = K_0 \exp(-\beta(E_0 - \hbar\omega)/kT)$	
E_0	3.23 eV	$E \perp c$	with $E_0 = E_\infty - CT$	
	3.66 eV	$E \parallel c$		
C	$6.2 \cdot 10^{-4}$ eV K^{-1}	$E \perp c$	for absorption spectrum near	
	$7.3 \cdot 10^{-4}$ eV K^{-1}	$E \parallel c$	threshold, see Fig. 6; for He I and He II spectra, see Fig. 7	
E_g	2.96 eV	$E \perp c$	analyzed from plot of K^2 vs. $\hbar\omega$ at RT	68D
	2.80 eV	$E \parallel c$		
	3.05 eV	RT		79H

Defects

MoO$_3$ is not a line phase, maximum values of γ in MoO$_{3-\gamma}$ before a second phase appears are

		T [°C]		
$10^4 \gamma$	1.4	607	in the range 607···739 °C	74Z
	1.6	637	γ is proportional to $p_{O_2}^{-1/6}$	
	2.5	670		
	3.1	691		
	3.6	714		
	5.2	739		

There is some controversy over the nature of the defects. The thermogravimetry data [74Z] is consistent with doubly ionized O vacancies. However further reduction apparently gives rise to (301)-shear planes (Fig. 8). The O-vacancy model was further developed [68D, 78S1] and epr data reported by [68D] is consistent with a small concentration of singly ionized O vacancies. This has been queried by [69I] and [72S]. [69I] found no epr signal in air-grown MoO$_3$ but in Ar-grown crystals a signal corresponding to the localization of the electron on two different Mo-sites with hyperfine structure associated with a proton – the defect being Mo^{5+}–OH$^-$–Mo^{6+}. [72S] also found a 4 d' signal associated with a proton provided the crystals were grown in an atmosphere containing traces of H$_2$. For Ar-grown crystals, or those reduced with H$_2$, a different epr signal is found but no evidence for O-vacancies. However the technically important colour centres in MoO$_3$ films [68D, 74T] are most naturally ascribed to Mo^{5+} localized near oxygen vacancies or OH$^-$ ions.

Lattice properties

infrared and Raman active phonon wavenumbers at RT (in cm^{-1}): (first, second and third column according to [69M], [71B] and [79B], respectively)

$\bar{\nu}_{IR}$		193			a_u [2]	the numbers in brackets refer
		229			b_u [1]	to the position of the O-ion,
	285	270		δ(O, MoO$_3$)	a_u [1]	δ are deformation modes,
	352	351			b_u [3]	ν are stretching modes
	362, 377	371		δ(O, MoO$_3$)	b_u [2]	
	485	505	490		b_u [3]	
	570	545	565	ν(MoO$_3$)	a_u [3]	Mo–O–Mo chain stretch
	821, 870	840	820, 860	ν(MoO$_2$, –O$_3$)	b_u [2]	ν(Mo–O$_2$)
	996	1004	986, 1000	ν(MoO$_1$)	b_u [1]	ν(Mo–O$_1$)
$\bar{\nu}_R$		84, 101			a_g [all 3]	
	116	118, 130			b_g [all 3]	
	158	160			a_g [all 3']	(continued)

Physical property	Numerical value	Experimental conditions	Experimental method, remarks	Ref.

Raman active phonon wavenumbers (continued)

$\bar{\nu}_R$	196 200		b_g [2]	
	220 218		a_g [1]	
	248 248		b_g [2]	
	285 286, 294	b_g (δO, MoO_3)	b_g [1]	
	340 338		a_g [3]	
	363 367	a_g (δO, MoO_3)	a_g [1, 2]	
	378 382 370	a_g (δO, MoO_3)	a_g [1, 2]	
	474 473 474		a_g [3]	
	665 668 668	b_g (νMoO_3)	b_g [3] Mo–O–Mo chains	
	807 820 822	a_g (νMoO_2)	a_g [2] $\nu(Mo-O_2)$	
	992 997 998	a_g (νMoO_1)	a_g [1] $\nu(Mo-O_1)$	

Transport properties

At high temperatures the conductivity of MoO_3 appears to be partially ionic.

ionic conductivity parameters:

$t_+ + t_-$	0.58	$T = 800$ K, 0.00025 atm $< p_{O_2} < 0.21$ atm	t_+ and t_- are transport numbers for cation (t_+) and anion (t_-)	74Z
	0.44	950 K		
	0.72	700···950 K		71E
D_0	$9.1 \cdot 10^{-8}$ cm^2 s^{-1}		parameters in the relation $D = D_0 \exp(-E/RT)$ where $D = kT(t_+ + t_-)\sigma/N_i z_i^2 e^2$ and N_i is the number concentration of ions of charge $z_i e$	
E	20.3 kcal mol^{-1}			

activation energy of electronic conductivity (in eV):

E_A	1.7	$p_{O_2} = 0.21$ atm, $T = 560\cdots739$ °C	typical data for a crystal grown and measured in Ar is given in Fig. 9; at temperatures below ≈ 500 °C reproducibility is found for samples pre-sintered at various temperatures: Fig. 10	74Z
	2.3	$2 \cdot 10^{-4}$ atm		
	3.6	$< 10^{-4}$ atm		
	0.60 (5)	$T < 500$ °C		69I
	0.29···0.70	$T < 500$ °C		68D
	1.83 (10)	$T > 500$ °C		68D
	0.30	≈ 25 °C		78S1

Seebeck coefficient: little reproducibility (Fig. 11), considerable hysteresis.

ac conductivity: proportional to ω^s, where s is a temperature dependent number near 0.8, Fig. 12. The ac conductivity can be written $\sigma \propto BT\omega^s \exp(-E_A/kT)$ with $E_A \approx 0.015\cdots0.020$ eV ($T = 90\cdots183$ K) [78S1].

photocurrent:

The (photo)current-voltage curve is shown in Fig. 16. A transition to a space-charge limited region ($I \propto U^{3/2}$) is apparent. Photocurrent spectrum (Fig. 17) shows a threshold at ca. 3.4 eV larger than the direct edge threshold (ca. 3.0 eV). The peculiar Urbach tail (3.0···3.4 eV) suggests that excitonic effects may be dominant at the threshold. If so the relatively low electric fields used in the photoconductivity experiment may be insufficient to separate the carriers. The photoconductivity shows complex time-dependence and the role of traps is very considerable [68D]. The activation energy for the photocurrent is about 0.24 eV.

Physical property	Numerical value	Experimental conditions	Experimental method, remarks	Ref.

Optical properties

absorption coefficient: see above under "Electronic properties". See also Figs. 14, 15.

refractive index:

n_α ($\|b$)	1.88	values obtained using Na "D" line	wavelength dependence: Fig. 13	68D
n_β ($\|a$)	2.29			
n_γ ($\|c$)	2.54			
n	2.30	$\lambda \to \infty$	extrapolated, thin film, close to average value (2.24)	

static dielectric constant:

$\varepsilon(0)$	18 (1)	$f = 1 \cdots 40$ kHz, RT	$\varepsilon(\infty) = 5.31$ [68D], $= 5.70$ [66D]	66D, 68D
	35 (5)	RT		78S1
	17 (1)	1 kHz		64A
	12 (1)	10 kHz, $T = 77$ K		78S1

The optical spectra of $Mo_{1-x}W_xO_3$ also show a direct edge (Figs. 14, 15). For a range of x values, the band gap of $Mo_{1-x}W_xO_3$ exceeds both that of WO_3 (2.77 eV) and pure MoO_3 (3.05 eV), an effect ascribed to Anderson localization [79H].

References for 9.15.2.7.1

63K	Kihlborg, L.: Ark. Kemi **21** (1963) 357.
64A	Ames, I., Gregor, L.V.: Electrochem. Technol. **2** (1964) 97.
64K	Kirkegaard, P.: Ark. Kemi **23** (1964) 223.
66D	Deb, S.K., Chopoovrian, J.A.: J. Appl. Phys. **37** (1966) 4818.
68B	Beattie, I.R., Gilson, T.R.: Proc. Roy. Soc. (London) Ser. A **307** (1968) 407.
68D	Deb, S.K.: Proc. Roy. Soc. **A304** (1968) 211.
69C	Chang, L.L.Y., Phillips, B.: J. Am. Ceram. Soc. **52** (1969) 530.
69I	Ioffe, V.A., Petrina, I.B., Zelenetskaya, E.V., Mikheeva, V.P.: Phys. Status Solidi **35** (1969) 535.
69M	Mattes, R., Schröder, F.: Z. Naturforsch. **24B** (1969) 1095.
71B	Beattie, I.R., Cheetham, N., Gardner, M., Rogers, D.E.: J. Chem. Soc. **A 1971**, 2240.
71E	Elyutin, V.P., Lenskaya, T.G., Povlov, Yu.A., Polyakov, V.P.: Dokl. Akad. Nauk SSSR **199** (1971) 62.
72S	Sperlich, G., Franck, G., Rhein, W.: Phys. Status Solidi (b) **54** (1972) 241.
74T	Tubbs, M.R.: Phys. Status Solidi (a) **21** (1974) 253.
74Z	Zhukovskii, V.M., Yanushkevich, T.M., Neimen, A.Ya., Lebedkin, V.P.: Zh. Fiz. Khim. **48** (1974) 715.
78B1	Beatham, N.: D. Phil. Thesis (Oxford) **1978**.
78B2	Broclawik, E., Foti, A.E., Smith, V.H.: J. Catal. **51** (1978) 380.
78S1	Sayer, M., Mansingh, A., Webb, J.B., Noad, J.: J. Phys. **C 11** (1978) 315.
78S2	Skorokhod, V.V., Solonin, Yu.M.: Kristallografiya **23** (1978) 653.
79B	Bart, J.C.J., Cariati, F., Sgiamdotti, A.: Inorg. Chim. Acta **36** (1979) 105.
79F	Fourcandot, G., Gourmada, M., Mercier, J.: J. Cryst. Growth **46** (1979) 132.
79H	Hoppmann, G., Salje, E.: Opt. Commun. **30** (1979) 199.

9.15.2.7.2 Tungsten oxides

The W–O phase diagram is complex (Fig. 1) with a number of oxides in the range WO_2–WO_3. The phase WO_2 has a narrow stoichiometry range. The neighbouring phase $W_{18}O_{49}$ ($WO_{2.72}$) also has an apparently very small homogeneity range [69M1, 76P]. The structure of $W_{18}O_{49}$ is not related to the homologous shear-plane structures, but is a compromise between the WO_2 and WO_3 structures [50M].

Homologous series of the general formula $W_nO_{3n-(m-1)}$ are derived from the ReO_3 structure by shearing along the {10m} planes (Fig. 2). The series with m=2 has been explored by electron microscopy [74B, 74S2]. It has not proved possible, however, to isolate any pure W_nO_{3n-1} phase. Weak reduction gives both m=2 and m=3 series and heavier reduction favours the latter. Sintering weakly reduced WO_3 at 1273 K gives $W_{30}O_{89}$ ($WO_{2.966}$) and $W_{26}O_{76}$ ($WO_{2.923}$) and values of n as low as 14 ($W_{14}O_{41}$, $WO_{2.929}$) have been observed. The series m=3 is better characterized; $W_{20}O_{58}$ and $W_{25}O_{73}$ are well known, though values of n from 12 to 28 have been found [69A].

The end-member of the homologous series, WO_3, has a very small homogeneity range. Pure WO_3 can only be made under high oxygen pressures [77G]. As usually prepared, WO_3 is slightly oxygen deficient, though true point defects appear only to be found for $x \lesssim 10^{-4}$ in WO_{3-x} [70S]. For values of $x > 10^{-4}$ condensation of these point defects to form shear planes is found [76T].

WO_3
(See also Landolt-Börnstein, Vol. III/16a.)

Crystal structure

The crystal chemistry of WO_3 is extremely complex. Below 130 K a transition has been detected by dielectric loss measurements [75L] but nothing is known of the structure. Above 130 K and below 223 K a monoclinic low-temperature modification α-WO_3 is found [60T, 76S1]. At higher temperatures a triclinic β-modification occurs, which converts to a second monoclinic γ-WO_3 form, stable between 290 K and 603 K. Between 603 K and ≈1013 K, WO_3 has orthorhombic symmetry and above 973 K a tetragonal form occurs. (For transition temperatures, see also below.)

structural data:

a) **α-WO_3**: space group $Pc - C_s^2$, $Z=4$. Coordination about W almost 5-fold, with zig-zag W strings.

lattice parameters (distances in Å):

a	5.275	$T=133$ K		76S1
b	5.155			
c	7.672			
β	91.7°			
$d(W-O)$	1.85		W–O distance in $a-b$ plane;	
	2.31		long sixth distance, see Fig. 3	

b) **β-WO_3**: space group $P\bar{1} - C_i^1$, $Z=8$. The detailed structure is derived from ReO_3 with the corner-shared octahedra tilted (Figs. 4, 5).

lattice parameters:

a	7.309 (2) Å	RT	data collected from a single	78D,
b	7.522 (2) Å		domain crystal found by repeated	60T
c	7.678 (2) Å		thermal cycling of WO_3	
α	88.81°			
β	90.92°			
γ	90.93°			

Bond lengths and angles	Numerical value	Bond lengths and angles	Numerical value

bond lengths (in Å) and angles (in °) of the independent WO_6 octahedra at RT [78D]:
(see Fig. 5 for numbering of atoms)

Bond lengths and angles	Numerical value	Bond lengths and angles	Numerical value
$W(1)O_6$ octahedron		$W(2)O_6$ octahedron	
d(W(1)–O(5))	1.751 (21)	d(W(2)–O(11))	1.775 (24)
O(12)	1.756 (22)	O(6)	1.787 (22)
O(4)	1.826 (21)	O(2)	1.853 (22)
O(1)	1.952 (23)	O(3)	1.915 (23)
O(6)	2.056 (23)	O(5)	2.116 (21)
O(9)	2.215 (23)	O(10)	2.199 (24)
∡(O(5)–W(1)–O(9))	87.62 (90)	∡(O(2)–W(2)–O(5))	84.10 (87)
O(1)	93.34 (96)	O(6)	96.02 (97)
O(4)	98.89 (96)	O(10)	81.24 (92)
O(12)	100.22 (100)	O(11)	99.53 (99)
O(9) O(1)	80.69 (90)	O(5) O(10)	78.41 (85)
O(4)	83.52 (90)	O(3)	79.57 (89)
O(6)	79.72 (85)	O(11)	93.70 (94)
O(1) O(6)	82.43 (93)	O(6) O(10)	84.96 (96)
O(12)	95.20 (98)	O(3)	95.66 (99)
O(4) O(6)	82.06 (92)	O(11)	102.85 (104)
O(12)	98.63 (98)	O(10) O(3)	81.96 (93)
O(6) O(12)	92.29 (95)	O(3) O(11)	95.30 (101)
O(5) O(6)	167.13 (96)	O(2) O(3)	158.57 (99)
O(9) O(12)	171.39 (88)	O(5) O(6)	163.17 (95)
O(1) O(4)	159.57 (102)	O(11) O(10)	170.97 (95)
$W(3)O_6$ octahedron		$W(4)O_6$ octahedron	
d(W(3)–O(9))	1.715 (23)	d(W(4)–O(10))	1.738 (25)
O(7)	1.796 (22)	O(8)	1.751 (21)
O(1)	1.856 (23)	O(3)	1.881 (25)
O(4)	1.974 (21)	O(2)	1.934 (21)
O(8)	2.085 (21)	O(7)	2.049 (22)
O(12)	2.176 (22)	O(11)	2.157 (24)
∡(O(7)–W(3)–O(12))	86.24 (92)	∡(O(7)–W(4)–O(10))	94.16 (100)
O(9)	101.08 (102)	O(11)	78.48 (88)
O(1)	99.89 (100)	O(2)	79.50 (88)
O(4)	91.83 (94)	O(3)	84.53 (93)
O(8) O(12)	77.41 (81)	O(8) O(10)	99.03 (104)
O(9)	94.85 (93)	O(11)	88.01 (93)
O(1)	82.73 (92)	O(2)	95.13 (93)
O(4)	81.27 (86)	O(3)	97.49 (98)
O(12) O(1)	83.40 (93)	O(10) O(2)	93.06 (101)
O(4)	80.46 (85)	O(3)	100.55 (106)
O(9) O(1)	100.16 (104)	O(7)	81.43 (88)
O(4)	94.11 (97)	O(8)	83.18 (93)
O(7) O(8)	163.08 (94)	O(7) O(8)	166.03 (95)
O(12) O(9)	171.09 (91)	O(10) O(11)	171.47 (98)
O(1) O(4)	159.37 (100)	O(2) O(3)	159.70 (97)

A second form has also been prepared differing from β-WO_3 in that there is no centre of symmetry (space group P1) [60T, 63C].

Physical property	Numerical value	Experimental conditions	Experimental method, remarks	Ref.

c) γ-WO$_3$: space group C_{2h}^5–P2$_1$/c, $Z=8$, distorted ReO$_3$-structure (Figs. 6, 7).

lattice parameters:

a	7.306 (1) Å	RT		69L2
b	7.540 (1) Å			
c	7.692 (1) Å			
β	90.88°			

interatomic distances (in Å) and fractional coordinates at RT:
(Atomic positions are shown in Fig. 6.)

$d(W(1)-O_{x1})$	1.89 (6)	$d(W(2)-O_{x1})$	1.92 (7)	66L
$d(W(1)-O_{x2})$	1.91 (6)	$d(W(2)-O_{x2})$	1.85 (8)	
$d(W(1)-O_{y1})$	1.72 (6)	$d(W(2)-O_{y2})$	1.85 (7)	
$d(W(1)-O_{y1})$	2.16 (6)	$d(W(2)-O_{y2})$	2.01 (8)	
$d(W(1)-O_{z1})$	2.13 (5)	$d(W(2)-O_{z1})$	1.75 (5)	
$d(W(1)-O_{z2})$	1.79 (5)	$d(W(2)-O_{z2})$	2.15 (5)	

Atom	x	y	z	
W(1)	0.254 (6)	0.037 (5)	0.282 (5)	69L2
W(2)	0.250 (9)	0.023 (6)	0.784 (5)	
O_{x1}	0.005 (5)	0.042 (5)	0.211 (4)	
O_{x2}	0.993 (6)	0.474 (5)	0.218 (5)	
O_{y1}	0.288 (3)	0.262 (6)	0.286 (3)	
O_{y2}	0.211 (3)	0.259 (8)	0.730 (3)	
O_{z1}	0.292 (6)	0.043 (5)	0.008 (4)	
O_{z2}	0.279 (7)	0.487 (6)	0.993 (3)	

density:

d	7.21 g cm^{-3}	$T=22$°C		66L

In this temperature regime, substoichiometric WO$_{2.98}$ has been indexed on space group C_{2h}^2–P2$_1$/m ($Z=4$) with:

a	7.354 (5) Å	RT		66G
b	7.569 (5) Å			
c	3.854 (5) Å			
β	90.6°			

d) orthorhombic WO$_3$: space group D_{2h}^{16}–Pmnb, $Z=8$, see Figs. 8 and 9.

lattice parameters:

a	7.341 (4) Å	$T=480$°C		77S
b	7.570 (4) Å			
c	7.754 (4) Å			

linear thermal expansion:

α_a	$1.7 \cdot 10^{-5}$ K^{-1}	$T=480$		77S
α_b	$-1.3 \cdot 10^{-6}$ K^{-1}	$\cdots 680$°C		
α_c	$1.8 \cdot 10^{-5}$ K^{-1}			

interatomic distances and fractional coordinates:

Atom	x	y	z	
W(1)	0.25	0.029	0.031	77S
W(2)	0.25	0.030	0.532	
O(1)	0	0	0	
O(1′)	0.5	0	0	

(continued)

Physical property	Numerical value	Experimental conditions	Experimental method, remarks	Ref.
interatomic distances and fractional coordinates (continued)				
O(2)	0	0	0.5		77S
O(2′)	0.5	0.5	0		
O(3)	0.25	0.269	0.027		
O(4)	0.25	0.278	0.471		
O(5)	0.25	0.004	0.262		
O(6)	0.25	0.015	0.776		
$d(W(1)-O(1))$	1.86	$d(W(2)-O(2))$	1.87	
$d(W(1)-O(1'))$	1.86	$d(W(2)-O(2'))$	1.87	
$d(W(1)-O(3))$	1.81	$d(W(2)-O(3))$	2.03	
$d(W(1)-O(4))$	1.91	$d(W(2)-O(4))$	1.94	
$d(W(1)-O(5))$	1.81	$d(W(2)-O(5))$	2.10	
$d(W(1)-O(6))$	1.98	$d(W(2)-O(6))$	1.89	

e) **tetragonal WO_3**: space group D_{4h}^7 – P4/nmm, $Z=2$. Four O-atoms lie in a plane and one each above and below the W-atoms, which are themselves displaced some $0.06\,c$ ($=0.23$ Å) out of the O_4 plane; see Fig. 10.

lattice parameters and linear thermal expansion:

a	5.191 (2) Å	$T=740\,°C$		75S1
c	3.858 Å			
α_a	$2.33 \cdot 10^{-5}\,K^{-1}$	$T=770\cdots950\,°C$		52K
α_c	$7.1 \cdot 10^{-6}\,K^{-1}$			

parameters of the phase transitions:

α-WO_3 to β-WO_3: first order transition showing considerable hysteresis

$T_{tr\downarrow}$	$-40\,°C$		X-ray	60T
	$-50\,°C$		domain structure	52H
	$-55\,°C$		dilatometry	49F
	$-40\,°C$		single crystal, conductivity	65C
	$-43\,°C$		single crystal, conductivity	70B1
$T_{tr\uparrow}$	$-25\,°C$		X-ray	60T
	$-10\,°C$		domain structure	52H
	$-15\,°C$		dilatometry	49F
	$-22\,°C$		single crystal, conductivity	65C
	$-23\,°C$		single crystal, conductivity	70B1
ΔH_{tr}	39 cal mol^{-1}			70B1
	38 (2) cal mol^{-1}			76C
ΔS_{tr}	0.16 cal K^{-1} mol^{-1}			70B1
	0.154 cal K^{-1} mol^{-1}			76C

β-WO_3 to γ-WO_3:

$T_{tr\downarrow}$	9 °C		single crystal, conductivity	70B1
	1 °C		polycrystalline, conductivity	
	10 °C		single crystal, conductivity + Hall effect	65C
	-15 °C		conductivity	72H
	-19 °C		microwave reflectivity	
$T_{tr\uparrow}$	25$\cdots$35 °C		single crystal, conductivity	70B1
	21$\cdots$40 °C		polycrystalline, conductivity	
	27 °C		single crystal, conductivity + Hall effect	65C
	17 °C		conductivity	72H
	0 °C		microwave reflectivity	
ΔH_{tr}	20 cal mol^{-1}			70B1
	19 (2) cal mol^{-1}			76C

(continued)

Physical property	Numerical value	Experimental conditions	Experimental method, remarks	Ref.
β-WO₃ to γ-WO₃ (continued)				
ΔS_{tr}	0.07 cal K^{-1} mol^{-1}			70B1
	0.064 (6) cal K^{-1} mol^{-1}			76C

The transition temperature for substoichiometric WO$_{3-x}$ is lower; for $x = 9 \cdot 10^{-5}$ T_{tr} is about 2 °C lower and $\Delta H_{tr} \approx 8.5$ cal mol^{-1}. For $x \gtrsim 2 \cdot 10^{-4}$ the transition from the γ-form to the β-form is suppressed.

γ-WO₃ to orthorhombic WO₃: the transition occurs at 330 °C [59S], though the monoclinic angle β only becomes 90° at a higher temperature (467 °C) [75S1].

T_{tr}	330 °C	dilatometry	the transition is at least of	51S
	330 °C	heat capacity	second order; for variation of	56S1
	339 °C	resistivity	lattice parameters, see Fig. 11	76C
	330 °C	resistivity		59S, 63C
	335 °C	resistivity		58F
ΔH_{tr}	330 cal mol^{-1}			56S1
	7 (1) cal mol^{-1}			76C
ΔS_{tr}	0.55 cal K^{-1} mol^{-1}			56S1
	0.011 cal K^{-1} mol^{-1}			76C

orthorhombic to tetragonal WO₃: the transition appears higher than second order and over a wide temperature range both forms coexist. For a structural model, see Fig. 12

T_{tr}	710···770 °C	X-ray diffraction		70A
	750 °C	X-ray diffraction		66R
	740 °C	X-ray diffraction		75S1
	755 °C	dilatometry		51S
	740 °C	single crystal expansion		56S2, 55H
	730 °C	heat capacity		51S, 56S1
ΔH_{tr}	450 cal mol^{-1}			51S, 56S1, 58F
ΔS_{tr}	0.45 cal K^{-1} mol^{-1}			

The systematics of these transitions have been explained in terms of the condensation of soft phonon modes at the points Γ, X, M and R of the Brillouin zone of the hypothetical cubic form (O_h^1–Pm3m) and the P, Z, M and A points in the tetragonal (D_{4h}^7–P4/nmm) prototype. The condensed modes are:

monoclinic low-temperature phase (C_s^2–Pc): 80H

W $\Gamma_{15}^z + M_3'^z + (X_5^z)^x + (X_5^z)^y$
O $(R_{25}^x + R_{25}^y + R_{25}^z)$

triclinic phase (C_i^1–P$\bar{1}$): 80H

W $M_3'^z + (X_5^x)^y$
O $R_{25}^x + R_{25}^y + R_{25}^z$

monoclinic high-temperature phase (C_{2h}^5–P2$_1$/c): 80H

W $M_3'^z + (X_5^x)^y$
O $R_{25}^x + M_3^y + R_{25}^z$

orthorhombic phase (D_{2h}^{16}–Pmnb): 80H

W $M_3'^z + (X_5^x)^y$
O $M_3'^z + (M_5'^y)^z + 2(X_5^x)^y$

tetragonal phase (D_{4h}^7–P4/nmm): 80H

W $M_3'^z$
O $M_3'^z$

Physical property	Numerical value	Experimental conditions	Experimental method, remarks	Ref.

Electronic properties

band structure: No calculation of monoclinic or triclinic WO_3 has been reported but a calculation of cubic $NaWO_3$ and idealized cubic WO_3 [77K] gives the band structure of Fig. 13. The $NaWO_3$ density of states is shown in Fig. 14. Strong covalent interaction is evident.

energy gap (due to $O(2p) - W(5d)$ transitions):

E_g	2.585 eV	$T=300$ K, $E \parallel [001]$	absorption, see Fig. 15 for absorption coefficient, Fig. 16 for temperature shift	74S1, 76H1
	3.1 eV	0 K (extrapolated)		
dE_g/dT	$-9.0 \cdot 10^{-4}$ eV K^{-1}	a-polarized light	in the temperature range $0 \cdots 700$ °C	60I
	$-6.5 \cdot 10^{-4}$ eV K^{-1}	c-polarized light		

For higher band-band transitions, see section on "Optical properties".

Defects

For WO_{3-x} electron microscopy, density and XRD have established that for $x \lesssim 10^{-4}$ point defects, primarily oxygen vacancies, predominate, whereas for $x \gtrsim 10^{-4}$ crystalline shear planes oriented in the {10m} directions of the ReO_3 phase are found. The oxygen deficit is accompanied by a marked colour change from yellow to green and finally to black. EPR data of slightly substoichiometric WO_3 after IR illumination showed a signal ascribed to W^{5+} [77G]. The change in resistivity with oxygen pressure at 750 °C is shown in Fig. 19 and there is clear evidence for seven shear-like phases [77B]. See also the section "Transport properties".

Lattice properties

wavenumbers of Raman active phonons at RT (in cm^{-1}):

$\bar{v}_R$	277, 330, 722, 810	intense peaks in Raman spectra	69B1
	275, 297, 328, 352, 718, 808		69R
	275, 330, 719, 808		75S2
	275, 718, 807		69M2
	224, 327, 349, 378, 406, 418, 445, 574, 642	weaker peaks in Raman spectra	69M2
	71, 83, 135	lines below 200 cm^{-1}. These low frequency phonons show marked changes with temperature, Fig. 17	69R
	33, 60, 73, 93, 133		75S2

wavenumbers of IR active phonons at RT (in cm^{-1}):

$\bar{v}_{IR}$	765, 825, 920	intense bands in IR spectra	75S2
	775, 835, 900		75K
	768, 825, 920		69R
	745, 815, 920		69M2
	226, 252, 280, 290, 307, 322, 340, 360, 386	lower frequency bands	75K
	230, 285, 310, 335, 370		75S1
	375, 390		70O
	218, 241, 275, 295.5, 314, 328, 350.5, 378		76A

Debye temperature:

Θ_D	380 (15) K	$T = 1 \cdots 55$ K	heat capacity	74B

Below -50 °C WO_3 is piezoelectric (Fig. 18). Antiferroelectric grain boundaries are evident at low temperatures [51M, 70B2, 72S2, 76S3].

Transport properties

The complexity of the crystallographic transitions in WO_3 is reflected in the behaviour of the *resistivity* (see Figs. 19, 20). Throughout the region 0···200 K the material behaves like a semiconductor but anomalies at 40, 70, 130 and 185 K have been observed. In the higher temperature region, about RT, there are several anomalies (Fig. 21). The discontinuous decrease of the resistivity on passing from the low temperature monoclinic phase to the triclinic phase at $\approx -50\,°C$ is due to a change of a factor of ≈ 200 in the carrier concentration [70B1]. The ac resistivity also shows a strong anomaly at $\approx -50\,°C$ (Fig. 22).

activation energies for conductivity (in eV):

		T [K]	
E_A	0.10	187···260	75L
	0.19	140···185	
	0.052	81···110	
	0.028	38···68	
	0.006	14···34	
	0	4.2···14	
	0.19···0.10	below 223	
	0.4	below 223	80H
	0.23	below 223	70B1
	0.039	triclinic phase, 233···RT	80H

Above RT the resistivity of the monoclinic phase *rises* with T (Fig. 23). Further studies of both the Hall and Seebeck coefficients of the monoclinic and triclinic phases have been reported [70B1]:

further transport properties:

triclinic phase ($\approx 223\,K < T < RT$):

Hall carrier concentration: Fig. 24, Hall mobility: Fig. 25, activation energy for carrier generation 0.031 eV. The Hall mobility has been analyzed in terms of a large polaron model [67L]. The best fit to the data give: Fröhlich coupling constant $\alpha = 4.2$, effective polaron mass $m^{**} = 1.4\,m_o$, phonon effective temperature $\hbar\omega_{LO}/k = 225\,(50)\,K$. These results are somewhat out of line with the data for the high T monoclinic phase (γ-WO_3) and some additional scattering mechanism may be operative [70B1].

Seebeck coefficient: Fig. 26. Also shown are calculated values. The origin of the poor fit may lie in phonon drag effects, but a straight-forward application of present-day phonon drag theories [54H, 66P] is not possible.

monoclinic phase (γ-WO_3):

Hall carrier concentration: Fig. 27, Hall mobility: Fig. 28. Best fit to the large polaron model: $\alpha = 3.00$, $m_n = 1.75\,m_o$, $m^{**} = 3.26\,m_o$, $\hbar\omega_{LO}/k = 565\,K$.

Seebeck effect (Fig. 29) shows a trend opposite to that expected from elementary theory and phonon-drag effects again seem implicated. Activation energies for carrier generation 0.01···0.016 eV. Energies of the donor states responsible are $E_d = 0.009\,eV$, assuming $n_d = 3.9 \cdot 10^{18}\,cm^{-3}$ and $E_d = 0.04\,eV$, assuming $n_d = 1.9 \cdot 10^{18}\,cm^{-3}$ (temperature range 230···290 K) [70B1].

At higher temperatures results are scanty, see Fig. 30 for results showing a slight anomaly at 330°C. ac measurements are shown in Fig. 31. Discontinuities are apparent at 750°C, 910°C and 1235°C. dc resistivity showing anomalies at high temperature: Fig. 32.

reduced WO_3:

Substantial qualitative changes occur in behaviour on reduction to WO_{3-x}: (a) The monoclinic-triclinic transformation at ca. $-10\,°C$ is suppressed for $x \gtrsim 10^{-4}$ (Fig. 33). (b) As x increases, the activation energy for conduction in the triclinic phase increases. (c) As the monoclinic-triclinic transition is lost, the conductivity in the high-temperature monoclinic phase becomes activated (Fig. 34). (d) In the low-temperature monoclinic phase (α-WO_3) the activation energy decreases with increasing x (Fig. 33). (e) The change in resistivity around the γ-WO_3-triclinic transition is mimicked by the carrier concentration (Fig. 35). As x increases the activation energy for the carrier concentration in the (low-T) monoclinic region increases from

0.01···0.016 eV in the stoichiometric material to 0.03···0.05 eV in the range $0.0002 \lesssim x \lesssim 0.0005$. (f) The Hall mobility in the high-temperature monoclinic phase also becomes activated as x increases (Fig. 36) and the distinct change at the high-temperature monoclinic-triclinic transition is lost. This activation has been ascribed to impurity scattering, though a quantitative fit to the theory could not be obtained. (g) The Seebeck coefficient decreases strongly with increasing x in the high-temperature monoclinic region (Fig. 37) and becomes temperature independent. Below 300 K the behaviour is complex (Fig. 38).

Optical properties

optical spectra: absorption in the visible: Fig. 15, in the near IR: Fig. 39, optical density of thin films: Fig. 40, derived absorption coefficient for amorphous films: Fig. 41.

photoemission spectra [76H1, 77A, 78B] show a valence band about 4 eV wide for RT (γ) phase (Fig. 42). On reduction the W(5d) conduction band becomes clearly visible (Fig. 43). EELS shows a peak at 1.3 eV which becomes strongly enhanced on reduction [77R].

higher energy peaks (in eV, RT values):

Physical property	Numerical value	Experimental conditions	Experimental method, remarks	Ref.
E	6.5	O(2p)–W(5d)	(probably γ-phase)	77R
	14.0	O(2p)–W(6s)		
	21.0	O(2s)–CB		
	34.7, 35.2	W(4f)–CB		
	35.7	W(5p)–CB		
	36.9, 37.4	W(4f)–CB		
	43.1, 46.1	W(5p)–CB		
	52.9, 56.6	W(5s)–CB (?)		

dielectric constant:

seems to depend quite critically on the nature of the sample (Fig. 44 for single crystals, Fig. 45 for sintered samples). Above room temperature ε has been reported as going through a maximum value of $2\cdots3\cdot10^5$ [69B2]. Much lower values have been reported:

$\varepsilon(0)$	20	RT	evaporated films	73D
$\varepsilon(\infty)$	5	RT		
$\varepsilon(0)$	90	RT	amorphous films	78M
$\varepsilon(0)$	12.1···14.0	RT	sintered samples (probably γ-phase)	76S2

refractive index:

n_a	2.703 (35)	tungsten lamp,	temperature dependence: Fig. 46;	59S
n_b	2.376 (35)	$T=23\,°C$	the material is double-refracting	
n_c	2.283 (35)		(probably γ-phase)	

Magnetic properties

magnetic susceptibility:

χ_m	$-1.7\cdot10^{-10}\,\mathrm{m^3\,mol^{-1}}$	RT	χ_m in SI units	78S

References for 9.15.2.7.2

49F Foëx, M.: C. R. Acad. Sci. **228** (1949) 1335.
50M Magnéli, A.: Arkiv Kemi **1** (1950) 223.
51M Matthias, B.T., Wood, E.A.: Phys. Rev. **84** (1951) 1255.
51S Sawada, S., Ando, R., Nomura, S.: Phys. Rev. **84** (1951) 1054.
52H Hirakawa, K.: J. Phys. Soc. Jpn. **7** (1952) 331.
52K Kehl, W.L., Hay, R.G., Wahl, D.: J. Appl. Phys. **23** (1952) 212.
53M Magnéli, A.: Acta Crystallogr. **6** (1953) 495.
54H Herring, C.: Phys. Rev. **96** (1954) 1163.
55H Hededüs, A.J., Millner, T., Neugebauer, J., Sasvári, K.: Z. Anorg. Allgem. Chem. **281** (1955) 64.

References for 9.15.2.7.2

56S1	Sawada, S.: J. Phys. Soc. Jpn. **11** (1956) 1237.
56S2	Sawada, S.: J. Phys. Soc. Jpn. **11** (1956) 1246.
57P	Perri, J.A., Banks, E., Post, B.: J. Appl. Phys. **28** (1957) 1272.
58F	Foëx, M.: J. Rech. C. N. R. S. Lab. Bellevue **9** (1958) 1.
59S	Sawada, S., Danielson, G.C.: Phys. Rev. **113** (1959) 1008.
60I	Iwai, T.: J. Phys. Soc. Jpn. **15** (1960) 1596.
60T	Tanisaki, S.: J. Phys. Soc. Jpn. **15** (1960) 566.
63A	Ackermann, R.J., Rauk, E.G.: J. Phys. Chem. **67** (1963) 2601.
63C	Crowder, B.L., Sienko, M.J.: J. Chem. Phys. **38** (1963) 1576.
65C	Crowder, B.L., Sienko, M.J.: Inorg. Chem. **4** (1965) 73.
66G	Gebert, E., Ackermann, R.J.: Inorg. Chem. **5** (1966) 136.
66L	Loopstra, B.O., Boldrini, P.: Acta Crystallogr. **21** (1966) 158.
66P	Plavitu, C.N.: Phys. Status Solidi **16** (1966) 69.
66R	Roth, R.S., Waring, J.L.: J. Res. Nat. Bur. Stand. Ser. A**70** (1966) 281, 289.
67L	Langreth, D.C.: Phys. Rev. **159** (1967) 717.
69A	Allpress, J.G., Godo, P.: Cryst. Lattice Defects **1** (1969) 331.
69B1	Beattie, I.R., Gilson, T.R.: J. Chem. Soc. A**1969**, 322.
69B2	Le Bihan, R., Vacherand, C.: Croissance Composes Miner. Monocryst. **2** (1969) 147.
69L1	van Landuyt, J., Amelinckx, S.: Cryst. Lattice Defects **1** (1969) 113.
69L2	Loopstra, B.O., Boldrini, P.: Acta Crystallogr. **B25** (1969) 1420.
69M1	Marucco, J.F., Gerdanian, P., Godó, M.: J. Chim. Phys. **66** (1969) 674.
69M2	Mattes, R., Schröder, F.: Z. Naturforsch. **24B** (1969) 1095.
69R	Rocchiccioli, C.: C. R. Acad. Sci. Ser. **B268** (1969) 45.
70A	Ackermann, R.J., Sorrell, C.A.: High Temp. Sci. **2** (1970) 119.
70B1	Berak, J.M., Sienko, M.J.: J. Solid State Chem. **2** (1970) 109.
70B2	Le Bihan, R., Vacherand, C.: J. Phys. Soc. Jpn. Suppl. **28** (1970) 159.
70O	Ohwada, K.: Spectrochim Acta **A26** (1970) 1035.
70S	Sienko, M.J., Berak, J.M.: Chem. Ext. Defect in Non-Metallic Solids **1970**, 541. O'Keeffe, M., Eyring, L. (eds.).
72B	Bursill, L.A., Hyde, B.G.: J. Solid State Chem. **4** (1972) 430.
72H	Hirose, T., Kawano, I., Niino, M.: J. Phys. Soc. Jpn. **33** (1972) 272.
72S1	Schröder, F.A., Felser, H.: Z. Kristallogr. **135** (1972) 391.
72S2	Schröder, F.A., Hartmann, P.: Z. Naturforsch. **B27** (1972) 902.
73D	Deb, S.K.: Philos. Mag. **27** (1973) 801.
74B	Bevolo, A.J., Shanks, H.R., Sidles, P.H., Danielson, G.C.: Phys. Rev. **B9** (1974) 3220.
74S1	Salje, E.: J. Appl. Crystallogr. **7** (1974) 615.
74S2	Sundberg, M., Tilley, R.J.D.: J. Solid State Chem. **11** (1974) 150.
75K	Kiss, A.B.: Acta Chem. Acad. Sci. Hung. **84** (1975) 393.
75L	Lefkowitz, M., Dowell, M.B., Shields, M.A.: J. Solid State Chem. **15** (1975) 24.
75S1	Salje, E., Viswanathan, K.: Acta Crystallogr. **A31** (1975) 356.
75S2	Salje, E.: Acta Crystallogr. **A31** (1975) 360.
76A	Anderson, A.: Spectrosc. Lett. **9** (1976) 809.
76C	Coey, J.M.D., Roux-Bouisson, H., Schlenker, C., Lakkis, S., Dumas, J.: Rev. Gen. Thermodyn. **15** (1976) 1013.
76H1	Hollinger, G., Tranh Minh Duc, Deneuville, A.: Phys. Rev. Lett. **37** (1976) 1564.
76H2	Hoppmann, G., Salje, E.: Phys. Status Solidi (a) **37** (1976) K187.
76P	Pickering, R., Tilley, R.J.D.: J. Solid State Chem. **16** (1976) 247.
76S1	Salje, E.: Ferroelectrics **12** (1976) 215.
76S2	Samsonov, G.V., Fen', E.K., Malakhov, Ya.S., Malakhov, V.Ya.: Izv. Akad. Nauk SSSR, Neorg. Mater. **12** (1976) 1404.
76S3	Schröder, F.A.: Acta Crystallogr. **A32** (1976) 342.
76T	Tilley, R.J.D.: J. Solid State Chem. **19** (1976) 53.
77A	de Angelis, B.A., Schiavello, M.: J. Solid State Chem. **21** (1977) 67.
77B	Bonnet, J.-P., Marquastant, E., Onillon, M.: Mater. Res. Bull. **12** (1977) 361.
77G	Grazzinelli, R., Schirmer, O.F.: J. Phys. C **10** (1977) L145.
77I	Iguchi, E., Tilley, R.J.D.: J. Solid State Chem. **21** (1977) 49.
77K	Kopp, L., Harmon, B.N., Liu, S.H.: Solid State Commun. **22** (1977) 677.
77R	Ritsko, J.J., Witzke, H., Deb, S.K.: Solid State Commun. **22** (1977) 455.

77S	Salje, E.: Acta Crystallogr. **B33** (1977) 574.
78B	Beatham, N.: D. Phil. Thesis (Oxford) 1978.
78D	Diehl, R., Brandt, G., Salje, E.: Acta Crystallogr. **B34** (1978) 1105.
78M	Mansingh, A., Sayer, M., Webb, J.B.: J. Non-Cryst. Solids **28** (1978) 123.
78S	Snare, J., Karner, R., Leftkowitz, J., Friedberg, C.B., Ruvalds, J.: Bull. Am. Phys. Soc. **23** (1978) 309.
80H	Hirose, T.: J. Phys. Soc. Jpn. **49** (1980) 562.

9.15.3 Binary transition-metal chalcogenides

Review papers: [68H, 69W, 72J, 75W2, 76R, 76W2, 82G]. For structural properties, see also Landolt-Börnstein, Vol. III/6 and III/14 (in preparation).

9.15.3.1 M_{IV} (= Ti, Zr, Hf)-chalcogenides

With the possible exception of Zr_2S_3, Hf_2S_3, and Hf_2Se_3, all the semiconducting chalcogenide phases with M_{IV} cations formally contain these cations in their quadrivalent state. The several phases in the Ti−S system are shown in Fig. 1.

The stoichiometric compounds $M_{IV}S_2$ and $M_{IV}Se_2$ crystallize in the C6 layer structure of CdI_2. The conduction bands (Figs. 6, 12, 18, 21, 24, 26) are primarily cation-d bands, the valence bands anion-p^6. The energy gap in TiS_2 is significantly smaller than those of ZrS_2 and HfS_2; in semimetallic $TiSe_2$ the gap has just vanished, and in metallic $TiTe_2$ the two bands overlap significantly. Stoichiometric TiS_2 has not been prepared; all samples are titanium-rich n-type $Ti_{1+x}S_2$, some interstitial titanium occupying the octahedral sites between TiS_2 layers. The lattice parameters increase linearly with x, Fig. 2. The room-temperature data in the table below are for samples approaching the stoichiometric limit. The optical data of Figs. 5, 11, 17, 20, 23, 25 have been interpreted in terms of the band diagrams of Figs. 6, 12, 18, 21, 24, 26. The anomalous tranport data of $Ti_{1+x}Se_2$ in the interval $100 < T < 200$ K and the change from n-type to p-type conductivity reflect a second-order phase transition at $T_{tr} \simeq 200$ K and the semimetallic character of the compound. Holes in the Se:$4p^6$ valence bands are more mobile than electrons in the Ti:3d conduction bands. ZrS_2 exhibits a pronounced decrease in conductivity with increasing temperature in the interval $250 < T < 500$ K where the donor centers for extrinsic conduction are exhausted and intrinsic conduction is not significant, Figs. 13···15.

The compounds $M_{IV}S_3$ crystallize in the $ZrSe_3$ structure illustrated in Fig. 16a. It contains linear chains of edge-shared trigonal prismatic sites held together by van der Waals forces. Each shaded triangle has a dimeric-anion base and a monomeric-anion apex to give a formal valence $Zr^{4+}Se^{2-}(Se_2)^{2-}$, and the empty antibonding orbitals of the $(Se_2)^{2-}$ unit have energies well above the bottom of the Zr:4d conduction band. Very little work has been done on the compounds Zr_2S_3, Hf_2S_3, and Hf_2Se_3. If these compounds are semiconductors, as suggested, then we must anticipate the formation of dimeric cations capturing two electrons per dimer in a metal-metal bond having an energy near the top of the anion-p^6 valence band. Such cationic dimeric units are found in Mo_2S_3 [61J1]. However, these compounds may prove to be metallic and without cationic dimerization.

In the tables below the following abbreviations have been used:

S: Structure (space group).
CG: Crystal growth (the numbers in parentheses correspond to T_1 and T_2, the temperatures (in °C) of the hot and cold end of the crystal growth tube, respectively).
C: Colour (colour of polycrystalline samples is given in parentheses).
$E_{g,th}$: Values of the thermal energy gap are twice the activation energy E_A obtained from a $\log \varrho$ vs. T^{-1} plot.

If not stated otherwise, all data in the tables below are at RT.

Physical property	Numerical value	Experimental conditions	Experimental method, remarks	Ref.
$Ti_{1+x}S_2$ (Figs. 1···6) ($0 \leq x \leq 0.1$)				
			at 1000°C: $0.04 \leq x \leq 0.1$	63B,
			at 800°C: $0.03 \leq x \leq 0.1$	65G,
			at 600°C: $0 \leq x \leq 0.1$	67G,
a	3.407 Å	$x \approx 0$	S: C6, $D_{3d}^3 - P\bar{3}m1$	68C2,
c	5.695 Å		CG: halogen transport (900/800)	69L,
d	3.22 g cm^{-3}		C: brassy	72T,
$\varrho_\perp$	$5 \cdot 10^{-3}$ Ω cm	n-type, synthetic single crystal	$d\varrho/dp < 0$	75C, 75T1, 75T2, 75W2,
$\varrho_a = \varrho_0 + \varrho_1 T^2 = 1.95 \cdot 10^{-4}$ Ω cm $+ (2.0 \cdot 10^{-8}/k^2 \, T^2)$ [Ω cm]			$\varrho_1 \propto n^{-5/3}$, where $n =$ carrier concentration; shows electron-electron scattering is dominant	77F, 78B, 83H1
$S_\perp$	$-265 \cdots -40$ μV K^{-1}	$0 \leq x \leq 0.1$		
$\mu_{H\perp}$	9.7 cm^2/Vs		$dR_H/dp = 0$	
n	$9 \cdot 10^{20}$ cm^{-3}			
m_n	1.5 m_0			
g	4.3			
E_g	0.7 eV		optical gap	
	0.7 eV		indirect gap, calculated	
α_a	$0.96 \cdot 10^{-5}$ K^{-1}			
α_c	$1.94 \cdot 10^{-5}$ K^{-1}			
χ_m	10^{-5} cm^3 mol^{-1}		χ_m in CGS-emu	
TiS_{3-x} (Figs. 1, 7) ($0 \leq x \leq ?$)				
a	4.958 Å		S: ZrSe$_3$-type, $C_{2h}^2 - P2_1/m$	61G,
b	3.401 Å		CG: sublimed from poly- crystalline mass with excess sulfur at 610°C	63H1, 75F
c	8.778 Å			
β	97.32°			
			C: graphite	
ϱ_b	4.0 Ω cm	n-type, synthetic single crystal		
S_b	-500 μV K^{-1}			
E_g	0.9 eV		optical gap	
$Ti_{1+x}Se_2$ (Figs. 8···12) ($0 \leq x \leq 0.018$)				
a	3.537 Å		S: C6, $D_{3d}^3 - P\bar{3}m1$	76D,
c	6.00 Å		second-order phase transition at $T_{tr} = 200$ K, $T < T_{tr}$: transverse atomic displacements with $q = (\frac{1}{2}, 0, \frac{1}{2})$	77F, 78B, 78Z
			CG: halogen transport (900/$\leq$800)	
			C: metallic	
$\varrho_\perp$	$1.32 \cdot 10^{-3}$ Ω cm	p-type, synthetic single crystal	$d\varrho/dp < 0$	
$\varrho_\parallel$	$3 \cdot 10^{-3}$ Ω cm			
$S_\perp$	20 μV K^{-1}		n-type for $T < T_{tr}$	
R_H ($B \parallel c$)	$3.25 \cdot 10^{-2}$ cm^3/C		$dR_H/dp < 0$	
p	10^{20} cm^{-3}			
E_g	≈ 0.0 eV		optical gap (semimetal)	
	0.0 eV		calculated	

Physical property	Numerical value	Experimental conditions	Experimental method, remarks	Ref.
Zr_2S_3				
a	10.253 Å		S: defective B1, random Zr vacancies	58M1, 58M2
			C: (black)	
ϱ	0.08 Ω cm	n-type, poly-		
S	-50 μV K^{-1}	crystalline sample		
ZrS_2 (Figs. 13···15, 17, 18)				
a	3.66 Å		S: C6, $D_{3d}^3 - P\bar{3}m1$	65G,
c	5.85 Å		CG: halogen transport (900/800)	68C2,
d	3.87 g cm^{-3}		C: violet metallic	69L,
$\varrho_\perp$	0.76 Ω cm	n-type,		78B,
$\mu_{H\perp}$	4.3 cm^2/Vs	synthetic		83H1
n	$1.1 \cdot 10^{18}$ cm^{-3}	single crystal		
E_g	1.7 eV		optical gap, $dE_g/dT = -4.2 \cdot 10^{-4}$ eV K^{-1}	
	1.7 eV		indirect gap, calculated	
S	-700 μV K^{-1}	n-type, poly-crystalline sample		
ZrS_{3-x} (Fig. 19) ($0 \leq x \leq 0.05$)				
a	5.123 Å		S: $ZrSe_3$-type, $C_{2h}^2 - P2_1/m$	61G,
b	3.627 Å		CG: halogen transport	63H1,
c	8.986 Å		(750···900/550···600)	73S,
β	97.15°		C: dark orange	75F
ϱ_b	10^4 Ω cm	n-type,		
S_b	-500 μV K^{-1}	synthetic single crystal		
E_g	2.8 eV		optical gap	
Zr_2Se_3				
a	3.757 Å		S: hexagonal	58M1,
c	18.63 Å		C: (black)	58M2
ϱ	$7 \cdot 10^{-3}$ Ω cm	n-type, poly-		
S	-50 μV K^{-1}	crystalline sample		
$Zr_{1+x}Se_2$ (Figs. 6b, 20, 21) ($0.05 \leq x \leq 0.08$)				
a	3.76 Å		S: C6, $D_{3d}^3 - P\bar{3}m1$	58M2,
c	6.15 Å		CG: halogen transport (900/800)	65G,
			C: metallic grey	69L,
ϱ	0.1 Ω cm	n-type, poly-		73B2,
S	-300 μV K^{-1}	crystalline sample		78B
E_g	0.87 eV	$Zr_{1.05}Se_2$	optical gap	
	1.2 eV	$ZrSe_2$	optical gap	
	1.0 eV		indirect gap, calculated	

Physical property	Numerical value	Experimental conditions	Experimental method, remarks	Ref.
ZrSe$_3$ (Fig. 19)				
a	5.411 Å		S: ZrSe$_3$ type, $C_{2h}^2 - P2_1/m$	61G,
b	3.749 Å		CG: from crystalline mass	65K2,
c	9.444 Å		C: dark grey	75F,
β	97.48°			79B
ϱ_b	10^3 Ω cm	n-type, synthetic single crystal		
S_b	-10^3 μV K^{-1}			
E_g	1.25 eV		optical gap	
	1.6 eV		direct gap, calculated	
	1.5 eV		indirect gap, calculated	
Hf$_2$S$_3$				
a	3.635 Å		S: hexagonal, cation vacancies randomly distributed on alternate (111) planes of B8 structure	58M1, 58M2
c	5.839 Å			
			C: (light yellow-brown)	
ϱ	150 Ω cm	n-type, poly-crystalline sample		
S	-200 μV K^{-1}			
HfS$_2$ (Figs. 6b, 22···24)				
a	3.622 Å		S: C6, $D_{3d}^3 - P\bar{3}m1$	65G,
c	5.88 Å		CG: halogen transport (900/800)	68C2,
			C: dark-red metallic	70W1,
$\varrho_\perp$	$3.3 \cdot 10^7$ ···$1.6 \cdot 10^8$ Ω cm	n-type, synthetic single crystal		78B
E_g	1.96 eV		optical gap, $dE_g/dT = -4.3 \cdot 10^{-4}$ eV/K	
	1.9 eV		indirect gap, calculated	
$E_{g,th}$	2.1 eV	$T > 500$ K		
HfS$_3$				
a	5.09 Å		S: ZrSe$_3$ type, $C_{2h}^2 - P2_1/m$	61G,
b	3.59 Å		CG: halogen transport (750···900/550···600)	63H1,
c	8.96 Å			71L,
β	98.4°		C: ochre	73S, 75F
ϱ_b	10^2 Ω cm	synthetic single crystal		
E_g	3.1 eV		optical gap	
Hf$_2$Se$_3$				
			C: black	58M2
ϱ	6 Ω cm	n-type, poly-crystalline sample		
S	-200 μV K^{-1}			

Physical property	Numerical value	Experimental conditions	Experimental method, remarks	Ref.
HfSe$_2$ (Figs. 6b, 25, 26)				
a	3.733 Å		S: C6, D_{3d}^3–P$\bar{3}$m1	65G,
c	6.146 Å		CG: halogen transport (900/800)	73B2,
			C: dark-red metallic	78B
E_g	1.08 eV	synthetic single crystal	$dE_g/dT = -6.8 \cdot 10^{-4}$ eV/K (E_g: optical gap)	
	1.1 eV		indirect gap, calculated	

9.15.3.2 M$_V$ (= V, Nb, Ta)-chalcogenides

Of the known binary vanadium chalcogenides, all are metallic with the probable exception of diamagnetic VS$_4$, which contains only dimeric anions and V–V pairing (2.83 vs. 3.22 Å) along the monoclinic c-axis [65K1]. NbS$_3$ crystallizes in a structure similar to the ZrSe$_3$ structure of Fig. 16a, but distorted by the formation of c-axis Nb–Nb dimers [78R]; TaS$_3$ has an analogous structure, Fig. 16b. With formal valences Nb^{4+} and Ta^{4+}, the compounds are semiconducting (see Figs. 27 and 34) because there are two crystallographically distinguishable cations within a chain, which splits the occupied d band in two. On the other hand, NbSe$_3$, which is isostructural with NbS$_3$, remains metallic at all temperatures although exhibiting phase transitions at 142 K and 58 K due to the formation of incommensurate charge-density waves in the linear chains. A conductivity strongly non-linear with temperature below these transition temperatures appears to reflect a contribution from sliding charge-density waves [77T2, 78F, 80F].

TaS$_2$, like several other dichalcogenides, crystallizes in a variety of polytypes; these are illustrated in Fig. 28 [69H]. Each consists of different stackings of X–M–X layers. Both octahedral and trigonal-prismatic coordination of the metal ions are found, and metallic behaviour is more common with trigonal-prismatic coordination. Of the several TaS$_2$ phases (1T, 2H, 3R and 4H), only 1T–TaS$_2$ (CdI$_2$ structure) becomes semiconducting at low temperatures, ($T < T_{d'} \simeq 350$ K), see Figs. 29···31. The 1T–TaS$_2$ phase is stable at high temperatures, but it can be retained at low temperatures by quenching from above 1050 K. Three low-temperature phases are identified [74W, 75S, 76W2]: 1T$_1$ stable in the interval $T_{d'} \simeq 350$ K $< T \lesssim 450$ K, (a transformation to the 2H structure is reported at about 450 K) contains incommensurate periodic distortions ($q = 0.283 a^*$) arising from softened LA modes (ICDW = incommensurate charge-density wave) and partially softened phonons in selected TA$_\parallel$ modes; 1T$_2$ stable in the interval $T_d = 190$ K $< T < T_{d'}$, with the ICDW having its q rotated through the transition by about 11.5° and continuing to rotate with decreasing temperature towards its commensurate position $q = 0.231 a^* + 0.077 c^*$ in the low-temperature phase 1T$_3$ stable below T_d. The commensurate charge-density wave (CCDW) has a superlattice edge $\sqrt{13}\, a$ in the basal plane and a repeat every 13 layers in the c-direction [80S]. At lowest temperatures, the CCDW amplitude, as measured by X-ray photoelectron spectroscopy, approaches one electron per atom, and the atomic displacements at room temperature are as large as 0.25 Å [76W1]. Band-structure calculations [73M1, 77W], Fig. 33, are consistent with the opening of a band gap at the Fermi surface due to CDW formation. Although 1T–TaSe$_2$ shows analogous structural changes at low temperature [75W2], it remains metallic (Fig. 29). The ICDW sets in below the onset temperature $T_0 \simeq 600$ K, and the first-order CCDW to ICDW transition occurs at $T_d = 473$ K. Calculation of the Fermi surface (Fig. 33a) is consistent with a CCDW having the measured $q = 0.28 a^* + c^*/3$, where a^* and c^* are the reciprocal-lattice vectors.

For the abbreviations used in the tables below, see the top of the tables in section 9.15.3.1

NbS$_3$ (Fig. 27)				
a	4.963 Å		S: distorted ZrSe$_3$ type, C_i^1–P$\bar{1}$	62G,
b	6.730 Å		(Nb displaced 0.16 Å to form	69K,
c	9.144 Å		Nb-Nb pairs)	71L,
α	90°		CG: halogen transport at 600 °C	78R
β	97.17°			
γ	90°			
d	4.143 g cm^{-3}			
ϱ	$6.6 \cdot 10^3$ Ω cm	p-type, poly- crystalline sample	diamagnetic	
S	500 μV K^{-1}			

Physical property	Numerical value	Experimental conditions	Experimental method, remarks	Ref.
1T – TaS$_2$ (Figs. 29···33)				
a	3.365 Å	(modified by	S: C6, D_{3d}^3–P$\bar{3}$m1. First-order	71C,
c	5.853 Å	formation	semiconductor-semiconductor	71T1,
		of charge	transition at T_d = 150···200 K,	72C2,
		density	first-order semiconductor-metal	73M2,
		waves)	transition at $T_{d'}$ = 325···350 K	74G,
			(see discussion).	74W,
ϱ_a	10^{-3} Ω cm	n-type	$dT_{d'}/dp = -3.48$ K/kbar,	75S,
$R_H (B\|c)$	-10^{-3} cm^3/C	synthetic	$dT_d(\uparrow)/dp = -5.65$ K/kbar,	75W1,
n	10^{22} cm^{-3}	single	$dT(\downarrow)/dp$ is non-linear, curve	75W2,
		crystal	fitted to $182(1-(p/5.7))^{1/2}$,	76W2,
S_a	-9.5 μV/K		p in kbar. $\uparrow$ and $\downarrow$ correspond	77D,
	$+3.1$ μV/K	$T = 363$ K	to heating and cooling,	77T1,
			respectively.	77W,
			CG: halogen transport (920/820)	77Y,
			C: golden yellow	79I2,
E_g	2.3 eV		calculated	80J,
$E_{g,th}$	$2 \cdot 10^{-4}$ eV	$8 < T < 140$ K		80S,
	$4 \cdot 10^{-2}$ eV	$250 < T < 340$ K		83H2
TaS$_3$ (Fig. 34)				
a	36.804 Å	$T < 210···218$ K	S: orthorhombic, D_2^5–C222$_1$.	63A,
b	15.173 Å		Semiconductor-metal tran-	64B2,
c	3.340 Å		sition at 210···218 K due to	77S,
			Peierls instability (Ta–Ta	78T,
			dimers in chains for $T < T_{tr}$).	79I1
			$dT_{tr}/dp = -1.3$ K/kbar	
			CG: from elements at 700 °C in	
			vacuum and slow cooling	
ϱ_c	$3 \cdot 10^{-4}$ Ω cm	p-type,		
S_c	21 μV K^{-1}	synthetic		
		single		
		crystal		
$E_{g,th}$	0.3 eV	$120 < T < 200$ K	$dE_{g,th}/dp = -8$ K/kbar	

9.15.3.3 M$_{VI}$ (= Cr, Mo, W)-chalcogenides

Of the several phases in the Cr–S system, Fig. 35, only CrS and Cr$_2$S$_3$ are semiconducting; the others are metallic, but retain spontaneous atomic moments and order magnetically at low temperatures.

Above about 550 K, stoichiometric CrS has the hexagonal B8 structure of NiAs. At lower temperatures the crystal symmetry is lowered to monoclinic [57J] by a cooperative Jahn-Teller distortion of the octahedral sites, which signals the presence of a high-spin d^4 configuration at the Cr^{2+} ions occupying these sites. The Cr^{2+}-ion atomic moments order antiferromagnetically below T_N = 450 K [69P]. The electrical conductivity increases sharply with temperature above T_N, particularly on passing through the monoclinic – hexagonal phase transition (≈ 550 K), see Fig. 37. The compounds tend to be Cr-deficient and p-type. Since the Cr^{2+} : 3d^4 configuration should have an energy in the gap between the filled S^{2-} : 3p^6 bands and the empty Cr : 4s band, the electrical conductivity should be due to electronic transport in the narrow, σ* (σ-antibonding) d bands of e$_g$ parentage. The twofold orbital degeneracy is removed by the Jahn-Teller distortion below $T_{tr} \simeq 550$ K, and intraatomic-exchange correlation splits the up-spin and down-spin energies at a given ionic position. However, the high conductivity in the hexagonal phase suggests that any correlation splitting of the σ* band is only comparable to the band width. Indeed, stoichiometric CrSe does not exhibit the cooperative Jahn-Teller distortion, and it is metallic above T_N; it is a small-gap semiconductor only below T_N [77J]. The reported T_N and semiconductor-metal transition (see Fig. 44) varies sensitively with composition from T_N = 575 K down to $T_N \simeq 300$ K [62M]. Below T_N, the magnetic order has the peculiar "umbrella" configuration of Fig. 46 [61C]. A semiconductor-metal transition occurs at room temperature under hydrosta-

Physical property	Numerical value	Experimental conditions	Experimental method, remarks	Ref.

tic pressure [77J]. The critical pressure for CrS is $p_{crit} \simeq 24$ kbar, for CrSe it is $p_{crit} \simeq 4$ kbar. All these data are consistent with high-spin configurations at the Cr^{2+} ions having three localized t_{2g} electrons and a single electron in orbitals of e_g parentage that are sufficiently localized in CrS to produce a cooperative Jahn-Teller distortion, but not in CrSe. Ferromagnetic CrTe is most probably metallic and mostly low-spin. The decrease in Curie temperature with increasing hydrostatic pressure indicates itinerant-electron ferromagnetism. The spontaneous ferromagnetism disappears at pressures $p > p_{crit} \simeq 28$ kbar [78L].

The two polymorphs of Cr_2S_3 consist of the hexagonal B8 structure with two-thirds of the octahedral sites vacant and ordered on every other layer. The ordering in rhombohedral Cr_2S_3 makes $c = 3c_0$, where c_0 is the c axis of the B8 unit cell, and in trigonal $Cr_{2+x}S_3$ makes $c = 2c_0$ [57J]. Trigonal $Cr_{2+x}S_3$ is metallic [73B1]. Rhombohedral Cr_2S_3 contains localized $^4A_{2g}$ (t_{2g}^3) d-electron configurations at each octahedral Cr^{3+} ion, and below a ferrimagnetic Curie temperature T_C ($= 120$ K), see Fig. 43, the spontaneous magnetization has a maximum at about 85 K, indicative of a noncollinear spin configuration at lowest temperatures. The n-type resistivity exhibits a maximum at T_C due to spin-disorder scattering and a broad maximum at 85 K, Fig. 40. Lack of any significant field dependence of the Hall coefficient both above and below T_C indicates that it represents a normal Hall effect. The Hall mobility $\mu_H = R_H/\varrho$ is shown in Fig. 41 along with the carrier concentration $n = 1/R_H e$. The transverse magnetoresistance $\Delta\varrho/\varrho_0$, where $\Delta\varrho$ is the change in resistivity with applied magnetic field relative to its value ϱ_0 in zero applied field, has the form $-\Delta\varrho/\varrho_0 =$ const $\cdot H^p$ with $1 < p \leq 2$ for $T \geq T_C$ and $0 < p < 1$ for $T < T_C$; the negative magnetoresistance exhibits a dramatic maximum between 85 K and $T_C = 120$ K, see Fig. 42, which may be attributed to spin-flip scattering. A maximum in μ_H at 85 K, where the spontaneous magnetism – and hence the spin splitting of the conduction band – has a maximum is also consistent with this interpretation.

Rhombohedral Cr_2Se_3 has been prepared at 3 kbar and at atmospheric pressure [67I, 73Y]. The former sample was an n-type semiconductor with a thermal energy gap at ≈ 300 K of 0.05 eV. The sample prepared at atmospheric pressure was metallic, but exhibited an apparent transition between two metallic states in the interval $70 < T < 225$ K. Trigonal $Cr_{2.04}Se_3$ prepared at atmospheric pressure was reported to be semiconducting with a thermal energy gap of 0.15 eV [73B1]. The magnetic data are consistent with a localized $Cr^{3+} : d^3$ configuration, which should have its energy below the top of the $Se^{2-} : 4p^6$ valence band.

Whereas Cr_3S_4 is reported to be metallic, isostructural Cr_3Se_4 is reported to be semiconducting, see Fig. 48. Semiconducting behaviour for this mixed-valent compound requires an ordering of Cr^{2+} in the half-occupied cation basal planes and Cr^{3+} in the totally occupied cation basal planes.

MoS_2 is found in the 2H and 3R polytypes of Fig. 28 [63J]. The valence band consists of overlapping $S : 3p^6$ and $Mo : d_{z^2}^2$ bands where the z axis is the crystallographic c axis; the conduction bands are the $Mo : d_{xy}^0 d_{x^2-y^2}^0$, and $\sigma_\perp/\sigma_\parallel \simeq 200$. $MoSe_2$ is analogous. $MoTe_{2-x}$, on the other hand, exhibits a semiconductor-metal transition at $T_{tr} = 820 \cdots 880$ °C going from the 2H (α-phase) to a distorted 1T (β-phase) polytype with increasing temperature [70V1]. The distortion consists of Mo-atom antiferroelectric displacements parallel to the c axis [66B]. The 1T polytype can be quenched to room temperature; this metastable phase is stable to nearly 750 K, see Fig. 55.

$Mo_{2.06}S_3$ exhibits two phase transitions, which makes its physical properties complex, see Fig. 36 [70D1]. The 3R and 2H polytypes of WS_2 and WSe_2 are similar to the molybdenum analogues. Orthorhombic WTe_2 is a layer structure differing from the 1T polytype by (1) hc stacking rather than hh stacking of the anion layers and (2) W displacement within the layers giving rise to zig-zag W chains having W–W separations of 2.861 Å [66B]. These displacements are not strong enough to split the 5d bands, so the compound is semimetallic with a band overlap of 0.05 eV and a conduction-band edge 0.075 eV above a broad $Te : 5p^6$ band [66K, 72A].

For the abbreviations used in the tables below, see the top of the tables in section 9.15.3.1.

CrS (Figs. 36⋯39)

a	3.826 Å	S: monoclinic, C2/c	57J,
b	5.913 Å	(hexagonal, space group	60K,
c	6.089 Å	$D_{3h}^4 - C\bar{6}2c$ with $a = 2\sqrt{3} \, a_0$	69P,
β	101° 36′	$= 12$ Å and $c = 2c_0 = 11.74$ Å,	76V,
d	4.85 g cm^{-3}	where a_0, c_0 are the lattice	77J,
		parameters of undistorted B8 cell,	77M1,
		also used for low temperature,	83H1
		semiconducting phase [77L].	

(continued)

Physical property	Numerical value	Experimental conditions	Experimental method, remarks	Ref.
CrS (continued)				
ϱ	$0.1 \cdots 1 \, \Omega$ cm	p-type, polycrystalline sample for CrS_x and $x \leq 1.12$	Semiconductor-metal transition at 550 K, $dT_{tr}/dp = -15$ K/kbar. Transition at 24 kbar (at RT) to metallic phase. Antiferromagnetic, $T_N = 450$ K, $p_A = 3.5 \, \mu_B$/Cr atom at 80 K, moments directed along c axis, see discussion.	
Cr_2S_3 (Figs. 40$\cdots$43)				
a	5.939 Å		S: trigonal, $C_{3i}^2 - R\bar{3}$,	57J,
c	16.65 Å		n- or p-type depending	69S,
d	3.77 g cm^{-3}		on preparation; ferrimagnetic, $T < T_C = 120$ K with three magnetic sublattices, Curie-Weiss paramagnetism, $T > 300$ K, $\Theta_p = -585 \, (3)$ K; $C_A = 2.12 \, (1)$ cm^3 K/g-atom	70V1, 73B1, 76B1, 83H1
ϱ	$2.6 \cdot 10^3 \, \Omega$ cm	n-type, polycrystal grown by halogen transport		
μ_H	4 cm^2/Vs			
n	$8 \cdot 10^{18}$ cm^{-3}			
$E_{g,th}$	1.1 eV	$T \approx 275$ K		
ϱ	$7 \cdot 10^3 \, \Omega$ cm	p-type, single crystal grown under 3 kbar pressure		
S	1600 μV K^{-1}			
$E_{g,th}$	0.1 eV			
CrSe (Figs. 44$\cdots$47)				
a	3.67 Å		S: B8, $D_{6h}^4 - P6_3/mmc$, see discussion	61C, 62M, 67I, 77J
c	6.01 Å			
$Cr_{2+x}Se_3$ ($x=0.04$)				
a	6.28 Å		S: trigonal, $P\bar{3}1c$ for Cr_2Se_3, antiferromagnetic with $T_N = 43$ K and $p_{eff} = 3.84 \, \mu_B$	67I, 73B1, 73Y
c	11.64 Å			
ϱ	0.32 Ω cm	polycrystal		
S	$-190 \, \mu$VK^{-1}			
$E_{g,th}$	0.148 eV	$T > 310$ K		
	0.05 eV	$T < 310$ K		
Cr_3Se_4 (Fig. 48)				
a	6.299 Å		S: monoclinic, $C_{2h}^3 - I2/m$, antiferromagnetic with $T_N = 82$ K and $p_{eff} = 4.49 \, \mu_B$	71L, 73B1, 73Y
b	3.607 Å			
c	11.731 Å			
β	91°52′			
ϱ	0.2 Ω cm	polycrystal		
$E_{g,th}$	0.14 eV			

Physical property	Numerical value	Experimental conditions	Experimental method, remarks	Ref.
$Cr_{1-x}Te$ ($0 \leq x \leq 0.25$)				
a	3.93 Å		S: B8, $D_{6h}^4 - P6_3/mmc$,	60H,
c	6.15 Å		ferromagnetic, $T_C = 343$ K,	66S2,
ϱ	10^{-3} Ω cm	n-type, poly-	$p_A = 2.39 \mu_B/Cr$ atom at 0 K.	76G
S	-24 μV K^{-1}	crystalline	T_C depends on x.	78L
$E_{g,th}$	0.02 eV	sample	$dT_C/dp = -6$ K/kbar,	
			$p > 28$ kbar:	
			no ferromagnetism observed	
			CG: Bridgman method at 1280 °C	
2H–MoS$_2$ (Figs. 6b, 49···51)				
a	3.16 Å		S: hexagonal, $D_{6h}^4 - P6_3/mmc$	60J,
c	12.29 Å		CG: from polycrystalline mass	69C,
			at 1050 °C	69W,
			C: black	72A,
$\varrho_\perp$	33 Ω cm	n-type,	$d\sigma_\perp/dp = 0.065$ Ω^{-1} cm^{-1}/kbar,	76N,
n	$6 \cdot 10^{15}$ cm^{-3}	synthetic	$p < 10$ kbar,	77E,
		single	$d\sigma_\perp/dp = 0.03$ Ω^{-1} cm^{-1}/kbar,	78B,
		crystal	$12 \leq p \leq 30$ kbar,	78E
			$\sigma_\perp/\sigma_\parallel \simeq 200$.	
			($\varrho_\perp = 0.27$ Ω cm [70W1].)	
E_g	1.971 eV	$T = 77$ K	optical gap,	
			$dE_g/dp = -3.8$ meV/kbar	
			($= -2.5$ meV/kbar for a natural	
			crystal)	
	0.7 eV		indirect gap, calculated	
	1.4 eV		direct gap, calculated	
$E_{g,th}$	0.47 eV	$T < 300$ K		
	1.27 eV	$T > 790$ K		
3R–MoS$_2$				
a	3.17 Å		S: trigonal, C_{3v}^5-R3m	67F,
c	18.38 Å		CG: halogen transport (947/890)	69C,
$\varrho_\perp$	0.3 Ω cm	synthetic	$\sigma_\perp/\sigma_\parallel \simeq 200$	70W1,
E_g	1.9 eV	single	optical gap,	72A
		crystal	$dE_g/dp = -2.5$ meV/kbar for a	
			natural crystal	
$E_{g,th}$	0.25 eV			
MoS$_2$ (data for non-specified types)				
$\varrho_\perp$	18 Ω cm	n-type,		75G
$\mu_{H\perp}$	57 cm^2/Vs	natural		
n	$6 \cdot 10^{15}$ cm^{-3}	crystal		
E_g	1.9 eV		optical gap	
$E_{g,th}$	1.6 eV			
$\varrho_\perp$	2.0 Ω cm	p-type,		
$\mu_{H\perp}$	20 cm^2/Vs	synthetic		
p	$1.6 \cdot 10^{17}$ cm^{-3}	single		
		crystal		
E_g	1.9 eV		optical gap	
$E_{g,th}$	1.6 eV			

Physical property	Numerical value	Experimental conditions	Experimental method, remarks	Ref.
2H−MoSe$_2$ (Figs. 50···52)				
a	3.288 Å		S: hexagonal, D_{6h}^4−P6$_3$/mmc	62B,
c	12.92 Å		(3R-MoSe$_2$ also exists with	69W,
			a=3.292 Å, c=19.392 Å)	71E
			CG: from polycrystalline mass at 1000°C	72A, 72B2
			C: black	75G
$\varrho_\perp$	≈1···5 Ω cm	n-type,	$\sigma_\perp/\sigma_\parallel \simeq 200$	77E
$S_\perp$	−900 μV K^{-1}	synthetic		
n	0.35···1.6·10^{17} cm^{-3}	single crystal		
E_g	1.6 eV		optical gap, dE_g/dp=1.2 meV/kbar	
	1.6 eV		direct gap, calculated	
$E_{g,th}$	0.28 eV	T<300 K		
	0.95 eV	T>760 K		
2H−MoTe$_{2-x}$ (Figs. 53···56) ($0.01 \leqq x \leqq 0.1$)				
a	3.519 Å	hexagonal	S: semiconductor-metal	66B,
c	13.964 Å	modification	transition at T_{tr}, where	70V2,
a	6.33 Å	monoclinic	T_{tr}=820°C for Te-rich	70W1,
b	3.469 Å	modification	MoTe$_{2-x}$ and T_{tr}=880°C	72A,
c	13.86 Å		for Mo-rich MoTe$_{2-x}$.	72B2,
β	93°55′		$T<T_{tr}$: hexagonal, D_{6h}^4−P6$_3$/mmc	72D, 75G,
			$T>T_{tr}$: monoclinic, C_{2h}^2−P2$_1$/m	79C2
			CG: halogen transport (900/700); high-temperature form halogen transport (1000/900) and water quenched	
$\varrho_\perp$	0.5···8.5 Ω cm	n-type,		
$S_\perp$	−422 μV K^{-1}	synthetic		
R_H ($B\parallel c$)	−24.17 cm^3/C	single		
n	2.6·10^{17} cm^{-3}	crystal		
E_g	1.1 eV		optical gap	
	1.2 eV		direct energy gap, calculated	
$E_{g,th}$	1.08 eV	T>660 K		
3R−WS$_2$ (Fig. 57)				
a	3.162 Å		S: trigonal, C_{3v}^5−R3m	64W3,
c	18.50 Å		(hexagonal P6$_3$/mmc also exists with a=3.155 Å, c=12.35 Å)	70W1, 72B2
			CG: (i) Na$_2$CO$_3$ flux at 900°C, (ii) halogen transport	
$\varrho_\perp$	0.04 Ω cm	synthetic single crystal		
	0.1 Ω cm	natural crystal		
E_g	2.1 eV		direct gap, calculated	
	2.07 eV		optical gap	

Physical property	Numerical value	Experimental conditions	Experimental method, remarks	Ref.
2H–WSe$_2$ (Figs. 50, 58, 59)				
a	3.282 Å		S: hexagonal, $D_{6h}^4 - P6_3/mmc$	62B,
c	12.937 Å		CG: (i) from polycrystalline mass,	67F,
			(ii) halogen transport (800/720)	69W,
			C: black	72A,
$\varrho_\perp$	0.167 Ω cm	p-type,	$\sigma_\perp/\sigma_\parallel \simeq 200$ ($\varrho_\perp = 1.3$ Ω cm [70W1])	72B2,
$S_\parallel$	990 μV K^{-1}	synthetic		72D,
p	$2.35 \cdot 10^{17}$ cm^{-3}	single		77E,
		crystal		78E,
E_g	1.78 eV		optical gap	79A
	1.73 eV		direct gap, calculated	
$E_{g,th}$	0.17 eV	$T < 300$ K		
	1.33 eV	$T > 890$ K		
m_p	0.01 m_o		optical effective hole mass	
$\mu_{H\perp}$	100 cm^2/Vs	n-type		
$R_{H\perp}$	-80 cm^3/C			
n	$1.25 \cdot 10^{16}$ cm^{-3}			
WTe$_2$ (Figs. 60···62)				
a	6.282 Å		S: orthorhombic, $C_{2v}^7 - Pnm2_1$	66B,
b	3.496 Å		CG: halogen transport (900/700)	66K,
c	14.07 Å			72A
$\varrho_\perp$	$0.75 \cdot 10^{-3}$ Ω cm	n-type,	p-type $T \leq 60$ K,	
$S_\parallel$	-30 μV K^{-1}	synthetic	n-type $60 < T \leq 600$ K	
$R_{H\perp}$	$-5 \cdot 10^{-2}$ cm^3/C	single crystal		
ϱ	$1.5 \cdot 10^{-3}$ Ω cm	poly-	p-type $T \leq 60$ K,	
S	-10 μV K^{-1}	crystalline	n-type $60 < T \leq 400$ K,	
R_H	$-2 \cdot 10^{-2}$ cm^3 C^{-1}	sample	p-type $400 < T \leq 600$ K	

9.15.3.4 M_{VII} (= Mn, Tc, Re)-chalcogenides

Cubic α-MnS appears to be stable above room temperature. The β- and γ-phases can be prepared at low temperatures, but they transform to the α-phase above 200°C [67S1]. The α-phase is retained at all temperatures. Photoelectron spectroscopy has confirmed a $Mn^{2+}:3d^5$ valence band [78H], but holes in this band have a mobility that is not activated [78H] even though the manganese atoms carry a localized, high-spin atomic moment [56C]. Reversal of the sign of the Hall effect with temperature, Fig. 65, is attributed to a much larger electron mobility in the Mn:4s conduction band than hole mobility in the valence band, (some authors assign the charge-transfer optical absorption edge to $Mn:3d^6$ rather than $Mn:3d^54s^1$). The properties of α-MnSe are similar. MnTe, on the other hand, has the B8 structure with the top of the $Te:5p^6$ band above the $Mn^{2+}:3d^5$ band [77A], and lattice hole mobilities are about 50 cm^2/Vs [56V]. However, drift mobilities are lowered markedly as the temperature is raised through $T_N = 310$ K because of a large magnon-drag contribution [64W2]. This magnon-drag contribution increases the Seebeck coefficient and the resistivity as shown in Fig. 69.

Antiferromagnetic MnS$_2$ has the pyrite structure with dimeric anions [59H]; the σ-antibonding S–S orbitals of a dimer are empty with energies above the top of the cationic conduction band. As in MnS, photoelectron spectroscopy [59H] indicates that the valence band is the high-spin $Mn^{2+}:3d^5$ band responsible for the localized atomic moments, which order below $T_N = 48.2$ K [68L]. The magnetic susceptibility data suggest that holes in the $Mn^{2+}:3d^5$ valence band have non-activated mobilities, as was found for α-MnS [78H]. The transport data for isostructural MnTe$_2$, Figs. 70 and 71, show evidence of a magnon drag, which should be even more pronounced in MnS$_2$ [79A2].

The technetium and rhenium disulfides, diselenides and ditellurides are layered compounds having distorted CdCl$_2$ structures [71W1] as a result of a metal-metal bonding that changes the translational symmetry so as to split the d bands into a filled valence band and empty conduction bands.

Physical property	Numerical value	Experimental conditions	Experimental method, remarks	Ref.
For the abbreviations used in the tables below see the top of the tables in section 9.15.3.1			
α-MnS (Figs. 63···66)			
a	5.223 Å	$T > T_N$	S: cubic, B1, $O_h^5 - Fm3m$; $T < T_N$:	56C,
a	5.198 Å	$T < T_N$	trigonal distortion; anti-	67H,
β	90.099°		ferromagnetic, $T_N = 130$ K,	67S1,
			$p_{eff} = 5.6\,\mu_B$, $\Theta_p = -465$ K.	70M,
			Ferromagnetic (111) sheets	70R1,
			coupled antiparallel; spins	70W2,
			in ferromagnetic (111) planes	71L,
			CG: (i) in silica gel between	78H
			4°C and 41°C, (ii) halogen	
			transport (1050/550)	
			C: green	
ϱ	10^3 Ω cm	p-type,		
S	850 μV K^{-1}	synthetic single crystal		
μ_H	10 cm^2/Vs	$T = 625$ K		
R_H	10^2 cm^3/C			
E_g	2.8 eV		optical gap	
	3.2 eV		direct gap, calculated	
$E_{g,th}$	0.62 eV	$T \leq 435$ K		
	3.0 eV	$T \geq 590$ K		
β-MnS			
a	5.606 Å		S: cubic, B3, $T_d^2 - F\bar{4}3m$, antiferromagnetic, $T_N = 160$ K, $p_{eff} = 5.82\,\mu_B$, $\Theta_p = -982$ K, fcc ordering of 3rd kind; spins along c axis of tetragonal magnetic cell	56C, 71L
γ-MnS			
a	3.99 Å		S: hexagonal, B4, $C_{6v}^4 - P6_3mc$, antiferromagnetic, $T_N = 100$ K, $p_{eff} = 6.1\,\mu_B$, $\Theta_p = -932$ K. Pairs of ferromagnetic (011) sheets of orthorhombic cell coupled antiparallel ($a_{orth} = \sqrt{3}\,a_{hex}$, $b_{orth} = 2\,a_{hex}$, $c_{orth} = c_{hex}$), spins along ortho-[011]	56C, 67S1, 71L
c	6.47 Å			
			CG: in silica gel between 4°C and 41°C	
			C: pink	
MnS$_2$			
a	6.097 Å		S: pyrite, C2, $T_h^6 - Pa3$, antiferromagnetic, $T_N = 48.2$ K, $p_{eff} = 6.16\,\mu_B$, $\Theta_p = -598$ K	59H, 68L, 71L

Physical property	Numerical value	Experimental conditions	Experimental method, remarks	Ref.
α-MnSe (Figs. 67, 68)				
a	5.464 Å		S: cubic, B1, O_h^5–Fm3m; B8 above $p=90$ kbar, antiferromagnetic, $T_N(\uparrow)=248$ K, $T_N(\downarrow)=197$ K, $p_{eff}=5.88\,\mu_B$, $\Theta_p=-373$ K. Ferromagnetic (111) sheets coupled antiparallel; spins in (111) planes. β-form ($a=5.82$ Å, space group T_d^2–F$\bar{4}$3m) and γ-form ($a=4.12$ Å, $c=6.72$ Å, space group C_{6v}^4–P6$_3$mc) are unstable	68C1, 70R1, 71D, 71L, 72C1, 78I
ϱ	10^4 Ω cm	p-type, synthetic single crystal		
S	60 μV K^{-1}			
			CG: chemical transport (1000/600)	
E_g	2.5 eV		optical gap	
$E_{g,th}$	0.48 eV	$283 \leq T \leq 197$ K (cooling)		
MnTe (Fig. 69)				
a	4.144 Å	B8 modification	S: B8, D_{6h}^4–P6$_3$/mmc, above 1040 °C B1 mod. (O_h^5–Fm3m), antiferromagnetic, $T_N=310$ K, $\Theta_p=-584$ K, $p_{eff}=5.91\,\mu_B$	54P, 56U, 61B, 61J2, 64W1, 64W2, 72M1, 77A
c	6.703 Å			
a	6.03 Å	B1 modification	CG: Bridgman-Stockbarger from Te-rich melt	
ϱ	10^{-3} Ω cm	$T=200$ K, p-type, polycrystalline sample	electron-spin scattering $\tau_{em}(T_N)=10^{-5}$ s; relaxation of magnon system $\tau_m(T_N)=4\cdot 10^{-12}$ s; large magnon drag for $T>200$ K; $T_N=310$ K	
S	50 μV K^{-1}	$T=200$ K		
p	$5\cdot 10^{19}$ cm^{-3}	$T<T_N$		
μ_H	50 cm^2/Vs		mobility excluding magnetic contributions ($m_p=0.5\,m_0$)	
μ_{dr}	6 cm^2/Vs			
μ_{opt}	115 cm^2/Vs		optical mobility	
E_g	1.3 eV		optical gap	
MnTe$_2$ (Figs. 70, 71)				
a	6.943 Å		S: pyrite, C2, T_h^6–Pa3, antiferromagnetic, $T_N=87.2$ K, $p_{eff}=5.84\,\mu_B$, $\Theta_p=-472$ K	65J, 68L, 71L, 74A2, 77O
ϱ	0.1 Ω cm	p-type, polycrystalline sample		
S	300 μV K^{-1}			
R_H	40 cm^3/C			
TcS$_2$				
a	6.415 Å		S: triclinic, primitive (unspecified)	71W1
b	6.375 Å			
c	6.659 Å		CG: halogen transport (1150/1080)	

(continued)

9.15.3.4 M_{VII} (= Mn, Tc, Re)-chalcogenides

Physical property	Numerical value	Experimental conditions	Experimental method, remarks	Ref.
TcS$_2$ (continued)				
α	103.61°			71W1
β	62.97°			
γ	118.96°			
d	5.066 g cm^{-3}		calculated	
E_g	1.0 eV	synthetic single crystal	optical gap	
TcSe$_2$				
			S: triclinic	71W1
			CG: halogen transport (1080/1000)	
E_g	0.88 eV	synthetic single crystal	optical gap	
TcTe$_2$				
a	12.522 Å	synthetic single crystal	S: monoclinic, Cc or C2/c	71W1
b	7.023 Å		CG: halogen transport (980/840)	
c	13.828 Å			
β	101.26°			
d	7.890 g cm^{-3}		calculated	
ReS$_2$				
a	6.455 Å		S: triclinic, $C_i^1 - P\bar{1}$	71W1,
b	6.362 Å		CG: halogen transport (1150/1080)	83H
c	6.401 Å			
α	105.04°			
β	91.6°			
γ	118.97°			
d	7.506 g cm^{-3}		experimental	
	7.613 g cm^{-3}		calculated	
$\varrho_\perp$	10^3 Ω cm	synthetic single crystal		
E_g	1.33 eV		optical gap	
$E_{g,th}$	0.4 eV	$T = 160 \cdots 570$ K		
	1.0 eV	$570 \cdots 920$ K		
ReSe$_2$				
a	6.727 Å		S: triclinic, $C_i^1 - P\bar{1}$	65A,
b	6.606 Å		CG: halogen transport (1080/1000)	71W1
c	6.72 Å			
α	118.93°			
β	91.82°			
γ	104.93°			
d	9.219 g cm^{-3}		calculated	
	9.237 g cm^{-3}		experimental	
E_g	1.15 eV	synthetic single crystal	optical gap	
ReTe$_2$				
a	12.987 Å		S: orthorhombic, $D_{2h}^{15} - Pcab$	65J,
b	13.055 Å		CG: halogen transport (980/840)	66F,
c	14.271 Å			71W1
d	8.50 g cm^{-3}		experimental	
	9.712 g cm^{-3}		calculated	
ϱ	>10^2 Ω cm	polycrystal, p-type		
S	490 μV K^{-1}			

Physical property	Numerical value	Experimental conditions	Experimental method, remarks	Ref.

9.15.3.5 M_{VIIIA} (= Fe, Ru, Os)-chalcogenides

In the Fe–S system, the mixed-valence compounds such as Fe_7S_8 and Fe_3S_4 are metallic. FeS tends to be iron-poor and crystallizes in three polymorphs: hexagonal troilite (B8 with $a = 5.962$ Å, $c = 11.75$ Å) [56B, 70E], tetragonal mackinawite (with $a = 3.68$ Å, $c = 5.04$ Å) [65B] and a cubic phase (B3 with $a = 5.42$ Å) [78W]. Troilite contains high-spin Fe^{2+} ions [62S], and the five parallel-spin electrons are localized with energies below the Fermi energy, which lies in the band of minority-spin 3d orbitals of $t_{2g} = a_1 + e_\pi$ parentage. There is one minority-spin 3d electron per iron atom, and at high temperatures these minority-spin electrons are itinerant, making troilite metallic above the magnetic-ordering temperature $T_N = 600$ K [82G]. Below T_N a $d\sigma/dT > 0$ is characteristic of a semiconductor, but a sharp drop in conductivity only occurs below a first-order transition at T_α. In the low-temperature 2C-phase below T_α, the iron atoms form triangular clusters within the basal planes of the B8 (1C) phase to give superlattice cell parameters $a = \sqrt{3}a_0$, $c = 2c_0$ (see Fig. 73). In this phase, the minority-spin electrons are trapped in the triangular clusters by splitting of the e_π minority-spin band. A change of the spin orientation from perpendicular to parallel to the c axis on lowering the temperature through $T_s \gtrsim T_\alpha$, see Fig. 72a, does not significantly change the conductivity. The structure changes smoothly from B8 to B31 on cooling through T_s [78K2, 82T]. The temperature T_α decreases and its associated hysteresis increases with increasing x in $Fe_{1-x}S$, see Figs. 75 and 76. The magnetic susceptibility data, Fig. 74, reveals the spin-flip transition at T_s as well as a sharp increase in magnetic anisotropy at T_α and the conventional maximum at T_N.

In both the tetragonal and cubic forms of FeS, the Fe(II) ions occupy tetrahedral sites [65B]. Tetragonal FeS is reported to be a semiconductor [65B]. Cubic FeS becomes antiferromagnetic and orthorhombic below $T_N = 238$ K [78W].

FeS_2, which contains diamagnetic, low-spin Fe(II) ions, has two polymorphs: pyrite and marcasite [55B, 70B1,2]. Of these, pyrite is the more easily synthesized; only natural marcasite crystals are available. Just the opposite situation occurs for $FeSe_2$ and $FeTe_2$. In both structures, the anions form dimers in which the σ^*-antibonding dimeric orbitals have energies above the bottom of the conduction band. The cubic component of the octahedral-site crystal field splits the iron 3d orbitals into filled bands of t_{2g} parentage and empty bands of e_g parentage. In FeS_2, $FeSe_2$ and $FeTe_2$, the conduction band of both marcasite and pyrite phases is associated with the empty 3d orbitals of e_g parentage at the low-spin Fe(II) [82G].

The Fe–Se phase diagram near FeSe, Fig. 78, indicates that FeSe disproportionates into an iron-rich $Fe_{1.04}Se$ α-phase with LiOH structure that is metallic, and an iron-poor β-phase with cation vacancies ordered onto alternate (001) planes of the B8 structure [66S1]. Vacancy ordering within these planes occurs at specific compositions: Fe_7Se_8 and Fe_3Se_4. In the range $0.12 \leq x \leq 0.125$, the $Fe_{1-x}Se$ phase orders ferrimagnetically below $T_C = (445 + 160x)$ K, ferromagnetic (001) planes of high-spin iron coupling antiparallel [56H, 66S1]. Although most authors report that these phases are metallic, Fig. 79 shows data indicative of a metal-semiconductor transition at $T_{tr} \simeq 350$ K $< T_N$ typical of a degenerate semiconductor. Moreover, a phase corresponding to Fe_2Se_3 is included.

Of the iron tellurides, only $FeTe_2$ appears to be semiconducting. However, pressed powders of $Fe_{1-x}Te$ were reported to exhibit semiconducting behaviour at higher temperatures ($T > T_C$) with a change from n to p-type conduction at Fe_2Te_3 [64A].

The RuX_2 and OsX_2 phases are analogous to the FeX_2 pyrite phases; the Ru(II) and Os(II) have low-spin d^6 configurations [63H2].

For the abbreviations used in the tables below, see the top of the tables in section 9.15.3.1.

$Fe_{1-x}S$ (Figs. 72···76) ($0 < x < 0.055$)

a	5.96 Å	$x \approx 0$,	S: hexagonal, $D_{3h}^4 - P\bar{6}2c$,	56B,
c	11.74 Å	300 K $< T_\alpha$	$T < T_\alpha = 413$ K ($x \approx 0$),	56K1,
			see Fig. 75; orthorhombic, B31,	56K2,
			$D_{2h}^{16} - Pmcn$, $T_\alpha < T < T_s$;	57H,
			hexagonal, B8, $D_{6h}^4 - P6_3mmc$,	58M3,
			$T > T_s$	60A,
			CG: Bridgman method	61F,
			C: golden yellow	61M,

(continued)

Physical property	Numerical value	Experimental conditions	Experimental method, remarks	Ref.
$Fe_{1-x}S$ (continued)				
$\varrho_\perp$	10^{-1} Ω cm x = 0.006	p-type,	antiferromagnetic, $T_N = 600$ K,	64B1,
$S_\perp$	15 μV K^{-1} x = 0.025	synthetic	$p_{eff} = 5.5\ \mu_B$, $\Theta_p = -1160$ K.	64K,
$S_\parallel$	50 μV K^{-1} x = 0.025	single	Spin-flip transition at $T_s = 420$ K.	65B,
	260 μV K^{-1} x = 0.006	crystal	$dT_N/dp = 3.2$ K/kbar	68M,
$R_{H\perp}$	$0.7 \cdot 10^{-4}$ cm^3/C	$R_{H\parallel} = R_{H\perp}$		70D2,
p	$8.5 \cdot 10^{22}$ cm^{-3}			71N,
$\mu_{H\parallel}$	$2.8 \cdot 10^{-4}$ cm^2/Vs			73T,
	0.21 cm^2/Vs	$T = 150$ °C		74A1,
$\mu_{H\perp}$	0.16 cm^2/Vs			76C,
$\mu_{p\perp}$	0.16 cm^2/Vs		from Seebeck and resistivity data	76H,
				76M,
				78W
FeS_2 (pyrite) (Fig. 77)				
a	5.418 Å		S: pyrite, C2, $T_h^6 - Pa3$	54M,
d	5.0 g cm^{-3}		CG: chemical transport (715/655)	68B1,
ϱ	1.74 Ω cm	n-type		68B2,
S	-500 μV K^{-1}	synthetic		76S,
μ_H	230 cm^2/Vs	single		79S,
R_H	-400 cm^3/C	crystal		83H1
$E_{g,th}$	0.02 eV	$T < 60$ K		
	0.4 eV	$T = 298$ K		
	0.92 eV	500 K		
ϱ	0.06 Ω cm	n-type		
μ_H	100 cm^2/Vs	natural		
R_H	-6.5 cm^3/C	crystal		
n	$1.1 \cdot 10^{18}$ cm^{-3}			
E_g	0.95 eV		optical gap	
ϱ	2.2 Ω cm	p-type		
μ_H	1.3 cm^2/Vs	natural		
R_H	2.8 cm^3/C	crystal		
p	$2.6 \cdot 10^{18}$ cm^{-3}			
$E_g(T)$	$E_g(0) + aT + bT^2$	$T < 425$ K	with $E_g(0) = 0.835$ eV, $a = 4 \cdot 10^{-5}$ eV/K, $b = -7.4 \cdot 10^{-7}$ eV/K^2	
FeS_2 (marcasite)				
a	4.44 Å		S: marcasite, C18, $D_{2h}^{12} - Pnnm$	70B1,
b	5.42 Å		CG: difficult to grow single	70B2,
c	3.39 Å		crystals, natural crystal	83H1
d	4.87 g cm^{-3}		available	
$Fe_{1-x}Se$ (Figs. 78, 79)				
a	3.548 Å	x = 0	S: hexagonal, B8, $D_{6h}^4 - P6_3/mmc$.	56H,
c	5.733 Å		Metal-semiconductor transition	66S1,
ϱ	10^3 Ω cm	p-type, poly-	at 350 K. Ferrimagnetic, $T_C = 445$ K.	71L,
S	10 μV K^{-1}	crystalline sample	See discussion in text above.	73A1, 73A2
$E_{g,th}$	0.14 eV	$T > 350$ K		
$FeSe_{2-x}$ ($0 \leq x \leq 0.05$)				
a	5.786 Å		S: pyrite, C2, $T_h^6 - Pa3$	68B1,
ϱ	$2 \cdot 10^{-2}$ Ω cm	n-type	CG: under 65 kbar pressure at 1200 °C	71L
S	-28 μV K^{-1}	synthetic		
$E_{g,th}$	0.2 eV	single crystal	C: silvery	

Physical property	Numerical value	Experimental conditions	Experimental method, remarks	Ref.
FeSe$_2$ (Fig. 80)				
a	4.789 Å		S: marcasite, C18, D_{2h}^{12}–Pnnm	58F, 61D, 71L
b	5.768 Å			
c	3.575 Å			
ϱ	1.0 Ω cm	p-type, poly-crystalline sample	n-type for $T > 300$ K	
S	62 μV K^{-1}			
$E_{g,\text{th}}$	0.3 (1) eV			
Fe$_{1-x}$Te				
See discussion in text above. See [55U, 59T, 64A, 66S2, 70R2].				
FeTe$_2$ (pyrite)				
a	6.293 Å		S: pyrite, C2, T_h^6–Pa3	68B1, 71L
ϱ	$2 \cdot 10^{-3}$ Ω cm	n-type synthetic single crystal	CG: under 65 kbar pressure at 1200 °C	
S	-25 μV K^{-1}		C: silvery	
FeTe$_2$ (marcasite) (Fig. 81)				
a	5.265 Å		S: marcasite, C18, D_{2h}^{12}–Pnnm	61D, 70B1, 70B2, 71L
b	6.268 Å		CG: halogen transport	
c	3.874 Å			
ϱ	$1.5 \cdot 10^{-2}$ Ω cm	p-type, poly-crystalline sample		
S	64 μV K^{-1}			
$E_{g,\text{th}}$	0.92 eV	$T > 600$ K		
RuS$_2$				
a	5.609 Å		S: pyrite, C2, T_h^6–Pa3	63H2, 71L
E_g	1.8 eV		optical gap diamagnetic	
RuSe$_2$				
a	5.934 Å		S: pyrite, C2, T_h^6–Pa3	63H2, 71L
S	-120 μV K^{-1}	n-type, poly-crystalline sample	diamagnetic optical gap	
E_g	1.0 eV			
$E_{g,\text{th}}$	≥ 0.6 eV			
RuTe$_2$				
a	6.39 Å		S: pyrite, C2, T_h^6–Pa3	63H2, 71L
a	5.271 Å		marcasite, C18, D_{2h}^{12}–Pnnm, $T = 77$ K	
b	6.387 Å			
c	4.038 Å			
S	-200 μV K^{-1}	n-type, poly-crystalline sample	diamagnetic	
$E_{g,\text{th}}$	0.25 eV			
OsS$_2$				
a	5.619 Å		S: pyrite, C2, T_h^6–Pa3	63H2, 71L, 83H1
d	9.47 g cm^{-3}		diamagnetic	
E_g	2.0 eV	polycrystalline sample	optical gap	

Physical property	Numerical value	Experimental conditions	Experimental method, remarks	Ref.
OsSe$_2$				
a	5.945 Å		S: pyrite, C2, T_h^6 – Pa3	63H2,
S	-80 μV K^{-1}	n-type, poly-crystalline sample	diamagnetic	71L
OsTe$_2$				
a	6.397 Å		S: pyrite, C2, T_h^6 – Pa3	
a	5.2804 Å		marcasite, C18, D_{2h}^{12} – Pnnm,	63H2,
b	6.4018 Å		$T = 77$ K	71L
c	4.0481 Å			
S	-100 μV K^{-1}	n-type, poly-crystalline sample	diamagnetic	
$E_{g,th}$	>0.2 eV			

9.15.3.6 M_{VIIIB} (= Co, Rh, Ir)-chalcogenides

All the known cobalt chalcogenides are metallic. The pyrite phases that form with CoX_2, RhX_2 or IrX_2 are metallic, but cation-defect pyrites $Rh_{2/3}X_2$ and $Ir_{2/3}X_2$ are semiconducting. These contain low-spin Rh(III) or Ir(III) in octahedral sites. – Similarly, the semiconducting phases RhX_2 and Ir_2 have an orthorhombic structure in which half the chalcogenides belong the polyanion $(X_2)^{2-}$ units and the cations are octahedral, low-spin Rh(III) or Ir(III). An alternate structure that stabilizes low-spin, octahedral-site Ir(III) is found in IrS_2 and $IrSe_2$; they contain two types of anion units, $(X_2)^{2-}$ and X^{2-}. Finally, octahedral-site, low-spin Rh(III) and Ir(III) are found in the semiconductors Rh_2S_3 and isostructural Rh_2Se_3 and Ir_2S_3 [65H]. In all these semiconductors, the conduction band is formed from cationic d orbitals of e_g parentage as in FeS_2. Any occupancy of this conduction band leads to metallic conductivity; $RhSe_{2-x}$ is a superconductor with $T_c = 6$ K for $x = 0.02$ [55G].

For the abbreviations used in the tables below, see the top of the tables in section 9.15.3.1

$Rh_{2/3}S_2$				
a	5.58 Å		S: defect pyrite, C2, T_h^6 – Pa3	64H,
S	-400 μVK^{-1}	n-type, poly-crystalline sample	optical gap	71L
E_g	>1.5 eV			
Rh_2S_3				
a	8.462 Å		S: orthorhombic, D_{2h}^{14} – Pbcn	64H,
b	5.985 Å			67P,
c	6.138 Å			83H1
d	6.40 g cm^{-3}			
S	-300 μV K^{-1}	n-type, poly-crystalline sample		
$E_{g,th}$	0.8 eV			
$Rh_{2/3}S_2$ (= RhS$_{\approx 3}$)				
a	5.58 Å		S: ~C2 (trigonal distortion)	64H
α	$>90°$			
S	-400 μV K^{-1}	n-type, poly-crystalline sample	diamagnetic optical gap	
E_g	≥ 1.5 eV			

Physical property	Numerical value	Experimental conditions	Experimental method, remarks	Ref.
$Rh_2Se_2(Se_2)$ $(=RhSe_2)$				
a	20.91 Å		S: $IrSe_2$ structure, D_{2h}^{16}–Pnam	64H
b	5.95 Å		(C2 at high T, metallic)	
c	3.709 Å		diamagnetic	
S	$-80\ \mu V\ K^{-1}$	n-type, poly-		
E_g	0.6 eV	crystalline sample	optical gap	
$Rh_{2/3}Se_2$ $(=RhSe_{\approx 3})$				
a	5.962 Å		S: ~C2 (trigonal distortion)	37B1,
α	90° 44′			64H
S	$-30\ \mu V\ K^{-1}$	n-type, poly-		
E_g	0.7 eV	crystalline sample	optical gap	
$Ir_2S_2(S_2)$ $(=IrS_2)$				
a	19.78 Å		S: D_{2h}^{16}–Pnam	64H,
b	5.624 Å			83H1
c	3.565 Å		diamagnetic	
d	8.43 g cm^{-3}			
S	250 $\mu V\ K^{-1}$	p-type, poly-		
E_g	0.9 eV	crystalline sample	optical gap	
$Ir_{2/3}S_2$ $(=IrS_{\approx 3})$				
a	5.5 Å		S: ~C2 (trigonal distortion)	37B1,
α	>90°			37B2,
E_g	2.0 eV	polycrystal	optical gap	64H
$Ir_2Se_2(Se_2)$ $(=IrSe_2)$				
a	20.94 Å		S: orthorhombic, D_{2h}^{16}–Pnam	58B,
b	5.93 Å			64H
c	3.74 Å		diamagnetic	
S	300 $\mu V\ K^{-1}$	p-type, poly-		
E_g	≈1.0 eV	crystalline sample	optical gap	
$Ir_{2/3}Se_2$ $(=IrSe_{\approx 3})$				
a	5.96 Å		S: ~C2 (trigonal distortion)	37B1,
α	>90°		diamagnetic	64H
S	$-150\ \mu V\ K^{-1}$	n-type, poly-		
$E_{g,\,th}$	≧0.45 eV	crystalline sample		

9.15.3.7 M_{VIII} (= Ni, Pd, Pt)-chalcogenides

The millerite forms of NiS and NiSe are metallic. However, the high-temperature B8 phase of NiS may be stabilized to low temperatures by rapid quenching to room temperature [63S, 70T] or by direct synthesis [35L]. This phase is metallic with temperature-independent (Pauli) paramagnetism for $T > T_N = $ 265 K [68S2]. As shown in Figs. 82 and 83, the $Ni_{1-x}S$ phase is stable only over a narrow range of x; and T_N decreases with increasing x, vanishing near x ≃ 0.03. The magnetic transition is characterized by a first-order dilatation of the structure (Fig. 84), and the magnetically ordered state is suppressed by the application of hydrostatic pressure (Fig. 85). The resistivity changes by more than one order of magnitude at T_N (Fig. 86), and the Hall coefficient changes by over two orders of magnitude (Fig. 87). The magnetic

Physical property	Numerical value	Experimental conditions	Experimental method, remarks	Ref.

moment of the nickel ions decreases sensitively with decreasing lattice parameter, and hence with x, changing from $1.5\,\mu_B$ at $T=0\,K$ for $x=0$ to $1.0\,\mu_B$ for $x=0.03$ [74C]. Below the transition the half-filled bands of e_g parentage at the ocrahedral-site Ni^{2+} ions are split in two by strong correlations, but it is not clear whether the correlation splitting is large enough to create a semiconducting state; the low-temperature phase is semimetallic [72M2]. Fig. 88 shows an APW band calculation obtained by introducing below T_N constant potentials $+V_s$ and $-V_s$ at the two nickel sublattices ($V_s=0.54\,eV$), but no correlation splitting. This calculation would make stoichiometric NiS a small-bandgap semiconductor for $T<T_N$.

Single-crystal NiS_2 with pyrite structure has been prepared by chlorine transport of powder samples obtained at 600 °C in sealed tubes [68B3] and by direct synthesis from elemental nickel and sulfur at 65 kbar [68B2]. Samples prepared at high pressure were sulfur-rich. From Fig. 89, it appears that the sulfur-rich samples with smaller lattice parameter can be distinguished from the nickel-rich samples. This conclusion is reinforced by their magnetic and electric properties [68B3, 73G]. The sulfur-rich samples are good conductors with low activation energy, as is illustrated in Fig. 90 for a sample prepared at high pressures. The nickel-rich samples exhibit a larger activation energy and hence a much smaller conductivity at low temperatures [68B3]. Measurements of conductivity vs. pressure reveal a semiconductor-metal transition near 30 kbar at room temperature; the magnitude of $\Delta\varrho$ across the transition decreases with increasing temperature (Fig. 91) and with increasing S/Ni ratio x. The semiconductor-metal phase boundary shown in Fig. 92 represents a maximum in the resistivity as a function of temperature for a fixed pressure, and this T_{max} vs. x is plotted in Fig. 93, where it can be seen that the magnetic susceptibility has a maximum at the same temperature in the sulfur-rich compounds. The nickel-rich compounds, on the other hand, exhibit no such maximum in the resistivity (Fig. 94 for $NiS_{1.99}$); they are semiconductors at all temperatures, and extrapolation of the temperature-dependent optical band gap of $NiS_{1.99}$, Fig. 95, indicates the gap would vanish at 650 °C, a temperature above the onset of decomposition at about 400 °C. Moreover, the nickel-rich compounds are antiferromagnetic with a Néel temperature $50 < T_N < 70\,K$ below which the nickel atomic moments order as shown in Fig. 96. Below $T_C \simeq 30\,K$, the more complex magnetic order of Fig. 97 is found; it has a weak ferromagnetic component that decreases with increasing sulfur content (Fig. 98). The proposed density-of-states model for nickel-rich $NiS_{1.99}$ is shown in Fig. 99; it contains a correlation splitting of half-filled 3d bands of e_g parentage at the nickel ions with impurity states presumably associated with anion vacancies.

The semiconducting PdX and PtX compounds with X=S or Se contain low-spin Pd(II) or Pt(II) in square-coplanar coordination. The tetragonal B17 structure of PtS may be derived from the hexagonal NiAs structure by diffusionless displacement of the atoms [57J]; the tetragonal B34 structure of PdS is more complex. In these compounds, the conduction bands are derived from the σ-bonding $d_{x^2-y^2}$ orbitals of the metal atoms; the valence bands are derived from the anionic p^6 configurations. The semiconducting PdX_2 compounds, X=S or Se, similarly contain low-spin Pd(II) in square-coplanar coordination; the orthorhombic Pbca structure can be derived by diffusionless displacements of the atoms from the cubic pyrite structure [65H]. The conduction and valence bands are similar to those in PdX and PtX compounds, the antibonding σ^* orbitals of a polyanion pair lying above the conduction-band edge. The $Pt_{1-x}S_2$ and $PtSe_2$ compounds, on the other hand, contain low-spin Pt(IV) in octahedral coordination, and the anions are monomeric. This makes the parentage of the conduction bands the metal d orbitals of e_g symmetry.

For the abbreviations used in the tables below, see the top of the tables in section 9.15.3.1.

$Ni_{1-x}S$ ($0 \leq x \leq 0.06$) (Figs. 82···88)

a, c	see Fig. 84	x=0, quenched from $T>620\,K$; n-type synthetic single crystal	S: B8, $D_{6h}^4 - P6_3/mmc$	59L,
			CG: halogen transport (720/650) slowly cooled to 550 °C and quenched	62K, 63S,
$\varrho_\perp$	$10^{-4}\,\Omega\,cm$			64K,
$S_\perp$	$-3\,\mu V\,K^{-1}$		C: golden yellow	65H, 67S2, 68S1, 68S2, 70O1, 70O2, 71H, 71T2, 72K2,

(continued)

Physical property	Numerical value	Experimental conditions	Experimental method, remarks	Ref.
$Ni_{1-x}S$ (continued)				
$\mu_{H\perp}$	$1 \cdots 5$ cm^2/Vs		Semiconductor-metal transition at $T_{tr} = 265$ K, accompanied by antiferromagnetic-paramagnetic transition. T_{tr} depends on x and p, it disappears for $x \geq 0.035$ and $p > 20$ kbar, see text above. Latent heat = 282 cal/mol	72M2
$R_{H\perp}$	-10^{-4} cm^3/C			73K
n	$4 \cdot 10^{22}$ cm^{-3}			74B,
p	$2 \cdot 10^{20}$ cm^{-3}	($T < T_{tr}$)		74C,
$\mu_{H\perp}$	$1 \cdots 5$ cm^2/Vs	($T < T_{tr}$)		74M,
				76B2,
				76C
NiS_2 (Figs. 89···99)				
a	5.69 Å		S: pyrite, C2, $T_h^6 - Pa3$	68B2,
ϱ	1 Ω cm	p-type	CG: halogen transport (740/710)	68B3,
S	311 μV K^{-1}	synthetic	C: shiny black	70H,
E_g	0.27 eV	single crystal	optical gap	71L,
$E_{g,th}$	0.64 eV	$T > 380$ K	Semiconductor-metal transition for $p \geq 30$ kbar at RT. $dT_{tr}/dp = 6$ (1) K/kbar E_g: optical gap $(=(1/2)(dE_{g,th}/dp)$ antiferromagnetic, $T_N \approx 50$ K, weak ferromagnetism below $T_C = 30$ K, see text above	71W2,
	0.14 eV	$T = 140 \cdots 380$ K		72G,
	0.009 eV	$T < 140$ K		72K1,
dE_g/dT	-0.4 meV/K			73G,
dE_A/dp	-4.3 (3) meV/kbar			73M2,
dT_N/dp	0.9 (1) K/kbar			75M,
dT_C/dp	0.4 (1) K/kbar			77M3,
				78K1,
				78M
PdS				
a	6.429 Å		S: tetragonal, B34, $C_{4h}^2 - P4_2/m$	57G,
c	6.608 Å			65H,
d	6.6 g cm^{-3}			83H1
S	-250 μV K^{-1}	n-type, poly-crystalline sample	diamagnetic	
$E_{g,th}$	0.5 eV			
PdS_2				
a	5.46 Å		S: orthorhombic, $D_{2h}^{15} - Pbca$	56G,
b	5.54 Å			57G,
c	7.53 Å			65H,
d	$4.7 \cdots 4.8$ g cm^{-3}			83H1
ϱ	100 Ω cm	n-type, poly-crystalline sample	diamagnetic	
S	-240 μV K^{-1}			
$E_{g,th}$	$0.7 \cdots 0.8$ eV			
PdSe				
a	6.73 Å		S: B34, $C_{4h}^2 - P4_2/m$	65H,
c	6.91 Å			71L
S	-120 μV K^{-1}	n-type, poly-crystalline sample		
$E_{g,th}$	0.2 eV			
$PdSe_2$				
a	5.74 Å		S: orthorhombic, $D_{2h}^{15} - Pbca$	57G,
b	5.87 Å			65H
c	7.69 Å			
ϱ	1 Ω cm	p-type, poly-crystalline sample	diamagnetic	
S	500 μV K^{-1}			
$E_{g,th}$	0.4 eV			

Figs. p. 620   9.15.3.8 References for 9.15.3

Physical property	Numerical value	Experimental conditions	Experimental method, remarks	Ref.
PtS (Fig. 100)				
a	3.469 Å		S: tetragonal, B17, $P4_2/mmc$	65H, 79C1, 83H1
c	6.109 Å			
d	10.04 g cm^{-3}			
a	6.409 Å		B34, $C_{4h}^2 - P4_2/m$, $p = 30$ kbar	
c	6.596 Å			
ϱ	10^2 Ω cm	p-type, poly- crystalline sample	tetragonal phase	
S	50 μV K^{-1}		diamagnetic	
E_g	0.8 eV		optical gap	
$E_{g,th}$	0.64 eV			
ϱ	10 Ω cm	polycrystalline sample	B34 high-pressure phase	
$E_{g,th}$	0.38 eV			
Pt$_{0.97}$S$_2$ (Fig. 101)				
a	3.542 Å		S: trigonal, $D_{3d}^3 - P\bar{3}m1$	74F, 77M2
c	5.043 Å		CG: halogen transport with traces of phosphorus (800/740)	
$\varrho_\perp$	$2 \cdot 10^{-2}$ Ω cm	p- or n-type, synthetic single crystal	diamagnetic	
$S_\perp$	500 μV K^{-1}			
E_g	0.95 eV		optical gap; indirect gap for both $E\perp$ and $E\parallel$ between d states of Pt with VB (max) at M (or L) and CB (min) at Γ.	
dE_g/dT	$-3.7 (2) \cdot 10^{-4}$ eV/K			
$E_{g,th}$	0.4 eV			
PtSe$_2$				
a	3.724 Å		S: trigonal, C6, $D_{3d}^3 - P\bar{3}m1$	65H
c	5.062 Å			
S	40 μV K^{-1}	p-type, poly- crystalline sample		
$E_{g,th}$	0.1 eV			

9.15.3.8 References for 9.15.3

35L	Levi, G.R., Baroni, A.: Z. Kristallogr. **92** (1935) 210.
37B1	Biltz, W.: Z. Anorg. Allg. Chem. **233** (1937) 282.
37B2	Biltz, W., Laar, J., Ehrlich, P., Meisel, K.: Z. Anorg. Allg. Chem. **233** (1937) 257.
54M	Marinace, J.C.: Phys. Rev. **96** (1954) 593.
54P	Palmer, W.: J. Appl. Phys. **25** (1954) 12 S.
55B	Benoit, R.: J. Chim. Phys. **52** (1955) 119.
55G	Geller, S., Cetlin, B.B.: Acta Crystallogr. **8** (1955) 272.
55U	Uchida, E., Kondoh, H.: J. Phys. Soc. Jpn. **10** (1955) 357.
56B	Bertaut, E.F.: Bull. Soc. Fr. Mineral. Cristallogr. **79** (1956) 276.
56C	Corliss, L.M., Elliott, N., Hastings, J.M.: Phys. Rev. **104** (1956) 924.
56G	Gronvold, F., Rost, E.: Acta Chem. Scand. **10** (1956) 1620.
56H	Hirone, T., Chiba, S.: J. Phys. Soc. Jpn. **11** (1956) 666.
56K1	Kamigaichi, T., Hihara, T., Tazaki, H., Hirahara, E.: J. Phys. Soc. Jpn. **11** (1956) 606.
56K2	Kamigaichi, T., Hihara, T., Tazaki, H., Hirahara, E.: J. Phys. Soc. Jpn. **11** (1956) 1123.
56U	Uchida, E., Kondoh, H., Fukuoka, N.: J. Phys. Soc. Jpn. **11** (1956) 27.
57G	Gronvold, F., Rost, E.: Acta Crystallogr. **10** (1957) 329.
57H	Hihara, T., Murakami, M., Hirahara, E.: J. Phys. Soc. Jpn. **12** (1957) 743.
57J	Jellinek, F.: Acta Crystallogr. **10** (1957) 620.
58B	Barricelli, L.B.: Acta Crystallogr. **11** (1958) 75.

9.15.3.8 References for 9.15.3

58F	Fischer, G.: Can. J. Phys. **36** (1958) 1435.
58M1	McTaggart, F.K., Wadsley, A.D.: Aust. J. Chem. **11** (1958) 445.
58M2	McTaggart, F.K.: Aust. J. Chem. **11** (1958) 471.
58M3	Murakami, M., Hirahara, E.: J. Phys. Soc. Jpn. **13** (1958) 1407.
59H	Hastings, J.M., Elliott, N., Corliss, L.M.: Phys. Rev. **115** (1959) 13.
59L	Laffitte, M.: Bull. Soc. Chim. Fr. **1959**, 1211.
59T	Tsubokawa, I., Chiba, S.: J. Phys. Soc. Jpn. **14** (1959) 1120.
60A	Andresen, A.F.: Acta Chem. Scand. **14** (1960) 919.
60H	Hirone, T., Chiba, S.: J. Phys. Soc. Jpn. **15** (1960) 1991.
60J	Jellinek, F., Brauer, G., Müller, H.: Nature (London) **185** (1960) 376.
60K	Kamigaichi, T., Masumoto, K., Hihara, T.: J. Phys. Soc. Jpn. **15** (1960) 1355.
61B	Banewicz, J.J., Heidelberg, R.F., Luxem, A.H.: J. Phys. Chem. **65** (1961) 615.
61C	Corliss, L.M., Elliott, N., Hastings, J.M., Sass, R.L.: Phys. Rev. **122** (1961) 1402.
61D	Dudkin, L.D., Vaidanich, V.I.: Sov. Phys. Solid State **2** (1961) 1384.
61F	Fujime, S., Murakami, M., Hirahara, E.: J. Phys. Soc. Jpn. **16** (1961) 183.
61G	Grimmeiss, H.G., Rabenau, A., Hahn, H., Ness, P.: Z. Elektrochem. **65** (1961) 776.
61J1	Jellinek, F.: Nature (London) **192** (1961) 1065.
61J2	Johnston, W.D., Sestrich, D.E.: J. Inorg. Nucl. Chem. **19** (1961) 229.
61M	Murakami, M.: J. Phys. Soc. Jpn. **16** (1961) 187.
62B	Brixner, L.H.: J. Inorg. Nucl. Chem. **24** (1962) 257.
62G	Grigoryan, L.A., Novoselova, A.V.: Dokl. Chem. **144** (1962) 496.
62K	Kullerud, G., Yund, R.A.: J. Petrol. **3** (1962) 126.
62M	Masumoto, K., Hihara, T., Kamigaichi, T.: J. Phys. Soc. Jpn. **17** (1962) 1209.
62S	Sparks, J.T., Mead, W., Komoto, T.: J. Phys. Soc. Jpn. **17**, Suppl. B-I (1962) 249.
63A	Aslanov, L.A., Simanov, Yu.P., Novoselova, A.V., Ukrainski, Yu.M.: Russ. J. Inorg. Chem. **8** (1963) 1381.
63B	Benard, J., Jeannin, Y.: Adv. Chem. Ser. **39** (1963) 191.
63H1	Haraldsen, H., Kjekshus, A., Rost, E., Steffensen, A.: Acta Chem. Scand. **17** (1963) 1283.
63H2	Hulliger, F.: Nature (London) **200** (1963) 1064.
63J	Jellinek, F.: Ark. Kemi **20** (1963) 447.
63S	Sparks, J.T., Komoto, T.: J. Appl. Phys. **34** (1963) 1191.
64A	Aramu, F., Manca, P.: Nuovo Cimento **34** (1964) 1025.
64B1	Bertaut, E.F.: Bull. Soc. Sci. Bretagne **39** (1964) 67.
64B2	Bjerkelund, E., Kjekshus, A.: Z. Anorg. Allg. Chem. **328** (1964) 235.
64H	Hulliger, F.: Nature (London) **204** (1964) 644.
64K	Kullerud, G.: Fortschr. Mineral. **41** (1964) 235.
64W1	Wasscher, J.D., Seuter, A.M.J.H., Haas, C.: Proc. Int. Conf. Phys. Semiconductors (1964) 1269.
64W2	Wasscher, J.D., Haas, C.: Phys. Lett. **8** (1964) 302.
64W3	Wildervanck, J.C., Jellinek, F.: Z. Anorg. Allg. Chem. **328** (1964) 309.
65A	Alcock, N.W., Kjekshus, A.: Acta Chem. Scand. **19** (1965) 79.
65B	Bertaut, E.F., Burlet, P., Chappert, J.: Solid State Commun. **3** (1965) 335.
65G	Greenaway, D.L., Nitsche, R.: J. Phys. Chem. Solids **26** (1965) 1445.
65H	Hulliger, F.: J. Phys. Chem. Solids **26** (1965) 639.
65J	Johnston, W.D., Miller, R.C., Damon, D.H.: J. Less-Common Met. **8** (1965) 272.
65K1	Klemm, W., Schnering, H.G.: Naturwissenschaften **52** (1965) 12.
65K2	Kronert, V.W., Plieth, K.: Z. Anorg, Allg. Chem. **336** (1965) 207.
66B	Brown, B.E.: Acta Crystallogr. **20** (1966) 268.
66F	Furusett, S., Kjekshus, A.: Acta Chem. Scand. **20** (1966) 245.
66K	Kabashima, S.: J. Phys. Soc. Jpn. **21** (1966) 945.
66S1	Serre, J., Druille, R.: Compt. Rend. Ser. B **262** (1966) 639.
66S2	Suchet, J., Druille, R., Loriers, J.: Inorg. Mater. (USSR) (English Transl.) **2** (1966) 679.
67F	Fivaz, R., Mooser, E.: Phys. Rev. **163** (1967) 743.
67G	Gilles, P.W.: Applications of fundamental Thermodynamics to Metallurgical processes, Fitterer, G.R. (ed.), New York: Gordon and Breach, Science Publishers Inc. **1967**, p. 281.
67H	Huffman, D.R., Wild, R.L.: Phys. Rev. **156** (1967) 989.
67I	Ivanova, V.A., Abdinov, D.Sh., Aliev, G.M.: Phys. Status. Solidi **24** (1967) K145.
67P	Parthé, E., Hohnke, D., Hulliger, F.: Acta Crystallogr. **23** (1967) 832.

9.15.3.8 References for 9.15.3

67S1	Schwartz, A., Tauber, A., Shappirio, J.R.: Mater. Res. Bull. **2** (1967) 375.
67S2	Sparks, J.T., Komoto, T.: Phys. Lett. **25A** (1967) 398.
68B1	Bertaut, E.F., Cohen, J., Lambert-Andron, B., Mollard, P.: J. Phys. (Paris) **29** (1968) 813.
68B2	Bither, T.A., Bouchard, R.J., Cloud, W.H., Donohue, P.C., Siemons, W.J.: Inorg. Chem. **7** (1968) 2208.
68B3	Bouchard, R.J.: J. Cryst. Growth **2** (1968) 40.
68C1	Carpay, F.M.A.: Philips Res. Rep., Suppl. **10** (1968) 1.
68C2	Conroy, L.E., Park, K.C.: Inorg. Chem. **7** (1968) 459.
68H	Hulliger, F.: Structure and Bonding, Berlin-Heidelberg-New York: Springer-Verlag **4** (1968) 83.
68L	Lin, M.S., Hacker Jr., H.: Solid State Commun. **6** (1968) 687.
68M	Masayuki, U.: Z. Anorg. Allg. Chem. **361** (1968) 94.
68S1	Sparks, J.T., Komoto, T.: Rev. Mod. Phys. **40** (1968) 752.
68S2	Sparks, J.T., Komoto, T.: J. Appl. Phys. **39** (1968) 715.
69C	Connell, G.A.N., Wilson, J.A., Yoffe, A.D.: J. Phys. Chem. Solids **30** (1969) 287.
69H	Huisman, R., Jellinek, F.: J. Less-Common Met. **17** (1969) 111.
69K	Kadijk, F., Jellinek, F.: J. Less-Common Met. **19** (1969) 421.
69L	Lee, P.A., Said, G., Davis, R., Lim, T.H.: J. Phys. Chem. Solids **30** (1969) 2719.
69P	Popma, T.J.A., Van Bruggen, C.F.: J. Inorg. Nucl. Chem. **31** (1969) 73.
69S	Sleight, A.W., Bither, T.A.: Inorg. Chem. **8** (1969) 566.
69W	Wilson, J.A., Yoffee, A.D.: Adv. Phys. **18** (1969) 193.
70B1	Brostigen, G., Kjekshus, A.: Acta Chem. Scand. **24** (1970) 1925.
70B2	Brostigen, G., Kjekshus, A.: Acta Chem. Scand. **24** (1970) 2993.
70D1	DeJonge, R., Popma, T.J.A., Wiegers, G.A., Jellinek, F.: J. Solid State Chem. **2** (1970) 188.
70D2	DeMedicis, R.: Rev. Chim. Miner. **7** (1970) 723.
70E	Evans, H.T.: Science **167** (1970) 621.
70H	Hastings, J.M., Corliss, L.M.: IBM J. Res. Dev. **14** (1970) 227.
70M	Morosin, B.: Phys. Rev. **1B** (1970) 236.
70O1	Ohtani, T., Kosuge, K., Kachi, S.: J. Phys. Soc. Jpn. **28** (1970) 1588.
70O2	Ohtani, T., Kosuge, K., Kachi, S.: J. Phys. Soc. Jpn. **29** (1970) 521.
70R1	Rustamov, A.G., Kerimov, I.G., Valiev, L.M., Babaev, S.Kh.: Inorg. Mater. **6** (1970) 1176.
70R2	Rustamov, A.G., Kerimov, I.G., Valiev, L.M., Babaev, S.Kh., Ibragimova, P.G.: Dokl. Akad. Nauk Az. SSR **26** (1970) 19.
70V1	Van Bruggen, C.F., Vellinga, M.B., Haas, C.: J. Solid State Chem. **2** (1970) 303.
70V2	Vellinga, M.B., de Jonge, R., Haas, C.: J. Solid State Chem. **2** (1970) 299.
70W1	Wieting, T.J.: J. Phys. Chem. Solids **31** (1970) 2148.
70W2	Wilson, T.M.: Int. J. Quantum. Chem. **3** (1970) 757.
70T	Trahan, J., Goodrich, R.G., Watkins, S.F.: Phys. Rev. **B2** (1970) 2859.
71C	Chu, C.W., Huang, S., Hambourger, P.D., Thompson, A.H.: Phys. Lett **36A** (1971) 93.
71D	Decker, D.L., Wild, R.L.: Phys. Rev. **B4** (1971) 3425.
71E	Evans, B.L., Hazelwood, R.A.: Phys. Status. Solidi **A4** (1971) 181.
71H	Horwood, J.L., Ripley, L.G., Townsend, M.G., Trembley, R.J.: J. Appl. Phys. **42** (1971) 1476.
71L	Landolt-Börnstein (New Series), ed.: K.-H. Hellwege, Vol. III/6, Springer Verlag: Berlin, Heidelberg, New York **1971**.
71N	Nakazawa, H., Morimoto, N.: Mater. Res. Bull. **6** (1971) 345.
71T1	Thompson, A.H., Gamble, F.R., Revelli, J.F.: Solid State Commun. **9** (1971) 981.
71T2	Townsend, M.G., Trembley, R., Horwood, J.L., Ripley, R.J.: J. Phys. **C4** (1971) 598.
71W1	Wildervanck, J.C., Jellinek, F.: J. Less-Common Met **24** (1971) 73.
71W2	Wilson, J.A., Pitt, G.D.: Phil. Mag. **23** (1971) 1297.
72A	Al-Hilli, A.A., Evans, B.L.: J. Cryst. Growth **15** (1972) 93.
72B1	Bradley, D.J., Katayama, Y., Evans, B.L.: Solid State Commun. **11** (1972) 1695.
72B2	Bromley, R.A., Murray, R.B., Yoffe, A.D.: J. Phys. **C5** (1972) 759.
72C1	Cemic, L., Neuhaus, A.: High Temp. – High Pressures **4** (1972) 97.
72C2	Conroy, L.E., Pisharody, K.R.: J. Solid State Chem. **4** (1972) 345.
72D	Davey, B., Evans, B.L.: Phys. Status. Solidi **13A** (1972) 483.
72G	Gautier, F., Krill, G., Lapierre, M.F., Robert, C.: Solid State Commun. **11** (1972) 1201.
72J	Jellinek, F.: MTP Int. Rev. Science, Inorg. Chem., Ser. one **5** (1972) 339.
72K1	Kautz, R.L., Dresselhaus, M.S., Adler, D., Linz, A.: Phys. Rev. **B6** (1972) 2078.

9.15.3.8 References for 9.15.3

72K2	Koehler Jr., R.F., Feigelson, R.S., Swarts, H.W., White, R.L.: J. Appl. Phys. **43** (1972) 3127.
72M1	Mateika, D.: J. Cryst. Growth **13/14** (1972) 698.
72M2	McWhan, D.B., Marezio, M., Remeika, J.P., Dernier, P.D.: Phys. Rev. **5B** (1972) 2552.
72T	Thompson, A.H., Pisharody, K.R., Koehler Jr., R.F.: Phys. Rev. Lett. **29** (1972) 163.
73A1	Abdullaev, G.B., Akhmedov, N.R., Yalilov, N.Z., Abdinov, D.Sh.: Phys. Status Solidi **A20** (1973) K29.
73A2	Akhmedov, N.R., Dzhalilov, N.Z., Abdinov, D.Sh.: Inorg. Mater. **9** (1973) 1271.
73B1	Babot, D., Chevreton, M.: J. Solid State Chem. **8** (1973) 166.
73B2	Brattas, L., Kjekshus, A.: Acta Chem. Scand. **27** (1973) 1290.
73G	Gautier, F., Krill, G., Lapierre, M.F., Robert, C.: J. Phys. C **6** (1973), L320.
73K	Koehler Jr., R.F., White, R.L.: J. Appl. Phys. **44** (1973) 1682.
73M1	Mattheis, L.F.: Phys. Rev. **B8** (1973) 3719.
73M2	Mori, N., Mitsui, T., Yomo, S.: Solid State Commun. **13** (1973) 1083.
73S	Schairer, W., Shafer, M.W.: Phys. Status. Solidi **17A** (1973) 181.
73T	Takahashi, T.: Solid State Commun. **13** (1973) 1335.
73Y	Yuzuri, M.: J. Phys. Soc. Jpn. **35** (1973) 1252.
74A1	Anzai, S., Ozawa, K.: Phys. Status. Solidi **A24** (1974) K31.
74A2	Avdeev, B.V., Krasheninin, Yu.P.: Sov. Phys. Solid State **15** (1974) 2028.
74B	Barker Jr., A.S., Remaika, J.P.: Phys. Rev. **B10** (1974) 987.
74C	Coey, J.M.D., Brusetti, R., Kallel, A., Schweizer, J., Fuess, H.: Phys. Rev. Lett. **32** (1974) 1257.
74F	Finley, A., Schleich, D., Ackerman, J., Soled, S., Wold, A.: Mater. Res. Bull. **9** (1974) 1655.
74G	Grant, A.J., Griffiths, T.M., Pitt, G.D., Yoffe, A.D.: J. Phys. C **7** (1974) L249.
74M	Mattheiss, L.F.: Phys. Rev. **B10** (1974) 995.
74W	Williams, P.M., Parry, G.S., Scruby, C.B.: Philos. Mag. **29** (1974) 695.
75C	Cianelli, R.R., Scanlon, J.C., Thompson, A.H.: Mater. Res. Bull. **10** (1975) 1379.
75F	Furuseth, S., Brattas, L., Kjekshus, A.: Acta Chem. Scand. **29A** (1975) 623.
75G	Grant, A.J., Griffiths, T.M., Pitt, G.D., Yoffe, A.D.: J. Phys. C **8** (1975) L17.
75M	Miyadai, T., Takizawa, K., Nagata, H., Ito, H., Miyahara, S., Hirakawa, K.: J. Phys. Soc. Jpn. **38** (1975) 115.
75S	Scruby, C.B., Williams, P.M., Parry, G.S.: Philos. Mag. **31** (1975) 255.
75T1	Thompson, A.H.: Phys. Rev. Lett. **35** (1975) 1786.
75T2	Thompson, A.H., Gamble, F.R., Symon, C.R.: Mater. Res. Bull. **10** (1975) 915.
75W1	Wertheim, G.K., Di Salvo, F.J., Chiang, S.: Phys. Lett. **A54** (1975) 304.
75W2	Wilson, J.A., Di Salvo, F.J., Mahajan, S.: Adv. Phys. **24** (1975) 117.
75W3	Whittingham, M.S., Thompson, A.H.: J. Chem. Phys. **62** (1975) 1588.
76B1	Babot, D., Peix, G., Chevreton, M.: J. Phys. (Paris) Colloq. **4** (1976) 111.
76B2	Barthelemy, E., Chavant, C., Collin, G., Gorochov, O.: J. Phys. (Paris) Colloq. **4** (1976) 17.
76C	Coey, J.M.D., Roux-Buisson, H., Brusetti, R.: J. Phys. (Paris) Colloq. **4** (1976) 1.
76D	DiSalvo, F.J., Moncton, D.E., Waszczak, J.V.: Phys. Rev. **B14** (1976) 4321.
76G	Grazhdankina, N.P., Bersenev, Yu.S.: High Temp. – High Pressures **8** (1976) 613.
76H	Horwood, J.L., Townsend, M.G., Webster, A.H.: J. Solid State Chem. **17** (1976) 35.
76M	Moldenhauer, W., Brückner, W.: Phys. Status. Solidi **A34** (1976) 565.
76N	Neville, R.A., Evans, B.L.: Phys. Status. Solidi **B73** (1976) 597.
76R	Rao, C.N.R., Pisharody, K.P.R.: Prog. Solid State Chem. **10** (1976) 207.
76S	Schlegel, A., Wachter, P.: J. Phys. C **9** (1976) 3363.
76V	Vaidya, S.N., Karunakaran, C., Joshi, D.K., Karkhanavala, M.D.: Proc. Nucl. Phys. Solid State Phys. Symp. **19C** (1976) 44.
76W1	Wertheim, G.K., Di Salvo, F.J., Chiang, S.: Phys. Rev. **B13** (1976) 5476.
76W2	Williams, P.M. in "Crystallography and Crystal Chemistry of Materials with Layered Structures", F. Levy (ed.) Dordrecht: Reidel, **1976**, 51.
77A	Allen, J.W., Lucovsky, G., Mikkelsen Jr., J.C.: Solid State Commun. **24** (1977) 367.
77D	DiSalvo, F.J., Graebner, J.E.: Solid State Commun. **23** (1977) 825.
77E	El-Mahalawy, S.H., Evans, B.L.: Phys. Status. Solidi **79B** (1977) 713.
77F	Friend, R.H., Jerome, D., Liang, W.Y., Mikkelsen, J.C., Yoffe, A.D.: J. Phys. C **10** (1977) L705.
77J	Joshi, D.K., Karunakaran, C., Vaidya, S.N., Karkhanavala, M.D.: Mater. Res. Bull. **12** (1977) 1111.
77K	Kjekahus, A., Rakke, T., Andresen, A.F.: Acta Chem. Scand. **A31** (1977) 253.

9.15.3.8 References for 9.15.3

77L	Loseva, G.V., Ovchinnikov, S.G., Sokolovich, V.V., Petukhov, E.P.: Sov. Phys. Solid State **19** (1977) 1713.
77M1	Makovetskii, G.I., Ryzhkovskii, V.M., Shakhlevich, G.M.: Phys. Status. Solidi **39A** (1977) K127.
77M2	Mankai, C., Martinez, G., Gorochov, O.: Phys. Rev. **16B** (1977) 4666.
77M3	Miyadai, T., Kikuchi, K., Ito, Y.: Physica B+C **86–88** (1977) 901.
77O	Okada, O., Miyadai, T.: J. Phys. Soc. Jpn. **43** (1977) L343.
77S	Sambongi, T., Tsutsumi, K., Shiozaki, Y., Yamamoto, M., Yamaya, K., Abe, Y.: Solid State Commun. **22** (1977) 729.
77T1	Tani, T., Osada, T., Tanaka, S.: Solid State Commun. **22** (1977) 269.
77T2	Tsutsumi, K., Takazaki, T., Yamamoto, M., Shiozaki, Y., Ido, M., Sambongi, T., Yamaya, K., Abe, Y.: Phys. Rev. Lett. **39** (1977) 1675.
77W	Woolley, A.M., Wexler, G.: J. Phys. C **10** (1977) 2601.
77Y	Yamada, Y., Takatera, H.: Solid State Commun. **21** (1977) 41.
78B	Bullett, D.W.: J. Phys. C **11** (1978) 4501.
78E	El-Mahalawy, S.H., Evans, B.L.: Phys. Status Solidi **B86** (1978) 151.
78F	Fleming, R.M., Moncton, D.E., McWhan, D.B.: Phys. Rev. **B18** (1978) 5560.
78H	Heikens, H.H., Van Bruggen, C.F., Haas, C.: J. Phys. Chem. Solids **39** (1978) 833.
78I	Ito, T., Ito, K., Oka, M.: Jpn. J. Appl. Phys. **17** (1978) 371.
78K1	Kikuchi, K., Miyadai, T., Fukui, T., Ito, H., Takizawa, K.: J. Phys. Soc. Jpn. **44** (1978) 410.
78K2	King, H.E., Prewitt, C.T.: Phys. Chem. Minerals **3** (1978) 72.
78L	Lambert-Andron, B., Grazhdankina, N.P., Vettier, C.: J. Phys. (Paris) Lett. **39** (1978) 43.
78M	Mori, N., Watanabe, T.: Solid State Commun. **27** (1978) 567.
78R	Rijnsdorp, J., Jellinek, F.: J. Solid State Chem. **25** (1978) 325.
78T	Tsutsumi, K., Sambongi, T., Kagoshima, S., Ishiguro, T.: J. Phys. Soc. Jpn. **44** (1978) 1735.
78W	Wintenberger, M., Buevoz, J.L.: Solid State Commun. **27** (1978) 511.
78Z	Zunger, A., Freeman, A.J.: Phys. Rev. **B17** (1978) 1839.
79A	Anedda, A., Fortin, E., Raga, F.: Can. J. Phys. **57** (1979) 368.
79B	Bullett, D.W.: J. Phys. C **12** (1979) 277.
79C1	Collins, R., Kaner, R., Russo, P., Wold, A., Avignant, D.: Inorg. Chem. **18** (1979) 727.
79C2	Conan, A., Delaunay, D., Bonnet, A., Moustafa, A.G., Spiesser, M.: Phys. Status. Solidi **B94** (1979) 279.
79I1	Ido, M., Tsutsumi, K., Sambongi, T., Mori, N.: Solid State Commun. **29** (1979) 399.
79I2	Inada, R., Onuki, Y., Tanuma, S.: Phys. Lett. **69A** (1979) 453.
79S	Seehra, M.S., Seehra, S.S.: Phys. Rev. **B19** (1979) 6620.
80F	Fung, K.K., Steeds, J.W.: Phys. Rev. Lett. **45** (1980) 1696.
80J	Jericho, M.H., Simpson, A.H., Frindt, R.F.: Phys. Rev. **B22** (1980) 4907.
80S	Sezermann, O., Simpson, A.M., Jericho, M.H.: Solid State Commun. **36** (1980) 737.
82G	Goodenough, J.B.: Ann. Chim. (Paris) **7** (1982) 489.
82T	Töpel-Schadt, J., Müller, W.F.: Phys. Chem. Minerals **8** (1982) 175.
83H1	Handbook of Chemistry and Physics, 64th ed. (ed.: R.C. Weast), CRC Press. Inc. **1983**.
83H2	Haza, T., Abe, Y., Okwamoto, Y.: Phys. Rev. Lett. **51** (1983) 678.

Physical property	Numerical value	Experimental conditions	Experimental method, remarks	Ref.

9.16 Binary rare earth compounds

This contribution concentrates on semiconducting properties of rare earth compounds. The magnetic properties of this class of compounds has been described in detail in volume III/12c of this series. In order to represent the compounds in a systematic manner we also list a few compounds from which only lattice data are known exactly but from which semiconducting properties are expected too. The tables and figures give room-temperature data if not otherwise specified. E_A is always the activation energy of the electrical conductivity.

9.16.1 Hydrides RH_x

LaH_x
Structure: cubic (O_h^5 – Fm3m)

			crystal structure: Fig. 1; lattice parameter vs. x: Fig. 2; heat capacity [79B1, 83I] and	55M, 57S, 59S
Semiconductor: x = 2.7		$T < 239$ K	semiconductor-metal transition: Fig. 3 (x = 3) and Fig. 4 (x = 2.7);	79B1, 79Z2
Semiconductor: x = 3.0		$T < 241$ K	band structure (x = 3): Figs. 5, 6;	83I
E_g	0.1 eV	$T < 210$ K	density of states (x = 3):	80B1
Θ_D	241.5 K	x = 3	Figs. 6, 7; proposed electronic level scheme (x = 3): Fig. 8; temperature dependence of electrical resistivity (x = 2.85): Fig. 9; XPS spectra (x = 3): Figs. 10, 11; EDC spectra (x = 2.89): Figs. 12, 13; hydrogen vibrations [82K]; optical phonon energies of 116 meV and 64 meV for x = 2.82 [71M]	83I

LaD_x
Structure: cubic (O_h^5 – Fm3m)

Semiconductor: $2.89 \leq x \leq 3$		$T < 210$ K	crystal structure: Fig. 1; lattice parameter vs. x: Fig. 2; temperature dependence of	79M 80B1
E_g	0.1 eV	$T < 210$ K, x = 3	heat capacity and metal-	
Θ_D	246.3 K	x = 3	semiconductor transition: Fig. 3	83I

CeH_x
Structure: cubic (O_h^5 – Fm3m)

			crystal structure: Fig. 1; lattice parameter vs. x: Fig. 14;	72L1, 72L2
a	5.539 Å	x = 3	metal-semiconductor transition	66H1
σ	$\approx 10 \, \Omega^{-1} \, cm^{-1}$	x = 2.81, n-type	for $x \geq 2.7$ [72L1, 79D, 81F1]; heat capacity [65B];	69L2
	$\approx 3 \, \Omega^{-1} \, cm^{-1}$	x = 2.85	temperature dependence of heat capacity (x = 2.7, 2.96): Fig. 4;	69L2, 72L1
E_A	0.104 eV	x = 2.81, $T = 200$ K	energy bands (x = 3): Fig. 15; density of states (x = 3, 2.75): Figs. 16, 17; XPS spectra (x = 2.9): Figs. 18, 19; Hall coefficient: Fig. 20; charge carrier concentration: Fig. 21; Seebeck coefficient: Fig. 22;	72L2

(continued)

Physical property	Numerical value	Experimental conditions	Experimental method, remarks	Ref.
CeH$_x$ (continued)			resistivity dependence on x: Fig. 23; temperature dependence of resistivity for x = 2.77, 2.81, 2.85: Figs. 24, 25, 26; optical phonon energies 110 meV and 65 meV for x = 2.72 [74V]; electron phonon interaction [81F2]	
CeD$_x$				
Structure: cubic (O_h^5 – Fm3m)			crystal structure: Fig. 1; lattice parameters vs. temperature for x = 2.75: Fig. 27, tetragonal structure below T = 252 K [72L2]	72L2
a	5.55253 (10)	x = 2.75		
PrH$_x$				
Structure: cubic (O_h^5 – Fm3m)			crystal structure: Fig. 1; lattice parameter vs. x: Fig. 28; heat capacity [79D]; metal-semiconductor transition [79D, 79B1] and molar heat capacity: Fig. 4	61W, 62P, 66K, 79D
a	5.486 Å	x = 3		
Θ_D	231 K	x = 2.65		
EuH$_2$				
Structure: orthorhombic (D_{2h}^{16} – Pnma)				75H, 56K1
a	6.254 Å			
b	3.806 Å			
c	7.212 Å			
DyH$_3$				
Structure: hexagonal (D_{6h}^4 – P6$_3$/mmc)			Mössbauer spectra [81S1]; magnetic ordering [81C]	62P, 63W
a	3.671 Å			
c	6.615 Å			
HoH$_3$				
Structure: hexagonal (D_{6h}^4 – P6$_3$/mmc)				62P, 63W
a	3.642 Å			
c	6.560 Å			
ErH$_3$				
Structure: hexagonal (D_{6h}^4 – P6$_3$/mmc)				62P, 76S
a	3.621 Å			
c	6.526 Å			
YbH$_x$				
Structure: orthorhombic (D_{2h}^{16} – Pnma) (x=2)				66W1
a	5.889 (5) Å	x = 2	optical phonon energies of 130 meV, 91 meV, 81 meV and 71 meV have been reported for x = 2	75H
b	3.576 (2) Å			
c	6.789 (5) Å			
σ	$10^{-7}\,\Omega^{-1}\,cm^{-1}$	x = 1.9	temperature dependence for x = 1.90: Fig. 29	69H

Physical property	Numerical value	Experimental conditions	Experimental method, remarks	Ref.

9.16.2 Borides RB_6

SmB_6
Structure: cubic (O_h^1 – Pm3m)

a	4.129 Å		lattice parameter: Fig. 30;	56P2
T_m	>2500 °C		crystal structure: Fig. 31;	80H1
σ	16 Ω^{-1} cm^{-1}		$L\gamma_1$ emission spectra [82T1];	70Y
R_H	$5 \cdot 10^{-4}$ cm^3 C^{-1}		temperature dependence of	71N
S	$8 \cdot 10^{-6}$ V K^{-1}		resistance: Fig. 32	62S

EuB_6
Structure: cubic (O_h^1 – Pm3m)

a	4.185 Å		lattice parameter: Fig. 30;	74M3
T_m	>2500 °C		crystal structure: Fig. 31;	80H1
σ	100 Ω^{-1} cm^{-1}		energy bands: Fig. 33; electron tunneling [82G2]	73M, 73G
	$2 \cdot 10^3$ Ω^{-1} cm^{-1}		semimetal	79F1
E_A	0.38 eV			73M, 73G
	0.3 eV			74M3
R_H	$-50 \cdot 10^{-4}$ cm^3 C^{-1}			62S
S	$-8 \cdot 10^{-5}$ V K^{-1}			73M, 73G

YbB_6
Structure: cubic (O_h^1 – Pm3m)

a	4.140 Å		lattice parameter: Fig. 30;	56P2
	4.160 Å		crystal structure: Fig. 31	80H1, 74M2
T_m	>2500 °C			80H1
σ	30 Ω^{-1} cm^{-1}			73M, 73G
E_A	0.14 eV			73M, 73G
$|R_H|$	$84 \cdot 10^{-4}$ cm^3 C^{-1}			62S
S	$-2 \cdot 10^{-4}$ V K^{-1}			73M, 73G

For further data of rare earth borides, see section 9.14.7.

9.16.3 Monochalcogenides RX

SmS
Structure: cubic (O_h^5 – Fm3m)

a	5.97 Å		lattice parameter of $Sm_{1-x}Gd_xS$: Fig. 34; for $Sm_{1-x}R_xS$ with R = Ce, Ho, Er, Tm, Lu: Fig. 35; p–T-diagram: Fig. 36; lattice parameter vs. pressure: Fig. 37;	73J2, 75J, 77M
T_m	2210 K			72G
E_g	2.3 eV	$3p^6$–5d, 6s trans.	band structure and density of	78S2
	0.06 eV	4f–cond. band	states: Figs. 38, 39; band	72K1
	≈0.2 eV	4f–5d trans.	structure under pressure [80F2];	70J
	0.24(1) eV	4f–cond. band	band structure of the metallic	78S2
dE_g/dp	-10 meV/kbar	opt. absorption (4f–5d trans.)	and semiconducting phase: Fig. 40; band structure and electron-electron interaction [81F3];	73J1

(continued)

9.16.3 Rare earth monochalcogenides RX

Physical property	Numerical value	Experimental conditions	Experimental method, remarks	Ref.
SmS (continued)				
B	150 kbar		s-type conduction band [78K1]; intermediate valence and semiconductor-metal transition [78S2, 80G2, 80K2, 81H, 81G1, 81L1, 81K1, 81L2, 81A, 81B3, 82G1, 82M3, 82T2, 82S4]; reflectivity spectrum: Fig. 41; $\varepsilon_2(\omega)$: Fig. 42; optical properties of thin films [78G1]; photoemission spectrum: Fig. 43; phonon dispersion curves: Fig. 44; Raman spectra [78G2]; photoelectron spectra [82M2]; inelastic electron scattering [82C]; $L\gamma_1$ emission spectra [82T1]; thermoreflectance spectra [81M]; stress dependent carrier density [81K2]; pressure dependence of resistivity: Fig. 45; temperature dependence of resistivity: Fig. 46; temperature dependence of Hall coefficient: Fig. 47; temperature dependence of carrier mobility: Fig. 48; electron tunneling [82G2]; relativistic ab initio band structure calculation [83N]; electron phonon coupling [81G2]	72C
	470 kbar			76D
	476 kbar			78M1
Θ_D	247 K			76M1
	155 (7) K			77S
c_{11}	1200 kbar			76D
	1270 kbar			82S3
c_{12}	110 kbar			76D
	120 kbar			82S3
c_{44}	250 kbar			76D
	269 kbar			82S3
dc_{11}/dp	10.4			82S3
dc_{12}/dp	-1.6			
dc_{44}/dp	-0.08			
σ	$10^2\cdots10^3\,\Omega^{-1}\,\text{cm}^{-1}$			70J
	$20\cdots30\,\Omega^{-1}\,\text{cm}^{-1}$			78S2
n	$\approx 10^{19}\,\text{cm}^{-3}$			78S2
μ_n	$20\cdots25\,\text{cm}^2\,\text{V}^{-1}\,\text{s}^{-1}$			78S2
SmSe				
Structure: cubic ($O_h^5 - Fm3m$)				
a	6.20 Å			70J
	6.1975 (3) Å			76B3
	6.192 (4) Å			71S2
E_g	0.46 (2) eV	4f–5d trans.	band structure: Fig. 49; absorption spectrum: Fig. 50; reflectivity spectrum: Fig. 41; photoemission spectrum: Fig. 43; photosensitivity [73S4]	70J
	0.55 eV	4f–5d trans.		73J1
	1.4 (2) eV	$4p^6$–5d, 6s trans.		74D
dE_g/dp	-11 meV/kbar	opt. spectra (4f–5d trans.)		72K3
B	400 kbar			72C
Θ_D	153 (14) K			77S
σ	$3\cdot10^{-4}\,\Omega^{-1}\,\text{cm}^{-1}$		pressure dependence of resistivity: Fig. 45; intermediate valence and metal-semiconductor transition [80F2, 78S2, 81H]; $L\gamma_1$ emission spectra [82T1]	70J
SmTe				
Structure: cubic ($O_h^5 - Fm3m$)				
a	6.594 Å			70J
	6.6600 (4) Å			71S2
E_g	0.63 eV	4f–5d trans.	band structure: Figs. 38, 51; photoemission spectrum: Fig. 43; reflectivity spectrum: Fig. 52	70J
dE_g/dp	-11.9 meV/kbar	opt. spectra		73J1
B	400 kbar			72C

(continued)

Physical property	Numerical value	Experimental conditions	Experimental method, remarks	Ref.
SmTe (continued)				
Θ_D	151 (20) K			77S
σ	$\approx 10^{-3}\,\Omega^{-1}\,cm^{-1}$		pressure dependence of resistivity Fig. 45; intermediate valence and metal-semiconductor transition [70J, 80F2, 78S2]	70J
EuO				
Structure: cubic (O_h^5–Fm3m)				
a	5.1435 (1) Å	$T = 298.15$ K	lattice parameter vs. T: Fig. 53; phase diagram: Fig. 54	74M1
	5.141 Å			72W1
T_m	2238 (10) K			72S1
	2249 (8) K			72R
d	8.197 g cm^{-1}			74M1
E_g	4.1 eV	$2p^6$–5d, 6s trans. (Fig. 58)	band structure: Figs. 55, 56, 57; density of states: Fig. 58; electronic structure and mixed valence [82N1, 81K3]; MO-calculation [81Z]; band structure and electron-electron interaction [81F4]	73E
	3.9 eV	$2p^6$–5d, 6s trans. (Fig. 58)		75M1
	1.12 eV	4f–5d trans. (Fig. 58)		72W1, 71G
dE_g/dp	-4.4 meV/kbar	optical absorption (4f–5d trans.)		69W
B	1070 kbar			66W2
	920 (60) kbar			69S
	910 (80) kbar	$T = 77$ K		71S3
	1100 (50) kbar			74J
γ_L	1.9		γ_L: lattice Grüneisen constant	67A
α	$13.2 \cdot 10^{-6}$ K^{-1}			69L1
c_{11}	$19.2 \cdot 10^{10}$ Pa	$T = 78$ K		69S, 72S3
c_{12}	$4.25 \cdot 10^{10}$ Pa	78 K		
c_{44}	$5.42 \cdot 10^{10}$ Pa	78 K		
κ	$1.1 \cdot 10^{-11}$ Pa^{-1}		κ: compressibility	70L
$v_{[100]}$	$4.83(6) \cdot 10^5$ cm s^{-1}	$T = 78$ K		69S
$v_{[110]}$	$4.6(1) \cdot 10^5$ cm s^{-1}	78 K		
Θ_D	353 K			72S3, 71P
	350 K			69S
C_p	48.74 J mol^{-1} K^{-1}			74M1
$\bar{\nu}_{TO}$	182 cm^{-1}		Raman scattering	74G
	199.3 cm^{-1}		IR measurement	69A
	350.1 cm^{-1}			
$\bar{\nu}_{LO}$	435 cm^{-1}			74G
	346.3 cm^{-1}			69A
	426.8 cm^{-1}			
$\varepsilon(0)$	26.5		dielectric constant ε_1: Fig. 61; dielectric constant ε_2: Figs. 62, 70	74G
	23.9			69A
$\varepsilon(\infty)$	4.6			74G
	5.0			69A
	3.85			68W
n	2.25	at 4f→5d (t_{2g}) absorption edge	reflectivity spectrum: Fig. 60	71G

(continued)

Physical property	Numerical value	Experimental conditions	Experimental method, remarks	Ref.
EuO (continued)				
$E(4f-5d)$	1.12 eV	edge	absorption spectrum: Fig. 59	72W1, 71G
$E(4f-5d)$	1.9 eV	maximum (Fig. 58)		71G
K	$1.23 \cdot 10^5$ cm^{-1}	max. 4f–5d trans.		75S2
f	0.14		oscillator strength of 4f–5d transition	71G
$W(4f-5d)$	1.1 eV		width of 4f–5d transition	71G
E_{thr}	3.6 eV		ionization energy (photothreshold)	75S3
	2.7 (2) eV			76M2
ϕ	1.8 eV		work function	73E
	1.80 (15) eV			76M2
			photosensitivity: Fig. 63	
σ	16···50 Ω^{-1} cm^{-1}		nonactivated	73S1
n	$5.5 \cdots 15 \cdot 10^{18}$ cm^{-3}			
μ_n	20 cm^2/Vs			
σ	$1.6 \cdots 4.5 \cdot 10^3$ Ω^{-1} cm^{-1}	$T=4$ K	nonactivated	73S1
n	$3.4 \cdots 7.5 \cdot 10^{19}$ cm^{-3}	4 K	metallic phase; temperature dependent resistivity and semiconductor-metal transition: Figs. 64, 65	
μ_n	290···370 cm^2/Vs	4 K		
σ	$50 \cdots 1.4 \cdot 10^{-5}$ Ω^{-1} cm^{-1}			73S1
E_A	0.3 eV			
n	$1.9 \cdots 2.9 \cdot 10^{13}$ cm^{-3}			
μ_n	12···29 cm^2/V s			
σ	250 Ω^{-1} cm^{-1}	$T=4$ K		73S1
E_A	0.3 eV			
n	$1.2 \cdots 1.7 \cdot 10^{19}$ cm^{-3}	$T=4$ K		
μ_n	9.2···130 cm^2/V s	4 K		
ϱ	$4 \cdot 10^6$ Ω cm	$T=280$ K	pressure dependent resistivity [78D1]; electric field dependence of conductivity [82G3]; carrier mobility [82S5]; magnetic field dependence of carrier mobility [81P2]; ac-conductivity [78K2]; electron-magnon interaction [81R1, 82A1, 83V]	77D
E_A	≈ 0.5 eV			72O
	0.360 eV			77D
dE_A/dp	-5 meV/kbar			72O
m_n	0.42 m_0			81P2
EuS				
Structure: cubic (O$_h^5$–Fm3m)				
a	5.968 Å		phase diagram: Fig. 66	72W1, 79W
	5.9679 (1) Å	$T=298.15$ K	lattice parameter vs. temperature: Fig. 67	74M1
T_m	2831 (16) K			72R
	1940 K			81Z
d	5.750 g cm^{-3}			74M1
E_g	2.3 eV	$3p^6$–5d, 6s trans. (Fig. 58)	band structure: Figs. 68, 69; density of states: Fig. 58; APW calculation [81F4]; MO calculation [81Z]	75M1
	1.65 eV	4f–5d trans. (Fig. 58)		71G, 72W1

(continued)

Physical property	Numerical value	Experimental conditions	Experimental method, remarks	Ref.
EuS (continued)				
dE_g/dp	-7.9 meV kbar^{-1}	opt. absorption (4f–5d) trans.		69W
B	560 (60) kbar			70L
	500 (75) kbar	$T = 77$ K		71S3
	610 (50) kbar			74J
α	$14.2 \cdot 10^{-6}$ K^{-1}			66D
	$12.6 \cdot 10^{-6}$ K^{-1}			69L1
c_{11}	$13.1 (5) \cdot 10^{10}$ Pa	$T = 77$ K		71S3,
c_{12}	$1.1 (8) \cdot 10^{10}$ Pa	77 K		72S3
c_{44}	$2.73 (11) \cdot 10^{10}$ Pa	77 K		
κ	$1.1 \cdot 10^{-11}$ Pa^{-1}		κ: compressibility	70L
$v_{[100]}$	$4.74 \cdot 10^5$ cm s^{-1}	$T = 77$ K		71S3
$v_{[110]}$	$4.09 \cdot 10^5$ cm s^{-1}	77 K		
Θ_D	280 K			66D
	276 K			71S3
	262 K			63M, 72S3
C_p	50.96 Jmol^{-1} K^{-1}			74M1
$\bar{v}_{TO}$	178.4 (20) cm^{-1}		IR measurements; for Raman scattering, see [79S, 80S1]	69A
	186.5 cm^{-1}	$T = 2$ K		73I
$\bar{v}_{LO}$	266.5 (18) cm^{-1}			69A
	278 cm^{-1}	2 K		73I
$\varepsilon(0)$	11.1 (15)		real and imaginary part of the dielectric constant: Figs. 61, 70	69A
	11.1			74G
$\varepsilon(\infty)$	4.9			68W
	4.7 (2)			69A
	4.7			74G
n	2.42	at 4f→5d absorption edge		71G
$E(4f-5d)$	1.65 eV	edge	absorption spectrum: Fig. 59	71G
$E(4f-5d)$	2.42 eV	maximum (Fig. 58)		75S2
K	$1.46 \cdot 10^5$ cm^{-1}	max. 4f–5d trans.		75S2
f	0.18		oscillator strength of 4f–5d transitions	71G
$W(4f-5d)$	0.7 eV		width of 4f–5d transition	71G
E_{thr}	4 eV		ionization energy (photothreshold)	76M2
ϕ	3.3 eV		work function photosensitivity: Figs. 71, 63; soft X-ray emission and absorption spectra: Fig. 72; bulk and surface states [82M2]	76M2
σ	$\approx 10^{-9}$ Ω^{-1} cm^{-1}		temperature dependence of resistivity: Fig. 73, temperature dependence of resistivity at different magnetic fields: Fig. 74; electron-magnon interaction [82A1]	71B1
	6···60 Ω^{-1} cm^{-1}			72S4
	60···700 Ω^{-1} cm^{-1}	$T = 4$ K		
n	$0.5 \cdots 1 \cdot 10^{19}$ cm^{-3}			72S4
μ_n	30 cm^2 V^{-1} s^{-1}		magnetic field dependence of Hall coefficient: Fig. 75; temperature dependence of mobility: Fig. 76	72S4

Physical property	Numerical value	Experimental conditions	Experimental method, remarks	Ref.
EuSe				
Structure: cubic (O_h^5 – Fm3m)				
a	6.195 Å			72W1, 79W
	6.1936 (2) Å	$T = 298.15$ K		74M1
T_m	2488 (8) K			72R
d	6.455 g cm^{-3}			74M1
E_g(4f→5d)	1.80 eV		optical spectroscopy	72W1
	1.78 eV		optical spectroscopy	71G
dE_g/dp	−8.4 meV kbar^{-1}	opt. absorption	band structure: Figs. 77, 78;	69W
B	530 (110) kbar		density of states: Fig. 58;	70L
	480 (50) kbar		temperature dependence of E_g:	71S3
	520 (50) kbar		Fig. 79; band structure and	74J
α	18.2 · 10^6 K^{-1}		electron-electron interaction	66D
	13.1 · 10^{-6} K^{-1}		[81F4]; MO calculation [81Z]	69L1
c_{11}	11.6 (4) · 10^{10} Pa	$T = 77$ K		71S3,
c_{12}	1.2 (6) · 10^{10} Pa	77 K		72S3
c_{44}	2.28 (9) · 10^{10} Pa	77 K		
κ	1.9 · 10^{-11} Pa^{-1}		κ: compressibilty	70L
$v_{[100]}$	4.22 · 10^5 cm s^{-1}	$T = 77$ K		71S3
Θ_D	153 (9) K			80S2
	180 K			64B
	220 K			66D
	232 K			71S3
	176 (9) K			72W2
C_p	51.34 Jmol^{-1} K^{-1}		heat capacity near the magnetic	74M1
$\bar{v}_{TO}$	130 cm^{-1}	$T = 4.2$ K ^{a})	phase transition [80J];	80S1
	127.8 (5) cm^{-1}	300 K ^{b})	phonon dispersion relations:	69A
	134 cm^{-1}	2 K ^{b})	Fig. 80; Raman spectra: Fig. 81;	73I
$\bar{v}_{LO}$	176 cm^{-1}	4.2 K ^{a})	phonon energies in various	80S1
	182 (3) cm^{-1}	300 K ^{b})	magnetic phases [80S1, 79S]	69A
	189 cm^{-1}	2 K ^{b})		73I
$\bar{v}_{TA}$	56 cm^{-1}	4.2 K ^{a})	^{a}) from Raman scattering	80S1
$\bar{v}_{LA}$	79 cm^{-1}	4.2 K ^{a})	^{b}) from IR measurements	80S1
$\varepsilon(0)$	9.4 (8)		real and imaginary parts of the	69A
	9.5		dielectric constant: Figs. 61, 70	74G
$\varepsilon(\infty)$	5.3			68W
	5.35			74G
n	2.43	at 4f−5d absorption edge		71G
E(4f−5d)	1.78 eV	edge	absorption spectrum: Fig. 59;	71G
E(4f−5d)	2.55 eV	maximum	luminescence and excitation	75S2
K	1.50 · 10^5 cm^{-1}	max. 4f−5d trans.	spectra: Figs. 82, 83	75S2
f	0.20		oscillator strength of 4f−5d transition	71G
W(4f−5d)	0.7 eV		width of 4f−5d transition	
E_{thr}	4.5 (1) eV		ionization energy (photothreshold) photosensitivity: Figs. 63, 84	76M2

(continued)

Physical property	Numerical value	Experimental conditions	Experimental method, remarks	Ref.
EuSe (continued)				
ϕ	3.90 (15) eV		work function	76M2
	4.00 (15) eV			
σ	$\approx 10^{-1} \, \Omega^{-1} \, cm^{-1}$	$T = 76$ K		78H
	$50 \, \Omega^{-1} \, cm^{-1}$			74S1
	$10^{-7} \, \Omega^{-1} \, cm^{-1}$		high resistivity sample	
	$1 \cdots 10 \, \Omega^{-1} \, cm^{-1}$	n-type	temperature dependence of ac- and	81Y
n	$0.42 \cdots 3.5 \cdot 10^{19} \, cm^{-3}$		dc-conductivity: Figs. 85, 86;	74S1
	$\approx 10^{18} \, cm^{-3}$		σ vs. magnetic field: Figs. 87, 88;	81Y
E_A	$11 \cdots 20$ meV	$T = 20 \cdots 50$ K	magnetically stimulated current	81Y
μ_H	$\approx 0.5 \, cm^2 \, V^{-1} \, s^{-1}$	$B = 5$ T, $T = 76$ K	vs. temperature: Fig. 89; dc- and microwave conductivity: Fig. 90; Hall voltage vs. magnetic field: Fig. 91; comparison of conductivity, carrier concentration and Hall mobility: Fig. 92	78H
EuTe				
Structure: cubic ($O_h^5 - Fm3m$)				
a	6.597 Å			69L1
	6.5984 (1) Å	$T = 298.15$ K		74M1
T_m	2456 (2) K			72R
d	6.461 g cm^{-3}			74M1
$E_g(4f \to 5d)$	2.0 eV		optical spectroscopy	71G
	2.062 (5) eV	$T = 1.7$ K	optical spectroscopy	78S3
	1.33 eV		optical spectroscopy density of states: Fig. 58 and [82L]	72S5
dE_g/dp	-12.0 meV kbar^{-1}	opt. absorption		69W
B	400 (30) kbar			70L
	357 (50) kbar	$T = 77.6$ K		72S6
	400 (50) kbar			74J
α	$13.6 \cdot 10^{-6}$ K^{-1}			69L1
c_{11}	$9.36 (40) \cdot 10^{10}$ Pa	$T = 77.6$ K		72S3, 72S6
c_{12}	$0.67 (60) \cdot 10^{10}$ Pa	77.6 K		
c_{44}	$1.63 (7) \cdot 10^{10}$ Pa	77.6 K		
κ	$2.5 \cdot 10^{-11}$ Pa^{-1}			70L
$v_{[100]}$	$3.78 \cdot 10^5$ cm s^{-1}	$T = 77.6$ K		72S6
$v_{[110]}$	$3.19 \cdot 10^5$ cm s^{-1}	77.6 K		
Θ_D	140 K			64B
	189 K			72S6
C_p	52.17 J mol^{-1} K^{-1}		heat capacity near the magnetic phase transition [80J]	74M1
$\bar{v}_{TO}$	102.3 (20) cm^{-1}		IR measurements [73H, 73I];	73H
	111.7 cm^{-1}	$T = 2$ K	Raman spectra [79G];	73I
$\bar{v}_{LO}$	141.5 (20) K		calculated Raman spectra	73H
	113 cm^{-1}	2 K	[83O1]	79G, 83O1
	111 cm^{-1}	0 K	theoretical value	83O1
$\varepsilon(0)$	6.9		real and imaginary parts of the dielectric constant: Figs. 61, 70	74G
	8.23			73H
$\varepsilon(\infty)$	5.75			74G
	5.92			68W
	4.18			73H

(continued)

Physical property	Numerical value	Experimental conditions	Experimental method, remarks	Ref.
EuTe (continued)				
n	2.71	at 4f–5d absorption edge		71G
$E(4f-5d)$	2.0 eV	edge	absorption spectrum: Fig. 59;	71G
	2.062 (5) eV	edge, $T=1.7$ K	absorption edge: Fig. 93;	78S3
$E(4f-5d)$	2.64 eV	maximum	shift of absorption edge [82M4];	75S2
K	$1.82 \cdot 10^5$ cm^{-1}	max. 4f–5d trans.	electronic excitation spectrum [82L]	75S2
f	0.25		oscillator strength of 4f–5d transition	71G
$W(4f-5d)$	0.6 eV		width of 4f–5d transition	71G
E_{thr}	4.6 (1) eV		ionization energy (photothreshold)	76M2
ϕ	3.6 (3) eV		work function	76M2
			photosensitivity: Fig. 63	
σ	10^{-8} Ω^{-1} cm^{-1}		high resistivity sample	71O
	40···250 Ω^{-1} cm^{-1}			72S2
	0.13···4 Ω^{-1} cm^{-1}	$T=4$ K		
μ_n	33···58 cm^2 V^{-1} s^{-1}		electron-phonon interaction [81U]	72S2
n	0.5···1.3 · 10^{19} cm^{-3}			72S2
E_A	33 meV			75V
TmTe				
Structure: cubic (O_h^5–Fm3m)				
a	6.049···6.364 Å		semiconductor-metal transition	64I
	6.353 Å		in TmSe$_{1-x}$Te$_x$ at x = 0.5 [80B2];	79K1
	6.26···6.35 Å		Tm-valency [83O2]; XPS	83O2
T_m	2143 (3) K		spectrum: Fig. 94, schematic energy level diagram: Fig. 95	78F
E_g	0.35 eV		optical determination	75S1
	0.22 eV		absorption	71B3
	0.32 (4) eV		optical determination	79B2
dE_g/dp	−10 meV kbar^{-1}			71B3, 79B2
B	460 (50) kbar		temperature dependence of thermal expansion for Tm$_{0.94}$Te: Fig. 96; molar heat capacity: Fig. 97	74J
c_{11}	1020 kbar			77O
c_{12}	60 kbar			
c_{44}	186 kbar			
$\bar{v}_{TO}$	115 (2) cm^{-1}		IR reflectivity	75W
$\bar{v}_{LO}$	146 (2) cm^{-1}			
$\varepsilon(0)$	7.2			76D
$\varepsilon(\infty)$	4.5			
			far-infrared reflectivity: Fig. 98; reflectivity: Fig. 99	
E_A	0.2 eV		temperature dependence of resistivity: Fig. 100	75B
ϱ	0.77 Ω cm		nonstoichiometric Tm$_{0.94}$Te	83O2

Physical property	Numerical value	Experimental conditions	Experimental method, remarks	Ref.
YbS				
Structure: cubic (O_h^5 – Fm3m)				
a	5.692 Å			71B3
	5.691 Å			78M2
E_g	0.98 eV		optical spectroscopy	71B3
	1.1 eV		optical absorption: Figs. 101, 102	74N
	1.325 (5) eV	4f→conduction band trans.		80G3
dE_g/dp	−6.5 meV kbar^{-1}		pressure dependence of the absorption spectrum: Fig. 103; photocurrent and photo-emf: Fig. 104	74N
B	720 (50) kbar			74J
$\bar{\nu}_{TO}$	199 (3) cm^{-1}		IR measurements; for Raman spectrum, see Fig. 105	78M2
	203 (3) cm^{-1}	$T = 77$ K		
$\bar{\nu}_{LO}$	272 (3) cm^{-1}			
	276 (3) cm^{-1}	77 K		
$\varepsilon(0)$	9.5 (5)			78M2
	9.6 (5)	$T = 77$ K		
$\varepsilon(\infty)$	5.0 (5)			
	5.0 (5)	77 K		
E	1.8 eV	$4f^{14}(^1S_0) \to 4f^{13}(^2F_{7/2})5d$ trans.;		80G3
	2.8 eV	$4f^{14}(^1S_0) \to 4f^{13}(^2F_{5/2})5d$ trans.		
E_A	0.3···0.4 eV	p-type	traps due to non-stoichiometry; bulk and surface states [82M2]	80G3
YbSe				
Structure: cubic (O_h^5 – Fm3m)				
a	5.934 (1) Å			74P
E_g	1.5 eV		optical absorption: Fig. 101	74N
dE_g/dp	−9.8 meV kbar^{-1}			74N
B	610 (50) kbar			74J
ϱ	100 Ω cm			64R1
YbTe				
Structure: cubic (O_h^5 – Fm3m)				
a	6.359 Å		variation of lattice parameter [81K4]; phase diagram: Fig. 106	79K1
T_m	2200 °C			81K4
E_g	1.8 eV		optical absorption: Fig. 101	74N
	2.0 eV			74S2
dE_g/dp	−11.2 meV kbar^{-1}			74N
B	460 (50) kbar			72C
$\bar{\nu}_{TO}$	98 (3) cm^{-1}		IR measurements	78M2
$\bar{\nu}_{LO}$	145 (3) cm^{-1}			
ϱ	$10^4 \cdots 10^6$ Ω cm			60B, 64R1

Physical property	Numerical value	Experimental conditions	Experimental method, remarks	Ref.

9.16.4 Chalcogenides R_xX_y with $x<y$

Sm_3S_4
Structure: cubic (Th_3P_4-type, T_d^6-$I\bar{4}3d$)

a	8.556 Å		mixed valence [80H2, 80W, 80M1, 80M2];	56P1
	8.5198 (3) Å			76B2
	8.543 (2) Å			79V
	8.549 Å			79C
T_m	1800 °C			72G
α	$11.8 \cdot 10^{-6}$ K^{-1}			72G
κ	$3.58 \cdot 10^{-2}$ W cm^{-1} K^{-1}		heat capacity: Fig. 107; thermal conductivity: Fig. 108	61H
$\bar{\nu}_{LO}$	≈ 300 cm^{-1}		Raman scattering	77W
$\bar{\nu}_{TO}$	≈ 250 cm^{-1}		schematic band structure: Fig. 109	76B2
σ	1.8 Ω^{-1} cm^{-1}			
ϱ	5.9 Ω cm		temperature dependence of conductivity: Fig. 110	61H
	≈ 10 Ω cm			73S5
W(5d)	≈ 1.3 eV		width of 5d-conduction band	80W
E_A	0.132 eV	$T > 125$ K	from electrical conductivity	76B2,
	0.142 eV	$T < 125$ K	from electrical conductivity	76B5
	0.13 eV		from electrical conductivity	76E
	0.1 ··· 0.2 eV			77W
	≈ 0.1 eV			77V1
dE_A/dp	3.0 (3) meV kbar^{-1}		from electrical conductivity	76B2
S	-150 μV K^{-1}			61H
			reflectivity spectra: Fig. 111; electronic Raman spectrum: Fig. 112	

Sm_3Se_4
Structure: cubic (Th_3P_4-type, T_d^6-$I\bar{4}3d$)

a	8.785 Å		heat capacity: Fig. 107	59B
	8.892 Å			79C

Sm_3Te_4
Structure: cubic (Th_3P_4-type, T_d^6-$I\bar{4}3d$)

a	9.506 Å			65F

Eu_3S_4
Structure: cubic (Th_3P_4-type, T_d^6-$I\bar{4}3d$)

a	8.533 (1) Å		phase transition at 182 K [83P, 82W]; thermal expansion: Fig. 113; photoluminescence: Figs. 114, 115; resistivity and Seebeck coefficient: Figs. 116, 117, 118, 119; Raman scattering: Figs. 120, 121; 122; heat capacity: Figs. 123, 124; mixed valence [80H2, 80W, 80M1, 81R2]	67B
	8.534 (3) Å			77E
	8.527 (5) Å			78P
T_m	1600 ··· 2500 °C			77E
d	6.27 g cm^{-3}			77E, 78P
ν_{LO}	≈ 300 cm^{-1}		Raman scattering	77W
ν_{TO}	≈ 250 cm^{-1}			

(continued)

Physical property	Numerical value	Experimental conditions	Experimental method, remarks	Ref.
Eu_3S_4 (continued)				
E_g	1.7 eV		optical absorption edge	76V
E_A	0.22 (1) eV		from conductivity	67B
	0.32 (1) eV	$T < 186$ K	from conductivity	83P
	0.160 (5) eV	$T > 186$ K		
	0.21 (1) eV	$T < 175$ K	from conductivity; endothermic DTA transition at 170 K	70B
	0.163 (4) eV	$T > 175$ K		
ϱ	100 Ω cm		$E_A = 0.18$ eV	76B5
Eu_3Se_4				
σ	$3 \cdot 10^{-6}$ Ω^{-1} cm^{-1}		conductivity ($E_A = 0.58$ eV)	73B2
S	900 μV K^{-1}	p-type		73B2
Eu_3Te_4				
Structure: tetragonal				
$a = b$	12.5 Å			73B2
c	6.4 Å			
σ	0.3 Ω^{-1} cm^{-1}	p-type	conductivity ($E_A = 1.1$ eV)	73B2
S	240 μV K^{-1}			73B2
α-La_2S_3				
Structure: orthorhombic (D_{2h}^{16} – Pnma)				
a	7.584 Å		heat capacity: Fig. 125	68S
	7.587 (7) Å			81G4
b	15.860 Å			68S
	15.863 (4) Å			81G4
c	4.144 Å			68S
	4.149 (2) Å			81G4
T_m	1950 °C			81G4
Θ_D	317 (1) K			81G4
β-La_2S_3				
Structure: tetragonal (D_{4h}^{20} – I4$_1$/acd)				
$a = b$	15.62 Å		phase transition β→γ at 1300 °C;	73B1
	15.65 (4) Å		heat capacity: Fig. 126;	81G4
c	20.62 Å		for heat capacity, enthalpy and	73B1
	20.64 (6) Å		entropy, see also [69P]	81G4
E_g	2.58 eV	direct	optical determination	80L
	2.6 eV	direct	fundamental absorption edge	81S2
	1.38 eV		indirect or impurity tail	80L
ϱ	$2 \cdot 10^{12} \ldots 4 \cdot 10^{14}$ Ω cm		optical spectra for Nd-doped crystals: Figs. 127, 128; optical spectra for Nd, Ce-codoped crystals: Figs. 129, 130, 131; photoconductivity of a Ce-doped crystal: Fig. 132; photoluminescence and cathodoluminescence of a Ce-doped crystal: Figs. 133, 134	82A2
β-$La_{10}S_{14}O_xS_{1-x}$				
Structure: tetragonal (D_{4h}^{20} – I4$_1$/acd)				
$a = b$	15.62 ⋯ 15.36 Å	$0 \leq x \leq 1$		73B1
c	20.62 ⋯ 20.41 Å			

Physical property	Numerical value	Experimental conditions	Experimental method, remarks	Ref.
β-La$_{10}$S$_{14}$O				
Structure: tetragonal (D_{4h}^{20} – I4$_1$/acd)				
$a=b$	15.365 (3) Å		crystal structure: Fig. 135	81B1
c	20.384 Å		cathodoluminescence of Cu-doped samples: Fig. 136	
E_g	2.69 (7) eV		optical determination	81B1
γ-La$_2$S$_3$				
Structure: cubic (Th$_3$P$_4$-defect structure, T_d^6 – I$\bar{4}$3d)				
a	8.726 (1) Å		phase diagram: Fig. 137; coordination polyhedra: Fig. 138	78K3
	8.731 Å			60P
	8.7253 (8) Å			76K
	8.725 Å			81K6
	8.69 Å			80K1
T_m	2080 °C			60P
	1980 (30) °C			76K, 81K6
	2100 °C			80K1
d	4.93 g cm^{-3}			60P
α	$9.9 \cdot 10^{-6}$ K^{-1}			66D
E_g	2.9 eV		E_g: optical gap	80K1
	2.4 eV	$T = 148$ K		82B, 81B2
	2.8 (1) eV			79Z1
κ	$2.3 \cdot 10^{-2}$ W cm^{-1} K^{-1}			72G
Θ_D	320 K		molar heat capacity: Fig. 151	72S7
	285 K			81G4
$\bar{\nu}_{TO}$	195 cm^{-1}		$\bar{\nu}$ from Raman spectrum; see also Figs. 140, 141; absorption and transmission spectra: Figs. 141, 142, 143; real and imaginary part of the dielectric constant: Fig. 144; reflectivity spectrum: Fig. 145; luminescence and absorption of Nd-doped crystals: Figs. 146, 147, 148, 149, 150	79A
	230 cm^{-1}			
$\bar{\nu}_{LO}$	301 cm^{-1}			
$\varepsilon(0)$	17.2			79A
$\varepsilon(\infty)$	7.0			79A, 79Z1
n	2.6	$\lambda = 1 \cdots 15$ μm		80K1
	2.8	$\lambda = 0.55$ μm		
σ	10^{-10} Ω^{-1} cm^{-1}			80K1
	$10^{-9} \cdots 10^{-10}$ Ω^{-1} cm^{-1}			79Z1
	$0.7 \cdot 10^{-12}$ Ω^{-1} cm^{-1}			81B2
	10^{-12} Ω^{-1} cm^{-1}	$T = 148$ K		82B
La$_2$Se$_3$				
Structure: cubic (Th$_3$P$_4$-defect structure, T_d^6 – I$\bar{4}$3d)				
a	9.055 Å		coordination polyhedra: Fig. 138; heat capacity: Fig. 151	65F
α	$11.5 \cdot 10^{-6}$ K^{-1}			66D
Θ_D	347 K			66D
	240 K			72S7

Physical property	Numerical value	Experimental conditions	Experimental method, remarks	Ref.
La$_2$Te$_3$				
Structure: cubic (Th$_3$P$_4$-defect structure, $T_d^6 - I\bar{4}3d$)			heat capacity: Figs. 139, 151; coordination polyhedra: Fig. 138	65F
a	9.627 Å			65R1
	9.619 (1) Å			73S3
	9.617 (3) Å			65H
d	6.6 g cm^{-3}			65R2
T_m	1485 (5) °C			73S3
	1465 (30) °C			72S7
Θ_D	190 K			
σ	$10^{-4}\,\Omega^{-1}\,cm^{-1}$	n-type	temperature dependence of conductivity: Fig. 152, of Seebeck coefficient: Fig. 153	65R2
E_A	0.43 eV			65R2
α-Ce$_2$S$_3$				
Structure: orthorhombic (D_{2h}^{16} – Pnma)				71B2
a	7.55 Å			
b	15.79 Å			
c	4.14 Å			
β-Ce$_2$S$_3$				
Structure: tetragonal (D_{4h}^{20} – I4$_1$/acd)			transition α-phase to β-phase at 1410 °C	66M1
a	15.30 Å			71B2
c	20.29 Å			
T_m	1700 °C			60P
γ-Ce$_2$S$_3$				
Structure: cubic (Th$_3$P$_4$-defect structure, $T_d^6 - I\bar{4}3d$)			coordination polyhedra: Fig. 138	
a	8.631 Å			81K6
	8.630 Å			65F
	8.6084 Å		from neutron diffraction	71A
T_m	2060 °C			65F
	1870 °C			81K6
	2296 °C			72G
d	5.31 g cm^{-3}			81K6
α	$13.2 \cdot 10^{-6}\,K^{-1}$			66D
κ	$3.52 \cdot 10^{-2}\,W\,cm^{-1}\,K^{-1}$		heat capacity: Fig. 154	72G
σ	$10^{-10}\,\Omega^{-1}\,cm^{-1}$		temperature dependence of resistivity: Fig. 155	71A
E_A	2.23 eV		from electrical conductivity	67M
Ce$_2$Se$_3$				
Structure: cubic (Th$_3$P$_4$-defect structure, $T_d^6 - I\bar{4}3d$)			coordination polyhedra: Fig. 138	65F, 66D
a	9.03 Å			
α	$12.2 \cdot 10^{-6}\,K^{-1}$			66D
Θ_D	342 K			66D
ϱ	$3.3 \cdot 10^{-3}\,\Omega\,cm$			64R2
Ce$_2$Te$_3$				
Structure: cubic (Th$_3$P$_4$-defect structure, $T_d^6 - I\bar{4}3d$)			coordination polyhedra: Fig. 138	65F
a	9.539 Å			

Physical property	Numerical value	Experimental conditions	Experimental method, remarks	Ref.
α-Pr$_2$S$_3$				
Structure: orthorhombic (D_{2h}^{16} – Pnma)				60P
a	7.49 Å			71B2
b	15.69 Å			
c	4.10 Å			
β-Pr$_2$S$_3$				
Structure: tetragonal (D_{4h}^{20} – I4$_1$/acd)				
a	15.21 Å			71B2
c	20.17 Å			
γ-Pr$_2$S$_3$				
Structure: cubic (Th$_3$P$_4$-defect structure, T_d^6 – I$\bar{4}$3d)				
a	8.573 Å		coordination polyhedra: Fig. 138	65F, 75B
	8.592 Å			60P
T_m	1758 °C			72G
	1795 °C			60P
α	$11.3 \cdot 10^{-6}$ K^{-1}			66D
κ	$2.26 \cdot 10^{-2}$ W cm^{-1} K^{-1}			72G
Pr$_2$Se$_3$				
Structure: cubic (Th$_3$P$_4$-defect structure, T_d^6 – I$\bar{4}$3d)				
a	8.927 Å		coordination polyhedra: Fig. 138	65F
Θ_D	338 K			66D
α	$12.6 \cdot 10^{-6}$ K^{-1}			66D
Pr$_2$Te$_3$				
Structure: cubic (Th$_3$P$_4$-defect structure, T_d^6 – I$\bar{4}$3d)				
a	9.479 Å		heat capacity: Fig. 139;	75B
	9.481 Å		coordination polyhedra: Fig. 138	65F
α-Nd$_2$S$_3$				
Structure: orthorhombic (D_{2h}^{16} – Pnma)				
a	7.42 Å		limit of stability < 900 °C;	71B2
b	15.95 Å		color black; thermal conductivity	
c	4.22 Å		data [73S2]	
a	7.442 (2) Å			68S
b	15.519 (3) Å			
c	4.029 (2) Å			
σ	35 Ω^{-1} cm^{-1}			68S
β-Nd$_2$S$_3$				
Structure: tetragonal (D_{4h}^{20} – I4$_1$/acd)				
a	14.99 Å		limit of stability 900···1300 °C;	71B2
c	19.90 Å		thermal conductivity [73S2]	
γ-Nd$_2$S$_3$				
Structure: cubic (Th$_3$P$_4$-defect structure, T_d^6 – I$\bar{4}$3d)				
a	8.527 Å		coordination polyhedra: Fig. 138	60P, 81K6
	8.533 (1) Å			82N2
T_m	2010 °C		heat capacity: Fig. 156; thermal conductivity: Fig. 157 and [73S2]; anomaly of temperature	60P, 72G
	1860 °C			81K6

(continued)

Physical property	Numerical value	Experimental conditions	Experimental method, remarks	Ref.
γ-Nd_2S_3 (continued)				
d	5.50 (1) g cm^{-3}		dependent expansion coefficient:	82N2
E_g	≈ 3 eV	optical absorption	Fig. 158	70H
	2.5 eV			80S3
$\bar{\nu}_{TO}$	192 cm^{-1}		Raman spectra; see also Fig. 159	79A
	235 cm^{-1}			
	243.9 cm^{-1}		IR measurements	80S3
	270.2 cm^{-1}			
	298.5 cm^{-1}			
$\bar{\nu}_{LO}$	305 cm^{-1}			79A
	250.9 cm^{-1}			80S3
	292.3 cm^{-1}			
	338.4 cm^{-1}			
$\varepsilon(0)$	13			80S3
	18.35		absorption and transmission	79A
$\varepsilon(\infty)$	7.0		spectra: Figs. 160, 161, 162;	80S3
	7.1		see also Fig. 142; reflectivity spectrum: Fig. 145	79A, 79Z1
E_A	2.7$\cdots$3.8 meV		electrical measurement	77T
σ	10^{-10} Ω^{-1} cm^{-1}	n-type	temperature dependence of electrical conductivity: Fig. 163;	70H, 80S3
	10^2 Ω^{-1} cm^{-1}	S-excess	temperature dependence of	70H
	1$\cdots$2$\cdot 10^{-3}$ Ω^{-1} cm^{-1}		thermoelectric power and Hall coefficient: Figs. 164, 165	74T
Nd_2Se_3				
Structure: cubic (Th_3P_4-defect structure, $T_d^6 - I\bar{4}3d$)				
a	8.859 Å		heat capacity: Fig. 156;	65F
α	13.1$\cdot 10^{-6}$ K^{-1}		coordination polyhedra: Fig. 138	66D
Θ_D	232 K			66D
Nd_2Te_3				
Structure: cubic (Th_3P_4-defect structure, $T_d^6 - I\bar{4}3d$)				
a	9.424 Å		coordination polyhedra, Fig. 138;	65L
	9.421 Å		phase diagram [65A];	65A
d	7.09 g cm^{-3}		heat capacity: Fig. 156	65H
T_m	1893 K			65L1
Structure: orthorhombic (U_2S_3-structure, D_{2h}^{16} – Pnma)				
a	11.93 Å			65F
b	12.16 Å			
c	4.37 Å			
α-Sm_2S_3				
Structure: orthorhombic (D_{2h}^{16} – Pnma)				
a	7.382 (2) Å		color: red	68S
b	15.378 (3) Å			
c	3.974 (2) Å			
σ	6 Ω^{-1} cm^{-1}		conductivity ($E_A = 0.001$ eV)	68S
S	-230 μV K^{-1}			68S

Physical property	Numerical value	Experimental conditions	Experimental method, remarks	Ref.
γ-Sm$_2$S$_3$				
Structure: cubic (Th$_3$P$_4$-defect structure, $T_d^6 - I\bar{4}3d$)				
a	8.448 Å		heat capacity [79C] and Fig. 107;	60P
	8.43 Å		coordination polyhedra: Fig. 138;	79C
T_m	1780 °C		photoconductivity spectra:	60P
	1750 °C		Figs. 166, 167, 168	61H
α	$11.4 \cdot 10^{-6}$ K^{-1}			66D
ϱ	$7 \cdot 10^5$ Ω cm			65L2
Sm$_2$Se$_3$				
Structure: cubic (Th$_3$P$_4$-defect structure, $T_d^6 - I\bar{4}3d$)				
a	8.785 Å		coordination polyhedra: Fig. 138	65F
α	$14.4 \cdot 10^{-6}$ K^{-1}			66D
Θ_D	350 K			66D
ϱ	$7 \cdot 10^7$ Ω cm			65L2
Eu$_2$Se$_3$				
ϱ	$5 \cdot 10^6$ Ω cm			73B2
E_A	1.02 eV			73B2
S	700 μV K^{-1}			73B2
α-Gd$_2$S$_3$				
Structure: orthorhombic (D_{2h}^{16} – Pnma)				
a	7.339 Å		color: red	68P
b	15.273 Å		work function [67P];	
c	3.932 Å		absorption and luminescence	
a	7.338 (2) Å		spectra of Nd^{3+}-doped samples	
b	15.273 (3) Å		[80L], see also Figs. 127, 128	68S
c	3.932 (2) Å			
σ	25 Ω^{-1} cm^{-1}		conductivity ($E_A = 0.007$ eV)	68S
	$2 \cdot 10^{-3}$ Ω^{-1} cm^{-1}	n-type		70P
S	-360 μV K^{-1}		(referred to lead)	68S
γ-Gd$_2$S$_3$				
Structure: cubic (Th$_3$P$_4$-defect structure, $T_d^6 - I\bar{4}3d$)				
a	8.387 Å		phase diagram: Fig. 137;	57F
	8.385 Å		coordination polyhedra: Fig. 138	81K6
	8.3723 (5) Å			73W
T_m	1885 °C		thermal conductivity: Fig. 157;	57F
	1850 °C		Raman and IR spectra: Fig. 141;	81K6
σ	250 Ω^{-1} cm^{-1}	n-type	temperature dependence of	74T
	10^{-8} Ω^{-1} cm^{-1}		resistivity, Hall coefficient and	82B
E_g	3.4 eV	optical gap	Seebeck coefficient:	82B
E_A	1.6···1.9 meV		Figs. 163, 164, 165	77T
Gd$_2$Se$_3$				
Structure: cubic (Th$_3$P$_4$-defect structure, $T_d^6 - I\bar{4}3d$)				
a	8.727 Å		coordination polyhedra: Fig. 138;	76L
	8.718 Å		magnetic properties [64H]	64H
σ	0.3 Ω^{-1} cm^{-1}			64H
	≈ 800 Ω^{-1} cm^{-1}			64R2

Physical property	Numerical value	Experimental conditions	Experimental method, remarks	Ref.
α-Tb$_2$S$_3$				
Structure: orthorhombic (D_{2h}^{16}–Pnma)				
a	7.303 (2) Å		color: red	68S
b	15.200 (3) Å			
c	3.901 (2) Å			
σ	0.3 Ω^{-1} cm^{-1}			68S
S	-180 μV K^{-1}		(referred to lead)	68S
γ-Tb$_2$S$_3$				
Structure: cubic (Th$_3$P$_4$-defect structure, T_d^6–I$\bar{4}$3d)				
a	8.344 Å		coordination polyhedra: Fig. 138	65F
α-Dy$_2$S$_3$				
Structure: orthorhombic (D_{2h}^{16}–Pnma)				
a	7.279 (2) Å			68S
b	15.136 (2) Å			
c	3.878 (2) Å			
δ-Dy$_2$S$_3$				
Structure: monoclinic (C_{2h}^2–P2$_1$/m)				
a	17.496 (4) Å		color: green	68S
b	4.022 (2) Å			
c	10.183 (3) Å			
β	98.67°			
T_m	1470 °C			60P
σ	500 Ω^{-1} cm^{-1}			68S
E_A	0.34 eV			68S
S	4000 μV K^{-1}		(referred to lead)	68S
γ-Dy$_2$S$_3$				
Structure: cubic (Th$_3$P$_4$-defect structure, T_d^6–I$\bar{4}$3d)				
a	8.292 Å		phase diagram: Fig. 137;	60P
	8.285 Å		coordination polyhedra: Fig. 138	81K6
T_m	1490 °C		thermal conductivity: Fig. 157	60P
	1780 °C			81K6
E_g	3 eV		optical absorption edge	70H
	> 3.8 eV		optical determination	82B
$\bar{\nu}_{TO}$	205 cm^{-1}		Raman spectrum; see also Fig. 169	79A
	250 cm^{-1}			
$\bar{\nu}_{LO}$	300 cm^{-1}			
$\varepsilon(0)$	17.1		absorption spectra: Figs. 142, 170;	79A
$\varepsilon(\infty)$	7.8		IR-reflection spectrum: Fig. 145	79A, 79Z1
E_A	1.5 meV		electrical measurement	77T
σ	250 Ω^{-1} cm^{-1}	n-type	electrical conductivity: Fig. 163	67H
	10^{-10} Ω^{-1} cm^{-1}			70H, 82B
S	50 μV K^{-1}		Seebeck coefficient: Fig. 164;	67H
	120 μV K^{-1}	$T = 1300$ K	magnetic properties [67H]	

Physical property	Numerical value	Experimental conditions	Experimental method, remarks	Ref.
Dy$_2$Se$_3$				
Structure: cubic (Th$_3$P$_4$-defect structure, T$_d^6$–I$\bar{4}$3d)				
a	8.65 Å		coordination polyhedra: Fig. 138 magnetic properties [76L]	76L
δ-Ho$_2$S$_3$				
Structure: monoclinic (C$_{2h}^2$–P2$_1$/m)				
a	17.50 (3) Å		structure: Fig. 171	67W
b	4.002 (5) Å			
c	10.15 (2) Å			
β	99.4 (2)°			
a	17.452 (2) Å		color: yellow orange	68S
b	4.001 (3) Å			
c	10.128 (2) Å			
β	98.69 (2)°			
σ	0.14 · 10^{-6} Ω^{-1} cm^{-1}			68S
E_A	0.5 eV			68S
δ-Er$_2$S$_3$				
Structure: monoclinic (C$_{2h}^2$–P2$_1$/m)				
a	17.404 (2) Å		color: beige	68S
b	3.978 (3) Å			
c	10.092 (2) Å			
β	98.67 (2)°			
T_m	1730 °C			60P
	1703 °C			72G
δ-Tm$_2$S$_3$				
Structure: monoclinic (C$_{2h}^2$–P2$_1$/m)				
a	17.363 (2) Å		color: yellow [75B]	68S
b	3.960 (3) Å			
c	10.039 (2) Å			
β	98.78 (2)°			
σ-Tm$_2$S$_3$				
Structure: cubic (Tl$_2$O$_3$-structure, T$_h^7$–Ia3)				
a	10.51 Å		color: yellow	75B
φ-Tm$_2$Se$_3$				
Structure: orthorhombic (D$_{2h}^{24}$–Fddd)				
a	11.31 Å		color: brown red	75B
b	8.06 Å		magnetic properties [75B]	
c	24.06 Å			
φ-Tm$_2$Te$_3$				
Structure: orthorhombic (D$_{2h}^{24}$–Fddd)				
a	12.09 Å		color: light green	75B
b	8.55 Å			
c	25.65 Å			

Physical property	Numerical value	Experimental conditions	Experimental method, remarks	Ref.
ε-Yb$_2$S$_3$				
Structure: rhombohedral (Al$_2$O$_3$-structure, $D_{3d}^6 - R\bar{3}c$)			color: yellow [68S]	64F
a	6.772 Å	(hexagonal setting)		
c	18.28 Å			
σ	$3 \cdot 10^{-5}\,\Omega^{-1}\,cm^{-1}$			68S
E_A	0.29 eV			68S
ε-Lu$_2$S$_3$				
Structure: rhombohedral (Al$_2$O$_3$-structure, $D_{3d}^6 - R\bar{3}c$)			color: yellow	75R
a	6.722 (2) Å	(hexagonal setting)		
c	18.160 (3) Å			
Θ_D	266 (1) K		heat capacity: Fig. 172	81G4
T_m	1850 °C			81G4
LaS$_2$				
Structure: tetragonal (pseudocubic, $D_{4h}^7 - P4/nmm$)			color: light brown [75B]	79F2
a	4.09 Å		IR and Raman spectra: Figs. 173, 174	
c	8.19 Å			
Structure: orthorhombic ($D_{2h}^{16} - Pnma$)				78D2
a	8.131 (5) Å			
b	16.34 (1) Å			
c	4.142 Å			
LaTe$_2$				
Structure: tetragonal (Fe$_2$As-structure, $D_{4h}^7 - P4/nmm$)			color: black	65R2
a	4.53 Å		conductivity: Fig. 175	
c	9.12 Å			
T_m	1450 °C			65R2
d	6.82 g cm^{-3}			64E
CeS$_2$				
Structure: tetragonal (pseudocubic, $D_{4h}^7 - P4/nmm$)			IR spectrum: Fig. 173	79F2
a	4.09 Å			
c	9.12 Å			
PrS$_2$				
Structure: tetragonal (pseudocubic, $D_{4h}^7 - P4/nmm$)			color: light brown [75B]	79F2
a	4.03 Å			
c	8.06 Å			
LaTe$_3$				
Structure: orthorhombic (pseudotetragonal, $D_{2h}^{17} - Bmmb$)			Seebeck coefficient: Fig. 176; conductivity: Fig. 177 (p-type, $E_A = 0.56$ eV)	65R2
a	4.41 Å			
c	26.14 Å			
T_m	835 °C			65R2
d	6.88 g cm^{-3}			65R2

9.16.5 References for 9.16

53Z	Ziegler, W.T., Young, R.A.: Phys. Rev. **90** (1953) 115.
55H	Holley C.E., Mulford, R.N.R., Ellinger, F.H., Koehler, W.C., Zachariasen, W.H.: J. Phys. Chem. **59** (1955) 1226.
55M	Mulford, R.N.R., Holley, C.E., Jr.: J. Phys. Chem. **59** (1955) 1222.
55S	Stalinski, B.: Bull. Acad. Pol. Sci., Cl III, **3** (1955) 613.
56K1	Korst, W.L., Warf, J.C.: Acta Cryst. **9** (1956) 452.
56K2	Korst, W.L.: Ph.D. Thesis, University of Southern California, **1956.**
56P1	Picon, M., Patrie, M.: C. R. Acad. Sci. Paris **243** (1956) 2074.
56P2	Post, B., Moskowitz, D., Glaser, F.W.: J. Am. Chem. Soc. **78** (1956) 1800.
57F	Flahaut, J., Guittard, M., Loriers, M.J., Patrie, M.: C. R. Acad. Sci. Paris **245** (1957) 2291.
57S	Stalinski, B.: Bull. Acad. Pol. Sci. **5** (1957) 1001.
59B	Benacerrat, S., Guittard, M.: C. R. Acad. Sci. Paris **248** (1959) 2012.
59G	Goon, E.J.: J. Phys. Chem. **63** (1959) 2018.
59S	Stalinski, B.: Bull. Acad. Pol., Sci., **7** (1959) 269.
60B	Brixner, L.H.: J. Inorg. Nucl. Chem. **15** (1960) 199.
60P	Picon, M., Domange, L., Flahaut, J., Guittard, M., Patrie, M.: Bull. Soc. Chim. Fr. **2** (1960) 221.
61H	Houston, M.D.: Rare Earth Research, Kleber, E.V. (ed.), New York: Mac Millan Comp. **1961**, p. 255.
61W	Warf, J.C., Hardcastle, K.: Final Report, Office of Naval Research, Contract **228** (1961) 15.
62P	Pebler, A., Wallace, W.E.: J. Phys. Chem. **66** (1962) 148.
62S	Samsonov, G.V., Vainshtein, E.Y., Paderno, Y.B.: Fiz. Metal. Metallov. **13** (1962) 764.
63M	Morwzzi, V.L., Teaney, D.T.: Solid State Commun. **1** (1963) 127.
63W	Wallace, W.E., Kubota, Y., Zanowick, R.L.: "Magnetic Characteristics of Gadolinium, Terbium and Ytterbium Hydrides in Relation to the Electronic Nature of the Lanthanide Hydrides" in "Advances in Chemistry Series" No. 39, p. 122, Washington, D.C.: American Chemical Society **1963**.
64B	Busch, G., Junod, P., Morris, R.G., Muheim, J., Stutius, W.: Phys. Lett. **11** (1964) 9.
64C	Cutler, M., Leavy, J.F.: Phys. Rev. **133** (1964) A 1153.
64E	Eliseev, A.A., Kuznecov, V.G., Jarembaš, E.I.: Zh. Strukt. Khim. **5** (1964) 641.
64F	Flahaut, J., Domange, L., Pardo, M.: C. R. Acad. Sci. Paris **258** (1964) 594.
64H	Holtzberg, F., McGuire, T.R., Methfessel, S., Suits, J.G.: J. Appl. Phys. **35** (1964) 1033.
64I	Iandelli, A.: Rend. Accad. Sci. Fis. Mat., Naples **37** (1964) 160.
64R1	Reid, F.J., Matson, L.K., Miller, J.F., Himes, R.C.: J. Phys. Chem. Solids **25** (1964) 969.
64R2	Reid, F.J., Matson, L.K., Miller, J.F., Himes, R.C.: J. Electrochem. Soc. **111** (1964) 943.
65A	Abrikosov, N.Kh., Zargaryan, V.Sh.: Izv. Acad. Nauk SSSR, Neorg. Mater. **9** (1965) 1462.
65B	Biegański, Z., Fesenko, W., Stalinski, B.: Bull. Acad. Pol. Sci., Ser. Sci. Chim. **3** (1965) 227.
65F	Flahaut, J., Guittard, M., Patrie, M., Pardo, M.P., Golabi, S.M., Domange, L.: Acta. Cryst. **19** (1965) 14.
65H	Haase, D.J., Steinfink, H., Weiss, E.J.: Rare Earth Research **3** (1965) 335; New York: Gordon and Breach, 1965.
65L1	Lin, W., Steinfink, H., Weiss, E.J.: Inorg. Chem **4** (1965) 877.
65L2	Lashkarev, G.V., Paderno, Yu.B.: Izv. Akad. Nauk SSSR, Neorg. Mater. **1** (1965) 1791.
65R1	Ramsey, T.H., Steinfink, H., Weiss, E.J.: Inorg. Chem. **4** (1965) 1154.
65R2	Ramsey, T.H., Steinfink, H., Weiss, E.J.: J. Appl. Phys. **36** (1965) 548.
66D	Dudnik, E.M., Lashkarev, G.V., Paderno, Y.B., Obolonchik, V.A.: Inorg. Mater. **2** (1966) 833.
66H1	Hardcastle, K.I., Warf, J.C.: Inorg. Chem. **5** (1966) 1728.
66H2	Holtzberg, F., Methfessel, S.: J. Appl. Phys. **37** (1966) 1433.
66K	Korst, W.L., Warf, J.C.: Inorg. Chem. **5** (1966) 1719.
66M1	Marcon, J.P., Pascard, R.: J. Inorg. Nucl. Chem. **28** (1966) 2551.
66M2	Messer, C.E., Miller, R.M., Barrante, J.R.: Inorg. Chem. **5** (1966) 1814.
66W1	Warf, J.C., Hardcastle, K.: Inorg. Chem. **5** (1966) 1736.
66W2	McWhan, D.B., Souers, P.C., Jura, G.: Phys. Rev. **143** (1966) 385.
67A	Argyle, B.E., Miyata, N., Schultz, T.D.: Phys. Rev. **160** (1967) 413.
67B	Berkooz, O., Malamud, M., Shtrikman, S.: Solid State Commun. **6** (1967) 185.
67H	Henderson, J.R., Muramato, M., Loh, E.: J. Chem. Phys. **47** (1967) 3347.

9.16.5 References for 9.16

67M	Marchenko, V.I., Samsonov, G.V.: Chem. Bond in Semicond. and Solids (ed. N.N. Sirota) New York: Consultants Bureau **1967**.
67P	Peshev, P., Bliznakov, G., Toshev, A.: J. Less-Common Metals **14** (1968) 379.
67W	White, J.G., Yocom, P.N., Lerner, S.: Inorg. Chem. **6** (1967) 1872.
68M	Mueller, W.M., in: Metal Hydrides, Mueller, W.M., Blackledge, J.P., Libowitz, G.G. (eds.), N.Y.: Academic Press, **1968**, p. 384.
68P	Prewitt, C.T., Sleight, A.W.: Inorg. Chem. **6** (1968) 1090.
68S	Sleight, A.W., Prewitt, C.T.: Inorg. Chem. **7** (1968) 2282.
68W	Wachter, P.: Phys. Kondens. Mater. **8** (1968) 80.
69A	Axe, J.D.: J. Phys. Chem. Solids **30** (1969) 1403.
69H	Heckman, R.C.: Electronic Properties of Rare Earth Hydrides, Report No. SC-RR-69-571, Sandia Laboratories, Alburquerque, N.M., Nov. **1969**.
69L1	Levy, F.: Phys. Kondens. Mater. **10** (1969) 71.
69L2	Libowitz, G.G., Pack, J.G.: J. Chem. Phys. **50** (1969) 3557.
69M	Mehth, A., Buehler, E., Geballe, T.H.: Phys. Rev. Lett. **22** (1969) 295.
69P	Pankov, I.E., Nogteva, V.V., Yarembash, E.I.: Russ. J. Phys. Chem. **43** (1969) 1316.
69S	Shapira, Y., Reed, T.B.: J. Appl. Phys. **40** (1969) 1197.
69W	Wachter, P.: Solid State Commun. **7** (1969) 693.
69Z	Zhuze, V.P., Goncharova, E.V., Lukirskü, D.P.: Sov. Phys. – Solid State **10** (1969) 1052.
70B	Bransky, I., Tallan, N.M., Hed, A.Z.: J. Appl. Phys. **41** (1960) 1787.
70C	Cho, S.J.: Phys. Rev. **B 1** (1970) 4589.
70H	Henderson, J.R. Muramato, M., Gruber, J.B., Menzel, R.: J. Chem. Phys. **52** (1970) 2311.
70J	Jayaraman, A., Narayanamurti, V., Bucher, E., Maines, R.G.: Phys. Rev. Lett. **25** (1970) 1430.
70L	Levy, F., Wachter, P.: Solid State Commun. **8** (1970) 183.
70P	Piekarczyk, W., Peshev, P.: J. Crystal Growth **6** (1970) 357.
70Y	Yajima, S., Niihara, K.: Proc. 9th Rare Earth Research Conference, U.S. Dept. of Commerce, **1970**, p. 598.
71A	Atoji, M.: J. Chem. Phys. **54** (1971) 3226.
71B1	Bayer, E., Zinn, W.: Z. Angew. Phys. **32** (1971) 83.
71B2	Besancon, P.: Doctor Thesis, Univ. René Descartes, Paris **1971**.
71B3	Bucher, E., Narayanamurti, V., Jayaraman, A.: J. Appl. Phys. **42** (1971) 1741.
71D	Davis, H.L.: Proc. 9th RE Conference, p. 3, Virginia Polytechnic Inst., Blacksbury, Va., **1971**.
71G	Günterodt, G., Wachter, B., Imboden, D.M.: Phys. Kondens. Mater. **12** (1971) 292.
71M	Maeland, A.J., Holmes, D.E.: J. Chem. Phys. **54** (1971) 3979.
71O	Oliveira, N.F., Foner, S., Shapira, Y., Reed, T.B.: J. Appl. Phys. **42** (1971) 1783.
71N	Nickerson, J.C., White, R.M., Lee, K.N., Bachmann, R., Geballe, T.H., Hull, G.W.: Phys. Rev. **B 3** (1971) 2030.
71P	Petrich, G., von Molnar, S., Penney, T.: Phys. Rev. Lett. **26** (1971) 885.
71S1	Shapira, Y., Oliveira, N.F., Foner, S., Reed, T.B.: J. Appl. Phys. **42** (1971) 1783.
71S2	Suryanarayanan, R., Paparoditis, C.: Int. Conf. on Rare Earths and Actinides, Durham p. 210, **1971**.
71S3	Shapira, Y., Reed, T.B.: AIP Conf. Proc. **5** (1971) 837.
72C	Chatterjee, A., Singh, A.K., Jayaraman, A.: Phys. Rev. **B 6** (1972) 2285.
72G	Goryachev, Yu.M., Kutsenok, T.G.: High Temp.-High Press. **4** (1972) 663.
72K1	Kaldis, E., Wachter, P.: Solid State Commun. **11** (1972) 907.
72K2	Kasuya, T.: CRC Crit. Rev. Solid State Sci. **3** (1972) 131.
72K3	Kirk, J.L., Vedam, K., Narayanamurti, V., Jayaraman, A., Bucher, E.: Phys. Rev. **B 6** (1972) 3023.
72L1	Libowitz, G.G.: Ber. Bunsenges. Phys. Chem. **8** (1972) 837.
72L2	Libowitz, G.G., Pack, J.G., Binnie, W.P.: Phys. Rev. **B 6** (1972) 4540.
72O	Oliver, M.R., Dimmock, J.D., McWhorter, A.L., Reed, T.B.: Phys. Rev. **B 5** (1972) 1078,
72R	Reed, T.B., Fahly, R.E., Strauss, A.J.: J. Cryst. Growth **15** (1972) 174.
72S1	Shafer, M.W., Torrance, J.B., Penney, T.: J. Phys. Chem. Solids **33** (1972) 2251.
72S2	Shapira, Y., Foner, S., Oliveira, N.F., Reed, T.B.: Phys. Rev. **B 5** (1972) 2647.
72S3	Shapira, Y., Reed, T.B.: AIP Conf. Proc. **5** (1972) 837.
72S4	Shapira Y., Reed, T.B.: Phys. Rev. **B 5** (1972) 4877.
72S5	Sadovskaya, O.A., Stepanov, E.P., Khrapov, V.V., Yarembash, E.I.: Inorg. Mater. (USSR) **8** (1972) 708.

9.16.5 References for 9.16

72S6	Shapiro, Y., Reed, T.B.: Phys. Rev. **B 5** (1972) 2657.
72S7	Smirnov, I.A.: Phys. Status Solidi (a) **14** (1972) 363.
72W1	Wachter, P.: CRC Crit. Rev. Solid State Sci. **3**/12 (1972) 189.
72W2	White, H.W., Mc Collum, D.C.: J. Appl. Phys. **43** (1972) 1225.
73B1	Besancon, P.: J. Solid State Chem. **7** (1973) 232.
73B2	Butusov, O.B., Checkernikov, V.I., Pechennikov, A.V., Yarembash, E.I., Sadsovskaya, O.A., Tyurin, E.G.: Izv. Akad. Nauk. Neorg. Mater. **9** (1973) 1339.
73E	Eastman, D.E.: Phys. Rev. **B 8** (1973) 6027.
73G	Goodenough, J.B., Mercurio, J.P., Etourneau, J., Naslain, R., Hagenmuller, P.: C. R. Acad. Sci. Paris **277C** (1973) 1239.
73H	Holah, G.D., Webb, J.S., Dennis, R.B., Pidgeon, C.R.: Solid State Commun. **13** (1973) 209.
73I	Ikezwa, H., Suzuki, T.: J. Phys. Soc. Jpn. **35** (1973) 1556.
73J1	Jayaraman, A.: IV. Int. Conf. on Solid Compounds of Transition Elements, Geneva, **1973**, p. 148.
73J2	Jayaraman, A., Bucher, E., Dernier, P.D., Longinotti, L.D.: Phys. Rev. Lett. **31** (1973) 700.
73M	Mercurio, J.P., Etourneau, J., Naslain, R., Hagenmuller, P.: Mater. Res. Bull. **8** (1973) 837.
73S1	Shapira, Y., Foner, S., Peed, T.B.: Phys. Rev. **B 8** (1973) 2299.
73S2	Smirnov, I.A., Parfeneva, L.S., Sergeeva, V.M., Zhukova, T.B.: Sov. Phys. Solid State **14** (1973) 2142.
73S3	Sokolov, V.V., Kravchenko, L.Kh., Kamarzin, A.A.: Inorg. Mater. **9** (1973) 944.
73S4	Suryanarayanan, R., Paparoditis, C.: Phys. Lett. **42A** (1973) 373.
73S5	Smirnov, I.A., Parfen'eva, L.S., Khusnutdinova, V.Y., Sergeeva, V.M.: Sov. Phys. Solid State **14** (1973) 2412.
73W	Wu, R., Gilles, P.W.: J. Chem. Phys. **59** (1973) 6136.
74A	Ananth, K.P., Gielisse, P.J., Rockett, T.J.: Mater. Res. Bull. **9** (1974) 1167.
74C	Campagna, M., Bucher, E., Wertheim, G.K., Longinotti, L.D.: Phys. Rev. Lett. **33** (1974) 165.
74D	Dumas, J., Schlenker, C.: Phys. Status Solidi **(a) 22** (1974) 89.
74G	Günterodt, G.: Phys. Condens. Matter **18** (1974) 37.
74J	Jayaraman, A., Singh, A.K., Chatterjee, A., Usha Devi, S.: Phys. Rev. **B 9** (1974) 2513.
74M1	Mc Masters, O.D., Gschneidner, K.A., Kaldis, E., Sampietro, G.: J. Chem. Thermodyn. **6** (1974) 845.
74M2	Mercurio, J.P.: "Synthesis Thermodynamic Stability and Electrical and Magentic Properties of Rare Earth Hexaborides", thesis Univ. Bordeaux, Order No. 434, **1974**.
74M3	Mercurio, J.P., Etournean, J., Naslain, R., Hagenmuller, P., Goodenough, J.B.: J. Solid State Chem. **9** (1974) 37.
74N	Narayanamurti, V., Jayaraman, A., Bucher, E.: Phys. Rev. **B 9** (1974) 2521.
74P	Petzel, T.: Inorg. Nucl. Chem. Lett. **10** (1974) 119.
74S1	Shapira, Y., Foner, S., Oliveira, N.F., Reed, T.B.: Phys. Rev. **B 10** (1974) 4765.
74S2	Suryanarayanan, R., Ferré, J., Briat, B.: Phys. Rev. **89** (1974) 554.
74T	Taher, S.M.A., Gruber, J.B., Olsen, L.C.: J. Chem. Phys. **60** (1974) 2050.
74V	Vorderwisch, P., Hautecler, S.: Phys. Stat. Solidi **(b) 64** (1964) 495.
75B	Bucher, E., Andres, K., die Salvo, F. J., Maita, J.P., Gossard, A.C., Cooper, A.S., Hull jr., G.W.: Phys. Rev. **B 11** (1975) 500.
75G	Golovin, Yu.M., Petrov, K.I., Loginova, E.M., Grizik, A.A., Ponomarev, N.M.: Zh. Neorg. Khim. **20** (1975) 283.
75H	Haschke, J.M., Clark, M.R.: High Temp. Sci. **7** (1975) 152.
75J	Jayaraman, A., Dernier, P.D., Longinotti, L.D.: Phys. Rev. **B 11** (1975) 2783.
75M1	Mariot, J.M., Karnatak, R.C.: Solid State Commun. **16** (1975) 611.
75M2	Mitarov, R.G., Tikhonov, V.V., Vasilev, L.N., Golubkov, A.V., Smirnov, I.A.: Phys. Status Solidi **(a) 30** (1975) 457.
75R	Range, K.-J., Leeb, R.: Z. Naturforsch. **30b** (1975) 637.
75S1	Suryanarayanan, R., Güntherod, G., Freeouf, J.L., Holtzberg, F.: Phys. Rev. **B12** (1975) 4215.
75S2	Schoenes, J.: Z. Phys. **B 20** (1975) 345.
75S3	Sattler, K., Siegmann, H.C.: Z. Phys. **B 20** (1975) 289.
75V	Vitins, J., Wachter, P.: Phys. Rev. **B 12** (1975) 3829.
75W	Ward, R.W., Clayman, P.B., Rice, T.M.: Solid State Commun. **17** (1975) 1297.
76B1	Barnier, S., Lucazeau, G.: J. Chim. Phys. **5** (1976) 73.
76B2	Batlogg, B., Kaldis, E., Schlegel, A., v. Schulthiess, G., Wachter, P.: Solid State Commun. **19** (1976) 673.

9.16.5 References for 9.16

76B3	Batlogg, B., Kaldis, E., Schlegel, A., Wachter, P.: Phys. Rev. **B 14** (1976) 5503.
76B4	Batlogg, B., Kaldis, E., Schlegel, A., Wachter, P.: Phys. Lett. **56A** (1976) 122.
76B5	Batlogg, B., Kaldis, E., Wachter, P.: J. Magn. Magn. Mater. **3** (1976) 96.
76C	Campagna, M., Wertheim, G.K., Bucher, E.: In Structure and Bonding, Vol. 30, Springer: Berlin **1976**.
76D	Dernier, P.D., Weber, W., Longinotti, L.D.: Phys. Rev. **B 14** (1976) 3635.
76E	Escorne, M., Ghazali, A., Leroux-Hugon, P., Smirnov, I.A.: Phys. Lett. **56A** (1976) 475.
76G	Günterodt, G.: In Festkörperprobleme XVI, Advances in Solid State Physics, Treusch, J. (ed.), Braunschweig: Vieweg **1976**.
76K	Kamarzin, A.A., Sokolov, V.V., Mironov, K.E., Maloviikii, N, Vasileva, I.G.: Mater. Res. Bull. **11** (1976) 695.
76L	Lashkarev, G.V., Fedorchenko, V.P., Obolonchik, V.A., Skripka, I.P.: Sov. Powder Metall. Met. Ceram. **15** (1976) 593.
76M1	von Molnar, S., Holtzberg, F.: AIP Conf. Proc. **29** (1976) 394.
76M2	Munz, P.: Helv. Phys. Acta **49** (1976) 281.
76M3	Massenet, O., Coey, J.M.D., Holtzberg, F.: J. Phys. (Paris) **37** Collq. (1976) C4-297.
76S	Singh, B., Surplice, N.A., Müller, J.: J. Phys. D: Appl. Phys. **9** (1976) 2087.
76V	Vitins, J., Wachter, P.: Phys. Lett. **58A** (1976) 275.
77D	Desfours, J.P., Lascaray, J.P., Llinares, C., Averous, M.: Solid State Commun. **21** (1977) 441.
77E	Eliseeu, A.A., Sadovskaya, O.A.: Izv. Akad. Nauk SSSR, Neorg. Mater. **13** (1977) 1394.
77I	Iisikawa, Y., Bajaj, M.M., Kasaya, M., Tanaka, T., Bannai, E.: Solid State Commun. **22** (1977) 573.
77M	Mal'kova, A.A., Shul'man, S.G., Ivanov-Omskii, V.I., Smirnov, I.A.: Sov. Phys. Solid State **19** (1977) 1668.
77O	Ott, H.R., Luethi, B., Wang, P.S.: Proc. Conf. on Valence Instabilities and Related Narrow-Band Phenomena, Parks, R.D. (ed.), Rochester **1977**, p. 289.
77S	Subhadra, K.G., Sirdeshmukh, D.B.: Pramana **9** (1977) 223.
77T	Taher, M.A., Gruber, J.B.: Phys. Rev. **B 16** (1977) 1624.
77V1	Vitins, J., Wachter, P.: Physica **89 B** (1977) 234.
77V2	Vitins, J., Wachter, P.: Phys. Rev. **B 15** (1977) 3225.
77W	Wachter, P.: See ref. 77O, p. 337.
78D1	Desfours, J.P., Godart, C., Weill, G., Averous, M., Llinares, C.: Phys. Rev. **B 18** (1978) 2750.
78D2	Dugué, J., Carré, D., Guittard, M.: Acta Cryst. **B 34** (1978) 403.
78F	Fischer, K.J.: (KFA Jülich) private communication.
78G1	Glurdzhidze, L.N., Gigineishvili, A.V., Bzhalava, T.L., Dzhabua, Z.U., Pagava, T.A., Sanadze, V.V., Oskotskii, V.S.: Sov. Phys. Solid State **20 (9)** (1978) 1573.
78G2	Güntherodt, G., Merlin, R., Frey, A., Cardona, M.: Solid State Commun. **27** (1978) 551.
78H	Heleskivi, J., Mäenpää, M.: Phys. Scr. **18** (1978) 441.
78K1	Kaminskii, V.V., Vinogradov, A.A., Kapustin, V.A., Smirnov, I.A.: Sov. Phys. Solid State **20 (9)** (1978) 1571.
78K2	Kuivalainen, P., Kaski, K., Sinkkonen, J., Stubb, T.: Phys. Scr. **18** (1978) 433.
78K3	Konstantinov, V.L., Skornyakov, G.P., Kamarzin, A.A., Sokolov, V.V.: Inorg. Mater. **14** (1978) 659.
78M1	Mook, H.A., Nicklow, R.M., Penney, T., Holtzberg, F., Shafer, M.W.: Phys. Rev. **B 18** (1978) 2925.
78M2	Merlin, R., Güntherodt, G., Humphreys, R., Cardona, M., Suryanarayanan, R., Holtzberg, F.: Phys. Rev. **B 17** (1978) 4951.
78P	Palazzi, M., Jaulmes, S.: Mater. Res. Bull. **13** (1978) 1153.
78S1	Shubha, V., Ramesh, T.G., Ramaseshan, S.: Solid State Commun. **26** (1978) 173.
78S2	Smirnov, I.A., Oskotskii, V.S.: Sov. Phys. Usp. **21 (2)** (1978) 117.
78S3	Schmutz, L.E., Dresselhaus, G., Dresselhaus, M.S.: Solid State Commun. **28** (1978) 597.
79A	Arkatova, T.G., Zhuze, V.P., Karin, M.G., Kamarzin, A.A., Kukharskii, A.A., Mikhailov, B.A., Shelykh, A.I.: Sov. Phys. Solid State **21** (1979) 1979.
79B1	Bieganski, Z., Stalinski, B.: Z. Phys. Chem. Neue Folge **116** (1979) 109.
79B2	Battlog, B.J.R.: Thesis, ETH Zürich, **1979**.
79C	Coey, J.M.D., Cornut, B., Holtzberg, F., von Molnar, S.: J. Appl. Phys. **50** (1979) 1923.
79D	Drulis, M., Bieganski, Z.: Phys. Status Solidi (a) **53** (1979) 277.
79E	Escorne, M., Mauger, A., Godart, C., Achard, J.C.: J. Phys. (Paris) **40** (1979) 315.

9.16.5 References for 9.16

79F	Fisk, Z., Johnston, D.C., Cornut B., von Molnar, S., Osseroff, S., Calvo, R.: J. Appl. Phys. **50** (1979) 1911.
79F2	Flahaut, J.: In Handbook on the Physics and Chemistry of Rare Carths; Gschneidner, K.A., and Eyring, L.R. (eds.), Amsterdam: North Holland **1979**.
79G	Güntherodt, G.: J. Magn. Magn. Mater. **11** (1979) 394.
79K1	Kaldis, E., Fritzler, B., Peteler, W.: Z. Naturforsch. **34a** (1979) 55.
79K2	Keller, R., Güntherodt, G., Holzapfel, W.B., Dietrich, M., Holtzberg, F.: Solid State Commun. **29** (1979) 753.
79M	Müller, H., Knappe, P., Greis, O.: Z. Phys. Chem. **114** (1979) 45.
79S	Safran, S.A., Silberstein, R.P., Dresselhaus, G., Lax, B.: Solid State Commun. **29** (1979) 339.
79V	Volkonskaya, T.I., Kizhaev, S.A., Smirnov, I.A.: Sov. Phys. Solid State **21** (1979) 726.
79W	Wachter, P.: Handbook on the Physics and Chemistry of Rare Earths, Vol. II, Gschneidner, K.A. and Eyring, L.R. (eds.), Amsterdam: North-Holland, **1979**.
79Z1	Zhuze, V.P., Kamarzin, A.A., Karin, M.G., Sidorin, K.K., Shelykh, A.I.: Sov, Phys. Solid State **21** (1979) 1968.
79Z2	Zogal, O.J.: Phys. Status Solidi **(a) 53** (1979) K203.
80B1	Barnes, R.G., Beaudry, B.J., Creel, R.B., Torgeson, D.R., de Groot, D.G.: Solid State Commun. **36** (1980) 105.
80B2	Battlog, B., Wachter, P.: J. Phys. (Paris) **41** (1980) C5-59.
80F1	Fujimori, A., Minami, F., Tsuda, N.: Phys. Rev. **B 22** (1980) 3573.
80F2	Farberovich, O.V.: Sov. Phys. Solid State **22 (3)** (1980) 393.
80G1	Gupta, M., Burger, J.P.: Phys. Rev. **B 22** (1980) 6074.
80G2	Glurdzhidze, L.N., Gigineishvili, A.V., Zurabishvili, N.G., Bzhalava, T.L., Sanadze, V.V.: Sov. Phys. Solid State **22 (6)** (1980) 1098.
80G3	Glurdzhidze, L.N., Kekhainov, T.D., Gzirishvili, D.G., Bzhalava, T.L., Sanadze, V.V.: Sov. Phys. Solid State **22 (3)** (1980) 388.
80H1	Holtzberg, F., von Molnar, S., Coly, J.M.D.: Handbook on Semiconductors Vol. 3, S.P. Keller (ed.), Amsterdam: North Holland, **1980**, p. 803.
80H2	Holtzberg, F.: Phil. Mag. **B 42** (1980) 491.
80J	Johanson, W.R., McCollum, D.C.: Phys. Rev. **B 22** (1980) 2435.
80K1	Kaminskii, A.A., Sarkisov, C.É., Chan Niok, Denisenko, G.A., Kamarzin, A.A., Sokolov, V.V., Klȳpin, V.V., Maloviikii, Yu.N.: Izv. Akad. Nauk SSSR, Neorg. Mater. **16** (1980) 1333.
80K2	Kaminskii, V.V., Kapustin, V.A., Smirnov, I.A.: Sov. Phys. Solid State **22 (12)** (1980) 2091.
80L	Leiss, M.: J. Phys. **C 13** (1980) 151.
80M1	Morán-López, J.L., Schlottmann, P.: Phys. Rev. **B 22** (1980) 1912.
80M2	Mott, N.F.: Phil. Mag. **B 42** (1980) 327.
80S1	Silberstein, R.P.: Phys. Rev. **B 22** (1980) 4791.
80S2	Subhadra, K.G., Sirdeshmukh, D.B.: Natl. Acad. Sci. Lett. (India) **3** (1980) 126.
80S3	Skornyakov, G.P., Ponosov, Y.S., Surov, M.E., Darienko, E.P., Kamarzin, A.A., Sokolov, V.V.: Sov. Phys. Solid State **22** (1980) 613.
80W	Wachter, P.: Phil. Mag. **B 42** (1980) 497.
81A	Aveline, I., Iglesias-Sicardi, J.R.: Solid State Commun. **37** (1981) 749.
81B1	Bludau, W., Wichelhaus, W.: J. Appl. Phys. **52** (1981) 2750.
81B2	Batirov, T.M., Fridkin, V.M., Kamarzin, A.A., Malovitskii, Y.I., Verkhovskaya, K.A.: Phys. Status Solidi (a) **65** (1981) K163.
81B3	Brasileiro, C., Fisicas, P.: Solid State Commun. **38** (1981) 903.
81C	Carlin, R.L., Krause, L.J.: Chem. Phys. Lett. **82** (1981) 323.
81F1	Fujimori, A., Tsuda, N.: J. Phys. **C 14** (1981) 1427.
81F2	Fujimori, A., Tsuda, N.: J. Phys. **C 14** (1981) L69.
81F3	Farberovich, O.V.: Phys. Status Solidi (b) **104** (1981) 365.
81F4	Farberovich, O.V., Vlasov, S.V.: Phys. Status Solidi (b) **105** (1981) 755.
81G1	Güntherodt, G., Jayaraman, A., Kress, W., Bilz, H.: Phys. Lett. **A 82** (1981) 26.
81G2	Güntherodt, G., Jayaraman, A., Bilz, H., Kress, W.: Valence Fluctuations in Solids, Falicov, L.M., Hanke, W., Maple, M.B. (eds.) Amsterdam: North Holland **1981**, p. 121.
81G3	Güntherodt, G., Wichelhaus, W.: Solid State Commun. **39** (1981) 1147.
81G4	Gschneidner, K.A., Beaudry, B.J., Takeshita, T., Eucker, S.S., Taher, S.M.A., Ho, J.C., Gruber, J.B.: Phys. Rev. **B 24** (1981) 7187.
81H	Holtzberg, F., Wittig, J.: Solid State Commun. **40** (1981) 315.

9.16.5 References for 9.16

81K1	Konczykowski, M., Morillo, J., Senateur, J.P.: Solid State Commun. **40** (1981) 517.
81K2	Kaminskii, V.V., Ryabov, A.V., Stepanov, N.N.: Sov. Phys. Solid State **23 (6)** (1981) 1052.
81K3	Kurganskii, S.I., Farberovich, O.V.: Phys. Stat. Sol. (b) **106** (1981) 437.
81K4	Kaldis, E., Peteler, W.: J. Cryst. Growth **52** (1982) 125.
81K5	Kaski, K., Kuivalainen, P., Eränen, S., Stubb, T.: Phys. Scr. **24** (1981) 472.
81K6	Kamarzin, A.A., Mironov, K.E., Sokolov, V.V., Malovitskii, Y.N., Vasil'yeva, I.G.: J. Cryst. Growth **52** (1981) 619.
81L1	Lapierre, F., Ribault, M., Holtzberg, F., Flouquet, J.: Solid State Commun. **40** (1981) 347.
81L2	Lin-Liu, Y.R., Falicov, L.M., Kohn, W.: Phys. Rev. **B 24** (1981) 5664.
81M	Mariani, C., Modesti, S., Rosei, R., Simoni, F., Tosatti, E.: Solid State Commun. **38** (1981) 833.
81P1	Peterman, D.J., Weaver, J.H., Peterson, D.T.: Phys. Rev. **B23** (1981) 3903.
81P2	Patil, C.G., Krishnamurthy, B.S.: Phys. Status Solidi (b) **105** (1981) 391.
81R1	Rozhkov, S.S., Semchuk, A.Y.: Sov. Phys. Solid State **23** (7) (1981) 1118.
81R2	Röhler, J., Kaindl, G.: Solid State Commun. **37** (1981) 737.
81S1	Stewart, G.A., Wortmann, G.: Phys. Lett. **A85** (1981) 185.
81S2	Scharmer, E.-G., Leiß, M., Huber, G.: J. Lumin. **24/25** (1981) 751.
81U	Umehara, M.: J. Phys. Soc. Jpn. **50** (1981) 1082.
81W	Wegrzyn, A., Lubecka, M.: Thin Solid Films **80** (1981) 343.
81Y	Yamada, K., Heleskivi, J., Salin, A.: Solid State Commun. **37** (1981) 957.
81Z	Zhukov, V.P., Gubanov, V.A., Weber, J.: J. Phys. Chem. Solids **42** (1981) 641.
82A1	Allan, S.R., Edwards, D.M.: J. Phys. **C 15** (1982) 2151.
82A2	Astaf'eva, L.V., Skornyakov, G.P., Kamarzin, A.A., Malovitskii, Y.N.: Sov. Phys. Solid State **24** (1982) 367.
82B	Batirov, T.M., Verkhovskaya, K.A., Kamarzin, A.A., Malovitskii, Y.N., Lisoivan, V.I., Fridkin, V.M.: Sov. Phys. Solid State **24** (1982) 746.
82C	Chen. C.H., Meixner, A.E., Schluter, M., Varma, C.M., Schmidt, P.H.: Phys. Rev. **B 25** (1982) 2036.
82G1	Grebinskii, S.I., Kaminskii, V.V., Ryabov, A.V., Stepanov, N.N.: Sov. Phys. Solid State **24 (6)** (1982) 1069.
82G2	Güntherodt, G., Thompson, W.A., Holtzberg, F., Fisk, Z.: Phys. Rev. Lett. **49** (1982) 1030.
82G3	Gal'dikas, A.P., Matulenene, I.B., Samokhvalov, A.A., Osipov, V.V.: Sov. Phys. Solid State **24 (6)** (1982) 939.
82G4	Glurdzhidze, L.N., Gzirishvili, D.G., Koshoridze, S.I., Dzhabua, Z.U., Sanadze, V.V.: Sov. Phys. Solid State **24** (1982) 795.
82G5	Glurdhidze, L.N., Gigineishvili, A.V., Bzhalava, T.L., Sanadze, V.V.: Sov. Phys. Solid State **24** (1982) 1050.
82K	Kamitakahara, W.A., Crawford, R.K.: Solid State Commun. **41** (1982) 843.
82L	v. d. Linden, W., Nolting, W.: Z. Phys. B – Condens. Matter **48** (1982) 191.
82M1	Misemer, D.K., Harmon, B.N.: Phys. Rev. **B 26** (1982) 5634.
82M2	Mårtensson, N., Reihl, B., Pollak, R.A., Holtzberg, F., Kaindl, G.: Phys. Rev. **B 25** (1982) 6522.
82M3	Mook, H.A., McWhan, D.B., Holtzberg, F.: Phys. Rev. **B 25** (1982) 4321.
82M4	Matlak, M., Ramakanth, A., Skrobiś, K.: Z. Phys. B – Condens. Matter **48** (1982) 227.
82N1	Nolting, W.: Z. Phys. B – Condens. Matter **49** (1982) 87.
82N2	Nabutovskaya, O.A., Nogteva, V.V., Sokolov, V.V., Kamarzin, A.A.: Sov. Phys. Solid State **24** (1982) 834.
82S1	Schlapbach, L., Scherrer, H.R.: Solid State Commun. **41** (1982) 893.
82S2	Schlapbach, L., Osterwalder, J.: Solid State Commun. **42** (1982) 271.
82S3	Saunders, G.A., Lambson, W.A., Hailing, T., Bullet, D.W., Bach, H., Methfessel, S.: J. Phys. **C 15** (1982) L 551.
82S4	Spronken, G., Avignon, M.: J. Phys. **F 12** (1982) 2541.
82S5	Sendorek, D., Lubecka, M., Wegrzyn, A.: Phys. Status Solidi (b) **113** (1982) K19.
82S6	Scharmer, E.-G., Leiß, M., Huber, G.: J. Phys. **C 15** (1982) 1071.
82T1	Tsutsumi, K., Aita, O., Watanabe, T.: Phys. Rev. **B 25** (1982) 5415.
82T2	Trzaskoma, P., Zielinski, J., Szopa, M.: Phys. Status Solidi (b) **109** (1982) K129.
82W	Wichelhaus, W., Simon, A., Stevens, K.W.H., Brown, P.J., Ziebeck, K.R.A.: Philos. Mag. **B 46** (1982) 115.
83E	Eränen, S., Sinkkonen, J.: Phys. Status Solidi (b) **115** (1983) 519.

9.16.5 References for 9.16

83I	Ito, T., Beaudry, B.J.: Gschneidner, K.A., Takeshita, T.: Phys. Rev. **B 27** (1983) 2830.
83K1	Kamarzin, A.A., Mamedov, A.A., Smirnov, V.A., Sobol', A.A., Sokolov, V.V., Shcherbakov, I.A.: Sov. J. Quantum Electron. **13** (1983) 1027.
83K2	Kamarzin, A.A., Mamedov, A.A., Smirnov, V.A., Sokolov, V.V., Shcherbakov, I.A.: Sov. J. Quantum Electron. **13** (1983) 328.
83K3	Kamarzin, A.A., Mamedov, A.A., Smirnov, V.A., Sokolov, V.V., Shcherbakov, I.A.: Sov. J. Quantum Electron. **13** (1983) 336.
83N	Nazareno, H.N., Caldas, A., Taft, C.A.: J. Phys. **C 16** (1983) 887.
83O1	Ousaka, Y., Sakai, O., Tachiki, M.: J. Phys. Soc. Jpn. **52** (1983) 1034.
83O2	Ott, H.R., Hulliger, F.: Z. Phys. B – Condens. Matter **49** (1983) 323.
83P	Pott, R., Güntherodt, G., Wichelhaus, W., Ohl, M., Bach, H.: Phys. Rev. **B 27** (1983) 359.
83V	Vonsovsky, S.V., Samokhvalov, A.A., Osipov, V.V., Kostylev, V.A.: J. Mag. Magn. Mater **31–34** (1983) 165.

Figures

B Physical data of semiconductors V

9 Non-tetrahedrally bonded binary compounds

(Sect. 9.1···9.6, see subvolume III/17e; Sect. 9.7···9.13, see subvolume III/17f)

9.14 Boron compounds (For tables, see p. 9ff.)

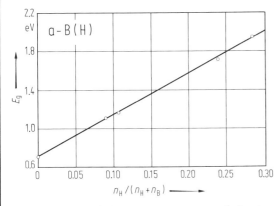

Fig. 1. a-B(H). Optical energy gap E_g vs. atomic fraction of hydrogen [80B].

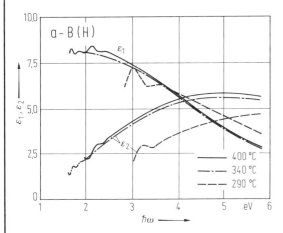

Fig. 2. a-B(H). Real and imaginary parts ε_1 and ε_2 of the dielectric function vs. photon energy for samples deposited at three different reaction temperatures (290 °C, 340 °C, 400 °C). The low energy structure of the curves is due to the onset of back reflection from the substrate [80B].

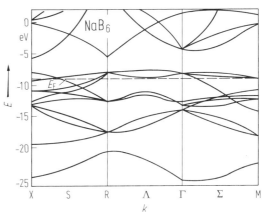

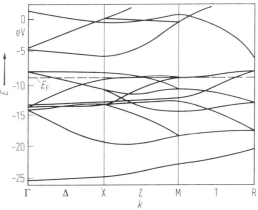

Fig. 3. NaB$_6$. Calculated band structure [76P1].

9.14 Boron compounds

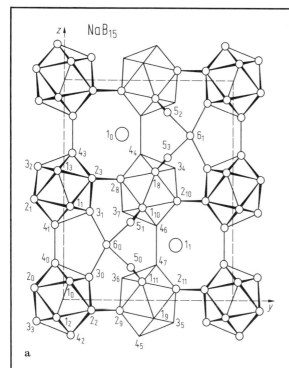

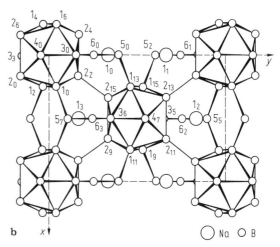

Fig. 4. NaB$_{15}$. Projection of the crystal structure (a) on the yz plane, (b) on the xy plane (only atoms with $-1/4 \leq x \leq 1/4$ are represented) [76N, 77N1]. The numbers denote crystallographically different atomic positions in the lattice.

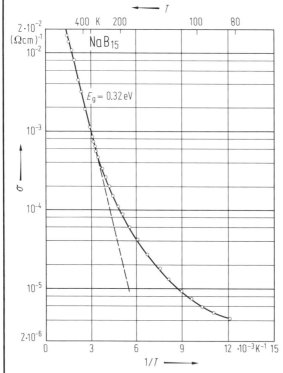

Fig. 5. NaB$_{15}$. Electrical conductivity vs. reciprocal temperature for polycrystalline sample [77N1].

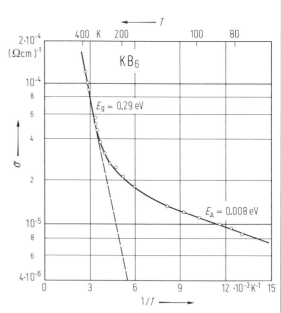

Fig. 6. KB$_6$. Electrical conductivity of sintered material vs. reciprocal temperature [77N1].

9.14 Boron compounds

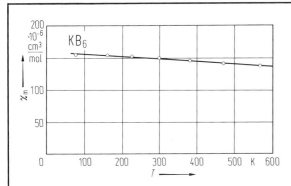

Fig. 7. KB_6 Magnetic susceptibility χ_m vs. temperature [77N1, 66N]. χ_m in CGS-emu.

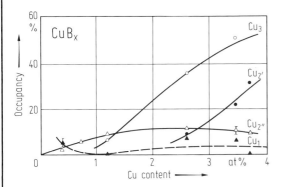

Fig. 8. Cu in β-rhombohedral boron. Occupancy of the four different crystallographic positions vs. Cu content [76C].

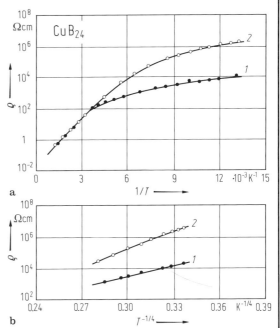

Fig. 10. CuB_{24} (Cu in β-rhombohedral boron). Resistivity of two different samples (a) vs. reciprocal temperature and (b) vs. $T^{-1/4}$ (according to Mott's law) [76G1, 79G2].

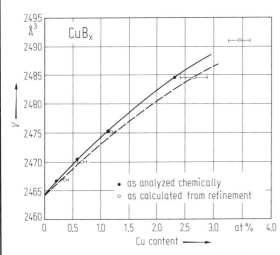

Fig. 9. Cu in β-rhombohedral boron. Unit cell volume V vs. Cu content [76C].

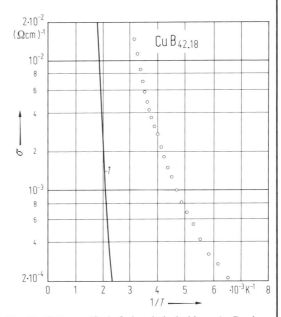

Fig. 11. $CuB_{42.18}$ (Cu in β-rhombohedral boron). Conductivity of a polycrystalline sample vs. reciprocal temperature (circles) [81W3]. Pure β-rhombohedral boron for comparison (curve 1).

9.14 Boron compounds

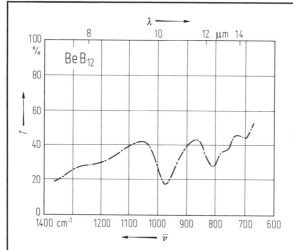

Fig. 12. BeB$_{12}$. Transmission vs. wavenumber in the lattice vibration range [77B1, 64B].

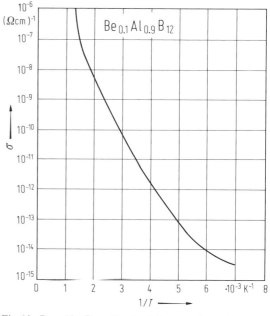

Fig. 13. Be$_{0.1}$Al$_{0.9}$B$_{12}$. Conductivity vs. reciprocal temperature [79G1].

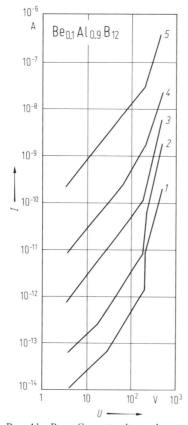

Fig. 14. Be$_{0.1}$Al$_{0.9}$B$_{12}$. Current-voltage characteristics at various temperatures. Curve 1: $T = 113$ K; curve 2: $T = 145$ K; curve 3: $T = 185$ K; curve 4: $T = 232$ K; curve 5: $T = 293$ K [79G1].

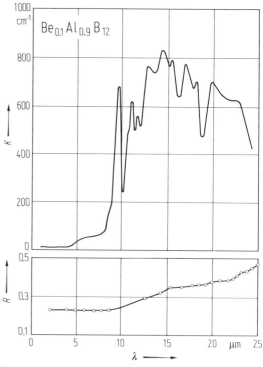

Fig. 15. Be$_{0.1}$Al$_{0.9}$B$_{12}$. Absorption coefficient K and reflectivity R vs. wavelength. (Remark: the strong increase of the reflectivity with increasing wavelength cannot be explained by lattice vibrations of such low oscillator strengths. Hence $\varepsilon(0)$, which has been derived from R should be put in question, too.) [79G1].

9.14 Boron compounds

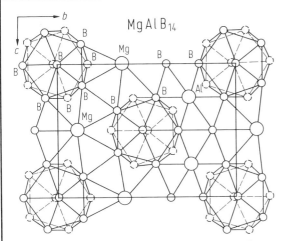

Fig. 16. MgAlB$_{14}$. A layer of icosahedra with extra-icosahedral atoms. The atoms with $-1/4 < x < +1/4$ are represented [77M3].

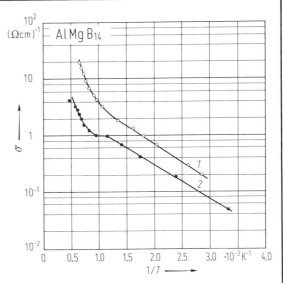

Fig. 17. MgAlB$_{14}$. Electrical conductivity vs. reciprocal temperature for polycrystalline samples. Curve 1: pure MgAlB$_{14}$; curve 2: MgAlB$_{14}$ (nickel alloyed) [79B1].

For Fig. 19, see next page.

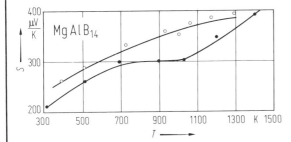

Fig. 18. MgAlB$_{14}$. Thermoelectric power vs. temperature for polycrystalline samples Curve 1: pure MgAlB$_{14}$; curve 2: AlMgB$_{14}$ (nickel alloyed) [79B1].

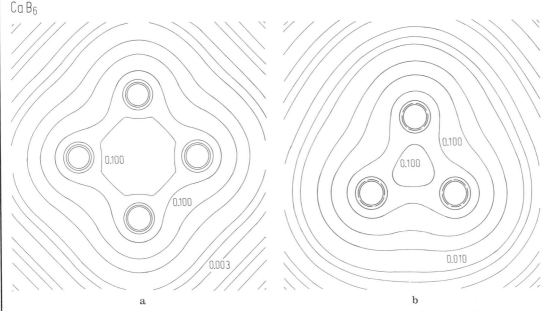

Fig. 20. CaB$_6$ a) Electron density contour across the fourfold axis of the octahedron. b) Electron density contour across the eight trigonal faces of B$_6^{2-}$ cage [77P1]. Electron density in e Å^{-3}.

9.14 Boron compounds

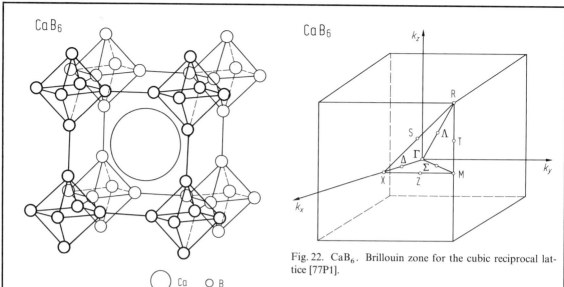

Fig. 19. CaB$_6$. Lattice structure [77E].

Fig. 22. CaB$_6$. Brillouin zone for the cubic reciprocal lattice [77P1].

For Fig. 21 α, see next page.

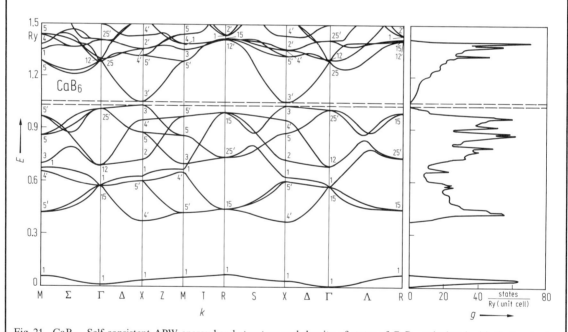

Fig. 21. CaB$_6$. Self consistent APW energy band structure and density of states of CaB$_6$, calculated with the local-spin-density approximation and with the non-muffin-tin corrections. The symmetry labels denote irreducible representations about the center of the B$_6$ octahedron. The two broken lines show the energy gap between the valence and the conduction bands. The density of states is the total density of states for the lowest 13 bands [79H1]. (Cp. also [77P1]). 1 Ry = 13.62 eV.

9.14 Boron compounds

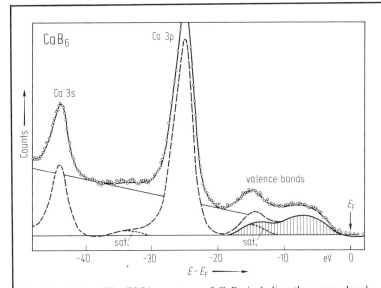

Fig. 21α. CaB$_6$. The ESCA spectrum of CaB$_6$ including the energy band region. The charging-up effect has been corrected for and hence the position $E_b^F = 0$ coincides with the Fermi level within experimental error (± 0.2 eV). The dotted curves denote the $\alpha_{3,4}$ satellite peaks of the Ca 3s and 3p main peaks [76A1].

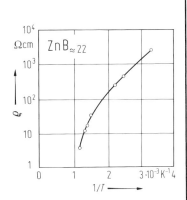

Fig. 24. ZnB$_{\approx 22}$. Electrical resistivity vs. reciprocal temperature for polycrystalline samples [74K].

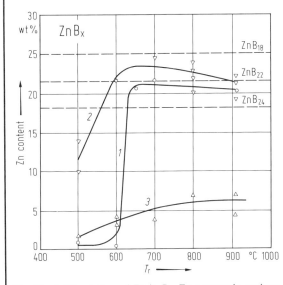

Fig. 23. Solid solution of Zn in B. Zn content in various boron modifications vs. reaction temperature. T_r. Curves: 1: β-rhombohedral boron; 2: amorphous boron; 3: α-rhombohedral boron [77G3].

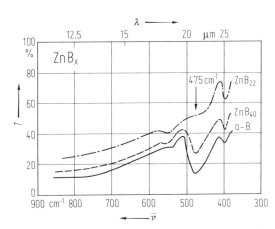

Fig. 25. ZnB$_{22}$/ZnB$_{\approx 40}$. IR transmission vs. wavenumber (transmission of amorphous boron for comparison) [77G3].

9.14 Boron compounds

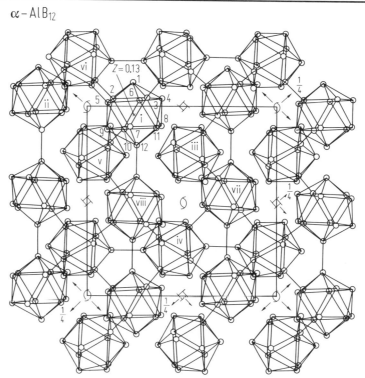

Fig. 26. α-AlB$_{12}$. B$_{12}$ icosahedral arrangement as seen along the c axis. The icosahedra are marked with the following symmetry codes: i(x, y, z); ii$(-x, -y, \frac{1}{2}+z)$; iii$(\frac{1}{2}-y, \frac{1}{2}+x, \frac{3}{4}+z)$; iv$(\frac{1}{2}+y, \frac{1}{2}-x, \frac{1}{4}+z)$; v$(y, x, -z)$; vi$(-y, -x, \frac{1}{2}-z)$; vii$(\frac{1}{2}-x, \frac{1}{2}+y, \frac{3}{4}-z)$; viii$(\frac{1}{2}+x, \frac{1}{2}-y, \frac{1}{4}-z)$ [77H1].

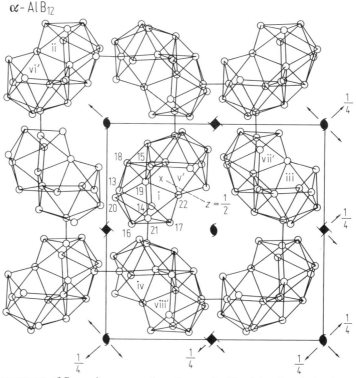

Fig. 27. α-AlB$_{12}$. Arrangement of B$_{19}$ units as seen along the c axis. The defect icosahedra involved in the B$_{19}$ units are marked with the following symmetry codes: i, ii, iii, iv (see Fig. 26), v'$(y, x, 1-z)$, vi'$(-y, -x, \frac{3}{2}-z)$, vii'$(\frac{1}{2}-x, \frac{1}{2}+y, \frac{7}{4}-z)$, viii'$(\frac{1}{2}+x, \frac{1}{2}-y, \frac{3}{4}-z)$ [77H1].

9.14 Boron compounds

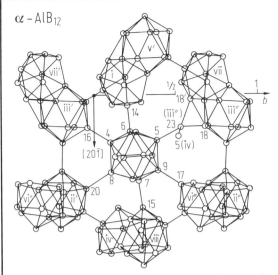

Fig. 28. α-AlB$_{12}$. The nature of the linkages between B$_{12}$ icosahedra and B$_{19}$ units as projected on the (102) plane along the [101]-axis. The icosahedron defect icosahedra involved in B$_{19}$ units and the B atoms, B(23) and B(5), are marked with following symmetry codes, where i, iv, vii and viii and v' are identical to those used in Figs. 26 and 27, respectively: i, ii'$(1-x, -y, -\frac{1}{2}+z)$, ii''$(1-x, 1-y, -\frac{1}{2}+z)$, iii'$(\frac{1}{2}-y, -\frac{1}{2}+x, -\frac{1}{4}+z)$, iii''$(\frac{1}{2}-y, \frac{1}{2}+x, -\frac{1}{4}+z)$, iv, iv'$(\frac{1}{2}+y, \frac{1}{2}-x, -\frac{3}{4}+z)$, v', vi''$(1-y, -x, \frac{1}{2}-z)$, vi'''$(1-y, 1-x, \frac{1}{2}-z)$, vii, vii'$(\frac{1}{2}-x, -\frac{1}{2}+y, \frac{3}{4}-z)$, viii [77H1].

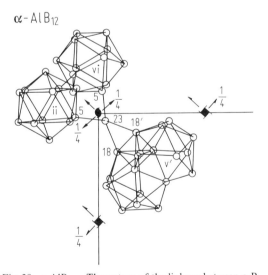

Fig. 29. α-AlB$_{12}$. The nature of the linkage between a B$_{12}$ icosahedron and a B$_{19}$ unit through an intermediary single B atom as seen along the c axis. These units are marked with the symmetry codes which are identical to those used in Figs. 27 and 28 [77H1].

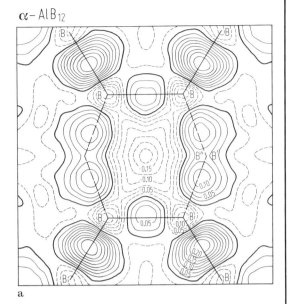

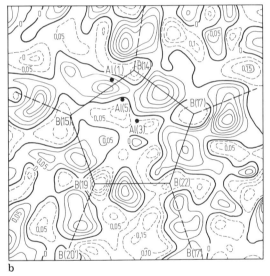

Fig. 30. α-AlB$_{12}$. a) Averaged difference Fourier map through the bisecting plane of the icosahedron. Contours as in Fig. 27. Refer to Fig. 26 for the geometry. The four external B–B bonds lie on the plane. b) Difference Fourier map through the open pentagonal face of the B$_{19}$ unit. Al(1)–Al(5)–Al(3) sites with the occupancies 0.717, 0.021 and 0.240, respectively are approximately 1.8 Å above the face. Contours as in Fig. 27 [77H1]. Electron density in eÅ^{-3}.

9.14 Boron compounds

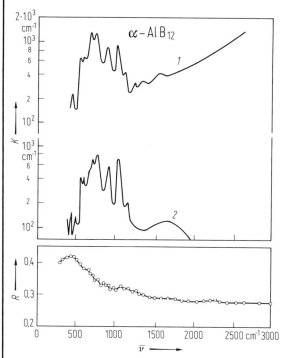

Fig. 31. α-AlB$_{12}$. Absorption coefficient and reflectivity vs. wavenumber. Curve *1*. Sample with 10^2 Ωcm resistivity; curve *2*: Sample with 10^6 Ωcm resistivity (Remark: the strong increase of the reflectivity with increasing wavelength cannot be explained by lattice vibrations of such low oscillator strengths. Hence $\varepsilon(0)$, which has been derived from R should be put in question, too) [74B3, 73B1, 77B1].

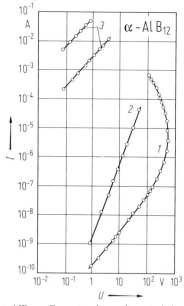

Fig. 35. α-AlB$_{12}$. Current-voltage characteristics obtained at 77 K in different states of a sample: Curves *1*: high-resistivity state, $I \propto \exp\sqrt{U}$; *2*: intermediate state with a quadratic dependence; $I \propto U^2$; *3*: low-resistivity memory state, $I \propto U$ [75Z].

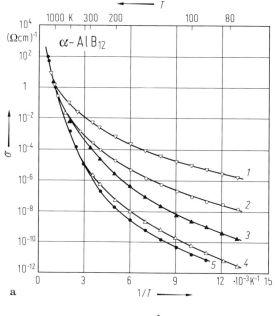

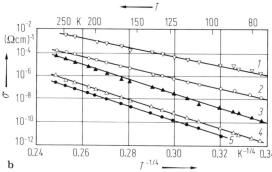

Fig. 32. α-AlB$_{12}$. Conductivity of unspecified different polycristalline samples a) vs. reciprocal temperature, b) vs. $T^{-1/4}$ (according to Mott's law) [74B3, 77B1].

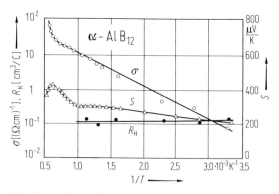

Fig. 33. α-AlB$_{12}$. Conductivity, Hall coefficient and thermoelectric power of polycrystalline samples vs. reciprocal temperature [74B3].

For Fig. 34, see next page.

9.14 Boron compounds

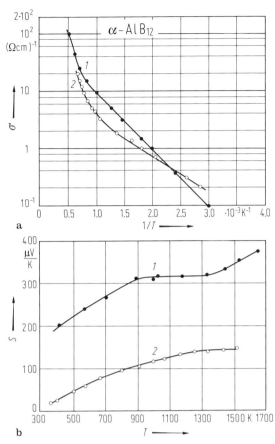

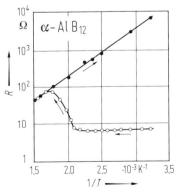

Fig. 34. α-AlB$_{12}$. a) Conductivity vs. reciprocal temperature. b) Thermoelectric power vs. temperature. Curve 1: single crystal (orientation unknown); curve 2: pressed sample [79B1].

Fig. 36. α-AlB$_{12}$. Resistance vs. reciprocal temperature in the low-resistivity memory state [75Z].

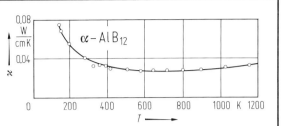

Fig. 37. α-AlB$_{12}$. Thermal conductivity of polycrystalline samples vs. temperature [74B3].

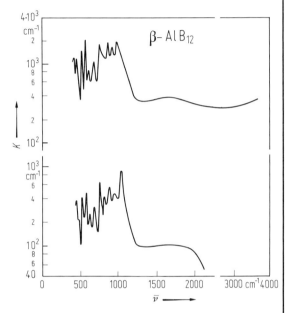

Fig. 38. β-AlB$_{12}$. Absorption coefficient of different samples vs. wavenumber [74G1].

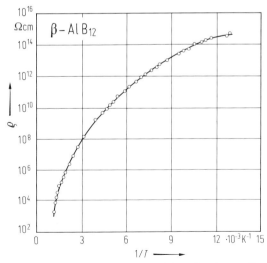

Fig. 39. β-AlB$_{12}$. Electrical resistivity of polycrystalline sample vs. reciprocal temperature [76G1, 79G2].

9.14 Boron compounds

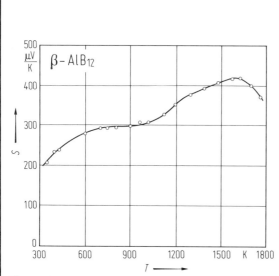

Fig. 40. β-AlB$_{12}$. Thermoelectric power of polycrystalline sample vs. temperature [79G2].

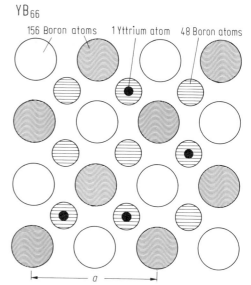

Fig. 41. YB$_{66}$. Simplified schematic representation of the crystal structure looking down along the [100] axis. The horizontal and vertical directions are [010] and [001]. Different shading of the spheres representing B$_{156}$ groups indicate different orientation. The B$_{48}$ groups are situated at $a/4$, the Y atoms at $0.054\ a$ below the surface. The Y sites are believed to be statistically occupied [77S4].

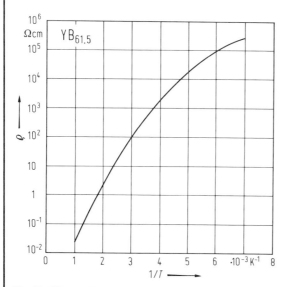

Fig. 42. YB$_{61.5}$. Electrical resistivity vs. reciprocal temperature [77S4].

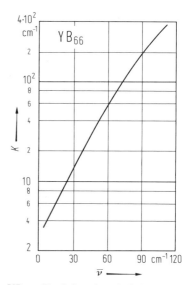

Fig. 43. YB$_{66}$. Far infrared optical absorption coefficient at 4.2 K vs. wavenumber [77S4].

9.14 Boron compounds

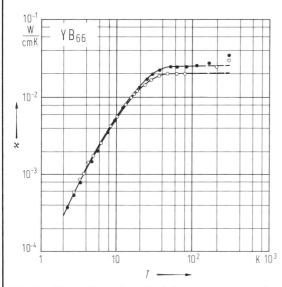

Fig. 44. YB$_{66}$. Thermal conductivity vs. temperature for two samples [71S3].

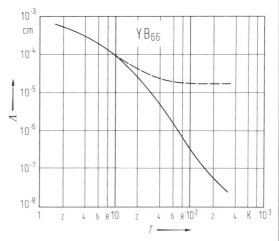

Fig. 45. YB$_{66}$. Phonon mean free path vs. temperature [71S3]. The dashed line represents the behavior of the acoustic phonons only and corresponds to the mean free path of the actual carriers with thermal energy.

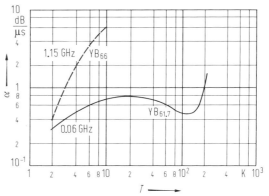

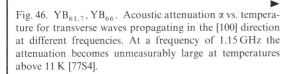

Fig. 46. YB$_{61.7}$, YB$_{66}$. Acoustic attenuation α vs. temperature for transverse waves propagating in the [100] direction at different frequencies. At a frequency of 1.15 GHz the attenuation becomes unmeasurably large at temperatures above 11 K [77S4].

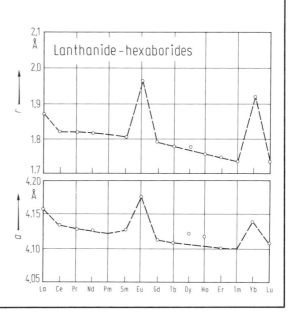

Fig. 47. Lanthanide hexaborides. Lattice parameters and atomic radii of the metals [75S].

9.14 Boron compounds

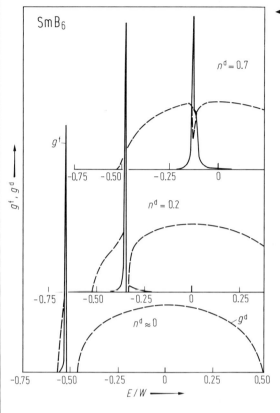

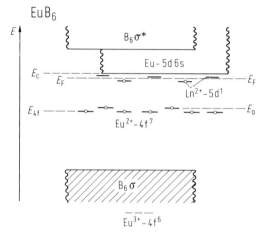

◀ Fig. 48. SmB$_6$. Calculated densities of d and f states g^d, g^f from the two-band Hubbard model with parabolic density of states for the d band. The bottom curve is for a filled non-degenerate f state with a gap to the d states and the upper curves represent partial f occupation $1 \leq n^f \leq 2$. The gap disappears for $n^d \approx 0.6$. The results are based on a solution of the single site coherent potential approximation in terms of the site Green's function. Parameters: $V/W = 0.05$ (V = one electron matrix element, W = band width of the d band); $g = j = 0$ (g = direct Coulomb interaction, j = exchange interaction); n^d, n^f = occupation density of the d and f states [79M3].

Fig. 49. EuB$_6$. Energy diagram according to the description in section 9.14.7 [77E, 73G].

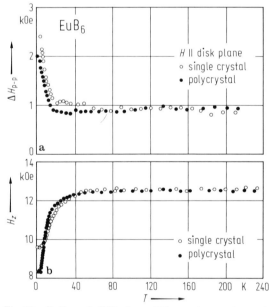

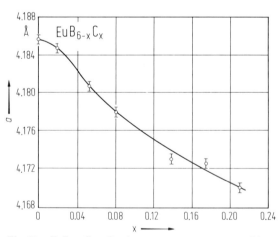

Fig. 50. EuB$_6$. a) ESR line width vs. temperature at 35.47 GHz. b) Temperature dependence of the ESR resonance points at 35.47 GHz [79K2]. ΔH_{p-p}: peak to peak line width; H_z: resonance field in z-direction. (For qualitatively similar results at 24.58 GHz, see original paper.)

Fig. 51. EuB$_{6-x}$C$_x$. Lattice parameter vs. composition [78K1, 79K3].

9.14 Boron compounds

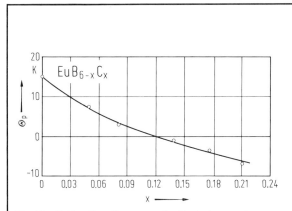

Fig. 52. EuB$_{6-x}$C$_x$. Paramagnetic Curie temperature vs. composition [78K1, 79K3].

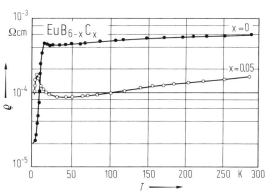

Fig. 53. EuB$_{6-x}$C$_x$. Electrical resistivity for $x=0$ and $x=0.05$ vs. temperature [78K1, 79K3].

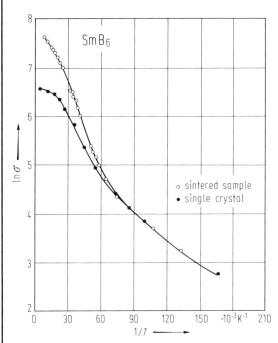

Fig. 54. SmB$_6$. Electrical conductivity vs. reciprocal temperature [77E, 73M1]; see also [77B5]. σ in Ω^{-1}cm^{-1}.

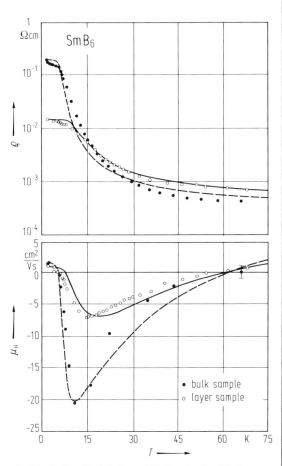

Fig. 55. SmB$_6$. Resistivity and Hall mobility of SmB$_6$ samples vs. temperature [71N2]. The calculated curves are discussed in the original paper.

9.14 Boron compounds

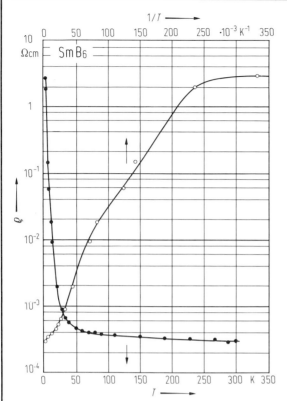

Fig. 55 α. SmB$_6$. Temperature dependence of resistivity measured at 230 Hz; same samples as in Fig. 55 β [79A6].

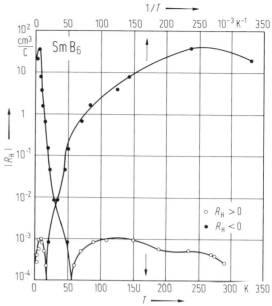

Fig. 55 β. SmB$_6$. Temperature dependence of the Hall coefficient, measured at 230 Hz with $B=1.7$ T; same samples as in Fig. 55 α [79A6].

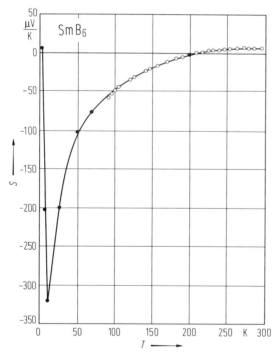

Fig. 56. SmB$_6$. Thermoelectric power vs. temperature; data on two samples [69P2, 73M1, 77E].

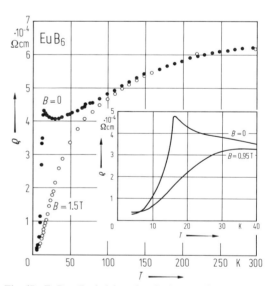

Fig. 57. EuB$_6$. Resistivity of a single crystal at $B=0$ and $B=1.5$ T vs. temperature [80G]. Insert: same plot at $B=0$ and $B=0.95$ T. $\Delta\varrho/\varrho$ changes sign at about 5.5 K [79W2] (cp. also [77I]).

9.14 Boron compounds

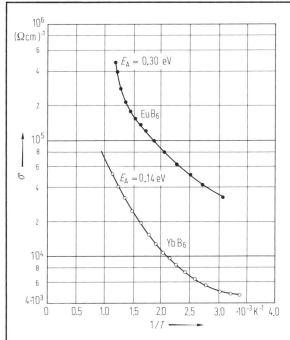

Fig. 58. EuB_6, YbB_6. Electrical conductivity vs. reciprocal temperature [77E, 73M1, 73B2]. Cp. also [69F, 75S].

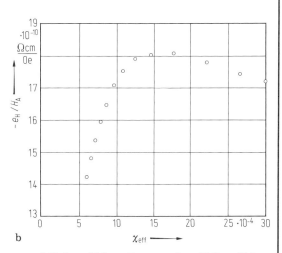

Fig. 60. EuB_6. Electrical resistivity ratios $\varrho(p)/\varrho(0)$ vs. hydrostatic pressure [79G3]. Curve I: EuB_6 prepared with a stoichiometric ratio Eu to B in the starting charge; curves II and III: with B surplus.

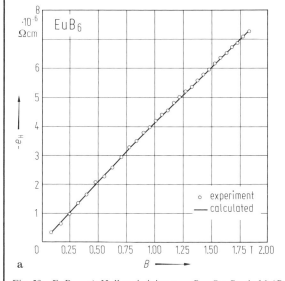

Fig. 59. EuB_6. a) Hall resistivity $e_H = R_H \cdot B + R_s \cdot 4\pi M$ (R_H = normal Hall coefficient, R_s = anomalous Hall coefficient, B = magnetic induction, M = magnetization of the sample) at 4.2 K vs. magnetic induction [80G]. Solid curve calculated with $R_H = -3.7 \cdot 10^{-2}$ cm^{3}C^{-1} and $R_s = -4.8 \cdot 10^{-3}$ cm^{3}C^{-1}. b) Relative Hall resistivity in the paramagnetic region $e_H/H_A = R_H + \{(1-N) \cdot R_H + R_s\} \cdot 4\pi\chi_{eff}$ (H_A = applied field; $N = 0.633$ (demagnetizing factor)) vs. the effective susceptibility $\chi_{eff} = \chi/(1+4\pi N\chi)$ [80G].

9.14 Boron compounds

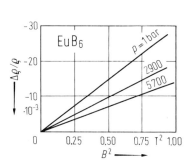

Fig. 61. EuB$_6$. Magnetoresistance vs. magnetic field (squared) at 77 K for different pressures [79W2].

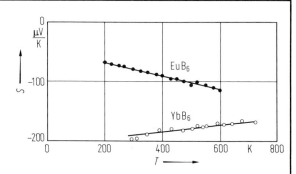

Fig. 62. EuB$_6$, YbB$_6$. Thermoelectric power vs. temperature [77E, 73M1, 73B2]. Cp. also [75S].

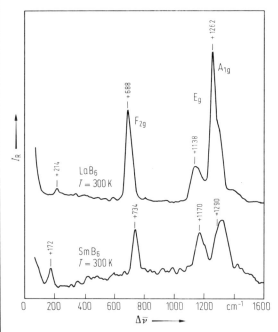

Fig. 62 a. LaB$_6$, SmB$_6$. Room temperature Raman spectra of single crystals. Experimental conditions: oblique backscattering, laser 50 mW, $\lambda = 514$ nm, resolution 8 cm^{-1} [81M].

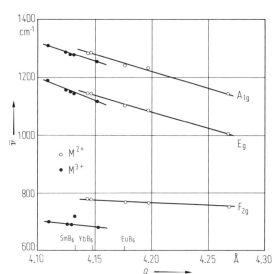

Fig. 63. SmB$_6$, EuB$_6$, YbB$_6$ (and other MB$_6$). Raman wavenumbers of metal hexaborides with bivalent and trivalent metals vs. lattice constant [76I].

9.14 Boron compounds

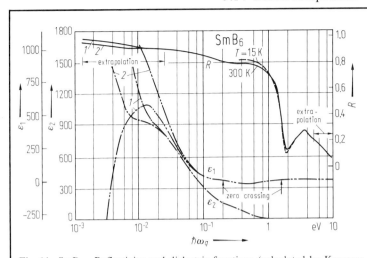

Fig. 64. SmB_6. Reflectivity and dielectric functions (calculated by Kramers-Kronig analysis) vs. phonon energy [77A, 78A]. The dielectric functions which are calculated according to two different extrapolations (*1*) (*2*) of the experimental reflectivity towards low photon energies, show that there are two limiting types of behavior, which cannot be distinguished at present because of the experimental uncertainty. Zero crossings of ε_1 are indicated by arrows.

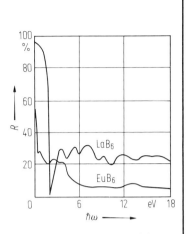

Fig. 64 a. EuB_6, LaB_6. Reflectivity vs. photon energy [80G2]. For permittivity spectra derived from the reflectivity, see [80G2].

Fig. 65. EuB_6. Raman spectrum. Scattering intensity vs. wavenumber (488 nm excitation) [76I].

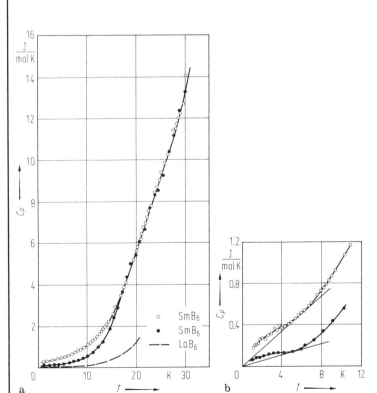

Fig. 66. SmB_6. Heat capacity vs. temperature for two samples (dashed line: LaB_6 for comparison). Fig. b shows the low temperature range on an expanded scale [71N2].

9.14 Boron compounds

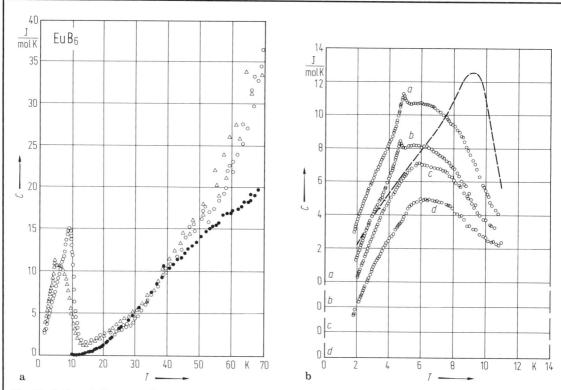

Fig. 67. EuB$_6$. a) Heat capacity vs. temperature. ○ ferromagnetic samples; △ antiferromagnetic samples, ● LaB$_6$ for comparison [80F]. b) Low temperature heat capacity of antiferromagnetic EuB$_6$ in different fields applied along a [100] axis: a: $H=0$, b: $H=0.2$ T, c: $H=0.6$ T, d: $H=1.2$ T. Data for the ferromagnetic EuB$_6$ at $H=0$ are also shown by the dashed curve [80F].

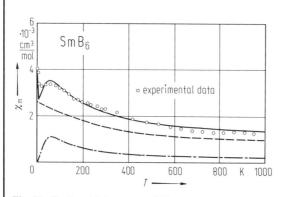

Fig. 68. SmB$_6$. Molar susceptibility vs. temperature. Experimental data: [69M1]; theoretical fit: [71N2] (dashed curve: sum of the susceptibilities of the 4f^6 and f^{5}s∥ configurations; dot – dashed curve: contribution due to the f^{5}s⊥ term). χ_m in CGS units.

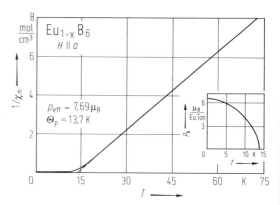

Fig. 69. Eu$_{1-x}$B$_6$. Low field magnetic susceptibility of an Al – grown Eu$_{1-x}$B$_6$ single crystal vs. temperature (the units of χ assume an exact EuB$_6$ composition) Insert: Saturation magnetic moment per Eu ion, p_A, vs. temperature derived from Arrot plots (accuracy 20%) [79F1].

9.14 Boron compounds

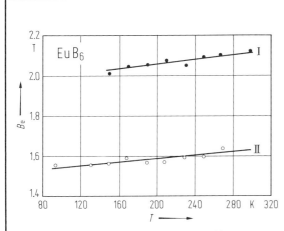

Fig. 70. EuB$_6$. Exchange field in bulk (I) and low-electron-density (II) material vs. temperature. Results obtained by EPR [76G2].

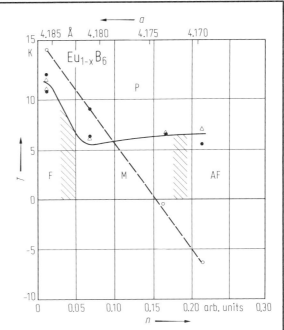

Fig. 71. EuB$_6$. Magnetic phase diagram derived from Mössbauer (●) and magnetic (△) data. Ordering temperatures vs. lattice parameter (P: paramgnetic; F: ferromagnetic; M: micromagnetic; AF: antiferromagnetic). The second scale relates a to the approximate conduction electron concentration. (○: paramagnetic Curie temperatures according to [78K1]). [79C1]. See also [80G1, 80T].

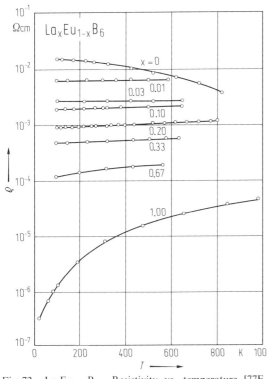

Fig. 72. La$_x$Eu$_{1-x}$B$_6$. Resistivity vs. temperature [77E, 74M1].

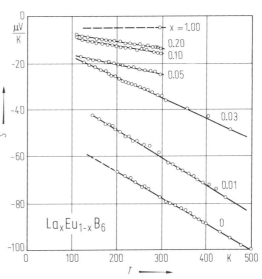

Fig. 73. La$_x$Eu$_{1-x}$B$_6$. Thermoelectric power vs. temperature [77E, 74M1].

9.14 Boron compounds

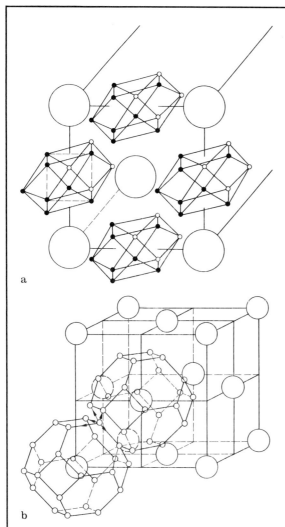

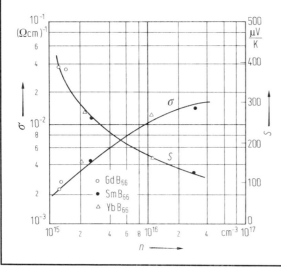

Fig. 74. Lanthanide dodecaborides. a) Structure model; metal atoms and cubo-octahedral B_{12} units arranged in a NaCl – like cubic face – centered lattice. b) Structure model; metal atoms arranged in a cubic lattice, each metal atom surrounded by a B_{24} cage [72S4, 81S2].

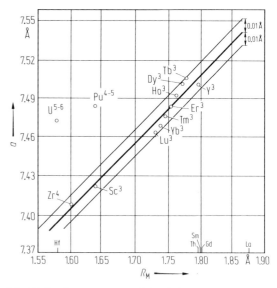

Fig. 75. Lattice parameters of the dodecaboride phases vs. the atomic radii of the Me atoms, R_M [72S4].

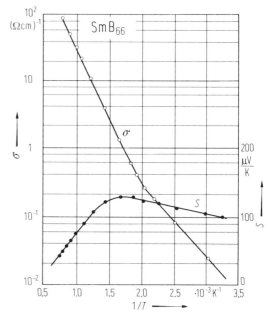

Fig. 76. SmB_{66}. Electrical conductivity and thermoelectric power vs. reciprocal temperature [81G3].

Fig. 77. SmB_{66}, GdB_{66}, YbB_{66}. Electrical conductivity and thermoelectric power vs. carrier concentration [81G3].

9.14 Boron compounds

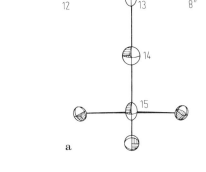

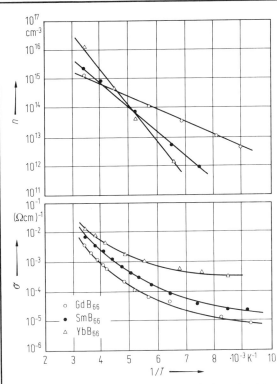

Fig. 78. SmB$_{66}$, GdB$_{66}$, YbB$_{66}$. Electrical conductivity and carrier concentration (derived from Hall effect) vs. reciprocal temperature [81G3].

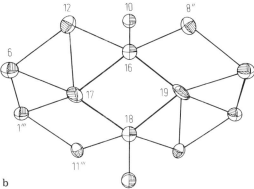

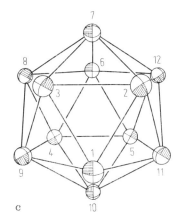

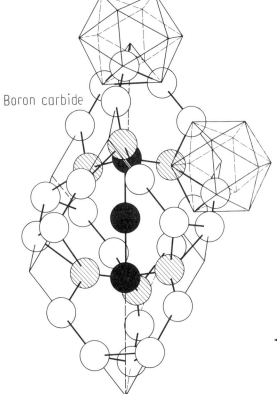

Fig. 80. Boron carbide. a) C—B—C chain in the unit cell and the equatorial icosahedral B atoms to which it bonds. b) Planar B$_4$ group and the icosahedral atoms to which it bonds in one of six possible locations for the pair of B(6) atoms. c) Icosahedron with the atom positions to which the C—B—C chain (Fig. a) or the planar B$_4$ group (Fig. b) bonds [75Y].

◀ Fig. 79. Boron carbide. Arrangement of the atoms in the unit cell. ○ atoms of the icosahedra located on the edges of the unit cell; ◐ equatorial atoms of the icosahedra forming multiple-center bonds; ● atoms of the three – atomic linear C—B—C chain; (cp. e.g. [66S, 79B4, 81S2]).

9.14 Boron compounds

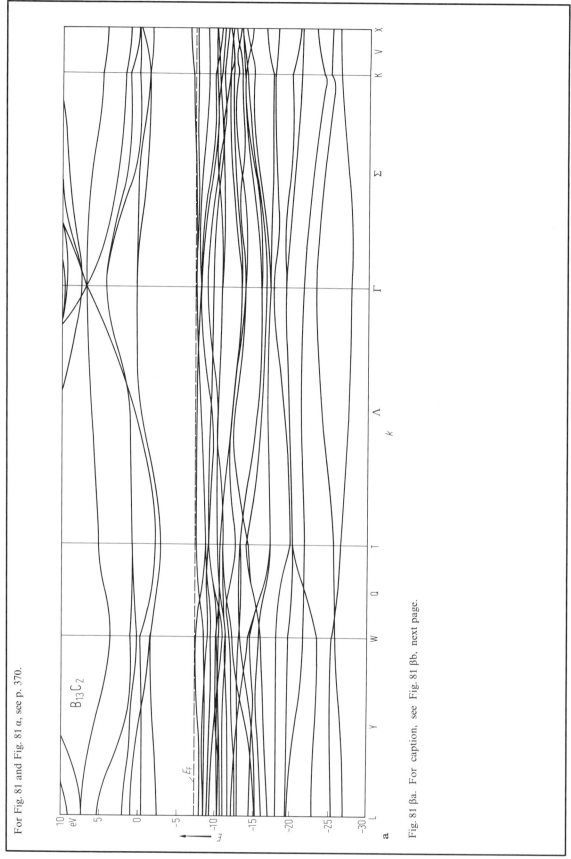

For Fig. 81 and Fig. 81 α, see p. 370.

Fig. 81 βa. For caption, see Fig. 81 βb, next page.

9.14 Boron compounds

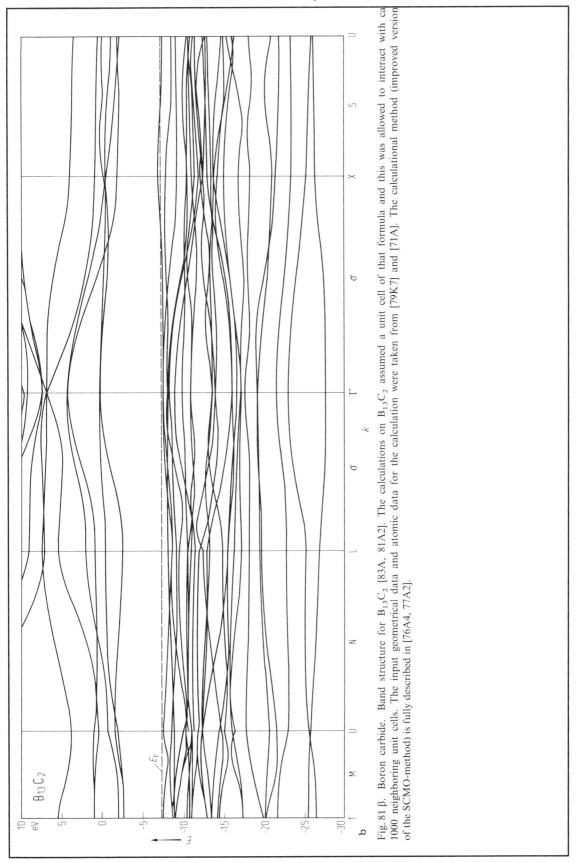

Fig. 81 β. Boron carbide. Band structure for $B_{13}C_2$ [83A, 81A2]. The calculations on $B_{13}C_2$ assumed a unit cell of that formula and this was allowed to interact with ca. 1000 neighboring unit cells. The input geometrical data and atomic data for the calculation were taken from [79K7] and [71A]. The calculational method (improved version of the SCMO-method) is fully described in [76A4, 77A2].

9.14 Boron compounds

Boron carbide

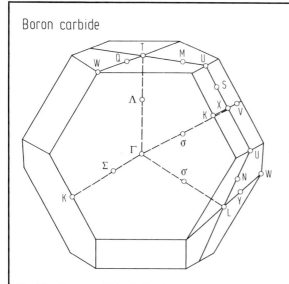

Fig. 81 α. Boron carbide. Brillouin zone (rhombohedral lattice) [81A2, 83A].

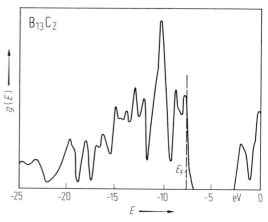

Fig. 81 γ. Boron carbide. Total electron density of states for $B_{13}C_2$ [81A2, 83A].

For Fig. 82, see next page.

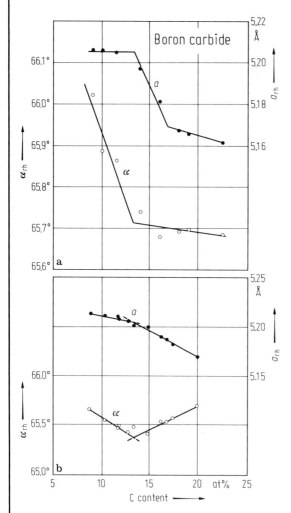

Fig. 83. Boron carbide. Valence electron density distribution in the B(3), C, B(1), B(2) plane (B(1) and B(2) are icosahedral atoms, B(3) and C are atoms of the C–B–C chain. Levels are at $0.1\ e\ \text{Å}^{-3}$, negative contours are dotted [79K8].

◀

Fig. 81. Boron carbide. Room temperature rhombohedral lattice parameters vs. carbon content a) according to [75Y] (see also [71K1]) (expected errors in the compositions: ±1%); b) according to [81B1].

9.14 Boron compounds

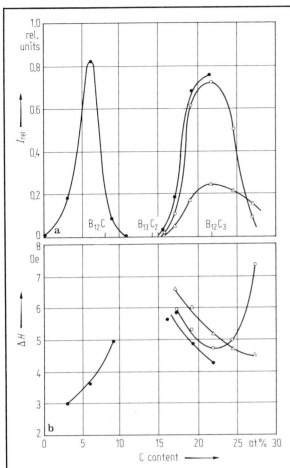

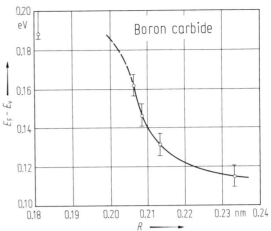

Fig. 82. Boron-carbon system. EPR results of powders (particle size < 50 μm) [80V]. a) relative line intensity vs. C content. b) line width vs. C content. Data of different samples.

Fig. 84. Boron carbide. Distance between the Fermi level and the valence band edge respectively the mobility edge vs. the medium distance R of the C atoms [80W; 81W6]. A possible contribution of a thermally activated mobility is included into the energy scale. $R = \{(4\pi/3)N\}^{1/3}$ where N is the number of C atoms per cm^3.

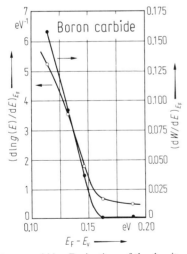

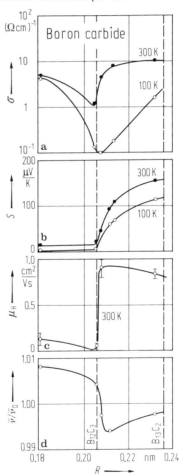

Fig. 85. Boron carbide. Derivatives of the density of states $g(E)$ and of the hopping energy W at the Fermi level vs. the shift of the Fermi level relative to the valence band edge E_v (resp. the mobility edge). This shift includes that the C content affects the density of states and the occupation density as well [80W, 81W6].

Fig. 86. Boron carbide. Transport properties and resonance warenumber of the 1080 cm^{-1} phonon vs. medium distance of the C atoms. The compositions $B_{12}C_3$ and $B_{13}C_2$ are outlined [81W1].

9.14 Boron compounds

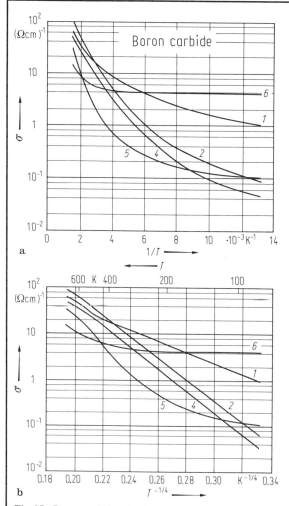

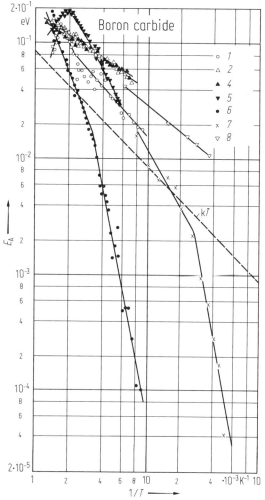

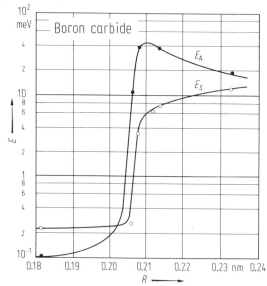

Fig. 87. Boron carbide. a) Electrical conductivity vs. reciprocal temperature. For sample composition, see Fig. 89. b) Electrical conductivity vs. $T^{-1/4}$ (according to Mott's law) [80W, 81W3].

Fig. 89. Boron carbide. Activation energy of the electrical conductivity of boron carbide of different composition vs. reciprocal temperature [80W, 81W6]. (Sample compositions: *1*) $B_{12.94}C_{2.06}$; *2*) $B_{12.35}C_{2.65}$; *3*) $B_{12.28}C_{2.72}$; *4*) $B_{12.13}C_{2.87}$; *5*) $B_{12.01}C_{2.99}$; *6*) $B_{10.56}C_{4.41}$; *7, 8*) [70G2] unknown composition, accordingly analyzed measurements.)

Fig. 88. Boron carbide. Activation energy E_A of the electrical conductivity and the Peltier energy $E_S = e \cdot S \cdot T$ at 100 K vs. medium distance of the C atoms [80W, 81W6].

9.14 Boron compounds

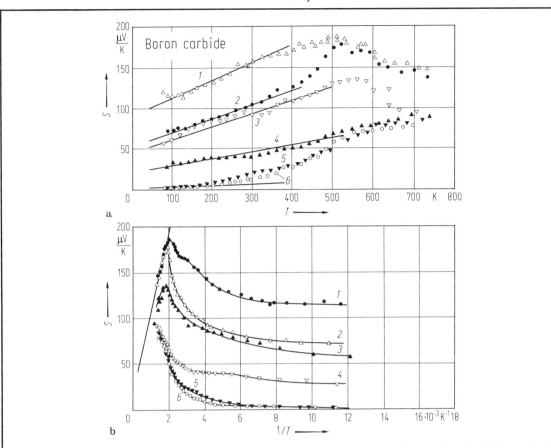

Fig. 90. Boron carbide. a) Thermoelectric power vs. temperature. For sample composition, see Fig. 89. b) Thermoelectric power vs. reciprocal temperature [80W, 81W3].

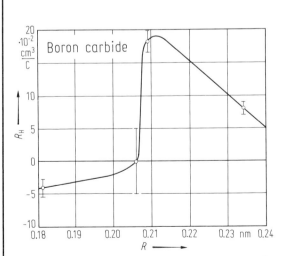

Fig. 91. Boron carbide. Hall coefficient vs. medium distance of the C atoms [81W3].

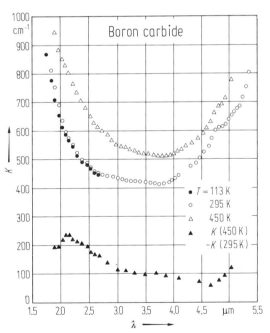

Fig. 92. Boron carbide. (unknown composition; approximately $B_{12}C_3$, polycrystalline samples) Absorption coefficient vs. wavelength [71W2, 74W].

9.14 Boron compounds

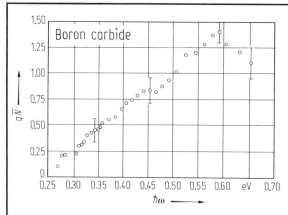

Fig. 93. Boron carbide. (unknown composition; approximately $B_{12}C_3$) Photoabsorption at 113 K. Optical excitation spectrum of optically excited and trapped electrons into the conduction band, absorption cross section q, times medium concentration of absorption centers per unit area, $\bar{N}$, vs. photon energy [71W2, 74W].

Fig. 94. Boron carbide. (unknown composition; approximately $B_{12}C_3$) Concentration of unoccupied traps at 113 K; dependence on the time of photo-excitation, measured by optical absorption at $\lambda = 3.25$ μm [71W2].

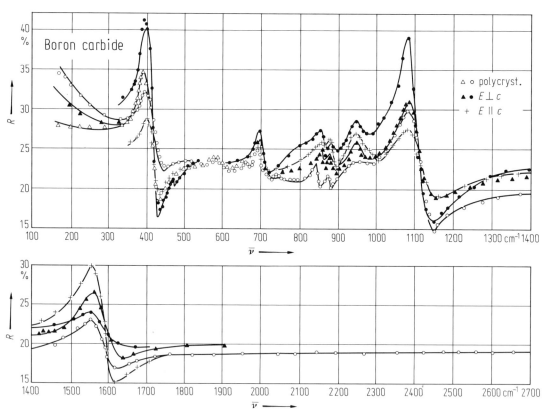

Fig. 95. Boron carbide. (unknown composition; approximately $B_{12}C_3$) IR reflectivity vs. wavenumber. The sample dependent increase of R to wavenumbers < 350 cm^{-1} is attributed to the plasma resonance of free carriers [79B4].

9.14 Boron compounds

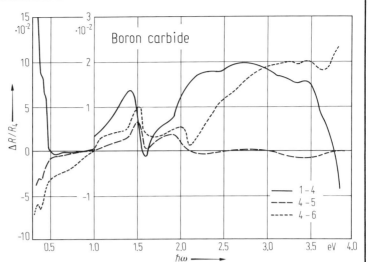

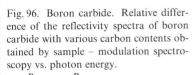

Fig. 96. Boron carbide. Relative difference of the reflectivity spectra of boron carbide with various carbon contents obtained by sample – modulation spectroscopy vs. photon energy.
—: $R_{sample\,1} - R_{sample\,4}$,
– – –: $R_{sample\,4} - R_{sample\,5}$,
·····: $R_{sample\,4} - R_{sample\,6}$.
For the composition of samples 1,4,5,6, see Fig. 86. The zero point is arbitrarily chosen, hence a relatively higher reflectivity of the more boron-rich samples is indicated by more positive values, a relatively higher reflectivity of more carbon-rich samples by more negative values [79W3].

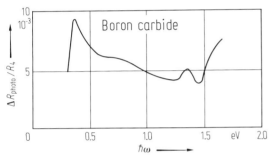

Fig. 97. Boron carbide. Light-induced change of the differential reflectivity of two samples with different C contents $\left(\dfrac{(R_1 - R_4)_{photo}}{R_4}\right)$ vs. photon energy. For composition of the samples 1 and 4, see Fig. 89 [79W3]. R_1: $R_{sample\,1}$, R_4: $R_{sample\,4}$.

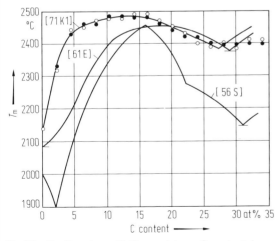

Fig. 99. B–C system. Melting point vs. C content (●○: different experimental arrangements) [71K1]. Data from [61E1] and [56S1] for comparison.

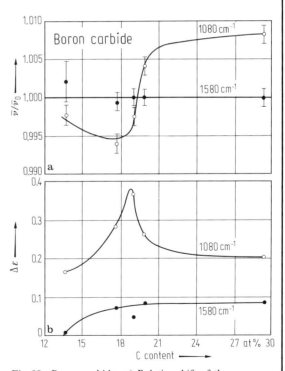

Fig. 98. Boron carbide. a) Relative shift of the resonance wavenumber of IR-active phonons at RT ($1080\,\text{cm}^{-1}$ phonon: split icosahedral mode; $1580\,\text{cm}^{-1}$ phonon: vibration of the C–B–C chain) vs. carbon content [82W] (cp. also [81B1, 81B2]). b) Oscillor strength of the phonons vs. carbon content [82W].

9.14 Boron compounds

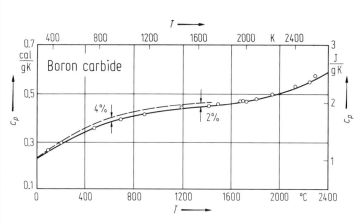

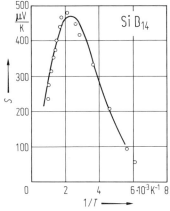

Fig. 100. Boron carbide. Specific heat capacity vs. temperature [72V]. Dashed line [60K2].

Fig. 101. SiB$_{14}$. Electrical conductivity vs. reciprocal temperature [79P1].

Fig. 102. SiB$_{14}$. Thermoelectric power vs. reciprocal temperature [79P1].

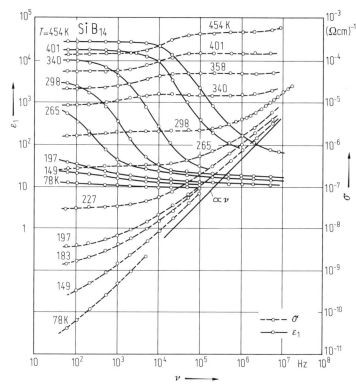

Fig. 103. SiB$_{14}$. Conductivity and dielectric constant of polycrystals at various temperatures vs. frequency [78P1].

For Fig. 104, see next page.

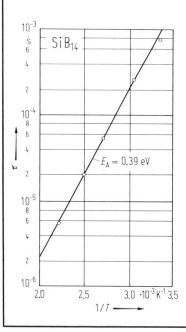

Fig. 105. SiB$_{14}$. Relaxation time derived from Cole-Cole diagrams vs. reciprocal temperature [79P1].

9.14 Boron compounds

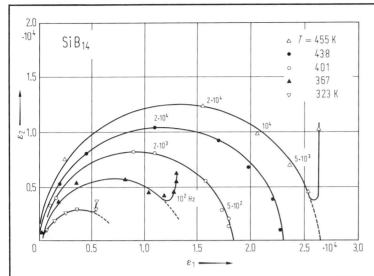

Fig. 104. SiB$_{14}$. Cole-Cole diagram. ε_2 vs. ε_1 at various temperatures [79P1]. The imaginary part ε_2 of the dielectric constant is plotted versus the real part ε_1 with the frequency as parameter. In case of only one relaxation time a half-circle results with its center at $(\varepsilon(0)+\varepsilon(\infty))/2$ on the ε_1 axis. $\varepsilon(0)$ and $\varepsilon(\infty)$ can be taken from the zero crossings of the circle. The maximum value of the half-circle is reached at $\omega=1/\tau$ (see Fig. 105).

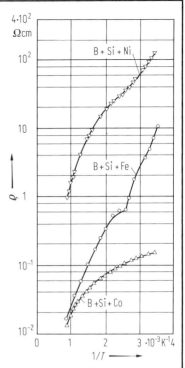

Fig. 106. Doped B–Si compounds. Resistivity vs. reciprocal temperature (Fe, Co, Ni as doping materials) [81D].

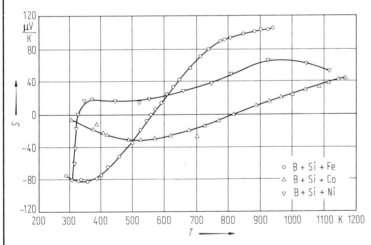

Fig. 107. Doped B–Si compounds. Thermoelectric power vs. temperature (Fe, Co, Ni as doping materials) [81D].

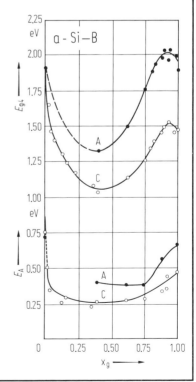

Fig. 108. a-Si–B alloys. Optical energy gap E_{g4} (photon energy at $K=10^4$ cm^{-1}) and electrical activation energy of anode (A) and cathode (C) deposited films (deposition temperature: 270°C) vs. the B fraction in plasma x_g [79T1].

9.14 Boron compounds

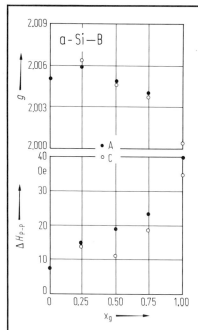

Fig. 109. a-Si—B alloys. g-value and peak-to-peak linewidth of ESR signals of samples deposited at 270°C vs. atomic B fraction in plasma x_g [79T1].

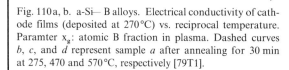

Fig. 110a, b. a-Si—B alloys. Electrical conductivity of cathode films (deposited at 270°C) vs. reciprocal temperature. Paramter x_g: atomic B fraction in plasma. Dashed curves b, c, and d represent sample a after annealing for 30 min at 275, 470 and 570°C, respectively [79T1].

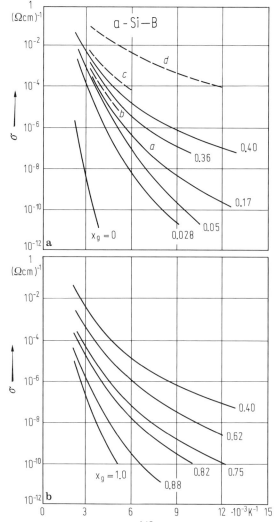

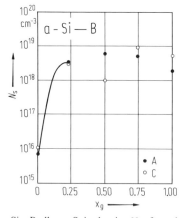

Fig. 111. a-Si—B alloys. Spin density N_s of anode (A) and cathode (C) deposited films (deposition temperature: 270°C) vs. B fraction in plasma x_g [79T1].

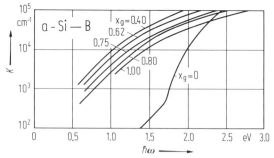

Fig. 112. a-Si—B alloys. Optical absorption edge of cathode deposited films (deposition temperature: 270°C) with different B fraction in plasma, x_g. Absorption coefficient vs. photon energy. For further results, see [79T1].

9.14 Boron compounds

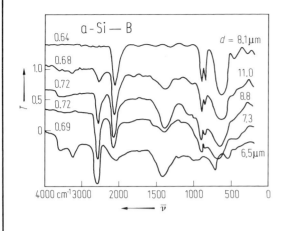

Fig. 113. a-Si—B alloys. IR transmission of anode deposited films (deposition temperature 25°C) vs. wavenumber. Atomic B fractions $x_g = 0$; 0.25; 0.5; 0.75; 1. Film thickness d and the transmission measured at $\bar{\nu} = 4000$ cm^{-1} are given for each curve. For further results, see [79T1].

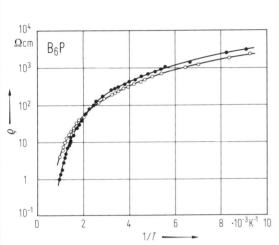

Fig. 114. $B_6P(B_{5.8}P)$. Electrical resistivity vs. reciprocal temperature for two polycrystalline samples with a phosphorus content of 14.7 at % [59G2].

Fig. 116. $B_{12}P_2$. Thermal conductivity of polycrystalline samples vs. temperature [71S3].

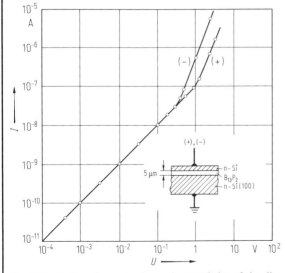

Fig. 115. $B_{13}P_2$. Current-voltage characteristics of the diode with n-type Si/$B_{13}P_2$/n-type Si structure [73T].

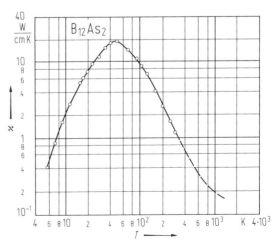

Fig. 117. $B_{12}As_2$. Thermal conductivity vs. temperature [71S3]. Crystal orientation not reported.

9.14 Boron compounds

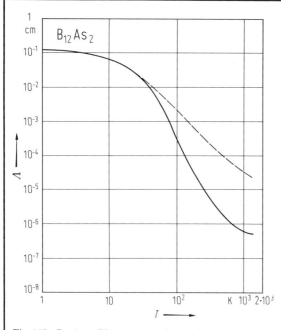

Fig. 118. $B_{12}As_2$. Phonon mean free path vs. temperature derived from the heat capacity. The dashed line represents the behavior of the acoustic phonons only and corresponds to the mean free path of the actual carriers with thermal energy [71S3].

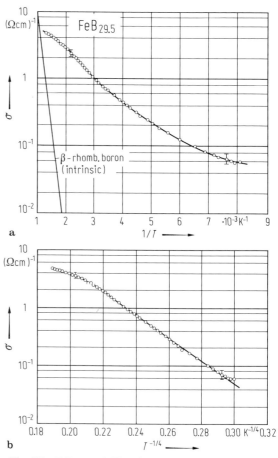

Fig. 120. $FeB_{29.5}$. a) Electrical conductivity vs. reciprocal temperature. Curve of pure β-rhombohedral boron is given for comparison. b) Electrical conductivity vs. $T^{-1/4}$ (according to Mott's law). From [81W1].

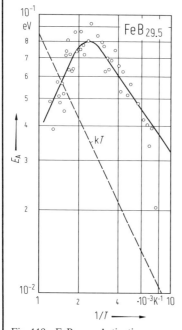

Fig. 119. $FeB_{29.5}$. Activation energy of the electrical conductivity vs. reciprocal temperature [81W1].

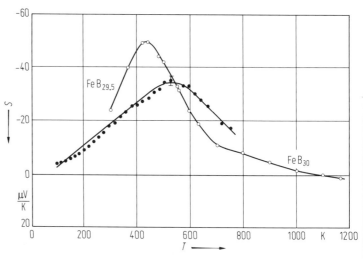

Fig. 121. FeB_x. Thermoelectric power vs. temperature. $FeB_{29.5}$: [81W1]. FeB_{30} (no hint on the structure is given by the authors): [81D].

9.15.1.1 $T_n(III)_{2n-m}$ and $T_n(IV)_{2n-m}$ compounds

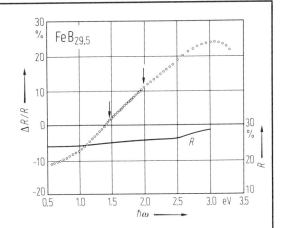

Fig. 122. FeB$_{29.5}$. Relative difference of the reflectivity spectra of pure β-rhombohedral boron and FeB$_{29.5}$ obtained by sample-modulation spectroscopy vs. photon energy (solid curve: reflectivity of pure β-rhombohedral boron for comparison). The zero point is arbitrarily chosen, hence a relatively higher reflectivity of pure boron is indicated by a more positive value, a relatively higher reflectivity of FeB$_{29.5}$ by a more negative value. – The arrows indicate kinks in the curve [81W1].

9.15 Binary transition-metal compounds
9.15.1 Compounds with elements of the IIId and IVth groups
9.15.1.1 $T_n(III)_{2n-m}$ and $T_n(IV)_{2n-m}$ compounds (For tables, see p. 63 ff.)

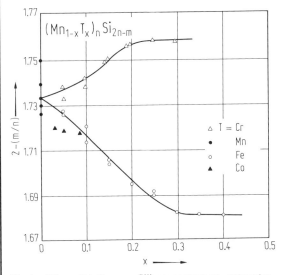

Fig. 1. $(Mn_{1-x}T_x)_nSi_{2n-m}$. Silicon content vs. concentration of substitutional transition metals [68F].

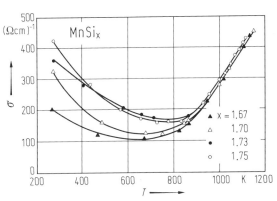

Fig. 2. MnSi$_x$. Electrical conductivity vs. temperature for polycrystalline samples [72U].

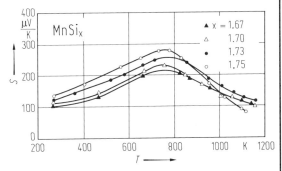

Fig. 3. MnSi$_x$. Thermoelectric power vs. temperature for polycrystalline samples [72U].

9.15.1.1 $T_n(III)_{2n-m}$ and $T_n(IV)_{2n-m}$ compounds

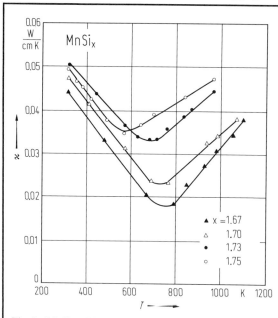

Fig. 4. $MnSi_x$. Thermal conductivity vs. temperature for polycrystalline samples [72U].

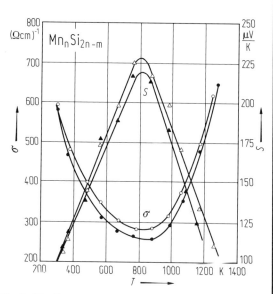

Fig. 5. Mn_nSi_{2n-m}. Electrical conductivity and Seebeck coefficient vs. temperature for polycrystalline samples [61K]. Open symbols: $MnSi_{1.735}$ (47 wt% Si); full symbols: $MnSi_{1.770}$ (47.5 wt% Si).

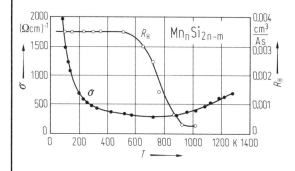

Fig. 6. Mn_nSi_{2n-m} ($Mn_{11}Si_{19}$?). Electrical conductivity σ and Hall coefficient R_H vs. temperature for p-type single crystals of unknown orientation [69N1, 69N2].

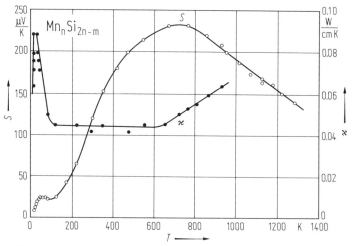

Fig. 7. Mn_nSi_{2n-m} ($Mn_{11}Si_{19}$?). Thermoelectric power S and thermal conductivity κ vs. temperature for p-type single crystals of unknown orientation [69N1, 69N2].

9.15.1.1 $T_n(III)_{2n-m}$ and $T_n(IV)_{2n-m}$ compounds

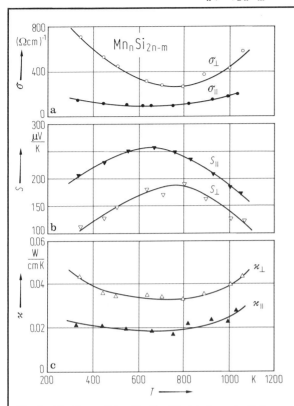

Fig. 8. Mn_nSi_{2n-m} ($Mn_{11}Si_{19}$?). Electrical conductivity σ (a), thermoelectric power S (b) and thermal conductivity κ (c) [69I]. $\parallel$: measured along [001], $\perp$: measured along [100].

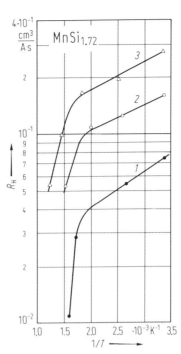

Fig. 10. $MnSi_{1.72}$. Hall coefficient vs. reciprocal temperature [69I]. *1*: magnetic field strength H parallel to [001], *2*: current I along [001], *3*: H and I perpendicular to [001].

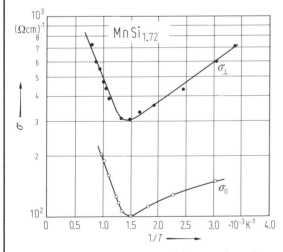

Fig. 9. $MnSi_{1.72}$. Electrical conductivity vs. reciprocal temperature [69I]. $\sigma_\parallel$: current I along [001], $\sigma_\perp$: current I along [100].

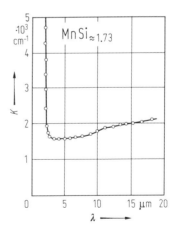

Fig. 11. $MnSi_{\approx 1.73}$. Absorption coefficient vs. wavelength [79Z]. Reflection spectra parallel to a fourfold axis.

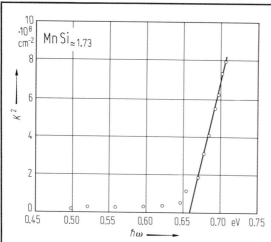

Fig. 12. $MnSi_{\approx 1.73}$. Square of the absorption coefficient vs. photon energy [79Z].

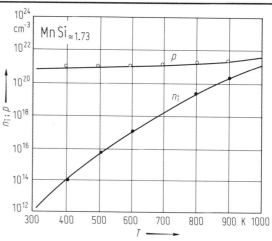

Fig. 13. $MnSi_{\approx 1.73}$. Concentrations of holes ($p_0 + p_i$) and electrons (n_i) vs. temperature [71N], n_i, p_i: intrinsic concentrations, $p_0 = 1/eR_H$ (impurity region).

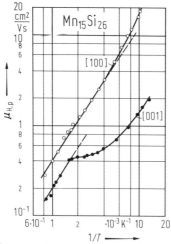

Fig. 14. $Mn_{15}Si_{26}$. Hall mobilities of holes $\mu_{H,p} = R_H/\varrho$ in [100] and [001] direction vs. reciprocal temperature in a doubly-logarithmic scale [81K]. The straight lines correspond to a $T^{-3/2}$ law (see tables).

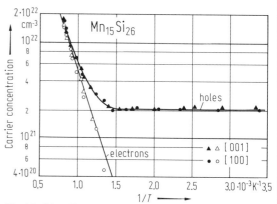

Fig. 15. $Mn_{15}Si_{26}$. Charge-carrier concentrations from resistivity and Hall coefficient vs. reciprocal temperature [81K].

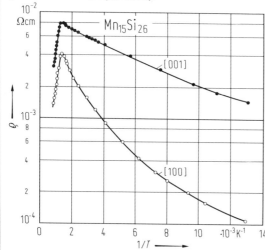

Fig. 16. $Mn_{15}Si_{26}$. Resistivity vs. reciprocal temperature [81K]. $\varrho^{[100]}$ and $\varrho^{[001]}$: measured along the a- and along the c-direction, respectively.

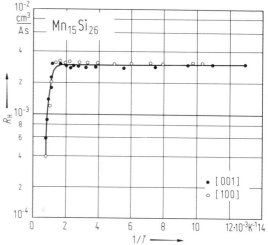

Fig. 17. $Mn_{15}Si_{26}$. Hall coefficient vs. reciprocal temperature [81K].

9.15.1.1 $T_n(III)_{2n-m}$ and $T_n(IV)_{2n-m}$ compounds

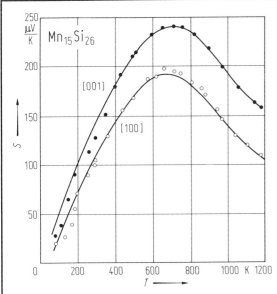

Fig. 18. $Mn_{15}Si_{26}$. Thermoelectric power vs. temperature [81K]. $S^{[100]}$ and $S^{[001]}$: measured with the temperature gradient along a and c, respectively. Solid curves are calculated with the formula for degenerate p-type semiconductors.

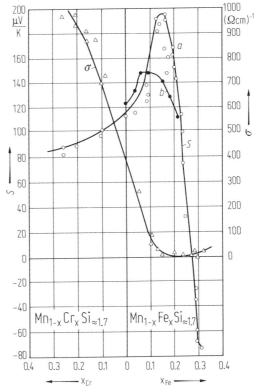

Fig. 19. $Mn_{1-x}Cr_xSi_{\approx1.7}$ and $Mn_{1-x}Fe_xSi_{\approx1.7}$. Electrical conductivity σ and thermoelectric power S vs. concentration x [71N]. a: S for samples with variable Si contents, b: S for $Mn_{1-x}Fe_xSi_{1.73}$.

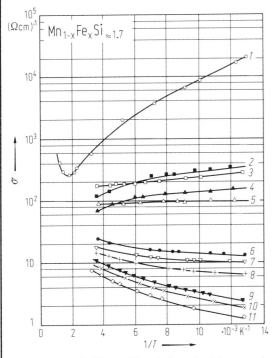

Fig. 20. $Mn_{1-x}Fe_xSi_{\approx1.7}$. Electrical conductivity vs. reciprocal temperature [75Z]. 1: x=0, 2: x=0.1, 3: x=0.15, 4: x=0.17, 5: x=0.21, 6: x=0.23, 7: x=0.24, 8: x=0.25, 9: x=0.28, 10: x=0.29, 11: x=0.30.

Fig. 21. $Mn_{1-x}Fe_xSi_{\approx1.72}$. Electrical conductivity vs. reciprocal temperature for oriented single crystals [74A1].

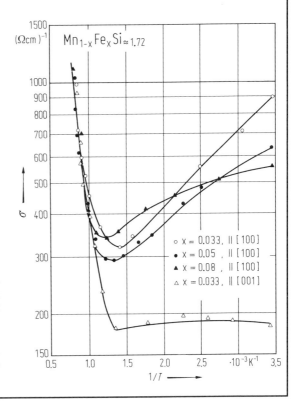

9.15.1.2 T(IV)$_2$ compounds

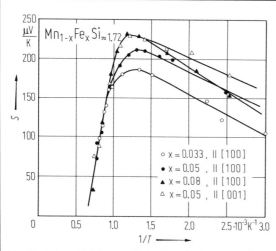

Fig. 22. Mn$_{1-x}$Fe$_x$Si$_{\approx 1.72}$. Seebeck coefficient vs. reciprocal temperature for oriented single crystals [74A1].

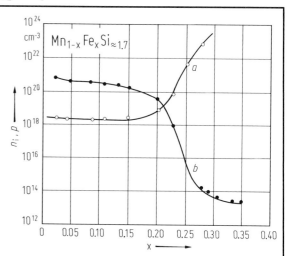

Fig. 23. Mn$_{1-x}$Fe$_x$Si$_{\approx 1.7}$. Concentrations of electrons (a) and holes (b) vs. concentration x, as derived from the data of Fig. 19 [71N].

9.15.1.2 T(IV)$_2$ compounds (For tables, see p. 72ff.)

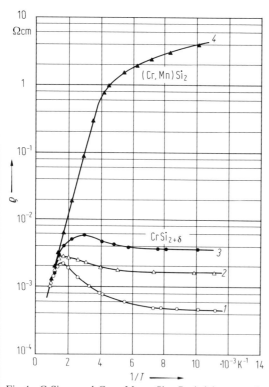

Fig. 1. CrSi$_{2+\delta}$ and Cr$_{0.88}$Mn$_{0.12}$Si$_2$. Resistivity vs. reciprocal temperature [64S]. 1: CrSi$_2$ (single crystal of unknown orientation), 2: CrSi$_{2.01}$ (polycrystalline), 3: CrSi$_{2.02}$ (polycrystalline), 4: Cr$_{0.88}$Mn$_{0.12}$Si$_2$ (crystal grown by the Bridgman technique).

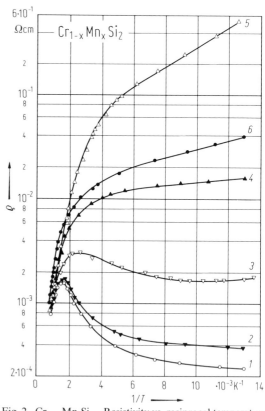

Fig. 2. Cr$_{1-x}$Mn$_x$Si$_2$. Resistivity vs. reciprocal temperature for hot-pressed samples annealed at 1373 K [78N]. Mn concentration (in parentheses) in 10^{21} cm^{-3}. 1: x=0, 2: x=0.059 (1.65), 3: x=0.100 (2.80), 4: x=0.115 (3.23), 5: x=0.142 (3.99), 6: x=0.182 (5.12).

9.15.1.2 T(IV)$_2$ compounds

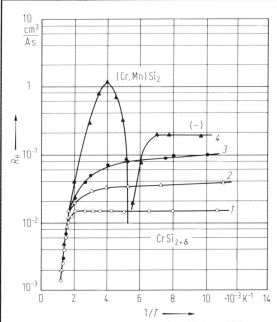

Fig. 3. CrSi$_{2+\delta}$ and Cr$_{0.88}$Mn$_{0.12}$Si$_2$. Hall coefficient vs. reciprocal temperature [64S]. Symbols as in Fig. 1.

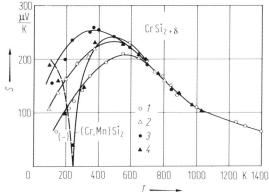

Fig. 5. CrSi$_{2+\delta}$ and Cr$_{0.88}$Mn$_{0.12}$Si$_2$. Thermoelectric power vs. temperature [64S]. The solid curves are calculated. Symbols as in Fig. 1.

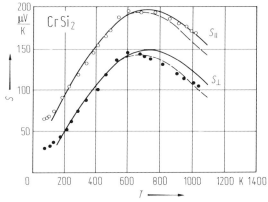

Fig. 7. CrSi$_2$. Anisotropic Seebeck coefficient vs. temperature [72N]. The solid curves are calculated with the mobility ratio $b=0$, while the dotted curves are calculated with $b=0.01$. Compare [72V].

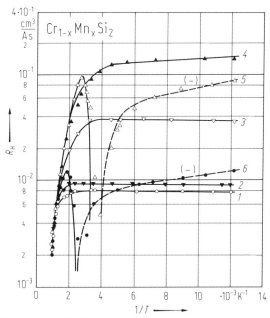

Fig. 4. Cr$_{1-x}$Mn$_x$Si$_2$. Hall coefficient vs. reciprocal temperature [78N]. Mn concentrations as in Fig. 2.

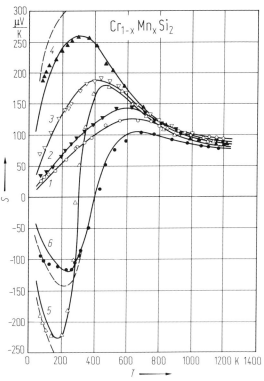

Fig. 6. Cr$_{1-x}$Mn$_x$Si$_2$. Seebeck coefficient S vs. temperature [78N]. Mn concentrations as in Fig. 2. Solid and dashed curves are calculated for acoustic lattice scattering and for impurity ion scattering, respectively.

9.15.1.2 T(IV)$_2$ compounds

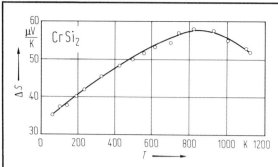

Fig. 8. CrSi$_2$. Anisotropy of the thermoelectric power vs. temperature [72N]. The sample contained 66.43% Si. $\Delta S = S_\parallel - S_\perp$.

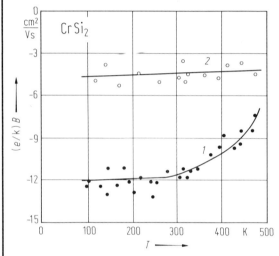

Fig. 9. CrSi$_2$. Nernst coefficient B vs. temperature [70K]. Curve 1: B_{aca}, 2: B_{aac}.

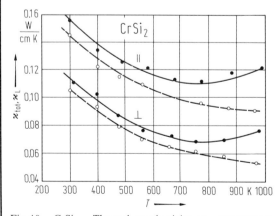

Fig. 10. CrSi$_2$. Thermal conductivity vs. temperature [72V]. Full curves: total thermal conductivity; broken curves: lattice contribution. $\parallel$: parallel to the c axis; $\perp$: perpendicular to the c axis.

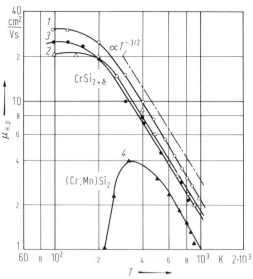

Fig. 11. CrSi$_{2+\delta}$ and Cr$_{0.88}$Mn$_{0.12}$Si$_2$. Hall mobility of holes vs. temperature in a doubly-logarithmic scale [64S]. Symbols as in Fig. 1.

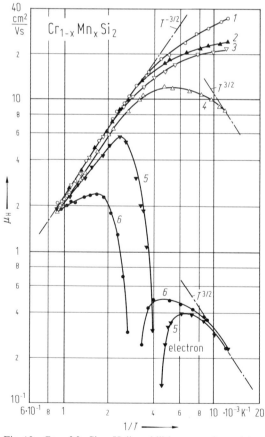

Fig. 12. Cr$_{1-x}$Mn$_x$Si$_2$. Hall mobilities vs. reciprocal temperature in a doubly-logarithmic scale [78N]. Mn concentrations as in Fig. 2. Samples are hot-pressed at 1473 K and annealed at 1373 K.

9.15.1.2 T(IV)$_2$ compounds

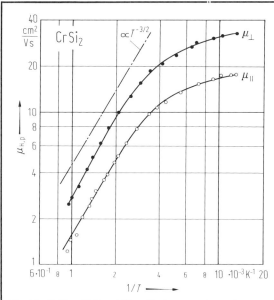

Fig. 13. CrSi$_2$. Hall mobility of holes vs. reciprocal temperature in a doubly-logarithmic scale [72N]. $\mu_\parallel$: along the c axis, $\mu_\perp$: along the a axis.

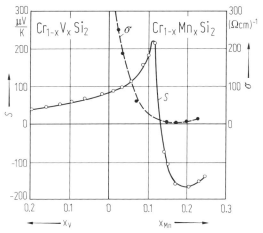

Fig. 15. Cr$_{1-x}$V$_x$Si$_2$ and Cr$_{1-x}$Mn$_x$Si$_2$. Electrical conductivity σ and thermoelectric power S vs. concentration [73N1]. $T = 295$ K, polycrystalline samples.

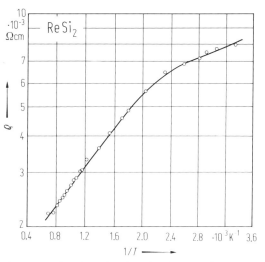

Fig. 16. ReSi$_2$. Logarithmic resistivity vs. reciprocal temperature for a polycrystalline sample [65N].

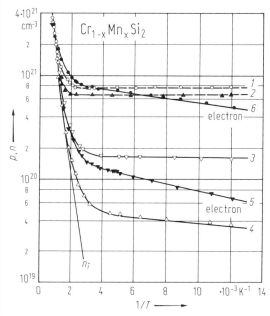

Fig. 14. Cr$_{1-x}$Mn$_x$Si$_2$. Carrier concentrations vs. reciprocal temperature [78N]. Mn concentrations as in Fig. 2. Curves 1, 2, 3, 4: holes; 5, 6: electrons.

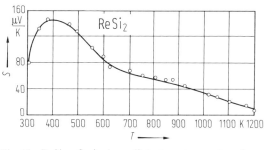

Fig. 17. ReSi$_2$. Seebeck coefficient vs. temperature for a polycrystalline sample [65N].

9.15.1.2 T(IV)$_2$ compounds

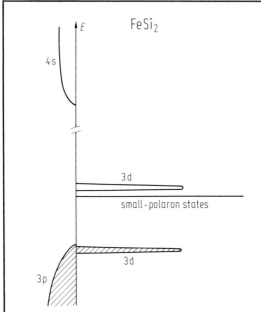

Fig. 18. FeSi$_2$. Energy-band scheme derived from optical measurements [70B1].

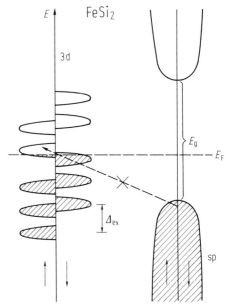

Fig. 19. FeSi$_2$. Schematic representation of the valence and conduction band and the localized 3d levels [73B]. Δ_{ex} is the exchange-splitting parameter. Arrows indicate spin-up and spin-down levels. $E_g \approx 0.85$ eV for the "vertical transition". The transition from the (s, p) valence band to the lowest unoccupied 3d level is crossed because of its negligible probability.

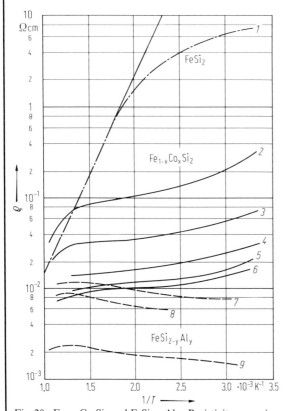

Fig. 20. Fe$_{1-x}$Co$_x$Si$_2$ and FeSi$_{2-y}$Al$_y$. Resistivity vs. reciprocal temperature [64W]. Samples quenched from the liquid state and annealed at 1173 K. 1: x=y=0, 2: x=0.005, 3: x=0.01, 4: x=0.02, 5: x=0.03, 6: x=0.04, 7: y=0.02, 8: y=0.04, 9: y=0.08.

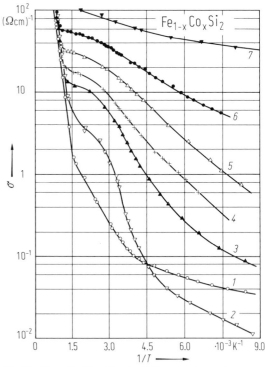

Fig. 21. Fe$_{1-x}$Co$_x$Si$_2$. Conductivity vs. reciprocal temperature [68B1]. Samples prepared by sintering at 1453 K and annealing at 1153 K. All samples n-type. Co concentration (in parentheses) in 10^{20} cm^{-3}. 1: x=y=0, 2: x=0.001, 3: x=0.0028, 4: x=0.005 (1.31), 5: x=0.01 (2.63), 6: x=0.029 (7.6), 7: x=0.060 (15.8).

9.15.1.2 T(IV)$_2$ compounds

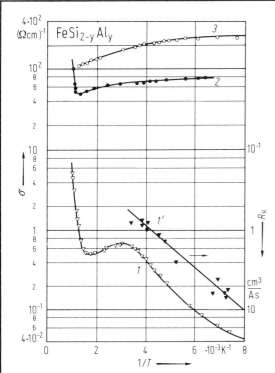

Fig. 22. FeSi$_{2-y}$Al$_y$. Conductivity (curves *1*, *2*, *3*) and Hall coefficient (curve *1'*) vs. reciprocal temperature [68B1]. All samples p-type; prepared by sintering at 1453 K and annealing at 1153 K *1*, *1'*: y = 0.04 (1.04·10^{21} Al atoms/cm^3), *2*: y = 0.08 (2.08·10^{21} Al atoms/cm^3), *3*: y = 0.08.

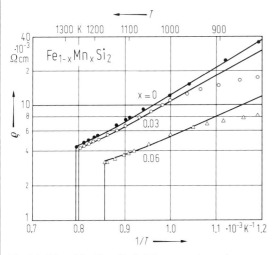

Fig. 24. Fe$_{1-x}$Mn$_x$Si$_2$. Resistivity vs. reciprocal temperature in the intrinsic region for hot-pressed annealed samples [73N2]. The solid curves represent the fitted calculated values.

Fig. 25. Fe$_{1-x}$Co$_x$Si$_2$. Hall coefficient and carrier concentration vs. reciprocal temperature [68B1]. All samples n-type; prepared by sintering at 1453 K and annealing at 1153 K.

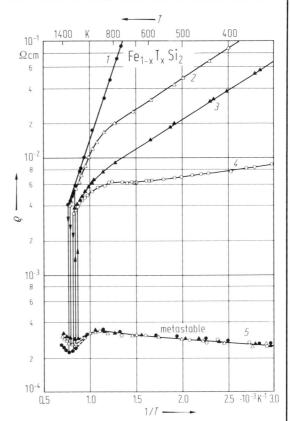

Fig. 23. Fe$_{1-x}$Mn$_x$Si$_2$ and Fe$_{0.95}$Co$_{0.05}$Si$_2$. Resistivity vs. reciprocal temperature [73N2]. The heating and cooling transition directions are indicated by arrows. *1*: FeSi$_2$ (n-type), *2*: Fe$_{0.97}$Mn$_{0.03}$Si$_2$ (p-type), *3*: Fe$_{0.94}$Mn$_{0.06}$Si$_2$ (p-type), *4*: Fe$_{0.95}$Co$_{0.05}$Si$_2$ (n-type), *5*: metastable inhomogeneous phase. The samples were prepared by hot-pressing at 1423 K and annealing at 1073 K. Curve *5* was measured after quenching the samples from ≈1400 K. These metastable and inhomogeneous samples were composed of cubic FeSi and tetragonal Fe$_{≈0.8}$Si$_2$ (incorrectly named α-FeSi$_2$, as if it represented the high-temperature modification of orthorhombic "β-FeSi$_2$". Obviously the transformation back to FeSi$_2$ is rather fast).

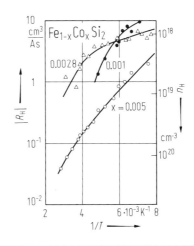

9.15.1.2 T(IV)₂ compounds

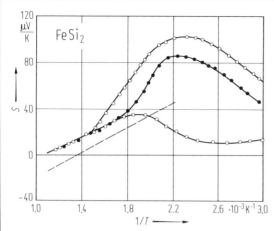

Fig. 26. FeSi$_2$. Thermoelectric power vs. reciprocal temperature [64W]. Reference element: platinum. The three samples differed slightly in impurity content. The samples were prepared by rapidly freezing the melt in 3 mm silica capillaries (to ensure a fine dispersion of FeSi in the FeSi–α-FeSi$_2$ eutectic) and annealing at 1173 K. $|R_H| < 1 \, \text{cm}^3 \, \text{C}^{-1}$. Unknown impurity content (compare Fig. 29). Corrected absolute values for the intrinsic region are indicated by a dashed line.

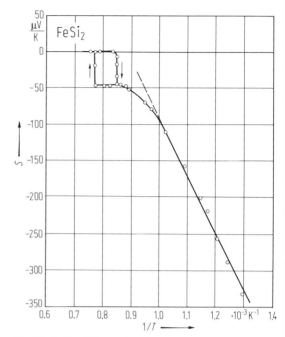

Fig. 27. FeSi$_2$. Thermoelectric power vs. reciprocal temperature in the intrinsic and phase transition range for sintered material [70B1].

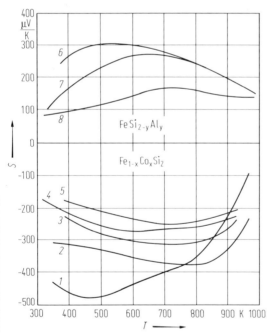

Fig. 28. Fe$_{1-x}$Co$_x$Si$_2$ and FeSi$_{2-y}$Al$_y$. Thermoelectric power vs. temperature for quenched samples annealed at 1173 K [64W]. 1: x=0.005, 2: x=0.01, 3: x=0.02, 4: x=0.03, 5: x=0.04, 6: y=0.02, 7: y=0.04, 8: y=0.08.

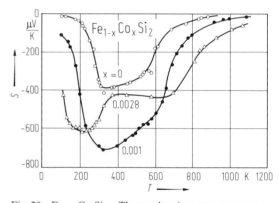

Fig. 29. Fe$_{1-x}$Co$_x$Si$_2$. Thermoelectric power vs. temperature for weakly doped n-type samples [68B1]. Samples sintered at 1453 K and annealed at 1153 K.

Fig. 30. Fe$_{1-x}$Co$_x$Si$_2$. Thermoelectric power vs. temperature for heavily doped n-type samples [68B1]. Samples sintered at 1453 K and annealed at 1153 K.

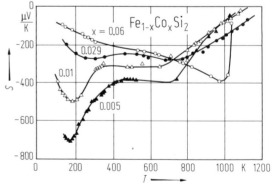

9.15.1.2 T(IV)$_2$ compounds

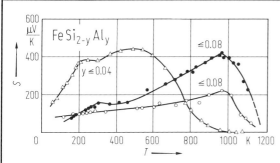

Fig. 31. FeSi$_{2-y}$Al$_y$. Thermoelectric power vs. temperature for p-type samples [68B1]. Samples sintered at 1453 K and annealed at 1153 K. Al loss due to evaporation during melting and sintering unknown.

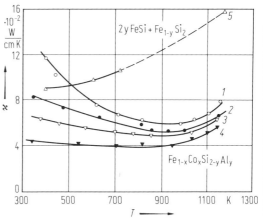

Fig. 33. Fe$_{1-x}$Co$_x$Si$_{2-y}$Al$_y$. Thermal conductivity vs. temperature for sintered samples [73W]. Data of the metastable inhomogeneous high-temperature phase for comparison (Curve 5). Curve 1: x=0, y=0; 2: x=0, y=0.04; 3: x=0.03, y=0; 4: x=0.06, y=0.

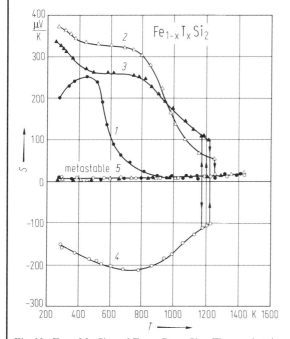

Fig. 32. Fe$_{1-x}$Mn$_x$Si$_2$ and Fe$_{0.95}$Co$_{0.05}$Si$_2$. Thermoelectric power vs. temperature for hot-pressed and annealed samples [73N2]. Heating and cooling transition directions are indicated by arrows. 1: x=0 (p-type), 2: x=0.03 (p-type), 3: x=0.06 (p-type), 4: Fe$_{0.95}$Co$_{0.05}$Si$_2$ (n-type), 5: metastable FeSi/Fe$_{\approx 0.8}$Si$_2$ mixture.

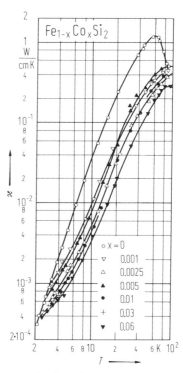

Fig. 34. Fe$_{1-x}$Co$_x$Si$_2$. Thermal conductivity vs. temperature in a doubly-logarithmic scale; sintered samples [73W].

9.15.1.2 T(IV)₂ compounds

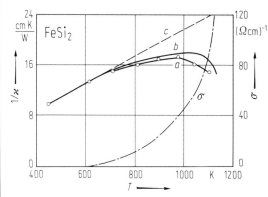

Fig. 35. FeSi₂. Thermal resistivity vs. temperature for sintered samples [73W]. Curve a measured, b calculated with the formula of Price for the ambipolar part.

$$\kappa_{\text{ambip.}} = \left(\frac{k}{e}\right)^2 T \left\{ \sigma_n \delta_n + \sigma_p \delta_p + \frac{\sigma_n \sigma_p}{\sigma} \left(\frac{E_g(0) - \beta T}{kT} + \delta_n + \delta_p \right)^2 \right\},$$

where δ_n, δ_p are scattering parameters. Polar optical scattering with $\delta_n = \delta_p \approx 2.5$ was assumed. Since $\mu_n \approx \mu_p$ in FeSi₂, $\sigma_n \approx \sigma_p$ in the intrinsic range. Best fit (curve b) with $E_g(0) = 0.9$ eV and $\beta = 4.5 \cdot 10^{-4}$ eV K⁻¹. c lattice thermal resistivity. Electrical conductivity is also shown.

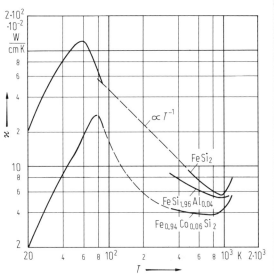

Fig. 36. FeSi₂. Thermal conductivity vs. temperature in a doubly-logarithmic scale [73W]. Sintered samples. Curve a: undoped, b: $2.1 \cdot 10^{21}$ Al atoms/cm³, c: $1.6 \cdot 10^{21}$ Co atoms/cm³.

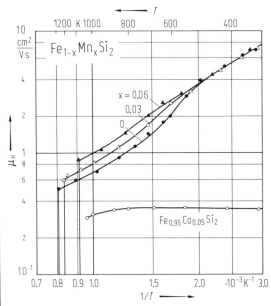

Fig. 37. Fe₁₋ₓMnₓSi₂ and Fe₀.₉₅Co₀.₀₅Si₂. Hall mobility $\mu_H = R_H/\varrho$ vs. reciprocal temperature in a doubly-logarithmic scale [73N2]. Mn-doped samples: $\mu_{H,p}$; Co-doped sample: $\mu_{H,n}$; all samples hot-pressed at ≈ 1400 K and annealed at 1073 K.

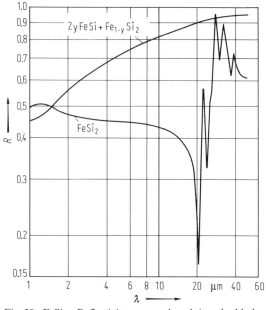

Fig. 38. FeSi₂. Reflectivity vs. wavelength in a doubly-logarithmic scale at RT [70B1, 70B2]. Comparison with the metallic metastable high-temperature phase mixture $2y$FeSi + Fe₁₋ᵧSi₂.

9.15.1.2 T(IV)₂ compounds

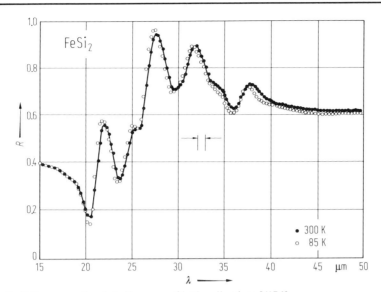

Fig. 39. FeSi₂. Reflectivity vs. wavelength in the range of lattice vibrations [68B2].

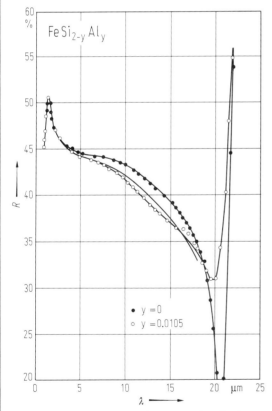

Fig. 40. FeSi$_{2-y}$Al$_y$. Reflectivity vs. wavelength of p-type samples, showing the effect of additional free charge carriers on the absorption and the shift of the reflectivity minimum towards shorter wavelengths (20 μm) [70B2]. Full circles: undoped, open circles: y=0.0105, $\sigma_{300K}=50\,\Omega^{-1}\,cm^{-1}$ (and $S_{300K}=+280\,\mu V\,K^{-1}$). Full line calculated for the Al-doped sample with the Drude theory assuming a relaxation time $\tau_D=6\cdot 10^{-15}$ s.

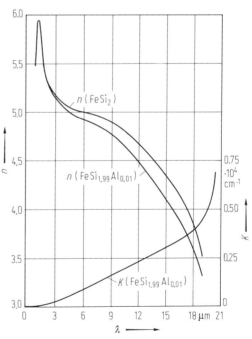

Fig. 41. FeSi₂ and FeSi$_{2-y}$Al$_y$. Refractive index n and absorption coefficient K vs. wavelength [70B2]. Upper curve: Measurements on an undoped p-type FeSi₂ sample ($S=+280\,\mu V\,K^{-1}$). Other curves: Calculated according to Drude with $\tau_D=6\cdot 10^{-15}$ s for a sample with y=0.0105, $\sigma_{300K}=50\,\Omega^{-1}\,cm^{-1}$.

9.15.1.2 T(IV)$_2$ compounds

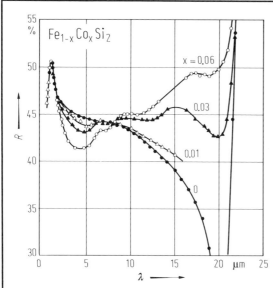

Fig. 42. Fe$_{1-x}$Co$_x$Si$_2$. Reflectivity vs. wavelength [70B2]. No dopant-dependent shift of the 1.3 μm maximum (which corresponds to the fundamental absorption). x=0: σ_{300K} [Ω^{-1} cm^{-1}]=0.3, S_{300K} [μV K^{-1}]= −460; x=0.01: σ_{300K} [Ω^{-1} cm^{-1}]=18, S_{300K} [μV K^{-1}]= −380; x=0.03: σ_{300K} [Ω^{-1} cm^{-1}]=53, S_{300K} [μV K^{-1}]= −230; x=0.06: σ_{300K} [Ω^{-1} cm^{-1}]=110, S_{300K} [μV K^{-1}]= −150.

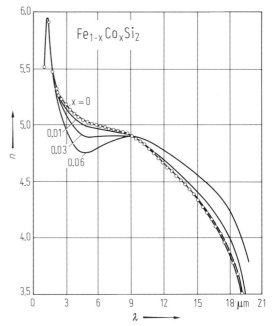

Fig. 43. Fe$_{1-x}$Co$_x$Si$_2$. Refractive index vs. wavelength of n-type samples calculated with Reik's small-polaron theory [70B2]. The same samples as in Fig. 42.

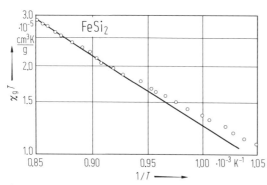

Fig. 44. FeSi$_2$. Product of magnetic mass susceptibility and temperature vs. reciprocal temperature in the intrinsic-conduction range [69B1]. The solid curve is calculated as

$$\chi_g T = \frac{2g\mu_B^2}{d k} e^{-E_g/2kT},$$

with g=density of states, d=density and μ$_B$=Bohr magneton. χ in CGS-emu.

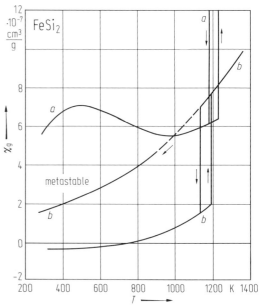

Fig. 45. FeSi$_2$. Mass susceptibility (χ_g in CGS-emu) vs. temperature [69B1, 70B1]. Sintered samples. a: nominal composition 50.15 wt% Si corresponding to FeSi$_2$, b: 55.0 wt% Si corresponding to Fe$_{0.82}$Si$_2$ or FeSi$_2$+0.43 Si. Sample a still contained traces of FeSi even after prolonged annealing. Its high-temperature part of $\chi_g(T)$ as well as the curve for the quenched state are cut off by the frame of the figure. The upper part of curve b represents the intrinsic property of Fe$_{0.8}$Si$_2$ (α-FeSi$_2$) while the lower curve b results from β-FeSi$_2$+0.43 Si, the contribution of the diamagnetic Si precipitations being small.

9.15.1.2 T(IV)$_2$ compounds

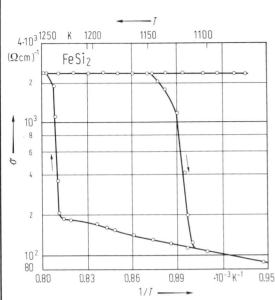

Fig. 46. FeSi$_2$ Electrical conductivity vs. reciprocal temperature [69B2]. Semiconductor→metal transition in a nominally stoichiometric polycrystalline sample (50.15 wt% Si). Comparison with the decomposition product below T_{tr}.

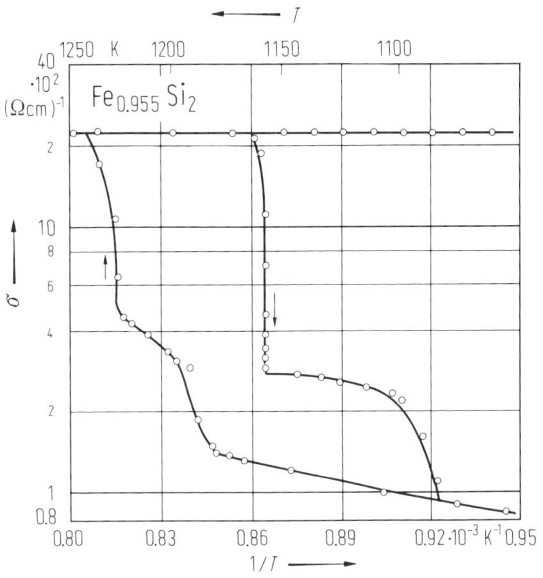

Fig. 47. Fe$_{0.955}$Si$_2$. Electrical conductivity vs. reciprocal temperature [69B2]. Semiconductor→metal transition in an iron-deficient polycrystalline sample (51.3 wt% Si). Comparison with the decomposition product below T_{tr}.

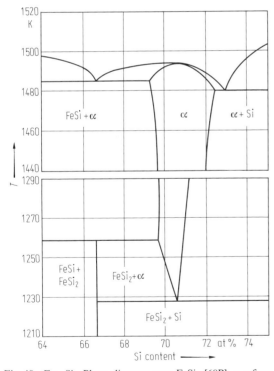

Fig. 48. Fe–Si. Phase diagram near FeSi$_2$ [68P]. α refers to the high-temperature phase Fe$_{\approx 0.8}$Si$_2$ often designated α-FeSi$_2$.

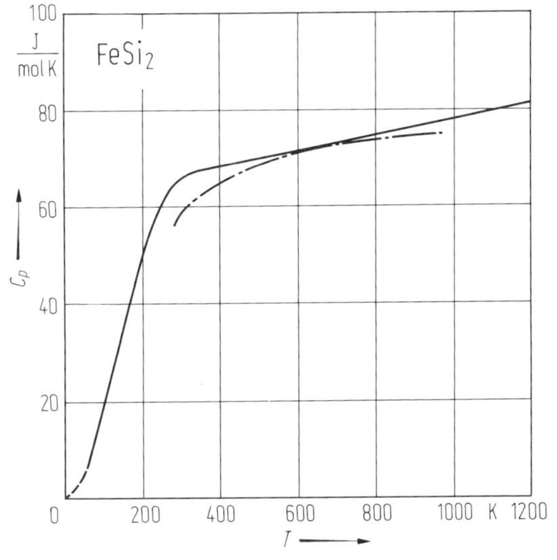

Fig. 49. FeSi$_2$. Heat capacity vs. temperature [73W]. Full curve: measurements of Krentsis, Gel'd and Kalishevich (1963); Broken curve: measurements of Maglic and Parrot (1970).

9.15.1.3 T(V)₂ compounds (For tables, see p. 80ff.)

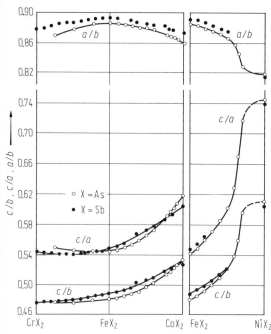

Fig. 1. CrX_2-FeX_2, FeX_2-CoX_2 and FeX_2-NiX_2; X = As, Sb. Axial ratios of the marcasite-type cells vs. composition [74K3].

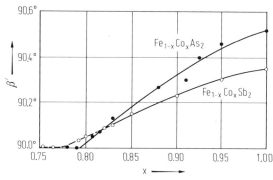

Fig. 2. $Fe_{1-x}Co_xAs_2$ and $Fe_{1-x}Co_xSb_2$. Monoclinic angle β' of the pseudo-marcasite cell vs. compositional parameter x [74K3].

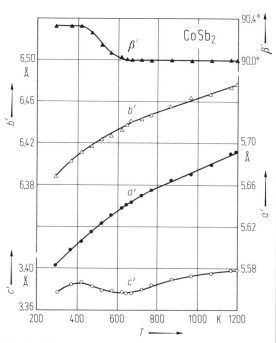

Fig. 4. $CoSb_2$. Lattice parameters vs. temperature [77K3]. Pseudo-marcasite cell (cf. caption of Fig. 3). $CoSb_2$ decomposes at 1204 K.

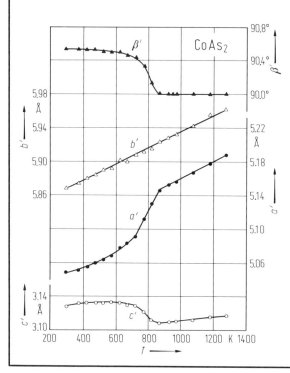

Fig. 3. $CoAs_2$. Lattice parameters vs. temperature [77K3]. Pseudo-marcasite cell a', b', c', β', with $\mathbf{a'}=(\mathbf{a}-\mathbf{c})/2$, $\mathbf{b'}=\mathbf{b}$, and $\mathbf{c'}=(\mathbf{a}+\mathbf{c})/2$, where a, b, c, β refer to the true arsenopyrite-type cell. Linear temperature dependence of the unit cell volume from 300 to 1300 K.

9.15.1.3 T(V)$_2$ compounds

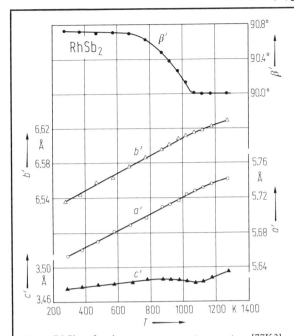

Fig. 5. RhSb$_2$. Lattice parameters vs. temperature [77K3]. The unit-cell volume varies linearly with temperature. Cf. caption of Fig. 3 for parameters a', b', c', β'.

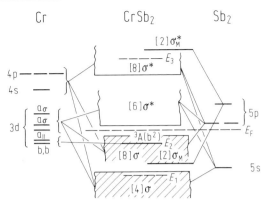

Fig. 7. CrSb$_2$. Energy-level scheme for the loellingite-type antiferromagnet. The two non-bonding d electrons per Cr^{4+} ion occupy a localized ^{3}A state [72G]. [n]: number of states per formula unit.

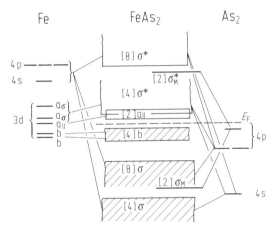

Fig. 8. FeAs$_2$. One-electron energy levels for the valence electrons in loellingite [72F1, 72G].

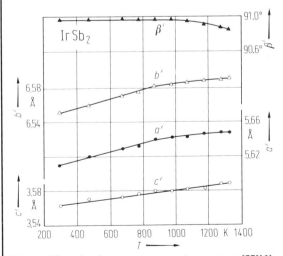

Fig. 6. IrSb$_2$. Lattice parameters vs. temperature [77K3]. Cf. caption of Fig. 3 for parameters a', b', c', β'.

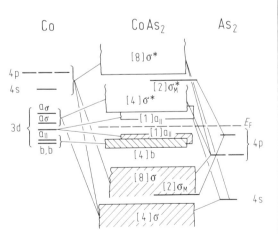

Fig. 9. CoAs$_2$. Energy-level scheme for the valence electrons in the arsenopyrite-type phase with formal d^5 configuration [72G].

9.15.1.3 T(V)$_2$ compounds

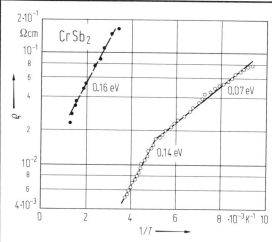

Fig. 10. CrSb$_2$. Electrical resistivity vs. reciprocal temperature for a sample quenched, crushed and annealed at 923 K [69A]. The broken curve represents the data of [57A].

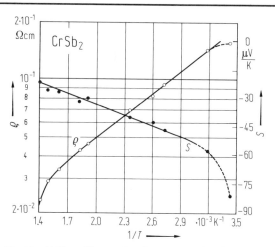

Fig. 11. CrSb$_2$. Thermoelectric power and resistivity vs. reciprocal temperature of polycrystalline samples [56A].

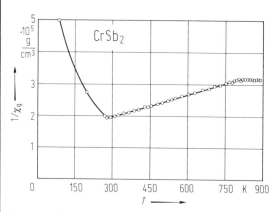

Fig. 12. CrSb$_2$. Reciprocal magnetic susceptibility vs. temperature [79K1]. Measurements on powdered single crystals grown by a chlorine transport reaction. χ in CGS-emu.

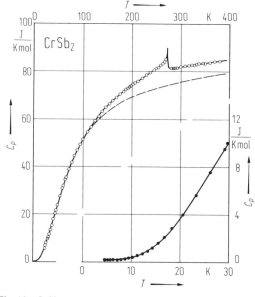

Fig. 13. CrSb$_2$. Heat capacity vs. temperature [78A2]. For explanation of dashed line, see Fig. 14.

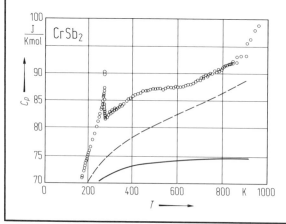

Fig. 14. CrSb$_2$. Heat capacity vs. temperature [78A2]. Full curve: estimated vibrational contribution $C_{\text{vibr.}}$. Broken curve: vibrational + dilatational (Nernst-Lindemann) contribution $C_{\text{vibr.}} + C_{\text{dil.}}$; $C_{\text{dil.}} = A\, C_p^2\, T_m/T$, with $A = 1.703 \cdot 10^{-3}$ K mol J^{-1} and $T_m = 991.3$ K, the peritectic temperature.

9.15.1.3 T(V)₂ compounds

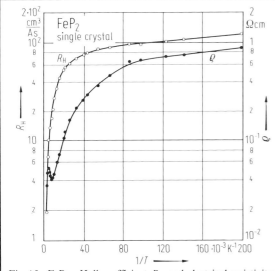

Fig. 15. FeP₂. Hall coefficient R_H and electrical resistivity ϱ vs. reciprocal temperature [71B]. Single crystal $1 \times 0.4 \times 0.4$ mm³ (largest dimension in c-direction?).

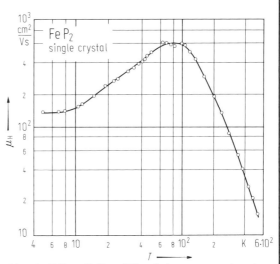

Fig. 16. FeP₂. Hall mobility R_H/ϱ vs. temperature in a doubly-logarithmic scale [71B]. Measurements on single crystals of unknown orientation.

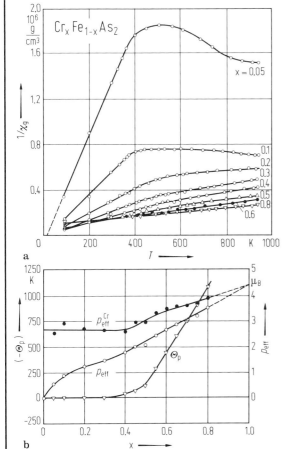

Fig. 17. $Cr_xFe_{1-x}As_2$. a) Reciprocal susceptibility vs. temperature [79K1]. b) gives the corresponding Curie-Weiss values for the paramagnetic moment p_{eff} and the paramagnetic Curie temperature Θ_p. Assuming Fe in the diamagnetic d⁴ state as in pure FeAs₂ a Cr moment is derived as $p_{eff}^{Cr} = p_{eff}/x^{1/2}$. χ_g in CGS-emu.

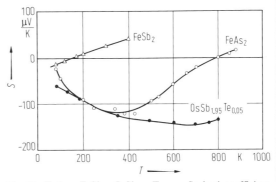

Fig. 18. FeAs₂, FeSb₂, OsSb₁.₉₅Te₀.₀₅. Seebeck coefficient vs. temperature [65J]. FeAs₂: polycrystalline natural sample containing traces of calcite; FeSb₂ and OsSb₁.₉₅Te₀.₀₅: pressed samples sintered at 873 K and 973 K, respectively.

For Fig. 19, see next page.

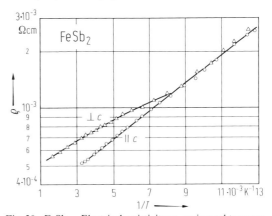

Fig. 20. FeSb₂. Electrical resistivity vs. reciprocal temperature of an n-type single crystal [72F1]. Triangles: measurement made in the (a, b) plane; circles: measurement made in the (a, c) or in the (b, c) plane.

9.15.1.3 T(V)$_2$ compounds

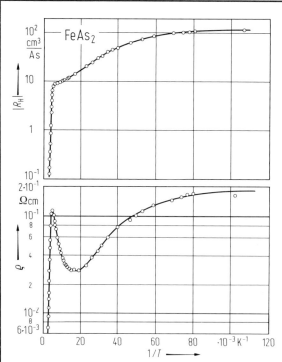

Fig. 19. FeAs$_2$. Hall coefficient and resistivity vs. reciprocal temperature of an n-type single crystal of unknown orientation [72F1].

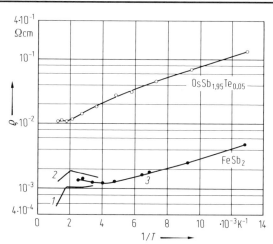

Fig. 21. FeSb$_2$ and OsSb$_{1.95}$Te$_{0.05}$. Electrical resistivity vs. reciprocal temperature [65J]. Sintered samples. FeSb$_2$: *1* [60D], *2* [59H], *3* [65J].

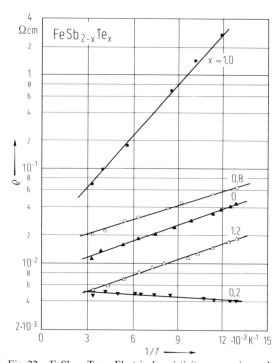

Fig. 22. FeSb$_{2-x}$Te$_x$. Electrical resistivity vs. reciprocal temperature [80Y]. Compounds with x = 0.8, 1 and 1.2 crystallize in the monoclinic arsenopyrite structure.

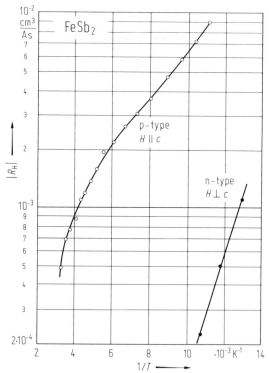

Fig. 23. FeSb$_2$. Hall coefficient vs. reciprocal temperature [72F1]. Open circles: Measurement made with the magnetic field perpendicular to the (a, b) plane; full circles: magnetic field parallel to the (a, b) plane.

9.15.1.3 T(V)₂ compounds

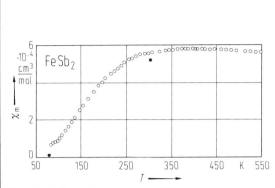

Fig. 24. FeSb₂. Magnetic susceptibility vs. temperature of non-oriented single crystals grown by a chlorine transport reaction [72F1]. Full circles represent the data of [80Y], see table. χ_m in CGS-emu.

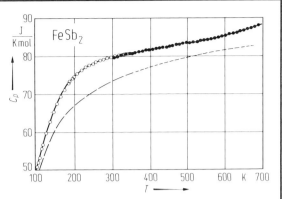

Fig. 25. FeSb₂. Molar heat capacity vs. temperature [77G]; 190 g sample. For comparison the experimental and (above 350 K) extrapolated heat capacity of FeTe₂ is added (broken curve).

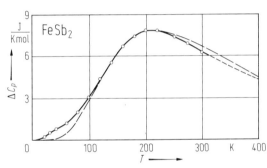

Fig. 26. FeSb₂. Excess molar heat capacity vs. temperature [77G]. Broken curve: Calculated Schottky anomaly for electron excitations from a ground level to a first level at $E_1/hc = 300$ cm^{-1} and to a second level $E_2/hc = 500$ cm^{-1}, with the corresponding degeneracies $g_0 : g_1 : g_2 = 1 : 1 : 2$.

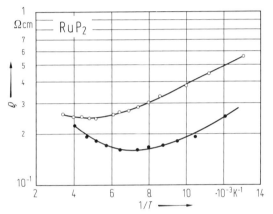

Fig. 27. RuP₂. Resistivity vs. reciprocal temperature [77K1]. Measurements on two different single crystals containing up to 0.85 wt% Sn from the tin flux; orientation unknown.

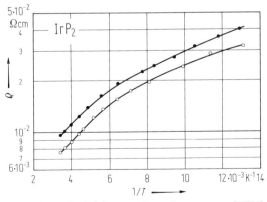

Fig. 28. IrP₂. Resistivity vs. reciprocal temperature [77K1]. Measurements on two different single crystals containing up to 0.24 wt% Sn and 0.006 wt% Cu from tin flux; orientation unknown.

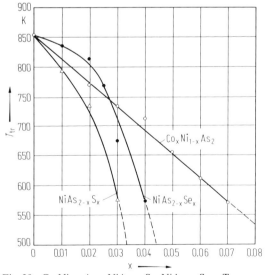

Fig. 29. Co$_x$Ni$_{1-x}$As₂, NiAs$_{2-x}$S$_x$, NiAs$_{2-x}$Se$_x$. Temperature of the pararammelsbergite (α) ⇄ marcasite (β) transformation vs. concentration of substituent [79K2].

9.15.1.3 T(V)₂ compounds

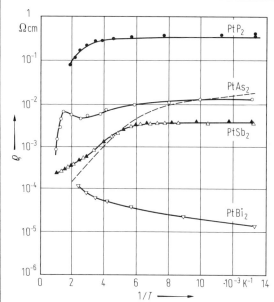

Fig. 30. PtP₂, PtAs₂, PtSb₂, PtBi₂. Resistivity vs. reciprocal temperature [65J]. The data for semimetallic pyrite-type PtBi₂(h) are added for comparison. PtP₂: hot-pressed sample; PtAs₂ and PtBi₂: sintered samples; PtSb₂: open triangles – polycrystalline sample, full triangles – single crystals. The broken curve shows the measurements of [66B] on a sintered PtAs₂ sample.

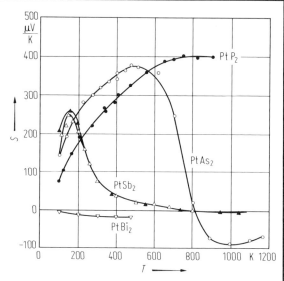

Fig. 31. PtP₂, PtAs₂, PtSb₂, PtBi₂. Thermoelectric power vs. temperature [65J]. PtSb₂: open triangles–polycrystalline sample, full triangles–single crystal; PtBi₂: pyrite-type high-temperature modification.

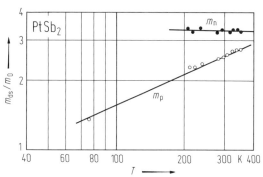

Fig. 32. PtSb₂. Density of states effective masses of electrons and holes vs. temperature in a doubly-logarithmic scale [73A].

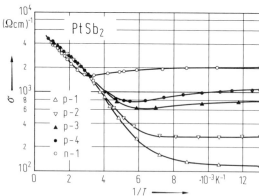

Fig. 33. PtSb₂. Conductivity vs. reciprocal temperature [65D]. Single crystal specimens as in Fig. 37.

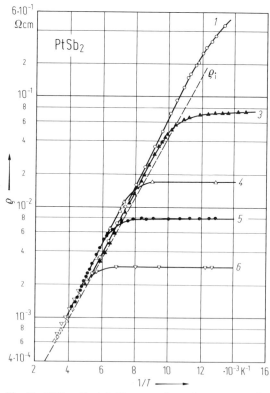

Fig. 34. PtSb₂. Resistivities vs. reciprocal temperature for n-type samples [68R2]. Carrier concentrations given in Fig. 38. ϱ_i: intrinsic resistivity.

9.15.1.3 T(V)$_2$ compounds

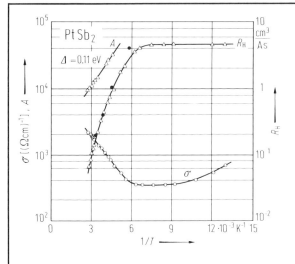

Fig. 36. PtSb$_2$. Electrical conductivity and Hall coefficient vs. reciprocal temperature for a p-type single crystal [73A]. The energy gap is derived from

$$A = R_H T^{3/2} \left(\frac{m_n \cdot m_p}{m_0^2} \right)^{3/4} \frac{1+b}{1-b},$$

where the density of states masses $m_{n,p} = n_{C,V}^{2/3} m_\parallel^{1/3} m_\perp^{2/3}$, with $n_{C,V}$ = number of energy ellipsoids, $m_\parallel$ and $m_\perp$ = longitudinal and transverse effective masses, respectively. A in $\frac{cm^3 \, K^{3/2}}{As}$.

Fig. 35. PtSb$_2$. Electrical conductivity vs. reciprocal temperature [76A]. The specimens were fragments of a Te-doped single crystal with a Te concentration close to the concentration of residual acceptor impurities in "undoped n-type PtSb$_2$". The donor (Te) concentration was approximately constant ($n_d = 4 \cdot 10^{16}$ cm^{-3}) along the ingot and only the acceptor concentration varied. The samples were cut from the n-type part at right angles to the growth direction [110]. The sample numbers increase with the degree of compensation (as estimated from R_H in the depletion region). $n_d - n_a$ in 10^{16} cm^{-3}: 1: 3.2, 2: 1.9, 3: 1.4, 4: 1.2, 5: 1.0, 6: 0.7, 7: 0.3.

Fig. 37. PtSb$_2$. Absolute values of the Hall coefficient vs. reciprocal temperature of single crystals [65D].

Sample	$1/(e R_H^{77K})$ [$\cdot 10^{18}$ cm^{-3}]	$R_H^{77K} \cdot \sigma_{77K}$ [cm^2 V^{-1} s^{-1}]	$R_H^{77K} \cdot \sigma_{4.2K}$ [cm^2 V^{-1} s^{-1}]
p−1 Te doped	1.3	610	780
p−2 Te doped	2.1	860	1170
p−3	3.2	1610	2160
p−4	5.2	1390	1680
n−1 Te doped	62	222	226

For sample n−1, R_H is negative throughout the temperature range. For samples p−1, p−2, p−3 and p−4, R_H is positive below about 500 K and negative at higher temperatures.

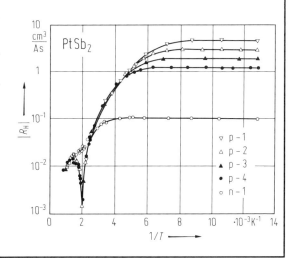

9.15.1.3 T(V)$_2$ compounds

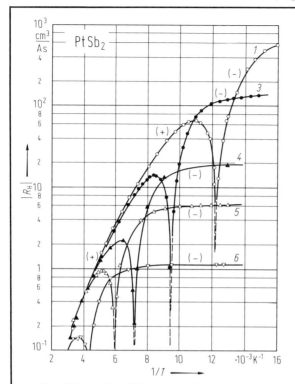

Fig. 38. PtSb$_2$. Hall coefficients vs. reciprocal temperature for n-type samples [68R2]. Extrinsic carrier concentrations $N_{\text{extr.}} = n - p = 1/eR_H$ (77 K); 1: $5.68 \cdot 10^{15}$ cm^{-3}, 3: $5.00 \cdot 10^{16}$ cm^{-3}, 4: $3.30 \cdot 10^{17}$ cm^{-3}, 5: $9.93 \cdot 10^{17}$ cm^{-3}, 6: $5.48 \cdot 10^{18}$ cm^{-3}.

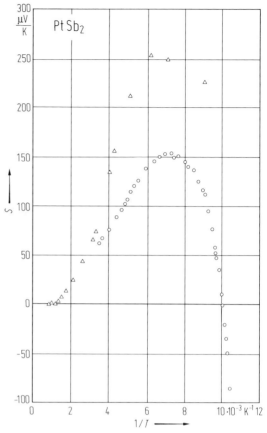

Fig. 40α. PtSb$_2$. Seebeck coefficient vs. reciprocal temperature in the intrinsic region. Circles: Te doped single crystal with $n_d - n_a \approx 10^{17}$ cm^{-3}, $n_a/n_d \approx 0.8$ and $(-eR_H)^{-1} = 6.6 \cdot 10^{16}$ cm^{-3} [83D]. Triangles: Sample p–3 of [65D]; compare Fig. 37.

For Fig. 40, see next page.

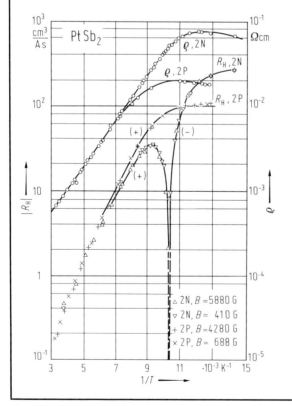

Fig. 39. PtSb$_2$. Hall coefficient and resistivity vs. reciprocal temperature in the mixed-conduction range for an n-type sample of low impurity contents [68R2]. 2N: $N_{\text{extr.}} = n - p \approx 2.4 \cdot 10^{16}$ cm^{-3}, $\mu_H(77\ \text{K}) = 3140$ cm^2/Vs; 2P: $P_{\text{extr.}} = p - n \approx 6.06 \cdot 10^{16}$ cm^{-3}, $\mu_H(77\ \text{K}) = 5840$ cm^2/Vs.

9.15.1.3 T(V)$_2$ compounds

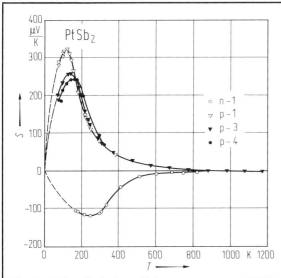

Fig. 40. PtSb$_2$. Seebeck coefficient vs. temperature [65D]. All samples were single crystals. For details, see Fig. 37. Above ≈ 1000 K the Seebeck coefficient of sample p−3 is negative ($\approx -1\,\mu$V/K).

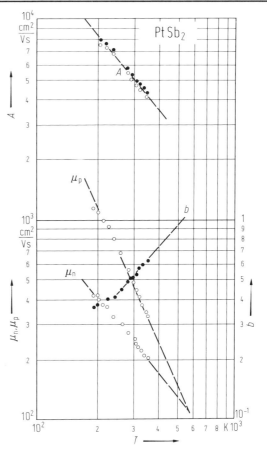

Fig. 41. PtSb$_2$. Mobilities of electrons and holes and mobility ratio $b = \mu_n/\mu_p$ vs. temperature in the intrinsic region; b derived from $S(T)$ [73A]. $A = \mu_p \left(\dfrac{m_p}{m_0}\right)^{5/2}$ (open circles) and $\mu_n \left(\dfrac{m_n}{m_0}\right)^{5/2}$ (full circles) varies as $\propto T^{-1.3}$.

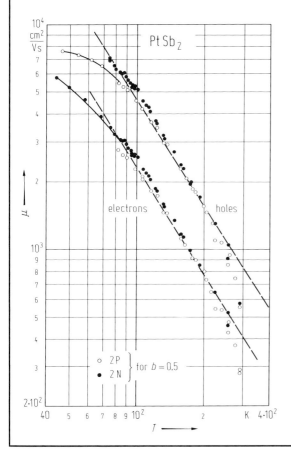

Fig. 42. PtSb$_2$. Electron and hole mobilities vs. temperature [68R2]. The broken lines represent values calculated for the intrinsic range: $\mu_p = 4.15 \cdot 10^6\,T^{-3/2}\,e^{+12.6\,\text{K}/T} \approx 6.55 \cdot 10^6\,T^{-1.57}$. $\mu_n = b\,\mu_p = 2.07 \cdot 10^6\,T^{-3/2}\,e^{+12.6\,\text{K}/T}$ (T in K), obtained from experimental data of Hall effect and resistivity. 2P, 2N: sample numbers.

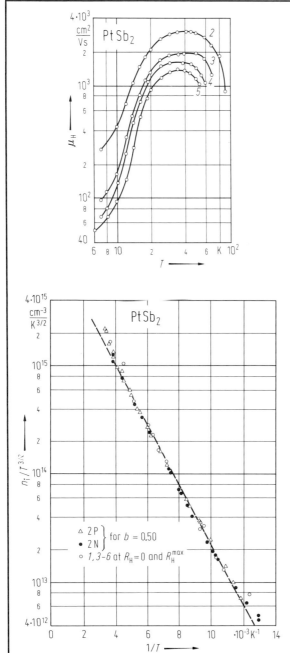

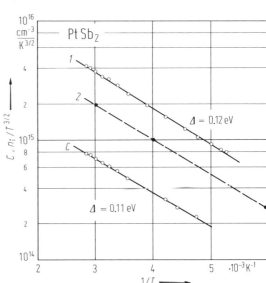

Fig. 43. PtSb$_2$. Hall mobility vs. temperature in a doubly-logarithmic scale [76A]. Sample numbers as in Fig. 35. At low temperatures the slopes are considerably steeper than for compensated semiconductors when electrons are scattered by ionized impurities i.e. $R\sigma$ ceases to represent the mobility at $T < 20$ K [76A].

Fig. 45. PtSb$_2$. Intrinsic carrier concentration vs. reciprocal temperature [73A]. 1: $n_i T^{-3/2}$ of [73A] calculated from Hall effect and Nernst-Ettingshausen effect, 2: $n_i T^{-3/2}$ data taken from [68R2], $C = n_i T^{-3/2} \left(\dfrac{m_n m_p}{m_0^2}\right)^{-3/4}$.

Fig. 44. PtSb$_2$. $n_i T^{-3/2}$ vs. reciprocal temperature [68R2]. For samples 2P and 2N the intrinsic carrier concentration n_i was calculated from $\sigma(T)$, $R_H(T)$ using the relations $n_i^2 = np$ and $n - p = N_{\text{extr.}} = (eR_H)^{-1}$ at 77 K, and assuming $b = \mu_n/\mu_p = 0.50$. For samples 1, 3···6 n_i was calculated at T where $R_H = 0$, using $nb^2 = p = n - N_{\text{extr.}}$, as well as at the temperature at which R_H reaches a maximum, using $nb = p = n - N_{\text{extr.}}$. The mobility ratio b was calculated at the positive maximum of R_H for each n-type sample using the relation $R_H^{\text{max.}}/R_H^{\text{extr.}} = (b-1)^2/4b$. The broken line represents $n_i T^{-3/2} = 4.84 \cdot 10^{15} A\,e^{-E_g(0)/2kT}\,[\text{cm}^{-3}\,\text{K}^{-3/2}]$ where $E_g(0) = 0.110$ eV and $A = (m_n m_p/m_0^2)^{3/4} e^{-\beta/2k} = 2.72$ ($m_{n,p}$: density of states masses), β: temperature coefficient of the band gap ($E_g = E_g(0) + \beta T$).

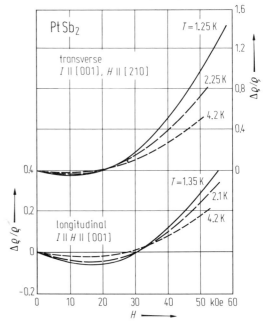

Fig. 46. PtSb$_2$. Transverse and longitudinal magnetoresistance vs. magnetic field strength [72D1]. Monocrystalline p-type sample of low carrier concentration; $R_H = 150$ cm^3/As, $\varrho = 0.169\,\Omega$ cm, $\mu_H = 890$ cm^2/Vs at 4.2 K.

9.15.1.3 T(V)₂ compounds

Fig. 47. $PtSb_2$. Transverse and longitudinal magnetoresistance vs. magnetic field strength [72D1]. Monocrystalline p-type samples with moderate to high carrier concentrations. Characteristic data at 4.2 K:

		R_H [cm³/As]	ϱ [Ωcm]	μ_H [cm²/Vs]
2:	Sn-doped	13.0	$6.4 \cdot 10^{-3}$	2030
3:	Rh-doped	7.87	$3.05 \cdot 10^{-3}$	2600
5:	Rh-doped	1.17	$6.4 \cdot 10^{-4}$	1830
6:	Rh-doped	0.422	$3.3 \cdot 10^{-4}$	1300

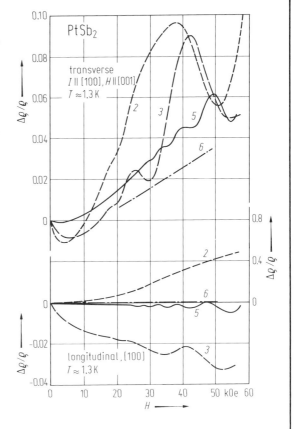

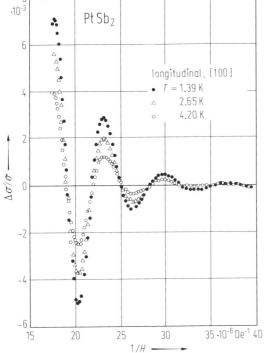

Fig. 48. $PtSb_2$. Oscillatory part of the longitudinal magnetoconductivity $\Delta\sigma'/\sigma$ vs. reciprocal magnetic field [72D1]. Longitudinal effect parallel to [100]. Rh-doped p-type single crystal with $R_H(4.2\,K) = 1.87$ cm³/As, $\varrho(4.2\,K) = 9.14 \cdot 10^{-4}$ Ωcm. Period $P = 5.77 \cdot 10^{-6}$ Oe^{-1}; $m_{\omega_c} = 0.175\, m_0$ (corrected).

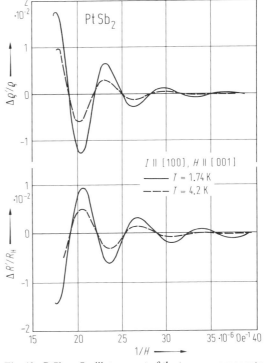

Fig. 49. $PtSb_2$. Oscillatory part of the transverse magnetoresistance $\Delta\varrho'/\varrho$ and Hall coefficient $\Delta R'/R$ vs. reciprocal magnetic field strength [72D1]. For sample data, see Fig. 48.

9.15.1.3 T(V)$_2$ compounds

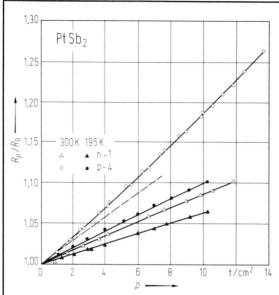

Fig. 50. PtSb$_2$. Resistance ratio R_p/R_0 vs. hydrostatic pressure [65D]. (1 t/cm^2 ≙ 9.807 · 10^7 Pa). Triangles and circles: n-type crystal n−1 and p-type crystal p−4, respectively, measured at 195 K (full symbols) and 300 K (open symbols). (Hydrostatic) piezoresistance coefficient $\pi_h = \pi_{11} + \pi_{12} + \pi_{13}$ (in 10^{-12} cm^2/dyn) derived from the slope d ln R/dp at $p \to 0$, where

$$d \ln R/dp = K'' + \pi_{11} + \pi_{12} + \pi_{13} \approx \pi_{11} + \pi_{12} + \pi_{13}$$

(K'': correction term):

	$T = 195$ K	$T = 300$ K
n−1	−6.3	−8.7
p−4	−10	−14.7

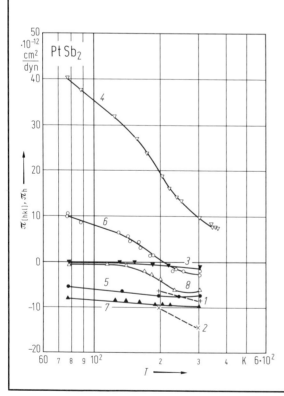

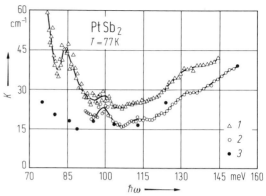

Fig. 52. PtSb$_2$. Absorption coefficient vs. photon energy at 77 K [70O]. *1* and *2*: n-type samples with extrinsic carrier concentration at 77 K ≈ 6 · 10^{16} cm^{-3}. *1* thickness 457 µm, *2* thickness 306 µm. *3*: data of [68R2] from an n-type specimen with carrier concentration 9.8 · 10^{17} cm^{-3} at 77 K.

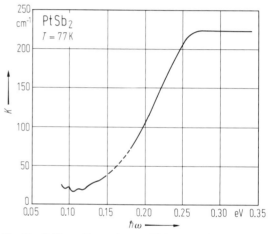

Fig. 53. PtSb$_2$. Absorption coefficient vs. photon energy at 77 K [70O]. Specimen 2 of Fig. 52.

◀

Fig. 51. PtSb$_2$. Piezoresistance coefficients vs. temperature [65D].

n−1	p−4	(samples characterized on Fig. 37)
1	2	$\pi_h = \pi_{11} + \pi_{12} + \pi_{13}$
3	4	$\pi_{[100]} = 1/3\,\pi_h + 2/3\,(\pi_{11} - \pi_{12} - \pi_{13})$
5	6	$\pi_{[110]} = 1/3\,\pi_h + 1/6\,(\pi_{11} - \pi_{12} - \pi_{13}) + 1/2\,\pi_{44}$
7	8	$\pi_{[111]} = 1/3\,\pi_h + 2/3\,\pi_{44}$

The piezoresistance tensor π, defined for a cubic crystal by $\Delta \varrho_i/\varrho_0 = \sum_j \pi_{ij} x_j$, where **X** is the stress tensor, has for the symmetry class $T_h - m3$ of the pyrite structure four independent coefficients

$$\pi_{lmn} = \frac{1}{X}\frac{\Delta R}{R} = K' + \pi_{11} - 2(l^2 m^2 + m^2 n^2 + n^2 l^2)\{\pi_{11} - 1/2\,(\pi_{12} - \pi_{13}) - \pi_{44}\},$$

where l, m, n, are the direction cosines of the specimen axis along which the stress is applied and K' is the neglected correction term due to the dimensional changes of the sample.

9.15.1.4 T(V)$_3$ compounds

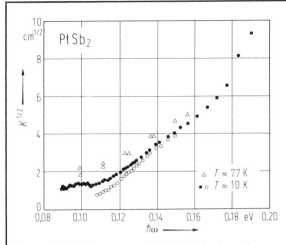

Fig. 54. PtSb$_2$. Square root of the intrinsic absorption coefficient vs. photon energy [68R2]. Open circles show the 10 K data (filled circles) after a wavelength-independent correction of 1.4 cm^{-1} has been subtracted from the experimental data, i.e. $(K-1.4)^{1/2}$. To obtain the intrinsic band-to-band absorption at 77 K, the free-carrier absorption was extrapolated to shorter wavelengths and subtracted from the total absorption.

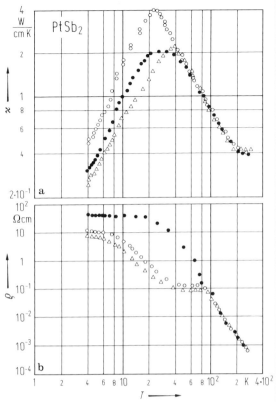

Fig. 55. PtSb$_2$. Thermal conductivity vs. temperature (a) and electrical resistivity vs. temperature (b), both in doubly-logarithmic scale, measured on single crystals of unknown orientation. The resistivity curves serve to characterize the different samples [82K].

9.15.1.4 T(V)$_3$ compounds (For tables, see p. 109 ff.)

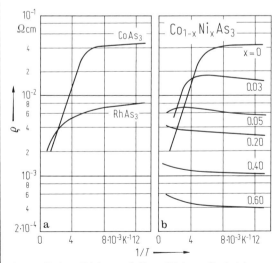

Fig. 1. CoAs$_3$, RhAs$_3$ and Co$_{1-x}$Ni$_x$As$_3$. Resistivity vs. reciprocal temperature [62P].

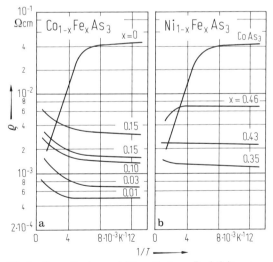

Fig. 2. Co$_{1-x}$Fe$_x$As$_3$ and Ni$_{1-x}$Fe$_x$As$_3$. Resistivity vs. reciprocal temperature [62P].

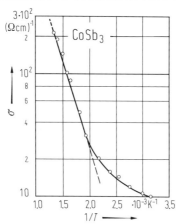

Fig. 3. CoSb$_3$. Electrical conductivity vs. reciprocal temperature [56D].

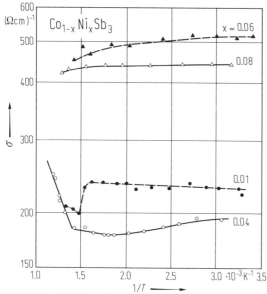

Fig. 4. Co$_{1-x}$Ni$_x$Sb$_3$. Conductivity vs. reciprocal temperature [57D].

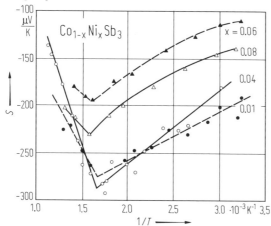

Fig. 6. Co$_{1-x}$Ni$_x$Sb$_3$. Thermoelectric power vs. reciprocal temperature [57D]. Concluding from Fig. 5 the curves for $x=0.06$ and $x=0.08$ should be interchanged.

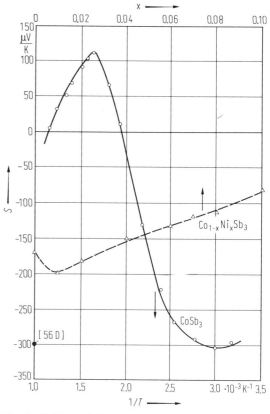

Fig. 5. CoSb$_3$ and Co$_{1-x}$Ni$_x$Sb$_3$. Thermoelectric power (circles) vs. reciprocal temperature for CoSb$_3$ [56D] and its room-temperature value (triangles) vs. concentration x for Co$_{1-x}$Ni$_x$Sb$_3$ [57D]. Compare Table 3 for $S(x)$.

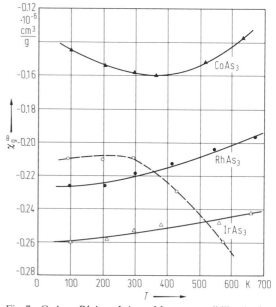

Fig. 7. CoAs$_3$, RhAs$_3$, IrAs$_3$. Mass susceptibility (at infinite field strength) vs. temperature [62P]. Open circles represent the data for IrAs$_3$ of [61K]. χ_g in CGS-emu.

9.15.2.1.1 Titanium oxide (TiO$_2$)

9.15.2 Binary transition-metal oxides
9.15.2.1 Titanium oxides
9.15.2.1.1 Titanium oxide (TiO$_2$) (For tables, see p. 129 ff.)

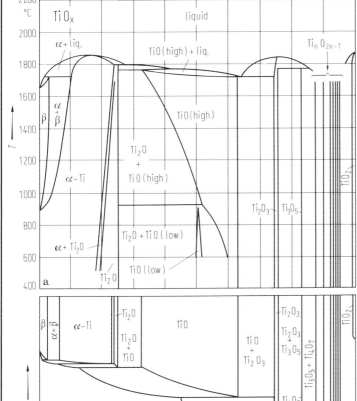

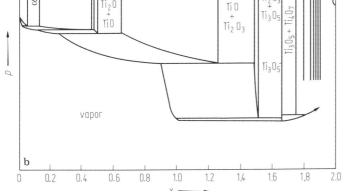

Fig. 1. Ti−O. Phase diagram. (a) [66W2], (b) [67G]. α and β refer to the hexagonal and cubic form of Ti metal.

Fig. 2. TiO$_2$, anatase. Relationship of anatase to NaCl. Metal ions removed from the NaCl structure are shown as dotted circles [75W].

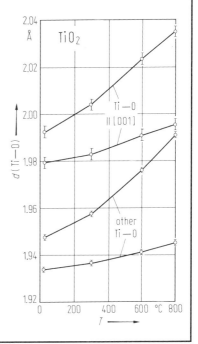

Fig. 3. TiO$_2$, anatase. Upper and lower bounds to the two nearest Ti−O bond lengths vs. temperature [72H2].

9.15.2.1.1 Titanium oxide (TiO$_2$)

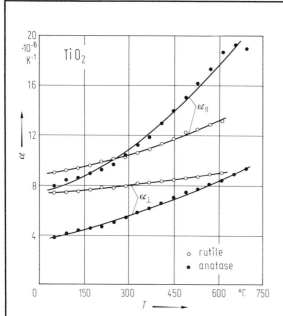

Fig. 4. TiO$_2$, anatase and rutile. Thermal expansion coefficients vs. temperature [70R].

Fig. 5. TiO$_2$, rutile. Crystal structure [79M1]. See also text in section 9.15.2.1.0.

For Figs. 6, 7, see next page.

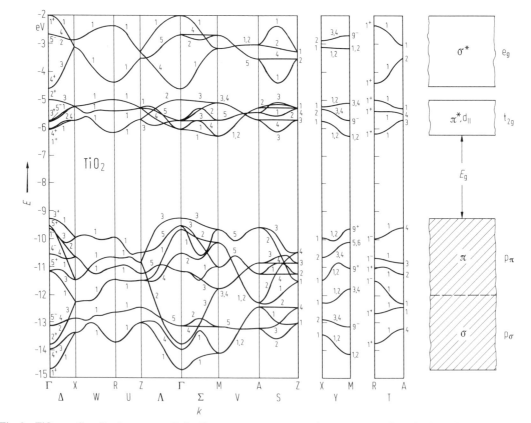

Fig. 8. TiO$_2$, rutile. Band structure. Left: Energy vs. wave vector along symmetry lines in the Brillouin zone, right: schematic bands with nomenclature used in the tables [77V1].

9.15.2.1.1 Titanium oxide (TiO$_2$)

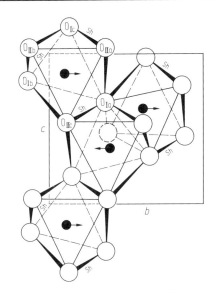

Fig. 6. TiO$_2$, brookite. Crystal structure. Shared edges are labelled Sh [79M1].

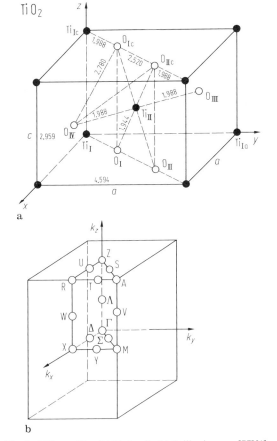

Fig. 9. TiO$_2$, rutile. (a) Unit cell, (b) Brillouin zone [77V1]. Distances in Å.

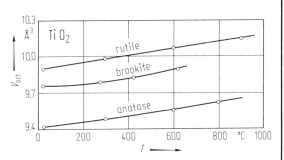

Fig. 7. TiO$_2$. Octahedral volumes in the three polymorphs vs. temperature [79M1].

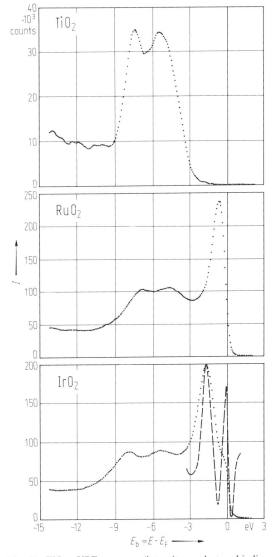

Fig. 10. TiO$_2$. XPE spectrum (intensity vs. electron binding energy $E_b = E - E_F$). For comparison the spectra for RuO$_2$ and IrO$_2$ are also shown [77R]. The dashed line for IrO$_2$ refers to the second derivative of the XPE spectrum.

9.15.2.1.1 Titanium oxide (TiO$_2$)

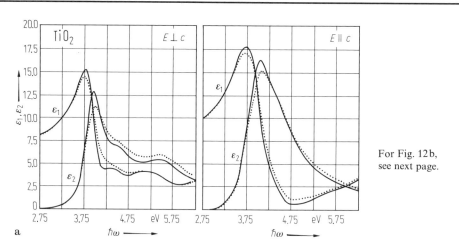

Fig. 12. TiO$_2$. (a) Real and imaginary parts of the dielectric constant. Left: $E \perp c$, right: $E \parallel c$; dotted curves $T = 300$ K, solid curves $T = 80$ K; (b) $\Delta\varepsilon_1$ and $\Delta\varepsilon_2$ as calculated from the electroreflectance spectrum at 300 K. Heavily drawn curves correspond to electrolyte configuration and lightly drawn curves to dry configuration. Broken curves $\Delta\varepsilon_1$, full curves $\Delta\varepsilon_2$. Numbers refer to estimated gold thickness in Å (measurements with gold electrodes evaporated onto crystal surface) [77V2].

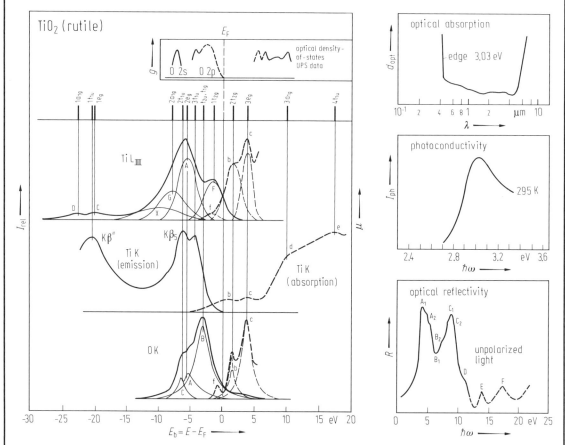

Fig. 11. TiO$_2$. Empirical energy level scheme (relative intensity of analyzed spectra vs. electron binding energy). For comments to the insets and the designations, compare the tables [72F].

9.15.2.1.1 Titanium oxide (TiO$_2$)

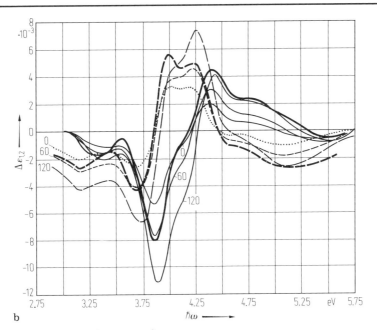

Fig. 12b. For caption, see previous page.

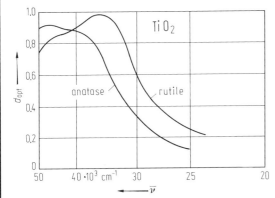

Fig. 14. TiO$_2$, anatase and rutile. Low resolution spectra (optical density vs. wavenumber) [58B].

For Fig. 13, see next page.

Fig. 15. TiO$_2$. Electron energy loss ($-\mathrm{Im}(\varepsilon^{-1})$) and derived optical constants vs. photon energy for a thin film [77B5], r is the reflection coefficient, n_{eff} the effective number of electrons per molecular unit involved in the optical absorption.

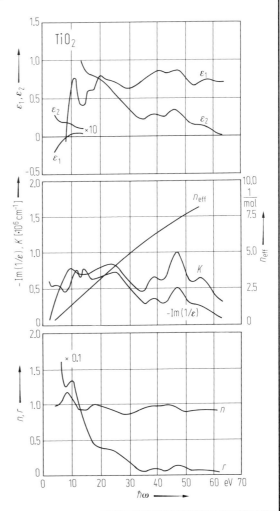

9.15.2.1.1 Titanium oxide (TiO$_2$)

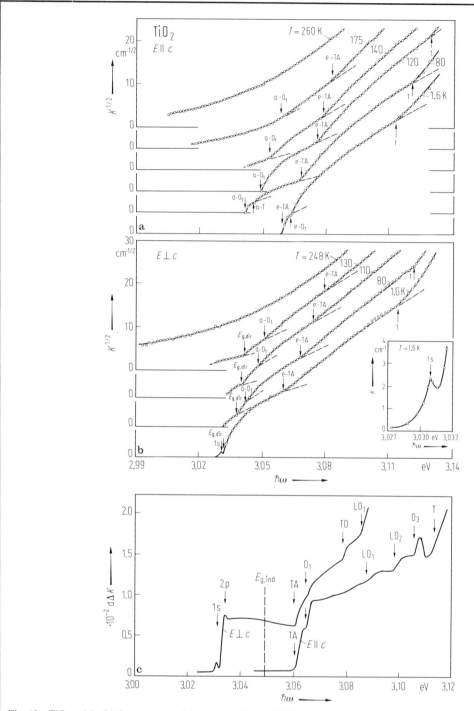

Fig. 13. TiO$_2$. (a), (b) Square root of the absorption coefficient vs. photon energy near the absorption edge for $E \parallel c$ (a) and $E \perp c$ (b), inset in (b) shows the 1s exciton; (c) wavelength modulated spectrum (differential of ΔK vs. photon energy) at 1.6 K [78P]. e−TA, a−O$_1$: emission of TA phonons, absorption of O$_1$ phonons.

9.15.2.1.1 Titanium oxide (TiO$_2$)

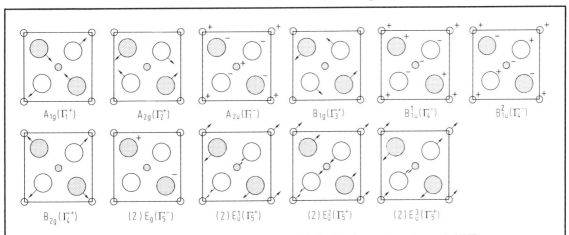

Fig. 16. TiO$_2$. Symmetry of the optical modes in rutile at $q=0$. Projection drawn down the c-axis [71T].

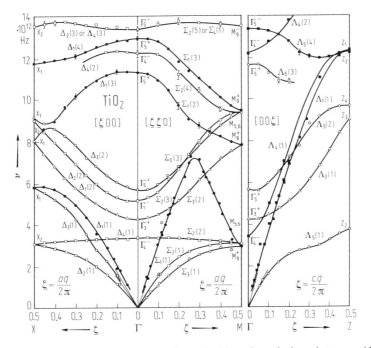

Fig. 17. TiO$_2$. Low-frequency phonon dispersion curves for rutile. Lines through the points are guide lines only. Solid (open) points represent measurements made under predominantly longitudinal (transverse) scattering conditions. Circular (square) points indicate that the scattering vector q lay predominantly normal to (parallel to) the z-axis. Triangular points are modes with displacements normal to z, but which may not be classified as longitudinal or transverse [71T].

9.15.2.1.1 Titanium oxide (TiO$_2$)

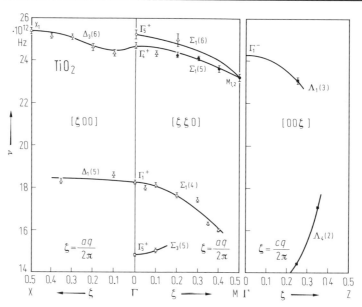

Fig. 18. TiO$_2$. High-frequency phonon dispersion spectrum. Notation as in Fig. 17 [71T].

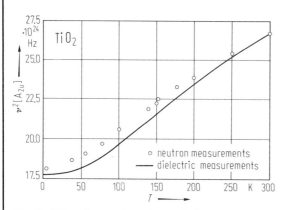

Fig. 19. TiO$_2$. Square of the A$_{2u}$-mode frequency vs. temperature. Solid line calculated from dielectric measurements [61P], circles from neutron scattering [71T].

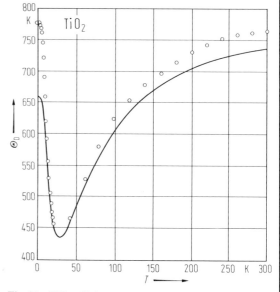

Fig. 20. TiO$_2$. Debye temperature vs. temperature. Solid line: calculated with shell model. Experimental points from heat capacity, 0 to 20 K from [69S1], 20 to 300 K from [69P]; [71T].

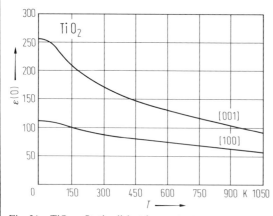

Fig. 21. TiO$_2$. Static dielectric constant vs. temperature for the [100] and [001] directions [61P].

9.15.2.1.1 Titanium oxide (TiO$_2$)

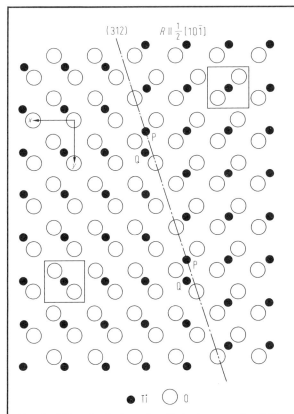

Fig. 22. TiO$_2$. Projection of rutile lattice onto the (001) plane intersected by a (312) fault. Ti-ion pairs PQ about the fault, and a single electron trapped at this site produces a W-centre EPR spectrum [77Y2]. R: Ti–Ti displacement.

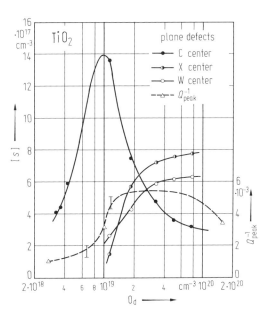

Fig. 24. TiO$_2$. Concentration of various lattice defects in slightly reduced rutile vs. oxygen deficit O_d (left ordinate spin concentration, right ordinate internal friction peak) [72H].

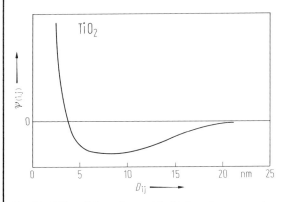

Fig. 23. TiO$_2$. Schematic empirical interaction energy between a pair of (312) CS-planes: wave function $\Psi(i,j)$ vs. plane separation D_{ij} [72B].

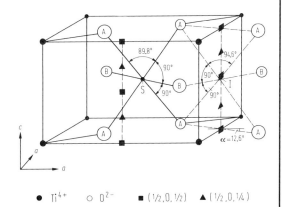

Fig. 25. TiO$_2$. Crystal structure of rutile showing some $[\frac{1}{2}, 0, \frac{1}{2}]$ and $[\frac{1}{2}, 0, \frac{1}{4}]$ types of interstitial site. Substitutional (S) and interstitial (I) site distances to oxygen atoms are: AS = 1.944 Å, BS = 1.988 Å, AI = 2.184 Å, BI = 1.603 Å (a = 4.594 Å, c = 2.959 Å) [73K1].

9.15.2.1.1 Titanium oxide (TiO$_2$)

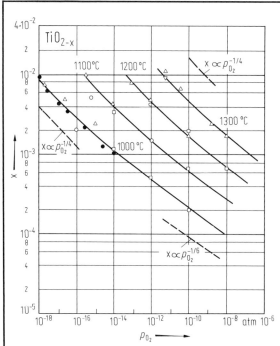

Fig. 26. TiO$_{2-x}$. Composition parameter x vs. oxygen partial pressure in the temperature range 1000···1350°C. (●) [67A1], (□) [64F], (△) [65M], (○) [62K], (▽) [65M]. 1 atm = 101 325 Pa.

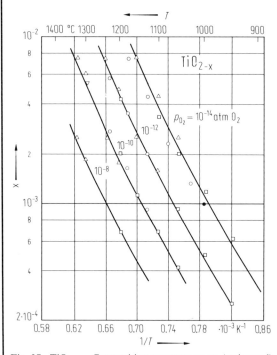

Fig. 27. TiO$_{2-x}$. Composition parameter x vs. (reciprocal) temperature at various oxygen partial pressures. Symbols as in Fig. 26.

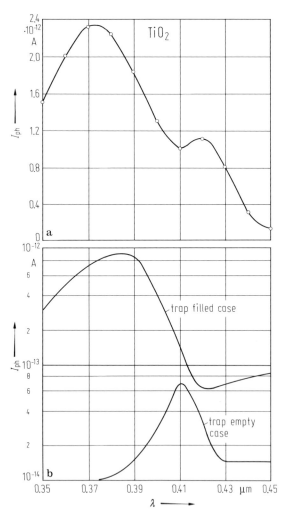

Fig. 28. TiO$_2$. Photoconductivity vs. wavelength of slightly reduced material (a) and comparison of photoconductivity in highly oxidized TiO$_2$ with and without empty shallow traps (b) [69G].

9.15.2.1.1 Titanium oxide (TiO$_2$)

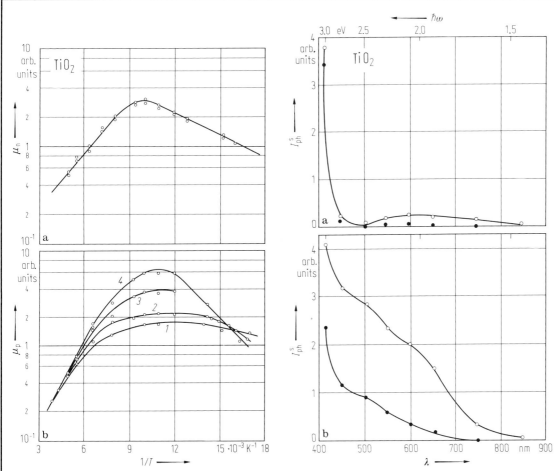

Fig. 29. TiO$_2$. (a) Electron and (b) hole mobility along c-axis vs. reciprocal temperature for photogenerated carriers. For electrons, μ_n is independent of field strength but (b) shows that μ_p increases with field strength; 1: $2.2 \cdot 10^3$ V/cm; 2: $4.4 \cdot 10^3$ V/cm; 3: $8.8 \cdot 10^3$ V/cm and 4: $17.6 \cdot 10^3$ V/cm [74T].

Fig. 30. TiO$_2$. Surface photocurrent vs. wavelength at 77 K (●) before, (○) after exposure to intense white light for (a) as sliced, (b) substitutionally doped with V, Cr, Mn or Fe and oxidized single crystals [80G2].

For Fig. 31, see next page.

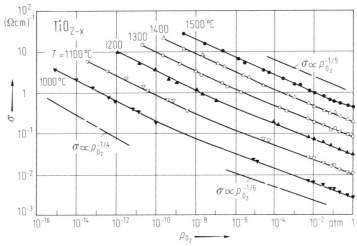

Fig. 32. TiO$_{2-x}$. Conductivity vs. oxygen partial pressure for various temperatures; current in [001] direction, single crystals [66B1,2].

9.15.2.1.1 Titanium oxide (TiO$_2$)

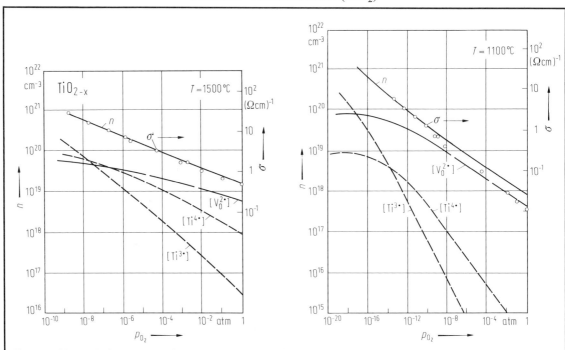

Fig. 31. TiO$_{2-x}$. Defect concentration vs. oxygen pressure from Kofstad model [67K] and comparison with conductivity data of [66B] for 1500°C (left hand figure) and 1100°C (right hand figure). n: concentration of defects, σ along [001] direction. 1 atm = 101325 Pa.

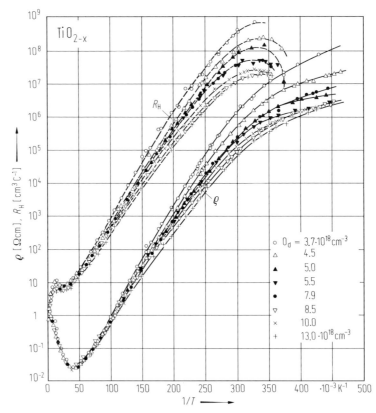

Fig. 33. TiO$_{2-x}$. Resistivity and Hall coefficient vs. reciprocal temperature for slightly reduced TiO$_2$. Oxygen deficit range: $3.7 \cdot 10^{18}$ to $13 \cdot 10^{18}$ cm^{-3} [72H1, 70H]. (Orientation not specified.)

9.15.2.1.1 Titanium oxide (TiO$_2$)

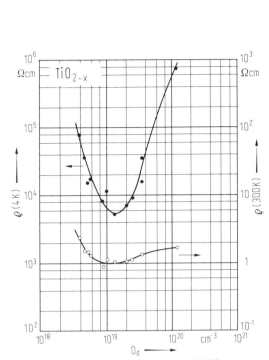

Fig. 34. TiO$_{2-x}$. Resistivities at 4 K and 300 K vs. oxygen deficiency of slightly reduced material [72H1, 70H]. (Orientation not specified.)

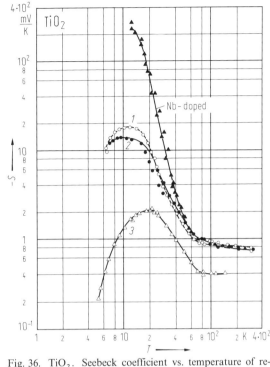

Fig. 36. TiO$_2$. Seebeck coefficient vs. temperature of reduced or Nb-doped rutile. Upper curve 0.1% Nb-doped. Curves *1, 2* heated at 1175 °C at 10^{-5} mm Hg for 27 h, curve *3* hydrogen reduced sample [65T]. Curve *1*: $S\|a$, curve *2*: $S\|c$, curve *3*: $S\|a$.

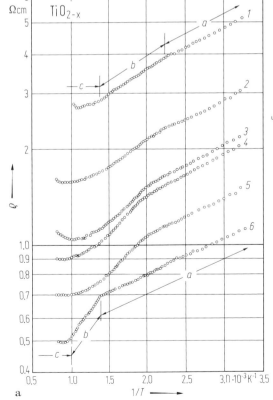

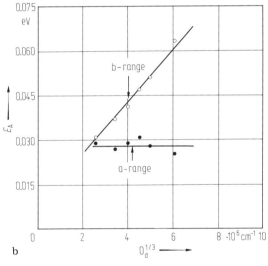

Fig. 35. TiO$_{2-x}$. (a) Resistivity in the [110] direction of moderately reduced TiO$_2$ vs. reciprocal temperature, (b) dependence of activation energy in ranges a and b on oxygen deficiency. For data on the samples 1···6, see the tables [59G, 74I].

9.15.2.1.1 Titanium oxide (TiO$_2$)

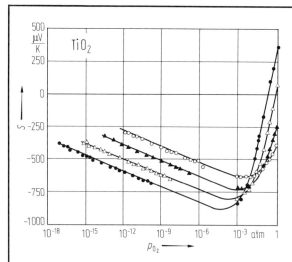

Fig. 37. TiO$_2$. Seebeck coefficient vs. oxygen partial pressure for polycrystalline rutile containing small quantities of Al and Fe. Temperature: (○) 1573 K, (▲) 1473 K, (△) 1393 K, (●) 1273 K. Lines are curves fitted to a suitable point-defect model [77B3].

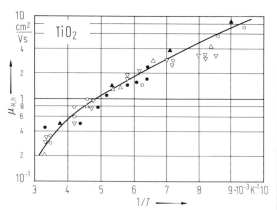

Fig. 38. TiO$_2$. Hall mobility in [001] direction vs. reciprocal temperature below 300 K [64A2]. Different symbols refer to samples of differing resistivities.

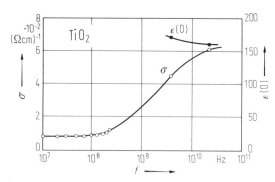

Fig. 40. TiO$_2$. Conductivity and static dielectric constant vs. frequency of a reduced rutile single crystal at 300 K along [001] direction [68G].

For Fig. 41, see next page.

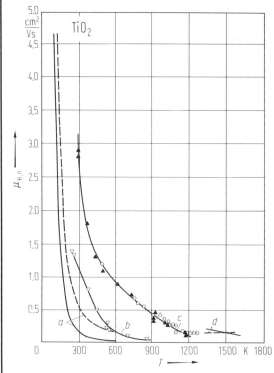

Fig. 39. TiO$_2$. Hall mobility vs. temperature at high temperatures for different samples. Curves a: —— $\perp c$, --- $\parallel c$ [67B1], curve b: $\perp c$ [69B2], curve c: $\parallel c$ [69B2], curve d: $\parallel c$ [66B1,2].

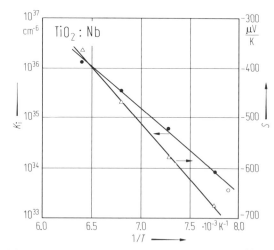

Fig. 42. TiO$_2$:Nb. Seebeck coefficient at $p_{O_2} = 10^{-10}$ atm and K_i ($= K_i^0 \exp(-E_g/kT)$) vs. reciprocal temperature for Nb-doped polycrystalline TiO$_2$ [77B3].

9.15.2.1.1 Titanium oxide (TiO$_2$)

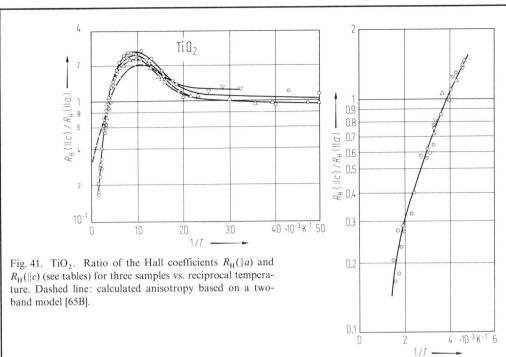

Fig. 41. TiO$_2$. Ratio of the Hall coefficients $R_H(\|a)$ and $R_H(\|c)$ (see tables) for three samples vs. reciprocal temperature. Dashed line: calculated anisotropy based on a two-band model [65B].

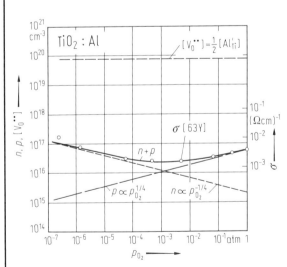

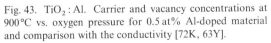

Fig. 43. TiO$_2$:Al. Carrier and vacancy concentrations at 900°C vs. oxygen pressure for 0.5 at% Al-doped material and comparison with the conductivity [72K, 63Y].

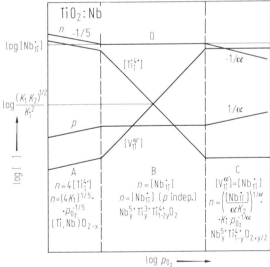

Fig. 44. TiO$_2$:Nb. Defect equilibria. The parameter α is the charge on Ti-vacancies and is assumed equal to 4. It is also assumed that $[Nb_{Ti}^{\cdot}] \gg K_i^{1/2} \gg (K_1 K_2)^{1/2} K_i^{-2}$ [77B4]. Region A: low p_{O_2}, concentration of native defects prevails over doping effect; B: electron concentration n is not dependent on p_{O_2}. C: high p_{O_2}, an overstoichiometric $Nb_y Ti_{1-y} O_{2+y/2}$ is expected owing to the rapid variation of $[Ti_i^{4+}]$ and $[V_{Ti}^{\alpha'}]$ in B. See also Fig. 49.

9.15.2.1.1 Titanium oxide (TiO$_2$)

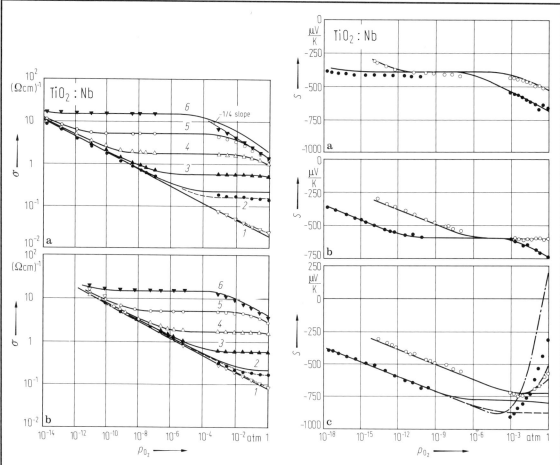

Fig. 45. TiO$_2$:Nb. Conductivity vs. oxygen partial pressure for polycrystalline Nb-doped material. (a) 1473 K, (b) 1623 K. *1*: pure TiO$_2$, Nb/Ti ratio: 0.0004 (*2*), 0.001 (*3*), 0.003 (*4*), 0.01 (*5*), 0.03 (*6*). Full lines: calculated curves using a point defect model. For (*2*), dashed curve is calculated with 100 ppm counterdopant. Dotted line: influence of intrinsic effects [77B4].

Fig. 47. TiO$_2$:Nb. Seebeck coefficient vs. oxygen partial pressure for three polycrystalline samples. $T = 1473$ K (o), 1273 K (●); (a) 1 at% Nb, (b) 0.1 at% Nb, (c) 0.01 at% Nb. In (c) the influence of trivalent impurities is clearly seen. Calculated curves: —— zero, --- 75 ppm, —·— 125 ppm trivalent impurities [77B3].

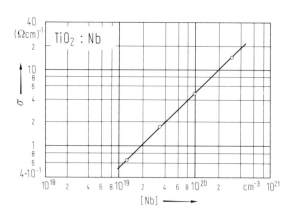

Fig. 46. TiO$_2$:Nb. Conductivity vs. Nb-concentration at 1273 K, $p_{O_2} = 10^{-10}$ atm [77B4]. Polycrystalline sample.

9.15.2.1.1 Titanium oxide (TiO$_2$)

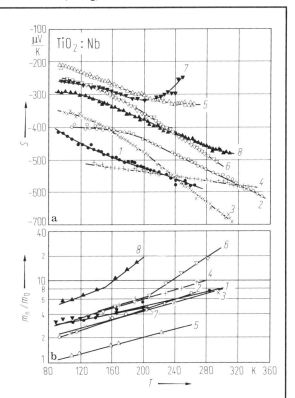

Fig. 48. TiO$_2$:Nb. a) Seebeck coefficient vs. temperature for eight Nb-doped single crystal TiO$_2$ samples measured along the c-axis. These samples had the characteristics:

Specimen	wt% Nb	σ_{150K} Ω^{-1} cm^{-1}	n_{150K} 10^{18} cm^{-3}
1	0.005	1.4	2.2
2	0.01	2.1	3.8
3	0.02	2.6	4.6
4	0.05	0.9	2.0
5	0.1	4.2	6.7
6	0.5	15.6	19.3
7	0.7	13.5	22.0
8	1.0	17.2	47.0

b) Effective masses vs. temperature for the same samples [78C].

For Fig. 49, see next page.

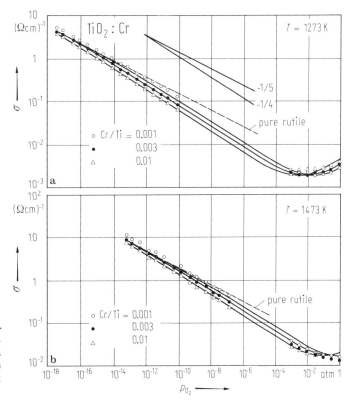

Fig. 49a. TiO$_2$:Cr. Conductivity of polycrystalline Cr:TiO$_2$ of various atomic ratios vs. oxygen partial pressure at (a) 1273 K (b) 1473 K. Full lines are best fit from defect model described in the text [80T].

9.15.2.1.1 Titanium oxide (TiO$_2$)

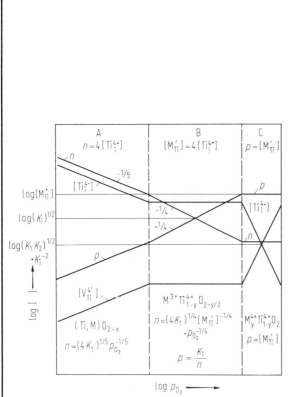

Fig. 49. TiO$_2$:M. Variation of the defect concentrations n, p, [Ti$_i^{4\cdot}$] and [V$_{Ti}^{4'}$] vs. oxygen partial pressure for TiO$_2$ doped with M^{3+}. The regions A···C correspond to ranges of different behaviour as indicated [80T].

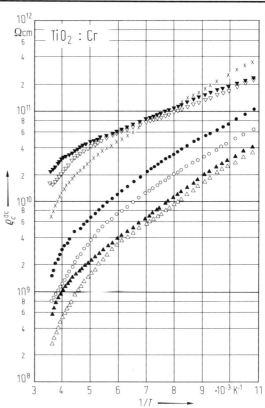

Fig. 50. TiO$_2$:Cr. dc electrical resistivity of single crystal slices in the c-direction vs. reciprocal temperature. The slices were doped by heating with powdered Cr$_2$O$_3$. The heating times were as follows: ▼ 4 weeks, ▽ 2 weeks, × 1 week, ● 2 days, ○ 1 day, ▲ 12 h, △ 3 h [79I].

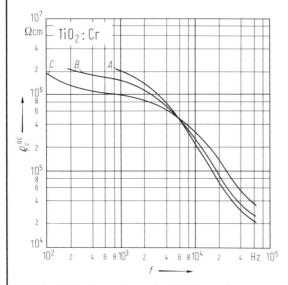

Fig. 51. TiO$_2$:Cr. ac electrical resistivity vs. frequency at 293 K in the c-direction. Samples prepared as in Fig. 50; heating times: A 1 day, B 2 days, C 2 weeks [79I].

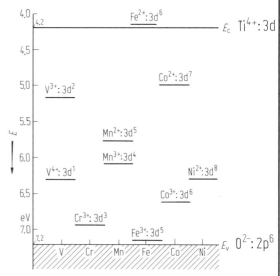

Fig. 52. TiO$_2$. Binding energies of transition-metal impurities in TiO$_2$ from epr and photoconductivity measurements [79M3].

9.15.2.1.1 Titanium oxide (TiO$_2$)

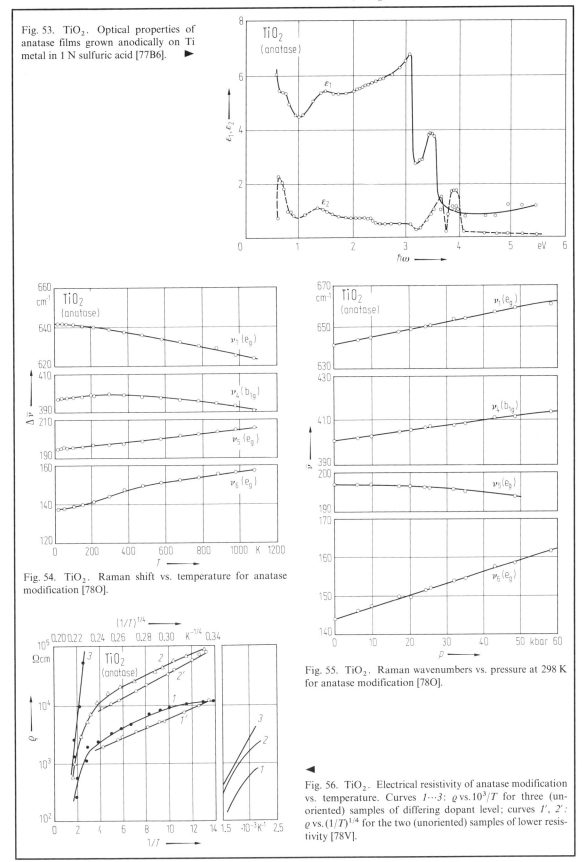

Fig. 53. TiO$_2$. Optical properties of anatase films grown anodically on Ti metal in 1 N sulfuric acid [77B6].

Fig. 54. TiO$_2$. Raman shift vs. temperature for anatase modification [78O].

Fig. 55. TiO$_2$. Raman wavenumbers vs. pressure at 298 K for anatase modification [78O].

Fig. 56. TiO$_2$. Electrical resistivity of anatase modification vs. temperature. Curves 1···3: ϱ vs. $10^3/T$ for three (unoriented) samples of differing dopant level; curves 1′, 2′: ϱ vs. $(1/T)^{1/4}$ for the two (unoriented) samples of lower resistivity [78V].

431

9.15.2.1.2 Titanium oxide (Ti$_2$O$_3$)

9.15.2.1.2 Titanium oxide (Ti$_2$O$_3$) (For tables, see p. 151 ff.)

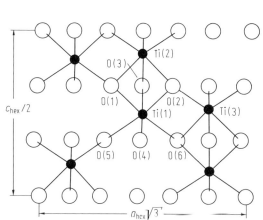

Fig. 1. Ti$_2$O$_3$. Projection of the structure on {110} [77R1].

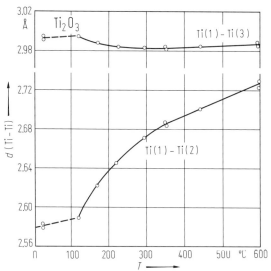

Fig. 2. Ti$_2$O$_3$. Interatomic distances vs. temperature [77R1].

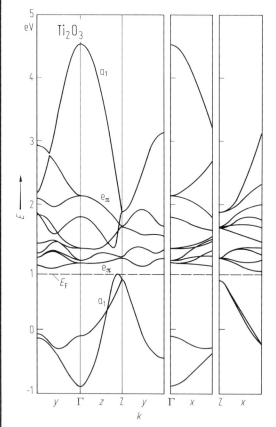

Fig. 3. Ti$_2$O$_3$. Band structure in the region of the t_{2g} levels [75A].

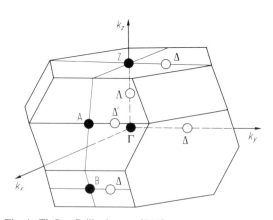

Fig. 4. Ti$_2$O$_3$. Brillouin zone [75A].

9.15.2.1.2 Titanium oxide (Ti_2O_3)

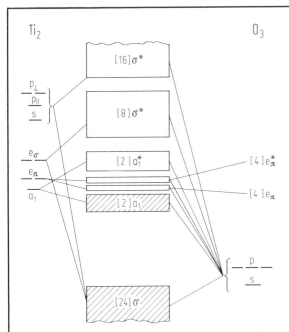

Fig. 5. Ti_2O_3. Phenomenological band structure [75H].

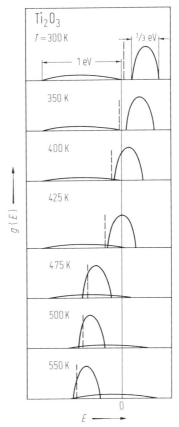

Fig. 6. Ti_2O_3. Density of states vs. energy showing band crossing with raising temperature [73B]. The vertical dashed line is the calculated Fermi energy.

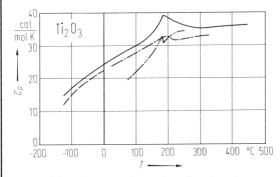

Fig. 7. Ti_2O_3. Heat capacity per formula unit vs. temperature. Different curves: data of various authors [73B].

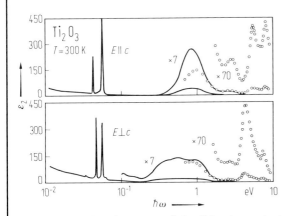

Fig. 8. Ti_2O_3. Imaginary part of the dielectric constant vs. photon energy in the near infrared for both polarization directions at 300 K. Solid lines from [78L1], dots from [78L2].

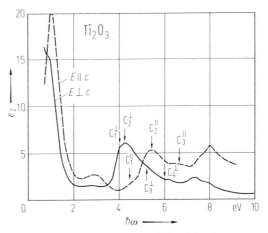

Fig. 9. Ti_2O_3. Imaginary part of the dielectric constant vs. photon energy in the visible and uv region for both polarization directions at 300 K [78L2].

9.15.2.1.2 Titanium oxide (Ti$_2$O$_3$)

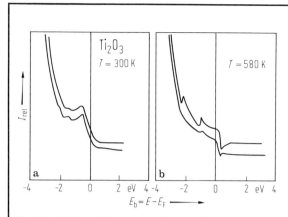

Fig. 10. Ti$_2$O$_3$. XPE spectrum (relative intensity vs. electron binding energy) at (a) 300 K and (b) 580 K [76S]. Two lines are upper and lower limits to the spectral noise.

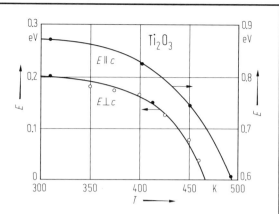

Fig. 11. Ti$_2$O$_3$. Energy of the $a_1 \to e_\pi$ transition ($E \perp c$) and the mixed $a_1 \to e_\pi$, $e_\pi \to e_\pi^*$ transition ($E \| c$). ● optical data, ○ dc conductivity [78L1].

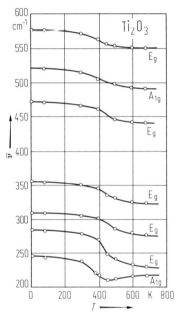

Fig. 12. Ti$_2$O$_3$. Wavenumbers of Raman frequencies vs. temperature [74S].

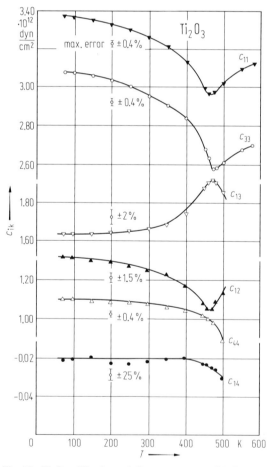

Fig. 13. Ti$_2$O$_3$. Elastic moduli vs. temperature [73C].

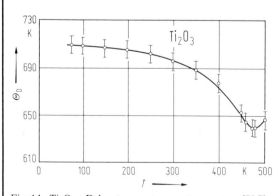

Fig. 14. Ti$_2$O$_3$. Debye temperature vs. temperature [73C].

9.15.2.1.2 Titanium oxide (Ti$_2$O$_3$)

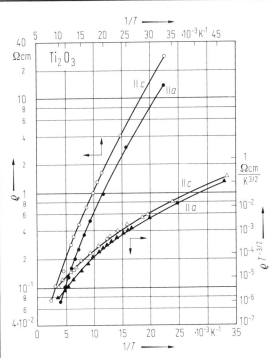

Fig. 15. Ti$_2$O$_3$. Resistivity vs. reciprocal temperature (right scale: resistivity times $T^{-3/2}$) at low temperatures [73S1].

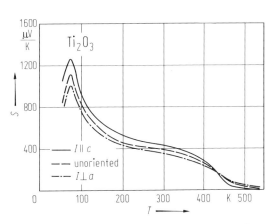

Fig. 16. Ti$_2$O$_3$. Seebeck coefficient vs. temperature for an unoriented and two oriented samples [73S1].

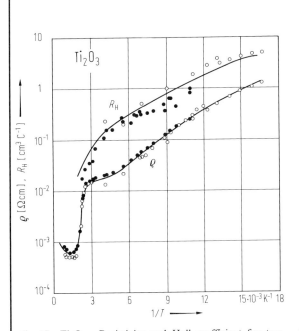

Fig. 17. Ti$_2$O$_3$. Resistivity and Hall coefficient for two crystals vs. reciprocal temperature [61Y]. Orientation not specified.

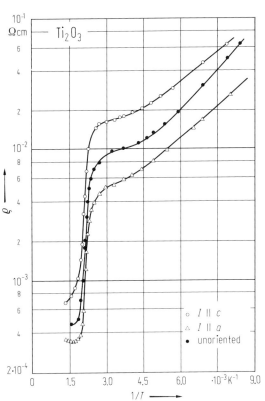

Fig. 18. Ti$_2$O$_3$. Resistivity vs. reciprocal temperature for an unoriented and two oriented samples [73S1].

9.15.2.1.2 Titanium oxide (Ti$_2$O$_3$)

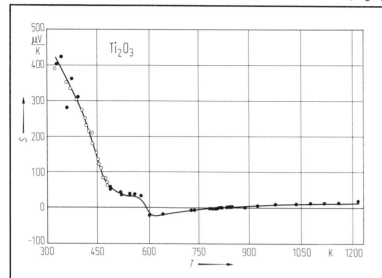

Fig. 19. Ti$_2$O$_3$. Seebeck coefficient vs. temperature above room temperature [61Y]. Orientation not specified.

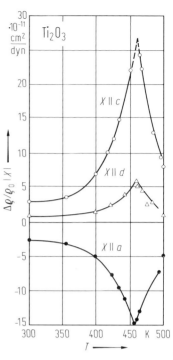

Fig. 22. Ti$_2$O$_3$. Piezoresistance vs. temperature for three directions of the stress along crystallographic directions [78C1].

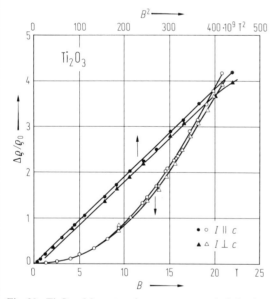

Fig. 20. Ti$_2$O$_3$. Magnetoresistance vs. magnetic induction (lower scale) and square of the magnetic induction (upper scale) for two directions of the current [68H].

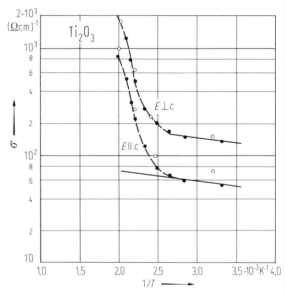

Fig. 21. Ti$_2$O$_3$. Conductivity vs. reciprocal temperature. Full circles: [76C2], open circles: [78L1].

9.15.2.1.2 Titanium oxide (Ti$_2$O$_3$)

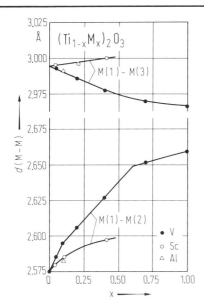

Fig. 23. (Ti$_{1-x}$M$_x$)$_2$O$_3$, M = V, Sc, Al. Interatomic distances vs. composition parameter x [77R2].

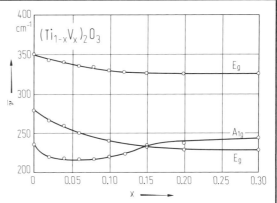

Fig. 24. (Ti$_{1-x}$V$_x$)$_2$O$_3$. Phonon wavenumbers vs. composition parameter x at 300 K [74S].

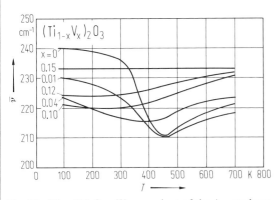

Fig. 25. (Ti$_{1-x}$V$_x$)$_2$O$_3$. Wavenumbers of the A$_{1g}$-mode vs. temperature for samples of different compositions [74S].

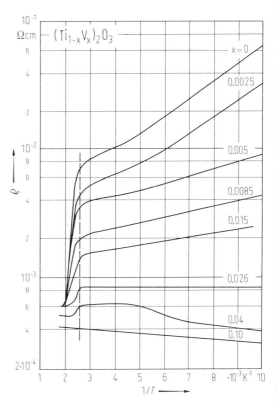

Fig. 26. (Ti$_{1-x}$V$_x$)$_2$O$_3$. Resistivity vs. reciprocal temperature for polycrystalline samples of different compositions [70C]. Dashed line: transition temperature from semiconductor to metal.

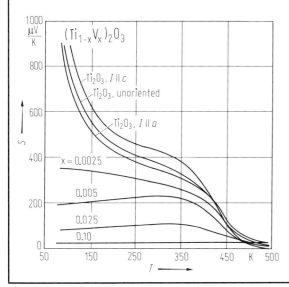

Fig. 27. (Ti$_{1-x}$V$_x$)$_2$O$_3$. Seebeck coefficient vs. temperature for polycrystalline doped samples of different compositions and for single crystals of Ti$_2$O$_3$ [73S1].

9.15.2.1.3 Phases between Ti_2O_3 and TiO_2

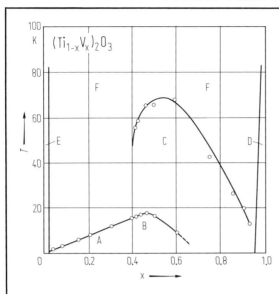

Fig. 28. $(Ti_{1-x}V_x)_2O_3$. Phase diagram. A: spin glass, B: mictomagnetic, C: antiferromagnetic metal, D: antiferromagnetic semiconductor, E: paramagnetic semiconductor, F: paramagnetic metal [79D2].

9.15.2.1.3 Phases between Ti_2O_3 and TiO_2 (For tables, see p. 155ff.)

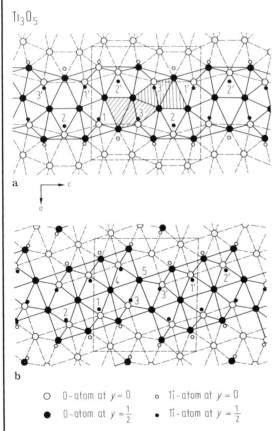

Fig. 1. Ti_3O_5. Crystal structures of the high temperature (a) and low temperature (b) phases. Unit cells are indicated by dotted lines and primes refer to symmetry related atomic sites [59A].

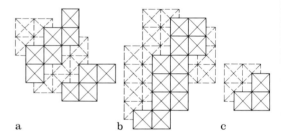

Fig. 2. Ti_3O_5. Arrangements of TiO_6 octahedra in anatase (010) plane (a) and low-temperature M-type Ti_3O_5 (010) plane (b). (c) shows the common structural unit, present in (a) and (b) [69I].

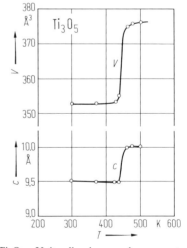

Fig. 3. Ti_3O_5. Unit cell volume and c-parameter vs. temperature [77R].

9.15.2.1.3 Phases between Ti_2O_3 and TiO_2

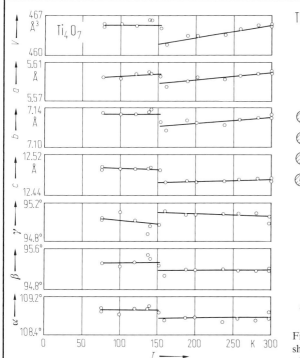

Fig. 4. Ti_4O_7. Lattice parameters vs. temperature [70M].

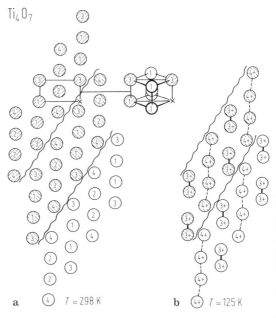

Fig. 6. Ti_4O_7. (a) Sheets of Ti ions at 298 K. Wavy lines show the shear planes which intersect the paper at an angle of 64°. Inset shows the rutile-derived cell. (b) Same but showing the situation at 125 K [73M].

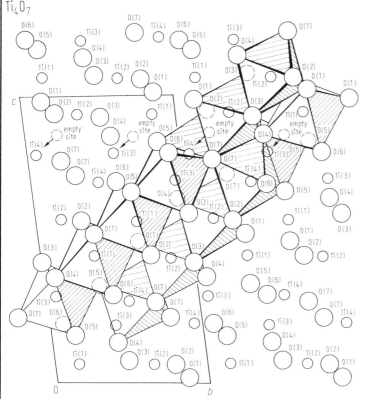

Fig. 5. Ti_4O_7. Projection of the structure looking down the triclinic a-axis at RT [71M]

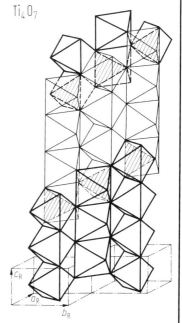

Fig. 7. Ti_4O_7. Stacking of oxygen octahedra containing Ti in each rutile block. Successive rutile blocks are represented by heavy and light lines. a_R, b_R, c_R represent the pseudorutile cell parameters. Shaded octahedra faces are common between 3-1-1-3 and 4-2-2-4 strings in neighbouring blocks [79H].

9.15.2.1.3 Phases between Ti$_2$O$_3$ and TiO$_2$

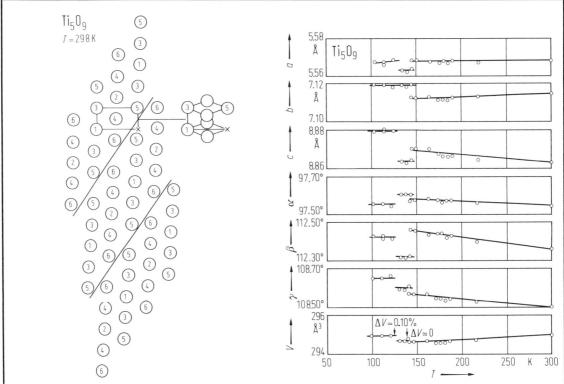

Fig. 8. Ti$_5$O$_9$. Structure showing two distinguishable strings [77M1]. Inset shows the rutile-derived cell.

Fig. 9. Ti$_5$O$_9$. Lattice parameters vs. temperature [77M1].

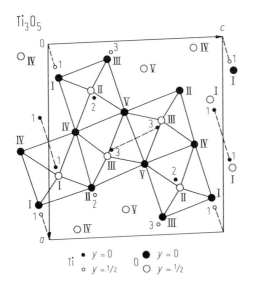

Fig. 10. Ti$_3$O$_5$. Detail of structure in the $a-c$ plane of the LT phase [74H3]. See also Fig. 1.

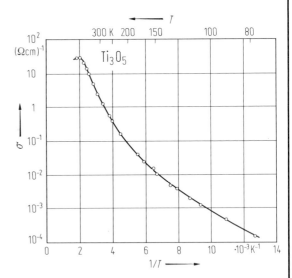

Fig. 11. Ti$_3$O$_5$. Conductivity along b-axis vs. (reciprocal) temperature of single crystal material [69B].

9.15.2.1.3 Phases between Ti_2O_3 and TiO_2

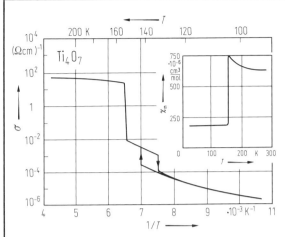

Fig. 12. Ti_4O_7. Conductivity along [031] pseudorutile direction vs. (reciprocal) temperature of a single crystal. The inset shows the molar magnetic susceptibility vs. temperature, (χ_m in CGS-emu) [76L].

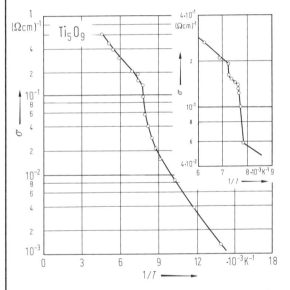

Fig. 13. Ti_5O_9. Conductivity vs. reciprocal temperature for a single crystal [77M1]. Inset shows transition range on an enlarged scale. Orientation not specified.

For Fig. 13a, see next page.

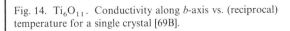

Fig. 14. Ti_6O_{11}. Conductivity along b-axis vs. (reciprocal) temperature for a single crystal [69B].

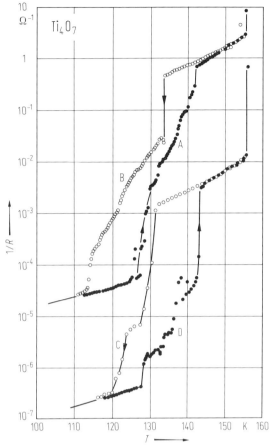

Fig. 12a. Ti_4O_7. Conductance $(1/R)$ vs. temperature, showing fine structure in the hysteresis range. Sample was warmed through the 154 K transition between runs B and C which accounts for the ordinate displacement between the two hysteresis loops. In the as-grown state this sample had an exceptionally low ϱ ($\approx 10^{-3}\,\Omega\,cm$) among the samples studied [83I].

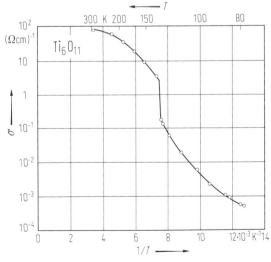

9.15.2.1.3 Phases between Ti_2O_3 and TiO_2

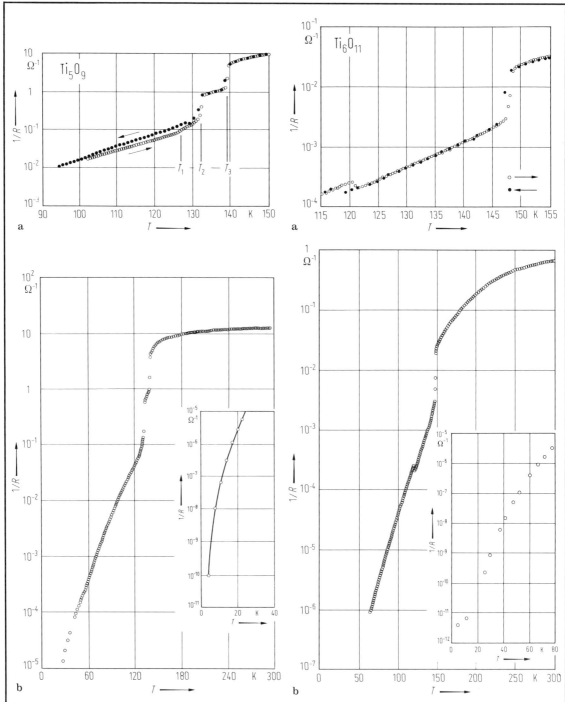

Fig. 13a. Ti_5O_9. Conductance vs. temperature for a single crystal. (a) shows detail of conductance in the transition region with clearly marked transition temperatures T_1, T_2, T_3 and (b) shows conductance over a wide temperature range; inset: low temperature range on an expanded scale [83I].

Fig. 14a. Ti_6O_{11}. Conductance vs. temperature. (a) shows detail of conductance in the transition region and (b) shows conductance over a wide temperature range; inset: low temperature range on an expanded scale [83I].

9.15.2.1.3 Phases between Ti_2O_3 and TiO_2

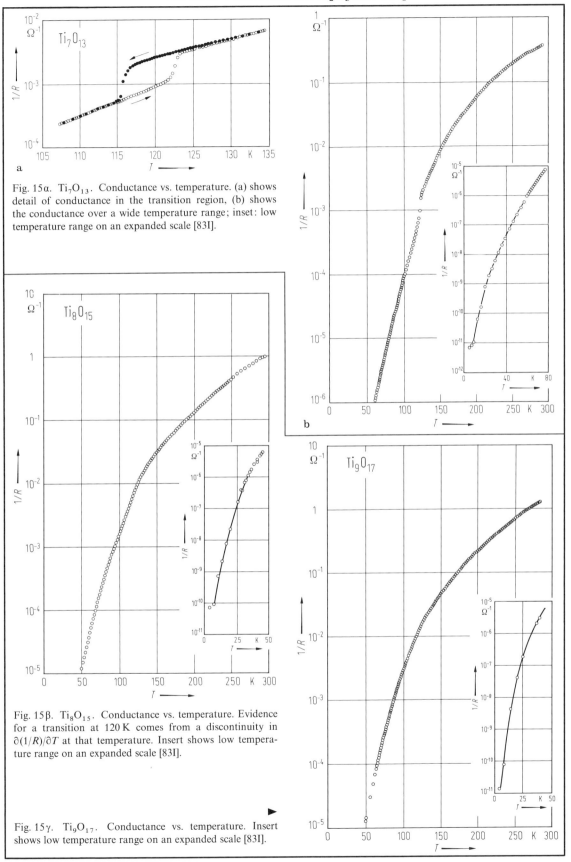

Fig. 15α. Ti_7O_{13}. Conductance vs. temperature. (a) shows detail of conductance in the transition region, (b) shows the conductance over a wide temperature range; inset: low temperature range on an expanded scale [83I].

Fig. 15β. Ti_8O_{15}. Conductance vs. temperature. Evidence for a transition at 120 K comes from a discontinuity in $\partial(1/R)/\partial T$ at that temperature. Insert shows low temperature range on an expanded scale [83I].

Fig. 15γ. Ti_9O_{17}. Conductance vs. temperature. Insert shows low temperature range on an expanded scale [83I].

9.15.2.1.3 Phases between Ti_2O_3 and TiO_2

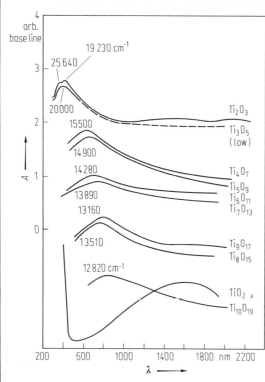

Fig. 16. Ti_nO_{2n-1}. Absorbance vs. wavelength for several intermediate oxides at RT [72P]. Wavenumbers of the maxima are also given.

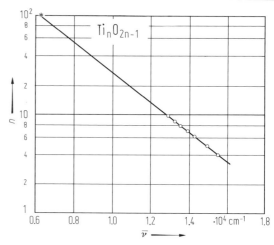

Fig. 17. Ti_nO_{2n-1}. Parameter n vs. wavenumber of absorption maximum at RT in diffuse reflectance spectrum. *: 1.6 μm band in "reduced rutile" [72P].

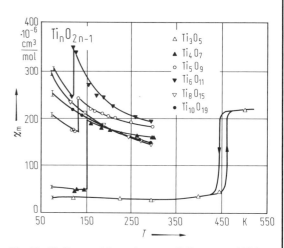

Fig. 18. Ti_nO_{2n-1}. Magnetic susceptibility per mol Ti ions vs. temperature for various microcrystalline samples; χ_m in CGS-emu [72P].

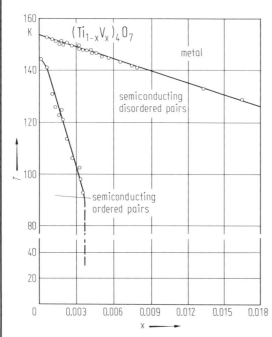

Fig. 20. $(Ti_{1-x}V_x)_4O_7$. Phase diagram showing regions of order/disorder of Ti-Ti pairs [79S].

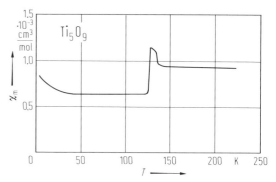

Fig. 19. Ti_5O_9. Molar magnetic susceptibility vs. temperature, single crystal sample. χ_m in CGS-emu [77M1].

9.15.2.1.3 Phases between Ti$_2$O$_3$ and TiO$_2$

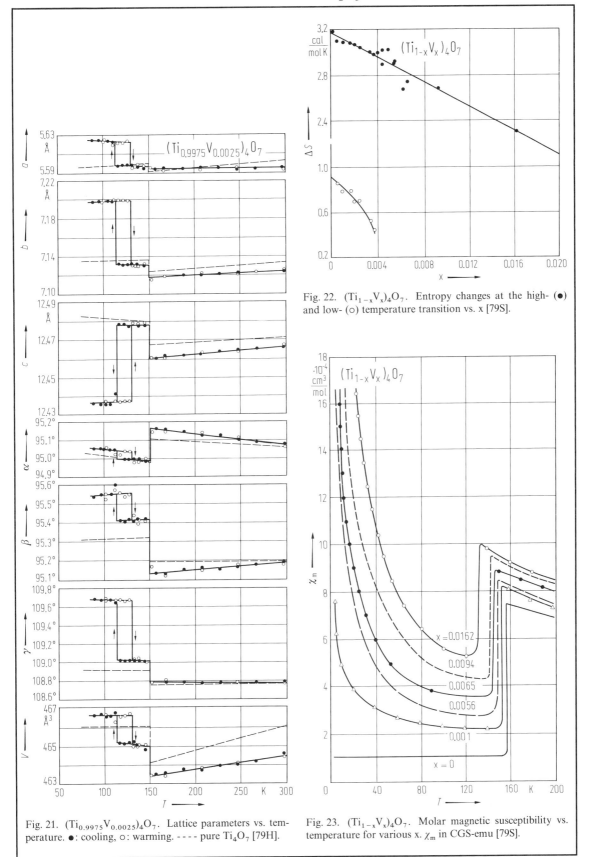

Fig. 21. (Ti$_{0.9975}$V$_{0.0025}$)$_4$O$_7$. Lattice parameters vs. temperature. ●: cooling, ○: warming. ---- pure Ti$_4$O$_7$ [79H].

Fig. 22. (Ti$_{1-x}$V$_x$)$_4$O$_7$. Entropy changes at the high- (●) and low- (○) temperature transition vs. x [79S].

Fig. 23. (Ti$_{1-x}$V$_x$)$_4$O$_7$. Molar magnetic susceptibility vs. temperature for various x. χ_m in CGS-emu [79S].

9.15.2.2.0 Introduction

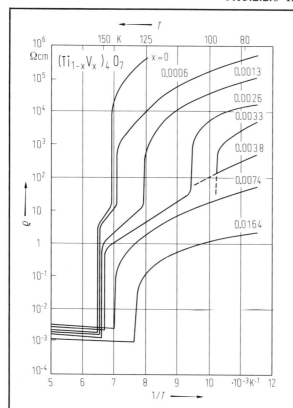

Fig. 24. $(Ti_{1-x}V_x)_4O_7$. Resistivity vs. (reciprocal) temperature for samples of different composition [79S]. Orientation not specified.

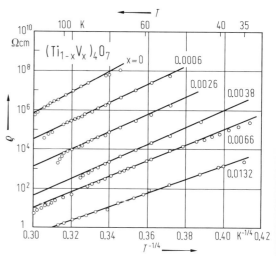

Fig. 25. $(Ti_{1-x}V_x)_4O_7$. Resistivity vs. $T^{-1/4}$ for various x [79S]. Orientation not specified.

9.15.2.2 Vanadium oxides

9.15.2.2.0 Introduction (For tables, see p. 167)

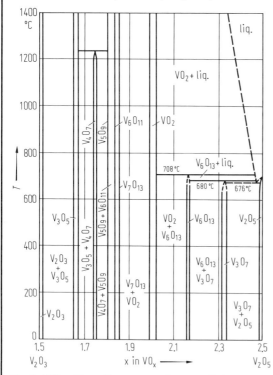

Fig. 1. VO_x. Phase diagram over the range V_2O_3 to V_2O_5 [63K, 67K, 73K].

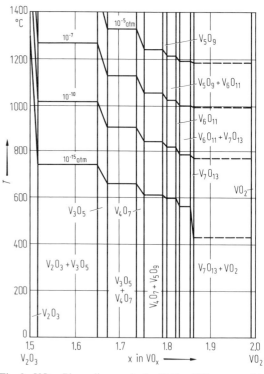

Fig. 2. VO_x. Phase diagram in the $V_2O_3-VO_2$ range with oxygen isobars [63K, 67K, 73K].

9.15.2.2.1 The shear phases V_nO_{2n-1} (For tables, see p. 167 ff.)

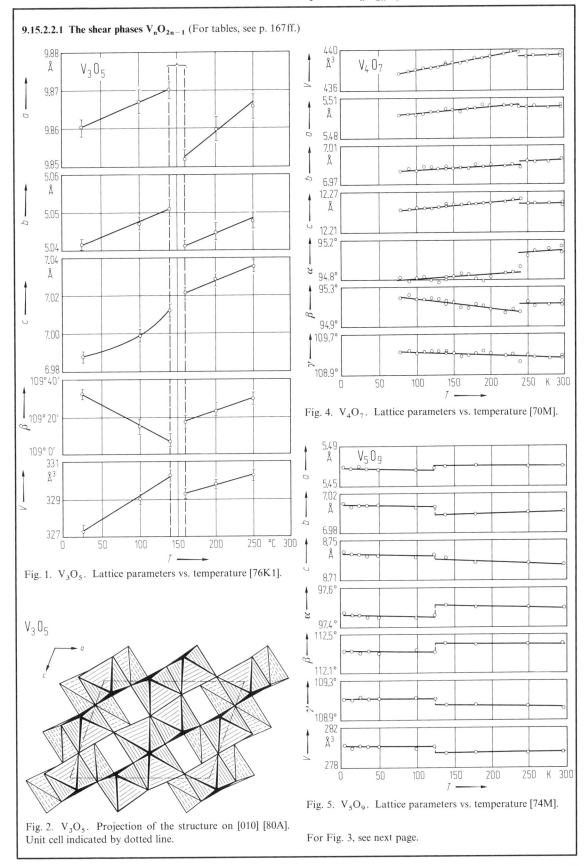

Fig. 1. V_3O_5. Lattice parameters vs. temperature [76K1].

Fig. 2. V_3O_5. Projection of the structure on [010] [80A]. Unit cell indicated by dotted line.

Fig. 4. V_4O_7. Lattice parameters vs. temperature [70M].

Fig. 5. V_5O_9. Lattice parameters vs. temperature [74M].

For Fig. 3, see next page.

9.15.2.2.1 The shear phases V_nO_{2n-1}

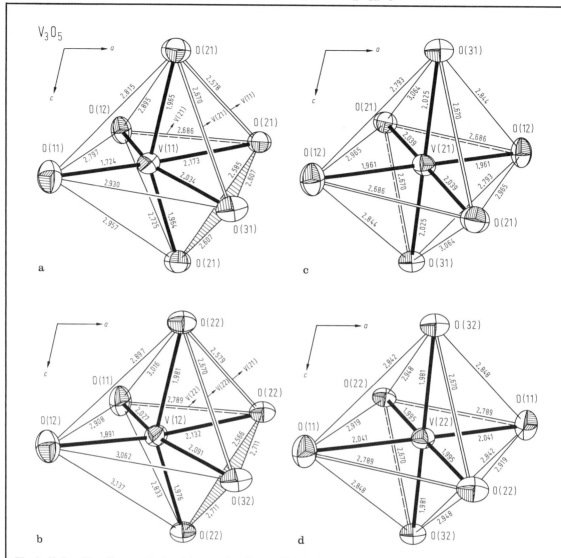

Fig. 3. V_3O_5. Vanadium octahedra: (a) tetravalent V-atom V(11), (b) trivalent V-atom V(12), (c) trivalent V-atom V(21), (d) trivalent V-atom V(22) [80A]. Distances in Å.

9.15.2.2.1 The shear phases V_nO_{2n-1}

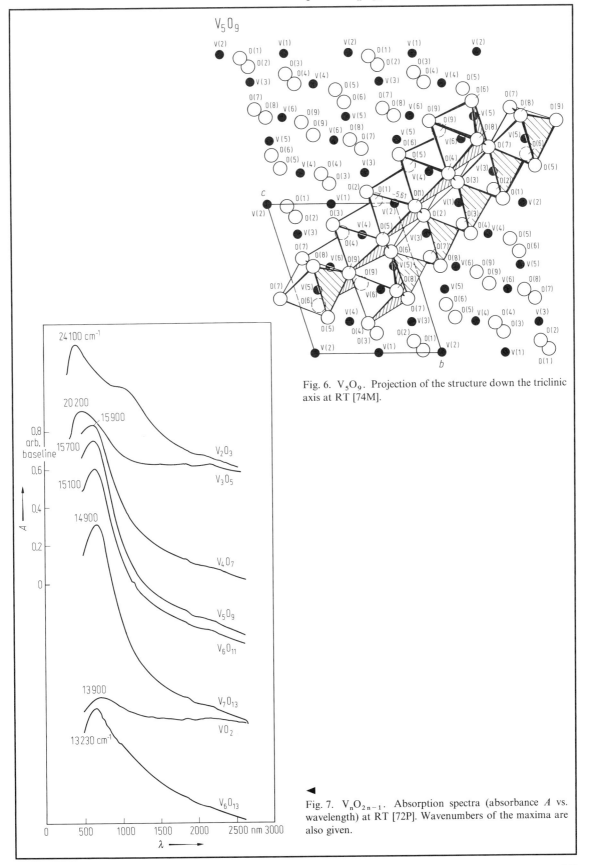

Fig. 6. V_5O_9. Projection of the structure down the triclinic axis at RT [74M].

Fig. 7. V_nO_{2n-1}. Absorption spectra (absorbance A vs. wavelength) at RT [72P]. Wavenumbers of the maxima are also given.

9.15.2.2.1 The shear phases V_nO_{2n-1}

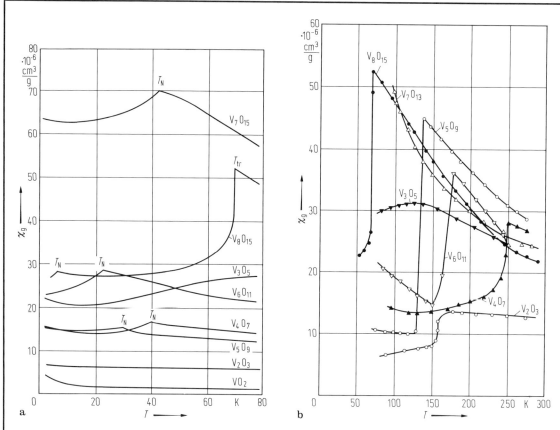

Fig. 8. V_nO_{2n-1}. Magnetic mass susceptibility of polycrystals vs. temperature [72K]. Fig. b shows high temperature range χ_g in CGS-emu.

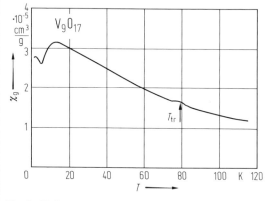

Fig. 9. V_9O_{17}. Low-field magnetic mass susceptibility of polycrystals vs. temperature. The metal-insulator transition is arrowed [81N].

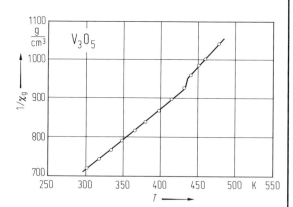

Fig. 10. V_3O_5. Reciprocal magnetic susceptibility of monocrystals vs. temperature [78C]. χ_g in CGS-emu. Orientation not specified.

9.15.2.2.1 The shear phases V_nO_{2n-1}

Fig. 11. V_3O_5. Resistivity and thermoelectric power vs. reciprocal temperature of stoichiometric and Mo-doped single crystals. (*1*) V_3O_5, (*2*) $V_3O_5 + 0.02$ at% Mo, (*3*) $V_3O_5 + 0.04$ at% Mo, (*4*) $V_3O_5 + 0.61$ at% Mo [78C]. Orientation not specified. ▶

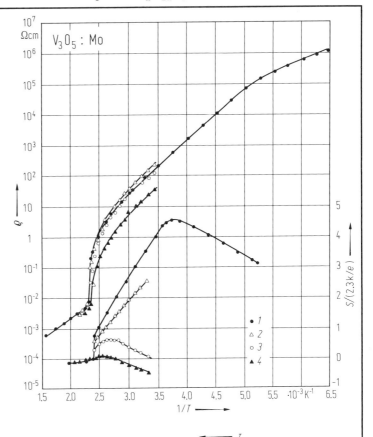

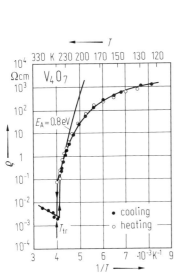

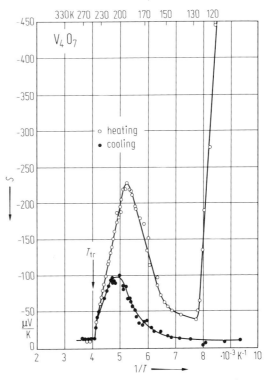

Fig. 12. V_4O_7. Resistivity vs. (reciprocal) temperature [70O1]. Orientation not given.

Fig. 13. V_4O_7. Thermoelectric power vs. (reciprocal) temperature [70O1]. Orientation not given.

9.15.2.2.1 The shear phases V_nO_{2n-1}

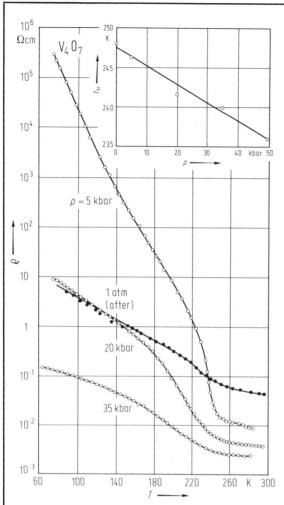

Fig. 14. V_4O_7. Resistivity vs. temperature at different pressures. Inset shows metal-semiconductor transition temperature vs. pressure [73M2]. The curve at 1 atm was obtained after pressurisation. Orientation not specified.

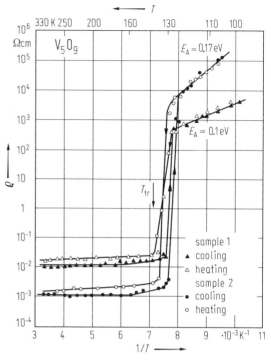

Fig. 15. V_5O_9. Resistivity vs. (reciprocal) temperature for two samples [70O4]. Activation energies are also shown. Orientation not given.

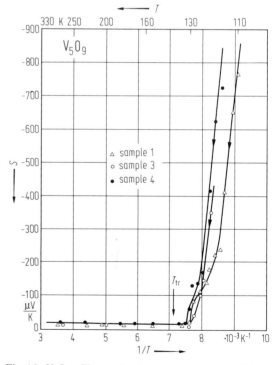

Fig. 16. V_5O_9. Thermoelectric power vs. (reciprocal) temperature for 3 samples [70O4]. Orientation not given.

9.15.2.2.1 The shear phases V_nO_{2n-1}

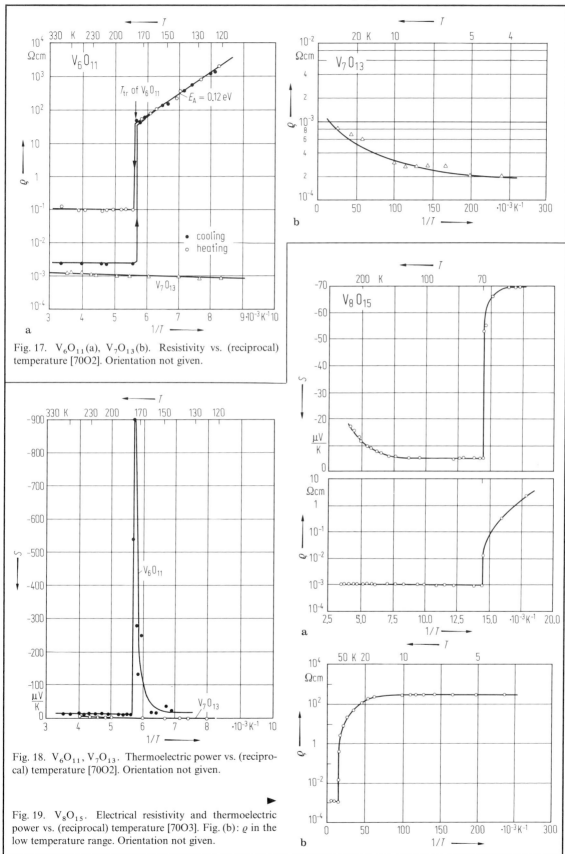

Fig. 17. V_6O_{11}(a), V_7O_{13}(b). Resistivity vs. (reciprocal) temperature [70O2]. Orientation not given.

Fig. 18. V_6O_{11}, V_7O_{13}. Thermoelectric power vs. (reciprocal) temperature [70O2]. Orientation not given.

Fig. 19. V_8O_{15}. Electrical resistivity and thermoelectric power vs. (reciprocal) temperature [70O3]. Fig. (b): ϱ in the low temperature range. Orientation not given.

9.15.2.2.1 The shear phases V_nO_{2n-1}

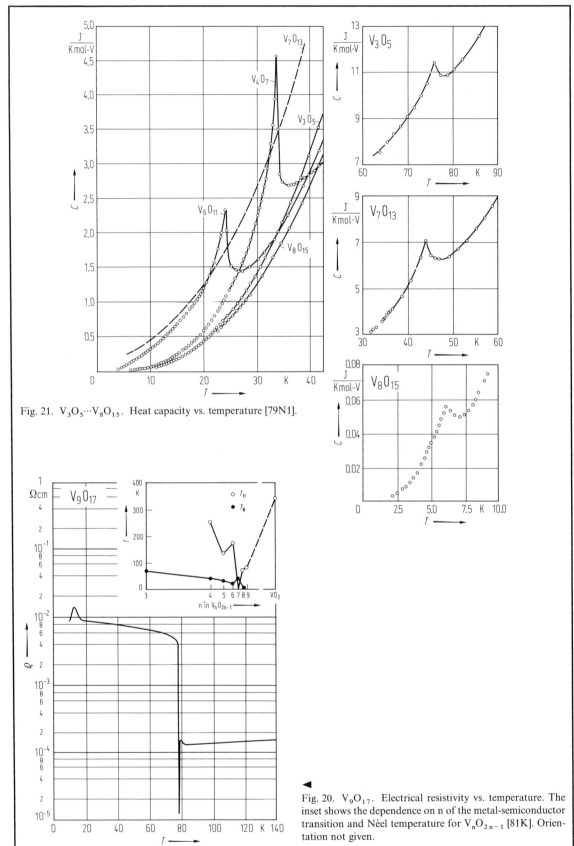

Fig. 21. $V_3O_5 \cdots V_8O_{15}$. Heat capacity vs. temperature [79N1].

Fig. 20. V_9O_{17}. Electrical resistivity vs. temperature. The inset shows the dependence on n of the metal-semiconductor transition and Néel temperature for V_nO_{2n-1} [81K]. Orientation not given.

9.15.2.2.2 The phases V_nO_{2n+1}

9.15.2.2.2 The phases V_nO_{2n+1} (For tables, see p. 173f.)

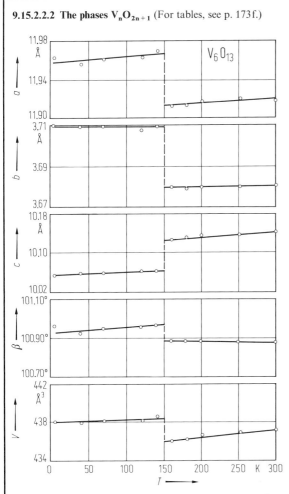

Fig. 1. V_6O_{13}. Lattice parameters vs. temperature [74D].

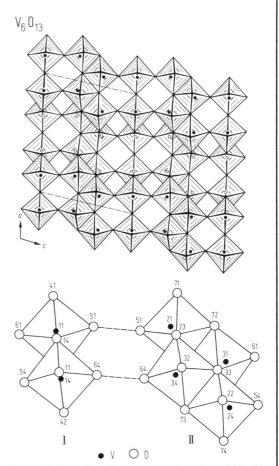

Fig. 2. V_6O_{13}. Crystal structure projected on (010). The two types of structure elements that can be distinguished are shown. The numbering of atoms refers to the tables [71W]. Dashed line in upper figure shows unit cell.
○ V atom above ac plane, ● V atom below ac plane.

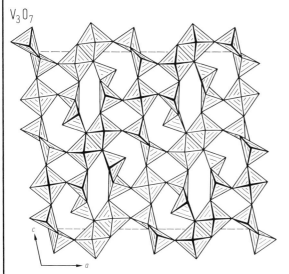

Fig. 3. V_3O_7. Unit cell projected on (010) [74W].

For Fig. 4, see next page.

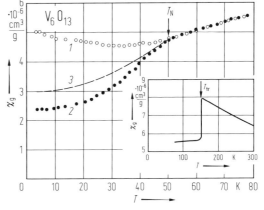

Fig. 5. V_6O_{13}. Magnetic susceptibility vs. temperature of a single crystal (1) along the a-axis, (2) along the b-axis, (3) along the c-axis. The inset illustrates the magnetic susceptibility in the temperature range from 77 K to 300 K [76U]. χ_g in CGS-emu.

9.15.2.2.2 The phases V_nO_{2n+1}

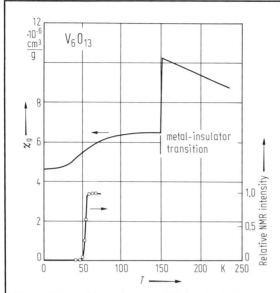

Fig. 4. V_6O_{13}. Magnetic susceptibility and relative NMR intensity vs. temperature for polycrystals. The decrease in NMR intensity between 50 and 60 K is associated with the onset of magnetic order and $T_N \approx 55$ K [74G]. χ_g in CGS-emu.

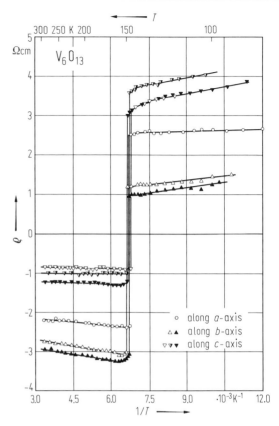

Fig. 6. V_6O_{13}. Electrical resistivity vs. (reciprocal) temperature along different crystallographic axes [74K]. Different symbols for different samples.

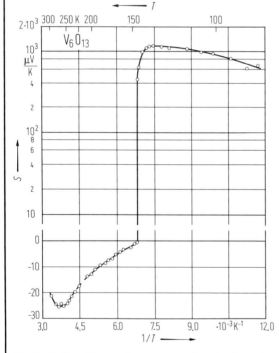

Fig. 7. V_6O_{13}. Thermoelectric power along b axis vs. (reciprocal) temperature [74K].

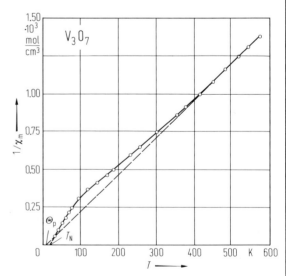

Fig. 8. V_3O_7. Molar magnetic susceptibility vs. temperature [74B]. Orientation not specified. χ_m in CGS-emu.

9.15.2.2.3 Vanadium oxide (V$_2$O$_3$) (For tables, see p. 175ff.)

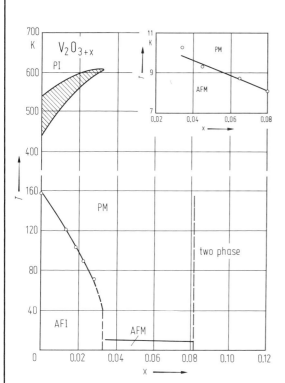

Fig. 1. V$_2$O$_{3+x}$. (Magnetic) phase diagram obtained from experimental results [80U]. PI: paramagnetic insulator, PM: paramagnetic metal, AFI: antiferromagnetic insulator, AFM: antiferromagnetic metal. Hatched area shows region where the upper transition is seen (see tables).

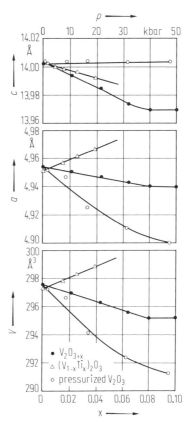

Fig. 3. V$_2$O$_{3+x}$. Comparison of the lattice parameters vs. composition x for V$_2$O$_{3+x}$, (V$_{1-x}$Ti$_x$)$_2$O$_3$ and vs. pressure for V$_2$O$_3$ [70M1].

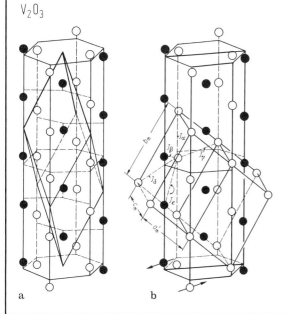

Fig. 2. V$_2$O$_3$. Relationships between V-atoms in hexagonal, rhombohedral and monoclinic unit cells. The primitive rhombohedral axes are shown as bold lines in (a). The magnetic and crystallographic unit cells are shown in (b). Spins on the filled circles are opposite to those on the open circles. Assuming equivalent metal atoms, the magnetic monoclinic cell is half ($a'_m = \frac{1}{2}a_m$) of the crystallographic cell. [81W]. $J_\alpha \cdots J_\varepsilon$ are magnetic coupling constants.

9.15.2.2.3 Vanadium oxide (V_2O_3)

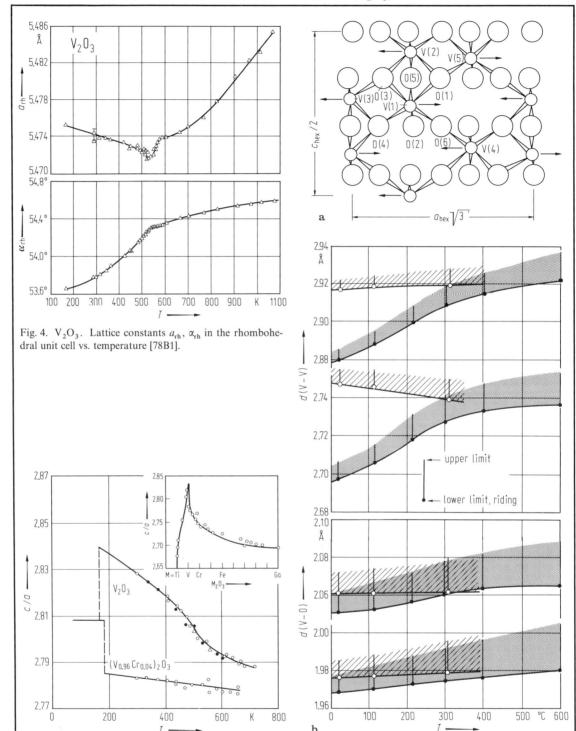

Fig. 4. V_2O_3. Lattice constants a_{rh}, α_{rh} in the rhombohedral unit cell vs. temperature [78B1].

Fig. 5. V_2O_3, $(V_{0.96}Cr_{0.04})_2O_3$. c/a ratios vs. temperature. Inset: c/a ratios for metal oxides of the corundum structure. ○ first heating cycle, △ second heating cycle, ● cooling cycle [69M2].

Fig. 6. V_2O_3. (a) Projection of the structure onto a plane perpendicular to $[\bar{1}\bar{1}0]$. The V-atoms are at zero height and the O are above and below the plane. The arrows indicate the direction of translation of V in the M−AF transition [70D1], (b) variation of V−V and V−O distances for V_2O_3 (●), β-Cr−V_2O_3 (○), and ranges of thermal corrections for V_2O_3 (dot shaded) and β-Cr−V_2O_3 (line shaded) [75R].

9.15.2.2.3 Vanadium oxide (V$_2$O$_3$)

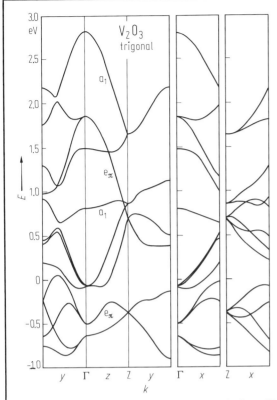

Fig. 7. V$_2$O$_3$. Energy dispersion in the t$_{2g}$-band of metallic V$_2$O$_3$ [76A2].

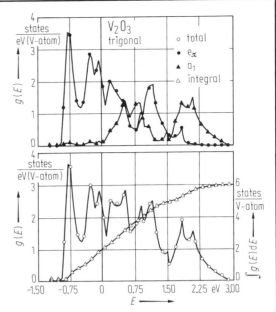

Fig. 8. V$_2$O$_3$. Density of states in the t$_{2g}$-band of metallic V$_2$O$_3$ [76A2].

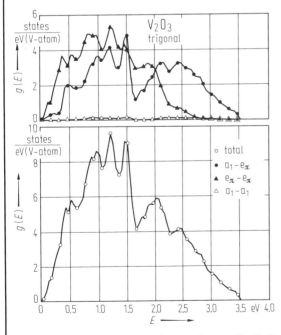

Fig. 9. V$_2$O$_3$. Joint density of states in metallic V$_2$O$_3$ [76A2].

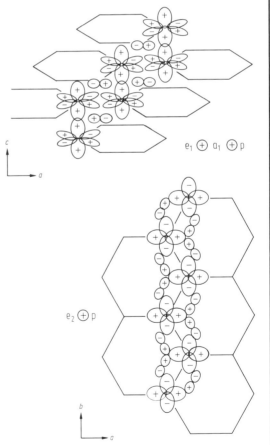

Fig. 10. V$_2$O$_3$. Ferromagnetic and antiferromagnetic chains [76A2].

9.15.2.2.3 Vanadium oxide (V_2O_3)

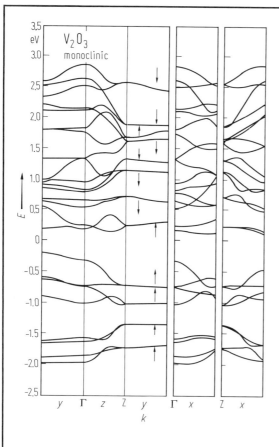

Fig. 11. V_2O_3. Energy dispersion in the t_{2g}-band of monoclinic V_2O_3; the arrows indicate the direction of the magnetic moments on sites where a spin-up state of the band is concentrated [76A2].

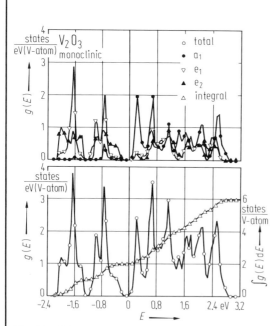

Fig. 12. V_2O_3. Density of states in monoclinic V_2O_3 [76A2].

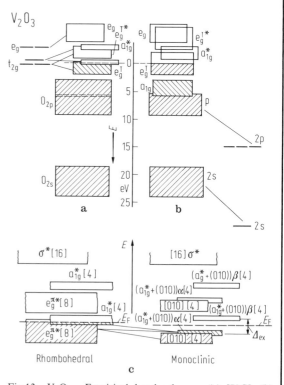

Fig. 13. V_2O_3. Empirical band schemes: (a) [70G], (b) [72H], (c) HT–LT transition [70G, 76S]. Numbers in brackets indicate number of band states per molecule.

9.15.2.2.3 Vanadium oxide (V$_2$O$_3$)

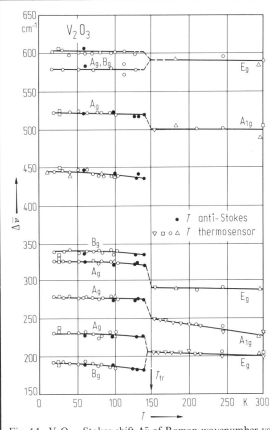

Fig. 14. V$_2$O$_3$. Stokes shift $\Delta\bar{\nu}$ of Raman wavenumber vs. temperature [77K2]. T was estimated by a thermosensor, if needed a better estimate was obtained from the intensity ratios of the Stokes/anti-Stokes scatterings.

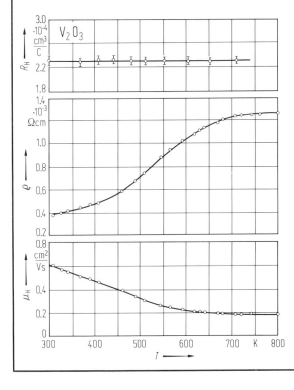

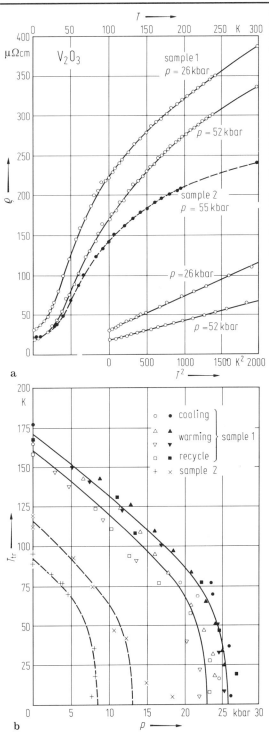

Fig. 15. V$_2$O$_3$. (a) Resistivity vs. temperature in the metallic phase above the critical pressure for two samples. Sample 2 is apparently V deficient. (b) AF−M transition temperature vs. pressure [69M1].

◀

Fig. 16. V$_2$O$_3$. Resistivity, Hall coefficient and Hall mobility vs. temperature for metallic V$_2$O$_3$. Magnetic field along c-axis, current in basal plane [69A].

9.15.2.2.3 Vanadium oxide (V$_2$O$_3$)

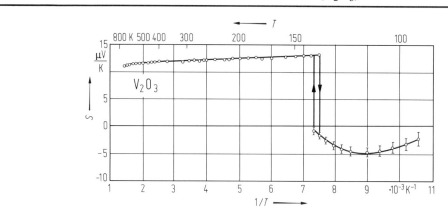

Fig. 17. V$_2$O$_3$. Seebeck coefficient vs. temperature above and below T_{tr} [69A]. Orientation not given.

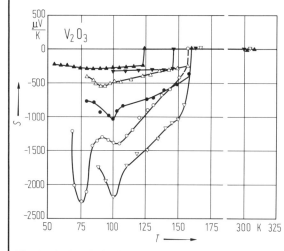

Fig. 18. V$_2$O$_3$. Seebeck coefficient vs. temperature for pure and Ti-doped V$_2$O$_3$. (▽) first cooling, (●) second cooling, (△) second warming, (▼) 0.1 at % Ti, (▲) 1 at % Ti. A second sample was also investigated, (○) cooling [75K]. Orientation not given.

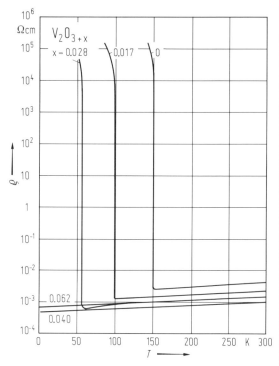

Fig. 19. V$_2$O$_{3+x}$. Resistivity vs. temperature on cooling [80U]. Orientation not given.

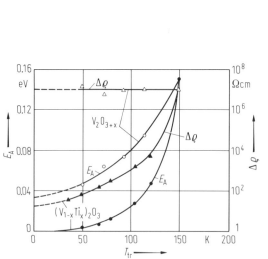

Fig. 20. V$_2$O$_{3+x}$, (V$_{1-x}$Ti$_x$)$_2$O$_3$. Comparison of activation energy E_A and discontinuity in resistivity ($\Delta\varrho$) vs. transition temperature T_{tr} [80U]. $\Delta\varrho = \varrho(T<T_{tr}) - \varrho(T>T_{tr})$.

9.15.2.2.3 Vanadium oxide (V_2O_3)

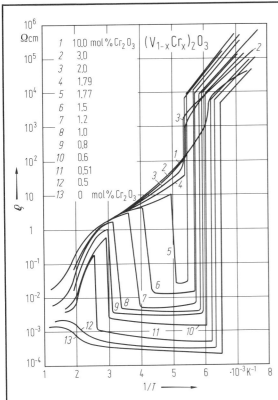

Fig. 21. $(V_{1-x}Cr_x)_2O_3$. Resistivity vs. reciprocal temperature [80K1]. Orientation not given.

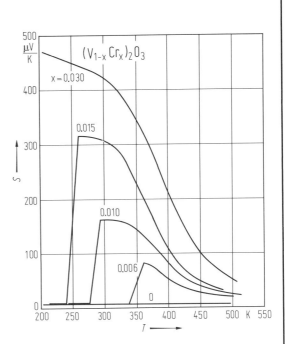

Fig. 22. $(V_{1-x}Cr_x)_2O_3$. Seebeck coefficient vs. temperature [80K1]. Orientation not given.

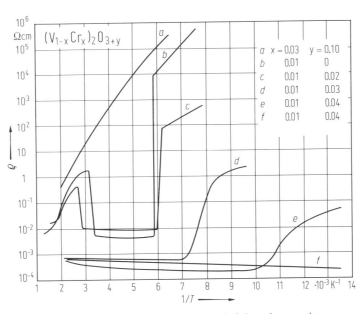

Fig. 23. $(V_{1-x}Cr_x)_2O_{3+y}$. Resistivity vs. reciprocal temperature [80K2]. Orientation not given.

9.15.2.2.3 Vanadium oxide (V_2O_3)

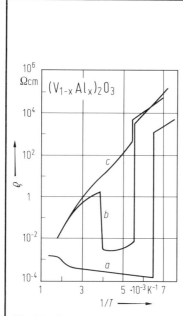

Fig. 24. $(V_{1-x}Al_x)_2O_3$. Electrical resistivity vs. reciprocal temperature for (curve a) x = 0.0033, (b) = 0.011, (c) = 0.02 [76K2]. Orientation not given.

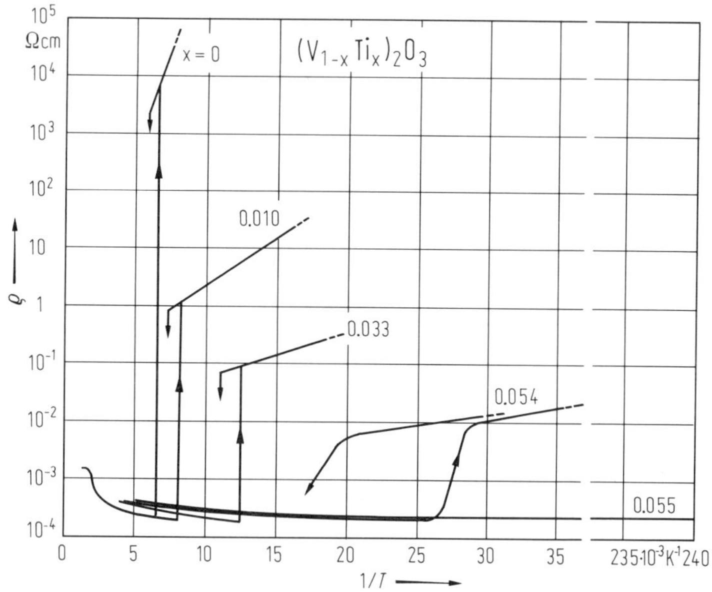

Fig. 25. $(V_{1-x}Ti_x)_2O_3$. Electrical resistivity vs. reciprocal temperature [76K3]. Orientation not given.

For Fig. 26, see next page.

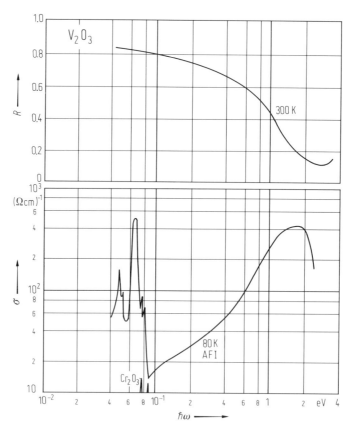

Fig. 27. V_2O_3. Reflectivity and optical conductivity of pure samples in the M and AF phases [70B, 71Z].

9.15.2.2.3 Vanadium oxide (V$_2$O$_3$)

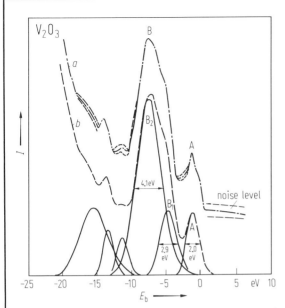

Fig. 26. V$_2$O$_3$. XPE spectra (intensity vs. binding energy) of a single crystal, relative to E_F for gold. (*a*) original data, (*b*) computer-based resolution into component Gaussians [72H].

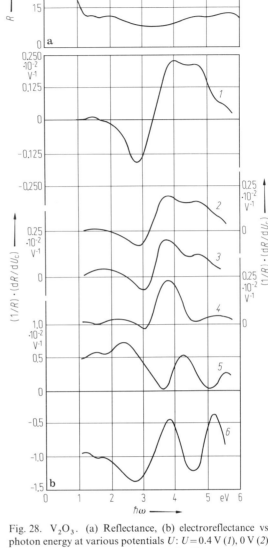

Fig. 28. V$_2$O$_3$. (a) Reflectance, (b) electroreflectance vs. photon energy at various potentials U: $U = 0.4$ V (*1*), 0 V (*2*), 1.5 V (*3*), 1.7 V (*4*), 2.2 V (*5*), 2.5 V (*6*) [71V].

For Fig. 29, see next page.

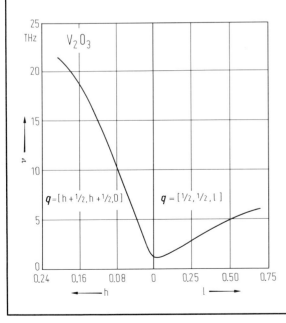

Fig. 30. V$_2$O$_3$. Computed magnon dispersion curves [81W].

9.15.2.2.3 Vanadium oxide (V_2O_3)

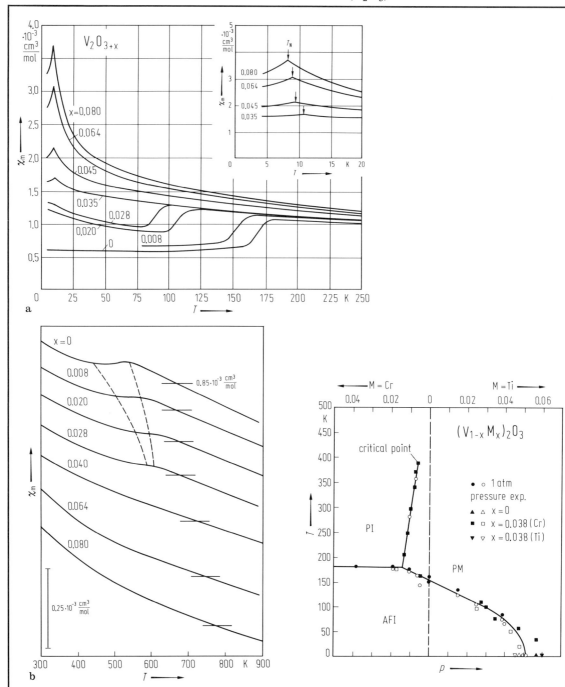

Fig. 29. V_2O_{3+x}. (a) Magnetic molar susceptibility below 250 K. For $x > 0.035$ no AF–M transition is seen. (b) Magnetic molar susceptibility above 300 K. The upper transition becomes smeared out at $x > 0.04$ [80U]. χ_m in CGS-emu. Orientation χ_m not given.

Fig. 31. $(V_{1-x}M_x)_2O_3$. Generalized phase diagram for the AF–M transition vs. doping with Cr and Ti and vs. pressure. Lower abscissa: increasing pressure 4 kbar/division, zero pressure point moves with top scale [80U].

9.15.2.2.4 Vanadium oxide (VO$_2$)

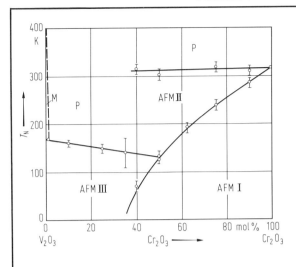

Fig. 32. (Cr$_x$V$_{1-x}$)$_2$O$_3$. Phase diagram defined by magnetic and crystallographic boundaries. Full lines by neutron diffraction, dashed line shows the metal-insulator boundary, which is also defined by crystallographic discontinuity. P: paramagnetic, AFM I: Cr$_2$O$_3$-type magnetic order, AFM II: new type of corundum magnetic ordering, AFM III: monoclinic V$_2$O$_3$, M: metallic conduction, corundum lattice [72R].

9.15.2.2.4 Vanadium oxide (VO$_2$) (For tables, see p. 185ff.)

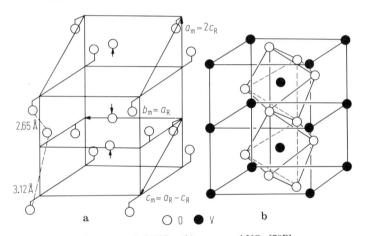

Fig. 1. VO$_2$. Structure and unit cell of (a) monoclinic VO$_2$, (b) tetragonal VO$_2$ [78P].

For Fig. 2, see next page.

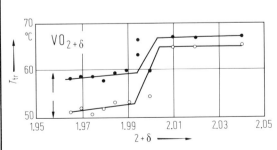

Fig. 3. VO$_{2\pm\delta}$. Transition temperature vs. stoichiometry [75B1]. Full circles: on heating, open circles: on cooling.

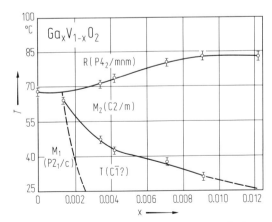

Fig. 4. Ga$_x$V$_{1-x}$O$_2$. Phase diagram [78P]. R: rutile (structure), M$_1$, M$_2$: monoclinic phases, T: triclinic phase.

9.15.2.2.4 Vanadium oxide (VO$_2$)

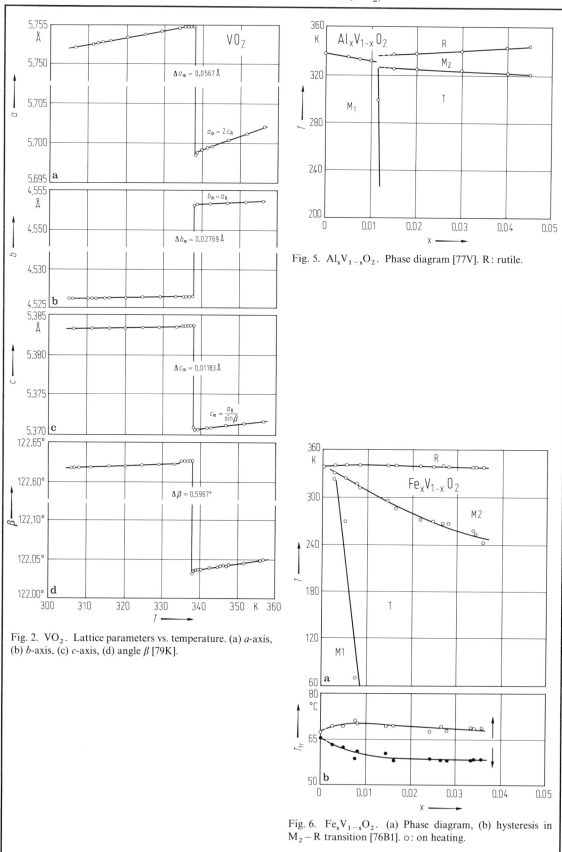

Fig. 2. VO$_2$. Lattice parameters vs. temperature. (a) a-axis, (b) b-axis, (c) c-axis, (d) angle β [79K].

Fig. 5. Al$_x$V$_{1-x}$O$_2$. Phase diagram [77V]. R: rutile.

Fig. 6. Fe$_x$V$_{1-x}$O$_2$. (a) Phase diagram, (b) hysteresis in M$_2$–R transition [76B1]. ○: on heating.

9.15.2.2.4 Vanadium oxide (VO$_2$)

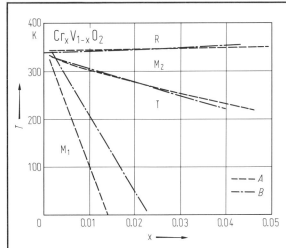

Fig. 7. $Cr_xV_{1-x}O_2$. Phase diagram (A acc. to [73V], B acc. to [72M], [74P1]).

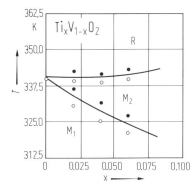

Fig. 8. $Ti_xV_{1-x}O_2$. Phase diagram for small x [76H2].

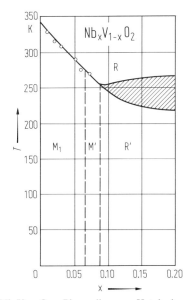

Fig. 9. $Nb_xV_{1-x}O_2$. Phase diagram. Hatched area represents uncertainty in the broad DTA peaks [76P]. M', R' phases: see tables.

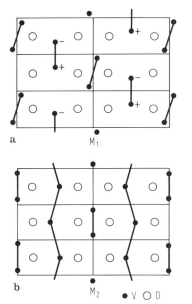

Fig. 10. VO$_2$. Schematic projection of the M_1 and M_2 structures onto $[110]_R$ [74D]. + and − signs refer to V atoms above and below the plane of paper.

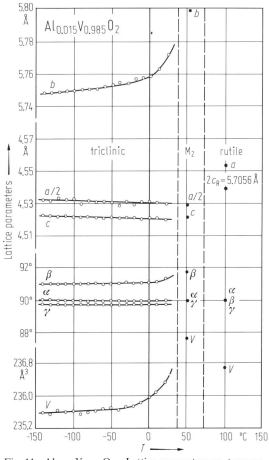

Fig. 11. $Al_{0.015}V_{0.985}O_2$. Lattice parameters vs. temperature [77G1]. R: rutile.

9.15.2.2.4 Vanadium oxide (VO$_2$)

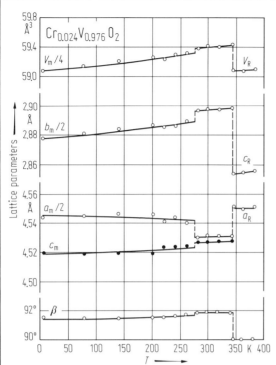

Fig. 12. $Cr_{0.024}V_{0.976}O_2$. Lattice parameters vs. temperature [72M]. Subscript m: monoclinic.

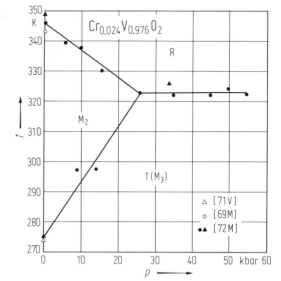

Fig. 13. $Cr_{0.024}V_{0.976}O_2$. Phase diagram. Experimental data according to [71V], [69M], [72M].

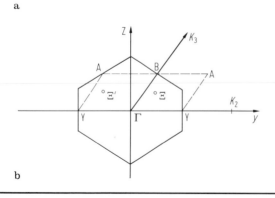

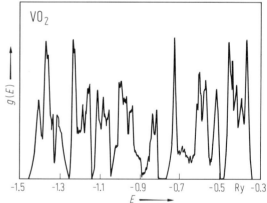

Fig. 16. VO$_2$. Density of states corresponding to the energy bands of Fig. 15 [73C2].

For Fig. 15, see next page.

Fig. 14. VO$_2$. a) Brillouin zone for monoclinic VO$_2$; b) the k_1 plane of the zone shown in a) [73C2]. $K_{1,2,3}$: reciprocal lattice vectors $\left(K_1 = \left(\frac{2\pi}{a_R}\right)\hat{x};\ K_2 = \left(\frac{2\pi}{a_R}\right)\hat{y};\ K_3 = \left(\frac{\pi}{a_R}\right)\hat{y} + \left(\frac{\pi}{c_R}\right)\hat{z};\ \hat{x},\ \hat{y},\ \hat{z}:\ \text{rectangular unit vectors}\right)$; k is an arbitrary vector with $k = 2\pi(k_1/a_R)\hat{x} + (k_2/a_R + \frac{1}{2}k_3/a_R)\hat{y} + \frac{1}{2}(k_3/c_R)\hat{z}$.

9.15.2.2.4 Vanadium oxide (VO$_2$)

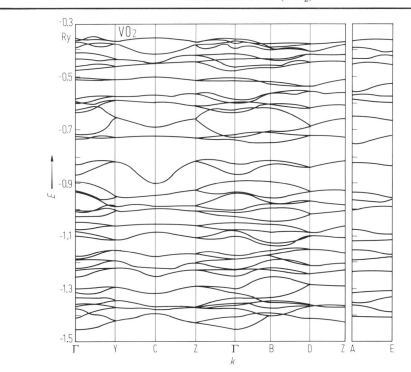

Fig. 15. VO$_2$. Energy bands for the d-band region for monoclinic VO$_2$ [73C2]. 1 Ry = 13.62 eV.

Fig. 17. VO$_2$. See Fig. 9 of section 9.15.2.1.1, p. 415.

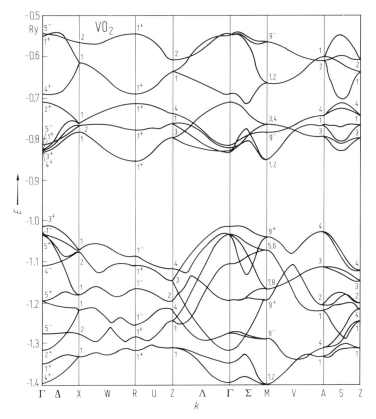

Fig. 18. VO$_2$. Energy bands for the d-band region for rutile (metallic) VO$_2$ [73C1].

9.15.2.2.4 Vanadium oxide (VO$_2$)

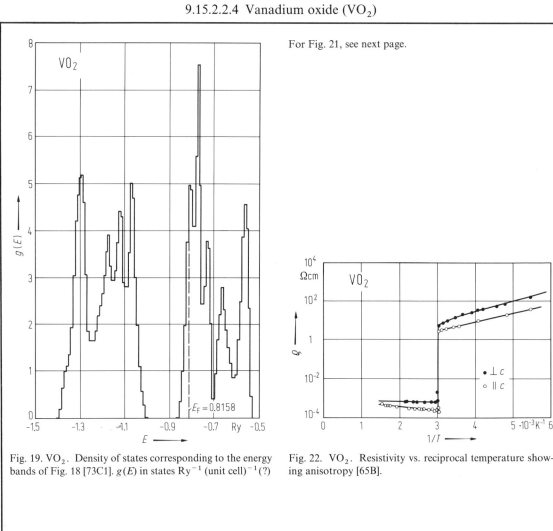

Fig. 19. VO$_2$. Density of states corresponding to the energy bands of Fig. 18 [73C1]. $g(E)$ in states Ry^{-1} (unit cell)$^{-1}$(?)

For Fig. 21, see next page.

Fig. 22. VO$_2$. Resistivity vs. reciprocal temperature showing anisotropy [65B].

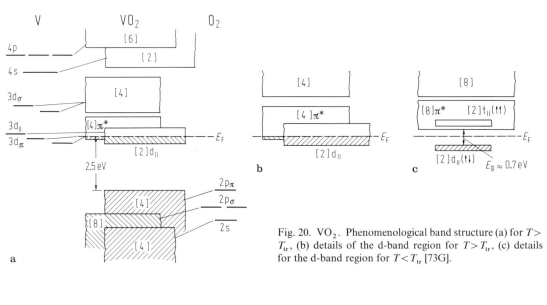

Fig. 20. VO$_2$. Phenomenological band structure (a) for $T > T_{tr}$, (b) details of the d-band region for $T > T_{tr}$, (c) details for the d-band region for $T < T_{tr}$ [73G].

9.15.2.2.4 Vanadium oxide (VO$_2$)

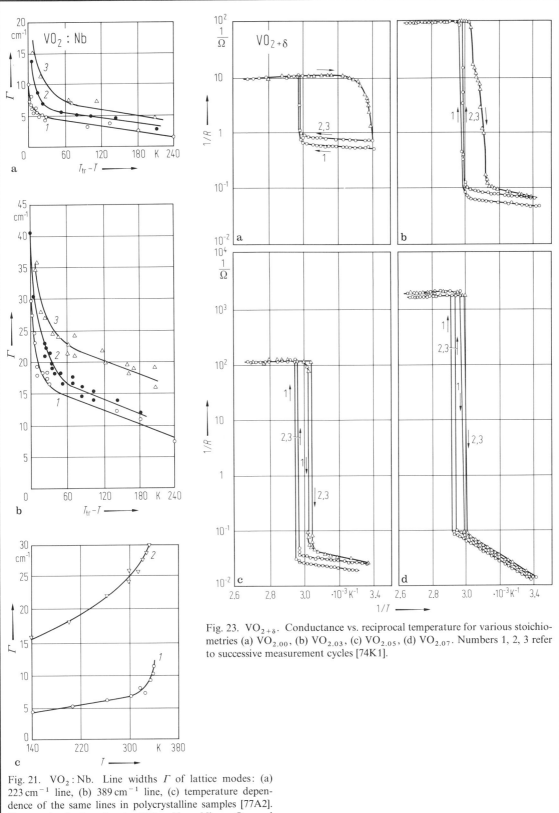

Fig. 23. VO$_{2+\delta}$. Conductance vs. reciprocal temperature for various stoichiometries (a) VO$_{2.00}$, (b) VO$_{2.03}$, (c) VO$_{2.05}$, (d) VO$_{2.07}$. Numbers 1, 2, 3 refer to successive measurement cycles [74K1].

Fig. 21. VO$_2$:Nb. Line widths Γ of lattice modes: (a) 223 cm^{-1} line, (b) 389 cm^{-1} line, (c) temperature dependence of the same lines in polycrystalline samples [77A2]. Curves 1, 2, 3 refer to VO$_2$, V$_{0.983}$Nb$_{0.017}$O$_2$ and V$_{0.976}$Nb$_{0.024}$O$_2$.

9.15.2.2.4 Vanadium oxide (VO$_2$)

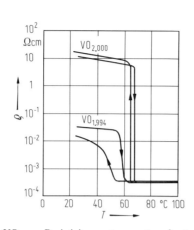

Fig. 24. VO$_{2-\delta}$. Resistivity vs. temperature for two samples of different stoichiometry in the tetragonal a direction [75B2].

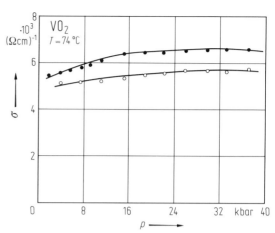

Fig. 25. VO$_2$. Conductivity vs. hydrostatic pressure for $T > T_{tr}$. The two curves are two different samples, measurement along c_R axis [69B2].

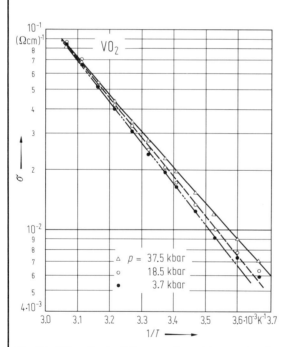

Fig. 26. VO$_2$. Conductivity vs. reciprocal temperature for three different pressures, $T < T_{tr}$, along c_R axis [69B2].

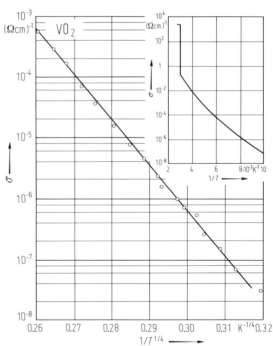

Fig. 27. VO$_2$. Conductivity vs. $T^{-1/4}$; inset: conductivity vs. reciprocal temperature [74S]. Orientation σ not given.

9.15.2.2.4 Vanadium oxide (VO$_2$)

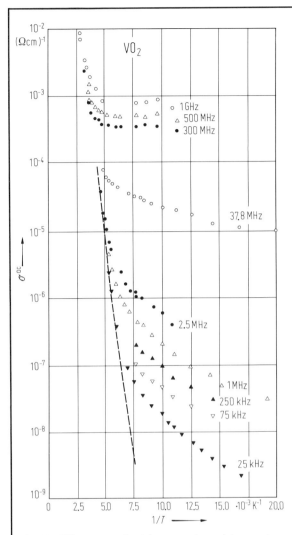

Fig. 28. VO$_2$. ac conductivity vs. reciprocal temperature [74P2]. Orientation not given.

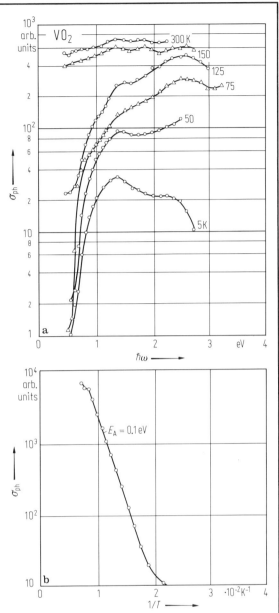

Fig. 29. VO$_2$. (a) Photoconductivity vs. photon energy at different temperatures. (b) shows photoconductivity for a fixed photon energy of 1.4 eV vs. reciprocal temperature [74S].

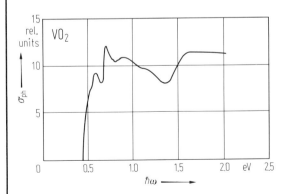

Fig. 29a. Photoconductivity vs. photon energy for a single crystal at 150 K [80K].

9.15.2.2.4 Vanadium oxide (VO$_2$)

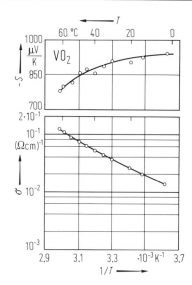

Fig. 30. VO$_2$. Thermoelectric power and conductivity vs. reciprocal temperature ($T < T_{tr}$) for a single-crystalline sample [66K2]. (Sign of S is assumed to be negative.)

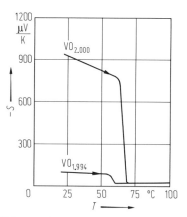

Fig. 31. VO$_{2-\delta}$. Thermoelectric power vs. temperature for samples of different stoichiometry [75B2]. Orientation S not specified.

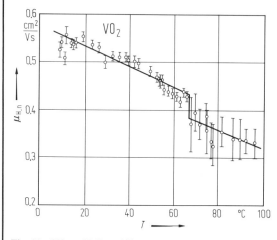

Fig. 32. VO$_2$. Hall mobility $\|c_R$ vs. temperature near T_{tr} [73R].

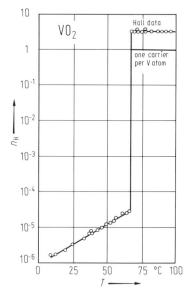

Fig. 33. VO$_2$. (Hall) carrier concentration (per V atom) vs. temperature near T_{tr} [73R].

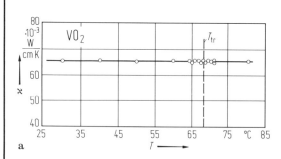
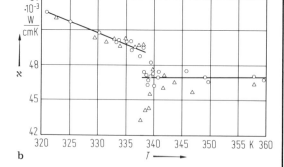

Fig. 34. VO$_2$. Thermal conductivity vs. temperature (a) according to [69B1] (b) according to [78A]. Orientation κ not given. Different symbols correspond to different samples.

9.15.2.2.4 Vanadium oxide (VO$_2$)

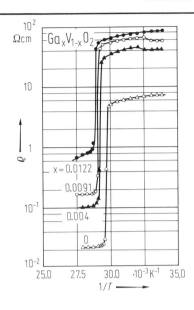

Fig. 35. $Ga_xV_{1-x}O_2$. Resistivity vs. reciprocal temperature for polycrystalline samples of different composition [78P].

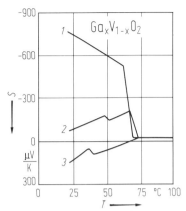

Fig. 36. $Ga_xV_{1-x}O_2$. Thermoelectric power vs. temperature for polycrystalline samples with x = 0 (1), 0.0039 (2) and 0.0078 (3) [76B2].

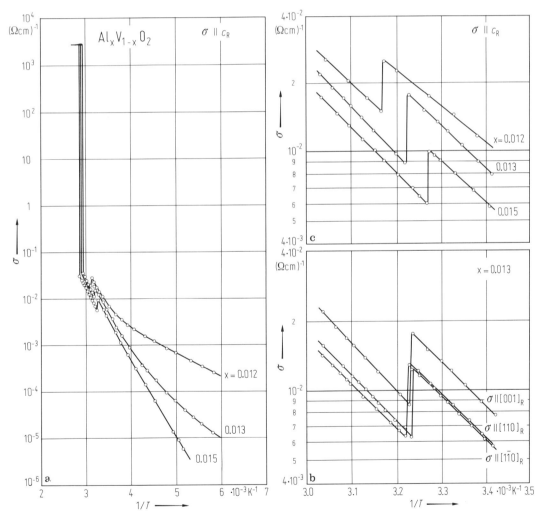

Fig. 37. $Al_xV_{1-x}O_2$. Conductivity $\parallel c_R$ vs. reciprocal temperature for samples of different composition, (b) conductivity vs. reciprocal temperature for different orientations and x = 0.013, (c) details of conductivity $\parallel c_R$ near the $T-M_2$ transition [75V].

9.15.2.2.4 Vanadium oxide (VO$_2$)

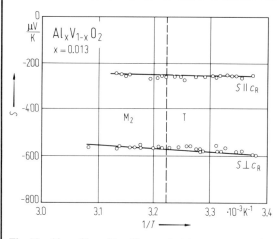

Fig. 38. Al$_{0.013}$V$_{0.987}$O$_2$. Thermoelectric power vs. reciprocal temperature near the T−M$_2$ transition [75V].

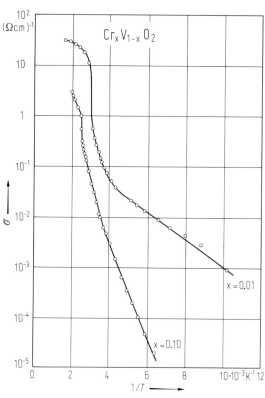

Fig. 39. Cr$_x$V$_{1-x}$O$_2$. Electrical conductivity vs. reciprocal temperature for two polycrystalline samples of different composition [71V].

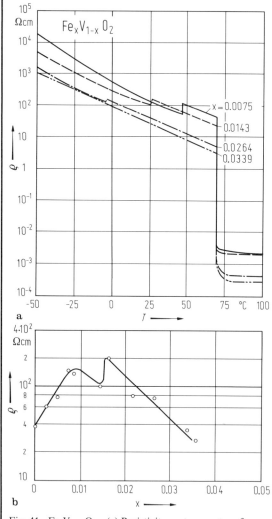

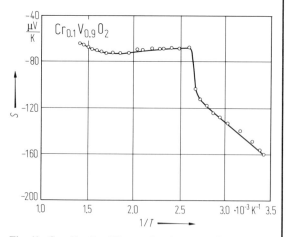

Fig. 41. Fe$_x$V$_{1-x}$O$_2$. (a) Resistivity vs. temperature for polycrystalline samples of different composition, (b) shows ϱ (20°C) vs. x [76B1].

Fig. 40. Cr$_{0.1}$V$_{0.9}$O$_2$. Thermoelectric power of polycrystalline sample vs. reciprocal temperature [71V].

9.15.2.2.4 Vanadium oxide (VO$_2$)

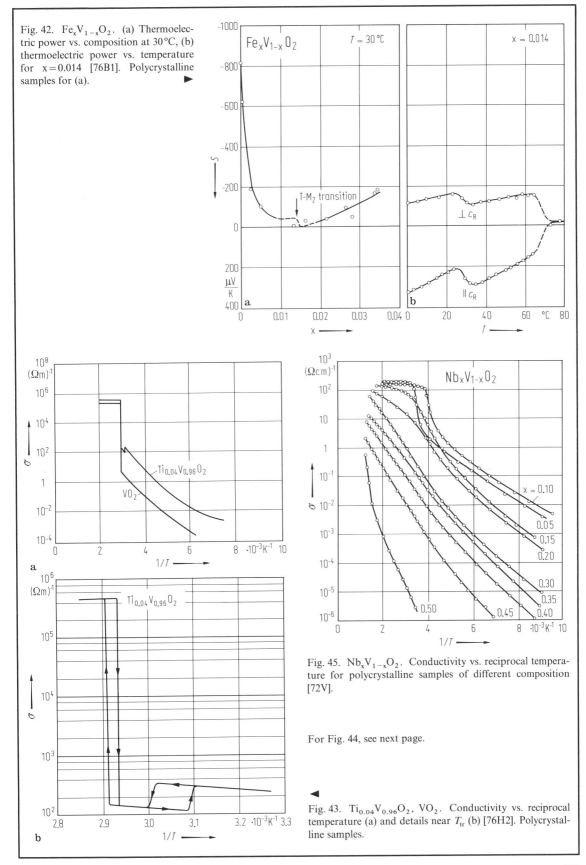

Fig. 42. Fe$_x$V$_{1-x}$O$_2$. (a) Thermoelectric power vs. composition at 30°C, (b) thermoelectric power vs. temperature for $x = 0.014$ [76B1]. Polycrystalline samples for (a).

Fig. 45. Nb$_x$V$_{1-x}$O$_2$. Conductivity vs. reciprocal temperature for polycrystalline samples of different composition [72V].

For Fig. 44, see next page.

Fig. 43. Ti$_{0.04}$V$_{0.96}$O$_2$, VO$_2$. Conductivity vs. reciprocal temperature (a) and details near T_{tr} (b) [76H2]. Polycrystalline samples.

9.15.2.2.4 Vanadium oxide (VO$_2$)

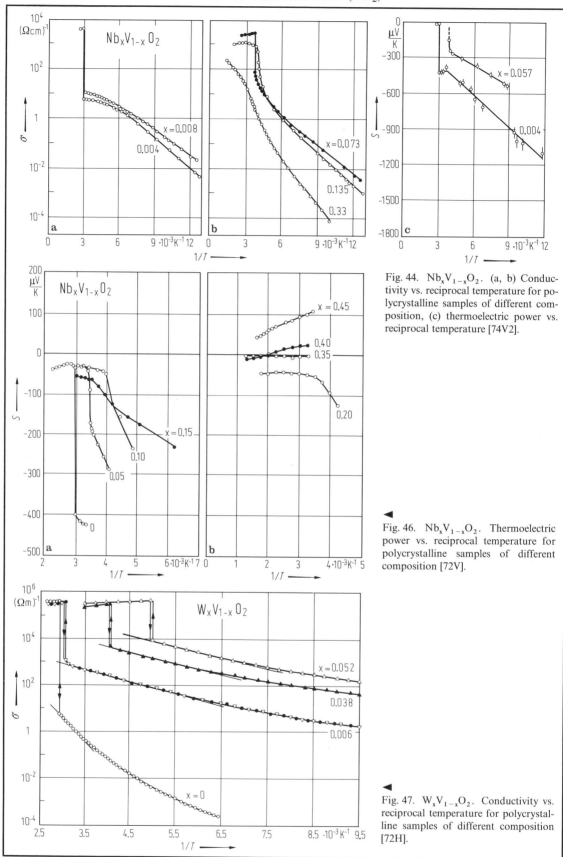

Fig. 44. Nb$_x$V$_{1-x}$O$_2$. (a, b) Conductivity vs. reciprocal temperature for polycrystalline samples of different composition, (c) thermoelectric power vs. reciprocal temperature [74V2].

Fig. 46. Nb$_x$V$_{1-x}$O$_2$. Thermoelectric power vs. reciprocal temperature for polycrystalline samples of different composition [72V].

Fig. 47. W$_x$V$_{1-x}$O$_2$. Conductivity vs. reciprocal temperature for polycrystalline samples of different composition [72H].

9.15.2.2.4 Vanadium oxide (VO$_2$)

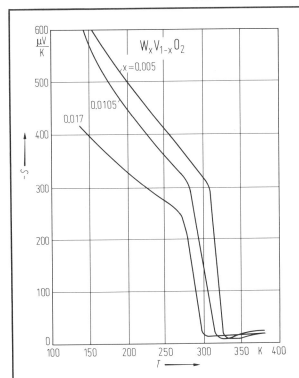

Fig. 48. $W_xV_{1-x}O_2$. Thermoelectric power vs. temperature for polycrystalline samples of different composition [73H2].

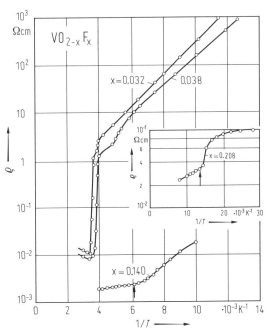

Fig. 51. $VO_{2-x}F_x$. Resistivity vs. reciprocal temperature for polycrystalline samples of different composition [71B]. Arrows indicate transition temperatures.

For Fig. 50, see next page.

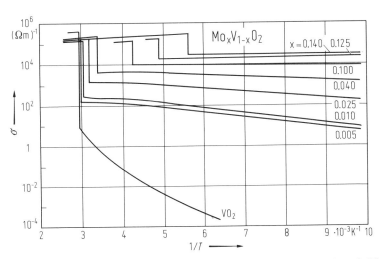

Fig. 49. $Mo_xV_{1-x}O_2$. Conductivity vs. reciprocal temperature for single-crystalline samples of different composition [73H2]. $\sigma \parallel c_R$.

9.15.2.2.4 Vanadium oxide (VO$_2$)

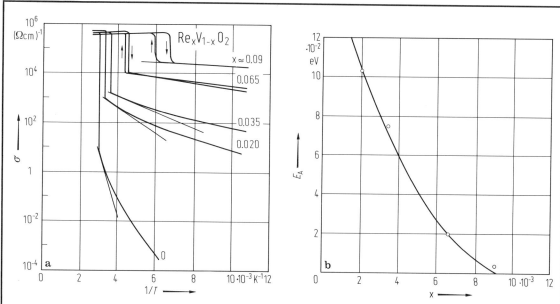

Fig. 50. Re$_x$V$_{1-x}$O$_2$. (a) Conductivity vs. reciprocal temperature for polycrystalline samples of different composition, (b) activation energy vs. composition [77S].

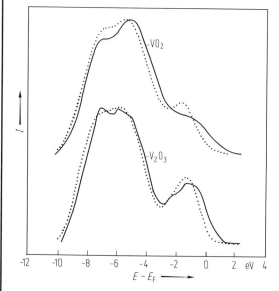

Fig. 52. VO$_2$, V$_2$O$_3$. XPE spectrum (intensity vs. binding energy). —— metallic phase, ··· semiconducting phase [79S].

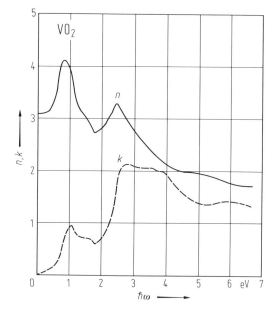

Fig. 53. VO$_2$. Real (n) and imaginary (k) parts of the refractive index vs. photon energy at RT for polycrystalline films [72G].

9.15.2.2.4 Vanadium oxide (VO$_2$)

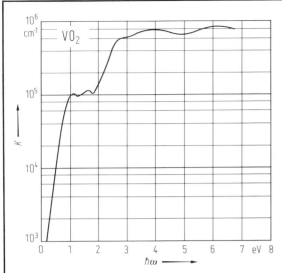

Fig. 54. VO$_2$. Absorption coefficient vs. photon energy in the semiconducting phase [72G].

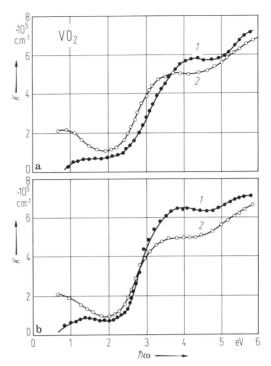

Fig. 55. VO$_2$. Absorption coefficient vs. photon energy of crystalline films for (a) $E \perp a_R$, (b) $E \parallel a_R$. Curves 1: $T < T_{tr}$, curves 2: $T > T_{tr}$ [76M2].

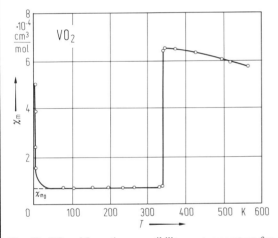

Fig. 56. VO$_2$. Magnetic susceptibility vs. temperature for pure polycrystalline material. The sharp rise at low temperature is thought to be due to magnetic impurities [72P]. χ_m in CGS-emu.

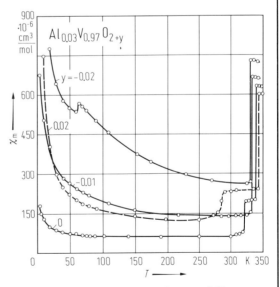

Fig. 58. Al$_{0.03}$V$_{0.97}$O$_{2+y}$. Magnetic susceptibility vs. temperature for polycrystalline samples of different composition [77V]. χ_m in CGS-emu.

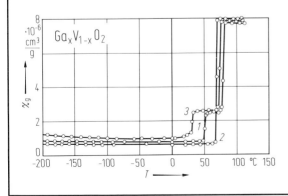

Fig. 57. Ga$_x$V$_{1-x}$O$_2$. Susceptibility vs. temperature for $x = 0$ (1), 0.0039 (2) and 0.0099 (3) [76B2]. χ_g in CGS-emu. Polycrystalline samples.

9.15.2.2.4 Vanadium oxide (VO$_2$)

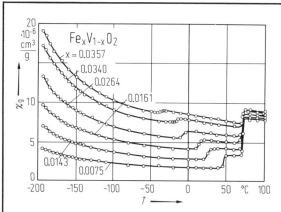

Fig. 59. Fe$_x$V$_{1-x}$O$_2$. Magnetic susceptibility vs. temperature for polycrystalline samples of different composition [76B1]. χ_g in CGS-emu.

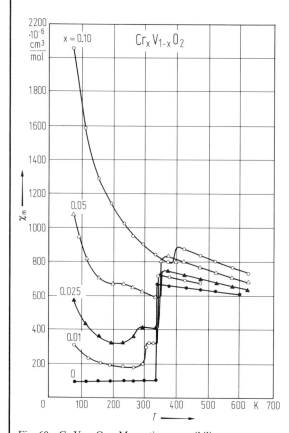

Fig. 60. Cr$_x$V$_{1-x}$O$_2$. Magnetic susceptibility vs. temperature for polycrystalline samples of different composition [77V]. χ_m in CGS-emu.

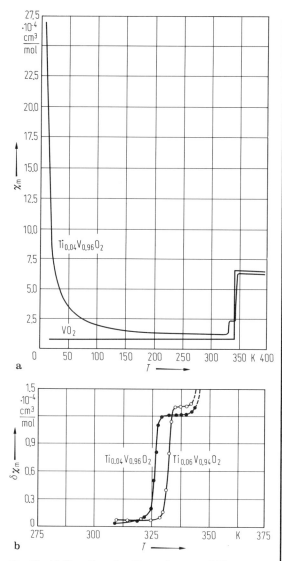

Fig. 61. (a) Ti$_{0.04}$V$_{0.96}$O$_2$. Magnetic susceptibility vs. temperature. (b) The deviation $\delta\chi_m$ of the susceptibility of Ti$_{0.04}$V$_{0.96}$O$_2$ and Ti$_{0.06}$V$_{0.94}$O$_2$ vs. T from the functional form $\frac{C_{obs}}{T}+\chi_o$ near the M$_1\rightarrow$M$_2$ transition [76H2]. χ_m in CGS-emu. Polycrystalline samples. χ_o is the T-independent contribution of χ_m.

9.15.2.2.4 Vanadium oxide (VO$_2$)

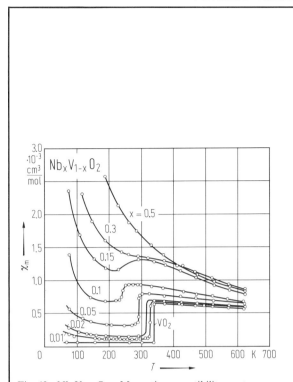

Fig. 62. $Nb_xV_{1-x}O_2$. Magnetic susceptibility vs. temperature for polycrystalline samples of different composition [72P]. χ_m in CGS-emu.

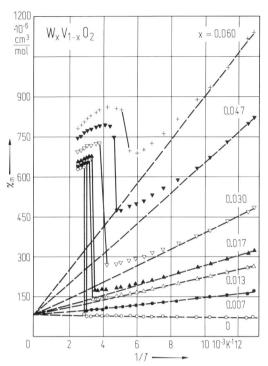

Fig. 63. $W_xV_{1-x}O_2$. Magnetic susceptibility vs. reciprocal temperature for polycrystalline samples of different composition [72H]. χ_m in CGS-emu.

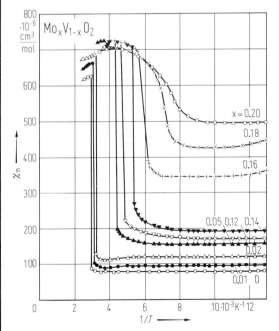

Fig. 64. $Mo_xV_{1-x}O_2$. Magnetic susceptibility vs. reciprocal temperature for polycrystalline samples of different composition [73H2]. χ_m in CGS-emu.

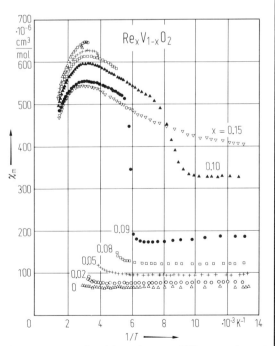

Fig. 65. $Re_xV_{1-x}O_2$. Magnetic susceptibility vs. reciprocal temperature for polycrystalline samples of different composition [77S]. χ_m in CGS-emu.

9.15.2.2.5 Vanadium oxide (V$_2$O$_5$)

9.15.2.2.5 Vanadium oxide (V$_2$O$_5$) (For tables, see p. 194ff.)

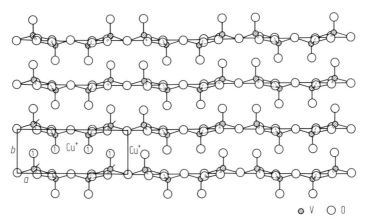

Fig. 1. V$_2$O$_5$. Structure projected along [001]. Shaded circles are vanadium, open circles are oxygen atoms. Inferred positions of monovalent impurities are shown [71P].

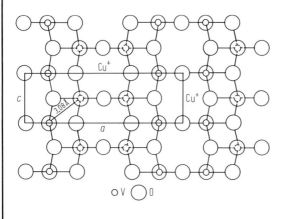

Fig. 2. V$_2$O$_5$. Structure projected along [010]. Small circles are vanadium with dotted ones below the $a-c$ plane and solid ones above, large circles are oxygen. Zig-zagging of vanadium separated by 3.08 Å along the c-axis is clearly seen [71P].

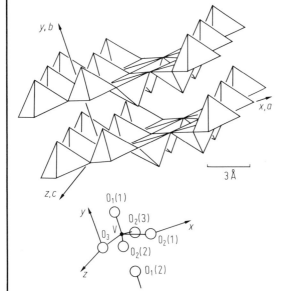

Fig. 3. V$_2$O$_5$. Crystal structure in perspective view [61B].

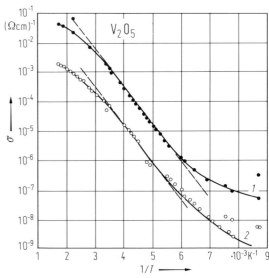

Fig. 4. V$_2$O$_5$. Conductivity vs. reciprocal temperature for (1) along c-axis (2) along b-axis [70I].

9.15.2.2.5 Vanadium oxide (V$_2$O$_5$)

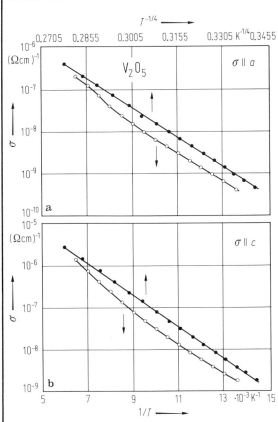

Fig. 5. V$_2$O$_5$. Conductivity vs. reciprocal temperature (open symbols) and $T^{-1/4}$ (full symbols) for (a) parallel a and (b) parallel c [73H].

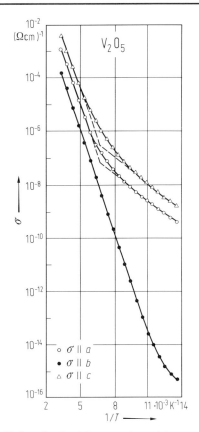

Fig. 6. V$_2$O$_5$. Conductivity vs. reciprocal temperature for low temperatures along the three crystallographic directions [73H].

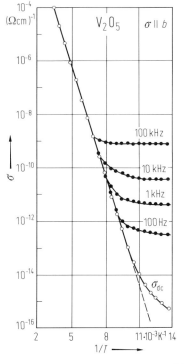

Fig. 7. V$_2$O$_5$. ac and dc conductivity vs. reciprocal temperature along the b-axis [73H].

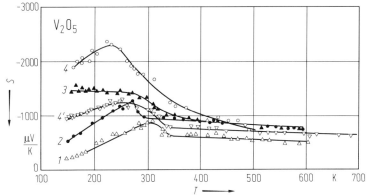

Fig. 8. V$_2$O$_5$. Seebeck coefficient vs. temperature for four crystals. (4′) is measured on the same crystal as (4) after annealing [70I]. Orientation not specified.

9.15.2.2.5 Vanadium oxide (V_2O_5)

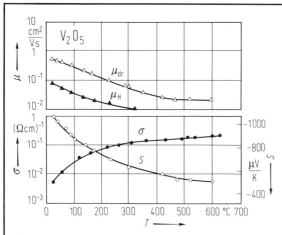

Fig. 9. V_2O_5. Seebeck coefficient, conductivity, drift and Hall mobility vs. temperature [72V]. Orientation not specified, assumed to be $\|a$.

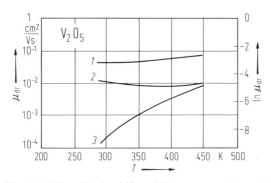

Fig. 10. V_2O_5. Carrier drift mobility vs. temperature (1) along c-axis, (2) along a-axis, (3) along b-axis [69V].

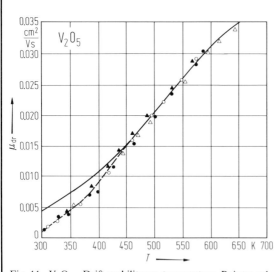

Fig. 11. V_2O_5. Drift mobility vs. temperature. Points: calculated values for five single crystals with different number of epr centres ($3 \cdot 10^{18} \cdots 2 \cdot 10^{19}$ cm^{-3}); full line: function $T^{-3/2} \exp(-0.16\,\text{eV}/kT)$ [70I]. Orientation not specified.

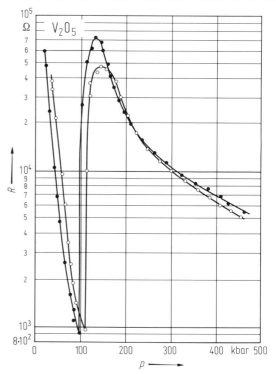

Fig. 12. V_2O_5. Resistance vs. pressure for two separate samples [63M].

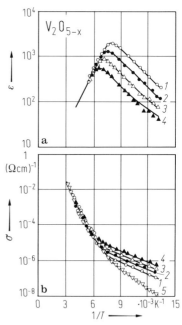

Fig. 12a V_2O_{5-x}. (a) Dielectric constant and (b) conductivity vs. reciprocal temperature for single crystal along [001] axis at different frequencies: 1: 10^3 Hz, 2: $3 \cdot 10^3$ Hz, 3: 10^4 Hz, 4: $3 \cdot 10^4$ Hz, 5: dc value [81I].

9.15.2.2.5 Vanadium oxide (V_2O_5)

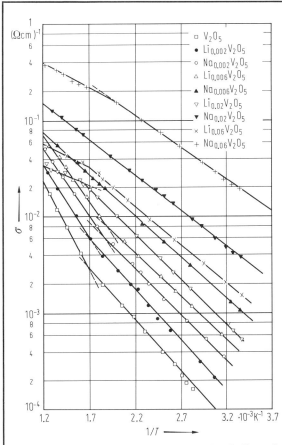

Fig. 13. V_2O_5. Conductivity of interstitially alkali-metal doped polycrystalline samples vs. reciprocal temperature [76C1].

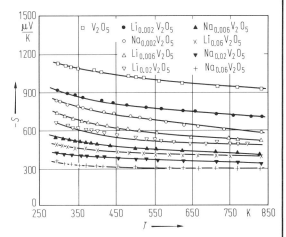

Fig. 14. V_2O_5. Seebeck coefficients vs. temperature for interstitially alkali-metal doped polycrystalline samples [76C1].

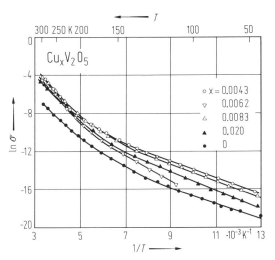

Fig. 15. V_2O_5. Natural logarithm of conductivity vs. (reciprocal) temperature for Cu-doped samples [71P]. σ in Ω^{-1} cm^{-1}; $\sigma \| c$.

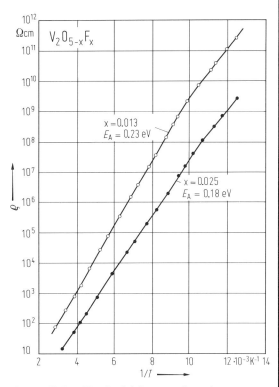

Fig. 16. $V_2O_{5-x}F_x$. Resistivity vs. reciprocal temperature for two samples of different composition [71B]. Orientation not specified.

9.15.2.2.5 Vanadium oxide (V$_2$O$_5$)

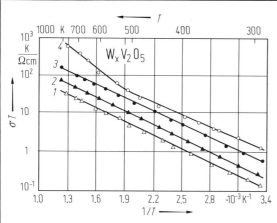

Fig. 16α. W$_x$V$_2$O$_5$. Conductivity times temperature vs. (reciprocal) temperature for polycrystalline samples. *1*: x = 0.005, *2*: x = 0.01, *3*: x = 0.02, *4*: x = 0.03 [79P].

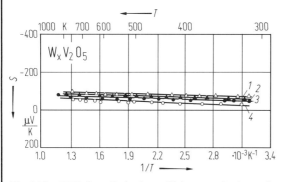

Fig. 16β. W$_x$V$_2$O$_5$. Seebeck coefficient vs. (reciprocal) temperature for polycrystalline samples [79P]. Curve numbering as for Fig. 16α.

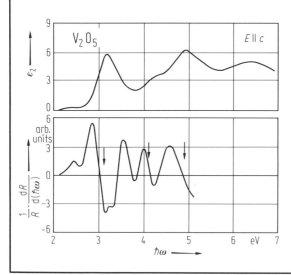

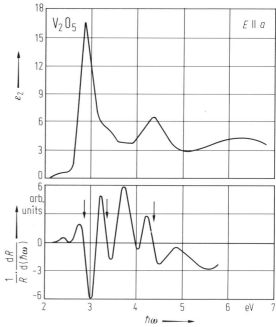

Fig. 17. V$_2$O$_5$. Electroreflectivity and imaginary part of the dielectric constant vs. photon energy for $E \| a$. Arrows mark peak positions in the reflectance spectrum [77L].

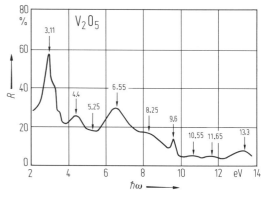

Fig. 19. V$_2$O$_5$. Reflectivity vs. photon energy [77L]. Peak positions are indicated.

◀ Fig. 18. V$_2$O$_5$. Electroreflectivity and imaginary part of the dielectric constant vs. photon energy for $E \| c$. Notation as in Fig. 17 [77L].

9.15.2.2.5 Vanadium oxide (V_2O_5)

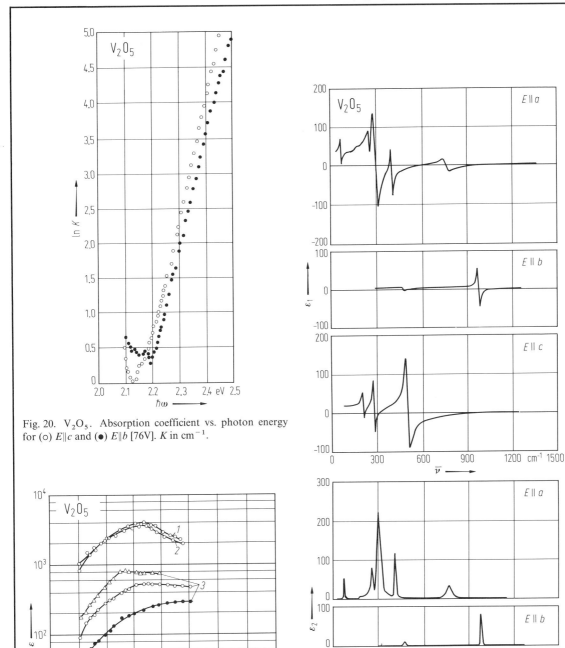

Fig. 20. V_2O_5. Absorption coefficient vs. photon energy for (○) $E\|c$ and (●) $E\|b$ [76V]. K in cm^{-1}.

Fig. 22. V_2O_5. Dielectric constant vs. temperature along the c-axis ($1\cdots 3$) and b-axis (4). Frequencies: I: 300 Hz, II: 1000 Hz, III: 10000 Hz [78I].

Fig. 21. V_2O_5. Real and imaginary parts of the dielectric constant vs. wavenumber for the three crystallographic directions [76C2].

9.15.2.3 Monoxides

9.15.2.3.1 Manganese oxide (MnO) (For tables, see p. 201 ff.)

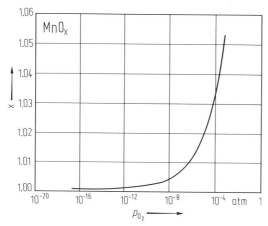

Fig. 1. MnO. Ratio of oxygen to manganese concentration vs. oxygen partial pressure in the range 1200···1650 °C [67S1].

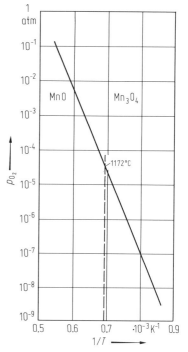

Fig. 2. MnO. Oxygen partial pressure vs. reciprocal temperature for the system MnO—Mn_3O_4 [67S1].

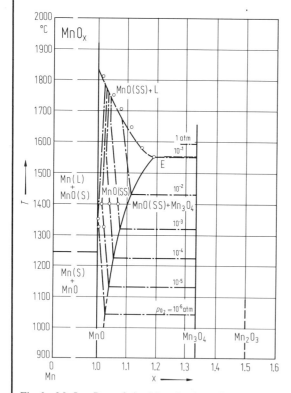

Fig. 3. MnO_x. Part of the Mn—O phase diagram. Dot-dashed lines are oxygen isobars [67H1]. L: liquid, S: solid, SS: solid solution, E: eutectic point; p_{O_2} in atm.

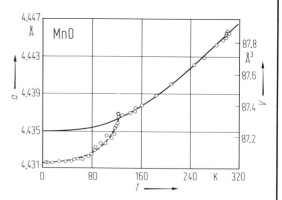

Fig. 4. MnO. Lattice constant and unit cell volume vs. temperature. Solid line: predicted from Debye Model; circles: experimental points [70M].

9.15.2.3.1 Manganese oxide (MnO)

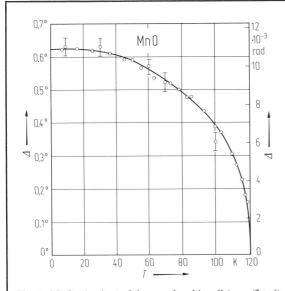

Fig. 5. MnO. Angle Δ of the pseudocubic cell ($\alpha = \pi/2 + \Delta$) vs. temperature [65R, 70M].

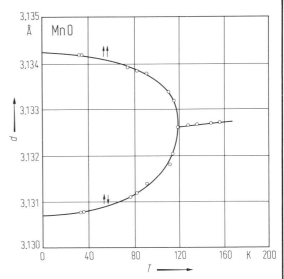

Fig. 6. MnO. Distance d between parallel and antiparallel spin neighbours vs. temperature for $T < T_N$ [68B].

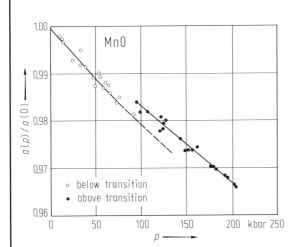

Fig. 7. MnO. Lattice constant vs. pressure [66C].

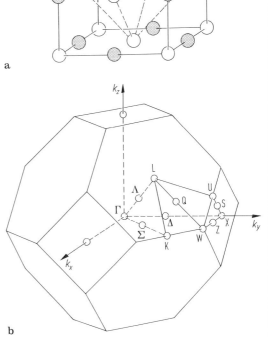

Fig. 8. MnO. Unit cell (a) and Brillouin zone (b) for the face centered cubic lattice.

9.15.2.3.1 Manganese oxide (MnO)

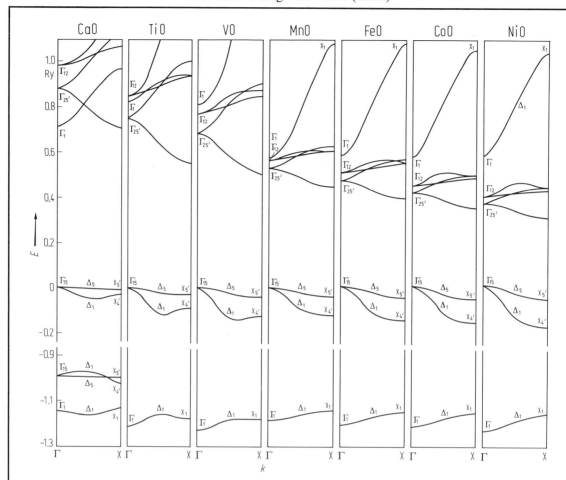

Fig. 9. 3d-monoxides. Energy band structure obtained by an APW calculation plotted along the Δ-direction [72M1].

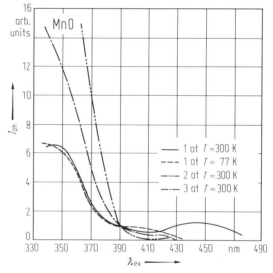

Fig. 10. MnO. Photocurrent vs. wavelength of exciting light for various crystals *1* as grown, *2* annealed in $p_{CO_2}/p_{H_2}=0.1$, *3* annealed in $p_{CO_2}/p_{H_2}=1.0$ [77U].

9.15.2.3.1 Manganese oxide (MnO)

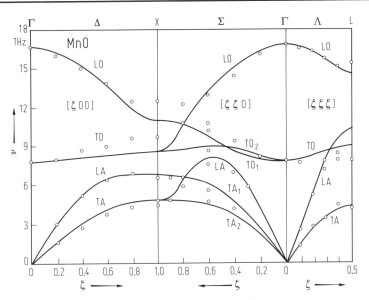

Fig. 11. MnO. Phonon dispersion curves along symmetry directions at 300 K (experimental data and theoretical curves) [77K, 79A].

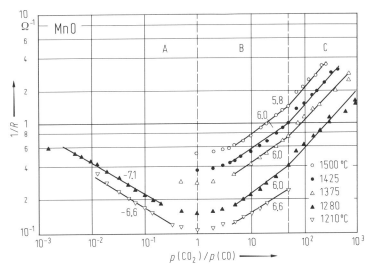

Fig. 12. MnO. Conductance vs. CO_2/CO pressure ratio of ambient atmosphere during measurement for single crystals. The regions A, B and C delineate regions of differing behaviour (see tables) and the numbers refer to the different exponents n as $1/R \propto p_{O_2}^{1/n}$ [67H2].

495

9.15.2.3.1 Manganese oxide (MnO)

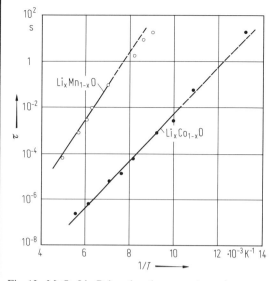

Fig. 13. MnO : Li. Relaxation time vs. reciprocal temperature for local hole-hopping in Li-doped MnO and CoO [70C2]. Li content $x = 10^{-3}$.

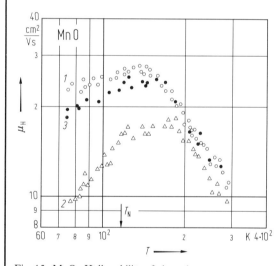

Fig. 15. MnO. Hall mobility of photoelectrons vs. temperature for various crystals 1 as grown, 2 annealed in $p_{CO_2}/p_{H_2} = 0.1$, 3 annealed in $p_{CO_2}/p_{H_2} = 1.0$ [77U].

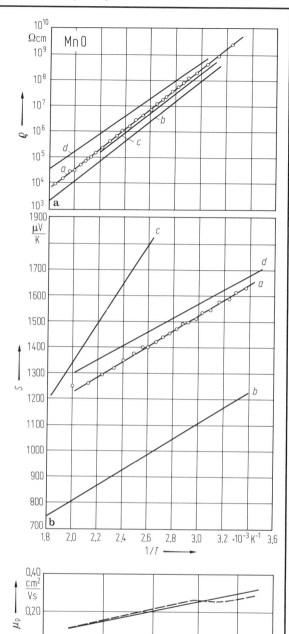

Fig. 14. MnO. (a) Resistivity vs. reciprocal temperature; a [81P], b [80J], c [76K1], d [71K]. (b) Thermoelectric power vs. reciprocal temperature; a···d as for Fig. (a). (c) Hole mobility (log scale) vs. temperature; dashed line: calculated directly from conductivity and thermogravimetric data, solid line: averaged [67H2].

9.15.2.3.1 Manganese oxide (MnO)

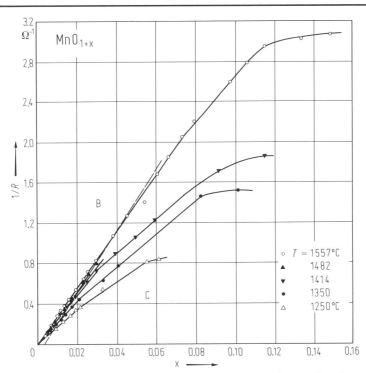

Fig. 16. MnO$_{1+x}$. Isotherms of the conductance $1/R$ of polycrystalline material for regions B and C of Fig. 12 vs. x in MnO$_{1+x}$ [67H2].

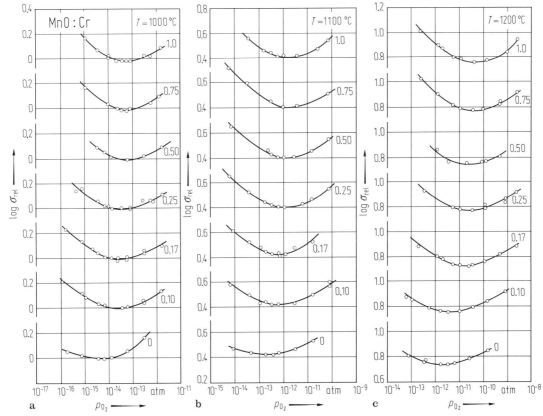

Fig. 17. MnO:Cr. Relative conductivity $\sigma_{rel} = \sigma/\sigma_{min}$ (1000°C) at three temperatures vs. oxygen partial pressure for pure and Cr doped samples. Numbers on curves correspond to at% Cr/Mn [70O].

9.15.2.3.1 Manganese oxide (MnO)

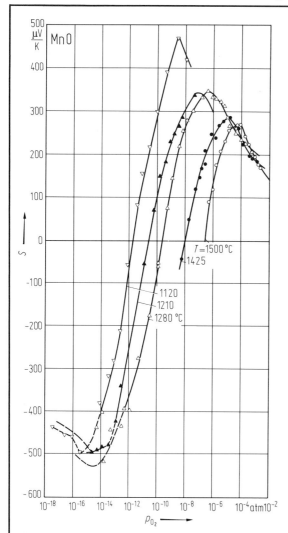

Fig. 18. MnO. Seebeck coefficient vs. oxygen partial pressure for single crystals at various temperatures [67H2].

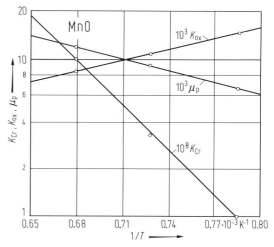

Fig. 19. MnO. Hole mobility (in $cm^2 V^{-1} s^{-1}$) and constants K_{ox} (in $atm^{1/2}$) and K_{Cr} vs. reciprocal temperature [70O].

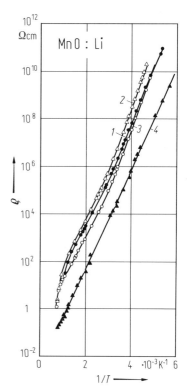

Fig. 20. MnO:Li. Resistivity vs. reciprocal temperature. Curve 1 pure single crystal, 2 0.03 at% Li doped, 3 0.1 at% Li doped, 4 5.0 at% Li doped [70C2].

For Fig. 21, see next page.

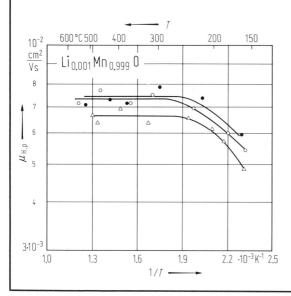

Fig. 22. $Li_{0.001}Mn_{0.999}O$. Hall mobility of holes for three single crystals vs. (reciprocal) temperature [67N].

9.15.2.3.1 Manganese oxide (MnO)

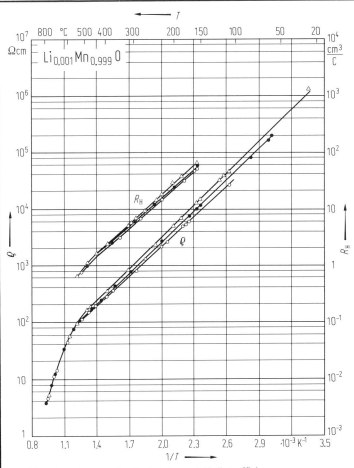

Fig. 21. $Li_{0.001}Mn_{0.999}O$. Resistivity and Hall coefficient vs. (reciprocal) temperature for three single crystals [67N].

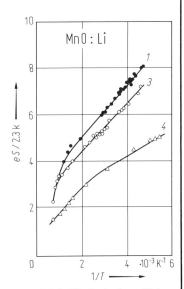

Fig. 23. MnO:Li. Seebeck coefficient (reduced) vs. reciprocal temperature for three samples of Fig. 20 [70C2].

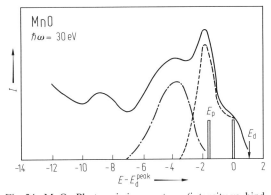

Fig. 24. MnO. Photoemission spectrum (intensity vs. binding energy relative to the E_d peak energy) showing estimated deconvolution into Mn 3d and O 2p; E_p and E_d denote the p- and d-band edges and the vertical lines are the calculated positions of the $3d^{n-1}$ states [75E].

For Fig. 25, see next page.

Fig. 26. MnO. Absorption coefficient vs. wavenumber for three temperatures near the absorption edge (arrow) [74C].

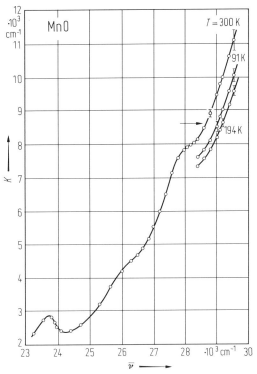

9.15.2.3.1 Manganese oxide (MnO)

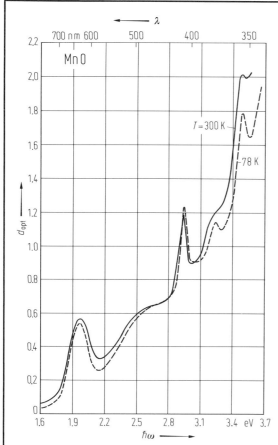

Fig. 25. MnO. Optical density vs. photon energy (wavelength) at 300 and 78 K. Film thickness 100 μm [69H1].

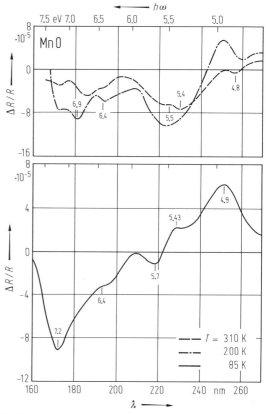

Fig. 27. MnO. Thermoreflectance vs. wavelength (photon energy) at 85 K (lower figure) and 200 and 310 K (upper figure) [72M3]. Peak positions (in eV) are indicated.

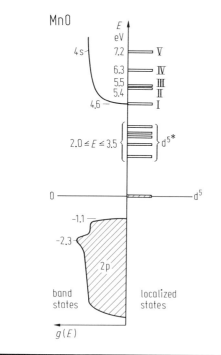

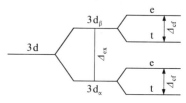

Fig. 28. MnO. Electronic states (at RT): one-electron bands are shown on the left and localized states on the right [72M3]. I: $d^4(t_\alpha^3 e_\alpha) + 4s$; II: $d^4(t_\alpha^3 e_\alpha) + d^6(t_\alpha^3 t_\beta e_\alpha^2)$; III: $d^4(t_\alpha^2 e_\alpha^2) + 4s$; IV: $d^4(t_\alpha^3 e_\alpha) + d^6(t_\alpha^3 e_\alpha^2 e_\beta)$ and $d^4(t_\alpha^2 e_\alpha^2) + d^6(t_\alpha^3 t_\beta e_\alpha^2)$; V: $d^4(t_\alpha^2 e_\alpha^2) + d^6(t_\alpha^3 e_\alpha^2 e_\beta)$. For the meaning of the t, e, α, β, see schematic figure:

9.15.2.3.1 Manganese oxide (MnO)

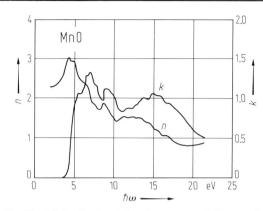

Fig. 29. MnO. Real and imaginary parts of the complex refractive index vs. photon energy [76K2].

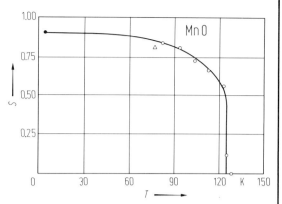

Fig. 31. MnO. Long range order parameter defined as $S = r_a - w_a = r_b - w_b$ where r_a is the fraction of spin-up sublattice sites occupied by spin-up Mn and w_a is the fraction of spin-up sublattice sites occupied by spin-down Mn and a similar definition holds for the b-sites vs. temperature. Measurements of various authors [66B].

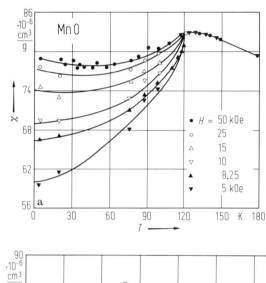

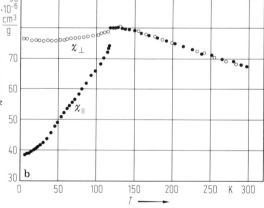

Fig. 30. MnO. Magnetic susceptibility vs. temperature for (a) MnO powder at various magnetic fields [67B], (b) MnO single crystals stressed at 52 bar along [111]; $\chi_\parallel$: $B \parallel$ (111), $\chi_\perp$: $B \perp$ (111) [81J]. χ in CGS-emu.

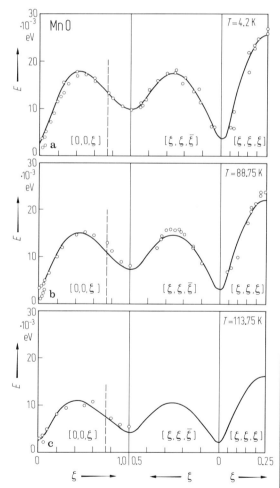

Fig. 32. MnO. Spin wave dispersion curves at (a) 4.2 K, (b) 88.75 K, (c) 113.75 K obtained by inelastic neutron scattering [74P].

9.15.2.3.2 Iron oxide (FeO and Fe$_{1-x}$O)

9.15.2.3.2 Iron oxide (FeO and Fe$_{1-x}$O) (For tables, see p. 208 ff.)

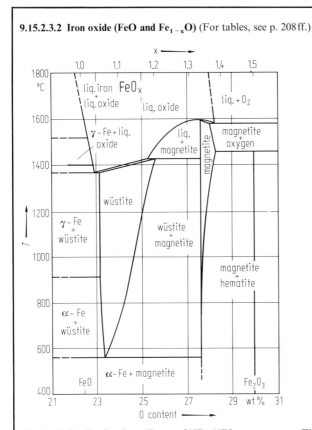

Fig. 1. FeO–Fe$_2$O$_3$ phase diagram [45D, 46D].

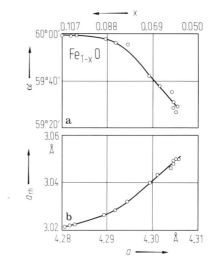

Fig. 2. Fe$_{1-x}$O. (a) Rhombohedral distortion angle and (b) lattice constant a_{rh} as a function of cubic lattice constant a at 300 K and stoichiometry [53W, 79B]. $T = 90$ K.

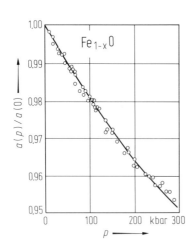

Fig. 3. Fe$_{1-x}$O. Lattice constant vs. pressure [66C].

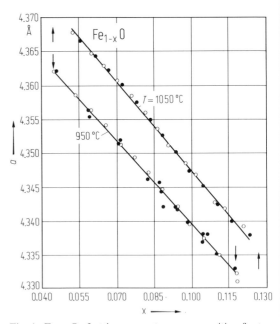

Fig. 4. Fe$_{1-x}$O. Lattice parameter vs. composition for two temperatures. Arrows show the phase boundaries at each temperature [72H].

9.15.2.3.2 Iron oxide (FeO and Fe$_{1-x}$O)

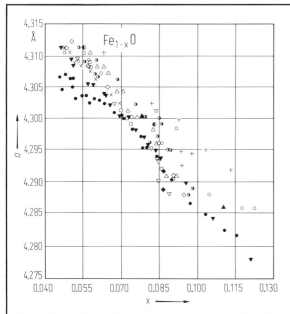

Fig. 5. Fe$_{1-x}$O. Lattice parameter vs. composition for quenched samples at 298 K. Results of different authors [80B].

Fig. 6. Fe$_{1-x}$O. Lattice constant vs. temperature for Fe$_{1-x}$O in 34% CO [72H]. $1-x = 0.905\cdots 0.908$.

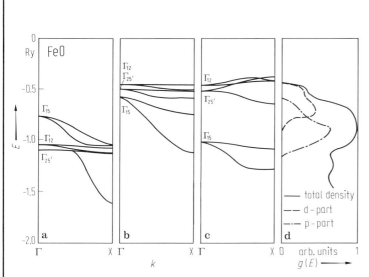

Fig. 7. FeO. Band structure showing (a) Hartree-Fock data, (b) self-consistent calculation with correlation, (c) non-self consistent calculation, (d) experimental density of states from photoemission spectroscopy data [75E].

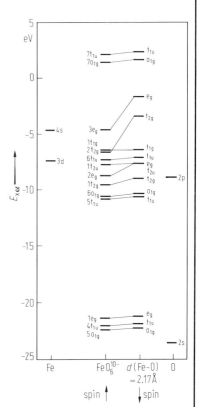

Fig. 8. FeO. One-electron unrestricted Xα cluster calculation on FeO$_6^{10-}$ [76T]. Shown are the electronic levels of an FeO$_6^{10-}$ cluster and the free-atom energy levels. $E_{x\alpha}$: orbital energy. For discussion of cluster method, see [72S].

9.15.2.3.2 Iron oxide (FeO and Fe$_{1-x}$O)

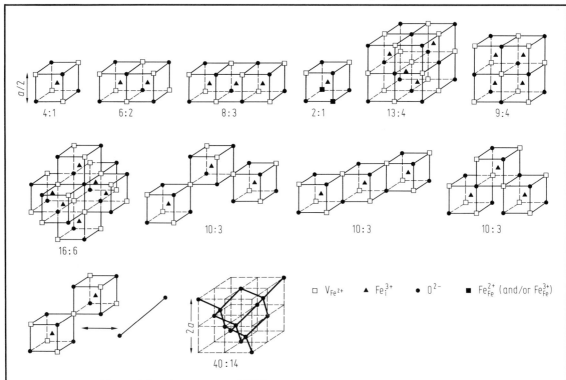

Fig. 9. Fe$_{1-x}$O. Different defect clusters proposed [80B].

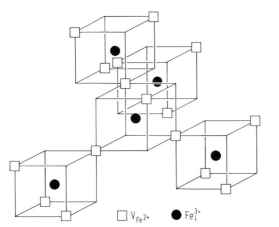

Fig. 10. Fe$_{1-x}$O. 16:5 spinel-like aggregate [79G, 77C].

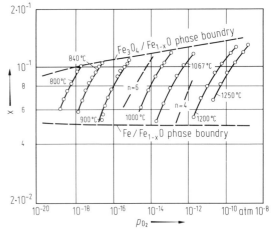

Fig. 11. Fe$_{1-x}$O. Non-stoichiometry vs. oxygen partial pressure for different temperatures. The curves marked $n=4$ and $n=6$ are theoretical slopes $x \propto p_{O_2}^{1/n}$ [65V2].

9.15.2.3.2 Iron oxide (FeO and $Fe_{1-x}O$)

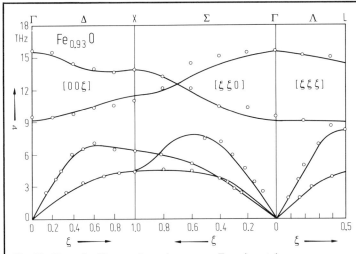

Fig. 12. $Fe_{0.93}O$. Phonon dispersion curves. Experimental neutron scattering data and shell-model best fit [77K].

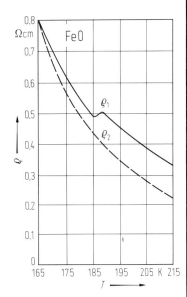

Fig. 17. FeO. Resistivity vs. temperature near T_N. Solid line: experimental value; dotted line: calculated assuming no critical behaviour [74B2].

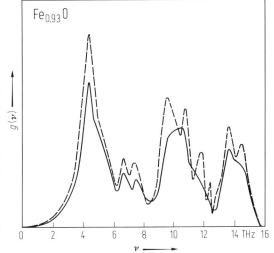

Fig. 13. $Fe_{0.93}O$. Density of phonon states vs. frequency according to the dispersion curves of Fig. 12. Solid line and dotted line are the results using two slightly different models [77K].

For Fig. 16, see next page.

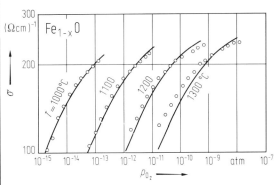

Fig. 14. $Fe_{1-x}O$. Electrical conductivity vs. oxygen partial pressure at various temperatures. Solid lines: fit to a complex defect model [70S].

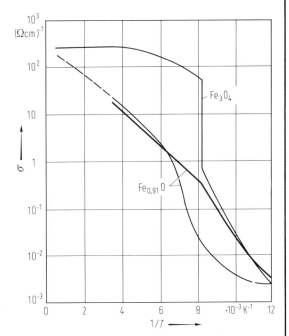

Fig. 15. $Fe_{0.91}O$. Conductivity vs. reciprocal temperature. Heavy line: [74B1], thin line: [62T]. Fe_3O_4 for comparison.

9.15.2.3.2 Iron oxide (FeO and $Fe_{1-x}O$)

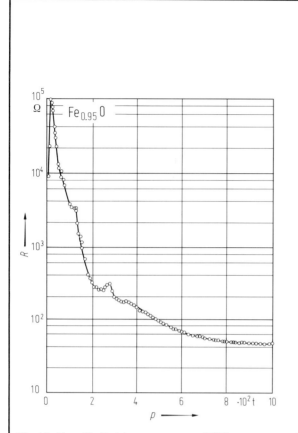

Fig. 16. $Fe_{0.95}O$. Resistance vs. pressure [74K].

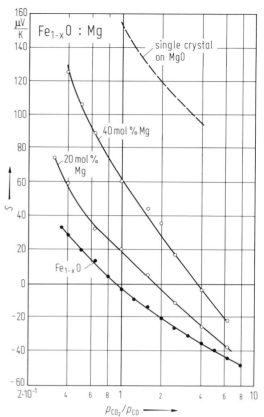

Fig. 19. $Fe_{1-x}O$:Mg. Seebeck coefficient vs. CO_2/CO pressure ratio of ambient atmosphere during measurement for various Mg doped samples [67H].

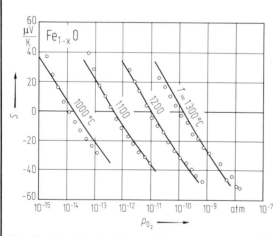

Fig. 18. $Fe_{1-x}O$. Seebeck coefficient vs. oxygen partial pressure at various temperatures [70S]. Solid lines: fit to the data (see tables).

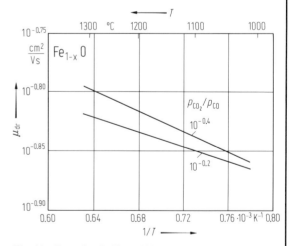

Fig. 20. $Fe_{1-x}O$. Drift mobility vs. (reciprocal) temperature for two values of p_{CO_2}/p_{CO} [67B1].

9.15.2.3.2 Iron oxide (FeO and Fe$_{1-x}$O)

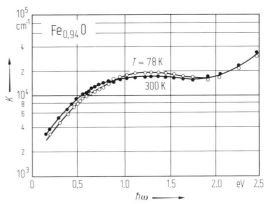

Fig. 22. Fe$_{0.94}$O. Absorption coefficient of a 3 μm thick film on MgO vs. photon energy at two temperatures [78B].

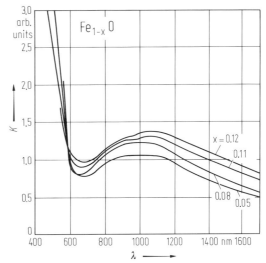

Fig. 23. Fe$_{1-x}$O. Absorption coefficient vs. wavelength at RT for single crystalline films of different composition [74B1].

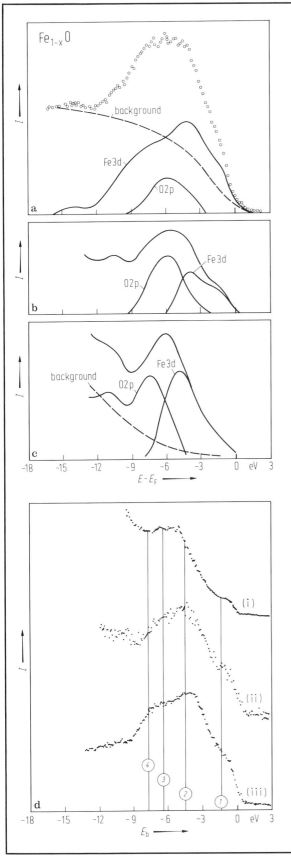

Fig. 21. Fe$_{1-x}$O. Photoemission spectrum (intensity vs. electron binding energy) in the valence band region.
(a) XPS spectrum [77B], (b) UPS spectrum, $\hbar\omega = 30$ eV,
(c) UPS spectrum, $\hbar\omega = 20$ eV [75E]. Figure from [77B].
(d) (i) HeI (21.2 eV), (ii) HeII (40.8 eV) and (iii) Mg K$_\alpha$(1253.6 eV) photoemission spectra for Fe$_{0.946}$O, peaks 1, 2 and 4 are assigned to Fe 3d and 3 to O 2p. E_b: binding energy relative to photoemission onset. See also Fig. 17 of 9.15.2.5.2 (Fe$_2$O$_3$) [77B].

9.15.2.3.3 Cobalt oxide (CoO)

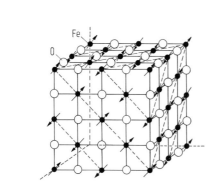

Fig. 24. $Fe_{1-x}O$. Crystalline and antiferromagnetic structure below the Néel temperature [58R].

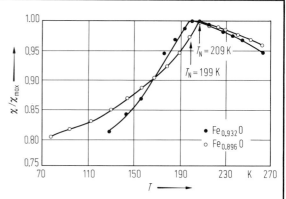

Fig. 25. $Fe_{1-x}O$. Magnetic susceptibility vs. temperature for two samples of different stoichiometry [67K1].

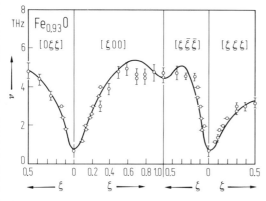

Fig. 26. $Fe_{0.93}O$. Spin-wave spectrum [78K1]. Solid line: calculated.

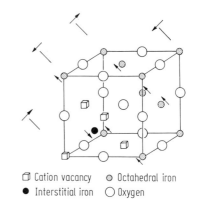

Fig. 27. $Fe_{1-x}O$. Local spin arrangements about a 4:1 cluster [79B].

9.15.2.3.3 Cobalt oxide (CoO) (For tables, see p. 213 ff.)

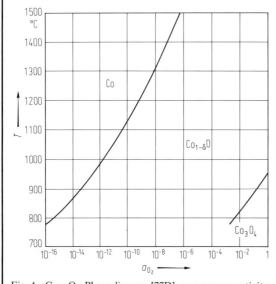

Fig. 1. Co–O. Phase diagram [77D]. a_{O_2}: oxygen activity.

For Fig. 2, see next page.

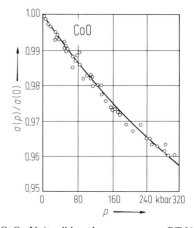

Fig. 3. CoO. Unit cell length vs. pressure at RT [66C1].

9.15.2.3.3 Cobalt oxide (CoO)

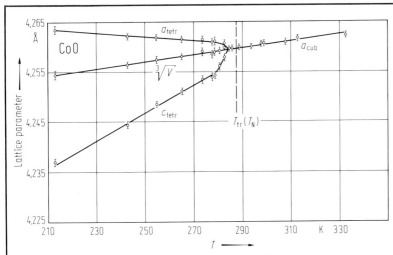

Fig. 2. CoO. Unit cell dimensions above and below the transition temperature. The dashed vertical line shows the transition temperature as determined from magnetic measurements [53G].

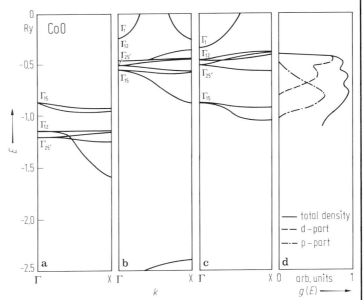

Fig. 4. CoO. (a) Hartree-Fock, (b) self-consistent calculation with correlation, (c) non-self consistent calculation of the band structure, (d) experimental photoemission spectrum of [75E]. [78K].

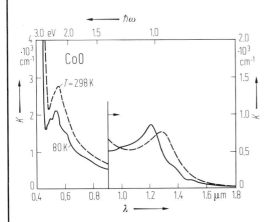

Fig. 5. CoO. Absorption coefficient vs. wavelength (photon energy) at 80 K and 298 K. Unpolarized light, sample (100) plane [59P].

9.15.2.3.3 Cobalt oxide (CoO)

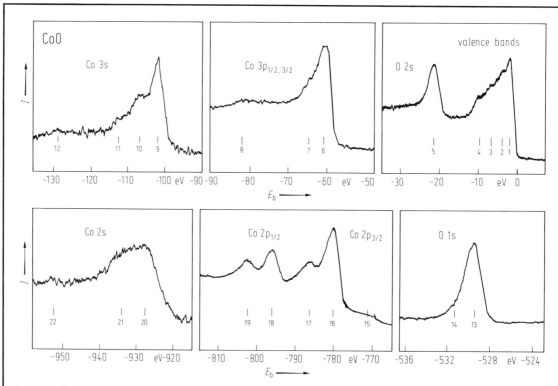

Fig. 6. CoO. XPS spectrum in the core region (intensity vs. binding energy) [75K]. For assignment of various peaks, see original paper. E_b: binding energy relative to the Fermi level.

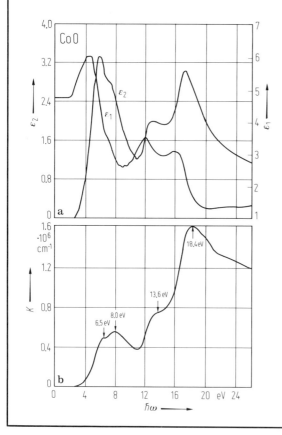

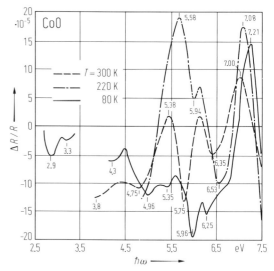

Fig. 8. CoO. Thermoreflectance vs. photon energy at various temperatures [72M2].

Fig. 7. CoO. (a) Real and imaginary parts of the dielectric function vs. photon energy, (b) absorption coefficient vs. photon energy at RT [70P].

9.15.2.3.3 Cobalt oxide (CoO)

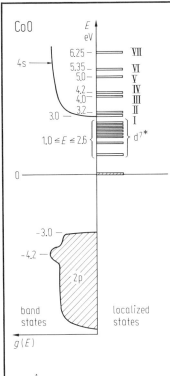

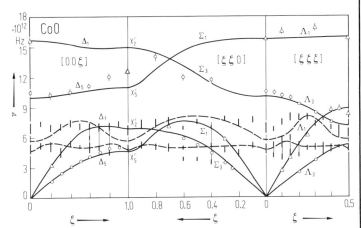

◄ Fig. 9. CoO. Empirical energy level diagram (RT data). Left: band states, right: localized states derived mostly from data at 80 K [72M2]. For assignment of states, see table.

Fig. 10. CoO. Phonon and magnon dispersion curves at 110 K. Circles represent transverse phonon modes and triangles are longitudinal phonon modes. The solid lines show a best fit shell model. The vertical bars represent broad peaks assigned to magnon dispersion and the dashed lines show a magnon model with next nearest neighbour interaction [68S1].

For Fig. 11, see next page.

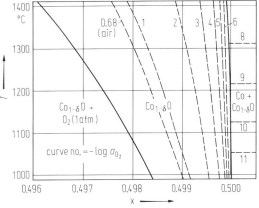

Fig. 13. $Co_{1-\delta}O$. $T-x$ diagram, where x is the mole fraction of Co, and the dotted lines are oxygen isobars, the numbers being $-\log_{10} a_{O_2}$ [77D].

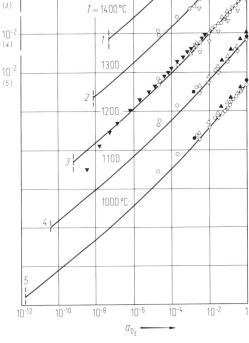

◄ Fig. 12. $Co_{1-\delta}O$. Deviation from stoichiometry δ vs. oxygen activity. Data from [54C, 64F, 72B, 68S2, 68E, 76F], [77D].

9.15.2.3.3 Cobalt oxide (CoO)

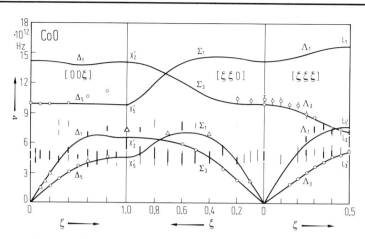

Fig. 11. CoO. Phonon and magnon dispersion curves at 330 K. Notation as in Fig. 10 for the phonon modes. The magnetic modes have split into the flat curves at 4.8 and $\approx 7.8 \cdot 10^{12}$ Hz [68S1].

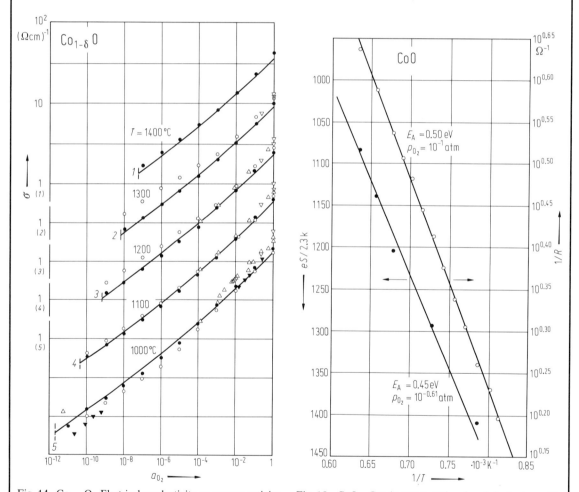

Fig. 14. $Co_{1-\delta}O$. Electrical conductivity vs. oxygen activity at various temperatures. Data from [68E, 66S2, 36W, 75M, 72G, 66F]. [77D]. σ in Ω^{-1} cm^{-1}.

Fig. 15. CoO. Conductance and reduced Seebeck coefficient vs. reciprocal temperature for a single crystal [71B].

9.15.2.3.3 Cobalt oxide (CoO)

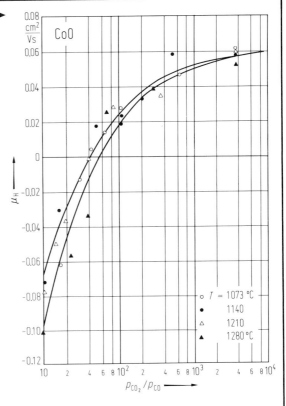

Fig. 17. CoO. Hall mobility for holes vs. CO_2/CO pressure ratio of ambient atmosphere during measurement at various temperatures for a single crystal. Two curves are calculated using a point defect model with two different values for the parameters [72G].

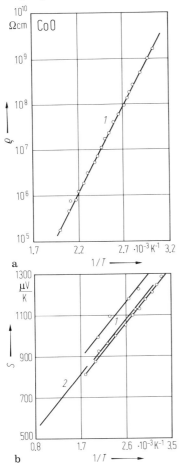

Fig. 15a. CoO. (a) Resistivity and (b) Seebeck coefficient vs. reciprocal temperature for various pure single crystals. 1 data from [80J], 2 data from [67B].

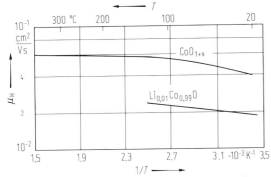

Fig. 16. CoO. Hall mobility for holes vs. (reciprocal) temperature for a non-stoichiometric and a Li doped sample [67N].

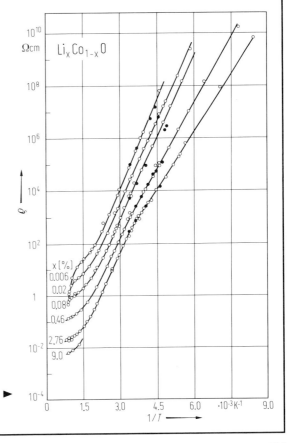

Fig. 18. $Li_xCo_{1-x}O$. Resistivity vs. reciprocal temperature for various Li doped samples. Open Symbols: from dc measurements, full symbols: from ac measurements [69B].

9.15.2.3.3 Cobalt oxide (CoO)

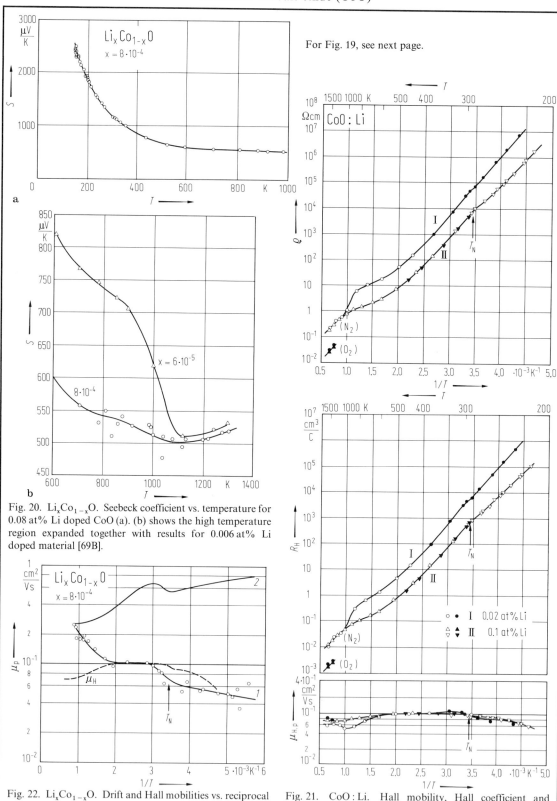

For Fig. 19, see next page.

Fig. 20. $Li_xCo_{1-x}O$. Seebeck coefficient vs. temperature for 0.08 at% Li doped CoO (a). (b) shows the high temperature region expanded together with results for 0.006 at% Li doped material [69B].

Fig. 22. $Li_xCo_{1-x}O$. Drift and Hall mobilities vs. reciprocal temperature for 0.08 at% Li doped p-type CoO calculated from Seebeck data assuming: 1 n_v = constant, 2 $n_v \propto T^{3/2}$, where n_v is the energy density at the valence band edge [69B].

Fig. 21. CoO:Li. Hall mobility, Hall coefficient and resistivity vs. (reciprocal) temperature for p-type Li doped ceramic samples measured in nitrogen (N_2) or oxygen (O_2) atmosphere [67D].

9.15.2.3.3 Cobalt oxide (CoO)

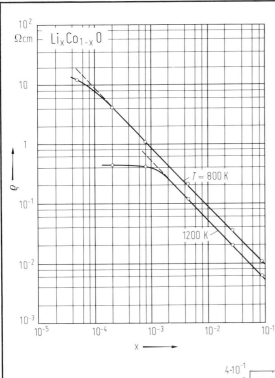

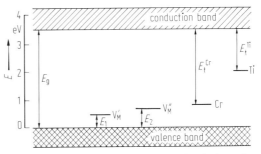

Fig. 19. $Li_xCo_{1-x}O$. Resistivity vs. Li content at 800 K and 1200 K [69B].

Fig. 23. CoO, Cr and Ti doped. Energy level scheme from transport data [72G]. V_M': metal vacancy, negatively charged.

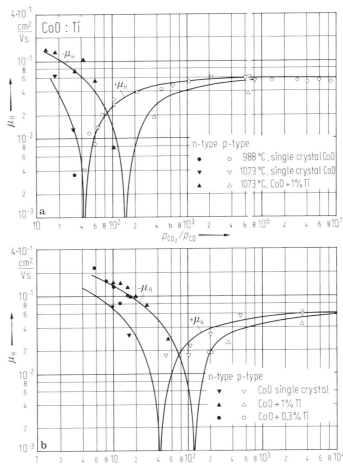

Fig. 24. CoO, pure and Ti doped. Hall mobilities vs. ratio of CO_2 and CO partial pressures of ambient atmosphere during measurement (a) at 988 °C and 1073 °C, (b) at 1140 °C; two curves correspond to a theoretical point defect model using different values for the parameters [72G]. See also Fig. 17.

9.15.2.3.3 Cobalt oxide (CoO)

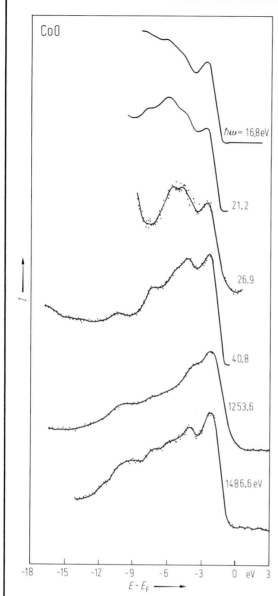

Fig. 25. CoO. Valence band spectra (intensity vs. electron binding energy at various excitation energies) [79J].

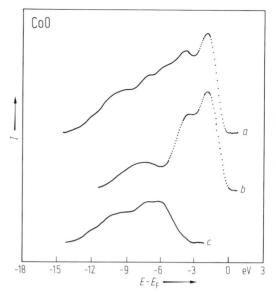

Fig. 26. CoO. Valence band XPS spectrum (intensity vs. electron binding energy) a experimental spectrum (after subtraction of background), b calculated spectrum after convoluting each expected d-excitation with a 1.8 eV Gaussian, c difference between a and b [79J].

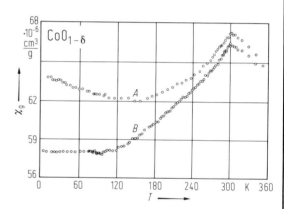

Fig. 27. $Co_{1-\delta}O$, powdered crystals. Magnetic susceptibility vs. temperature for two samples: A: $\delta = 0.01$, B: $\delta = 0.005$ [79S]. χ in CGS-emu.

9.15.2.3.4 Nickel oxide (NiO)

9.15.2.3.4 Nickel oxide (NiO) (For tables, see p. 219ff.)

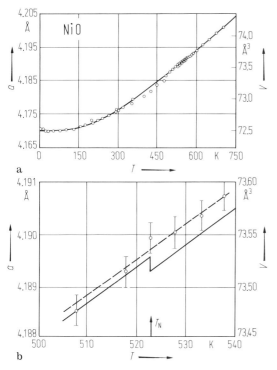

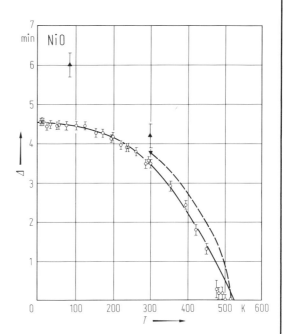

Fig. 3. NiO. Angle Δ of the pseudocubic cell, $\alpha = \pi/2 + \Delta$, vs. temperature for T below T_N. Symbols, dashed line: experimental data from [71B1, 60S, 54S, 70V1], solid line: theory [71B1].

Fig. 1. NiO. Lattice constant (volume of the unit cell) vs. temperature. Solid curves calculated. Fig. b: solid curve: calculated lattice constants with magnetic contribution, dashed line: without that contribution [71B1].

For Figs. 4 and 5, see next page.

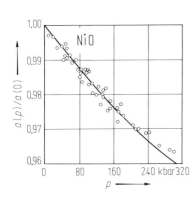

Fig. 2. NiO. Variation of the lattice constant with pressure [66C].

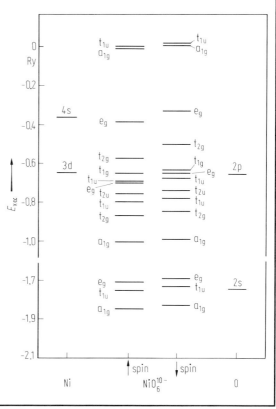

Fig. 6. NiO. Ground-state spin-polarized SCF–Xα electronic levels of an NiO$_6^{10-}$ cluster. Also shown are the corresponding spin-restricted SCF–Xα free-atom energy levels for O and Ni (in $3d^8 4s^2$ configuration) [73J]. For discussion of self consistent field cluster method, see [72S2].

9.15.2.3.4 Nickel oxide (NiO)

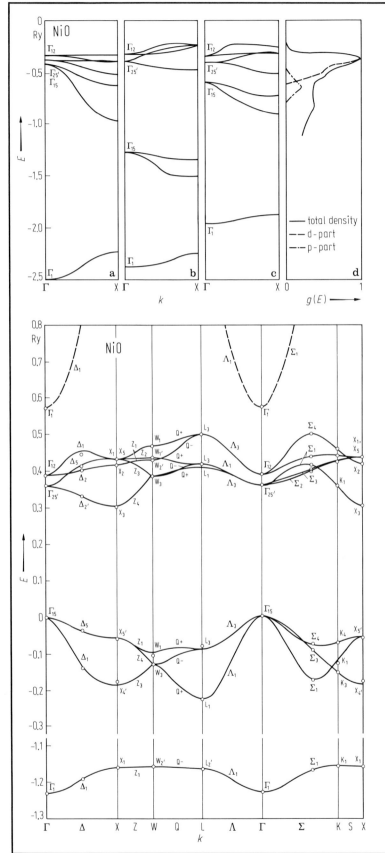

Fig. 4. NiO. Band calculations (a) [78K3], (b) [72M1], (c) [75C] along the ΓX-axis in k-space; (d) comparison with the experimental density of states from photoemission spectra [78K3].

Fig. 5. NiO. LCAO energy bands obtained by fitting APW results at symmetry points [72M1].

9.15.2.3.4 Nickel oxide (NiO)

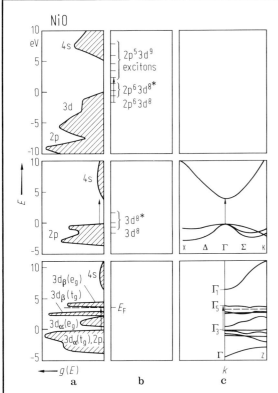

Fig. 7. NiO. Various models for the electronic structure of NiO (a) one electron density of states (b) many electron states (c) energy band structure [69M].

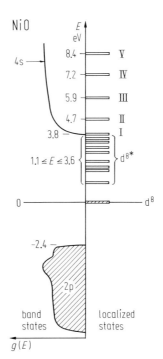

Fig. 8. NiO. Electronic states from RT data. Left: band states, right: localized states I: $d^7(t_\alpha^3 t_\beta^3 e_\alpha)+4s$, II: $d^7(t_\alpha^2 t_\beta^3 e_\alpha^2)+4s$, III: $d^7(t_\alpha^3 t_\beta^2 e_\alpha^2)+d^9(t_\alpha^3 t_\beta^2 e_\alpha^2 e_\beta)$, IV: $d^7(t_\alpha^3 t_\beta^3 e_\alpha)+d^9(t_\alpha^3 t_\beta^3 e_\alpha^2 e_\alpha)$, V: $d^7(t_\alpha^2 t_\beta^3 e_\alpha^2)+d^9(t_\alpha^3 t_\beta^2 e_\alpha^2 e_\beta)$ [72M2]. For the meaning of the t, e, α, β, see caption of Fig. 28 of 9.15.2.3.1

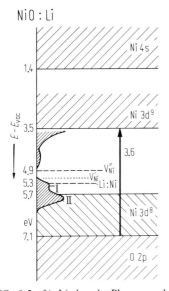

Fig. 9. NiO, 0.5 at% Li doped. Phenomenological band structure from photoelectrochemical measurements at 295 K; the ordinate is the band energies relative to vacuum; I and II correspond to surface states detected electrochemically [81D1].

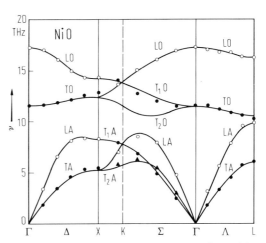

Fig. 10. NiO. Phonon dispersion curves, experimental data from neutron scattering, solid curves theoretical shell-model best fit [75R].

9.15.2.3.4 Nickel oxide (NiO)

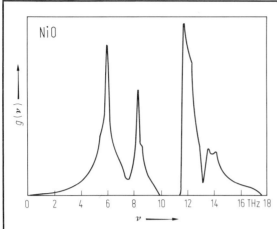

Fig. 11. NiO. Phonon density of states vs. frequency [76C].

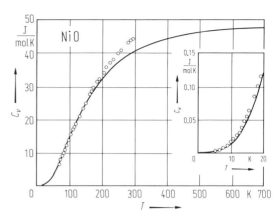

Fig. 12. NiO. Heat capacity vs. temperature [76C]. Dots: [40S]. Insert shows low temperature range on an expanded scale.

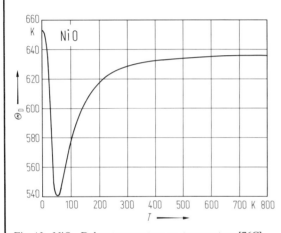

Fig. 13. NiO. Debye temperature vs. temperature [76C].

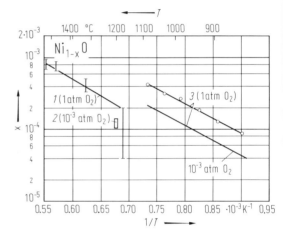

Fig. 14. $Ni_{1-x}O$. Stoichiometry x vs. (reciprocal) temperature at 1 and 10^{-3} atm O_2, curve 1 [61M], 2 [68S], 3 [69T].

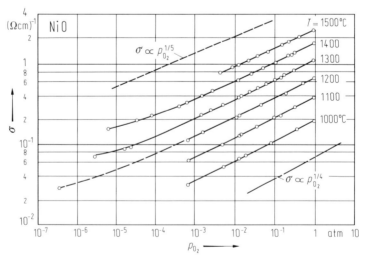

Fig. 15. NiO. Electrical conductivity vs. oxygen partial pressure for various temperatures [68B].

9.15.2.3.4 Nickel oxide (NiO)

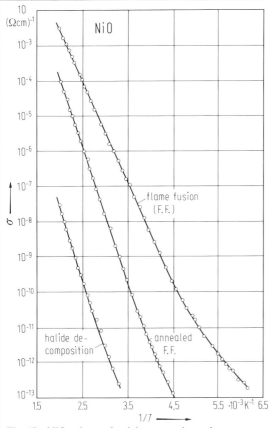

Fig. 17. NiO. dc conductivity vs. reciprocal temperature for single crystals containing excess oxygen [68A1]. Flame fusion: crystal grown by flame fusion techniques; annealed F.F.: flame fusion grown crystal annealed in O_2 at 1050 °C for two days; halide decomposition: crystal grown by decomposition of $NiBr_2$ in gas phase transport.

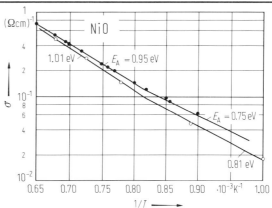

Fig. 16. NiO. High temperature conductivity vs. reciprocal temperature. Data from two authors [79D].

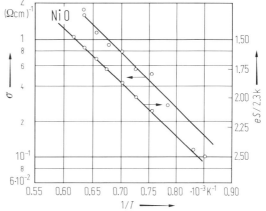

Fig. 18. NiO. Reduced Seebeck coefficient and conductivity vs. reciprocal temperature of polycrystalline material (prepared from spectroscopically pure powder) at 10^{-1} atm oxygen [71B2].

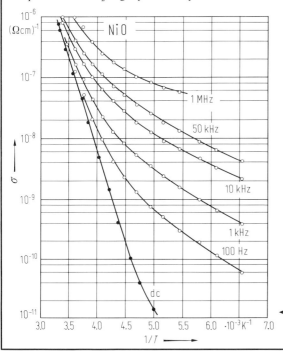

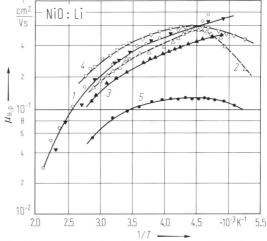

Fig. 20. NiO:Li. Hall mobility for holes vs. reciprocal temperature for various single crystal samples doped with Li. at% Li: 1 (○ 0.032; ▼ 0.029), 2 0.144, 3 0.211, 4 0.202, 5 0.537 [67A].

◀ Fig. 19. NiO. dc conductivity and real part of the ac conductivity vs. reciprocal temperature at fixed frequencies [68A2].

9.15.2.3.4 Nickel oxide (NiO)

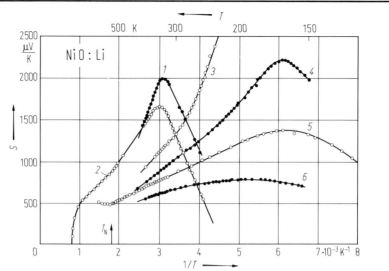

Fig. 21. NiO:Li. Seebeck coefficient vs. (reciprocal) temperature for several pure and Li doped crystals. at% Li: *1* 0, *2* 0, *3* 0.032, *4* 0.202, *5* 0.211, *6* 0.537 [67A].

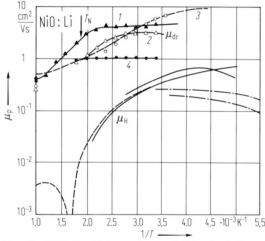

Fig. 22. NiO:Li. Hall and drift mobilities for holes vs. reciprocal temperature. —— single crystal data for Li doped samples [67A], – – – ceramics, 0.088···0.005 at% Li [66B], – · – single crystals pure or 0.002 at% Li [64R]; at% Li: *1* 0.03, *2* 0.21, *3* 0.202, *4* 0.537 [67A].

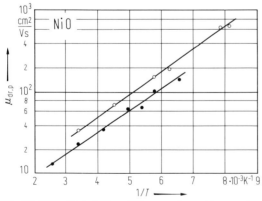

Fig. 23. NiO. Hole drift mobility vs. reciprocal temperature for two specimens [73S3].

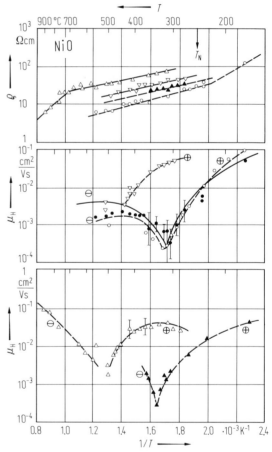

Fig. 24. NiO. Resistivity and Hall mobility vs. (reciprocal) temperature for different partial pressures of O_2. Two methods of analysing the signal were presented. p_{O_2} (atm): ○, ●: $3 \cdot 10^{-1}$, ▲: 10^{-2}, ▽: $4 \cdot 10^{-3}$, △: 10^{-3} [75F].

9.15.2.3.4 Nickel oxide (NiO)

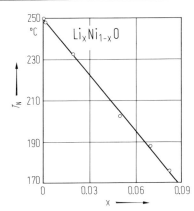

Fig. 25. $Li_xNi_{1-x}O$. Néel temperature vs. Li concentration [73N2].

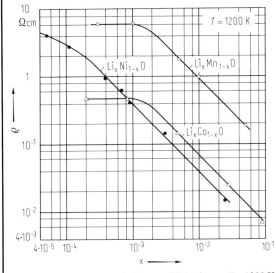

Fig. 27. $Li_xNi_{1-x}O$. Resistivity vs. Li doping at $T = 1200$ K for NiO, CoO and MnO. Deviation from the linear behaviour at low doping concentration are due to the pressure of cation vacancies [70B].

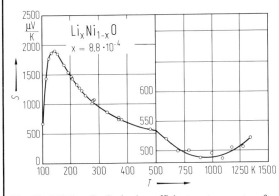

Fig. 28. $Li_xNi_{1-x}O$. Seebeck coefficient vs. temperature for a sample containing $8.8 \cdot 10^{-2}$ at% Li. Note the change in both scales at 500 K [66B].

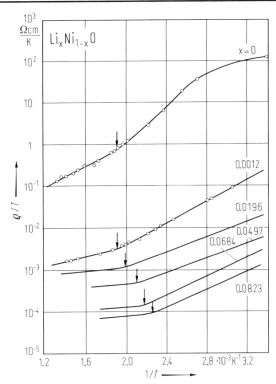

Fig. 26. $Li_xNi_{1-x}O$. Resistivity (over T) vs. reciprocal temperature for different doping levels. The arrows indicate T_N for each sample [73N2].

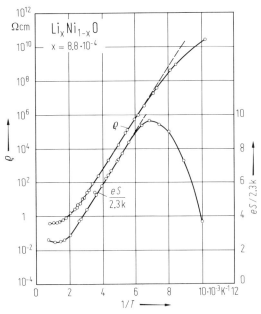

Fig. 29. $Li_xNi_{1-x}O$. Resistivity and reduced Seebeck coefficient vs. reciprocal temperature for a sample containing $8.8 \cdot 10^{-2}$ at% Li [66B].

9.15.2.3.4 Nickel oxide (NiO)

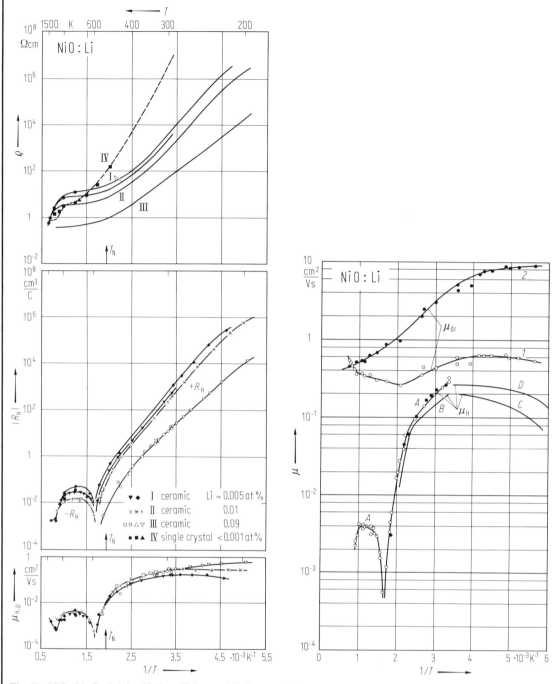

Fig. 30. NiO:Li. Resistivity, Hall coefficient and Hall mobility vs. (reciprocal) temperature for p-type ceramic and non-deliberately-doped single crystal material. Values of ϱ for ceramic samples have been indicated by solid lines solely. With regard to the resistivity of single crystals at high temperatures the black circles and squares represent different values for a thin disk measured when cooling the sample slowly and rapidly, respectively. The broken line results from a rapid run performed on a thick bar [67D].

Fig. 31. NiO:Li. Mobility 1, 2 and Hall mobility $A \cdots D$ for holes vs. reciprocal temperature. 1, 2 0.088 at% Li doped, calculated σ and S using different approximations. A Li doped crystals with 0.088⋯0.005 at%, B single crystal undoped and C, D single crystals 0.2 at% Li doped [66B].

9.15.2.3.4 Nickel oxide (NiO)

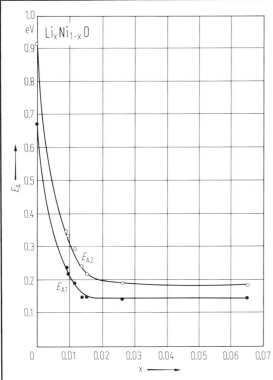

Fig. 32. $Li_xNi_{1-x}O$. Activation energy vs. Li content for two specimens. ●-●-●: E_{A1} for $T>T^*$ ○-○-○: E_{A2} for $T<T^*$ where T^* is the temperature at which a break occurs in the $\log\sigma$ vs. $1/T$ plot and E_{A1} and E_{A2} are the activation energies $T \gtrless T^*$ [65K1].

For Fig. 34, see next page.

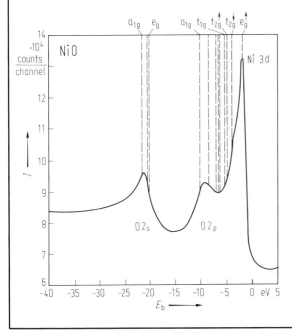

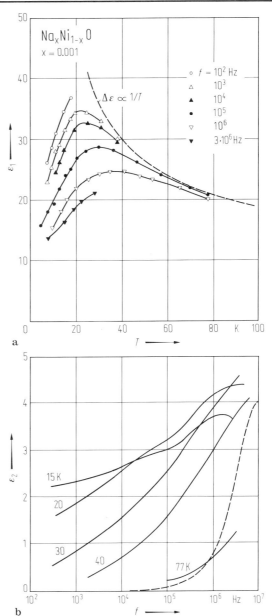

Fig. 33. NiO:Na. (a) Real part of the dielectric constant vs. temperature for different frequencies for a sample doped with 0.1 at% Na. The broken line represents the behaviour of $\varepsilon(0)$ if $\Delta\varepsilon$ (for definition, see the tables) is proportional to T^{-1}, (b) imaginary part of the dielectric constant vs. frequency for different temperatures measured on the same crystal as in (a). The broken line represents a Debye curve with its maximum at 10^7 Hz [70B].

◀

Fig. 35. NiO. SCF−Xα transition state energies of an NiO_6^{10-} cluster superimposed on the NiO XPS-data of [72W]. Only the spin polarization of predominantly 3d-like levels is indicated. E_b is relative to the experimental Fermi energy of the spectrometer [73J, 75E].

9.15.2.3.4 Nickel oxide (NiO)

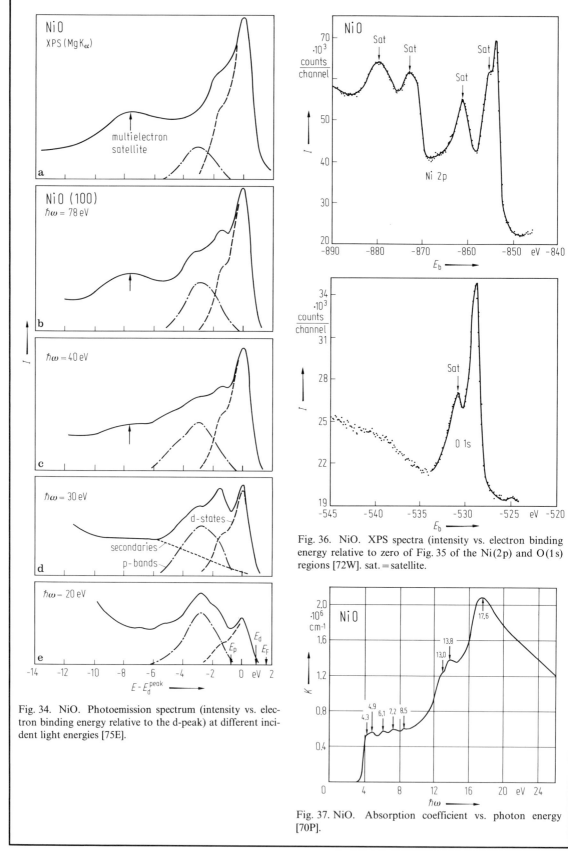

Fig. 34. NiO. Photoemission spectrum (intensity vs. electron binding energy relative to the d-peak) at different incident light energies [75E].

Fig. 36. NiO. XPS spectra (intensity vs. electron binding energy relative to zero of Fig. 35 of the Ni(2p) and O(1s) regions [72W]. sat. = satellite.

Fig. 37. NiO. Absorption coefficient vs. photon energy [70P].

9.15.2.3.4 Nickel oxide (NiO)

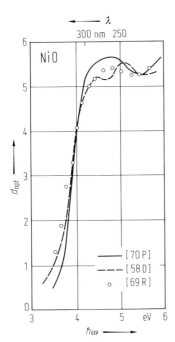

Fig. 38. NiO. Optical density vs. photon energy (wavelength) showing details of the absorption edge [70P].

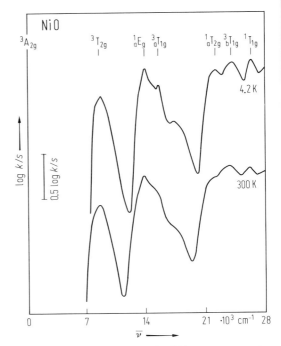

Fig. 39. NiO. Absorbance ($\log k/s$) vs. wavenumber. Assignment in terms of local exciton formation is given [78P]. k: absorbed radiant flux, s: incident radiant flux.

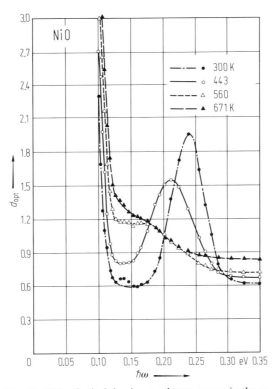

Fig. 40. NiO. Optical density vs. photon energy in the region of the 0.24 eV absorption peak for various temperatures. The peak has been assigned to a two-magnon and one phonon excitation [59N]. $d_{opt} = \log(I_0/I)$.

9.15.2.3.4 Nickel oxide (NiO)

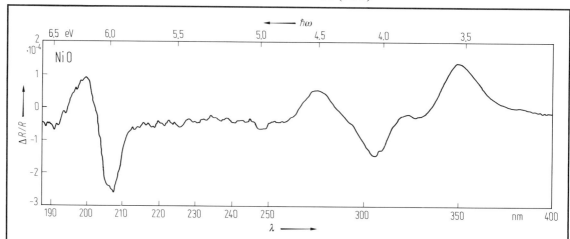

Fig. 41. NiO. Electroreflectance spectrum (electroreflectance vs. photon energy and wavelength) [71G, 69M].

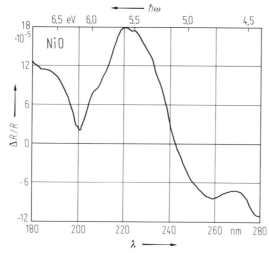

Fig. 42. NiO. Thermoreflectance vs. photon energy and wavelength [72M2].

Fig. 43. NiO. Real and imaginary parts of the complex refractive index vs. photon energy [70P].

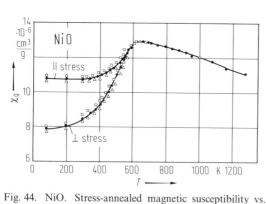

Fig. 44. NiO. Stress-annealed magnetic susceptibility vs. temperature measured parallel and perpendicular to the stress direction. The three symbols represent values found after annealing with stress along three different ⟨111⟩ directions [56S].

Fig. 45. NiO. Density of spin-wave states vs. energy. Solid line is calculated best fit, dashed line is from $J_2/k = 221$ K alone [72H].

9.15.2.3.5 Palladium oxide (PdO) (For tables, see p. 227 ff.)

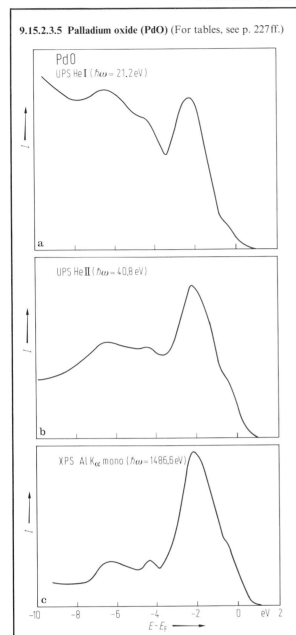

Fig. 1. PdO. Photoelectron spectra at different excitation energies (a) He I (b) He II (c) Al K$_\alpha$ monochromated to a half-width of 0.6 eV. Photoelectron intensity vs. energy relative to the Fermi level E_F [79H].

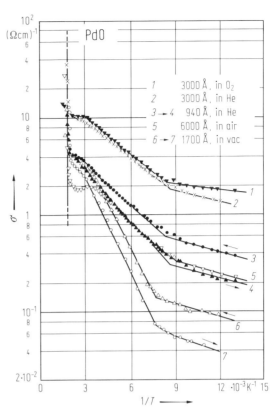

Fig. 2. PdO. Conductivity vs. reciprocal temperature for polycrystalline films of various thicknesses measured in various atmospheres [67O].

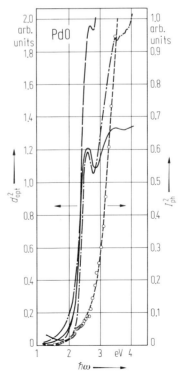

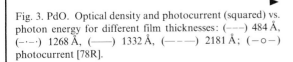

Fig. 3. PdO. Optical density and photocurrent (squared) vs. photon energy for different film thicknesses: (---) 484 Å, (-·-·) 1268 Å, (——) 1332 Å, (— — —) 2181 Å; (-o-) photocurrent [78R].

9.15.2.4.1 Magnetite (Fe$_3$O$_4$)

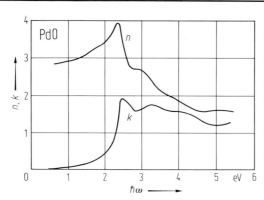

Fig. 4. PdO. Real and imaginary parts of the refractive index vs. photon energy [79N].

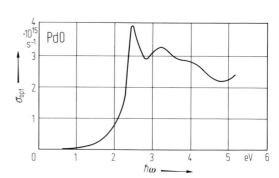

Fig. 5. PdO. Optical conductivity defined as $\sigma_{opt} = \omega \varepsilon_2(\omega)/4\pi$ vs. photon energy [79N].

9.15.2.4 Spinel oxides

9.15.2.4.1 Magnetite (Fe$_3$O$_4$) (For tables, see p. 229 ff.)

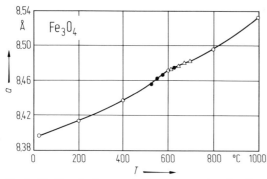

Fig. 1. Fe$_3$O$_4$. Lattice constant vs. temperature for three specimens [65G].

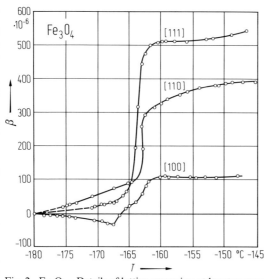

Fig. 2. Fe$_3$O$_4$. Details of lattice expansion at low temperatures. Coefficient β in $L(T) = L_0(1+\beta(T))$ (L_0 = length at $-180\,°C$) vs. temperature. Measurements in three crystallographic directions [50D].

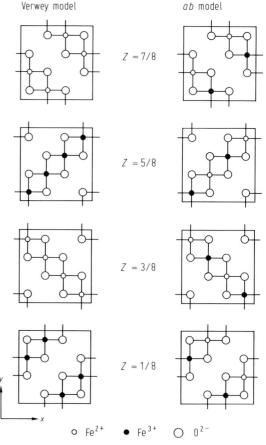

Fig. 3. Fe$_3$O$_4$. Comparison of the Verwey and ab models of ion ordering below the Verwey temperature [75S2]. $x \parallel [100]_{cub}$; $y \parallel [010]_{cub}$; $z \parallel [001]_{cub}$. Fe^{2+} and Fe^{3+} are arranged on octahedral (B) sites. Z in units of the lattice parameter.

9.15.2.4.1 Magnetite (Fe$_3$O$_4$)

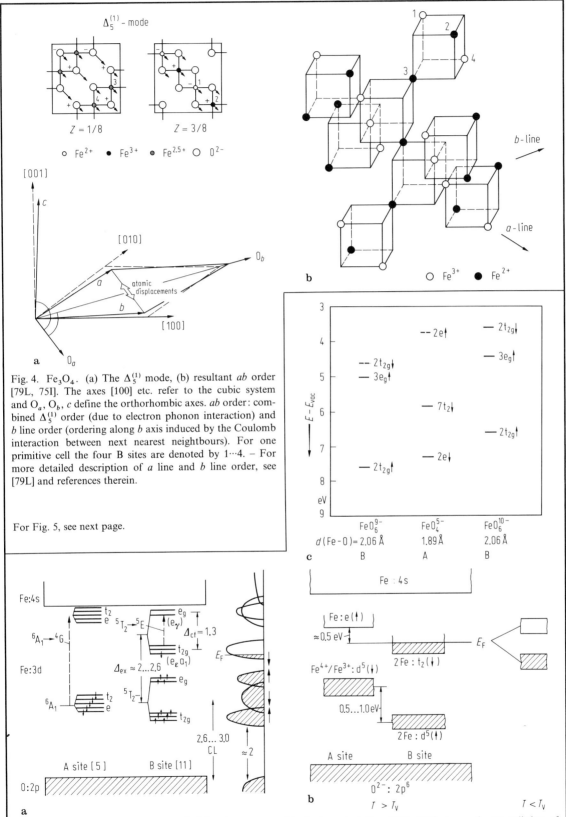

Fig. 4. Fe$_3$O$_4$. (a) The $\Delta_5^{(1)}$ mode, (b) resultant ab order [79L, 75I]. The axes [100] etc. refer to the cubic system and O_a, O_b, c define the orthorhombic axes. ab order: combined $\Delta_5^{(1)}$ order (due to electron phonon interaction) and b line order (ordering along b axis induced by the Coulomb interaction between next nearest neightbours). For one primitive cell the four B sites are denoted by 1···4. – For more detailed description of a line and b line order, see [79L] and references therein.

For Fig. 5, see next page.

Fig. 6. Fe$_3$O$_4$. (a, b) Phenomenological band structures, (c) cluster calculation [72C, 78T] Δ_{ex}: exchange splitting of the d-orbitals, Δ_{cf}: crystal field splitted at the B sites, CL: cathodoluminescence onset fixing the Fe 3d, O 2p separation.

9.15.2.4.1 Magnetite (Fe$_3$O$_4$)

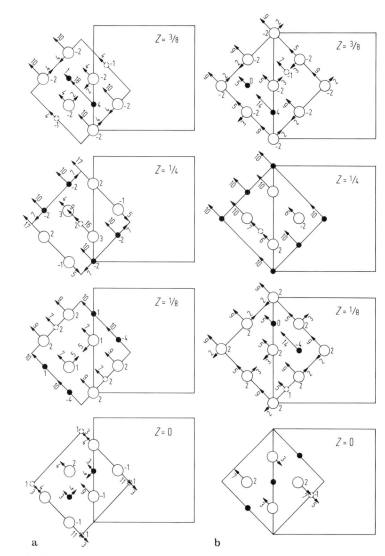

Fig. 5. Fe$_3$O$_4$. Atomic displacements from the cubic high temperature phase determined from neutron diffraction studies (a) refined in Pmc2$_1$ (b) refined in Pmca. Displacements in x and y are shown by arrows and z displacements are also indicated. All displacements are in fractional coordinates $\cdot 10^3$. O^{2-} ions are shown as open circles, B-site Fe-ions as full circles and A-site Fe ions as dashed circles [82I].

9.15.2.4.1 Magnetite (Fe$_3$O$_4$)

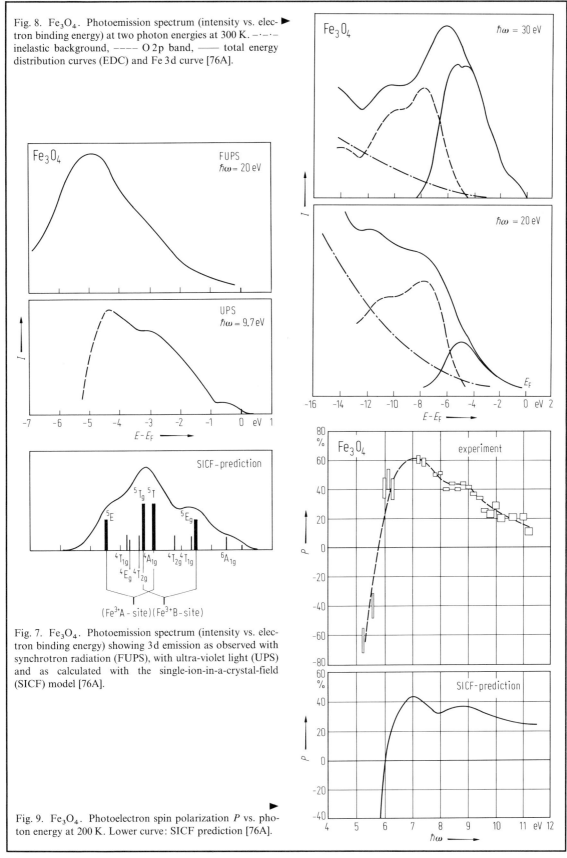

Fig. 8. Fe$_3$O$_4$. Photoemission spectrum (intensity vs. electron binding energy) at two photon energies at 300 K. –·–·– inelastic background, – – – – O 2p band, ——— total energy distribution curves (EDC) and Fe 3d curve [76A].

Fig. 7. Fe$_3$O$_4$. Photoemission spectrum (intensity vs. electron binding energy) showing 3d emission as observed with synchrotron radiation (FUPS), with ultra-violet light (UPS) and as calculated with the single-ion-in-a-crystal-field (SICF) model [76A].

Fig. 9. Fe$_3$O$_4$. Photoelectron spin polarization P vs. photon energy at 200 K. Lower curve: SICF prediction [76A].

9.15.2.4.1 Magnetite (Fe$_3$O$_4$)

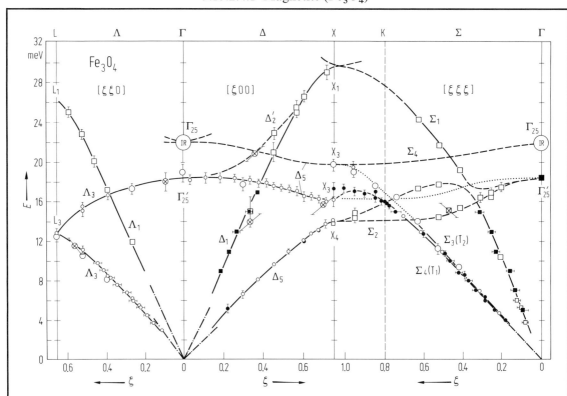

Fig. 10. Fe$_3$O$_4$. Phonon dispersion curves at 300 K for the principal symmetry directions [74S]. □, ■: longitudinal modes; ○, ●: transverse modes; open symbols: observations in [110] zone, filled symbols: observations in [100] zone; small symbols: observations with 13.8 meV incident energy neutrons, large symbols: observations with 38 meV incident energy neutrons. Circles with crosses: observations with time of flight. Solid lines: confidently assigned, dotted and dashed lines: tentatively assigned, dot-dash lines near the Γ point: ultrasonic measurements, IR: measured by [72G].

Fig. 11. Fe$_3$O$_4$. Brillouin zone of spinel [74S]. See Fig. 8 of 9.15.2.3.1 (MnO), p. 493.

For Fig. 12, see next page.

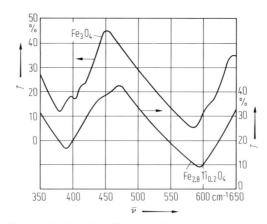

Fig. 13. Fe$_3$O$_4$, Fe$_{2.8}$Ti$_{0.2}$O$_4$. Transmission vs. wavenumber at 77 K [77K].

9.15.2.4.1 Magnetite (Fe$_3$O$_4$)

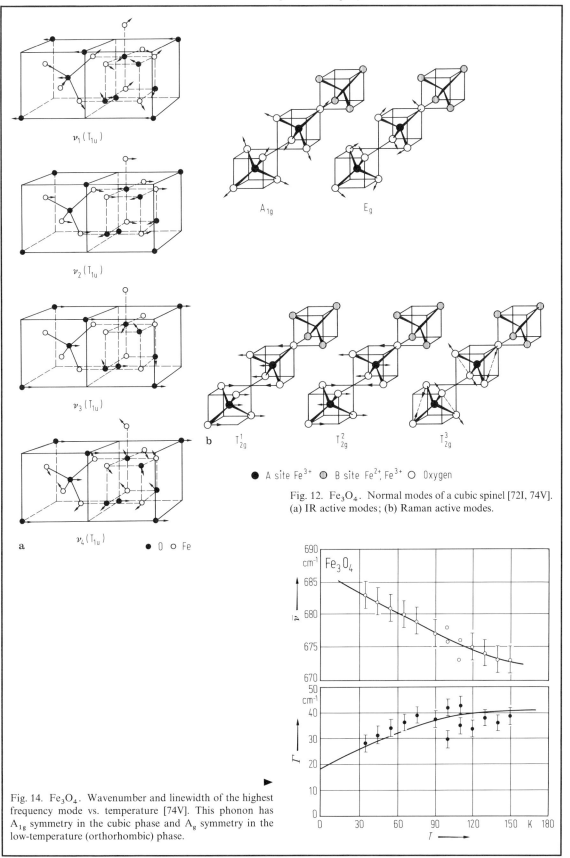

Fig. 12. Fe$_3$O$_4$. Normal modes of a cubic spinel [72I, 74V]. (a) IR active modes; (b) Raman active modes.

Fig. 14. Fe$_3$O$_4$. Wavenumber and linewidth of the highest frequency mode vs. temperature [74V]. This phonon has A$_{1g}$ symmetry in the cubic phase and A$_g$ symmetry in the low-temperature (orthorhombic) phase.

9.15.2.4.1 Magnetite (Fe$_3$O$_4$)

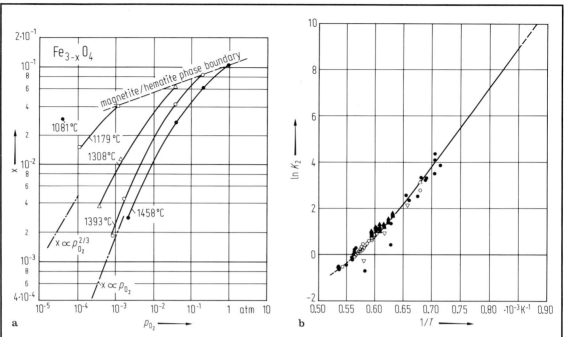

Fig. 15. Fe$_{3-x}$O$_4$. Composition parameter x vs. oxygen partial pressure at various temperatures (a) results from [72K] (b) results from several authors of ln K_2 (defined in text) vs. $1/T$ collected by [77D2].

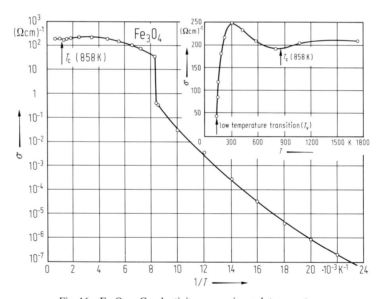

Fig. 16. Fe$_3$O$_4$. Conductivity vs. reciprocal temperature (in the inset: vs. temperature) for stoichiometric single crystal material [57M].

9.15.2.4.1 Magnetite (Fe$_3$O$_4$)

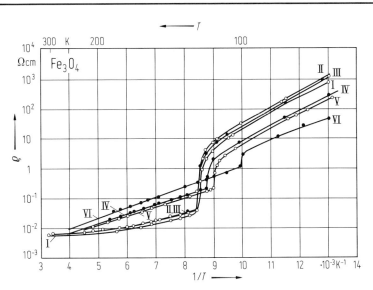

Fig. 17. Fe$_3$O$_4$. Resistivity vs. (reciprocal) temperature for polycrystalline sample. The curves I to VI represent increasing Fe-deficit [41V].

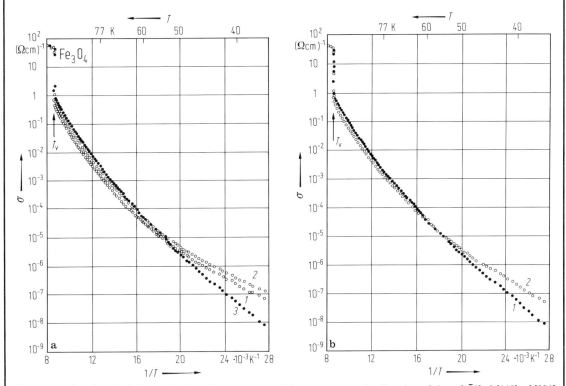

Fig. 18. Fe$_3$O$_4$. Conductivity vs. (reciprocal) temperature. (a) anisotropy in the direction of the 1 [1$\bar{1}$0], 2 [110], 3 [001] axes, (b) conductivity parallel to the [001] axis, while the direction of the magnetic field cooling is parallel to 1 [110] and 2 [001]+40° [77M2].

9.15.2.4.1 Magnetite (Fe$_3$O$_4$)

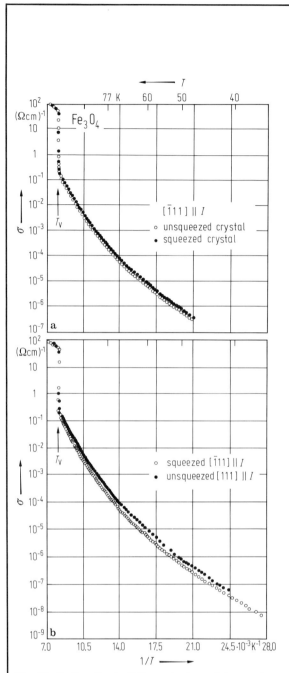

Fig. 19. Fe$_3$O$_4$. Conductivity vs. (reciprocal) temperature in the [$\bar{1}$11] direction for a squeezed and unsqueezed crystal (a), compared to the [111] direction for an unsqueezed crystal (b); field cooled in (a) [001] direction, (b) [001]+40° direction [77M2].

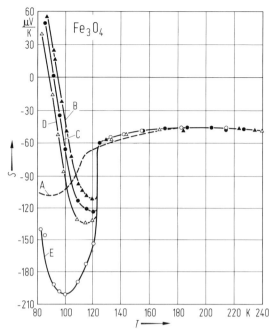

Fig. 20. Fe$_3$O$_4$. Seebeck coefficient vs. temperature for samples annealed under various oxygen pressures measured ∥ [110]. log p_{O_2} = (A) −4.2, (B) −9.0, (C) −9.7, (D) −9.9, (E) −10.2; p_{O_2} in atm [76K].

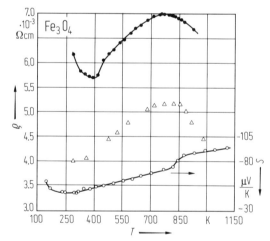

Fig. 21. Fe$_3$O$_4$. Resistivity (● [70G], △ [52S]) and Seebeck coefficient (○) vs. temperature at high temperatures ($T > T_V$) [70G].

9.15.2.4.1 Magnetite (Fe$_3$O$_4$)

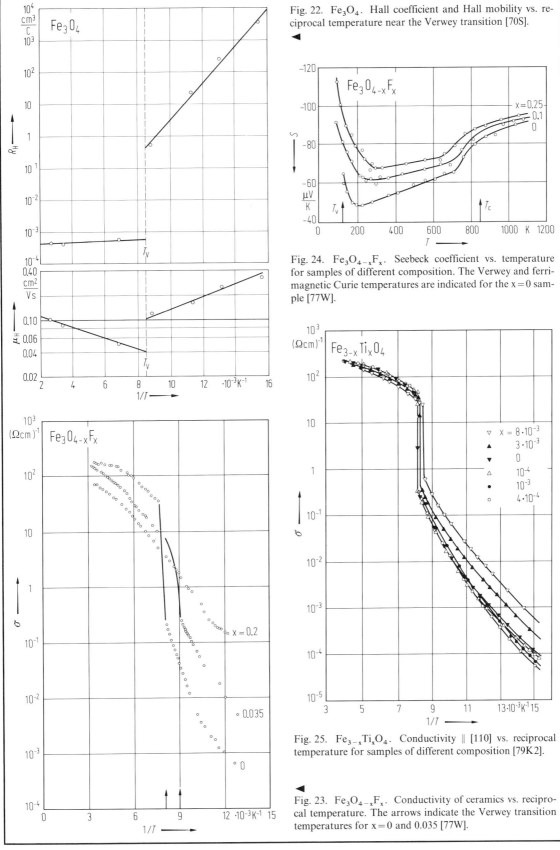

Fig. 22. Fe$_3$O$_4$. Hall coefficient and Hall mobility vs. reciprocal temperature near the Verwey transition [70S].

Fig. 24. Fe$_3$O$_{4-x}$F$_x$. Seebeck coefficient vs. temperature for samples of different composition. The Verwey and ferrimagnetic Curie temperatures are indicated for the x = 0 sample [77W].

Fig. 25. Fe$_{3-x}$Ti$_x$O$_4$. Conductivity ∥ [110] vs. reciprocal temperature for samples of different composition [79K2].

Fig. 23. Fe$_3$O$_{4-x}$F$_x$. Conductivity of ceramics vs. reciprocal temperature. The arrows indicate the Verwey transition temperatures for x = 0 and 0.035 [77W].

9.15.2.4.1 Magnetite (Fe$_3$O$_4$)

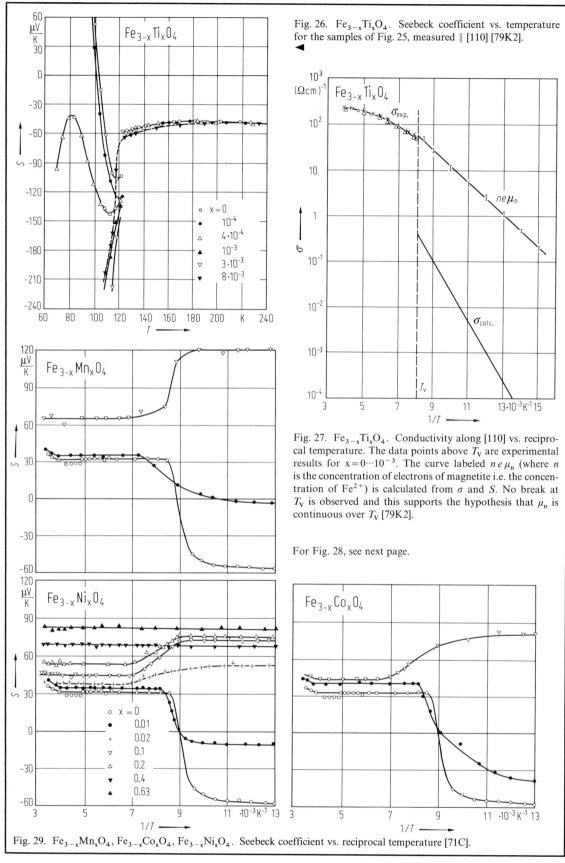

Fig. 26. Fe$_{3-x}$Ti$_x$O$_4$. Seebeck coefficient vs. temperature for the samples of Fig. 25, measured ∥ [110] [79K2].

Fig. 27. Fe$_{3-x}$Ti$_x$O$_4$. Conductivity along [110] vs. reciprocal temperature. The data points above T_v are experimental results for $x = 0 \cdots 10^{-3}$. The curve labeled $n e \mu_n$ (where n is the concentration of electrons of magnetite i.e. the concentration of Fe^{2+}) is calculated from σ and S. No break at T_v is observed and this supports the hypothesis that μ_n is continuous over T_v [79K2].

For Fig. 28, see next page.

Fig. 29. Fe$_{3-x}$Mn$_x$O$_4$, Fe$_{3-x}$Co$_x$O$_4$, Fe$_{3-x}$Ni$_x$O$_4$. Seebeck coefficient vs. reciprocal temperature [71C].

9.15.2.4.1 Magnetite (Fe$_3$O$_4$)

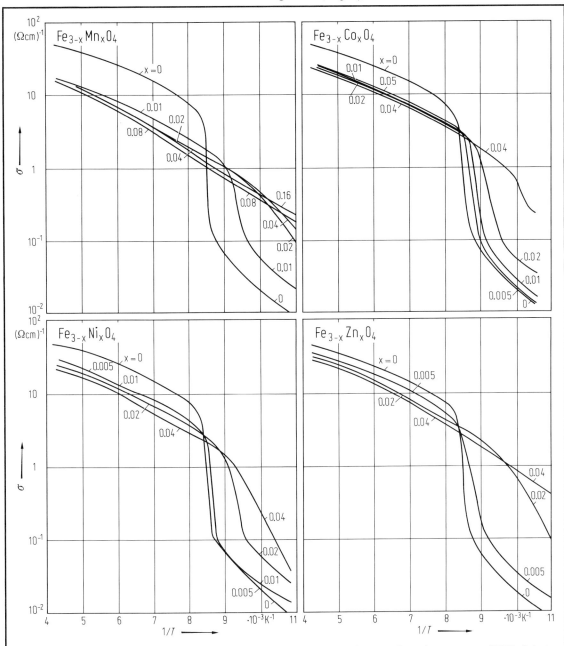

Fig. 28. Fe$_{3-x}$Mn$_x$O$_4$, Fe$_{3-x}$Co$_x$O$_4$, Fe$_{3-x}$Ni$_x$O$_4$, Fe$_{3-x}$Zn$_x$O$_4$. Conductivity vs. reciprocal temperature [57M]. Orientation not specified.

9.15.2.4.1 Magnetite (Fe$_3$O$_4$)

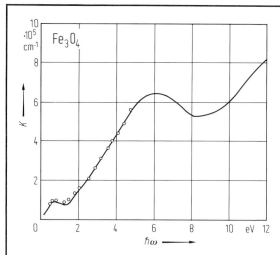

Fig. 30. Fe$_3$O$_4$. Absorption coefficient vs. photon energy in the UV at 300 K [79S2].

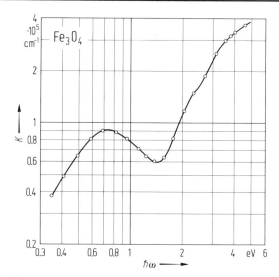

Fig. 31. Fe$_3$O$_4$. Absorption coefficient at RT vs. photon energy in the visible region for thin magnetite films [74M].

For Fig. 32, see next page.

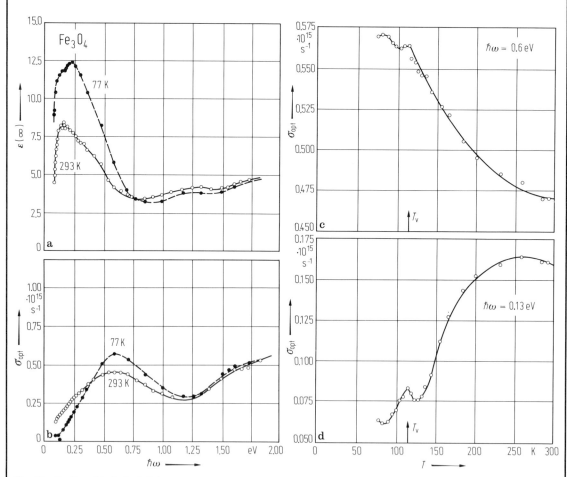

Fig. 33. Fe$_3$O$_4$. (a) Optical dielectric constant and (b) optical conductivity vs. photon energy at 77 K and 293 K; (c, d) optical conductivity vs. temperature for $\hbar\omega = 0.6$ and 0.13 eV, respectively [72B].

9.15.2.4.1 Magnetite (Fe$_3$O$_4$)

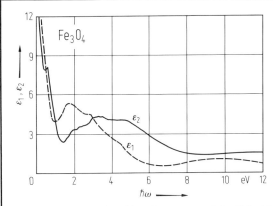

Fig. 32. Fe$_3$O$_4$. Real and imaginary parts of the dielectric function at 300 K vs. photon energy [79S2].

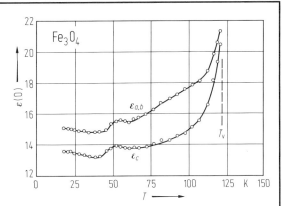

Fig. 34. Fe$_3$O$_4$. Dielectric constant vs. temperature in the microwave region for the low-temperature phase [78M4].

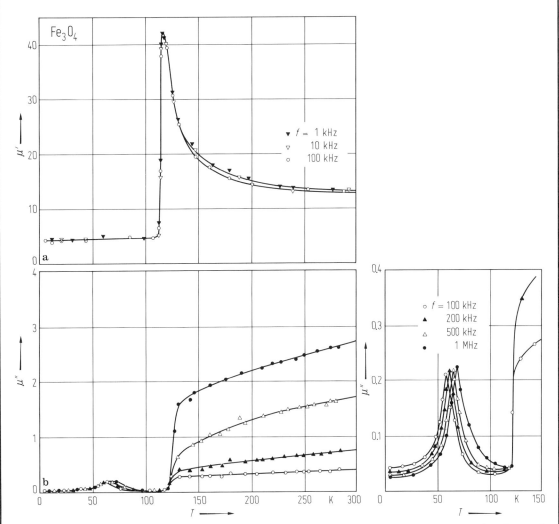

Fig. 35. Fe$_3$O$_4$. Real (μ') and imaginary (μ'') parts of the relative permeability vs. temperature at various frequencies [80I]. Lower right hand figure shows low-temperature range on an enlarged scale.

9.15.2.4.2 Cobalt oxide (Co$_3$O$_4$)

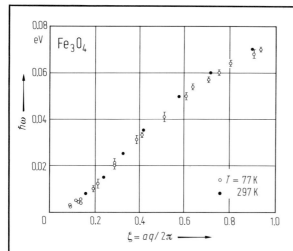

Fig. 36. Fe$_3$O$_4$. Acoustic spin-wave spectrum. Energy transfer $\hbar\omega$ vs. reduced wave vector. ○: data of [67T] at 77 K. ●: data of [62W] at 297 K.

9.15.2.4.2 Cobalt oxide (Co$_3$O$_4$) (For tables, see p. 236ff.)

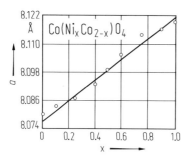

Fig. 1. Co(Ni$_x$Co$_{2-x}$)O$_4$. Lattice parameter vs. composition [78A].

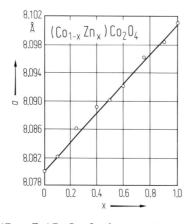

Fig. 2. (Co$_{1-x}$Zn$_x$)Co$_2$O$_4$. Lattice parameter vs. composition [79G].

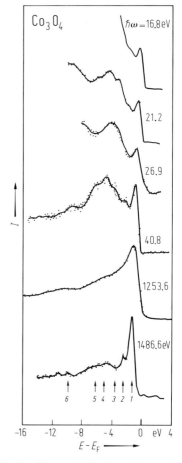

Fig. 3. Co$_3$O$_4$. Photoemission intensity vs. electron binding energy relative to Fermi energy for different photon energies showing valence band structure. 1, 2 Co 3d ionization, 3, 4 O 2p ionization, 5, 6 ionization from and to states of Co 3d by configuration mixing [79J].

9.15.2.4.2 Cobalt oxide (Co$_3$O$_4$)

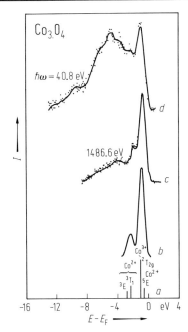

Fig. 4. Co$_3$O$_4$. Detailed valence region PE spectra for HeII and AlK$_\alpha$ exciting radiation: curves d and c respectively. Curve a shows the expected atomic energy levels split by crystal-field theory and b shows these levels convoluted by a 0.8 eV broad Gaussian to simulate the instrumental resolution [79J].

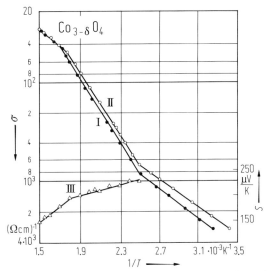

Fig. 6. Co$_{3-\delta}$O$_4$ film. I and II: conductivity vs. reciprocal temperature for films annealed at 350°C and 400°C, respectively. III: Seebeck coefficient vs. reciprocal temperature for a Co$_3$O$_4$ film annealed at 400°C [78B].

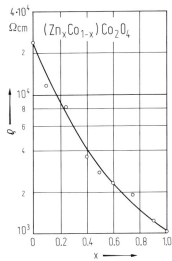

Fig. 7. (Zn$_x$Co$_{1-x}$)Co$_2$O$_4$ ceramic. Room temperature resistivity vs. composition [77V].

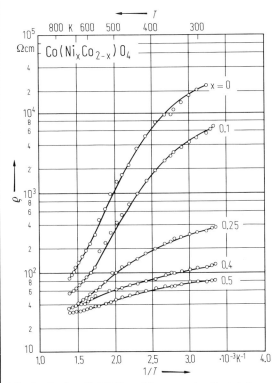

Fig. 5. Co$_3$O$_4$, Co(Ni$_x$Co$_{2-x}$)O$_4$, ceramic. Resistivity vs. reciprocal temperature for various x [78A].

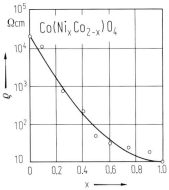

Fig. 8. Co(Ni$_x$Co$_{2-x}$)O$_4$ ceramic. Room temperature resistivity vs. composition [77V].

9.15.2.4.3 Manganese oxide (Mn_3O_4)

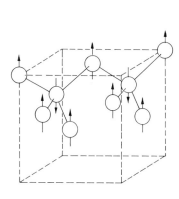

Fig. 9. Co_3O_4. Magnetic structure at low temperatures [64R].

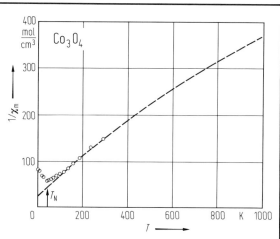

Fig. 10. Co_3O_4. Reciprocal paramagnetic susceptibility vs. temperature. Broken line: Curie-Weiss law using data listed in the tables [64R]. χ_m in CGS-emu.

9.15.2.4.3 Manganese oxide (Mn_3O_4) (For tables, see p. 238 ff.)

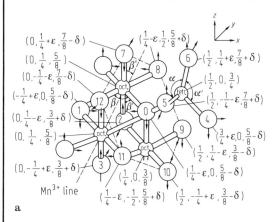

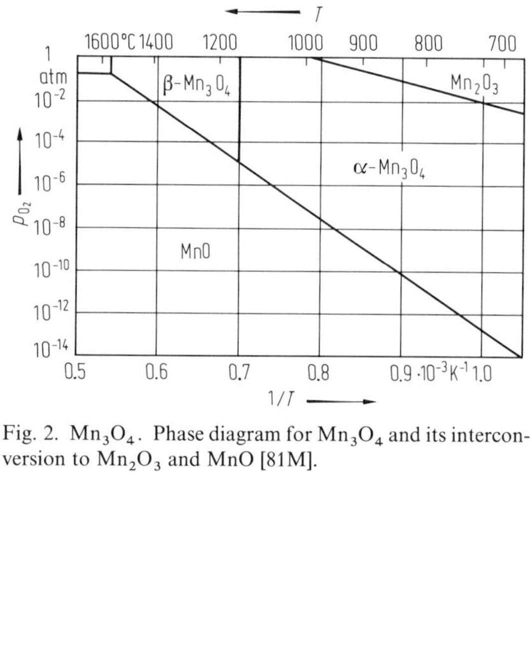

Fig. 2. Mn_3O_4. Phase diagram for Mn_3O_4 and its interconversion to Mn_2O_3 and MnO [81M].

○ ◯ Oxygen and octahedral site Mn^{3+} of the upper layer
◌ ◯ Oxygen and octahedral site Mn^{3+} of the lower layer
⊙ Tetrahedral site Mn^{2+}

Fig. 1. Mn_3O_4. Arrangement of the axis and its displacements in the tetragonal phase ($T < 1170$ °C) [61S].

9.15.2.4.3 Manganese oxide (Mn$_3$O$_4$)

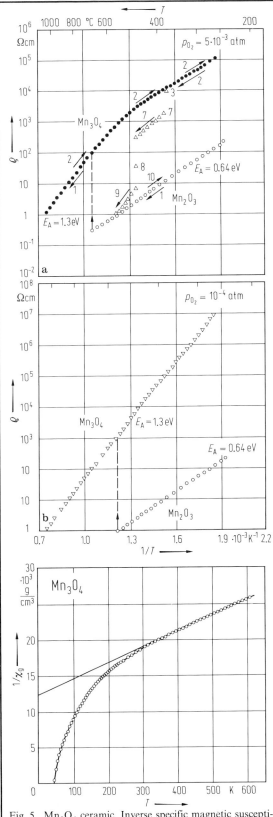

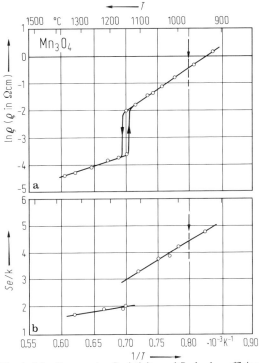

Fig. 3. Mn$_3$O$_4$ ceramic. Resistivity vs. (reciprocal) temperature of Mn$_3$O$_4$ and Mn$_2$O$_3$ in the low-temperature region for (a) $p_{O_2} = 5 \cdot 10^{-3}$ atm, (b) $p_{O_2} = 10^{-4}$ atm. Numbers in (a) refer to the thermal history of the sample [75L]. Vertical arrow indicates transition point between oxides.

Fig. 4. Mn$_3$O$_4$ ceramic. Resistivity and Seebeck coefficient vs. (reciprocal) temperature for sintered samples at 1 atm oxygen pressure. The arrows indicate the phase boundary for Mn$_3$O$_4$/Mn$_2$O$_3$, though this transformation is sufficiently slow to permit metastable measurements below this temperature on Mn$_3$O$_4$ [81M].

Fig. 5. Mn$_3$O$_4$ ceramic. Inverse specific magnetic susceptibility vs. temperature. Solid line shows best fit to Curie-Weiss law using parameters in text. χ in CGS-emu [71B2].

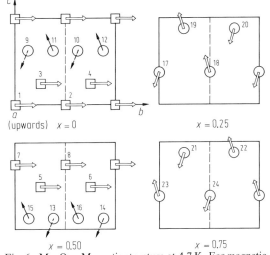

Fig. 6. Mn$_3$O$_4$. Magnetic structure at 4.7 K. For magnetic moment magnitudes and direction cosines, see the tables. □: tetragonal site, ○: octahedral site, ⇒: moment in the plane, →: moment directed upwards, --→ moment directed downwards [74J]. x refers to the value of x in tetragonal unit cell.

9.15.2.5 Sesquioxides and related oxides
9.15.2.5.1 Oxides of chromium (For tables, see p. 242ff.)

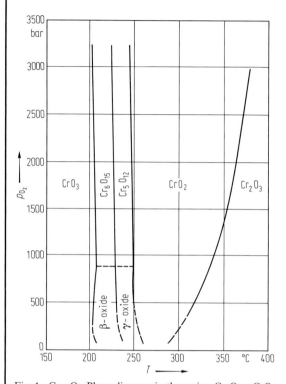

Fig. 1. Cr–O. Phase diagram in the region Cr_2O_3–CrO_3 [68W, 77C1].

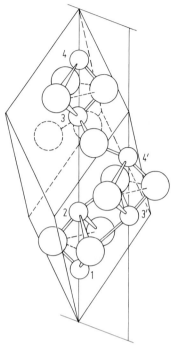

Fig. 2. Cr_2O_3. Rhombohedral cell of the corundum structure [70S]. Small spheres are metal ions.

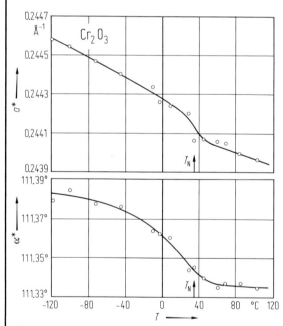

Fig. 3. Cr_2O_3. Reciprocal lattice parameters a^* and α^* of the rhombohedral setting vs. temperature [76A].

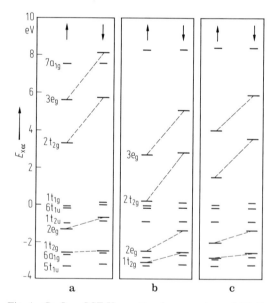

Fig. 4. Cr_2O_3. SCF-Xα scattered wave valence MO diagrams for (a) a CrO_6^{9-} cluster, (b) CrO_6^{8-} 2p hole state, (c) Cr_6^{8-} $2t_{2g}$ valence hole state [76T]. Arrows indicate spin direction.

9.15.2.5.1 Oxides of chromium

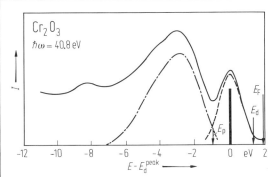

Fig. 5. Cr_2O_3. He II UPS spectrum (intensity vs. electron binding energy relative to the d-peak) at 40.8 eV photon energy deconvoluted into Cr 3d (dashed line) and O 2p (dashed-dotted line) signals; E_p, E_d are the onsets of the Cr 3d and O 2p photoemission signals and the vertical line is the unbroadened single-state d-ionization [75E].

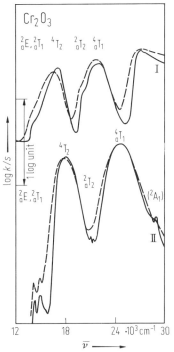

Fig. 6. Cr_2O_3. Absorbance vs. wavenumber in the visible region; I Cr_2O_3, II $Cr_{0.1}Al_{1.9}O_3$; Solid line: $T=4$ K, dashed line: $T=300$ K [74G].

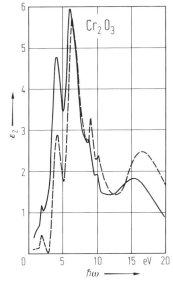

Fig. 7. Cr_2O_3. Imaginary part of the dielectric constant vs. photon energy for two different samples [77A].

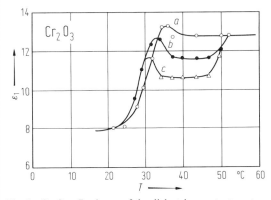

Fig. 8. Cr_2O_3. Real part of the dielectric constant vs. temperature for a single crystal with and without impressed magnetic fields; (a) without magnetic fields, (b) $H=3.5$ kOe, (c) $H=9$ kOe [67L].

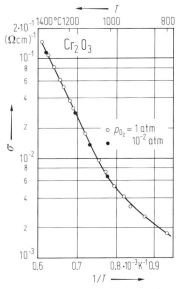

Fig. 9. Cr_2O_3. Conductivity vs. (reciprocal) temperature for unoriented single crystal in the intrinsic region for two values of p_{O_2} [64C].

9.15.2.5.1 Oxides of chromium

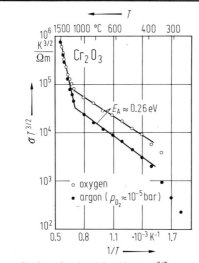

Fig. 10. Cr_2O_3. Conductivity (times $T^{3/2}$) vs. (reciprocal) temperature for hot-pressed ceramic samples under oxygen and argon atmosphere, respectively [72M].

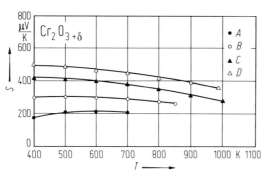

Fig. 12. $Cr_2O_{3+\delta}$. Seebeck coefficient vs. temperature for the cold-pressed ceramic samples of Fig. 11 [69C].

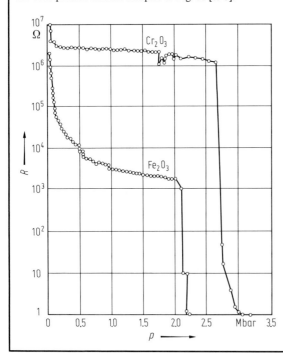

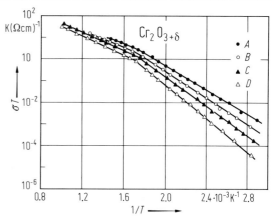

Fig. 11. $Cr_2O_{3+\delta}$. Conductivity (times temperature) vs. reciprocal temperature for cold-pressed ceramic samples of different stoichiometry. $\delta = 0.06$ (A), 0.02 (B), 0.01 (C), 0.004 (D) [69C].

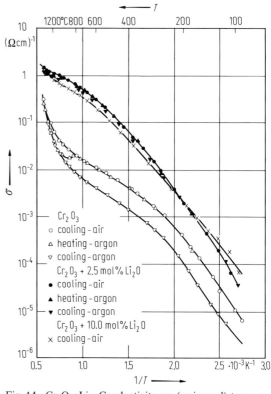

Fig. 14. Cr_2O_3:Li. Conductivity vs. (reciprocal) temperature for several hot-pressed ceramic samples showing the effect of Li doping [65H]. Measurements on cooling and heating under different atmospheres.

◄ Fig. 13. Cr_2O_3, Fe_2O_3. Resistance vs. pressure up to very high pressures [71K].

9.15.2.5.1 Oxides of chromium

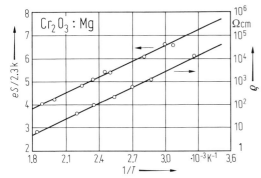

Fig. 15. Cr_2O_3: Mg. Resistivity and reduced Seebeck coefficient vs. reciprocal temperature [77C2]. Mg concentration: 1 at%.

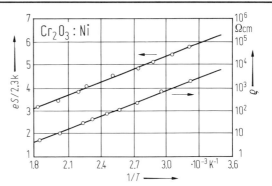

Fig. 16. Cr_2O_3: Ni. Resistivity and reduced Seebeck coefficient vs. reciprocal temperature [77C2]. Ni concentration: 1 at%.

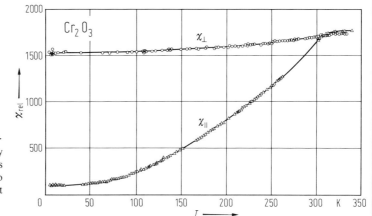

Fig. 17. Cr_2O_3. Magnetic susceptibility parallel and perpendicular to the c axis vs. temperature (susceptibility relative to a value of $\chi_\perp = 22.4\,(4) \cdot 10^{-6}\,cm^3/g$ at 4.2 K) [56M, 63F]. χ in CGS-emu.

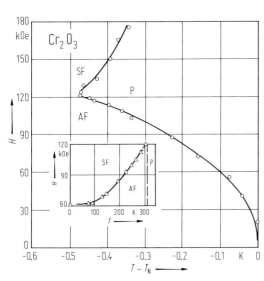

Fig. 18. Cr_2O_3. Magnetic phase diagram. $H\|c$ axis, $T_N = 307.3$ K [77Y]. AF: antiferromagnetic, P: paramagnetic, SF: spin flop region.

For Fig. 19, see next page.

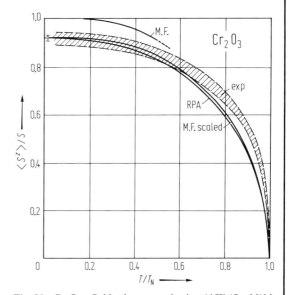

Fig. 20. Cr_2O_3. Sublattice magnetization ($\langle S^z\rangle/S = M/M_0$ where M_0 is the saturation magnetization) vs. reduced temperature (T/T_N). Hatched region gives the spread of experimental results from several authors and the solid lines give the results of calculations using Mean Field (M.F.), scaled M.F. and Random Phase Approximation (R.P.A.) methods [70S].

9.15.2.5.2 Hematite (α-Fe$_2$O$_3$)

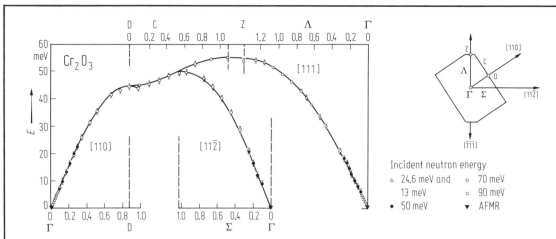

Fig. 19. Cr$_2$O$_3$. Spin wave dispersion at 78 K. A sketch of the Brillouin zone is shown in the upper right hand corner; upper curve relates to the upper scale and lower curve to the lower scale, where appropriate [70S].

9.15.2.5.2 Hematite (α-Fe$_2$O$_3$) (For tables, see p. 247ff.)

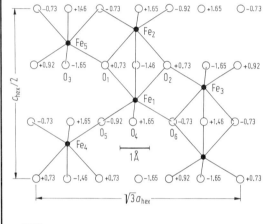

Fig. 2. Fe$_2$O$_3$. The rhombohedral unit cell of the corundum structure [70S]. See Fig. 2 of 9.15.2.5.1 (Oxides of chromium), p. 548.

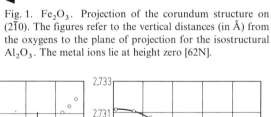

Fig. 1. Fe$_2$O$_3$. Projection of the corundum structure on $(2\bar{1}0)$. The figures refer to the vertical distances (in Å) from the oxygens to the plane of projection for the isostructural Al$_2$O$_3$. The metal ions lie at height zero [62N].

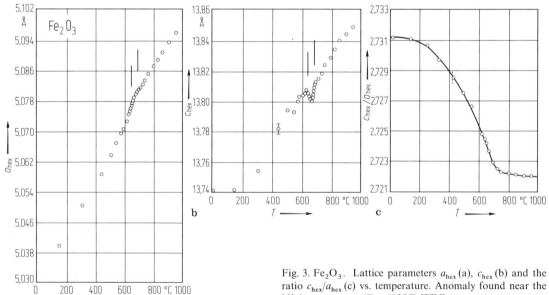

Fig. 3. Fe$_2$O$_3$. Lattice parameters a_{hex} (a), c_{hex} (b) and the ratio c_{hex}/a_{hex} (c) vs. temperature. Anomaly found near the Néel temperature ($T_N \approx 682$ °C) [77N].

9.15.2.5.2 Hematite (α-Fe$_2$O$_3$)

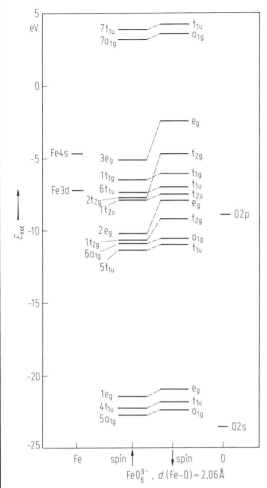

Fig. 4. Fe$_2$O$_3$. Xα-cluster calculation on FeO$_6^{9-}$. Energy for Fe, O and the spin up and down states of the cluster [74T].

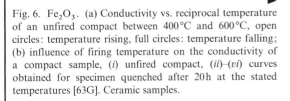

Fig. 6. Fe$_2$O$_3$. (a) Conductivity vs. reciprocal temperature of an unfired compact between 400°C and 600°C, open circles: temperature rising, full circles: temperature falling; (b) influence of firing temperature on the conductivity of a compact sample, (i) unfired compact, (ii)–(vi) curves obtained for specimen quenched after 20 h at the stated temperatures [63G]. Ceramic samples.

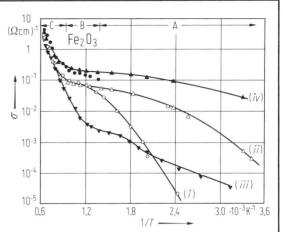

Fig. 5. Fe$_2$O$_3$. Conductivity vs. reciprocal temperature for high-purity samples (i) polycrystalline bar sintered at 1300°C in O$_2$ and cooled to 900°C in ca. 30 min., (ii) polycrystalline bar sintered at 1300°C in O$_2$ and quenched rapidly to 20°C, (iii) polycrystalline bar sintered at 1300°C and cooled slowly to RT, (iv) oxidized iron wire heated to 1300°C for 5 h and rapidly quenched to 20°C. A, B, C are the regions discussed in the tables. Solid circles: single crystal flux grown from sodium tetraborate; four-probe resistivity along the (111) plane of a specimen quenched after 13 h at 1300°C in air [63G].

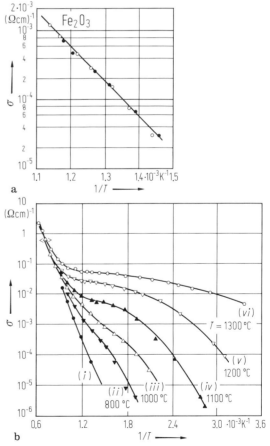

553

9.15.2.5.2 Hematite (α-Fe$_2$O$_3$)

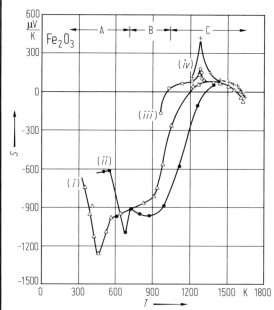

Fig. 7. Fe$_2$O$_3$. Seebeck coefficient vs. temperature for a high purity ceramic sample. (*i*) and (*ii*) fired specimens cooled (*i*) from 1364°C in air or (*ii*) quenched from 1300°C and heated, (*iii*) and (*iv*) unfired specimens, rising temperature [63G].

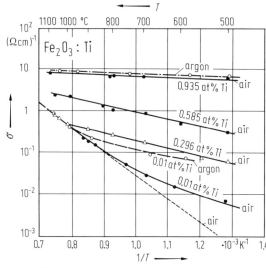

Fig. 8. Fe$_2$O$_3$:Ti. Conductivity vs. (reciprocal) temperature for Ti doped ceramic specimens in air and argon [65G]. For argon: p_{O_2} = equilibrium with Fe$_2$O$_3$ and Fe$_3$O$_4$. Short-dashed curve: pure Fe$_2$O$_3$ from [51M].

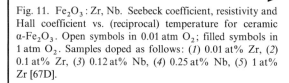

Fig. 11. Fe$_2$O$_3$:Zr, Nb. Seebeck coefficient, resistivity and Hall coefficient vs. (reciprocal) temperature for ceramic α-Fe$_2$O$_3$. Open symbols in 0.01 atm O$_2$; filled symbols in 1 atm O$_2$. Samples doped as follows: (*1*) 0.01 at% Zr, (*2*) 0.1 at% Zr, (*3*) 0.12 at% Nb, (*4*) 0.25 at% Nb, (*5*) 1 at% Zr [67D].

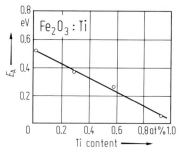

Fig. 9. Fe$_2$O$_3$:Ti. Activation energy for conductivity vs. Ti doping concentration in the temperature range where n_d (=[Ti]) is a constant [69G]. Ceramic sample.

For Fig. 10, see next page.

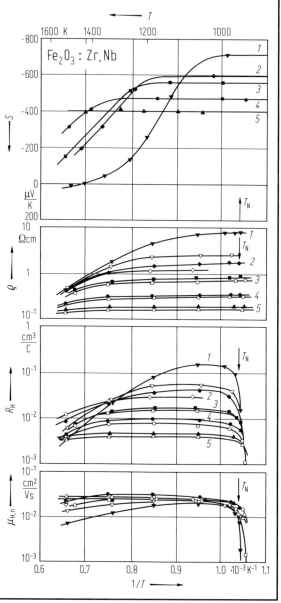

9.15.2.5.2 Hematite (α-Fe$_2$O$_3$)

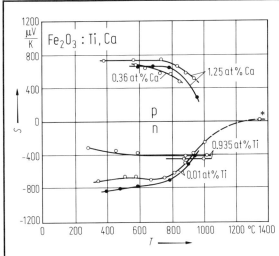

Fig. 10. Fe$_2$O$_3$: Ti, Ca. Seebeck coefficient vs. temperature for Ti and Ca doped ceramic samples in air (open circles) and argon (full circles) ($p_{O_2} \simeq 10^{-4}$ atm) (∗: data from [62T]) [65G].

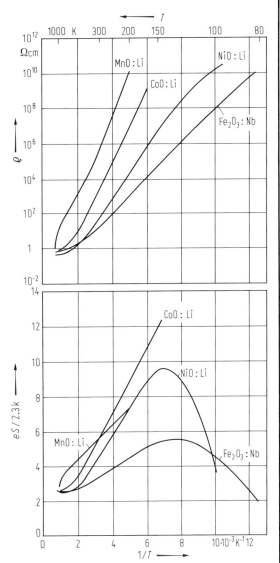

Fig. 13. Fe$_2$O$_3$: Nb. Resistivity and reduced Seebeck coefficient vs. (reciprocal) temperature for 0.1 at% Li doped p-type NiO, p-type CoO and p-type MnO and 0.1 at% Nb doped n-type Fe$_2$O$_3$ [70B]. Ceramic samples.

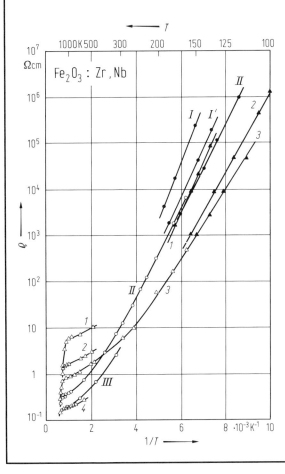

Fig. 12. Fe$_2$O$_3$: Zr, Nb. Resistivity vs. (reciprocal) temperature for Zr and Nb doped ceramic samples. Open symbols: dc results, filled symbols; ac results. Zr doped samples: (1) 0.01 at%, (2) 0.1 at%, (3) 0.25 at%, (4) 1.0 at%, Nb doped samples: (I, I') 0.1 at%, (II) 0.25 at%, (III) 0.5 at% [70B].

9.15.2.5.2 Hematite (α-Fe$_2$O$_3$)

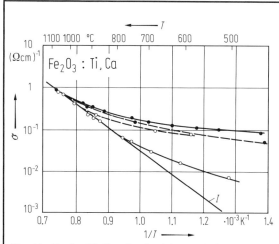

Fig. 14. Fe$_2$O$_3$:Ti, Ca. Conductivity vs. (reciprocal) temperature for 0.01 at% Ti doped Fe$_2$O$_3$ (open circles), 0.30 at% Ca doped Fe$_2$O$_3$ (full circles), in air (solid line), in argon (dashed line) (p_{O_2} in equilibrium with Fe$_2$O$_3$ and Fe$_3$O$_4$); *I*: data from [51M] on pure Fe$_2$O$_3$. [65G]. Ceramic samples.

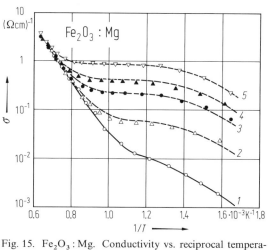

Fig. 15. Fe$_2$O$_3$:Mg. Conductivity vs. reciprocal temperature for Mg doped ceramic samples fired at 1300 °C. (*1*) pure sample, (*2*) 0.01 at% Mg, (*3*) 0.03 at% Mg, (*4*) 0.05 at% Mg, (*5*) 0.2 at% Mg [66G].

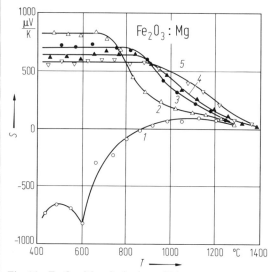

Fig. 16. Fe$_2$O$_3$:Mg. Seebeck coefficient vs. temperature. Samples as in Fig. 15 [66G].

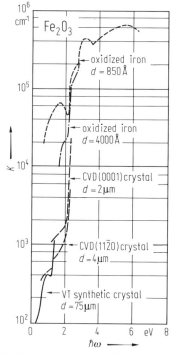

Fig. 18. Fe$_2$O$_3$. Absorption coefficient vs. photon energy for samples of varying thickness; CVD: chemically vapour deposited, VT: vapour transport grown [78B].

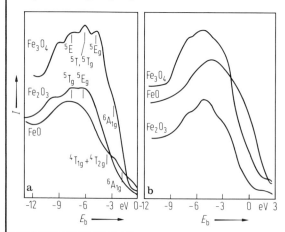

Fig. 17. Fe$_2$O$_3$, FeO, Fe$_3$O$_4$. (a) He II and (b) AlK$_\alpha$ XPE spectra (intensity vs. electron binding energy) showing valence bands in various iron oxides [79V].

9.15.2.5.2 Hematite (α-Fe$_2$O$_3$)

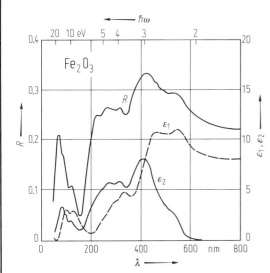

Fig. 19. Fe$_2$O$_3$. Reflectance (R) and real and imaginary parts of the dielectric function vs. wavelength (photon energy) [79G].

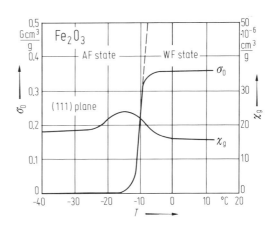

Fig. 20. Fe$_2$O$_3$. Magnetic susceptibility (χ_g) and weak ferromagnetic spontaneous specific magnetization (σ_0) vs. temperature [63T]. σ, $\chi_g \perp$ [111].

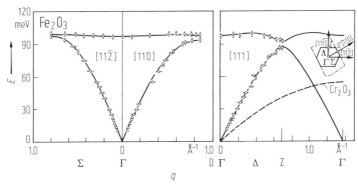

Fig. 21. Fe$_2$O$_3$. Spin wave spectrum at 240 K. A part of the Brillouin zone is shown in the upper right corner [70S].

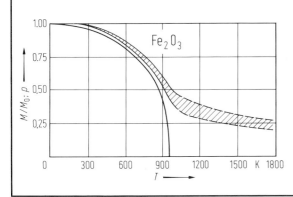

Fig. 22. Fe$_2$O$_3$. Short and long-range magnetic order for Fe$_2$O$_3$. Full curve: long-range order, defined as M/M_0 where M is the sublattice magnetization and M_0 the saturation sublattice magnetization (from neutron scattering). Hatched curve: short-range magnetic order parameter, p, defined as $\langle S_0^x S_1^x + S_0^y S_1^y + S_0^z S_1^z \rangle / S(S+1)$ where S_0 is the spin on the central ion and S_1 the spin on the ions in the first ion-shell surrounding the central ion. This parameter can be obtained from the heat capacity [75G].

9.15.2.5.3 Oxides of rhodium

(For tables, see p. 254f.)

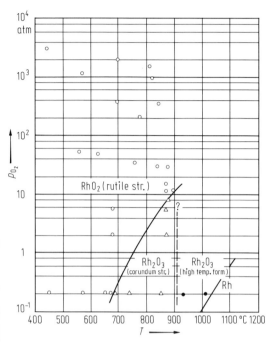

Fig. 1. Rh_2O_3. Phase diagram for the Rh−O system [68M].

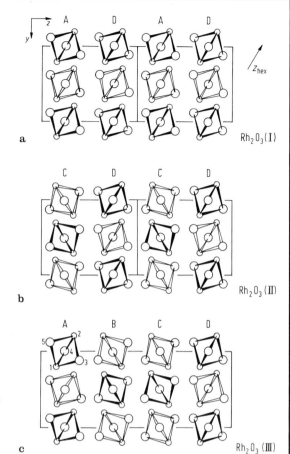

Fig. 2. Rh_2O_3. A schematic representation of the crystal structure of the polymorphs of Rh_2O_3. The small circles represent Rh atoms. The dark motifs are centred around the plane $x=0$, the light motifs around $x=\frac{1}{2}$. Rh_2O_3(I) is projected along the hexagonal x axis; Rh_2O_3(II) and Rh_2O_3(III) are projected along the orthorhombic x axis. Two unit cells of Rh_2O_3(I) and Rh_2O_3(II) and one unit cell of Rh_2O_3(III) have been drawn [73B].

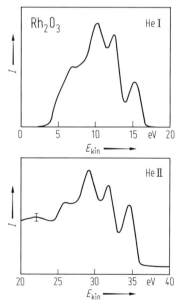

Fig. 3. Rh_2O_3. UPE spectra (intensity vs. kinetic electron energy) [78B].

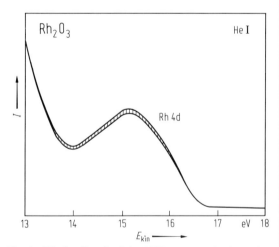

Fig. 4. Rh_2O_3. Detail of the UPE spectrum in the region of the Rh 4d peak [78B].

9.15.2.5.4 Oxides of manganese
(For tables, see p. 256 ff.)

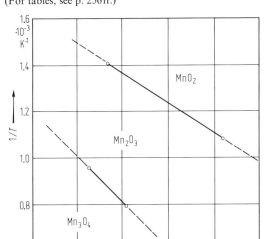

Fig. 1. Phase diagram for the systems MnO_2/Mn_2O_3 and Mn_2O_3/Mn_3O_4 drawn from the data of [65O, 67S]. See also Fig. 2 of 9.15.2.4.3 (Mn_3O_4). 1 atm ≈ 1 bar.

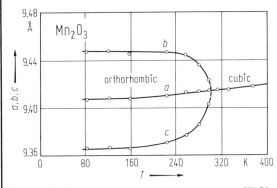

Fig. 3. Mn_2O_3. Lattice parameters vs. temperature [70G].

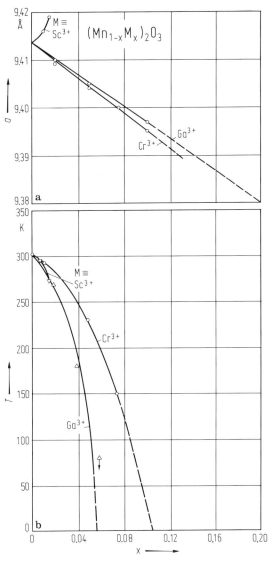

Fig. 4. $(Mn_{1-x}M_x)_2O_3$, M = Sc, Ga, Cr. (a) Lattice parameter vs. composition; (b) Orthorhombic-cubic transition temperature vs. composition [70G].

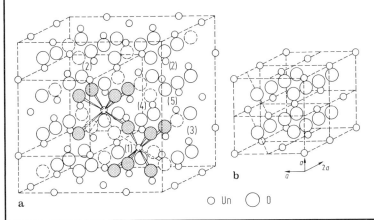

Fig. 2. Mn_2O_3. (a) The cubic bixbyite structure and its relationship to (b) fluorite. The dotted circles show the oxygen atoms that have been removed from the fluorite structure; the oxygen atoms forming the two types of the Mn site 8(a) and 24(d) described in the text, are shown shaded [75W].

9.15.2.5.4 Oxides of manganese

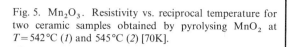

Fig. 5. Mn_2O_3. Resistivity vs. reciprocal temperature for two ceramic samples obtained by pyrolysing MnO_2 at $T = 542\,°C$ (1) and $545\,°C$ (2) [70K].

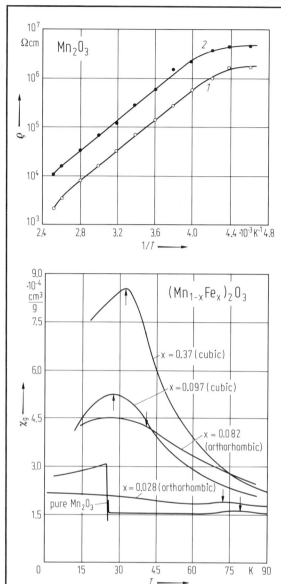

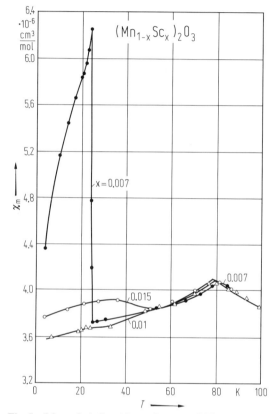

Fig. 6. $(Mn_{1-x}Fe_x)_2O_3$. Magnetic susceptibility vs. temperature for five differently doped ceramic samples [68G]. Arrows point to magnetic transition temperature. χ_g in CGS-emu.

Fig. 7. $(Mn_{1-x}Sc_x)_2O_3$. Magnetic susceptibility per mol of substance (gross formula) vs. temperature for four differently doped ceramic samples [70G]. χ_m in CGS-emu.

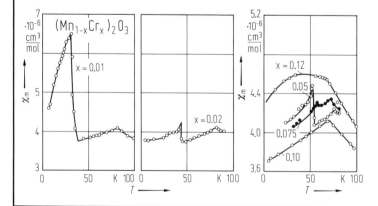

Fig. 8. $(Mn_{1-x}Cr_x)_2O_3$. Magnetic susceptibility per mol of substance (gross formula) vs. temperature for six differently doped ceramic samples [70G]. χ_m in CGS-emu.

9.15.2.5.4 Oxides of manganese

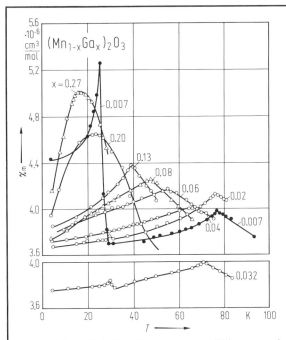

Fig. 9. $(Mn_{1-x}Ga_x)_2O_3$. Magnetic susceptibility per mol of substance (gross formula) vs. temperature for ten differently doped ceramic samples [70G]. χ_m in CGS-emu.

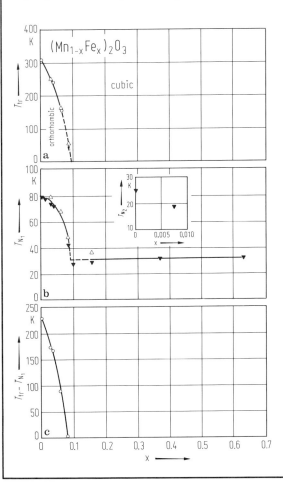

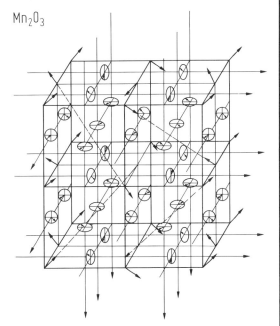

Fig. 11. Mn_2O_3. Proposed magnetic structure for cubic phases belonging to $T_h^7 - I\,a3$ [68G]. Shown are spin directions on b sites (along 3-fold axes) and d sites (in planes perpendicular to 2'-fold axes).

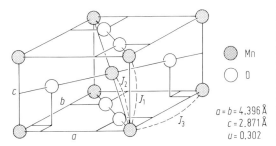

Fig. 12. MnO_2. Crystal structure and dominant exchange interactions [71O].

For Fig. 13, see next page.

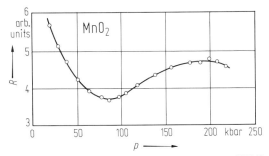

Fig. 14. MnO_2. Resistance (arb. units) vs. pressure [63M].

◄

Fig. 10. $(Mn_{1-x}Fe_x)_2O_3$. (a) Crystallographic transition temperature, (b) upper Néel temperature (inset: lower Néel temperature), (c) difference $T_{tr} - T_{N_1}$ vs. composition. Open circles: X-ray diffraction, open triangles: Moessbauer spectroscopy, full triangles: magnetic susceptibility [68G].

9.15.2.6.1 Oxides of niobium

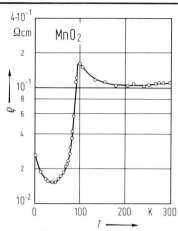

Fig. 13. MnO_2. Resistivity vs. temperature for a single crystal [69R]. Orientation not specified.

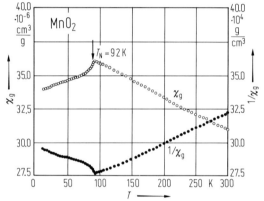

Fig. 16. MnO_2. Magnetic susceptibility and its inverse vs. temperature for powdered material [71O].

Fig. 15. MnO_2. (A) the screw-type magnetic structure with a pitch of $\frac{7}{2}$, (B) projection on the c plane [59Y]. $u_{1,2}$: real unit vectors perpendicular to each other.

9.15.2.6 Oxides of group V elements (Nb, Ta)

9.15.2.6.1 Oxides of niobium (For tables, see p. 260 ff.)

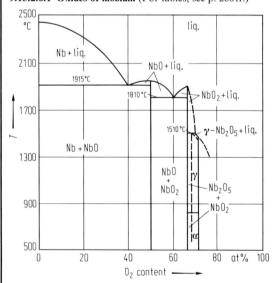

Fig. 1. Nb–O. Phase diagram [69C].

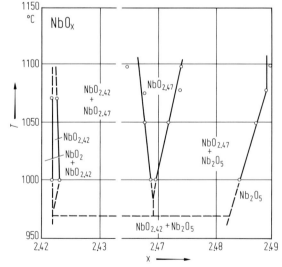

Fig. 2. NbO_x. Details of the phase diagram in the region $Nb_{12}O_{29} - Nb_2O_5$ [74M].

9.15.2.6.1 Oxides of niobium

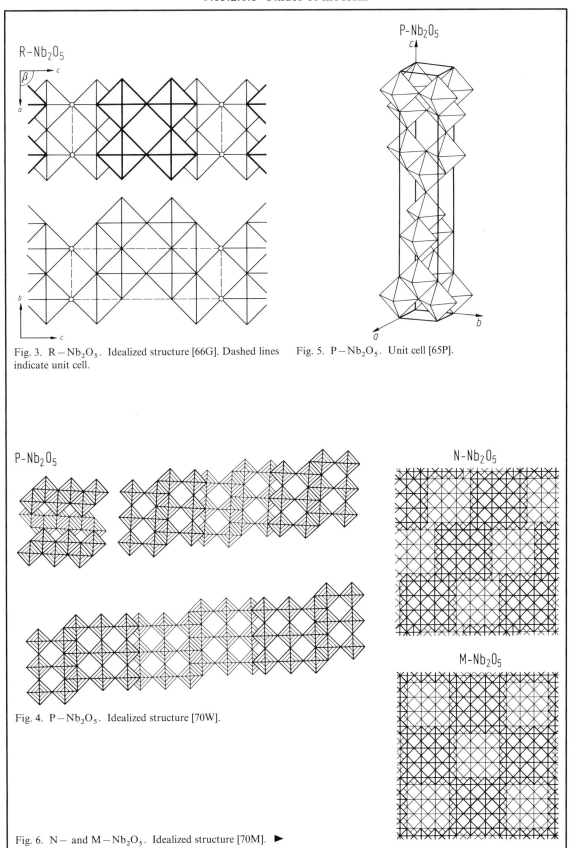

Fig. 3. R−Nb_2O_5. Idealized structure [66G]. Dashed lines indicate unit cell.

Fig. 5. P−Nb_2O_5. Unit cell [65P].

Fig. 4. P−Nb_2O_5. Idealized structure [70W].

Fig. 6. N− and M−Nb_2O_5. Idealized structure [70M]. ▶

9.15.2.6.1 Oxides of niobium

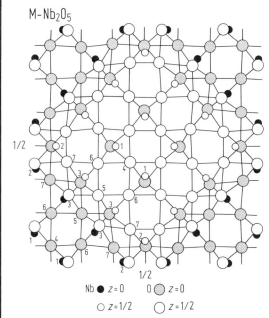

Fig. 7. M–Nb$_2$O$_5$. Projection of the unit cell in the ab plane [70M].

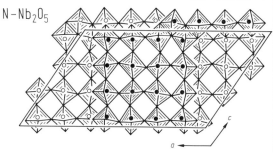

Fig. 8. N–Nb$_2$O$_5$. Unit cell [67A]. Open and full circles: Nb atoms at $y=0$ and $y=\frac{1}{2}$.

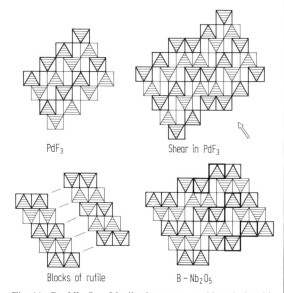

Fig. 11. B–Nb$_2$O$_5$. Idealized structure and its relationship to rutile and PdF$_3$ [70W].

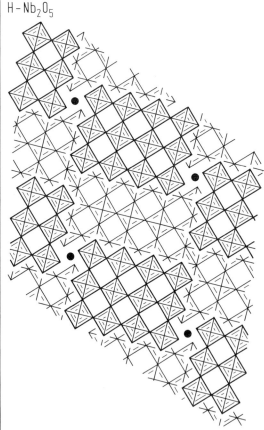

Fig. 9. H–Nb$_2$O$_5$. Idealized structure [70W]. Full circles are the Nb atoms in tetrahedral sites in the tunnels.

For Fig. 10, see next page.

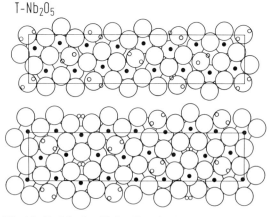

Fig. 12. T–Nb$_2$O$_5$. Unit cell projection parallel [001]. The small open circles represent possible positions for the nine-coordinate "additional" Nb; full circles: Nb atoms; large open circles: oxygen. Upper part: space group Pbam, lower part: space group Pbmn [75K1].

9.15.2.6.1 Oxides of niobium

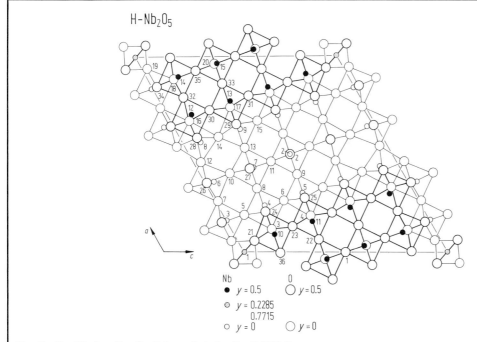

Fig. 10. H−Nb$_2$O$_5$. Details of the projected unit cell [76K1].

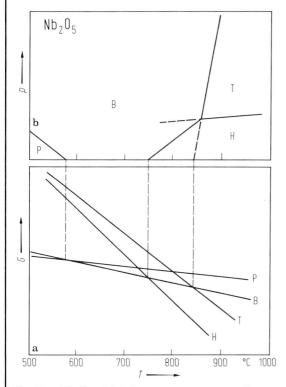

Fig. 12a. Nb$_2$O$_5$. (a) A free energy-temperature diagram for four modifications. (b) A pressure-temperature diagram for the four modifications [72K].

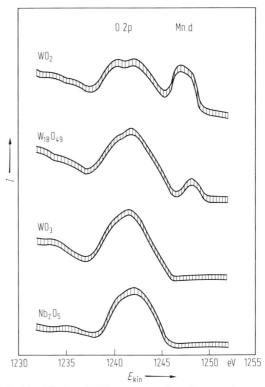

Fig. 13. Nb$_2$O$_5$. XPE spectrum (intensity vs. electron kinetic energy) compared to WO$_3$, W$_{18}$O$_{49}$ and WO$_2$ [78B1].

Fig. 15. Nb_2O_5. Raman spectra at RT (intensity vs. wavenumber) of the B- and H-modifications with $Ti_2Nb_{10}O_{29}$ and VNb_9O_{25} for comparison [76M2].

Fig. 14. Nb_2O_5. IR spectra (transmission vs. wavenumber) at RT for various modifications [68E].

Fig. 16. NbO_x. Composition parameter x vs. oxygen pressure. Solid lines: [72N2]; full and open circles: [65B]; dashed lines: [61K]; triangles: [68A]; 1300°C isotherm [69S3]. Fig. from [72N2]. 1 atm ≈ 1 bar.

Fig. 17. NbO_x. Oxygen partial pressure vs. composition at 1300°C and 1400°C [76K2].

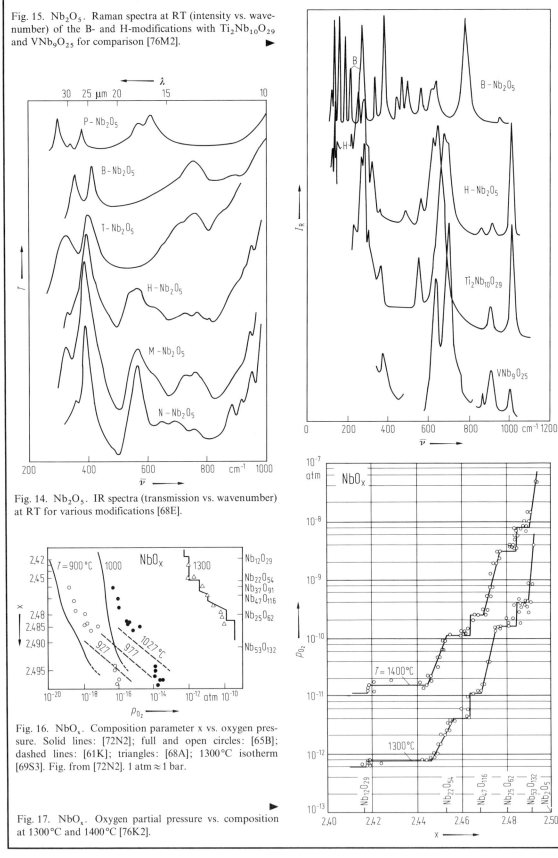

9.15.2.6.1 Oxides of niobium

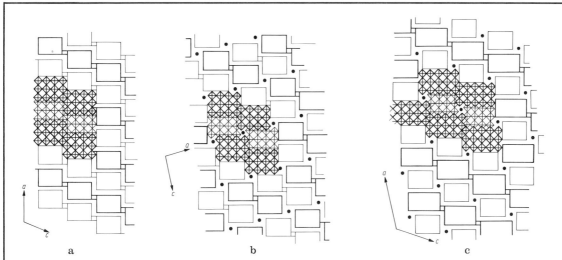

Fig. 18. Structural relationship between (a) $Nb_{12}O_{29}$, (b) $Nb_{25}O_{62}$ and (c) $H-Nb_2O_5$ in (010) projection [72N2].

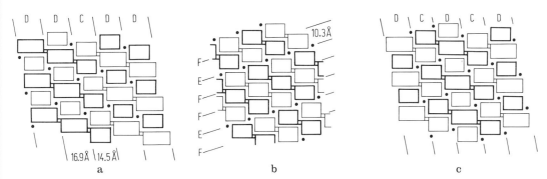

Fig. 19. Mechanisms whereby intergrowth phases may form at high temperatures. (a) coherent intergrowth of one row of $Nb_{25}O_{62}$ in $H-Nb_2O_5$; (b) coherent intergrowth of one row of $Nb_{12}O_{29}$ in $Nb_{25}O_{62}$; (c) regular coherent intergrowth corresponding to the phase $Nb_{53}O_{132}$ [72N2].

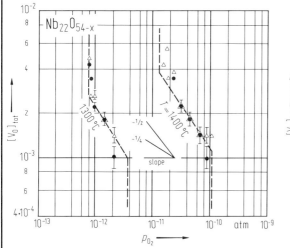

Fig. 20. $Nb_{22}O_{54-x}$. Total vacancy concentration vs. oxygen pressure. $[V_O]_{tot} = x/54$ [76K2]. Open triangles are data derived from Nb_2O_5 and dots are those derived from NbO_2.

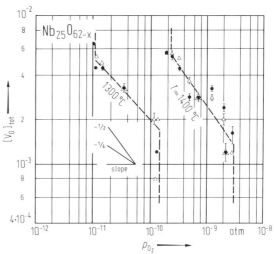

Fig. 21. $Nb_{25}O_{62-x}$. Total vacancy concentration vs. oxygen pressure. $[V_O]_{tot} = x/62$ [76K2]. For △ and ●, see caption of Fig. 20.

9.15.2.6.1 Oxides of niobium

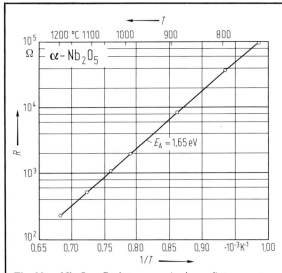

Fig. 22. α-Nb$_2$O$_5$. Resistance vs. (reciprocal) temperature at 1 atm ($p_{O_2}^{-1/4}$ region) [62K].

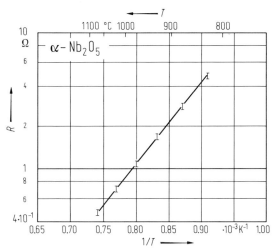

Fig. 23. α-Nb$_2$O$_5$. Resistance vs. (reciprocal) temperature at $p_{O_2} = 10^{-13}$ atm ($p_{O_2}^{-1/6}$ region) [62K].

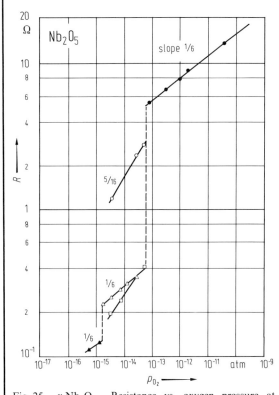

Fig. 25. α-Nb$_2$O$_5$. Resistance vs. oxygen pressure at 1100°C. Full circles: Nb$_2$O$_{5-x}$ (stable phase); open circles: metastable phase of Nb$_2$O$_{5-x}$, open triangles: stable phase of Nb$_{25}$O$_{62-x}$, full triangles: Nb$_{12}$O$_{29-x}$, squares: metastable phase of oxidation of Nb$_{12}$O$_{29-x}$ [74M].

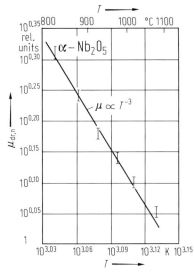

Fig. 24. α-Nb$_2$O$_5$. Electron drift mobility vs. temperature for ceramic sample at $p_{O_2} = 10^{-13}$ atm showing $\mu \propto T^{-3}$ [62K].

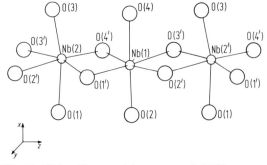

For Fig. 26, see next page.

Fig. 27. NbO$_2$. The essential structure unit [76C].

9.15.2.6.1 Oxides of niobium

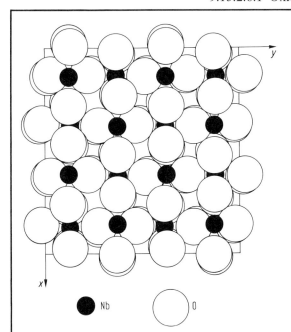

Fig. 26. NbO$_2$. Projection of the structure along the c axis [62M].

Fig. 28. NbO$_2$. Rutile or pseudorutile unit cell parameters vs. temperature [69S1].

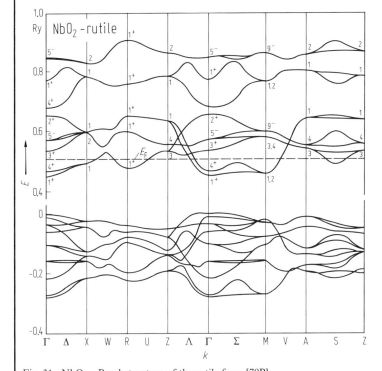

Fig. 31. NbO$_2$. Band structure of the rutile form [79P].

For Fig. 30, see next page.

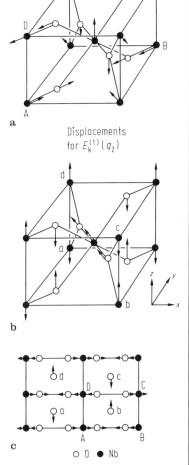

Fig. 29. NbO$_2$. Normal modes of the lattice involved in the structural transition [76P, 78P].

9.15.2.6.1 Oxides of niobium

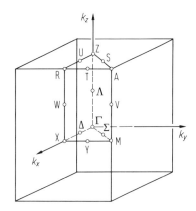

Fig. 30. NbO_2. Brillouin zone of the rutile form [79P].

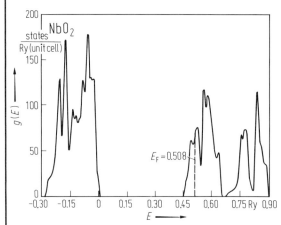

Fig. 32. NbO_2. Density of states for metallic rutile form NbO_2 [79P].

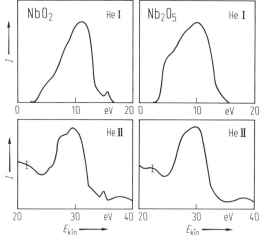

Fig. 34. NbO_2. He I and He II spectra (intensity vs. electron kinetic energy) of NbO_2 and Nb_2O_5 at RT [78B1].

Fig. 35. NbO_2. XPE spectrum (intensity vs. electron kinetic energy) of the valence band region (at RT) and comparison with MoO_2 and RuO_2 [78B1]. ▶

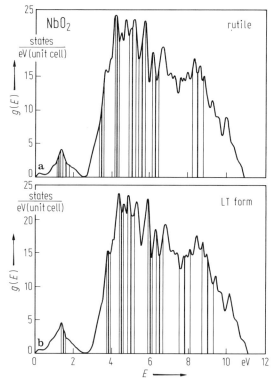

Fig. 33. NbO_2. Joint density of states for (a) the rutile and (b) the low-temperature forms compared to experimental values of ε_2 shown as perpendicular lines [76L].

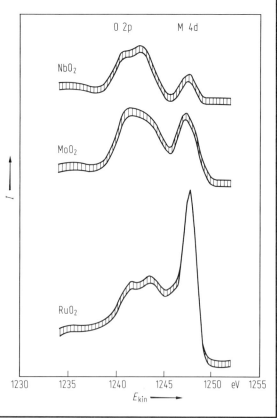

9.15.2.6.1 Oxides of niobium

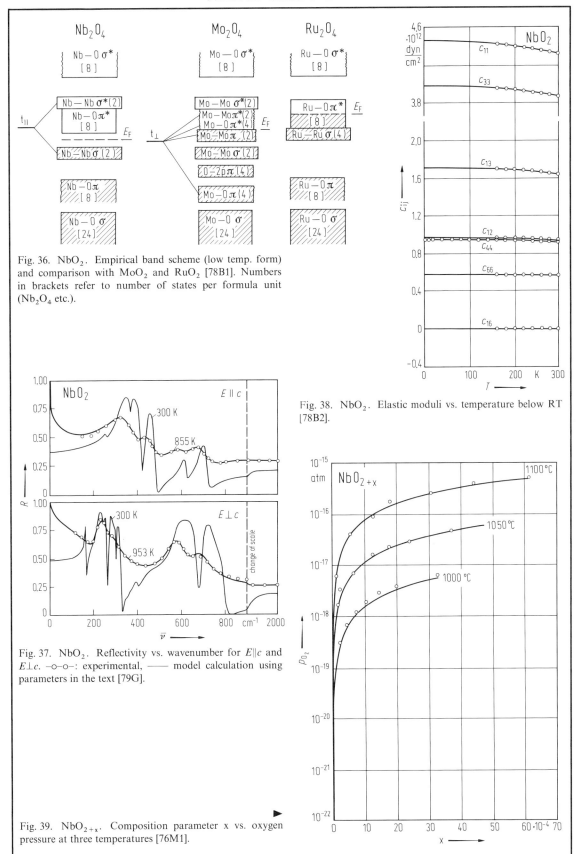

Fig. 36. NbO_2. Empirical band scheme (low temp. form) and comparison with MoO_2 and RuO_2 [78B1]. Numbers in brackets refer to number of states per formula unit (Nb_2O_4 etc.).

Fig. 38. NbO_2. Elastic moduli vs. temperature below RT [78B2].

Fig. 37. NbO_2. Reflectivity vs. wavenumber for $E \| c$ and $E \perp c$. –o–o–: experimental, —— model calculation using parameters in the text [79G].

Fig. 39. NbO_{2+x}. Composition parameter x vs. oxygen pressure at three temperatures [76M1].

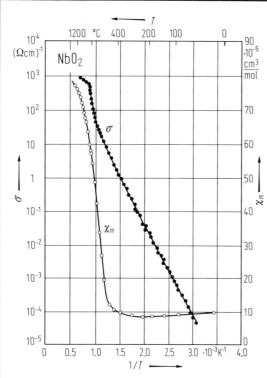

Fig. 40. NbO_2. Conductivity and magnetic susceptibility vs. (reciprocal) temperature [72S]. χ_m in CGS-emu.

Fig. 41. NbO_2. Resistivity vs. reciprocal temperature in a and c direction [74B].

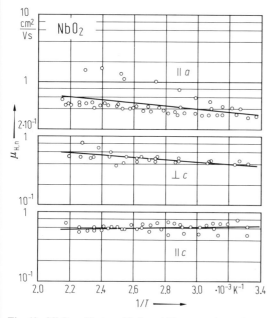

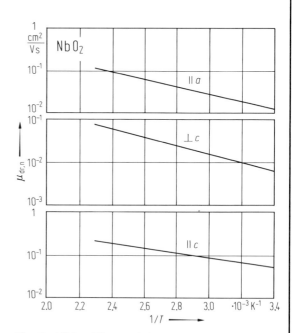

Fig. 42. NbO_2. Electron Hall mobility vs. reciprocal temperature along different directions [74B].

Fig. 43. NbO_2. Electron drift mobility vs. reciprocal temperature along different directions [74B].

9.15.2.6.2 Oxides of tantalum (Ta$_2$O$_5$)

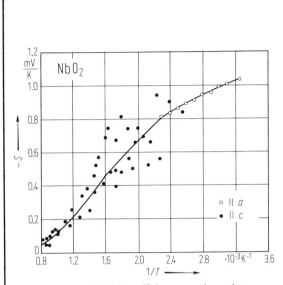

Fig. 44. NbO$_2$. Seebeck coefficient vs. reciprocal temperature in a and c direction [74B].

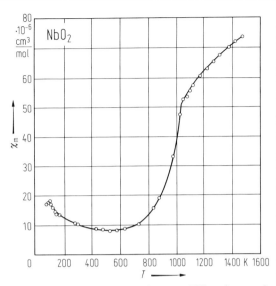

Fig. 45. NbO$_2$. Molar magnetic susceptibility of a ceramic sample vs. temperature [69S1]. χ_m in CGS-emu.

9.15.2.6.2 Oxides of tantalum (Ta$_2$O$_5$) (For tables, see p. 274 ff.)

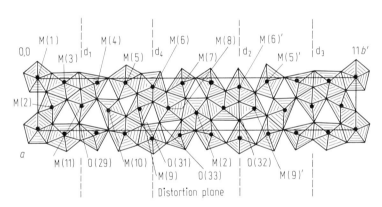

Fig. 1. L–Ta$_2$O$_5$. A (001) projection of the structure. There are three distortion planes in the unit cell located at d_1, d_2 and d_3. The fourth position at d_4 is related by symmetry but is not used in this unit cell. Black dots represent metal atoms and shaded areas oxygen coordination polyhedra [71S1]. 11b': unit cell having 11 UO$_3$ type subcell structure with b' = length of UO$_3$ type subcell.

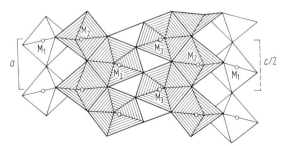

Fig. 2. H–Ta$_2$O$_5$. A (010) projection of the structure. Heavily shaded polyhedra are located at $y = 1/2$ and the remainder are at $y = 0$ [71S2].

9.15.2.6.2 Oxides of tantalum (Ta$_2$O$_5$)

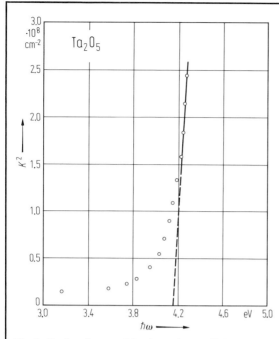

Fig. 3. Ta$_2$O$_5$. Square of the absorption coefficient vs. photon energy for a non-crystalline film of thickness 6960 Å at RT [73K].

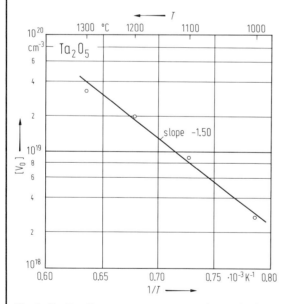

Fig. 4. Ta$_2$O$_5$. Oxygen vacancy concentration vs. (reciprocal) temperature at $p_{O_2} = 10^{-16}$ atm [74S].

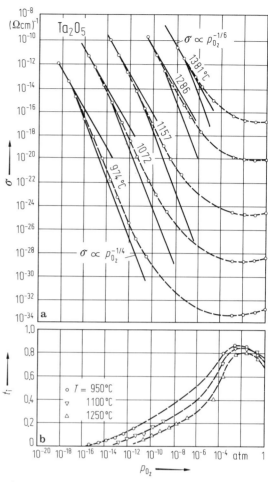

Fig. 5. Ta$_2$O$_5$. (a) Conductivity vs. oxygen partial pressure at 10 kHz at various temperatures, (b) ionic transfer number t_i $\left(\text{defined as } t_i = \frac{\sigma_i}{\sigma_{tot}}; \sigma_i = \text{ionic conductivity}\right)$ vs. oxygen partial pressure at various temperatures for polycrystalline samples [74S].

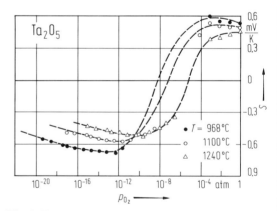

Fig. 6. Ta$_2$O$_5$. Seebeck coefficient vs. oxygen partial pressure at various temperatures for polycrystalline samples [74S].

9.15.2.6.2 Oxides of tantalum (Ta$_2$O$_5$)

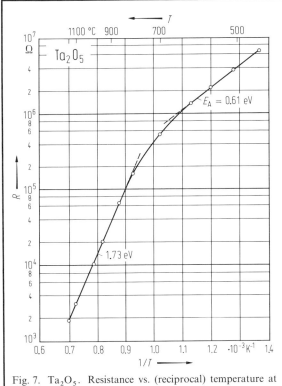

Fig. 7. Ta$_2$O$_5$. Resistance vs. (reciprocal) temperature at $p_{O_2} = 1$ atm [62K].

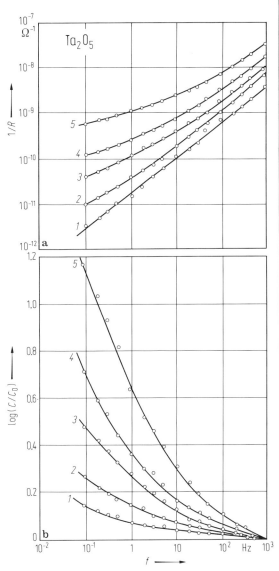

Fig. 8. Ta$_2$O$_5$. (a) Conductance vs. frequency of single crystal α-Ta$_2$O$_5$ at (1) 295 K, (2) 320 K, (3) 340 K, (4) 360 K, (5) 380 K, (b) capacitance (over C at 1 kHz) vs. frequency at various temperatures as in (a) [79S2].

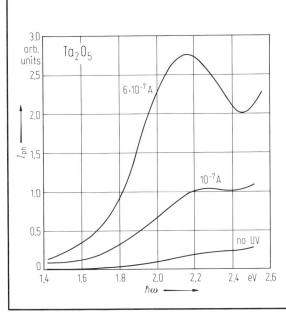

Fig. 9. Ta$_2$O$_5$. Photocurrent (corrected for lamp intensity) vs. photon energy; the sample was ω-irradiated by UV lamp giving the steady-state photocurrents shown [74T].

9.15.2.6.2 Oxides of tantalum (Ta$_2$O$_5$)

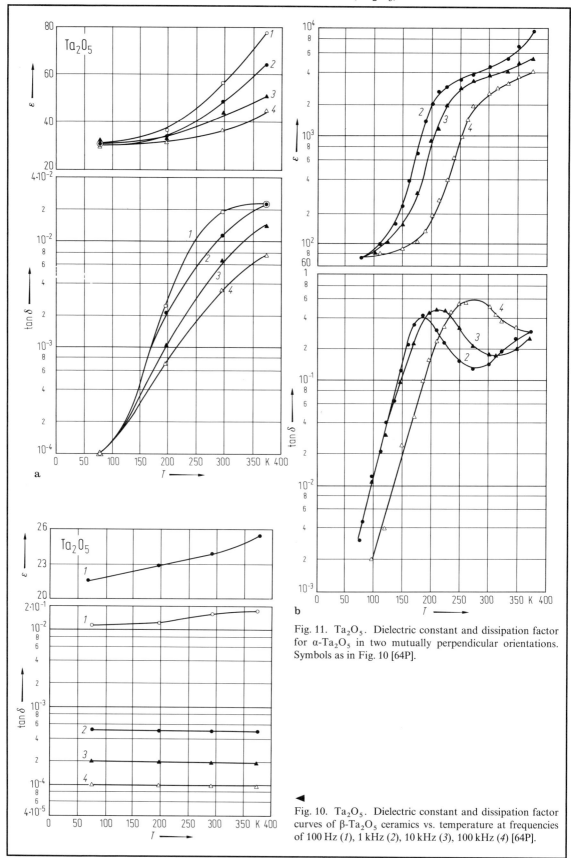

Fig. 11. Ta$_2$O$_5$. Dielectric constant and dissipation factor for α-Ta$_2$O$_5$ in two mutually perpendicular orientations. Symbols as in Fig. 10 [64P].

Fig. 10. Ta$_2$O$_5$. Dielectric constant and dissipation factor curves of β-Ta$_2$O$_5$ ceramics vs. temperature at frequencies of 100 Hz (*1*), 1 kHz (*2*), 10 kHz (*3*), 100 kHz (*4*) [64P].

9.15.2.7 Oxides of group VI elements (Mo, W)

9.15.2.7.1 Molybdenum oxides (For tables, see p. 276ff.)

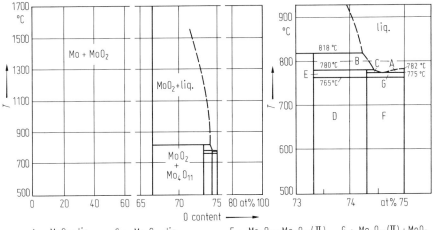

A: MoO_3 + liq. C: Mo_9O_{26} + liq. E: Mo_4O_{11} + Mo_9O_{26}(II) G: Mo_9O_{26}(II) + MoO_3
B: Mo_4O_{11} + liq. D: Mo_4O_{11} + Mo_9O_{26}(I) F: Mo_9O_{26}(I) + MoO_3

Fig. 1. Mo–O. Equilibrium phase diagram [69C].

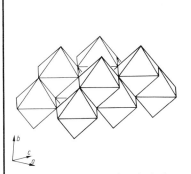

Fig. 2. MoO_3. Part of a single layer idealized as regular octahedra fused together [63K].

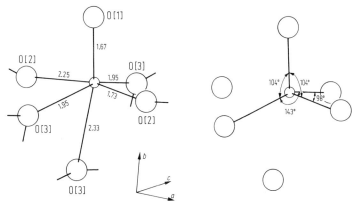

Fig. 4. MoO_3. Coordination about each molybdenum ion [63K].

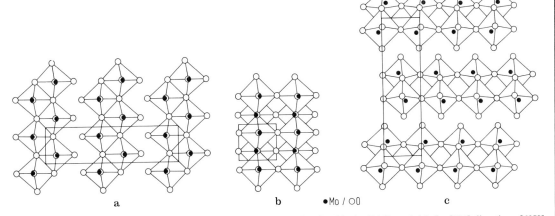

Fig. 3. MoO_3. Projections of the crystal structure along (a) the [100]-, (b) the [010]- and (c) the [001]-directions [63K]. Unit cell is indicated.

9.15.2.7.1 Molybdenum oxides

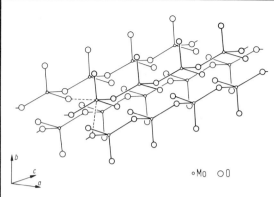

Fig. 5. MoO$_3$. Structure visualized as built up of 4 co-ordinated Mo atoms (small spheres) [63K].

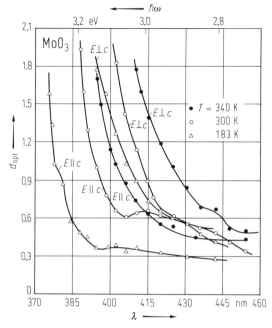

Fig. 6. MoO$_3$. Optical density vs. photon energy (wavelength) of a 44 μm thick single crystal for polarized light in the temperature range 150⋯340 K. The crystal was oriented with the (010) plane perpendicular to the incident light beam [68D]. $d_{opt} = \log(I_0/I)$.

For Fig. 8, see next page.

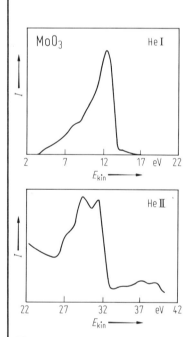

Fig. 7. MoO$_3$. He I and He II spectra (photoelectron intensity vs. electron kinetic energy) [78B1].

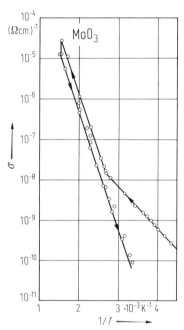

Fig. 9. MoO$_3$. Conductivity vs. reciprocal temperature for a single crystal grown and measured in argon; crystal heated from RT to 650 K, held for 20 h and then cooled as shown (orientation not given) [69I].

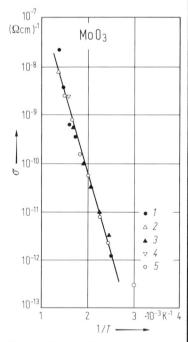

Fig. 10. MoO$_3$. Conductivity vs. reciprocal temperature for an unoriented single crystal sintered at (1) 715 K, (2) 710 K, (3) 649 K, (4) 629 K, (5) 653 K [69I].

9.15.2.7.1 Molybdenum oxides

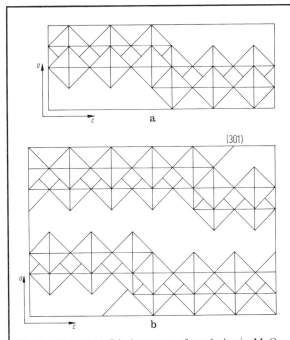

Fig. 8. MoO$_3$. (a) Displacements of octahedra in MoO$_3$ layers, (b) (301) shear planes through adjacent layers [78S2].

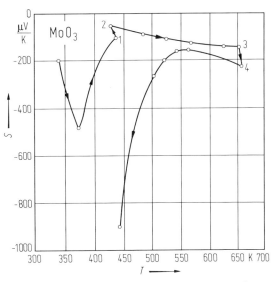

Fig. 11. MoO$_3$. Seebeck coefficient vs. temperature for an unoriented single crystal grown and measured in argon. Sample held between 1 and 2 for 17 h and between 3 and 4 for 23 h [69I].

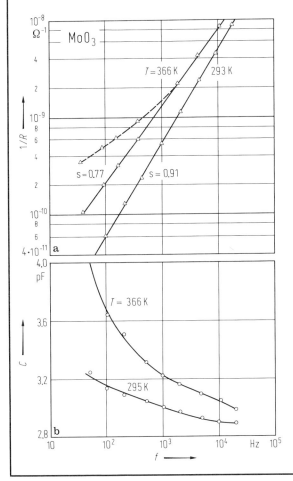

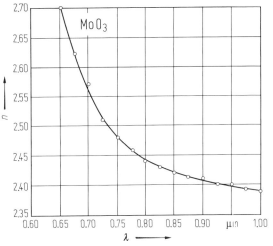

Fig. 13. MoO$_3$. Refractive index vs. wavelength of a thin film [68D].

Fig. 12. MoO$_3$. (a) Conductance and (b) capacitance vs. frequency for a single crystal. At 366 K the dashed line in Fig. (a) was obtained by subtracting the dc conductance from the solid curve [78S1]. s is a temperature dependent number in $\sigma_{ac} \propto \omega^s$.

9.15.2.7.1 Molybdenum oxides

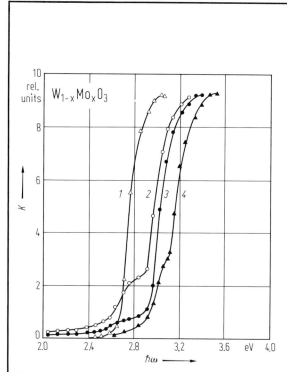

Fig. 14. $W_{1-x}Mo_xO_3$. Absorption coefficient vs. photon energy for (1) WO_3, (2) $W_{0.9}Mo_{0.1}O_3$, (3) $W_{0.63}Mo_{0.37}O_3$, (4) $W_{0.05}Mo_{0.95}O_3$ [79H].

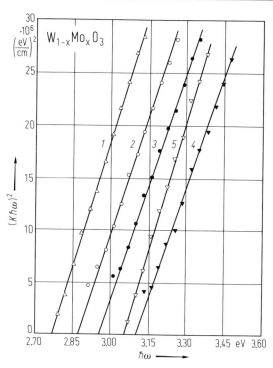

Fig. 15. $W_{1-x}Mo_xO_3$. Absorption coefficient times photon energy squared vs. photon energy indicating direct optical transitions for (1) WO_3, (2) $W_{0.9}Mo_{0.1}O_3$, (3) $W_{0.79}Mo_{0.21}O_3$, (4) MoO_3:W, (5) MoO_3 [79H].

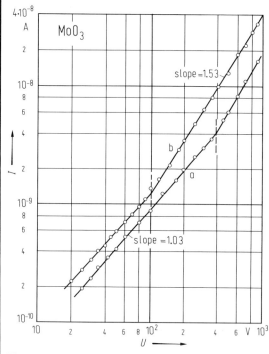

Fig. 16. MoO_3. Current-voltage characteristic of a single crystal at 300 K (a) in the dark, (b) during illumination with the full output of a mercury lamp [68D].

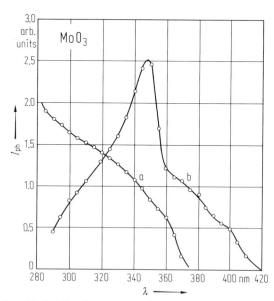

Fig. 17. MoO_3. Photocurrent intensity vs. wavelength in (a) a single crystal, (b) a thin film [68D].

9.15.2.7.2 Tungsten oxides (For tables, see p. 282 ff.)

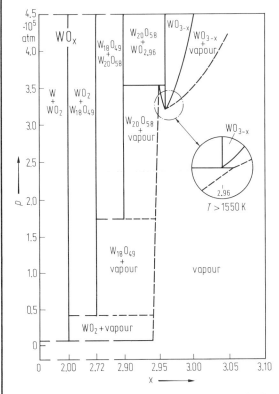

Fig. 1. WO_x. Pressure-composition diagram at 1450 K. Inset shows incongruent evaporation of WO_{3-x} above 1550 K [63A].

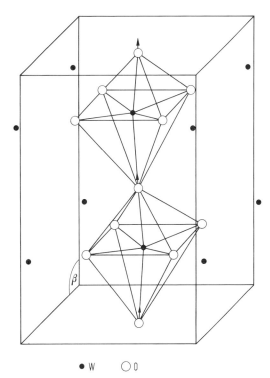

Fig. 3. α-WO_3. Structure at $-70\,°C$ [76S1].

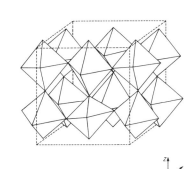

Fig. 4. β-WO_3 (triclinic). Corner sharing O-octahedra around one unit cell (dotted line) [78D].

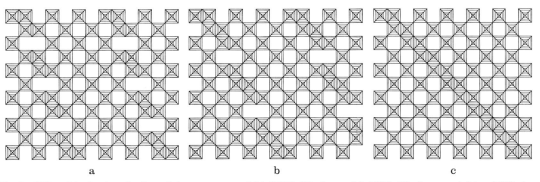

Fig. 2. WO_3. Idealized projection of the structures of (a) {102} CS planes, (b) {103} CS planes and (c) a {101} shear plane in an idealized matrix [77I].

9.15.2.7.2 Tungsten oxides

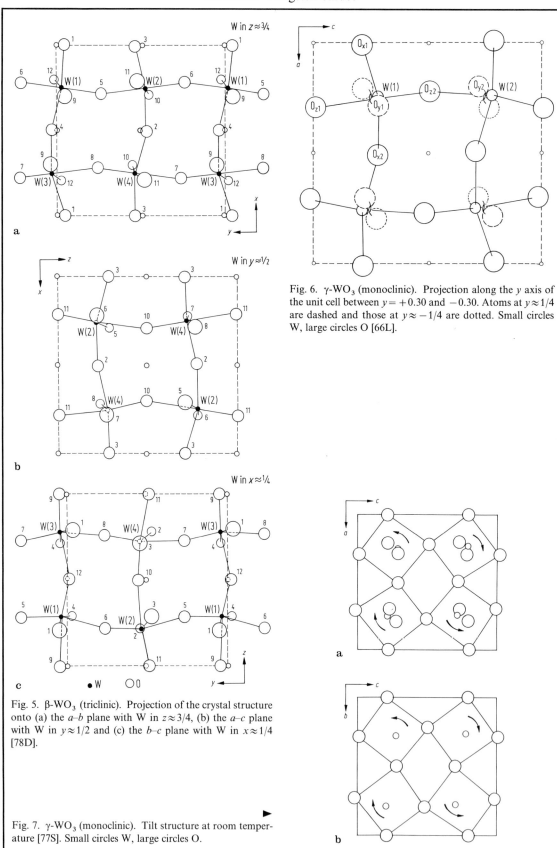

Fig. 5. β-WO$_3$ (triclinic). Projection of the crystal structure onto (a) the a–b plane with W in $z \approx 3/4$, (b) the a–c plane with W in $y \approx 1/2$ and (c) the b–c plane with W in $x \approx 1/4$ [78D].

Fig. 6. γ-WO$_3$ (monoclinic). Projection along the y axis of the unit cell between $y = +0.30$ and -0.30. Atoms at $y \approx 1/4$ are dashed and those at $y \approx -1/4$ are dotted. Small circles W, large circles O [66L].

Fig. 7. γ-WO$_3$ (monoclinic). Tilt structure at room temperature [77S]. Small circles W, large circles O.

9.15.2.7.2 Tungsten oxides

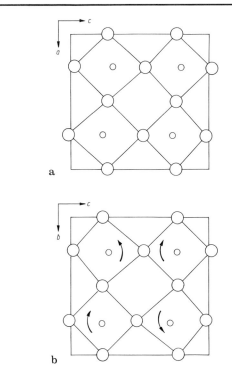

Fig. 8. WO_3 (orthorhombic). Projection of the structure on ac-plane (a) and bc-plane (b) [77S]. Small circles W, large circles O.

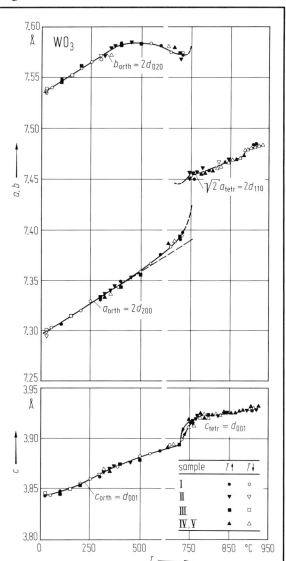

Fig. 11. WO_3. Lattice parameters and interplanar spacings vs. temperature for various samples [70A]. The crystal symmetry is monoclinic between 17 and 330 °C and orthorhombic between 330 and ≈740 °C (see tables). The unique angle β is, however close to 90° in the monoclinic phase (90.91° at RT), and a, b, c were calculated as if the crystal were orthorhombic even below 330 °C.

For Fig. 10, see next page.

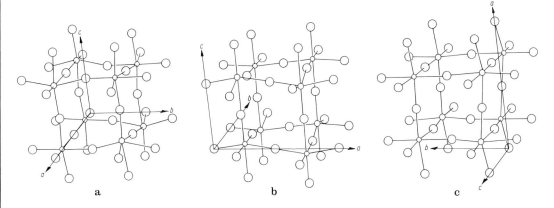

Fig. 9. WO_3 (orthorhombic). Structure [77S]. (a), (b) and (c) show three different projections indicated by the axes a, b, c.

9.15.2.7.2 Tungsten oxides

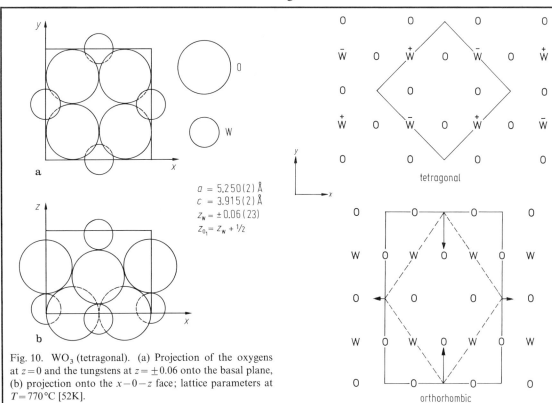

$a = 5.250(2)$ Å
$c = 3.915(2)$ Å
$z_W = \pm 0.06(23)$
$z_{O_1} = z_W + 1/2$

Fig. 10. WO$_3$ (tetragonal). (a) Projection of the oxygens at $z=0$ and the tungstens at $z=\pm 0.06$ onto the basal plane, (b) projection onto the $x-0-z$ face; lattice parameters at $T=770$°C [52K].

Fig. 12. WO$_3$. Model for the orthorhombic-tetragonal distortion [70A]. $\overset{+}{W}$, $\overset{-}{W}$ are W atoms above and below the O-plane.

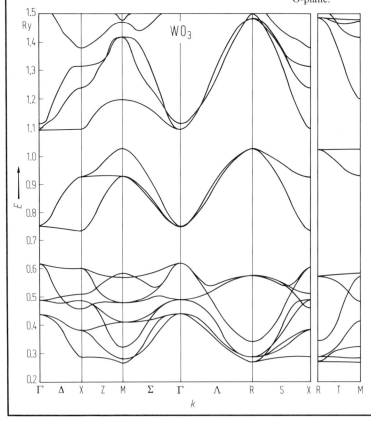

Fig. 13. WO$_3$. Energy bands of ideal cubic WO$_3$ along the major symmetry axes [77K].

9.15.2.7.2 Tungsten oxides

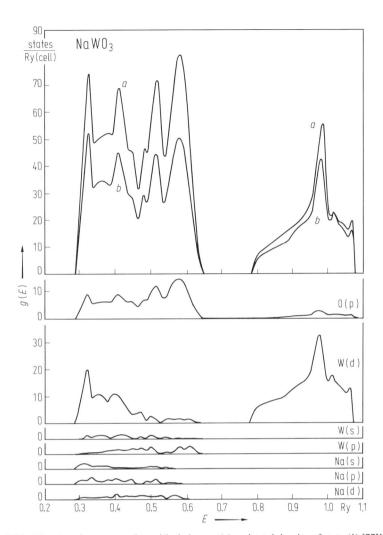

Fig. 14. NaWO$_3$. Orbital density of states together with their sum (*a*) and total density of state (*b*) [77K].

9.15.2.7.2 Tungsten oxides

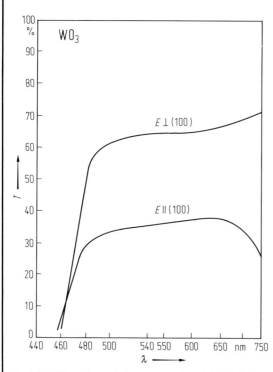

Fig. 15. WO_3. Transmission vs. wavelength in [001]-direction at RT [72S1].

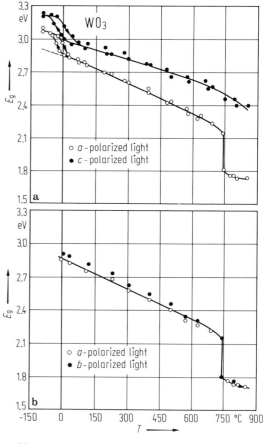

Fig. 16. WO_3. Optical absorption edge vs. temperature in the b plane (a), in the c plane (b), in the c plane (c) at low temperature [60I].

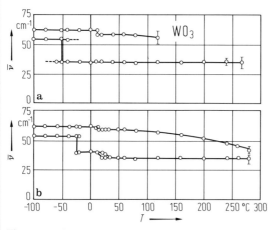

Fig. 17. WO_3. Wavenumber of low-frequency optical phonons vs. temperature. (a) and (b) refer to the monoclinic room temperature form and a triclinic form also stable at room temperature [75S1].

9.15.2.7.2 Tungsten oxides

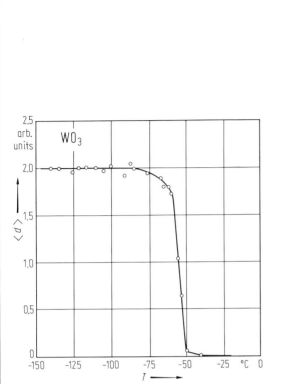

Fig. 18. α-WO$_3$. Average piezoelectric ocefficient vs. temperature [75S1].

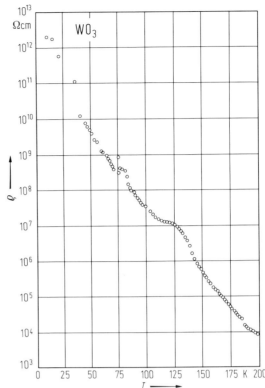

Fig. 20. WO$_3$. dc resistivity vs. temperature for a stoichiometric crystal [75L]. Bulk volume ϱ measured between (001) faces of a single crystal during cooling.

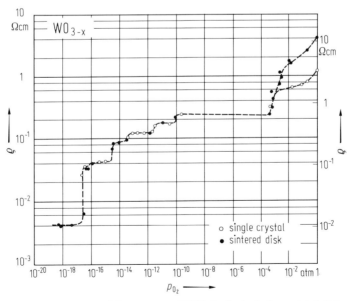

Fig. 19. WO$_{3-x}$. Resistivity vs. oxygen partial pressure at 750 °C for both single crystal (left hand scale) and sintered discs. Values of x in the range 0···0.04 [77B]. Orientation for single crystals not specified.

9.15.2.7.2 Tungsten oxides

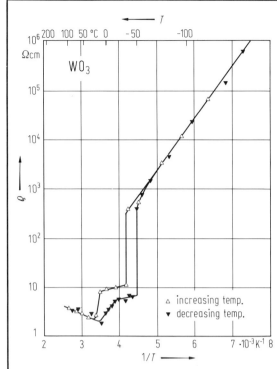

Fig. 21. WO_3. Resistivity vs. (reciprocal) temperature [80H].

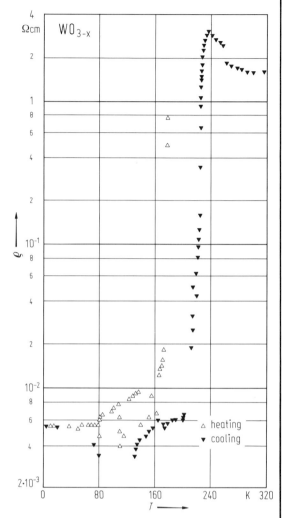

Fig. 22. WO_{3-x}. ac resistivity vs. temperature for a single crystal sample prepared by heating WO_3 in vacuum for 36 h at 1090 K [75L]. ϱ measured by 4-point probe method, points are collinear in an arb. direction on the (001) face.

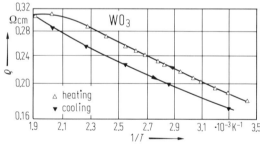

Fig. 23. WO_3. Resistivity along a axis vs. reciprocal temperature for $T > 300$ K [70B1].

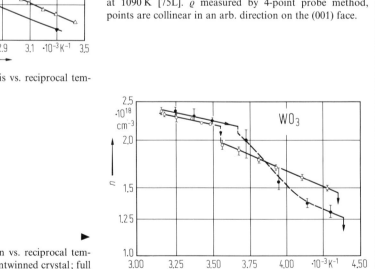

Fig. 24. WO_3. Electron concentration vs. reciprocal temperature for $T < 300$ K; open circles: untwinned crystal; full circles: crystal twinned (100) along its entire length [70B1].

9.15.2.7.2 Tungsten oxides

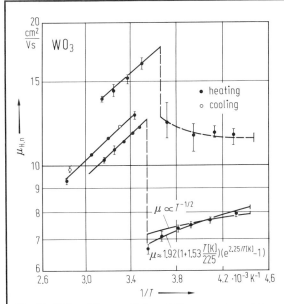

Fig. 25. WO_3. Electron Hall mobility vs. reciprocal temperature below 300 K showing the monoclinic-triclinic transition. Upper curve, twinned (100) crystal; lower curves, two untwinned crystals [70B1]. $\mu \parallel a$.

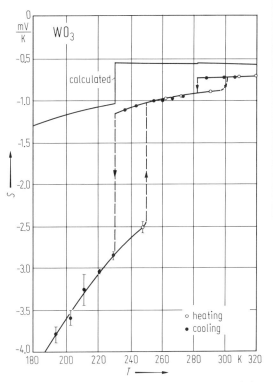

Fig. 26. WO_3. Seebeck coefficient vs. temperature below 320 K. Upper curve is calculated without assuming a phonon-drag contribution [70B1]. Specimen contains some (100) twin planes. $S \parallel a$.

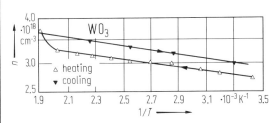

Fig. 27. WO_3. Electron concentration vs. reciprocal temperature above 300 K [70B1].

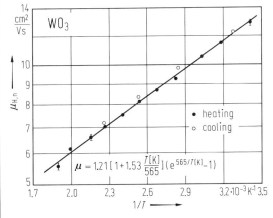

Fig. 28. WO_3. Hall mobility vs. reciprocal temperature above 300 K. Solid line is the best fit to a large-polaron theory [67L, 70B1]. $\mu \parallel a$.

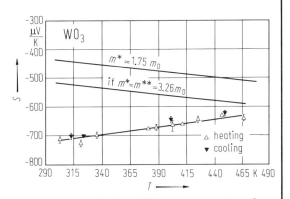

Fig. 29. WO_3. Seebeck coefficient vs. temperature above 300 K. The two curves are calculated using a rigid band mass and polaron mass, respectively [70B1]. $S \parallel a$.

9.15.2.7.2 Tungsten oxides

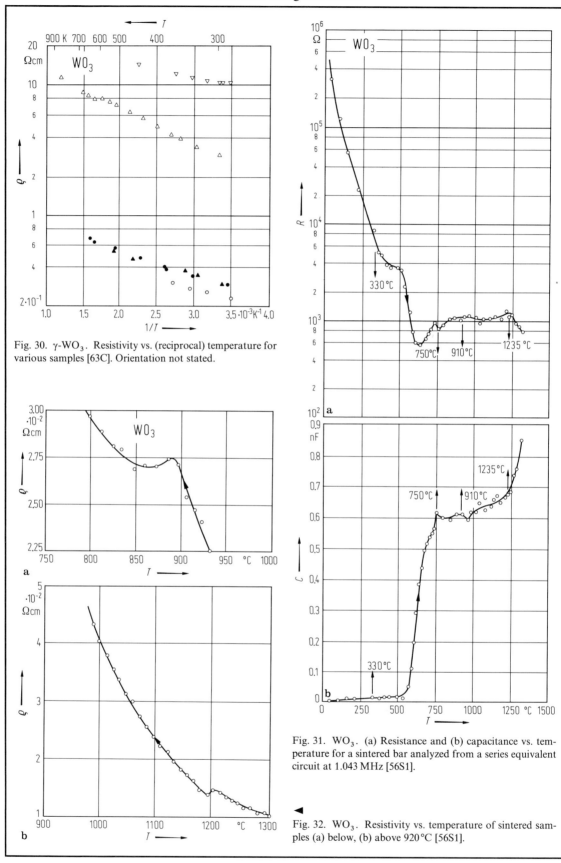

Fig. 30. γ-WO$_3$. Resistivity vs. (reciprocal) temperature for various samples [63C]. Orientation not stated.

Fig. 31. WO$_3$. (a) Resistance and (b) capacitance vs. temperature for a sintered bar analyzed from a series equivalent circuit at 1.043 MHz [56S1].

Fig. 32. WO$_3$. Resistivity vs. temperature of sintered samples (a) below, (b) above 920 °C [56S1].

9.15.2.7.2 Tungsten oxides

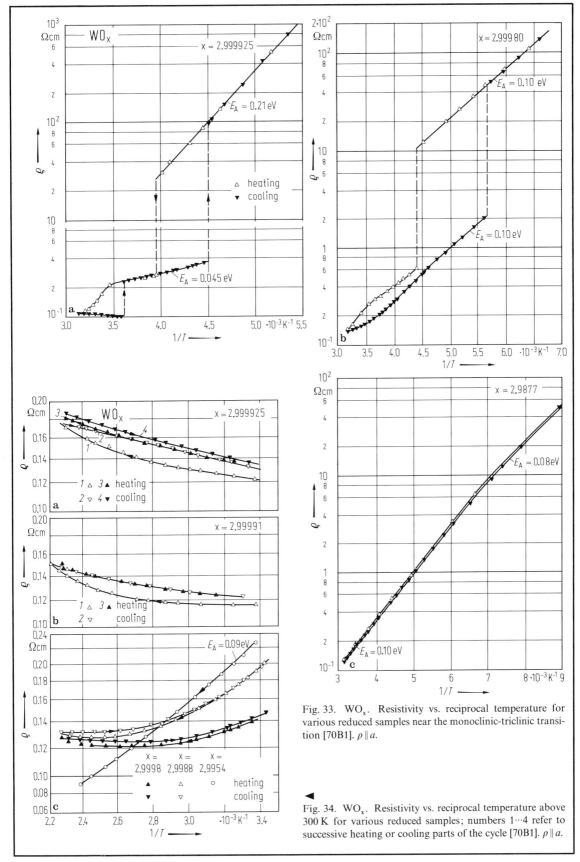

Fig. 33. WO_x. Resistivity vs. reciprocal temperature for various reduced samples near the monoclinic-triclinic transition [70B1]. $\rho \parallel a$.

Fig. 34. WO_x. Resistivity vs. reciprocal temperature above 300 K for various reduced samples; numbers 1···4 refer to successive heating or cooling parts of the cycle [70B1]. $\rho \parallel a$.

9.15.2.7.2 Tungsten oxides

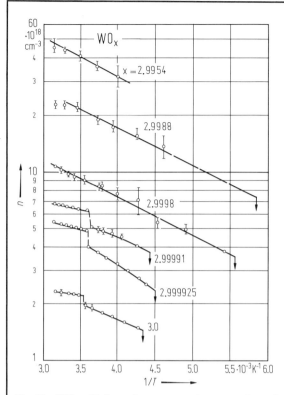

Fig. 35. WO_x. Hall carrier concentration vs. reciprocal temperature below 300 K for various reduced samples [70B1].

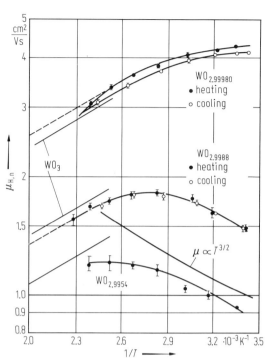

Fig. 36. WO_x. Hall mobility vs. reciprocal temperature above 300 K for various reduced samples [70B1]. Slope of WO_3 is shown for comparison (cf. Fig. 28). $\mu \parallel a$.

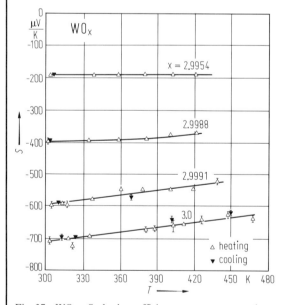

Fig. 37. WO_x. Seebeck coefficient vs. temperature above 300 K for various reduced samples [70B1]. $S \parallel a$.

For Fig. 38, see next page.

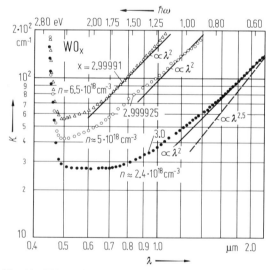

Fig. 39. WO_x. Absorption coefficient ($E \perp c$-axis) vs. wavelength (photon energy) for three crystals at RT [73B1].

9.15.2.7.2 Tungsten oxides

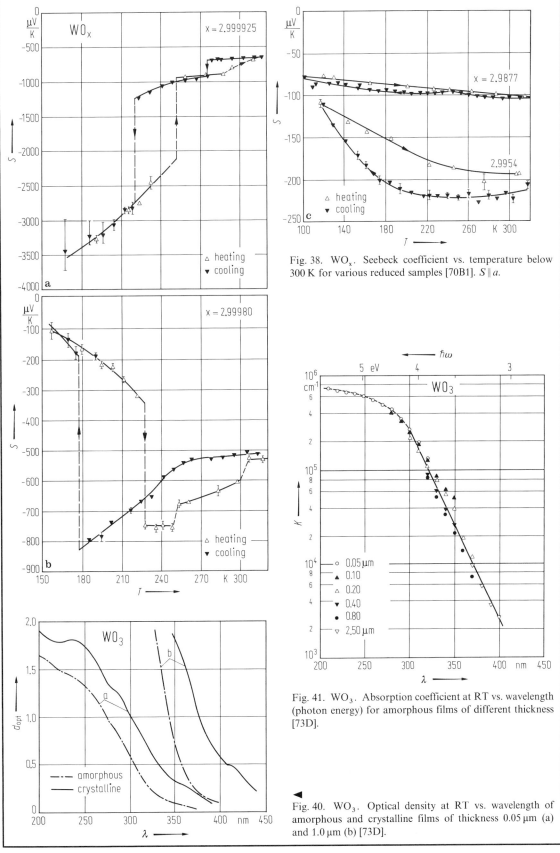

Fig. 38. WO_x. Seebeck coefficient vs. temperature below 300 K for various reduced samples [70B1]. $S \parallel a$.

Fig. 41. WO_3. Absorption coefficient at RT vs. wavelength (photon energy) for amorphous films of different thickness [73D].

Fig. 40. WO_3. Optical density at RT vs. wavelength of amorphous and crystalline films of thickness 0.05 μm (a) and 1.0 μm (b) [73D].

9.15.2.7.2 Tungsten oxides

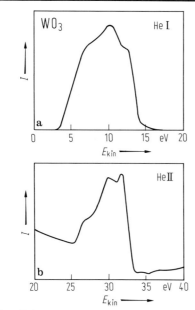

Fig. 42. γ-WO$_3$. UV-photoemission spectra at RT (intensity vs. electron kinetic energy), (a) He I, (b) He II [78B]. He I: light source −21.2 eV He line, He II: light source −40.8 eV He line.

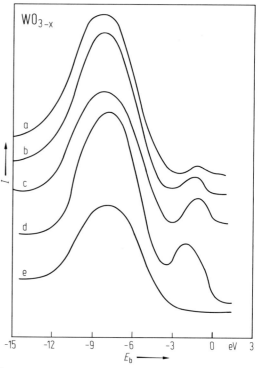

Fig. 43. WO$_{3-x}$. XPE spectra (intensity vs. binding energy) for samples with x = 0.04 (a), 0.10 (b), 0.28 (c), 1.00 (d), 0 (e) [77A].

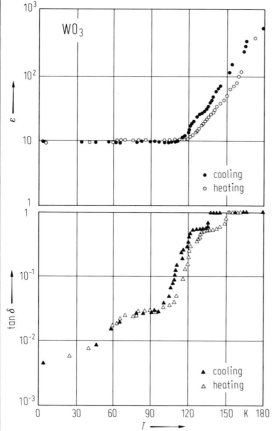

Fig. 44. WO$_3$. Dielectric constant and dissipation factor vs. temperature measured between nominal RT (001) surfaces of a stoichiometric single crystal at 1 kHz [75L].

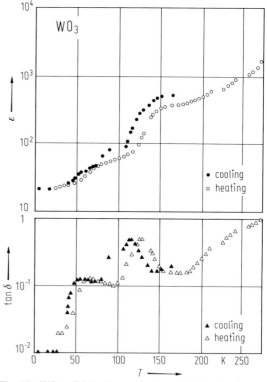

Fig. 45. WO$_3$. Dielectric constant and dissipation factor vs. temperature derived from capacitive measurements on sintered wafers of stoichiometric material at 1 kHz [75L].

9.15.3 Binary transition-metal chalcogenides

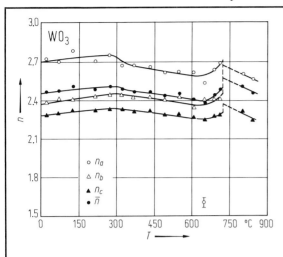

Fig. 46. WO_3. Indices of refraction vs. temperature [59S]. $\bar{n} = \frac{1}{3}(n_a + n_b + n_c)$.

9.15.3 Binary transition-metal chalcogenides (For tables, see p. 291 ff.)

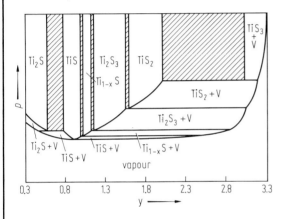

Fig. 1. Ti–S. Schematic pressure – composition diagram of the Ti–S system [67G]. y: atomic ratio S/Ti.

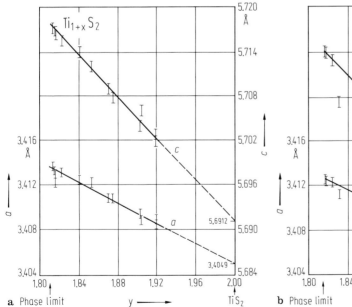

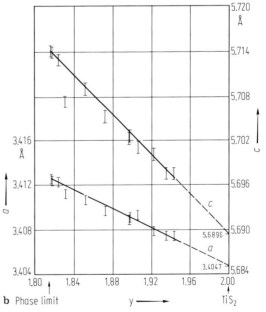

Fig. 2. $Ti_{1+x}S_2$. Cell parameters at RT vs. composition. Sulfides prepared at 1000 °C (a) and 800 °C (b) [63B]. y: atomic ratio S/Ti.

9.15.3 Binary transition-metal chalcogenides

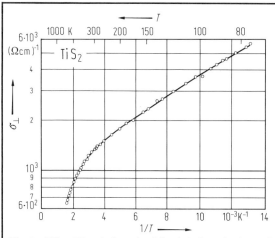

Fig. 3. TiS_2. Electrical conductivity ($\sigma \perp c$) vs. (reciprocal) temperature for a single crystal [68C2].

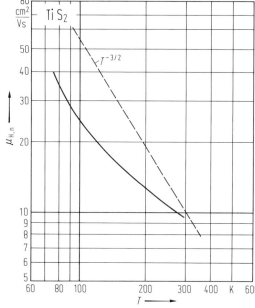

Fig. 4. TiS_2. Hall mobility vs. temperature [68C2]. $\mu \perp c$.

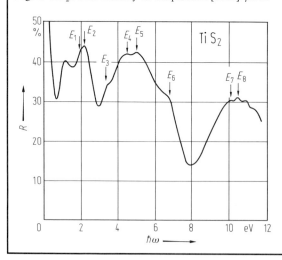

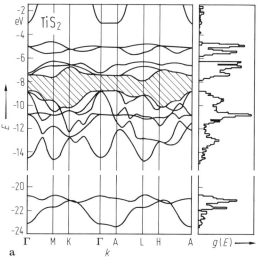

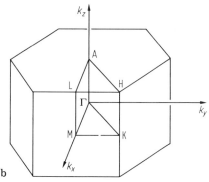

Fig. 6. TiS_2. (a) Band structure and density of states, (b) Brillouin zone [78B].

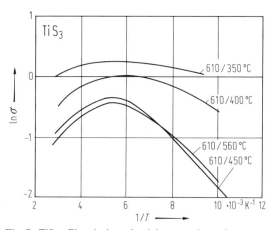

Fig. 7. TiS_3. Electrical conductivity vs. reciprocal temperature. The designated temperatures for each curve indicate the sublimation and condensation temperatures during crystal growth [61G]. σ in Ω^{-1} cm^{-1}, $\sigma \parallel b$.

◀

Fig. 5. TiS_2. Reflectivity vs. photon energy in the fundamental region at room temperature [65G].

9.15.3 Binary transition-metal chalcogenides

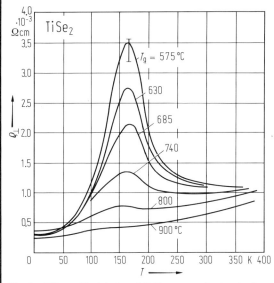

Fig. 8. TiSe$_2$. Electrical resistivity perpendicular to the c axis (parallel to the layers) for crystals grown by iodine vapor transport at different growth temperatures (T_g). The bar on the curve for $T_g = 575$ °C represents the spread in peak values observed in crystals from the same bath [76D].

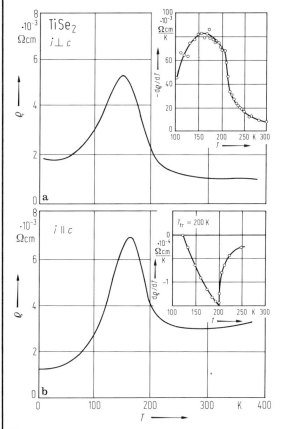

Fig. 10. TiSe$_2$. Electrical resistivity vs. temperature perpendicular to the c axis (a) and parallel to the c axis (b) for sublimation-grown (at 630 °C) TiSe$_2$. The insets show $d\varrho/dT$ for the same sample [76D].

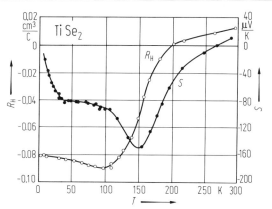

Fig. 9. TiSe$_2$. Hall coefficient and thermoelectric power perpendicular to c axis vs. temperature [76D].

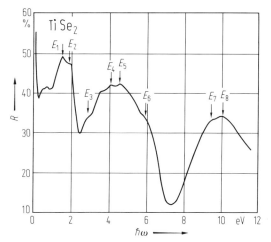

Fig. 11. TiSe$_2$. Reflectivity vs. photon energy in the fundamental region at room temperature [65G].

For Fig. 12, see next page.

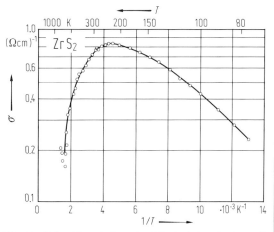

Fig. 13. ZrS$_2$. Electrical conductivity ($\sigma \perp c$) vs. (reciprocal) temperature for a single crystal [68C2].

9.15.3 Binary transition-metal chalcogenides

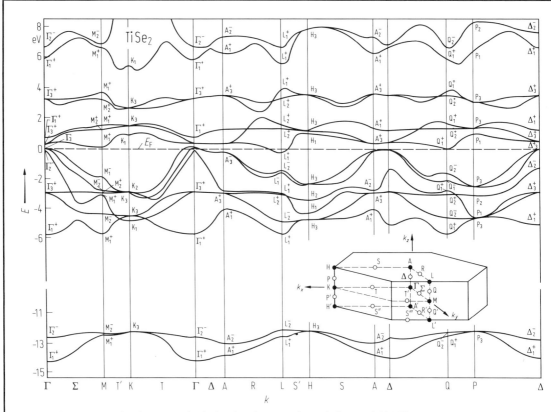

Fig. 12. TiSe$_2$. Energy-band structure in the local exchange and correlation model [78Z].

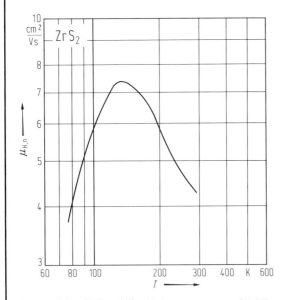

Fig. 14. ZrS$_2$. Hall mobility ($\perp c$) vs. temperature [68C2].

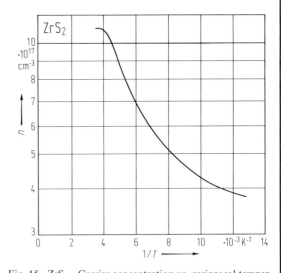

Fig. 15. ZrS$_2$. Carrier concentration vs. reciprocal temperature [68C2].

9.15.3 Binary transition-metal chalcogenides

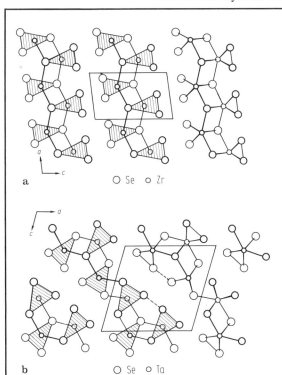

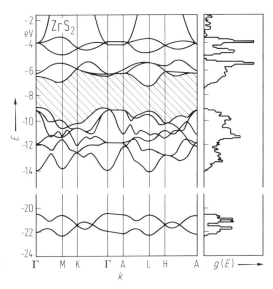

Fig. 16. The monoclinic structure of (a) ZrSe$_3$ and (b) TaSe$_3$ [68H].

Fig. 18. ZrS$_2$. Band structure and density of states [78B]. See Fig. 6b for Brillouin zone.

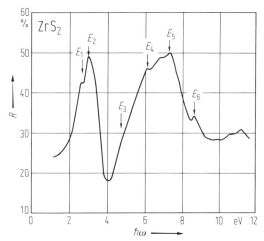

Fig. 17. ZrS$_2$. Reflectivity vs. photon energy in the fundamental region at room temperature [65G].

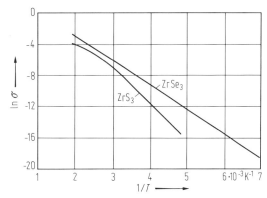

Fig. 19. ZrS$_3$, ZrSe$_3$. Conductivity vs. reciprocal temperature [61G]. σ in Ω^{-1} cm, $\sigma \parallel b$.

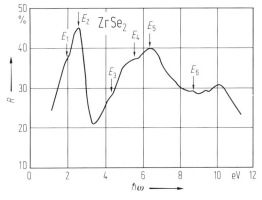

Fig. 20. ZrSe$_2$. Reflectivity vs. photon energy in the fundamental region at room temperature [65G].

9.15.3 Binary transition-metal chalcogenides

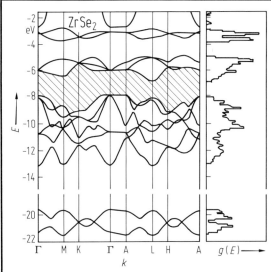

Fig. 21. $ZrSe_2$. Band structure and density of states [78B]. See Fig. 6b for Brillouin zone.

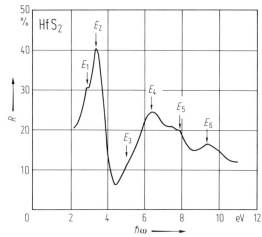

Fig. 23. HfS_2. Reflectivity vs. photon energy in the fundamental region at room temperature [65G].

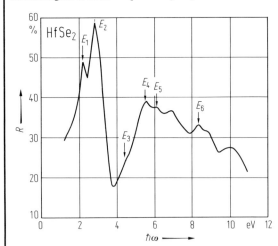

Fig. 25. $HfSe_2$. Reflectivity vs. photon energy in the fundamental region at room temperature [65G].

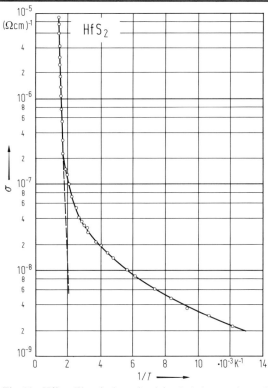

Fig. 22. HfS_2. Electrical conductivity ($\sigma \perp c$) vs. reciprocal temperature for a single crystal [68C2].

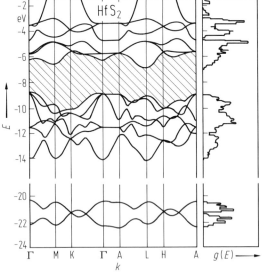

Fig. 24. HfS_2. Band structure and density of states [78B]. See Fig. 6b for Brillouin zone.

9.15.3 Binary transition-metal chalcogenides

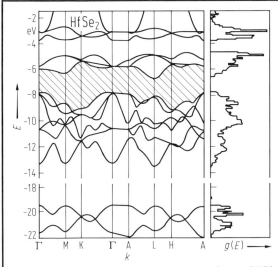

Fig. 26. HfSe$_2$. Band structure and density of states [78B]. See Fig. 6b for Brillouin zone.

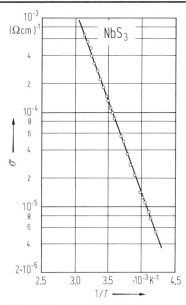

Fig. 27. NbS$_3$. Electrical conductivity vs. reciprocal temperature [62G].

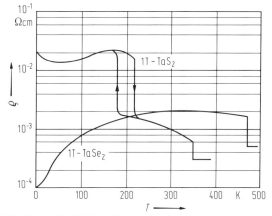

Fig. 29. TaS$_2$, TaSe$_2$. Electrical resistivity vs. temperature for 1T−TaS$_2$ and 1T−TaSe$_2$ [77D]. $\varrho \perp c$.

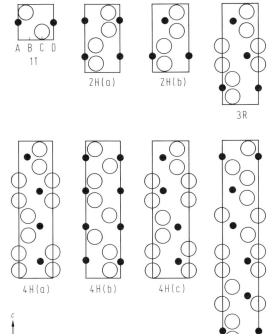

Fig. 28. Sections through (11$\bar{2}$0) planes of the hexagonal unit cells for the different kinds of polymorphic transition-metal disulfides, MS$_2$, of groups IV, V and VI. Metal atoms are in black circles and S atoms are in open circles [69H].

For Fig. 30, see next page.

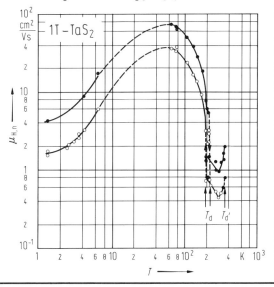

Fig. 31. TaS$_2$. Hall mobility vs. temperature in 1T−TaS$_2$ (two samples) [79I2]. $\mu \perp c$.

9.15.3 Binary transition-metal chalcogenides

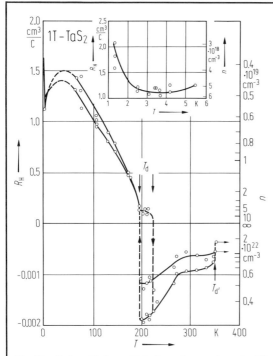

Fig. 30. TaS$_2$. Hall coefficient vs. temperature in 1T–TaS$_2$. The lower scale on the vertical axis is enlarged five hundred times as compared with the upper one. The right-hand scale on the vertical axis gives the carrier concentration assuming a one-carrier model. Measurements on two samples [79I2]. Inset shows low-temperature range on an expanded scale. $B \parallel c$.

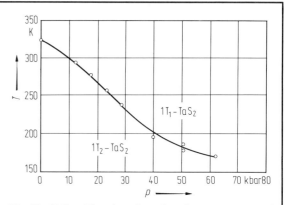

Fig. 32. TaS$_2$. The phase boundary in temperature and pressure between 1T$_1$– and 1T$_2$–TaS$_2$ [74G].

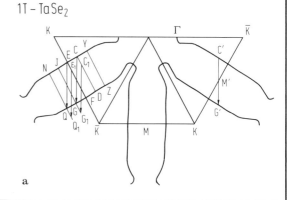

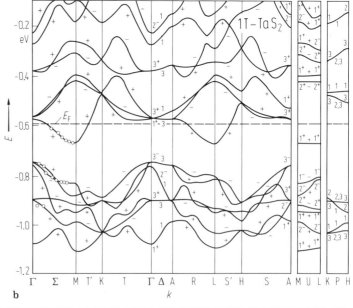

Fig. 33. TaS$_2$, TaSe$_2$. (a) The Fermi surface of 1T–TaSe$_2$ in the midplane ($k_z = 0$) showing the direction of the spanning vector, (b) band structure of 1T–TaS$_2$. Circles in Γ–M are from other literature (see original paper) [77W]. See Fig. 12 for Brillouin zone.

9.15.3 Binary transition-metal chalcogenides

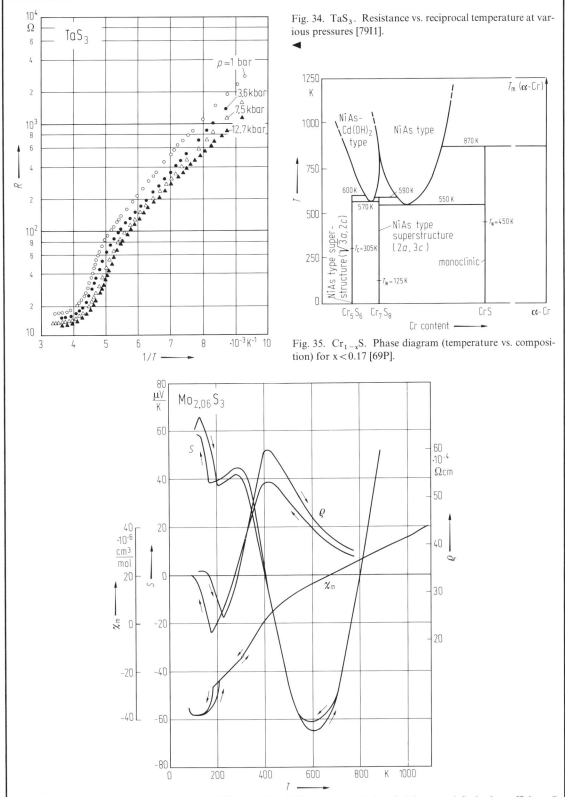

Fig. 34. TaS$_3$. Resistance vs. reciprocal temperature at various pressures [79I1].

Fig. 35. Cr$_{1-x}$S. Phase diagram (temperature vs. composition) for $x < 0.17$ [69P].

Fig. 36. Mo$_{2.06}$S$_3$. Molar magnetic susceptibility χ_m (in CGS-emu), electrical resistivity ϱ and Seebeck coefficient S vs. temperature. The effect of the phase transitions at 194 K and 310 K (when heating) is clearly seen: the low-temperature transition shows a pronounced hysteresis [70D1]. Polycrystalline sample.

9.15.3 Binary transition-metal chalcogenides

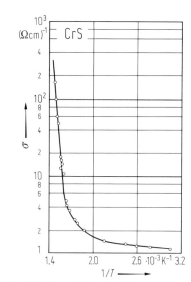

Fig. 37. CrS. Electrical conductivity vs. reciprocal temperature [77J].

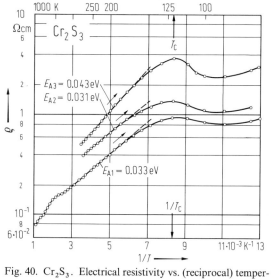

Fig. 40. Cr_2S_3. Electrical resistivity vs. (reciprocal) temperature for three samples of rhombohedral Cr_2S_3 [70V1]. Activation energies are indicated. Polycrystalline sample.

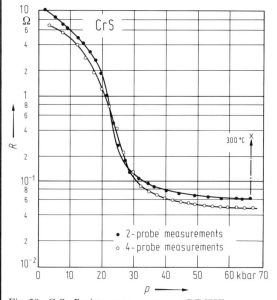

Fig. 38. CrS. Resistance vs. pressure at RT [77J].

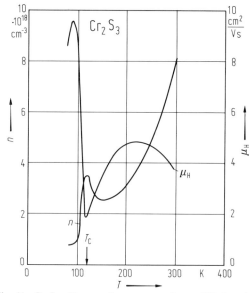

Fig. 41. Cr_2S_3. Free-carrier concentration and Hall mobility calculated from Hall and resistivity data vs. temperature [70V1]. Polycrystalline sample.

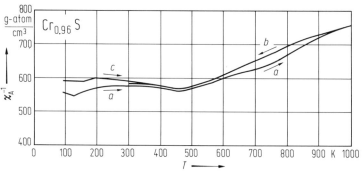

Fig. 39. $Cr_{0.96}S$. Reciprocal magnetic susceptibility per g-atom Cr (in CGS-emu) vs. temperature. Curve a: heating curve of "slowly cooled" specimen, b: cooling curve of "slowly cooled" specimen, c: heating curve of "quenched" specimen [69P]. Polycrystalline sample.

9.15.3 Binary transition-metal chalcogenides

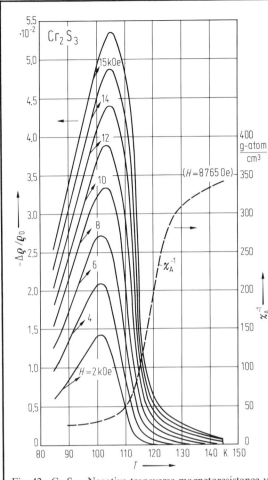

Fig. 42. Cr_2S_3. Negative transverse magnetoresistance vs. temperature; for comparison the temperature dependence of the reciprocal susceptibility (in CGS-emu) is included [70V1]. Polycrystalline sample.

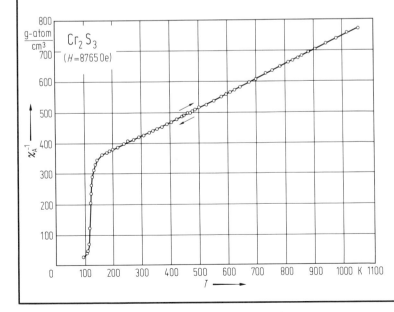

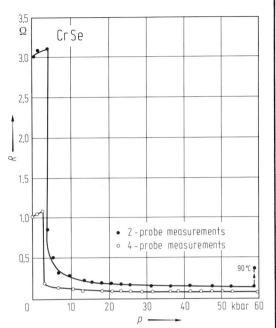

Fig. 45. CrSe. Resistance vs. pressure at RT [77J].

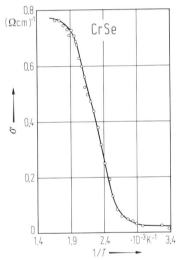

Fig. 44. CrSe. Electrical conductivity vs. reciprocal temperature [77J]. Polycrystalline pellets.

Fig. 43. Cr_2S_3. Reciprocal magnetic susceptibility per g-atom Cr (in CGS-emu) of rhombohedral Cr_2S_3 vs. temperature [70V1]. Polycrystalline sample.

9.15.3 Binary transition-metal chalcogenides

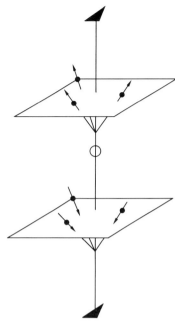

Fig. 46. CrSe. "Umbrella" model for the spin configuration. The magnetic moments in both the $z=0$ and $z=1/2$ level of the unit cell are shown in relation to the threefold axis [61C].

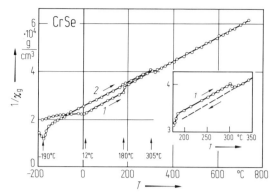

Fig. 47. CrSe. Inverse magnetic susceptibility per gram-atom (in CGS-emu) vs. temperature. Curves 1 and 2 are obtained for quenched and annealed specimens, respectivley [62M].

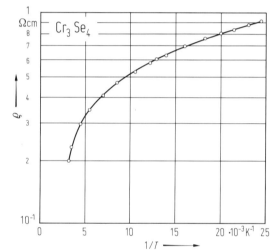

Fig. 48. Cr_3Se_4. Electrical resistivity vs. reciprocal temperature for a polycrystal [73B1], ϱ in Ω cm.

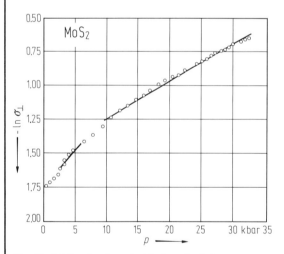

Fig. 49. MoS_2. In plane electrical conductivity vs. pressure in 2H-MoS_2 at room temperature [78E]. σ in Ω^{-1} cm^{-1}.

For Fig. 50, see next page.

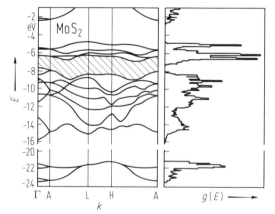

Fig. 51. MoS_2. Band structure and density of states of 2H modification [78B].

606

9.15.3 Binary transition-metal chalcogenides

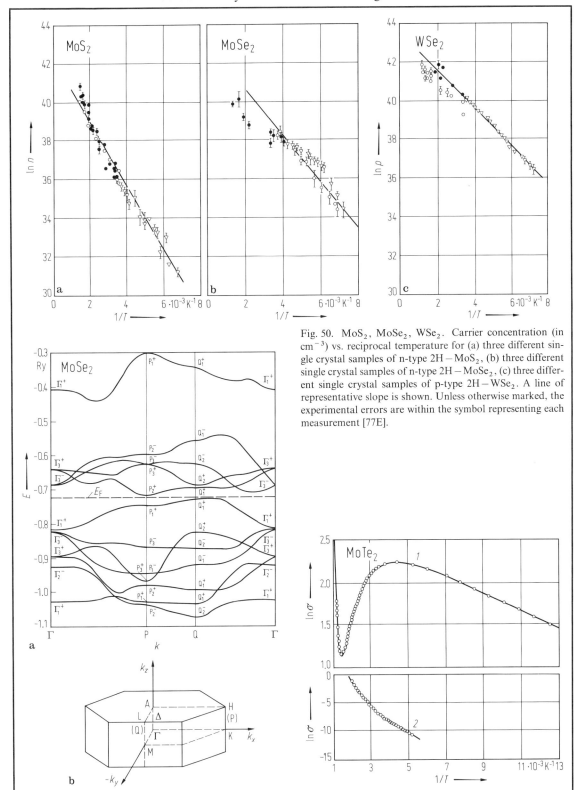

Fig. 50. MoS_2, $MoSe_2$, WSe_2. Carrier concentration (in cm^{-3}) vs. reciprocal temperature for (a) three different single crystal samples of n-type $2H-MoS_2$, (b) three different single crystal samples of n-type $2H-MoSe_2$, (c) three different single crystal samples of p-type $2H-WSe_2$. A line of representative slope is shown. Unless otherwise marked, the experimental errors are within the symbol representing each measurement [77E].

Fig. 52. $MoSe_2$. (a) Band structure and (b) Brillouin zone for $2H-MoSe_2$ [72B2]. Q and P are the middle of each edge and the vertex of the hexagon in the two-dimensional approximation used for the calculations.

Fig. 53. $MoTe_2$. Electrical conductivity vs. reciprocal temperature for 2H-modification. Theoretical results are drawn in full lines. 1: monocrystalline sample, 2: polycrystalline sample [79C2]. σ in $\Omega^{-1} cm^{-1}$, $\sigma \perp c$ (curve 1).

9.15.3 Binary transition-metal chalcogenides

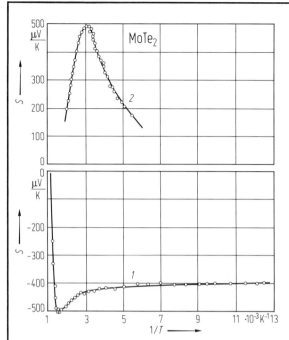

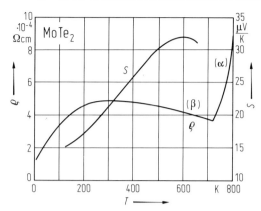

Fig. 55. MoTe$_2$. Resistivity and thermoelectric power vs. temperature for metastable β-MoTe$_2$. The rapid increase of the resistivity at 750 K corresponds to a transition to the α-phase [70V2]. Pressed-powder bar.

Fig. 54. MoTe$_2$. Seebeck coefficient vs. reciprocal temperature for 2H-modification. *1*: monocrystalline sample, *2*: polycrystalline sample [79C2]. $S \perp c$ (curve *1*).

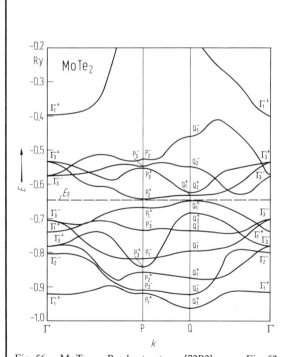

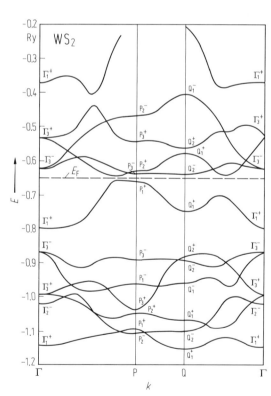

Fig. 56. MoTe$_2$. Band structure [72B2]; see Fig. 52 for Brillouin zone, and for the meaning of P and Q. 1 Ry = 13,6 eV.

Fig. 57. WS$_2$. Band structure [72B2]; see Fig. 52 for Brillouin zone and for the meaning of P and Q.

9.15.3 Binary transition-metal chalcogenides

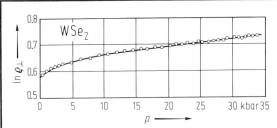

Fig. 58. WSe$_2$. Electrical conductivity vs. pressure at room temperature for 2H-modification [78E]. σ in Ω^{-1} cm^{-1}.

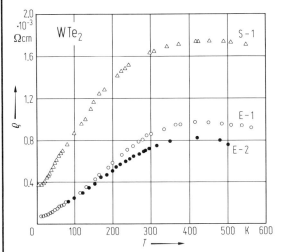

Fig. 60. WTe$_2$. Electrical resistivity vs. temperature. E-1, E-2: specimen of single crystals ($\varrho \parallel a$), S-1: sintered specimen [66K].

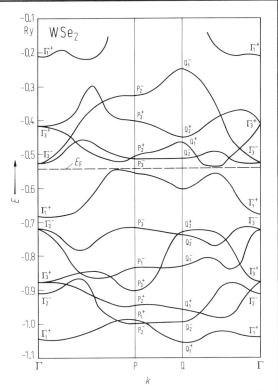

Fig. 59. WSe$_2$. Band structure for 2H-modificaion [72B2]; See Fig. 52 for Brillouin zone and for the meaning of P and Q.

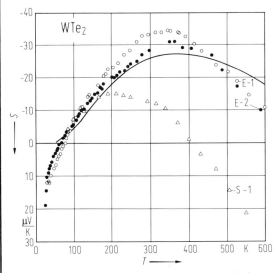

Fig. 61. WTe$_2$. Thermoelectric power of WTe$_2$ relative to Cu vs. temperature for the specimens of Fig. 60. $S \parallel a$. Solid curve calculated [66K].

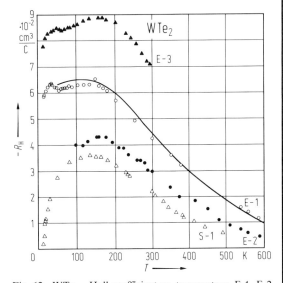

Fig. 62. WTe$_2$. Hall coefficient vs. temperature. E-1, E-2, E-3: single crystals, S-1: sintered specimen. The specimen E-1 is typical. Solid curve calculated [66K]. $B \parallel c$, $I \parallel a$.

9.15.3 Binary transition-metal chalcogenides

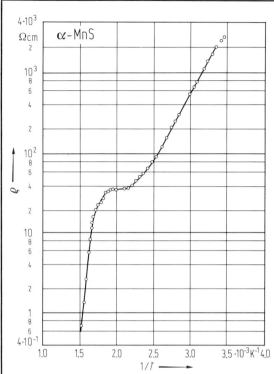

Fig. 63. α-MnS. Electrical resistivity vs. temperature [78H].

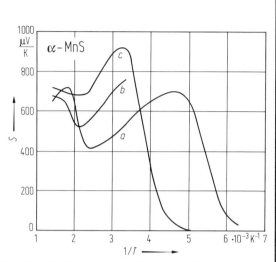

Fig. 64. α-MnS. Seebeck coefficient vs. reciprocal temperature for a single crystal grown by iodine-vapour transport and quenched from 600 K. First heating curve: a, second heating curve: b, ultimate behaviour: c [78H].

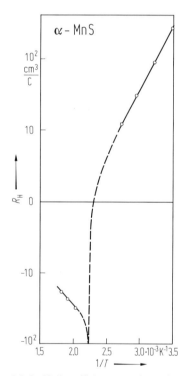

Fig. 65. α-MnS. Hall coefficient vs. reciprocal temperature for a sintered bar [78H].

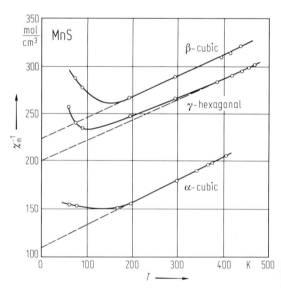

Fig. 66. MnS. Inverse magnetic molar susceptibility vs. temperature for 3 modifications [56C]. χ_m in CGS-emu.

9.15.3 Binary transition-metal chalcogenides

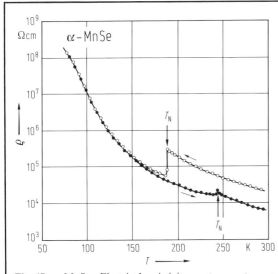

Fig. 67. α-MnSe. Electrical resistivity vs. temperature, ○: cooling curve, ●: warming curve [78I].

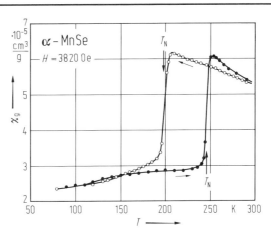

Fig. 68. α-MnSe. Magnetic susceptibility vs. temperature, ○: cooling curve, ●: warming curve [78I].

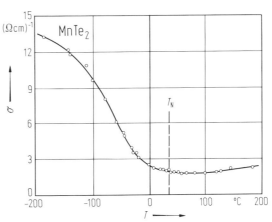

Fig. 70. MnTe$_2$. Electrical conductivity vs. temperature of a pure sample [74A2].

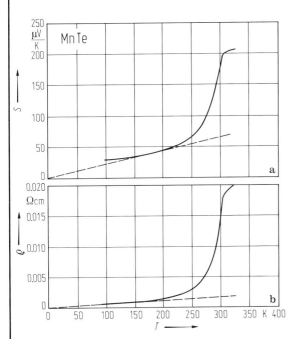

Fig. 69. MnTe. Thermoelectric power (a) and resistivity perpendicular to the c axis (b) vs. temperature for a degenerate p-type sample [64W2].

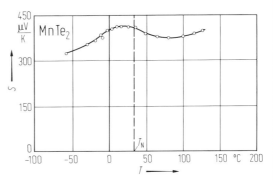

Fig. 71. MnTe$_2$. Thermoelectric power vs. temperature for a pure sample [74A2].

9.15.3 Binary transition-metal chalcogenides

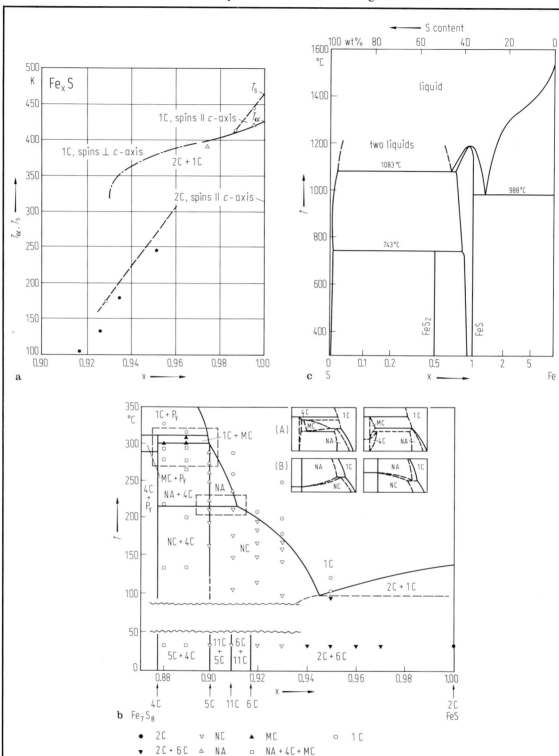

Fig. 72. Fe_xS. (a) T_α and T_s vs. composition (different symbols correspond to different measurements) [76H], (b) phase diagram of the $FeS-Fe_7S_8$ system (the inserts show two possible phase relations, see original paper) [71N], (c) phase diagram of the Fe−S system [64K].

9.15.3 Binary transition-metal chalcogenides

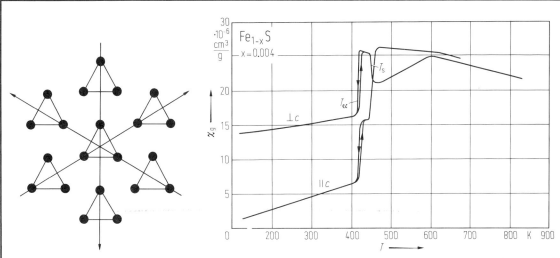

Fig. 73. FeS. Displacement of Fe atoms to form triangular clusters at low temperatures [64B1].

Fig. 74. Fe$_{0.996}$S. Magnetic susceptibility vs. temperature of a single crystal [76H].

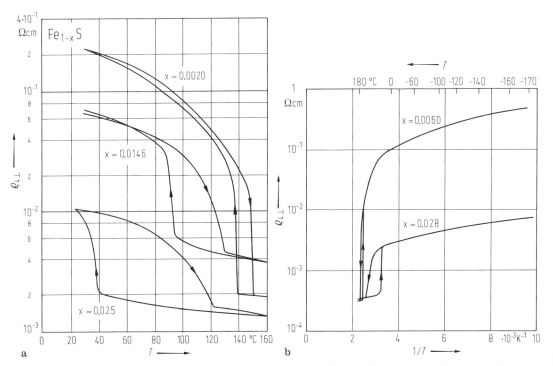

Fig. 75. Fe$_{1-x}$S. (a) Electrical resistivity $\varrho_{4\perp}$ vs. temperature (curve with x = 0.0146 represents the values of the second heating cycle), (b) electrical resistivity $\varrho_{4\perp}$ vs. reciprocal temperature [76M]. Four-probe measuring cell. The four electrodes are placed in the centre of a crystal surface $\perp c$. $R_{4\perp}$ then results from the voltage drop between the inner electrodes and current I.

9.15.3 Binary transition-metal chalcogenides

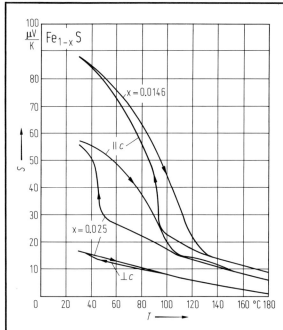

Fig. 76. $Fe_{1-x}S$. Seebeck coefficient vs. temperature [76M].

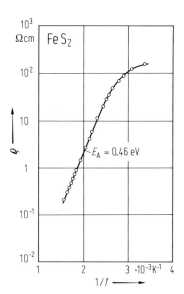

Fig. 77. FeS_2 (pyrite). Resistivity vs. reciprocal temperature [68B1].

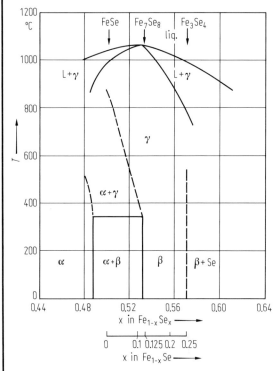

Fig. 78. Fe–Se. Phase diagram [66S1].

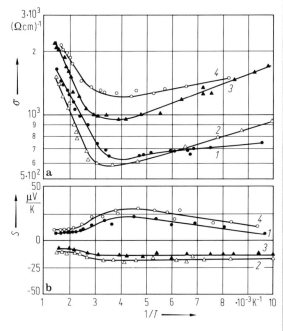

Fig. 79. Electrical conductivity (a) and thermoelectric power (b) vs. reciprocal temperature for 1: Fe_7Se_8, 2: Fe_3Se_4, 3: Fe_2Se_3, 4: FeSe [73A2]. Polycrystalline samples.

9.15.3 Binary transition-metal chalcogenides

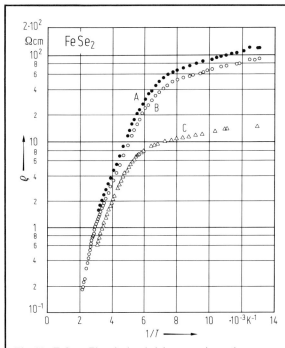

Fig. 80. FeSe$_2$. Electrical resistivity vs. reciprocal temperature for three polycrystalline samples. Sample A and B are stoichiometric, sample C is iron-rich and rather porous [58F].

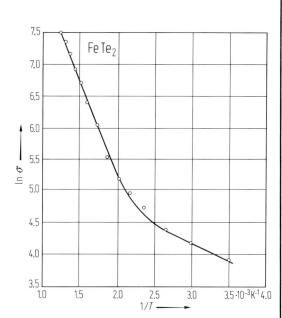

Fig. 81. FeTe$_2$. Electrical conductivity vs. reciprocal temperature [61D]. σ in Ω^{-1} cm^{-1}. Polycrystalline marcasite.

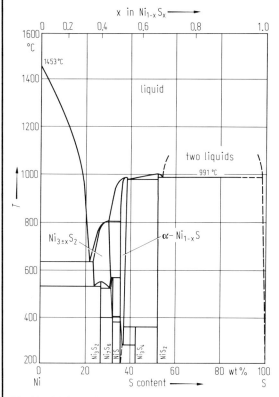

Fig. 82. Ni–S. Phase diagram. Vapour is present in all assemblages, and the pressure of the system is not constant [64K].

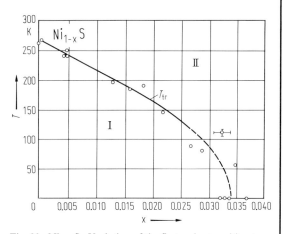

Fig. 83. Ni$_{1-x}$S. Variation of the first-order transition temperature with stoichiometry. Phase I is an antiferromagnetic semimetal, phase II is a Pauli-paramagnetic metal [76C].

9.15.3 Binary transition-metal chalcogenides

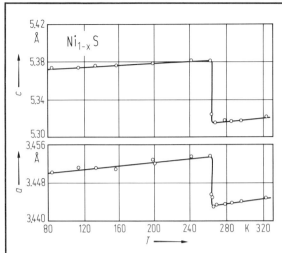

Fig. 84. $Ni_{1-x}S$ (x unspecified) and 0.2⋯0.02% Fe impurity. Lattice parameters vs. temperature [63S].

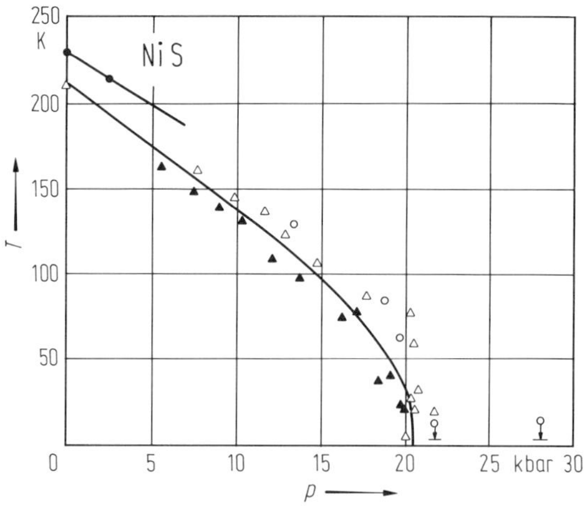

Fig. 85. NiS. Pressure-temperature phase diagram. Open and full symbols are for increasing and decreasing temperature or pressure. Circles are samples with $T_N = 230$ K at 1 atm and triangles are for samples with $T_N = 210$ K [72M2].

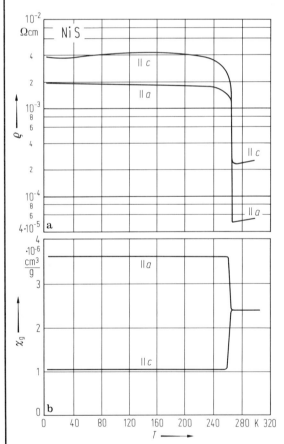

Fig. 86. NiS. Resistivity (a) and magnetic susceptibility (b) of a stoichiometric crystal of hexagonal NiS vs. temperature [74C].

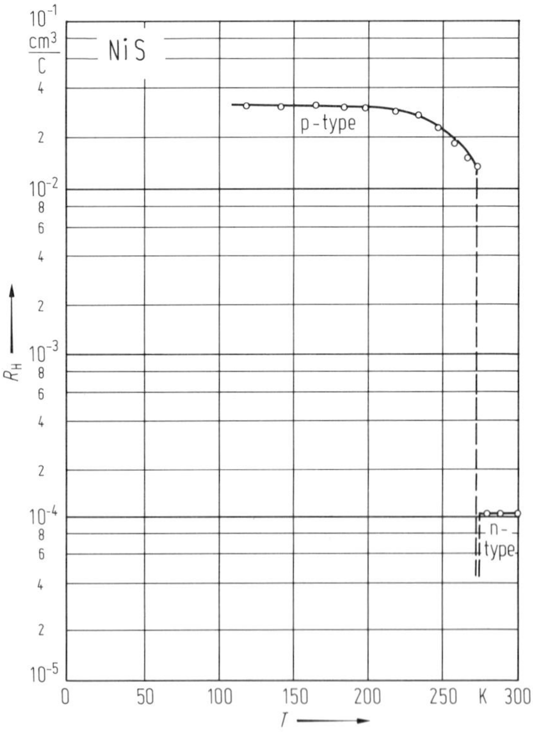

Fig. 87. NiS. Hall coefficient vs. temperature for a B8-phase single crystal [76B2]. $B \parallel c$.

9.15.3 Binary transition-metal chalcogenides

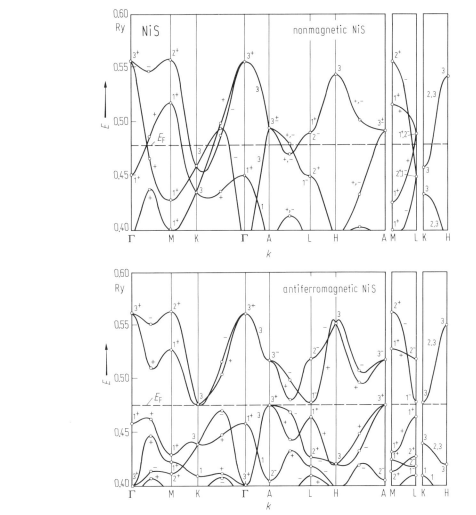

Fig. 88. NiS. Comparison of nonmagnetic and antiferromagnetic energy bands near the Fermi energy. The nonmagnetic states are labelled using the D_{3d}^3 space-group representations rather than those for D_{6h}^4 space group [74M].

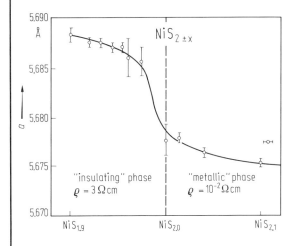

Fig. 89. $NiS_{2\pm x}$. Lattice parameters at RT vs. stoichiometry (stoichiometry modified according to [73G]) [72G].

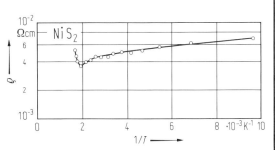

Fig. 90. NiS_2. Resistivity vs. reciprocal temperature for a sample prepared under high pressure [68B1].

9.15.3 Binary transition-metal chalcogenides

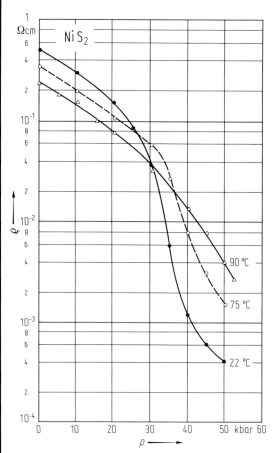

Fig. 91. NiS$_2$. Resistivity vs. pressure at different temperatures for a single crystal [71W2].

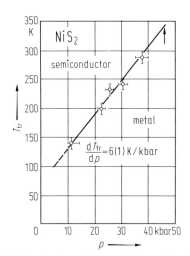

Fig. 92. NiS$_2$. Semiconductor-metal transition boundary in a temperature-pressure diagram. The critical point predicted by this resistance measurement is shown by an arrow [73M].

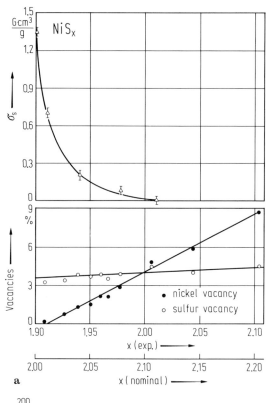

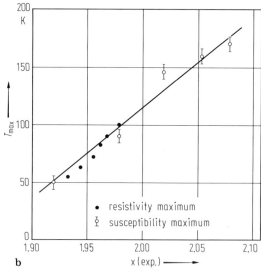

Fig. 93. NiS$_x$. (a) Variation of the saturation magnetization σ_s and the number of nickel and sulfur vacancies, with nominal and experimental x, (b) positions of the resistivity and susceptibility maxima against experimental x [73G].

9.15.3 Binary transition-metal chalcogenides

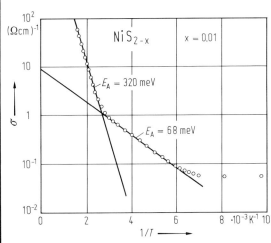

Fig. 94. $NiS_{1.99}$. Electrical conductivity vs. inverse temperature between 100 and 610 K [72K1].

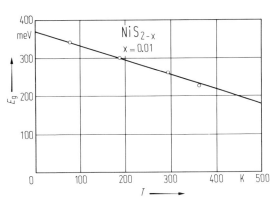

Fig. 95. $NiS_{1.99}$. Optical energy gap vs. temperature [72K1].

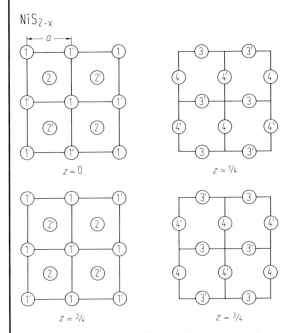

Fig. 96. NiS_{2-x}. Magnetic sublattices for a general ordering of second kind of fcc lattice [78K1]. Magnetic unit cells consist of 8 sublattices in which the spin directions of i and i'-th sites are antiparallel to each other.

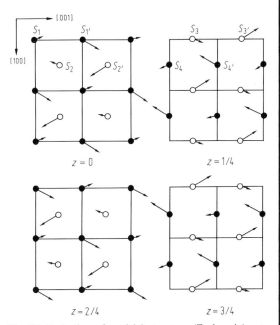

Fig. 97. Projection of model-3 structure (Z_a-domain) onto four adjacent (001) planes. The solid or open circle denotes the positive or negative sign of z component of the spin, respectively. The direction cosines of each spin are S_1: ($\bar{a}bc$), $S_{1'}$: (def), S_2: ($\bar{a}\bar{b}c$), $S_{2'}$: ($d\bar{e}\bar{f}$), S_3: ($ab\bar{c}$), $S_{3'}$: ($\bar{d}e\bar{f}$), S_4: ($a\bar{b}c$), $S_{4'}$: ($\bar{d}\bar{e}f$), with $a=0.04$, $b=0.35$, $c=0.94$, $d=0.47$, $e=0.87$, $f=0.17$ [78K1].

9.15.3 Binary transition-metal chalcogenides

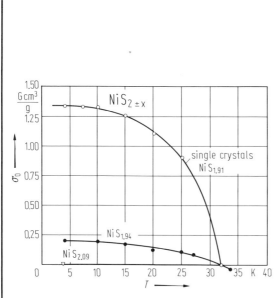

Fig. 98. $NiS_{2\pm x}$. Spontaneous magnetization σ_0 vs. temperature [73G].

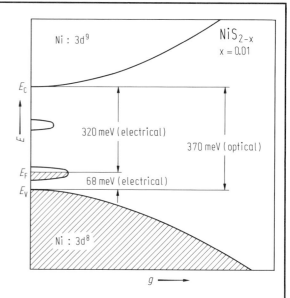

Fig. 99. $NiS_{1.99}$. A density of states model. The electron occupation (shaded area) and various energies are given for $T = 0$ K. The upper impurity band is a donor band and the lower one an acceptor band [72K1].

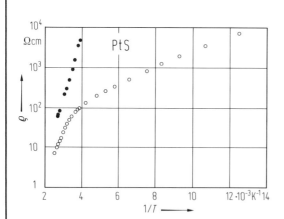

Fig. 100. PtS. Resistivity of ambient (●) and high-pressure (○) (50 kbar) phases vs. reciprocal temperature [79C1]. Polycrystalline sample.

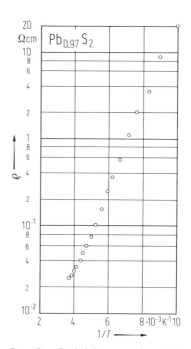

Fig. 101. $Pt_{0.97}S_2$. Resistivity vs. reciprocal temperature for a polycrystalline sample [74F].

9.16 Binary rare earth compounds
(For tables, see p. 317ff.)

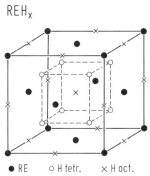

Fig. 1. REH_x. Unit cell of the CaF_2-type structure.

Fig. 3. LaH_3, LaD_3. Temperature dependence of heat capacities. I, II, III and IV denote phase transitions. α: semiconductor; β: metal; γ, δ, ε: phases with different H- and D-distribution on lattice sites [83I].

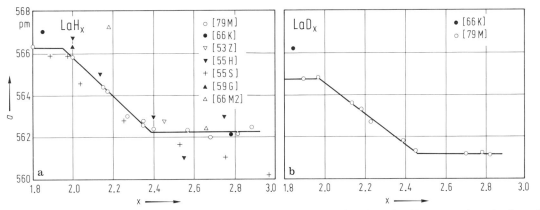

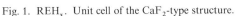

Fig. 2. LaH_x, LaD_x. Lattice parameters of LaH_x (a) and LaD_x (b) as a function of x. For comparison the data of previous works are also shown [79M].

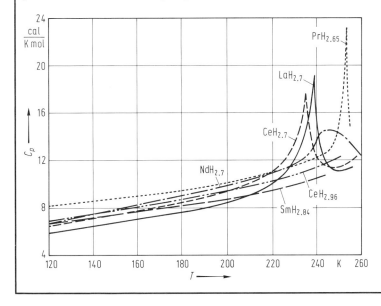

Fig. 4. RH_x, R = La, Ce, Pr, Nd, Sm. Molar heat capacities vs. temperature. The maxima at about 250 K are attributed to metal-semiconductor transitions. No anomaly was found for $CeH_{2.96}$ and $SmH_{2.84}$ in this temperature range [79B1].

9.16 Binary rare earth compounds

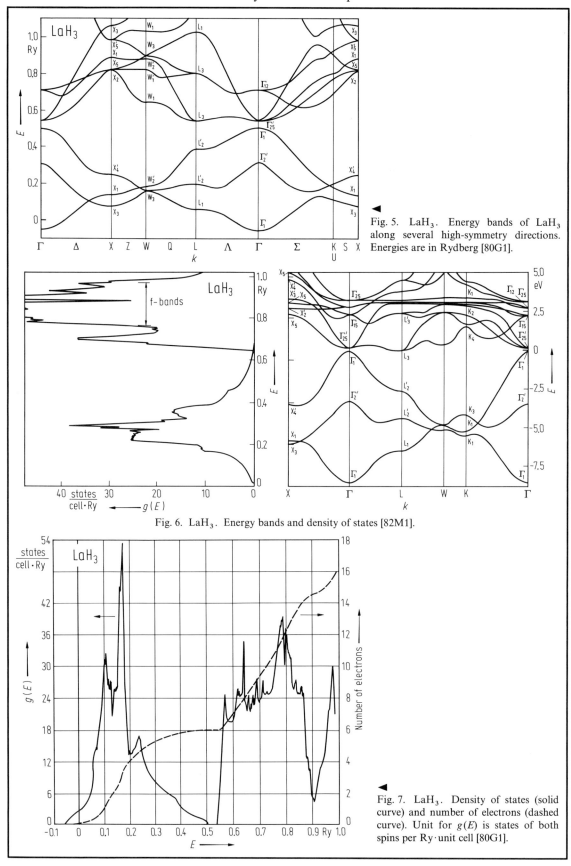

Fig. 5. LaH$_3$. Energy bands of LaH$_3$ along several high-symmetry directions. Energies are in Rydberg [80G1].

Fig. 6. LaH$_3$. Energy bands and density of states [82M1].

Fig. 7. LaH$_3$. Density of states (solid curve) and number of electrons (dashed curve). Unit for $g(E)$ is states of both spins per Ry·unit cell [80G1].

9.16 Binary rare earth compounds

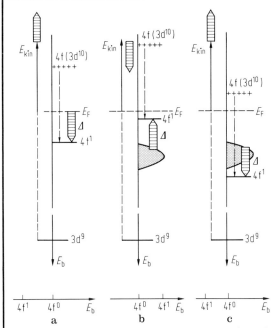

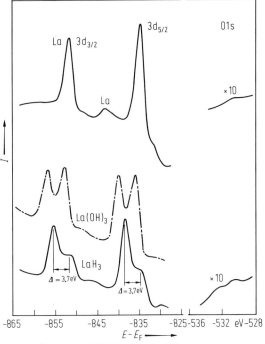

Fig. 8. LaH$_3$. Schematic electronic structure of A = La metal, B = up to now used for insulating La compounds, and C = proposed for LaH$_3$. Arrows indicate core hole screening mechanism [82S1]. For a detailed discussion, see original paper.

Fig. 11. LaH$_3$, La. XPS lanthanum 3d and oxygen 1s spectra (intensity vs. binding energy) [82S1]. La(OH)$_3$ for comparison.

Fig. 10. LaH$_3$. XPS valence band and 5p core level spectra ▶ (intensity vs. binding energy) [82S1].

For Fig. 12, see next page.

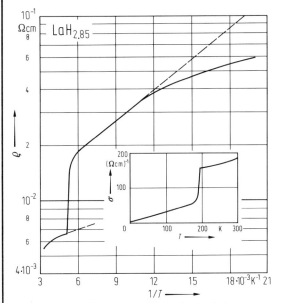

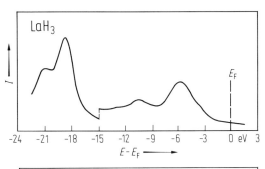

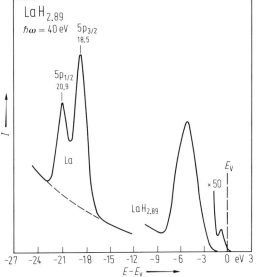

Fig. 9. LaH$_{2.85}$. Temperature dependence of electrical resistivity [82M1]. Inset shows conductivity vs. temperature.

Fig. 13. LaH$_{2.89}$. Photoelectron energy distribution curves ▶ (photoelectron emission intensity vs. initial state energy) for $\hbar\omega = 40$ eV normalized to the emission from the La 5p core levels [81P1].

623

9.16 Binary rare earth compounds

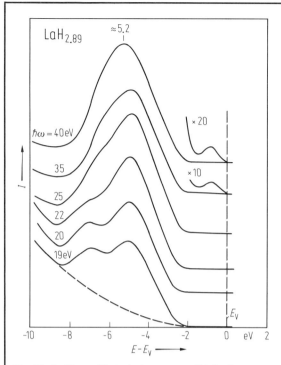

Fig. 12. $LaH_{2.89}$. Photoelectron energy distribution curves (photoelectron emission intensity vs. initial state energy) showing very weak valence band emission [81P1]. Parameter: photon excitation energy.

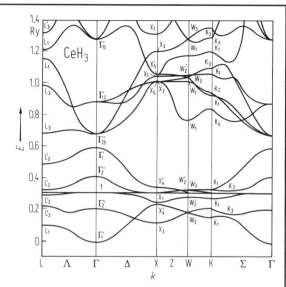

Fig. 15. CeH_3. Energy bands obtained from an APW calculation [80F1].

For Fig. 16, see next page.

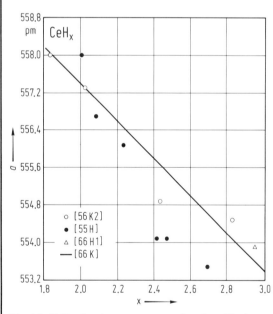

Fig. 14. CeH_x. Lattice parameter as a function of hydrogen concentration [68M].

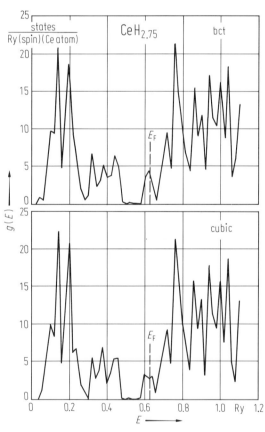

Fig. 17. $CeH_{2.75}$. Density of states for bct and cubic supercells [81F1].

9.16 Binary rare earth compounds

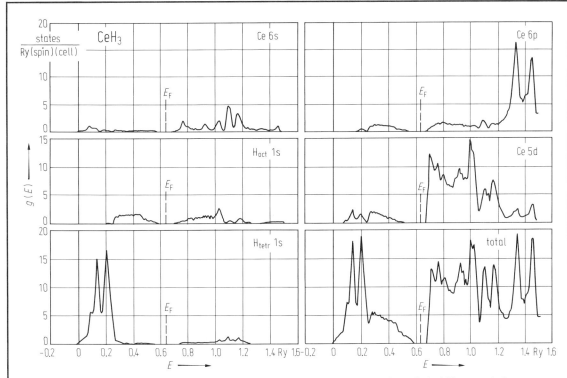

Fig. 16. CeH$_3$. Partial and total density of states [80F1]. H$_{oct}$, H$_{tetr}$: octahedral and tetrahedral H, respectively.

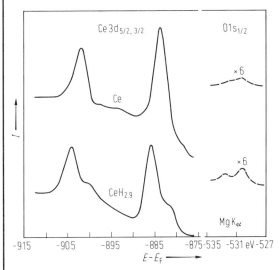

Fig. 18. CeH$_{2.9}$, Ce. XPS-spectra (intensity vs. binding energy). A comparison of 3d core level spectra [82S2].

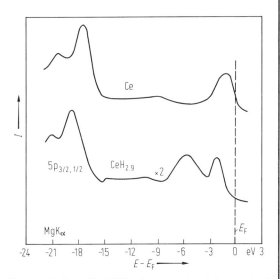

Fig. 19. CeH$_{2.9}$, Ce. XPS valence band and 5p core level spectra (intensity vs. binding energy) [82S2].

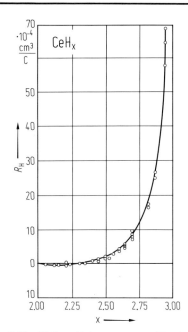

Fig. 20. CeH_x. Hall coefficient vs. composition in the range $2.0 \leq x \leq 2.8$ [69H].

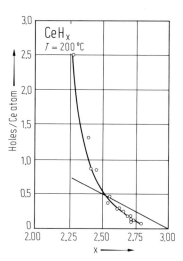

Fig. 21. CeH_x. Charge carrier concentration vs. composition for $2.3 \leq x \leq 2.8$. The straight line is predicted for the case of one hole removed for each hydrogen added [69H].

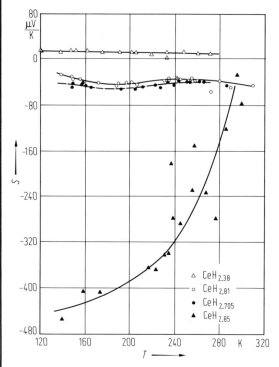

Fig. 22. CeH_x. Seebeck coefficient vs. temperature for several compositions [72L1].

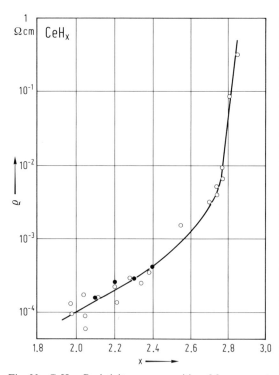

Fig. 23. CeH_x. Resistivity vs. composition. Measurements by various authors [72L1].

9.16 Binary rare earth compounds

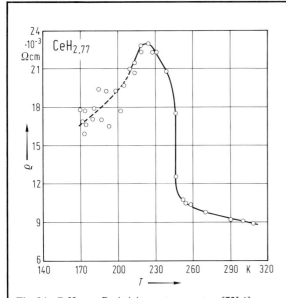

Fig. 24. $CeH_{2.77}$. Resistivity vs. temperature [72L1].

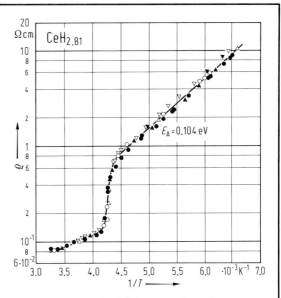

Fig. 25. $CeH_{2.81}$. Resistivity vs. reciprocal temperature [72L2].

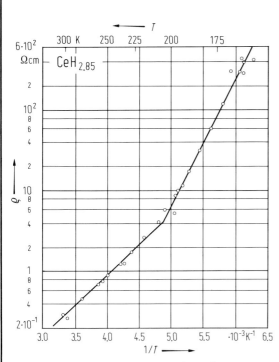

Fig. 26. $CeH_{2.85}$. Resistivity vs. (reciprocal) temperature [69L2, 72L1].

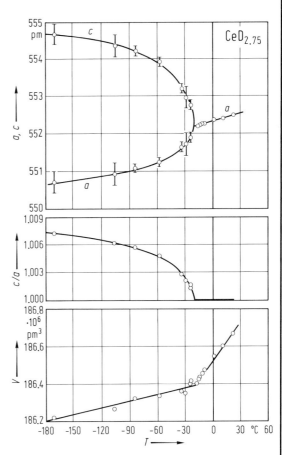

Fig. 27. $CeD_{2.75}$. Lattice parameters a, c, c/a ratio and unit cell volume V as a function of temperature [72L2].

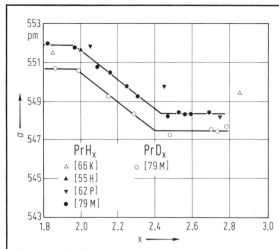

Fig. 28. PrH_x, PrD_x. Lattice parameters vs. x from various authors [79M].

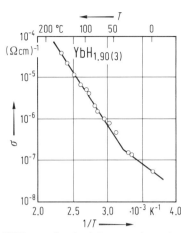

Fig. 29. $YbH_{1.90}$. Conductivity vs. reciprocal temperature [69H].

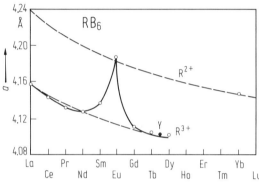

Fig. 30. RB_6. Lattice parameters for several rare earths hexaborides [80H1, 74M2].

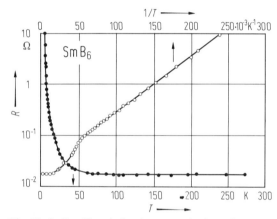

Fig. 32. SmB_6. Electrical resistance vs. (reciprocal) temperature [80H1, 69M].

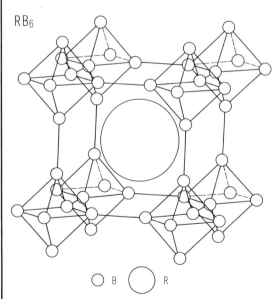

Fig. 31. RB_6. Crystal structure of rare earth hexaborides [80H1, 74M2].

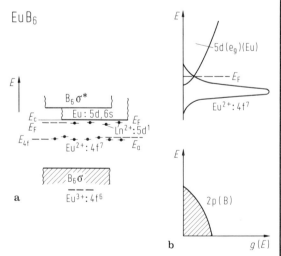

Fig. 33. EuB_6. Energy band schemes proposed by (a) [73M, 73G], (b) [77I], Ln: lanthanides.

9.16 Binary rare earth compounds

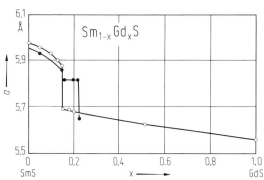

Fig. 34. $Sm_{1-x}Gd_xS$. Lattice parameters vs. composition at 298 K (open circles) and 4.2 K (full circles). The parameter is independent of concentration in the range $0.15 \leq x \leq 0.22$ [73J2]. See also [75J].

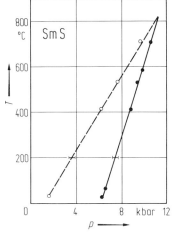

Fig. 36. SmS. The $p-T$ pase diagram. Solid line determined from thermopower data. Dashed line gives the pressure at which reverse transformation (semiconductor-metal) takes place. The intersection of the two lines gives the critical point (825 °C) [78S1].

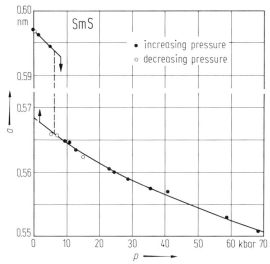

Fig. 37. SmS. Lattice parameter vs. hydrostatic pressure. Metallic phase for $p > 6.5$ kbar [79K2].

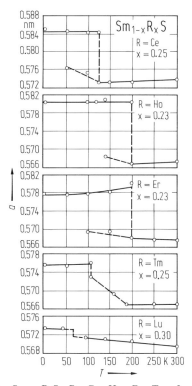

Fig. 35. $Sm_{1-x}R_xS$, R = Ce, Ho, Er, Tm, Lu. Lattice parameter vs. temperature for some compounds. Vertical broken lines mark the first order transitions (semiconductor-metal) [75J]. The compositions represent the nominal atomic percent R^{3+}.

For Fig. 38, see next page.

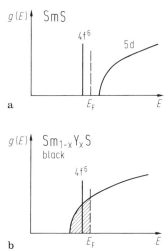

Fig. 39. SmS, $Sm_{1-x}Y_xS$. Schematic representation of the density of states of (a) semiconducting SmS, (b) black phase of $Sm_{1-x}Y_xS$ [76G].

9.16 Binary rare earth compounds

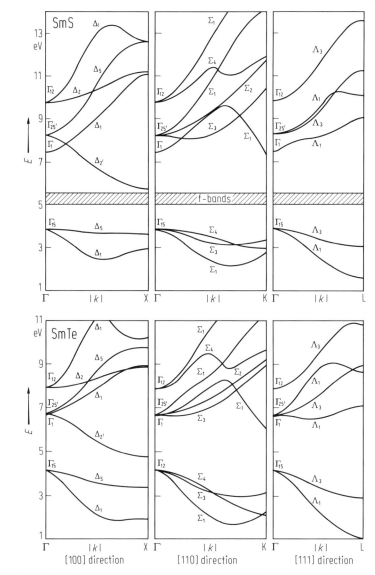

Fig. 38. SmS, SmTe. Calculated band structure for semiconducting SmS (top) and SmTe (bottom). Note the absence of 4f states in the case of the telluride [76C, 71D].

9.16 Binary rare earth compounds

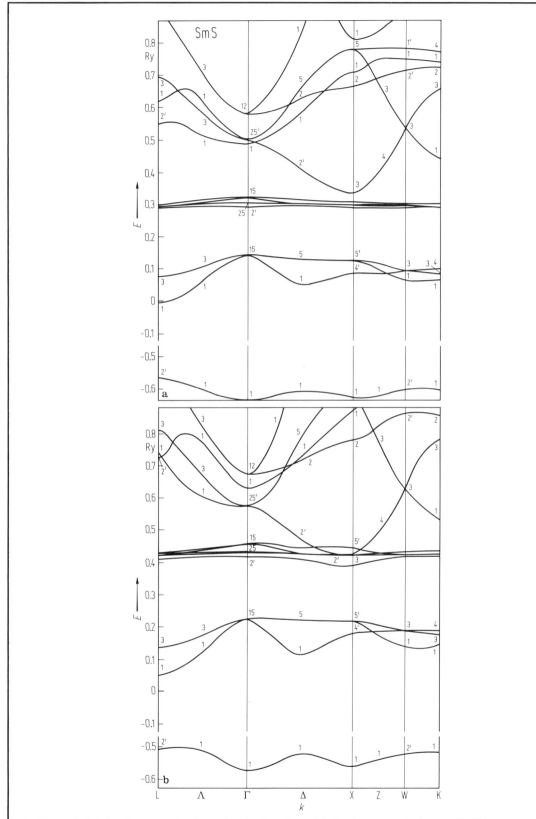

Fig. 40. SmS. (a): band structure in the semiconducting phase. (b): band structure in the metallic high pressure phase [81F3].

9.16 Binary rare earth compounds

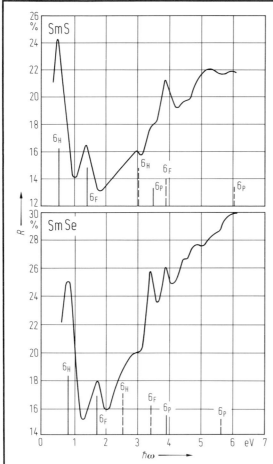

Fig. 41. SmS, SmSe. Reflectivity vs. photon energy for cleaved single crystals at 300 K. (———) 5d (t_{2g}) and (---) 5d (e_g) final states [76G].

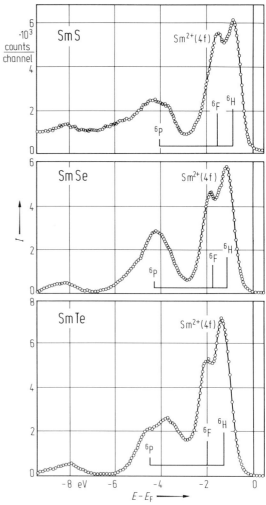

Fig. 43. SmS, SmSe, SmTe. XPS spectra of the valence band and the 4f region (electron intensitiy vs. binding energy) [76C].

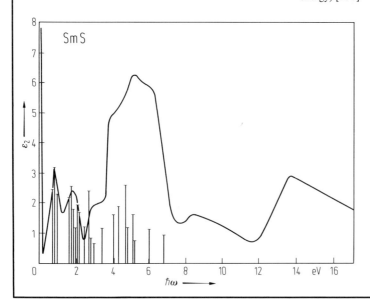

Fig. 42. SmS. Optical spectrum (imaginary part of dielectric constant vs. photon energy) and terms of excited f^5 configuration [81F3].

9.16 Binary rare earth compounds

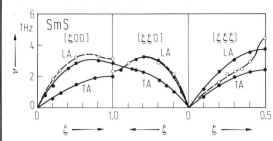

Fig. 44. SmS. Phonon dispersion (frequency vs. reduced wavevector coordinate) Solid points: measurements at normal pressure. Open points: measurements at 0.7 GPa [82M3].

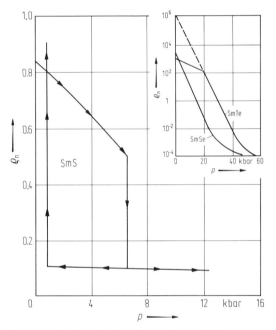

Fig. 45. SmS, SmSe, SmTe. Normalized resistivity vs. pressure for SmS. The actual resistivity at pressures greater than 6.5 kbar is $\approx 3 \cdots 4 \cdot 10^{-4}\,\Omega\,\mathrm{cm}$. The data for SmSe and SmTe are shown in the inset [70J].

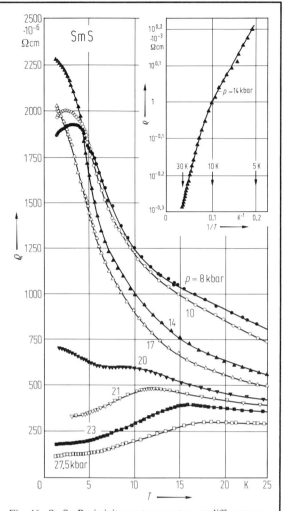

Fig. 46. SmS. Resistivity vs. temperature at different pressures [81L1]. Inset shows ϱ vs. $1/T$ at $p = 14$ kbar.

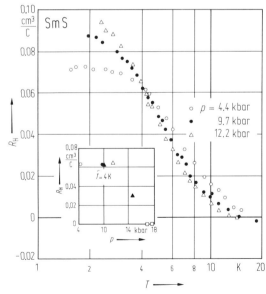

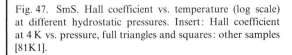

Fig. 47. SmS. Hall coefficient vs. temperature (log scale) at different hydrostatic pressures. Insert: Hall coefficient at 4 K vs. pressure, full triangles and squares: other samples [81K1].

9.16 Binary rare earth compounds

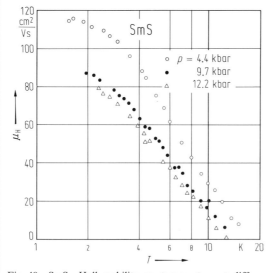

Fig. 48. SmS. Hall mobility vs. temperature at different pressures [81K1].

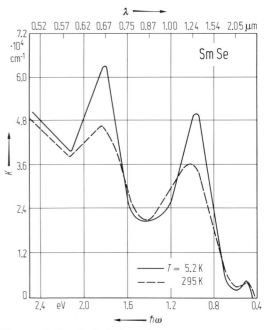

Fig. 50. SmSe. Optical absorption coefficient vs. photon energy for films of ≈ 8000 Å thickness [71B3].

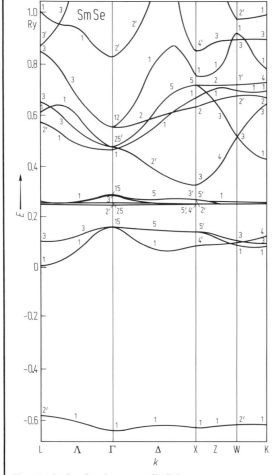

Fig. 49. SmSe. Band structure [80F2].

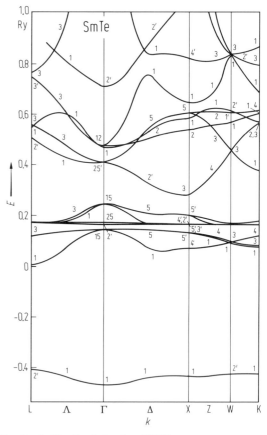

Fig. 51. SmTe. Band structure [80F2].

9.16 Binary rare earth compounds

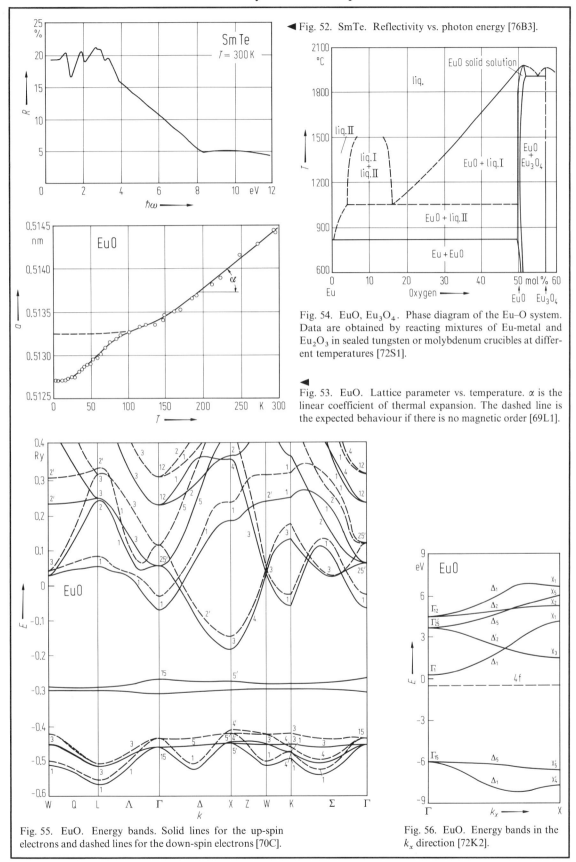

Fig. 52. SmTe. Reflectivity vs. photon energy [76B3].

Fig. 54. EuO, Eu_3O_4. Phase diagram of the Eu–O system. Data are obtained by reacting mixtures of Eu-metal and Eu_2O_3 in sealed tungsten or molybdenum crucibles at different temperatures [72S1].

Fig. 53. EuO. Lattice parameter vs. temperature. α is the linear coefficient of thermal expansion. The dashed line is the expected behaviour if there is no magnetic order [69L1].

Fig. 55. EuO. Energy bands. Solid lines for the up-spin electrons and dashed lines for the down-spin electrons [70C].

Fig. 56. EuO. Energy bands in the k_x direction [72K2].

9.16 Binary rare earth compounds

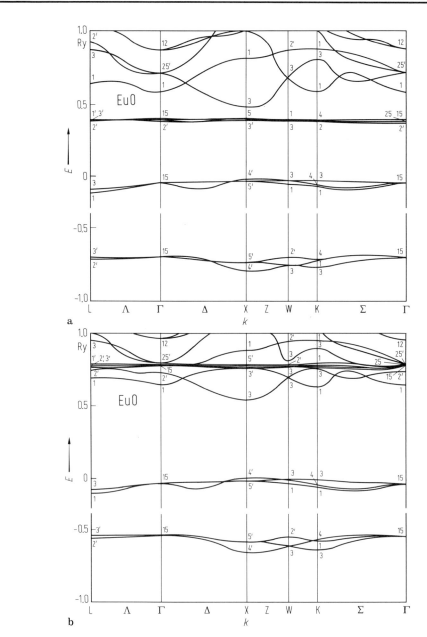

Fig. 57. EuO. Spin-polarized energy bands, a: spin-up, b: spin-down [81F4].

9.16 Binary rare earth compounds

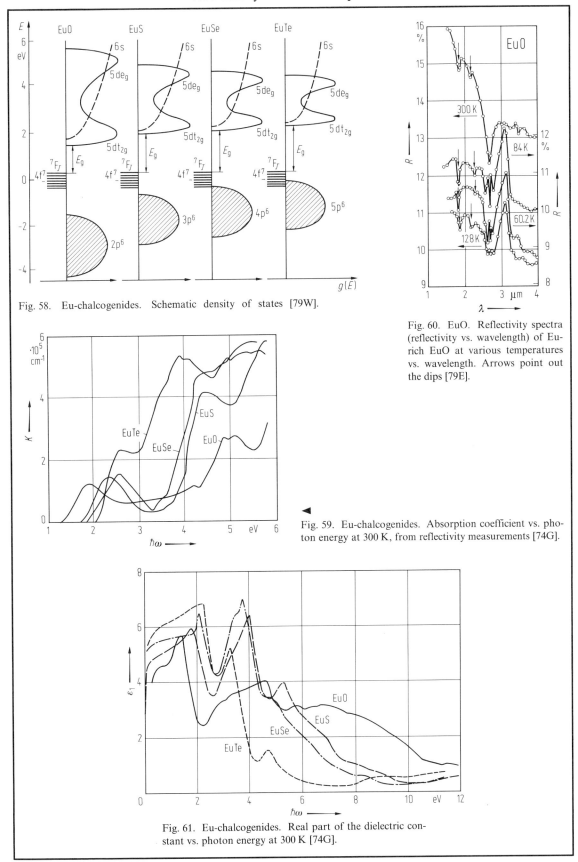

Fig. 58. Eu-chalcogenides. Schematic density of states [79W].

Fig. 60. EuO. Reflectivity spectra (reflectivity vs. wavelength) of Eu-rich EuO at various temperatures vs. wavelength. Arrows point out the dips [79E].

Fig. 59. Eu-chalcogenides. Absorption coefficient vs. photon energy at 300 K, from reflectivity measurements [74G].

Fig. 61. Eu-chalcogenides. Real part of the dielectric constant vs. photon energy at 300 K [74G].

9.16 Binary rare earth compounds

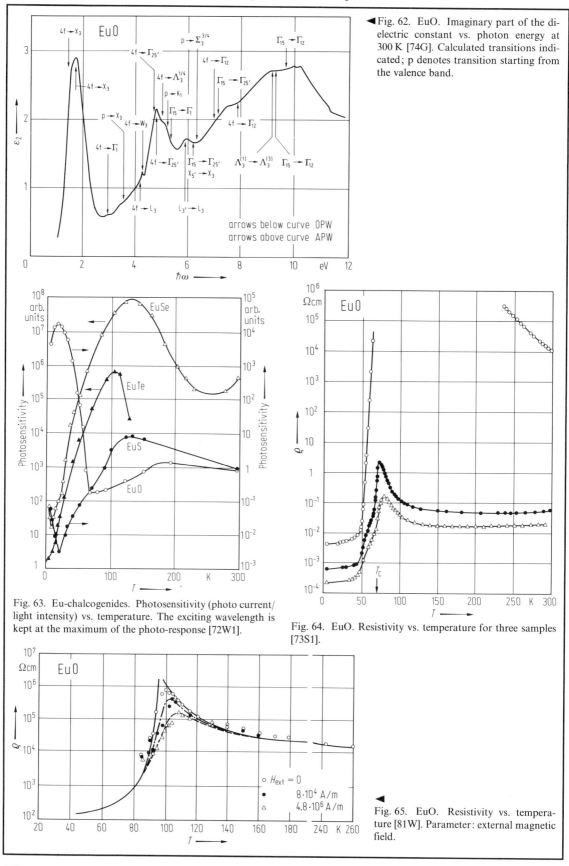

Fig. 62. EuO. Imaginary part of the dielectric constant vs. photon energy at 300 K [74G]. Calculated transitions indicated; p denotes transition starting from the valence band.

Fig. 63. Eu-chalcogenides. Photosensitivity (photo current/light intensity) vs. temperature. The exciting wavelength is kept at the maximum of the photo-response [72W1].

Fig. 64. EuO. Resistivity vs. temperature for three samples [73S1].

Fig. 65. EuO. Resistivity vs. temperature [81W]. Parameter: external magnetic field.

9.16 Binary rare earth compounds

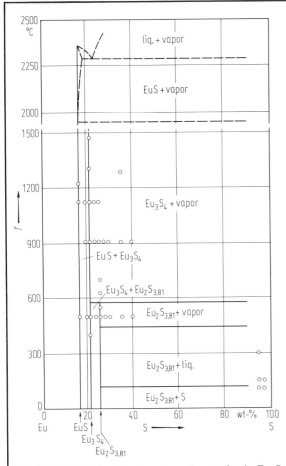

Fig. 66. EuS, Eu_3S_4. Tentative phase diagram for the Eu–S system determined by X-ray diffraction [74A].

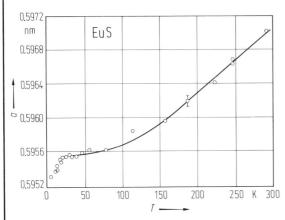

Fig. 67. EuS. Lattice parameter vs. temperature [69L1].

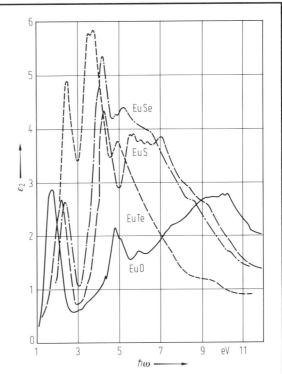

Fig. 70. Eu-chalcogenides. Imaginary part of the dielectric constant vs. photon energy at 300 K [74G].

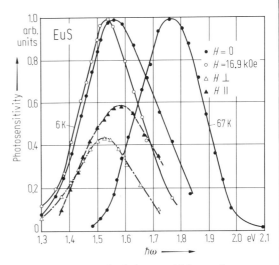

Fig. 71. EuS. Normalized photosensitivity vs. photon energy for two temperatures. Also shown is the photosensitivity at 6 K for light polarized parallel and perpendicular to the direction of the magnetic field [72W1].

For Figs. 68, 69, see next pages.

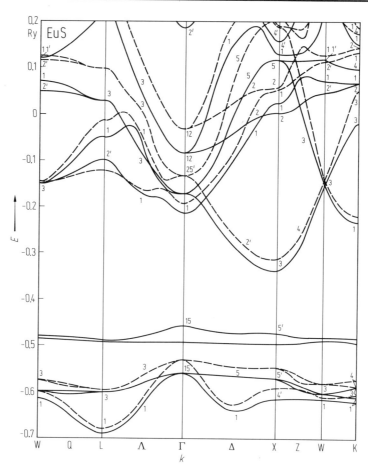

Fig. 68. EuS. Energy band structure. Solid lines for up-spin electrons, dashed lines for down-spin electrons [70C].

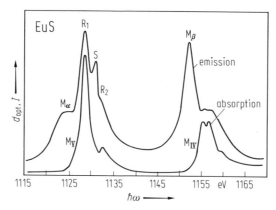

Fig. 72. EuS. Soft X-ray M_{IV} and M_v emission spectra (optical density (d_{opt}) and emission intensity (I) vs. photon energy) of a thin film ($3d^{10}4f^7 \leftrightarrow 3d^94f^8$) [75M1].

9.16 Binary rare earth compounds

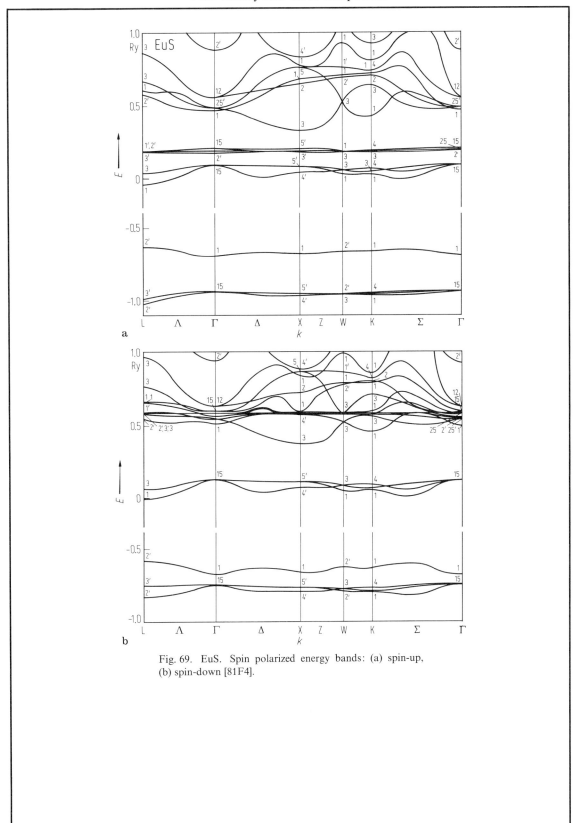

Fig. 69. EuS. Spin polarized energy bands: (a) spin-up, (b) spin-down [81F4].

9.16 Binary rare earth compounds

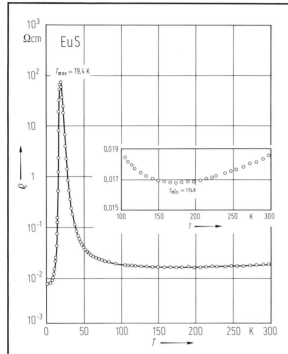

Fig. 73. EuS. Resistivity vs. temperature. The insert shows the resistivity vs. temperature in a linear scale near the resistivity minimum [72S4].

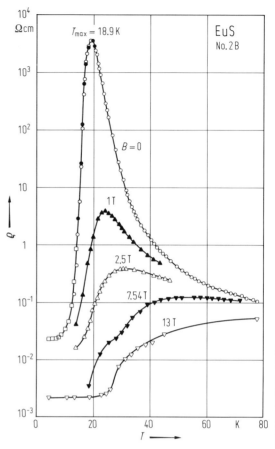

Fig. 74. EuS. Resistivity vs. temperature at several applied magnetic fields for an Eu-rich sample [72S4].

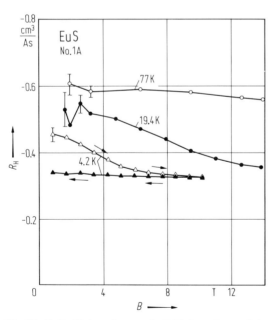

Fig. 75. EuS. (External) magnetic field dependence of the Hall coefficient for an n-type sample. Charge carrier density is of the order of 10^{19} cm^{-3}. Note the hysteresis at 4.2 K [72S4].

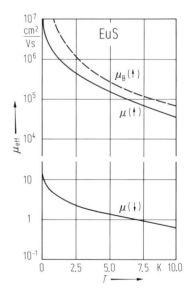

Fig. 76. EuS. Calculated spin-dependent effective mobility as a function of temperature. The dashed curve gives the Boltzmann result for spin-up electrons in the spin-polarized subband model [83E].

9.16 Binary rare earth compounds

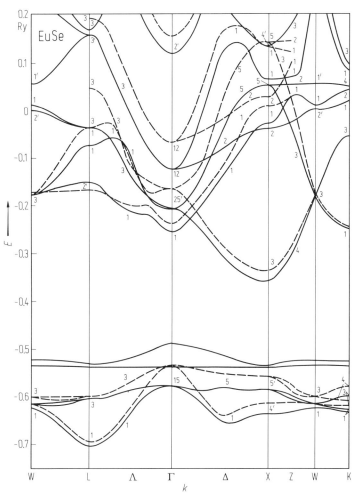

Fig. 77. EuSe. Energy band structure. Solid lines for up-spin electron, dashed lines for down-spin electrons [70C].

For Fig. 78, see next page.

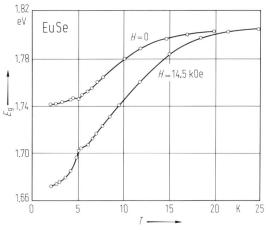

Fig. 79. EuSe. Absorption edge vs. temperature with and without an external magnetic field of 14.5 kOe [72W1].

9.16 Binary rare earth compounds

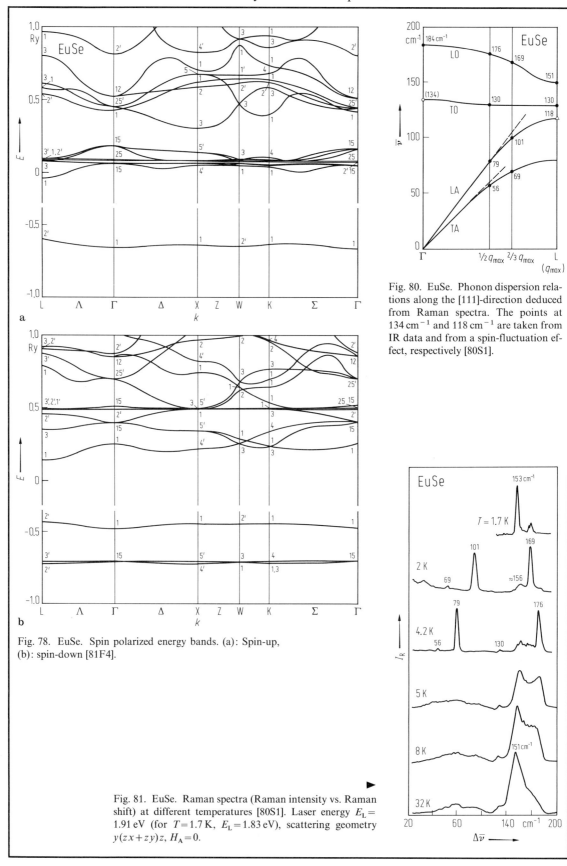

Fig. 78. EuSe. Spin polarized energy bands. (a): Spin-up, (b): spin-down [81F4].

Fig. 80. EuSe. Phonon dispersion relations along the [111]-direction deduced from Raman spectra. The points at 134 cm^{-1} and 118 cm^{-1} are taken from IR data and from a spin-fluctuation effect, respectively [80S1].

Fig. 81. EuSe. Raman spectra (Raman intensity vs. Raman shift) at different temperatures [80S1]. Laser energy $E_L = 1.91$ eV (for $T = 1.7$ K, $E_L = 1.83$ eV), scattering geometry $y(zx+zy)z$, $H_A = 0$.

9.16 Binary rare earth compounds

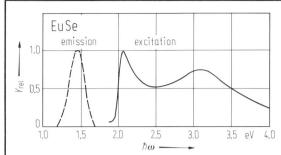

Fig. 82. EuSe. Relative quantum yield of the luminescence vs. photon energy at 4.3 K [72W1].

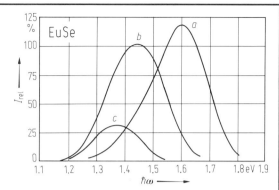

Fig. 83. EuSe. Emission spectrum (relative intensity vs. photon energy) (a) at 51 K, (b) at 4.3 K, (c) at 4.3 K with a magnetic field of 11 kOe [72W1].

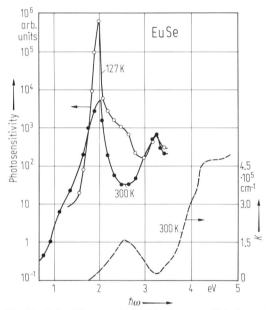

Fig. 84. EuSe. Photosensitivity (photo current/light intensity) vs. photon energy. Also shown is the absorption coefficient [72W1].

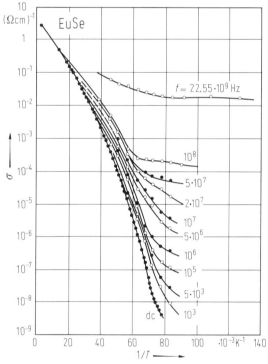

Fig. 85. EuSe. dc- and ac- conductivity vs. reciprocal temperature [78K2].

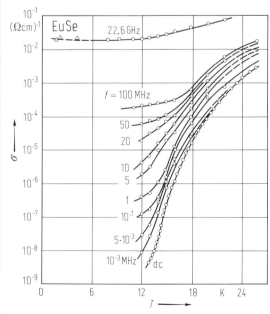

Fig. 86. EuSe. dc and ac- conductivity at low temperatures [78K2]. Triangles are results from other literature.

9.16 Binary rare earth compounds

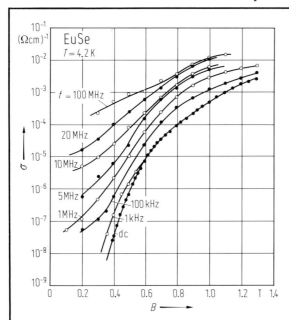

Fig. 87. EuSe. dc- and ac-conductivity vs. external magnetic field at 4.2 K [78K2].

Fig. 88. EuSe. dc- and ac-conductivity vs. external magnetic field at 10 K [78K2].

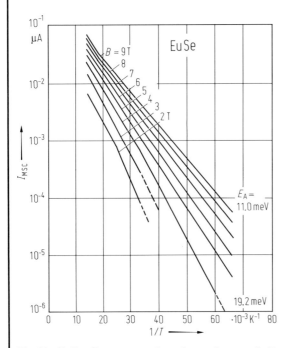

Fig. 89. EuSe. Temperature dependence of magnetically stimulated current I_{MSC} at different external magnetic fields [81Y].

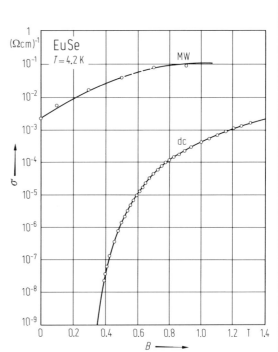

Fig. 90. EuSe. dc and microwave (MW) conductivities vs. external magnetic field at 4.2 K. The r.f. frequency was 24 GHz [81K5].

9.16 Binary rare earth compounds

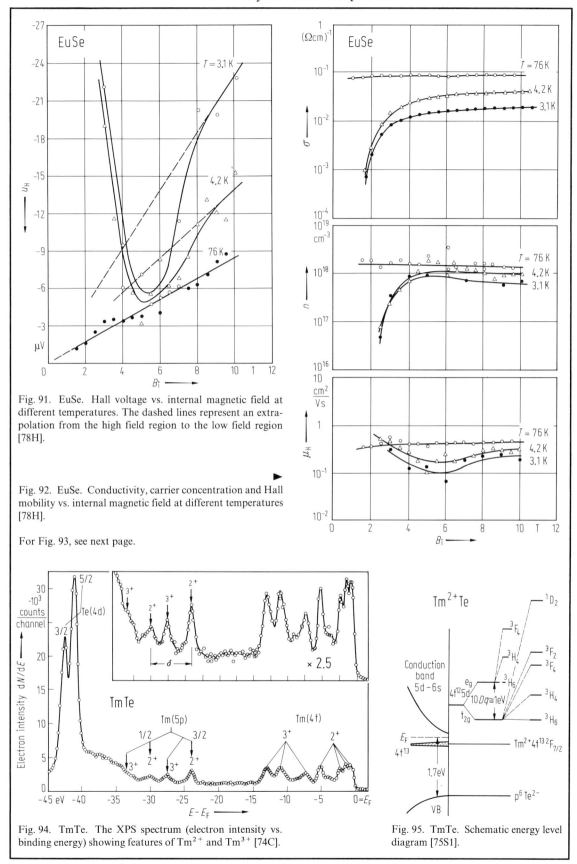

Fig. 91. EuSe. Hall voltage vs. internal magnetic field at different temperatures. The dashed lines represent an extrapolation from the high field region to the low field region [78H].

Fig. 92. EuSe. Conductivity, carrier concentration and Hall mobility vs. internal magnetic field at different temperatures [78H].

For Fig. 93, see next page.

Fig. 94. TmTe. The XPS spectrum (electron intensity vs. binding energy) showing features of Tm^{2+} and Tm^{3+} [74C].

Fig. 95. TmTe. Schematic energy level diagram [75S1].

9.16 Binary rare earth compounds

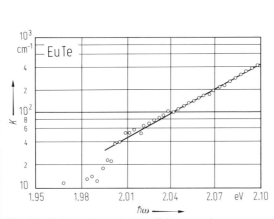

Fig. 93. EuTe. Absorption coefficient vs. photon energy near the absorption edge at $T=1.7$ K and $B=8.34$ T with $E \parallel B$ [78S3].

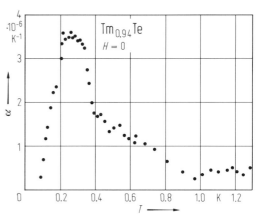

Fig. 96. $Tm_{0.94}Te$. Temperature dependence of the thermal expansion coefficient [83O2].

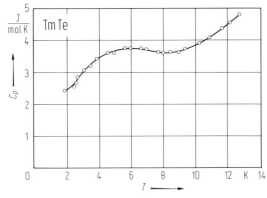

Fig. 97. TmTe. Molar heat capacity vs. temperature [75B].

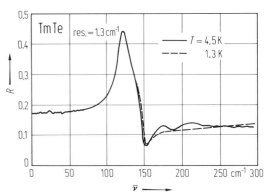

Fig. 98. TmTe. Far-infrared reflectivity spectrum (reflectivity vs. wavenumber) at 4.5 K and 1.3 K [75W].

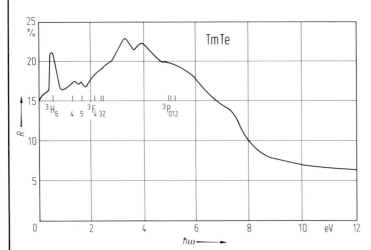

Fig. 99. TmTe. Reflectivity vs. photon energy [76B4]. Final states are indicated.

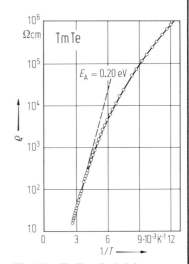

Fig. 100. TmTe. Resistivity vs. reciprocal temperature [75B].

9.16 Binary rare earth compounds

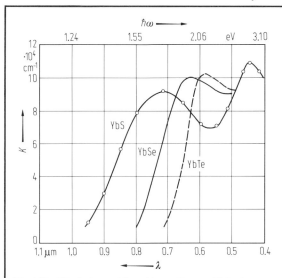

Fig. 101. Yb-chalcogenides. Absorption coefficient vs. wavelength (photon energy) for thin films on NaCl substrates at atmospheric pressure [74N].

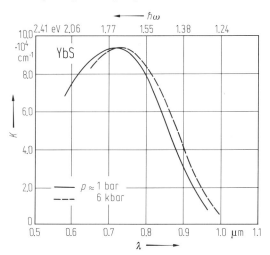

Fig. 103. YbS. Absorption edge at atmospheric pressure and at 6 kbar vs. wavelength (photon energy) [74N].

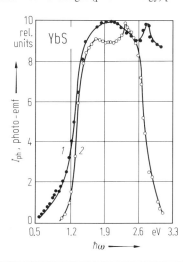

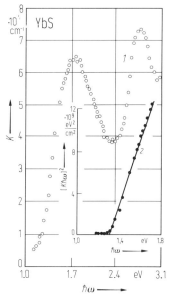

Fig. 102. YbS. Absorption coefficient (1) and absorption coefficient times photon energy (2) vs. photon energy for a 0.3 μm thick film [80G3].

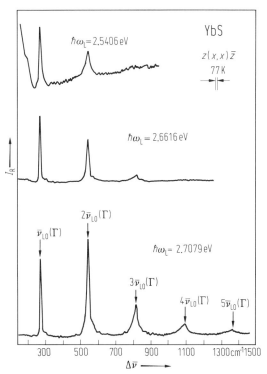

Fig. 105. YbS. Raman spectra (intensity vs. Raman shift) for different exciting laser energies [78M2].

Fig. 104. YbS. Photocurrent (1) and photo-emf (2) vs. photon energy of a 2.8 μm thick film [80G3].

9.16 Binary rare earth compounds

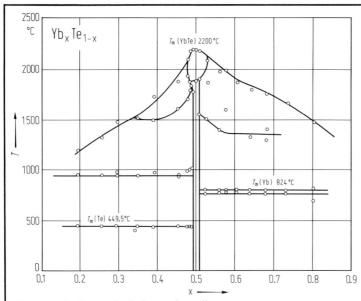

Fig. 106. Yb_xTe_{1-x}. Preliminary phase diagram as compiled from DTA measurements in sealed tungsten crucibles [81K4].

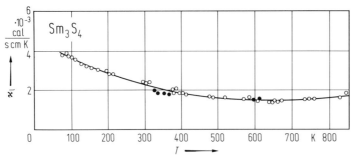

Fig. 108. Sm_3S_4. Temperature dependence of the thermal conductivity. Due to the small electronic contribution the total thermal conductivity κ_{tot} is assumed to be equal to the lattice contribution κ_l [73S5]. Full circles: on cooling, open circles: on heating.

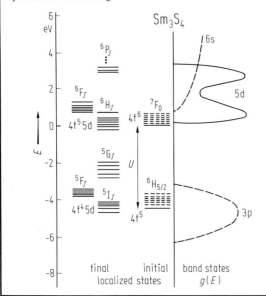

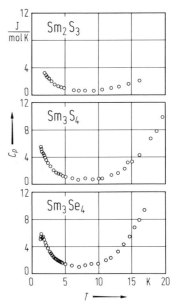

Fig. 107. γ-Sm_2S_3, Sm_3S_4, Sm_3Se_4. Temperature dependence of molar heat capacity. The increase below 7 K is explained by a Schottky anomaly due to the crystal field splitting ($\Delta_{cf} \triangleq 2.4$ K) of Sm^{3+} [79C].

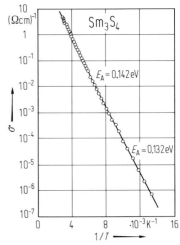

Fig. 110. Sm_3S_4. Electrical conductivity vs. reciprocal temperature for single crystals [76B2]. Change of activation energy near 125 K.

Fig. 109. Sm_3S_4. Energy levels and schematic density of states derived from optical data [76B2]. The energy of the $4f^6$ state of Sm^{2+} is chosen as zero point of energy scale. U is the Coulomb correlation energy.

9.16 Binary rare earth compounds

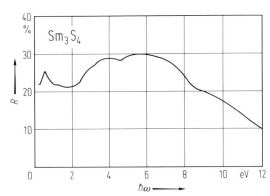

Fig. 111. Sm_3S_4. Reflectivity vs. photon energy for single crystals at 300 K [76B2].

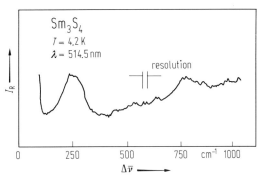

Fig. 112. Sm_3S_4. Electronic Raman scattering (relative intensity vs. Raman shift) at 4.2 K with 514.5 nm laser excitation. The peaks at 255 cm^{-1} and 760 cm^{-1} are assigned to the $^7F_0$–$^7F_1$ and $^7F_0$–$^7F_2$ transition of Sm^{2+}, respectively [77V1].

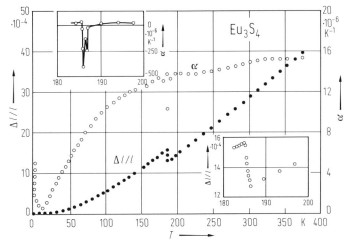

Fig. 113. Eu_3S_4. Relative length change and thermal expansion coefficient vs. temperature [83P].

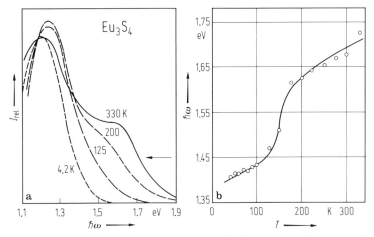

Fig. 114. Eu_3S_4. (a) 5d–4f interband photoluminescence spectra (relative intensity vs. photon energy) at different temperatures. All curves are normalized at 1.18 eV and have had the spectral sensitivity of the detector corrected. Fig. (b) shows the energies on the high energy side of the spectra (chosen at an arbitrary level marked by an arrow in Fig. (a)) vs. temperature [76V].

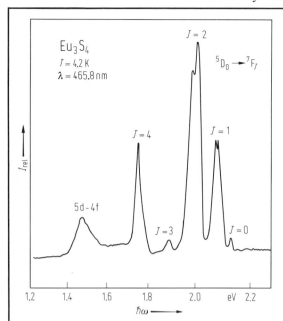

Fig. 115. Eu_3S_4. Intra-4f photoluminescence spectrum (relative intensity vs. photon energy) at 4.2 K resulting from the $^5D_0$–^7F_J transitions of Eu^{3+}. The optical excitation (465.8 nm) was into the $^5D_2$ multiplet level of Eu^{3+} [76V].

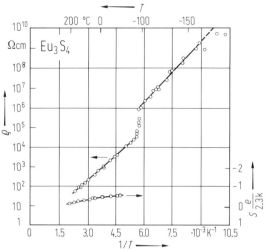

Fig. 116. Eu_3S_4. Resistivity and absolute Seebeck coefficient vs. (reciprocal) temperature [70B].

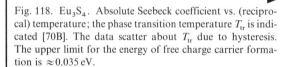

Fig. 118. Eu_3S_4. Absolute Seebeck coefficient vs. (reciprocal) temperature; the phase transition temperature T_{tr} is indicated [70B]. The data scatter about T_{tr} due to hysteresis. The upper limit for the energy of free charge carrier formation is ≈ 0.035 eV.

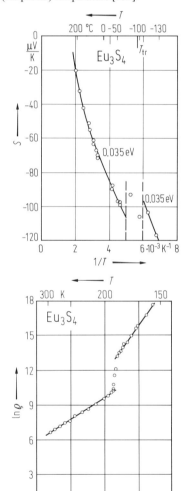

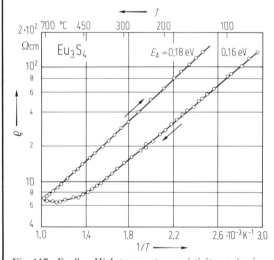

Fig. 117. Eu_3S_4. High temperature resistivity vs. (reciprocal) temperature; the arrows indicate data taken while heating and cooling [70B].

Fig. 119. Eu_3S_4. Electrical resistivity as a function of (reciprocal) temperature for a single crystal [83P].

9.16 Binary rare earth compounds

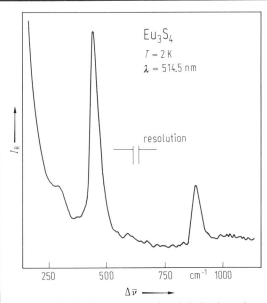

Fig. 120. Eu_3S_4. Raman scattering (relative intensity vs. Raman shift) of a polycrystalline sample at 2 K using 514.5 nm laser excitation. Phonon modes are located at 300 cm^{-1} and 425 cm^{-1} [77V2].

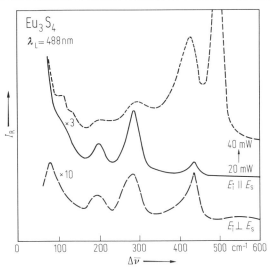

Fig. 121. Eu_3S_4. Raman spectrum (scattering intensity vs. Raman shift) of an unoriented single crystal at 300 K, measured at 20 mW (solid and long-dashed line) and 40 mW (short-dashed line) laser power. $E_{i(s)}$: electric field vector of incident (scattered) photon [81G3].

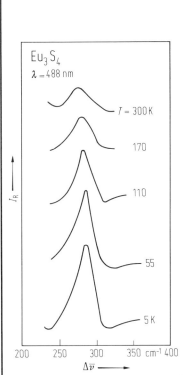

Fig. 122. Eu_3S_4. Temperature dependence of the 280 cm^{-1} Raman peak [81G3].

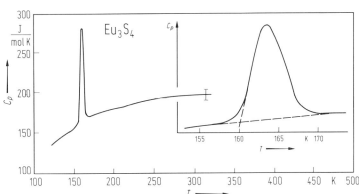

Fig. 123. Eu_3S_4. Molar heat capacity vs. temperature. The peak, which is also shown in the insert, indicates a change in the charge ordering of Eu^{2+} and Eu^{3+} ions [76M3].

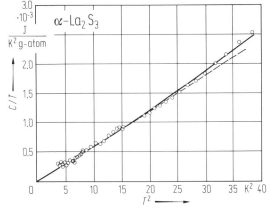

Fig. 125. α-La_2S_3. Temperature dependence of heat capacity [81G4].

For Fig. 124, see next page.

9.16 Binary rare earth compounds

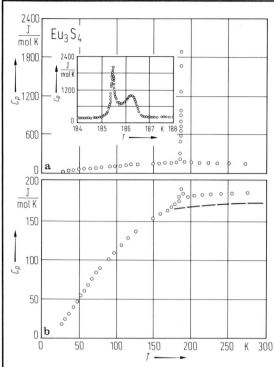

Fig. 124. Eu_3S_4. Molar heat capacity vs. temperature at full (a) and expanded scale (b). The dashed line in (b) is a Debye fit extrapolated for $T > 186$ K [83P].

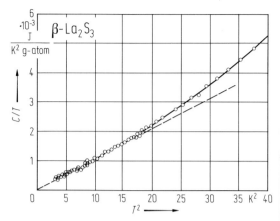

Fig. 126. β-La_2S_3. Temperature dependence of heat capacity [81G4].

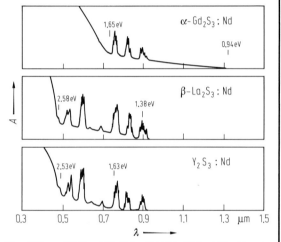

Fig. 127. α-Gd_2S_3:Nd, β-La_2S_3:Nd, Y_2S_3:Nd. Absorbance A vs. wavelength [80L].

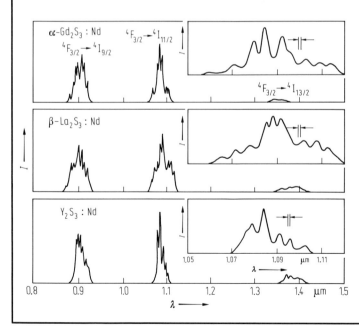

Fig. 128. α-Gd_2S_3:Nd, β-La_2S_3:Nd, Y_2S_3:Nd. Photoluminescence intensity vs. wavelength for α-Gd_2S_3:Nd excited at 752.5 nm, β-La_2S_3:Nd and Y_2S_3:Nd excited at 600 nm [80L].

9.16 Binary rare earth compounds

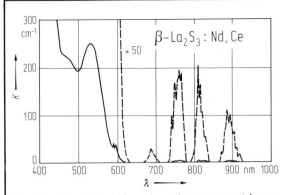

Fig. 129. β-La$_2$S$_3$:Nd, Ce. Absorption spectrum (absorption coefficient vs. wavelength) of Nd(1%) and Ce(1%) codoped crystals; unpolarized light [82S6].

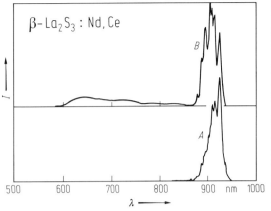

Fig. 130. β-La$_2$S$_3$:Nd, Ce. Fluorescence spectra (fluorescence intensity vs. wavelength) of (A) Nd (10%), Ce(1%) and (B) Nd(1%), Ce(0.2%) codoped crystals [82S6].

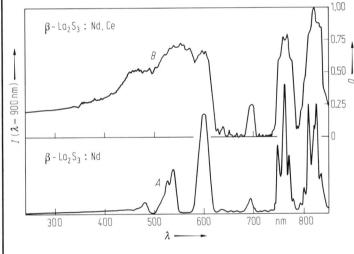

Fig. 131. β-La$_2$S$_3$: Nd, Ce; β-La$_2$S$_3$: Nd. Excitation spectra of Nd^{3+} (fluorescence intensity vs. excitation wavelength) registered at $\lambda = 900$ nm for (A) Nd(1%) and (B) Nd(10%), Ce(1%) doped crystals. The codoped crystal (B) shows efficient energy transfer from band excitation to Nd^{3+} [82S6]. Q: quantum efficiency of the transfer to Nd^{3+}.

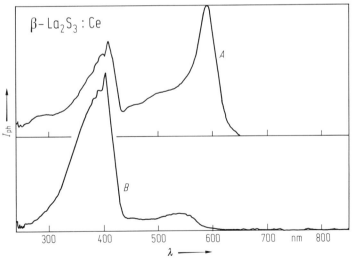

Fig. 132. β-La$_2$S$_3$:Ce. Photoconductivity vs. wavelength for two crystals. (A): 1% Ce, (B): 0.05% Ce doping levels [82S6].

9.16 Binary rare earth compounds

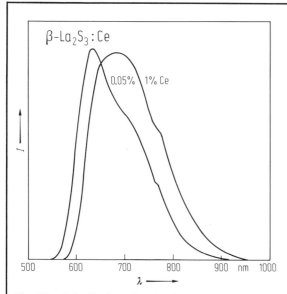

Fig. 133. β-La_2S_3:Ce. Photoluminescence intensity vs. wavelength for two Ce doping levels [81S2]. Excitation wavelength: 540 nm.

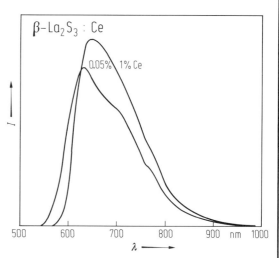

Fig. 134. β-La_2S_3:Ce. Cathodoluminescence (intensity vs. wavelength) at two Ce doping levels for 20 kV electron beam excitation [81S2].

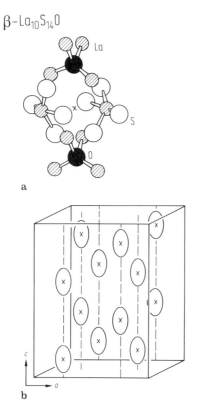

Fig. 135. β-$La_{10}S_{14}O$. Cavities in the $La_{10}S_{14}O$-structure. (a) Formation of a cavity, (b) arrangement of the cavities in the unit cell. The center of a cavity is marked by x [81B1].

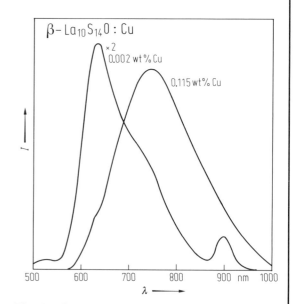

Fig. 136. β-$La_{10}S_{14}O$. Cathodoluminescence intensity vs. wavelength for samples with different Cu content (at 20 kV electron beam excitation) [81B1].

9.16 Binary rare earth compounds

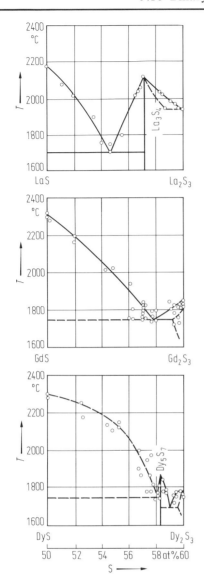

Fig. 137. LnS−Ln$_2$S$_3$ system, Ln = La, Gd, Dy. Phase diagrams [81K6].

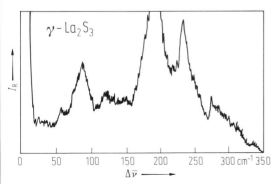

Fig. 140. γ-La$_2$S$_3$. Raman intensity vs. Raman shift (in wavenumbers) [79A].

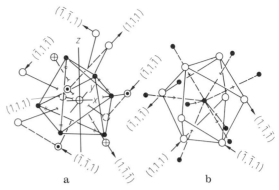

Fig. 138. Th$_3$P$_4$-type compounds. The coordination polyhedra of the cations and the anions with the lattice constant a for $u = 1/12$. (a) P○−6 Th●: 0.34a (distorted octahedra); −3 P⊙: 0.373a; −2 P⊕: 0.433a (along [1$\bar{1}\bar{1}$]); −6 P○: 0.493a (along [1$\bar{1}\bar{1}$]); (b) Th●−8 P○: 0.346a (octaverticon); −8 Th●: 0.467a (octaverticon) [66H2].

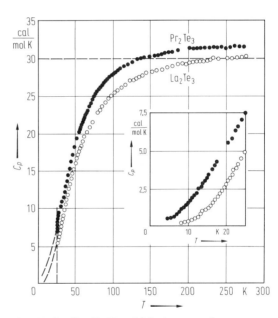

Fig. 139. La$_2$Te$_3$, Pr$_2$Te$_3$. Molar heat capacity vs. temperature with inserted low temperature region [75M2].

9.16 Binary rare earth compounds

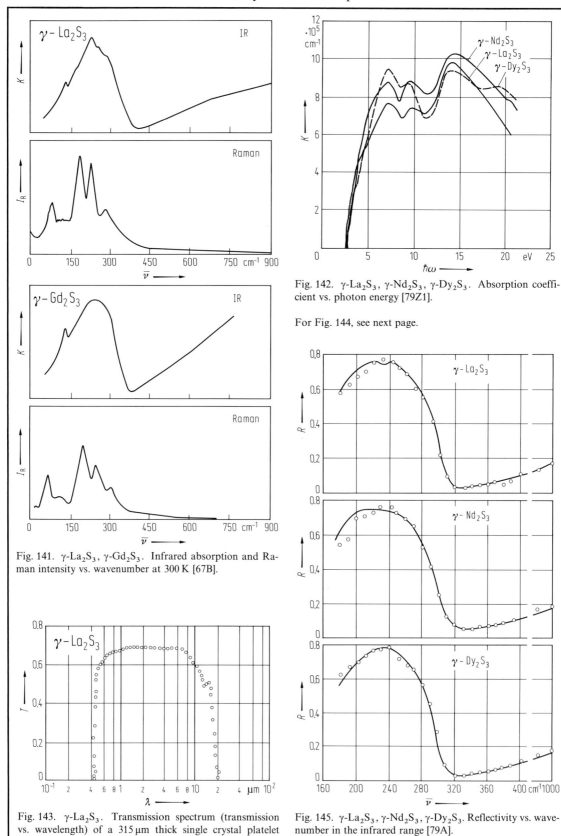

Fig. 141. γ-La_2S_3, γ-Gd_2S_3. Infrared absorption and Raman intensity vs. wavenumber at 300 K [67B].

Fig. 142. γ-La_2S_3, γ-Nd_2S_3, γ-Dy_2S_3. Absorption coefficient vs. photon energy [79Z1].

For Fig. 144, see next page.

Fig. 143. γ-La_2S_3. Transmission spectrum (transmission vs. wavelength) of a 315 µm thick single crystal platelet [81K6].

Fig. 145. γ-La_2S_3, γ-Nd_2S_3, γ-Dy_2S_3. Reflectivity vs. wavenumber in the infrared range [79A].

9.16 Binary rare earth compounds

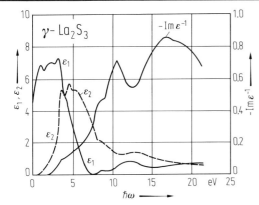

Fig. 144. γ-La$_2$S$_3$. Real and imaginary parts of the dielectric constant and energy loss function vs. photon energy [79Z1].

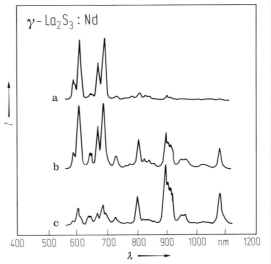

Fig. 147. γ-La$_2$S$_3$:Nd (0.9%). Luminescence spectra (intensity vs. wavelength) with 530 nm excitation taken 1 µs (a), 5 µs (b), and 10 µs (c) after excitation [83K2].

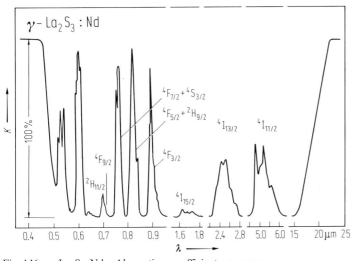

Fig. 146. γ-La$_2$S$_3$:Nd. Absorption coefficient vs. wavelength at 300 K [80K1]. Cf. Figs. 148, 149.

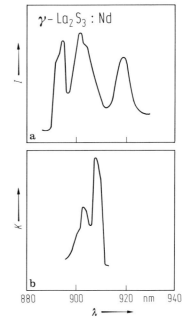

Fig. 148. γ-La$_2$S$_3$:Nd (0.9%). Luminescence intensity (a) and absorption coefficient (b) vs. wavelength at 4.2 K [83K2].

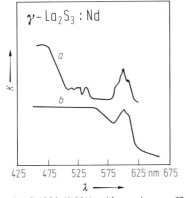

Fig. 149. γ-La$_2$S$_3$:Nd (0.9%). Absorption coefficient vs. wavelength at 300 K (a) and 600 K (b) [83K1]. Cf. Fig. 146.

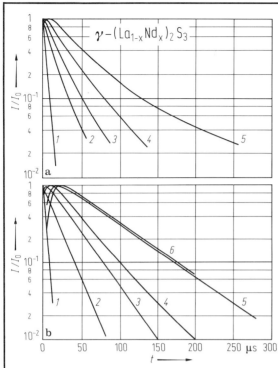

Fig. 150. γ-$(La_{1-x}Nd_x)_2S_3$. Temporal evolution of the $^4F_{3/2}$ neodymium population for $x = 0.27(1)$, $0.073(2)$, $0.046(3)$, $0.027(4)$, $0.009(5)$, and $0.003(6)$ after a short pulse excitation at 530 nm(a) and 600 nm(b). (Normalized fluorescence intensity I/I_0 vs. time t) [83K3].

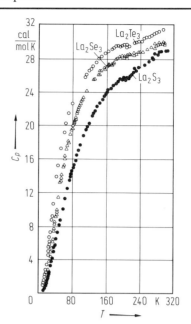

Fig. 151. γ-La_2S_3, La_2Se_3, La_2Te_3. Temperature dependence of the molar heat capacity [72S7].

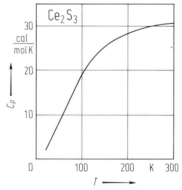

Fig. 154. γ-Ce_2S_3. Temperature dependence of the molar heat capacity [72S7].

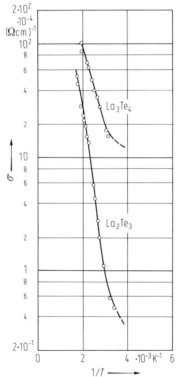

Fig. 152. La_2Te_3, La_3Te_4. Conductivity vs. reciprocal temperature [65R2].

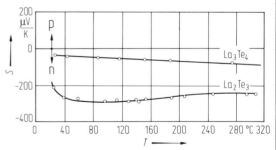

Fig. 153. La_2Te_3, La_3Te_4. Seebeck coefficient vs. temperature [65R2].

9.16 Binary rare earth compounds

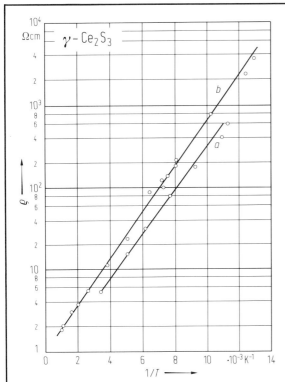

Fig. 155. γ-Ce$_2$S$_3$. Low temperature resistivity vs. reciprocal temperature. For curves a and b the electron concentrations are $8.7 \cdot 10^{17}$ cm^{-3} and $5 \cdot 10^{17}$ cm^{-3}, respectively [64C].

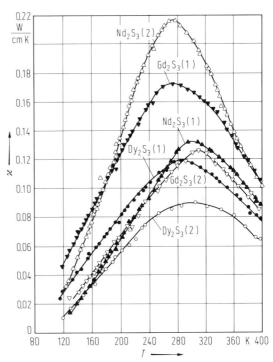

Fig. 157. γ-Nd$_2$S$_3$, γ-Gd$_2$S$_3$, γ-Dy$_2$S$_3$. Thermal conductivity vs. temperature of several single crystals. The samples show the following excess metal content: Nd$_2$S$_3$(1): 0.7% Nd, Nd$_2$S$_3$(2): 0.6% Nd, Gd$_2$S$_3$(1): 0.4% Gd, Gd$_2$S$_3$(2): 0.5% Gd, Dy$_2$S$_3$(1): 0.1% Dy, Dy$_2$S$_3$(2): 0.2% Dy [77T].

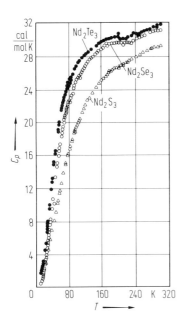

Fig. 156. γ-Nd$_2$S$_3$, Nd$_2$Se$_3$, Nd$_2$Te$_3$. Molar heat capacity vs. temperature [72S7].

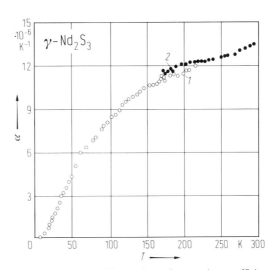

Fig. 158. γ-Nd$_2$S$_3$. Linear thermal expansion coefficient vs. temperature showing an anomaly. (*1*): values obtained after preliminary cooling below 169 K, (*2*): values obtained when the sample was not cooled below 169 K [82N2].

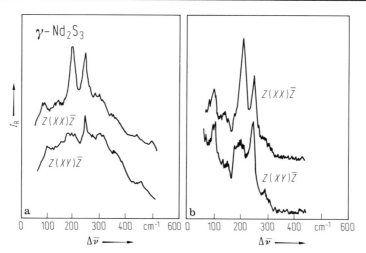

Fig. 159. γ-Nd$_2$S$_3$. Raman spectra (intensity vs. Raman shift) in the Stokes (a) and anti-Stokes (b) regions for different polarizations [80S3].

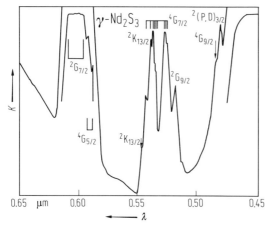

Fig. 160. γ-Nd$_2$S$_3$. Absorption coefficient vs. wavelength at 80 K, uv-region, Nd^{3+}-ground state multiplets are indicated [70H].

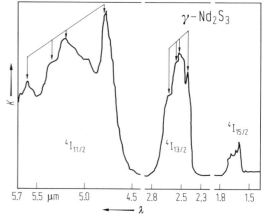

Fig. 161. γ-Nd$_2$S$_3$. Absorption coefficient vs. wavelength at 80 K, ir-region, Nd^{3+} ground state multiplets are indicated [70H].

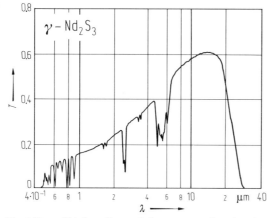

Fig. 162. γ-Nd$_2$S$_3$. Transmission vs. wavelength of a 0.07 mm thick crystal [80S3].

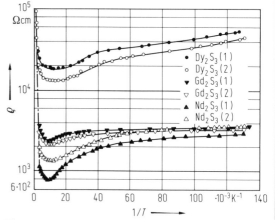

Fig. 163. γ-Nd$_2$S$_3$, γ-Gd$_2$S$_3$, γ-Dy$_2$S$_3$. Temperature dependence of the electrical conductivity. The sample compositions are given in Fig. 157 [77T].

9.16 Binary rare earth compounds

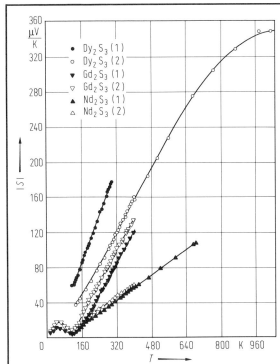

Fig. 164. γ-Nd$_2$S$_3$, γ-Gd$_2$S$_3$, γ-Dy$_2$S$_3$. Thermoelectric power vs. temperature. The sample compositions are given in Fig. 157 [77T].

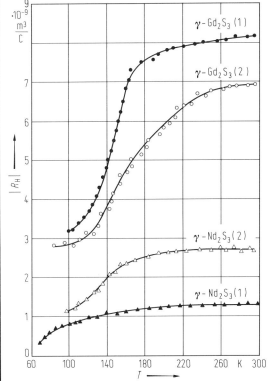

Fig. 165. γ-Nd$_2$S$_3$, γ-Gd$_2$S$_3$. Absolute value of the Hall coefficient vs. temperature. The excess metal content is as follows: Nd$_2$S$_3$(1): 0.7% Nd, Nd$_2$S$_3$(2): 0.65% Nd, Gd$_2$S$_3$(1): 0.6% Gd, Gd$_2$S$_3$(2): 0.6% Gd [74T].

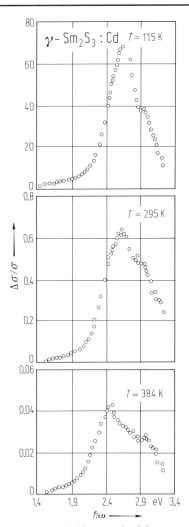

Fig. 166. γ-Sm$_2$S$_3$:Cd. Photoconductivity spectra (relative change $\Delta\sigma/\sigma$ vs. photon energy) of a Cd-doped polycrystalline film at 115 K, 295 K, 384 K [82G4].

9.16 Binary rare earth compounds

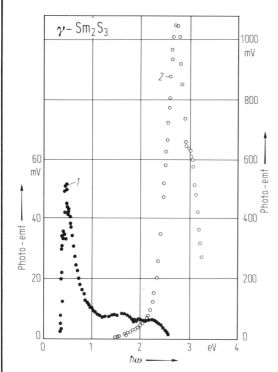

Fig. 167. γ-Sm$_2$S$_3$: Pb, Cd. Photoconductivity spectra (photo emf vs. photon energy) of Pb-doped (*1*) and Cd-doped (*2*) films [82G4].

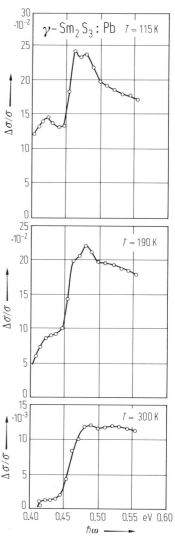

Fig. 168. γ-Sm$_2$S$_3$: Pb. Fine structure of the photoconductivity spectrum (relative change $\Delta\sigma/\sigma$ vs. photon energy) of a Pb-doped film at 115 K, 190 K, and 300 K [82G4].

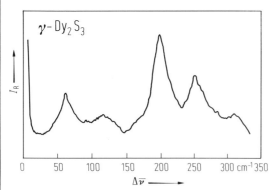

Fig. 169. γ-Dy$_2$S$_3$. Raman intensity vs. Raman shift (in wavenumbers) [79A].

9.16 Binary rare earth compounds

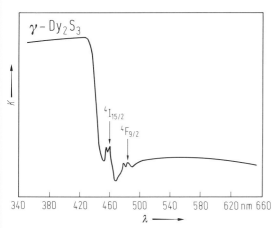

Fig. 170. γ-Dy$_2$S$_3$. Absorption coefficient vs. wavelength for single crystals at 300 K [67H].

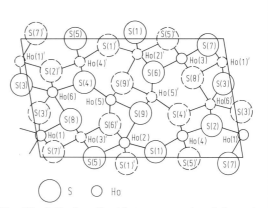

Fig. 171. δ-Ho$_2$S$_3$. Crystal structure projected down the crystal b axis. Solid circles represent atoms in the plane $y = 1/4$, dashed circles those in the plane $y = 3/4$ [67W].

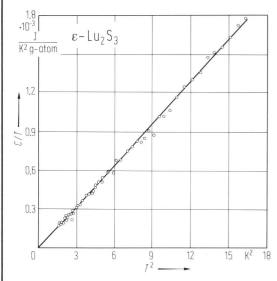

Fig. 172. ε-Lu$_2$S$_3$. Temperature dependence of heat capacity [81G4].

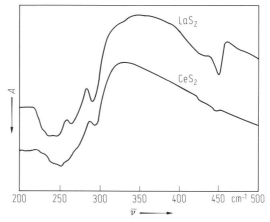

Fig. 173. LaS$_2$, CeS$_2$. Absorbance vs. wavenumber for a polycrystalline sample [75G].

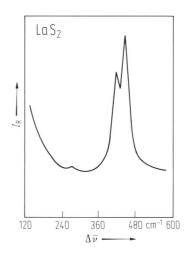

Fig. 174. LaS$_2$. Raman intensity vs. Raman shift (in wavenumbers) [75G].

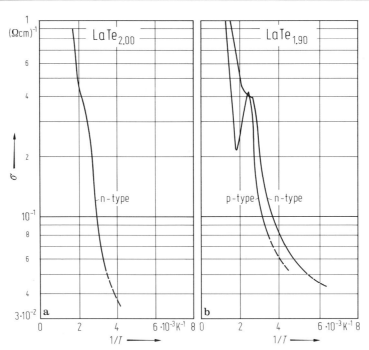

Fig. 175. LaTe$_{2.0}$, LaTe$_{1.90}$. Conductivity vs. reciprocal temperature for n-type LaTe$_{2.0}$ and n- and p-type LaTe$_{1.90}$, polycrystalline samples [65R2].

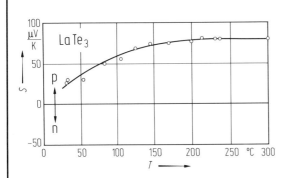

Fig. 176. LaTe$_3$. Seebeck coefficient vs. temperature, polycrystalline sample [65R2].

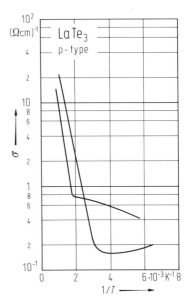

Fig. 177. LaTe$_3$. Conductivity vs. reciprocal temperature for two polycrystalline specimens [65R2].